Handbook of Industrial Hydrocarbon Processes

Handbook of Industrial Hydrocarbon Processes

Second Edition

James G. Speight, PhD, DSc, PhD
CD & W Inc. Laramie, WY, United States

ELSEVIER

Gulf Professional Publishing
An imprint of Elsevier

Gulf Professional Publishing is an imprint of Elsevier
50 Hampshire Street, 5th Floor, Cambridge, MA 02139, United States
The Boulevard, Langford Lane, Kidlington, Oxford, OX5 1GB, United Kingdom

Library of Congress Cataloging-in-Publication Data
A catalog record for this book is available from the Library of Congress

British Library Cataloguing-in-Publication Data
A catalogue record for this book is available from the British Library

ISBN: 978-0-12-809923-0

For information on all Gulf Professional Publishing publications visit our website at https://www.elsevier.com/books-and-journals

Publisher: Joe Hayton
Acquisition Editor: Katie Hammon
Editorial Project Manager: Aleksandra Packowska
Production Project Manager: Surya Narayanan Jayachandran
Cover Designer: Mark Rogers

Typeset by TNQ Technologies

Contents

About the Author

Dr James G. Speight has doctorate degrees in chemistry, geological sciences, and petroleum engineering and is the author of more than 80 books in petroleum science, petroleum engineering, environmental sciences, and ethics.

He has more than 50 years of experience in areas associated with (1) the properties, recovery, and refining of reservoir fluids, conventional petroleum, heavy oil, and tar and bitumen; (2) the properties and refining of natural gas, gaseous fuels, (3) the production and properties of petrochemicals; (4) the properties and refining of biomass, biofuels, biogas, and the generation of bioenergy; and (5) the environmental and toxicological effects of fuels. His work has also focused on safety issues, environmental effects, remediation, and reactors associated with the production and use of fuels and biofuels.

Although he has always worked in private industry with emphasis on contract-based work, Dr Speight was a visiting professor in the College of Science, University of Mosul, Iraq, and has also been a visiting professor in chemical engineering at the following universities: University of Missouri-Columbia, the Technical University of Denmark, and the University of Trinidad and Tobago.

In 1996, Dr Speight was elected to the Russian Academy of Sciences and awarded the Gold Medal of Honor that same year for outstanding contributions to the field of petroleum sciences. In 2001, he received the Scientists Without Borders Medal of Honor of the Russian Academy of Sciences and was also awarded the Einstein Medal for outstanding contributions and service in the field of Geological Sciences. In 2005, Dr Speight was awarded the Gold Medal—Scientists Without Frontiers, Russian Academy of Sciences, in recognition of Continuous Encouragement of Scientists to Work Together Across International Borders. In 2007, Dr. Speight received the Methanex Distinguished Professor award at the University of Trinidad and Tobago in recognition of excellence in research. In 2018, he received the American Excellence Award for Excellence in Client Solutions from the United States Institute of Trade and Commerce, Washington, DC.

Preface

The success of the First Edition of this text has been the primary factor in the decision to publish an updated Second Edition.

The hydrocarbon industry had its modern origins in the later years of the 19th century. By the time that the 19th century had dawned, it was known that kerosene, a fuel for heating and cooking, was the primary product of the crude oil industry in the 1800s. Rockefeller and other refinery owners considered gasoline a useless by-product of the distillation process. But all of that changed around 1900 when electric lights began to replace kerosene lamps and automobiles came on the scene. New hydrocarbon fuels were also needed to power the ships and airplanes used in World War I. After the war, an increasing number of farmers began to operate tractors and other equipment powered by oil. The growing demand for petrochemicals and the availability of crude oil and natural gas caused the industry to quickly expand in the 1920s and 1930s. During World War II, vast amounts of crude oil were produced and converted into hydrocarbon fuels and lubricants.

The term *hydrocarbons* (compounds which contain carbon and hydrogen *only*) represents a large group of chemicals manufactured from crude oil and natural gas as distinct from fuels and other products that are also derived from crude oil and natural gas by a variety of processes and used for a variety of commercial purposes. Hydrocarbons lead to products which include such items as plastics, soaps and detergents, solvents, drugs, fertilizers, pesticides, explosives, synthetic fibers and rubbers, paints, epoxy resins, and flooring and insulating materials. Hydrocarbons are also used to produce chemicals as diverse as aspirin, luggage, boats, automobiles, aircraft, polyester clothes, and recording discs and tapes. It is the changes in product demand that have been largely responsible for the evolution of the hydrocarbon industry from the demand for the hydrocarbonaceous asphalt mastic used in ancient times to the current high demand for gasoline and other hydrocarbon fuels as well as increasing demand for a wide variety of petrochemical products.

As a result, the hydrocarbon industry is a huge field that encompasses many commercial chemicals and polymers. The organic chemicals produced in large volumes are methanol, ethylene, propylene, butadiene, benzene, toluene, and the xylene isomers. Ethylene, propylene, and butadiene, along with butylenes, are collectively called olefins, which belong to a class of unsaturated aliphatic hydrocarbons, having the general formula C_nH_{2n}. Olefins contain one or more double bonds, which make them chemically reactive. Benzene, toluene, and the xylene isomers, commonly referred to as aromatics (BTX), are unsaturated cyclic hydrocarbons containing one or more rings. Olefins, aromatics, and methanol are precursors to a variety of chemical products and are generally referred to as primary petrochemicals. Furthermore, because ethylene and propylene are the major building blocks for petrochemicals, alternative ways for their production have always been sought. The main route for producing ethylene and propylene is steam cracking, which is an energy extensive process.

Basic hydrocarbon chemicals are the key building blocks for manufacture of a wide variety of durable and nondurable consumer goods. Considering the items that are encountered every day—clothes, construction materials used to build our homes and offices, a variety of household appliances and electronic equipment, food and beverage packaging, and many products used in various modes of transportation—hydrocarbons provide the fundamental building blocks that enable the manufacture of the vast majority of these goods. Demand for chemicals and plastics is driven by global economic conditions, which are directly linked to demand for consumer goods.

The search for alternative ways to produce monomers and chemicals from sources other than crude oil. In fact, Fisher Tropsch technology, which produces low molecular weight olefins in addition to hydrocarbon fuels, could enable non–crude oil feedstocks (such as extra heavy oil, tar and bitumen, coal, oil shale, and biomass) to be used as feedstocks for petrochemicals.

In addition, the continued high demand for hydrocarbon products, such as liquid fuels (gasoline and diesel fuel) and petrochemical feedstocks (such as aromatic derivatives and olefin derivatives), is increasing throughout the world. Traditional markets such as North America and Europe are experiencing a steady increase in demand, whereas emerging Asian markets, such as India and China, are witnessing a rapid surge in demand for hydrocarbons. This has resulted in a tendency for existing refineries to seek fresh refining approaches to optimize efficiency and throughput. Furthermore, the increasing use of the heavier feedstocks for refineries is forcing technology suppliers/licensors to revamp their refining technologies in an effort to cater to the growing customer base.

The evolution in product specifications caused by various environmental regulations plays a major role in the development of crude oil refining technologies. In many countries, especially in the United States and Europe, gasoline and diesel fuel specifications have changed radically in the past decades and will continue to do so in the future. Currently, reducing the sulfur levels of hydrocarbon fuels is the dominant objective of many refiners. There is also an increasing demand for hydrocarbon derivatives for other uses. This is pushing the technological limits of hydrocarbon production to the maximum, and the continuing issue is the processes that will increase hydrocarbon production and purity.

Refineries must, and indeed are eager to, adapt to changing circumstances and are amenable to trying new technologies that are radically different in character. Currently, refineries are also looking to exploit heavy (more viscous) crude oils and tar and bitumen (sometimes referred to as extra heavy crude oil) provided they have the refinery technology capable of handling such feedstocks. Transforming the higher boiling constituents of these feedstock components into single hydrocarbon derivative as well as hydrocarbon fuels is becoming a necessity.

The reader might also be surprised at the number of older references that are included. The purpose of this is to remind the reader that there is much valuable work cited in the older literature. Work that is still of value and, even though in some cases, there has been similar work performed with advanced equipment, the older work has stood the test of time. Many of the ideas are still pertinent and should not be forgotten in terms of the valuable contributions they have made to crude oil science and technology. However, many of the older references included in previous editions of this book have been deleted—unavailability of the source for the general scientific researcher and the current lack of a substantiated sources (other than the files collected by the author) have been the root cause of such omissions.

Therefore, it is the purpose of this book to provide the reader with a detailed overview of the production and properties of hydrocarbon derivatives and hydrocarbon fuels as the world evolves into the 21st century. With this in mind, many of the chapters that appeared in the First Edition have been rewritten to include the latest developments related to hydrocarbon products. Updates on the evolving processes and new processes as well as the various environmental regulations are presented. However, the text still maintains its initial premise, that is, to introduce the reader to the hydrocarbon science and technology as well as the production of a wide variety of products and petrochemical intermediates. However, the text will also prove useful for those scientists and engineers already engaged in the crude oil industry as well as in the catalyst manufacturing industry who wish to gain a general overview or update of the science of crude oil.

Thus, the book focuses on the interfaces between chemical technology and biotechnology especially where these impact on health and safety and the environment. Also, the book has been adjusted, polished, and improved for the benefit of new readers as well as for the benefit of readers of the four previous editions.

Dr. James G. Speight
Laramie, Wyoming, USA
June, 2019

Chapter 1

Chemistry and chemical technology

1. Introduction

Chemistry (from the Arabic *al khymia*) is the science of matter and is concerned with the composition, behavior, structure, and properties of matter, as well as the changes matter undergo during chemical reactions. Chemistry is a physical science and is used for the investigation of atoms, molecules, crystals, and other assemblages of matter whether in isolation or combination, which incorporates the concepts of energy and entropy in relation to the spontaneity or initiation of chemical reactions or chemical processes.

Disciplines within chemistry are traditionally grouped by the type of matter being studied or the kind of study and include the following (presented alphabetically): (i) analytical chemistry, which is the analysis of material samples to gain an understanding of their chemical composition and structure, (ii) biochemistry, which is the study of substances found in biological organisms, (iii) inorganic chemistry, which is the study of inorganic matter (inorganic chemicals, such as minerals), (iv) organic chemistry, which is the study of organic matter (inorganic chemicals, such as hydrocarbon derivatives), and (v) physical chemistry, which is the study of the energy relations of chemical systems at macro, molecular, and submolecular scales.

In fact, the history of human culture can be viewed as the progressive development of chemical technology through evolution of the scientific and engineering disciplines in which chemistry and chemical engineering have played major roles in producing a wide variety of industrial chemicals, especially industrial organic chemicals (Ali et al., 2005). Chemical technology, in the context of the present book, relies on chemical bonds of hydrocarbon derivatives. Nature has favored the storage of solar energy in the hydrocarbon bonds of plants and animals, and the evolution of chemical technology has exploited this hydrocarbon energy profitably.

The focus of this book is hydrocarbon derivatives and the chemistry associated with these organic compounds which will be used to explain the aspects of hydrocarbon properties, structure, and manufacture. The book will provide information relating to the structure and properties of hydrocarbon

Handbook of Industrial Hydrocarbon Processes. https://doi.org/10.1016/B978-0-12-809923-0.00001-1

derivatives and their production through process chemistry and chemical technology to their conversion into commercial products.

2. Organic chemistry

Organic chemistry is a discipline within chemistry that involves study of the structure, properties, composition, reactions, and preparation (by synthesis or by other means) of carbon-based compounds (in this context—hydrocarbon derivatives).

On the other hand, inorganic chemistry is the branch of chemistry concerned with the properties and behavior of inorganic compounds. This field covers all chemical compounds except the myriad of carbon-based compounds (such as the hydrocarbon derivatives), which are the subjects of organic chemistry. The distinction between the two disciplines is far from absolute, and there is much overlap, most importantly in the subdiscipline of organometallic chemistry in which organic compounds and metals form distinct and stable products. An example is tetraethyl lead which was formerly used in gasoline (until it was banned by various national environmental agencies) as an octane enhancer to prevent engine knocking or pinging during operation. Other than this clarification and brief mention here, neither inorganic chemistry nor organometallic chemistry will be described further in this text.

2.1 Organic chemicals

Organic compounds are structurally diverse, and the range of application of organic compounds is enormous. In addition, organic compounds may contain any number of other elements, including nitrogen, oxygen, sulfur, halogens, phosphorus, and silicon. They form the basis of, or are important constituents of, many products (such as plastics, drugs, petrochemicals, food, explosives, and paints) and, with very few exceptions, they form the basis of all life processes and many industrial processes.

Organic compounds—of which the hydrocarbon derivatives are a subgroup—are classified according to the presence of functional groups in the molecule (Table 1.1, Table 1.2, Table 1.3, Table 1.4). A *functional group* is a molecular moiety that typically dictates the behavior (reactivity) of the organic compound in the environment and the reactivity of that functional group is assumed to be the same in a variety of molecules, within some limits and if steric effects (that arise from the three-dimensional structure of the molecule) do not interfere. Thus, most organic functional groups feature heteroatoms (atoms other than carbon and hydrogen, such as: nitrogen, oxygen, and sulfur). The concept of functional groups is a major concept in organic chemistry, both to classify the structure of organic compounds and to predict the physical and chemical properties especially, in the context of this

TABLE 1.1 General classes of hydrocarbons.

Chemical class	Group	Formula	Structural formulae
Alkane	Alkyl	$R(CH_2)_nH$	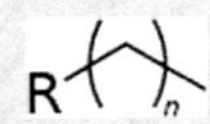
Alkene	Alkenyl	$R_2C = CR_2$	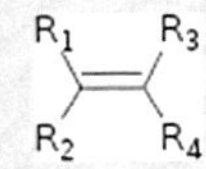
Alkyne	Alkynyl	$R_1C{\equiv}CR_2$	
Benzene derivative	Phenyl	RC_6H_5	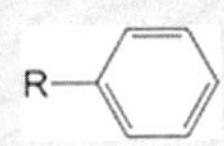

book, those properties that relate to behavior and reactivity in technological processes.

For example, when comparing the properties of ethane (CH_3CH_3) with the properties of propionic acid ($CH_3CH_2CO_2H$), which is a chemical that is formed due to the replacement of a hydrogen atom in the ethane molecule by a carboxylic acid functional group (CO_2H) the change in properties and behavior is spectacular. Alternatively, the replacement of a methyl group (CH_3) into the ethane molecule by the carboxylic acid function to produce acetic acid (CH_3CO_2H) (or the replacement of a hydrogen in the methane molecule (CH_4) by the carboxylic acid function) produces significant changes in the properties of the product vis-à-vis the original molecule.

Thus, organic chemicals from very simple compounds such as methane (CH_4) to organic chemicals that contain more than one carbon atom, as many as 10 carbon atoms to chemicals that contain hundreds or more carbon atoms that are linked in carbon—carbon bonds. Those that contain only carbon and hydrogen are called hydrocarbons; a simple example is $H_3C(CH_2)_3CH_3$ (pentane). Organic chemicals commonly contain other elements too, such as oxygen, nitrogen, or sulfur. An organometallic chemical has a carbon atom bonded to a metal as in tetraethyl lead. Some organometallic chemicals are found naturally.

Many organic chemicals are synthetic, i.e., produced not by living creatures, but manufactured by human beings. However, the feedstocks from which the chemicals are made come from nature. Commonly synthetic organic chemicals are made from crude oil or natural gas feedstocks, which are

TABLE 1.2 General classes of oxygen compounds.

Chemical class	Group	Formula	Structural formula
Alcohol	Hydroxyl	ROH	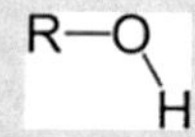
Ketone	Carbonyl	RCOR′	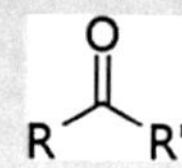
Aldehyde	Aldehyde	RCHO	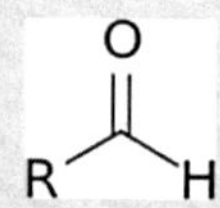
Acyl halide	Haloformyl	RCOX	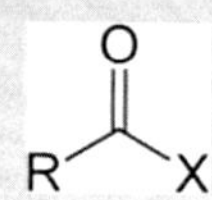
Carbonate	Carbonate ester	ROCOOR	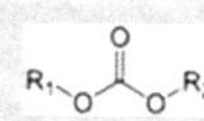
Carboxylate	Carboxylate	$RCOO^-$	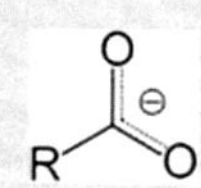
Carboxylic acid	Carboxyl	RCOOH	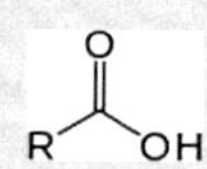
Ester	Ester	RCOOR′	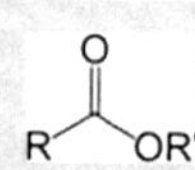
Methoxy	Methoxy	$ROCH_3$	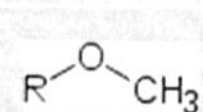

TABLE 1.3 General classes of nitrogen compounds.

Class	Group	Formula	Structural formula
Amide	Carboxamide	$RCONR_2$	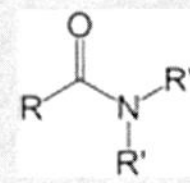
Amines	Primary amine	RNH_2	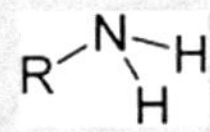
	Secondary amine	R_2NH	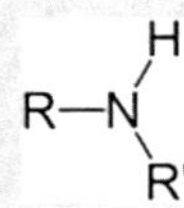
	Tertiary amine	R_3N	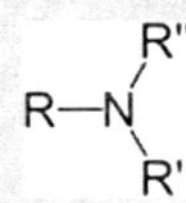
	4 degrees ammonium ion	R_4N^+	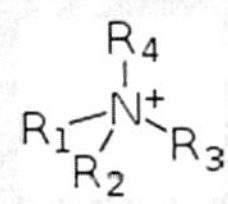
Imine	Primary ketimine	RC(=NH)R′	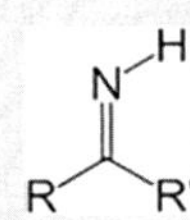
	Secondary ketimine	RC(=NR)R′	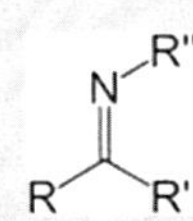
	Primary aldimine	RC(=NH)H	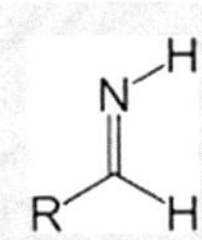

Continued

TABLE 1.3 General classes of nitrogen compounds.—cont'd

Class	Group	Formula	Structural formula
	Secondary aldimine	RC(=NR′)H	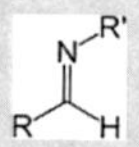
Imide	Imide	$(RCO)_2NR'$	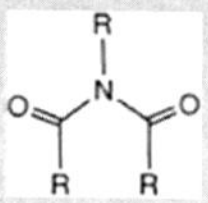
Azide	Azide	RN_3	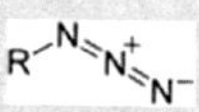
Azo compound	Azo (di-imide)	RN_2R'	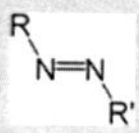
Cyanates	Cyanate	ROCN	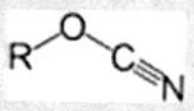
	Isocyanate	RNCO	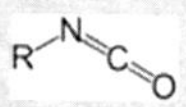
Nitrate	Nitrate	$RONO_2$	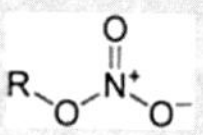
Nitrile	Nitrile	RCN	
	Isonitrile	RNC	
Nitrite	Nitroso-oxy	RONO	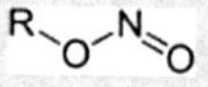
Nitro compound	Nitro	RNO_2	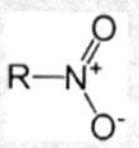

TABLE 1.4 General classes of sulfur compounds.

Chemical class	Group	Formula	Structural formula
Thiol	Sulfhydryl	RSH	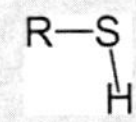
Sulfide (thioether)	Sulfide	RSR′	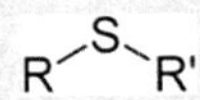
Disulfide	Disulfide	RSSR′	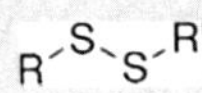
Sulfoxide	Sulfinyl	RSOR′	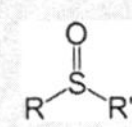
Sulfone	Sulfonyl	RSO_2R'	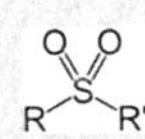
Sulfinic acid	Sulfino	RSO_2H	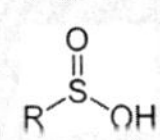
Sulfonic acid	Sulfo	RSO_3H	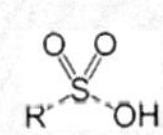
Thiocyanate	Thiocyanate	RSCN	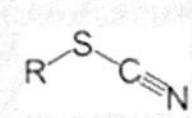
	Isothiocyanate	RNCS	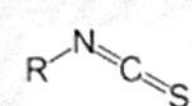
Thione	Carbonothioyl	RCSR′	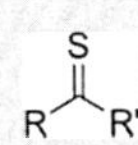
Thial	Carbonothioyl	RCSH	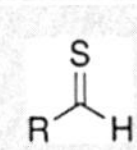

referred to as petrochemicals (Speight, 2014, 2017). Coal or wood also sometimes serve as feedstocks for organic chemicals. Plastics are synthetic organic chemicals and the so-called bioplastics, which humans produce from plant materials, involve some synthetic chemistry. Some commercial organic chemicals are produced too by cultures of molds or bacteria; such chemicals must then be purified from these cultures by human actions.

Finally, for clarification, a biochemical is an organic chemical synthesized by a living creature. Proteins, fats, and carbohydrates are biochemicals. Sucrose (table sugar) and the acetic acid (CH_3CO_2H) in vinegar are examples of simple biochemicals. Many biochemicals can also be made synthetically, not only simple chemicals such as vinegar or the sugars, sucrose, and xylose, but also complex ones. If the structure of a chemical made by synthetic means is exactly the same as that found in nature, it is indeed the same chemical—the body treats both in exactly the same way so there is no biological difference between them. Chemicals synthesized by living creatures can also be extensively manipulated during extraction and purification and still legally be called natural.

2.2 The chemical bond

The most basic concept in all of chemistry is the chemical bond. The chemical bond is essentially the sharing of electrons between two atoms, a sharing which holds or bonds the atoms together.

Atoms have three components: protons, neutrons, and electrons. Protons have a positive charge of +1, neutrons have 0 charge, and electrons have a negative charge of −1. The protons and neutrons occupy the center of the atom as a piece of solid matter called the nucleus. The electrons exist in *orbitals* surrounding the nucleus. In reality, it is impossible to tell the precise trajectory of an electron and the best that can be achieved is to describe the probability of locating the electron in a region of space.

The simplest case is when the nucleus is surrounded by just one electron (for example, the hydrogen atom). In this case, the probability of finding an electron in its lowest energy, or most stable, state is it being distributed in a spherically symmetric way around the nucleus. The probability of finding the electron is highest at the nucleus and decreases as the distance from the nucleus increases.

The spherically symmetric *1s orbital* is the lowest energy orbital that an electron can occupy, but several higher energy orbitals are significant in organic chemistry. The next lowest energy orbital that an electron can occupy is the *2s orbital*, which looks much like the *1s orbital* except that the electron is more likely to be found farther from the nucleus. The third lowest energy orbital is the *2p orbital*. The major and highly important difference between a *p orbital* and an *s orbital* is that the *p orbital* is not spherically symmetric and is oriented along a specific axis in space. There are three *p* orbitals, which are oriented along the x, y, and z axes.

2.3 Bonding in carbon-based systems

A chemical bond is essentially the sharing of electrons between two atoms. Since electrons are negatively charged and exert an attractive force on nuclei, they serve to hold the atoms together if they are located between two nuclei.

When two atoms approach each other, their *atomic orbitals* overlap. The overlapped atomic orbitals can add together to form a *molecular orbital* (*linear combination of atomic orbitals*, LCAO). The area of greatest overlap between the original atomic orbitals represents the chemical bond that is formed between them. Since the sharing of electrons is the basis of the chemical bond, the molecular orbitals formed represent chemical bonds.

For example, in the case of hydrogen, the two *1s* orbitals gradually come closer together until there is a good deal of overlap between them. At this point, the area in space of greatest electron density will be between the two nuclei, which themselves were at the center of the original atomic orbitals. This electron density, now part of a new molecular orbital, represents the chemical bond. When the area of greatest overlap occurs directly between the two nuclei on an axis containing the nuclei of both atoms (*internuclear axis*), the bond is a *sigma bond* (*σ bond*) (Fig. 1.1).

More than one atomic orbital from a single atom can be used to form new molecular orbitals. For example, a *2s* orbital and a *2p* orbital from one atom might add together and overlap with one or more orbitals from a second atom to form new molecular orbitals. Second, parts of orbitals can possess a sign (+or -). The *s* orbital has the same sign throughout, while in the *p* orbitals, one lobe is + and the other lobe is -. Signs do not matter with respect to electron density, but they must be considered when orbitals are added or subtracted. If two orbitals of the same sign are added, electron density will increase, while if two orbitals of opposite signs (charges) are added, the shared electron density will cancel out.

Carbon has six electrons—only two electrons can occupy an *s* orbital at a time. The first two electrons in carbon occupy the *1s* orbital and the next two occupy the higher-energy, but similarly shaped *2s* orbital while the final two electrons occupy the *2p* orbitals.

$$\underset{1s}{\underline{\uparrow\downarrow}}\ \underset{2s}{\underline{\uparrow\downarrow}}\ \underset{2p_x}{\underline{\ \uparrow\ }}\ \underset{2p_y}{\underline{\ \uparrow\ }}\ \underset{2p_z}{\underline{\quad}}$$

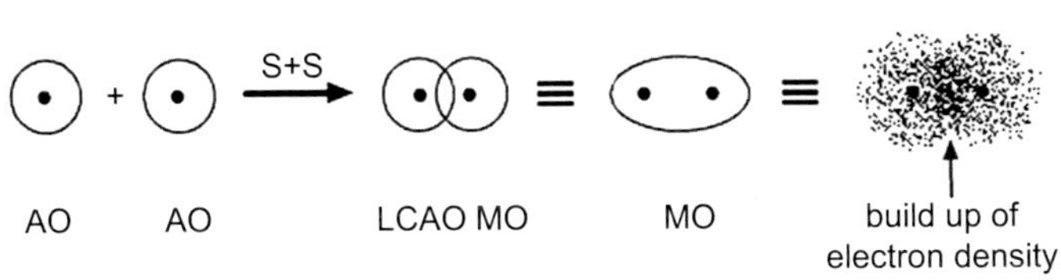

FIGURE 1.1 Two hydrogen *1s* atomic orbitals overlap to form a hydrogen molecular orbital.

In carbon, the electrons in the *1s* orbital are too low in energy to form bonds. Thus, electrons used to form bonds must come from the *2s* and *2p* orbitals. Carbon very often makes four bonds by redistribution of the *2p* electrons:

$$\underset{1s}{\underline{\uparrow\downarrow}}\ \underset{sp^3}{\underline{\uparrow}}\ \underset{sp^3}{\underline{\uparrow}}\ \underset{sp^3}{\underline{\uparrow}}\ \underset{sp^3}{\underline{\uparrow}}$$

When it does so, these bonds are arranged so that they are as far away from each other as possible. This arrangement is referred to as a *tetrahedral bond* (Fig. 1.2).

The individual *2s* orbital and the 2p orbital cannot form bonds in this arrangement due to their geometry. The *2s* orbital is completely symmetric, while the *2p* orbitals are aligned along specific axes. None of these orbitals is well-equipped to form bonds in the tetrahedral geometry alone.

Since a chemical bond does not have to be formed from individual atomic orbitals, but can be formed from a combination of several atomic orbitals from the same atom, each bond that is made in the tetrahedral geometry, a part of the *2s* and a part of each of the *2p* orbitals will contribute, resulting in a tetrahedral arrangement and there is a 109.5 degrees angle between each of the bonds (Fig. 1.2). To achieve this geometry, both the *2s* and all three of the *2p* orbitals ($2p_x$, $2p_y$, and $2p_z$) must contribute. The new bonds that are formed are called *sp^3 bonds*, since 1 *s* orbital and 3 *p* orbitals were used to form the bonds.

$$\underset{1s}{\underline{\uparrow\downarrow}}\ \underset{2s}{\underline{\uparrow\downarrow}}\ \underset{2p_x}{\underline{\uparrow\downarrow}}\ \underset{2p_y}{\underline{\uparrow\downarrow}}\ \underset{2p_z}{\underline{\uparrow\downarrow}}$$

Carbon sometimes makes three bonds instead of four. In this case, not all of the *2p* orbitals combine with the *2s* orbital to form bonds. Instead, a combination of the 2s orbital and two of the *2p* orbitals make three *sp^2 bonds*, while the other *p* orbital does not participate in this combination and can make a fourth bond on its own.

Like the sp^3 bonds, the sp^2 bonds are oriented such that they are as far away from each other as possible (*trigonal planar geometry*). Each of the bonds points to one of the vertices of a triangle, but all three bonds are located in the same plane. The other *2p* orbital, the one which did not add to make sp^2 bonds exists perpendicular to the plane in which the sp^2 bonds form. It too is able to form bonds, and it does so independently of the sp^2 bonds.

When two carbon atoms with sp^2 orbitals form a bond to each other using their sp^2 orbitals, a σ bond is formed between then. Moreover, the extra *p* orbitals, which exist above and below each carbon atom, also overlap with each other. This overlap between *p* orbitals leads to the formation of a second bond in addition to the σ bond formed between the sp^2 orbitals. This second bond which does not occur directly between the nuclei on the internuclear axis but above and below the internuclear axis is a π bond (pi bond). When a σ

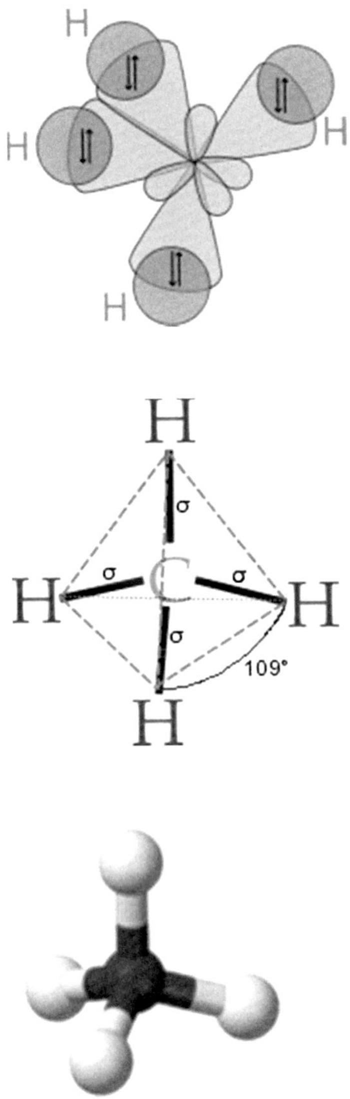

FIGURE 1.2 Tetrahedral geometry as exhibited by the carbon atom surrounded by four hydrogen atoms (methane).

bond and a π bond form together between two atoms, it leads to the formation of a double bond (Fig. 1.3).

Briefly, many inorganic compounds are ionic compounds, consisting of cations and anions joined by ionic bonding. Examples of salts (which are ionic compounds) are magnesium chloride ($MgCl_2$) which consists of magnesium cations (Mg^{2+}) and chloride anions (Cl^-). In any salt, the proportions of the

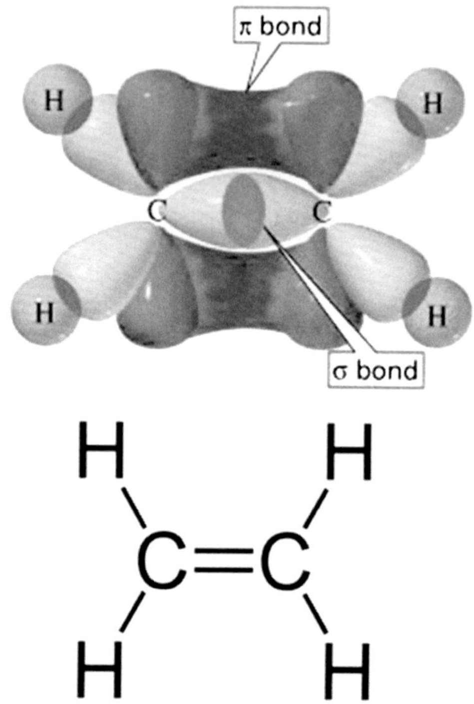

FIGURE 1.3 The molecule ethylene is formed from two carbon atoms and four hydrogen atoms— a σ bond is formed from two sp^2 orbitals and a π bond is formed from two *2p* orbitals to comprise a double bond.

ions are such that the electric charges cancel out, so that the bulk compound is electrically neutral. Important classes of inorganic compounds are the oxides, the carbonates, the sulfates, and the halides. Many inorganic compounds are characterized by high melting points, ease of crystallization, and solubility—where some salts (such as sodium chloride, NaCl) are highly soluble in water, others (such as silica, silicon oxide, SiO_2) are insoluble in water.

3. Chemical engineering

Chemical engineering is the branch of engineering that deals with the application of physical science (such as chemistry) to the process of converting raw materials (for example, crude oil) or chemicals into more useful or valuable forms.

Chemical engineering largely involves the design, improvement, and maintenance of processes involving chemical transformations for large-scale manufacture. Chemical engineers (process engineers) ensure the processes are operated safely, sustainably, and economically. Chemical engineering is applied in the manufacture of a wide variety of products. The chemical industry manufactures inorganic and organic industrial chemicals, ceramics, fuels and petrochemicals, agrochemicals (fertilizers, insecticides, herbicides), plastics and elastomers, oleo-chemicals, explosives, detergents and detergent products (soap, shampoo, cleaning fluids), fragrances and flavors, additives, dietary supplements, and pharmaceuticals. Closely allied or overlapping disciplines include wood processing, food processing, environmental technology, and the engineering of crude oil, glass, paints, and other coatings, inks, sealants, and adhesives.

Chemical engineers design processes to ensure the most economical operation in which the entire production chain must be planned and controlled for costs. A chemical engineer can both simplify and complicate showcase reactions for an economic advantage. Using a higher pressure or temperature makes several reactions easier; ammonia, for example, is simply produced from its component elements in a high-pressure reactor. On the other hand, reactions with a low yield can be recycled continuously (recycled to extinction in which no further product is made), which would be complex, arduous work if done by hand in the laboratory. It is not unusual to build 6-step, or even 12-step evaporators to reuse the vaporization energy for an economic advantage. In contrast, laboratory chemists evaporate samples in a single step.

The individual processes used by chemical engineers (e.g., distillation or filtration) are called unit operations and consist of chemical reactions, mass-transfer operations, and heat transfer operations. Unit operations are grouped together in various configurations for the purpose of chemical synthesis and/or chemical separation. Some processes are a combination of intertwined transport and separation unit operations, such as reactive distillation in which the product is formed as the still temperature is raised and the product distills from the reaction mixture.

Three basic physical laws that underlie chemical engineering design are (i) conservation of mass, (ii) conservation of energy, and (iii) conservation of momentum.

3.1 Conservation of mass

The law of conservation of mass (principle of mass/matter conservation) states that the mass of a closed system (in the sense of a completely isolated system) remains constant over time. The mass of an isolated system cannot be changed as a result of processes acting inside the system but while mass cannot be created or destroyed, it may be rearranged in space, and changed into different types of particles. Put simply, the law states that matter cannot be created or destroyed in a chemical reaction. This implies that for any chemical process in a closed system, the mass of the reactants must equal the mass of the products.

The law implies that mass can neither be created nor destroyed, although it may be rearranged in space, or the entities associated with it may be changed in form. This is illustrated in chemical reactions example in which the mass of the chemical components before the reaction is equal to the mass of the components after the reaction. Thus, during any chemical reaction and low-energy thermodynamic processes in an isolated system, the total mass of the reactants (the starting materials) will be equal to the mass of the reaction products. For example, using the molecular proportions as the weights of the reactants and the products:

$$CH_4 + 2O_2 \rightarrow CO_2 + 2H_2O$$

$$16 + (2 \times 32) \rightarrow 44 + (2 \times 18)$$

$$\text{Mass of the reactants} = \text{mass of the products}$$

The change in mass of certain kinds of open systems where atoms or massive particles are not allowed to escape, but other types of energy (such as light or heat) were allowed to enter or escape, went unnoticed during the 19th century, because the mass-change associated with addition or loss of the fractional amounts of heat and light associated with chemical reactions, was very small. Mass is also not generally conserved in open systems (even if only open to heat and work), when various forms of energy are allowed into, or out of, the system. Mass conservation for closed systems continues to be true exactly. The mass-energy equivalence theorem states that mass conservation is equivalent to energy conservation, which is the first law of thermodynamics. The mass-energy equivalence formula requires closed systems, since if energy is allowed to escape a system, mass will escape also.

3.2 Conservation of energy

The law of conservation of energy states that the total amount of energy in an isolated system remains constant over time. A consequence of this law is that energy can neither be created nor destroyed; it can only be transformed from one state to another. The only thing that can happen to energy in a closed system is that it can change form, such as a transformation of chemical energy to kinetic energy.

Conservation of energy refers to the conservation of the total system energy over time. This energy includes the energy associated with the mass of the reactants as well as all other forms of energy in the system. In an isolated system, although mass and energy (heat and light) can be converted to one another, both the total amount of energy and the total amount of mass of such systems remain constant over time. If energy in any form is allowed to escape such systems, the mass of the system will decrease in correspondence with the loss. The conservation of energy is a fundamental concept of physics along with the conservation of mass and the conservation of momentum. Within some problem domain, the amount of energy remains constant and energy is neither created nor destroyed. Energy can be converted from one form to another (potential energy can be converted to kinetic energy) but the total energy within the domain remains fixed.

Thus, energy conservation is the effort made to reduce the consumption of energy by using less of an energy service. This can be achieved either by using energy more efficiently (using less energy for a constant service) or by reducing the amount of service used (for example, by driving less). Energy can be conserved by reducing wastage and losses, improving efficiency through technological upgrades, and improving operation and maintenance. However, energy can only be transformed from one form to another.

3.3 Conservation of momentum

Momentum is the product of the mass and the velocity of an object. The conservation of momentum is a fundamental law of physics which states that the momentum of a system is constant if there are no external forces acting on the system. Momentum is a conserved quantity insofar as the total momentum of any closed system (a system not affected by external forces) cannot change. One of the consequences of the law is that the center of mass of any system of objects will always continue with the same velocity unless acted on by a force from outside the system. In an isolated system (one where external forces are absent) the total momentum will be constant, which dictates that the forces acting between systems are equal in magnitude, but opposite in sign, is due to the conservation of momentum.

Thus, for two objects (A and B) colliding in an isolated system, the total momentum before and after the collision is equal—the momentum lost by one object is equal to the momentum gained by the other. By Newton's third law of thermodynamics, for every action there is an equal but opposite reaction. Hence, the force exerted by object A on object B is equal but opposite to the force that object B exerts on object A. Then by Newton's second law of thermodynamics, this force is equal to the product of the mass and the acceleration of the objects. Hence, the product of the mass and acceleration of object A is equal but opposite to the product between the mass and acceleration of object B. in other words, momentum is conserved.

4. Chemical technology

The definitions of technology vary from one discipline to another. However, in the current context, chemical technology is the practical application of chemistry and chemical engineering to the needs of commerce or industry and is a multicomponent discipline which, in this context, deals with the application of chemical knowledge to the solution of practical problems (Badger and Baker, 1941; Henglein, 1969; Jess and Wasserscheid, 2013). Chemical technology is also a human action that involves the generation of knowledge and (usually innovative) processes to develop systems that solve problems and extend human capabilities.

In chemical technology, when a reaction not studied before is planned, it is of the utmost importance to know and to calculate the equilibrium, that is, the equilibrium constants at various temperatures, before expensive equipment for experimental studies is installed. It is the aim of chemical thermodynamics to answer technically important questions, such as the final state of chemical transformations, reaction mechanisms (passing intermediate stages), and reaction rates.

4.1 Historical aspects

Historically, the word *technology* is a modern term and rose to prominence during the *industrial revolution* when it became associated with science and engineering. The word *technology* can also be used to refer to a collection of techniques, which refers to the current state of humanity's knowledge of how to combine resources to produce desired products, to solve problems, fulfill needs, or satisfy wants; it includes technical methods, skills, processes, techniques, tools, and raw materials. The distinction between science, engineering, and technology is not always clear. However, technologies are not usually exclusively products of science because they have to satisfy requirements, such as utility.

In the context of technology as a technical endeavor, engineering technology is the process of designing and making tools and systems to exploit natural phenomena for practical human means, often (but not always) using results and techniques from chemistry and other sciences. Thus, the development of technology may draw upon many fields of knowledge from the scientific and engineering disciplines in order to achieve a practical result.

To some, technology is often a consequence of science and engineering—in this sense, scientists and engineers may both be considered technologists; the three fields are often considered as one for the purposes of research and reference.

Chemical technology is the study of technology related to chemistry. To be more specific, chemical technology takes chemistry beyond the laboratory and into the industrial world where products are made through knowledge of chemistry. Thus, chemical technology also involves various aspects of chemical engineering such as reactor design and performance. This differs from chemistry itself because the focus is also on the means by which chemistry can be employed to make useful products. Chemical technologists are more likely than technicians to participate in the actual design of experiments, and may be involved in the interpretation of experimental data. They may also be responsible for the operation of chemical processes in large plants, and may even assist chemical engineers in the design of the same.

Within technology falls the concept of innovation, which is the change in the thought process for performing a scientific or engineering task that will lead to (i) a new process, (ii) a new product, or (iii) a new use for an old product. In fact, innovation may refer to incremental or radical changes in products and/or processes and the goal of innovation is a positive change in a product or process. Innovation is considered to be a major driver of the economy, especially when it leads to new product categories or increasing productivity.

For example, using the crude oil industry as an example, innovative use of crude oil and its derivatives (particularly as an asphalt mastic) started 6000 years ago; current innovations can be considered to have commenced in the 1860s and continue to this day (Table 1.5) to the point where heavy oil (once considered a difficult-to-refine feedstock) is now refined on a very regular basis (Ancheyta and Speight, 2007; Speight, 2014).

4.2 Technology and human culture

The use of technology in the form of the development of tools and harnessing the energy of fire has often been regarded as the defining characteristic of *Homo sapiens,* and is a means of defining the species. Furthermore, the history of human culture can be viewed as the progressive development of new energy sources and their associated conversion technologies (Hall et al., 2003). Most of these energy technologies rely on the properties (i.e., the chemical bonds) of hydrocarbon derivatives.

Technology, the systematic application of scientific and engineering knowledge in developing and applying technology, has grown immensely. Technological knowledge provides a means of estimating what the behavior of things will be even before they are made or observed in service. Moreover, technology often suggests new kinds of behavior that had not even been imagined before, and so leads to strategies of design, to solve practical problems.

Although the development of hunting weapons can be considered a key event in the evolution of human culture, harnessing the energy of fire was probably the most seminal event of human history. This, more than any other event, assisted humans in their exploitation of colder, more northerly ecosystems.

The principal energy sources of antiquity were all derived directly from the sun: human and animal muscle power, wood, flowing water, and wind. In the mid-to-late 18th century the industrial revolution began with stationary wind-powered and water-powered technologies, which were essentially replaced by fossil hydrocarbon derivatives: coal in the 19th century, oil since the 20th century, and now, increasingly, natural gas (Speight, 2014, 2017). Furthermore, hydrocarbon-based energy has a strong connection with economic activity for industrialized and developing economies (Hall et al., 2001; Tharakan et al., 2001).

Technology provides the *raison d'être* of science and engineering. Technology is essential to science and engineering for purposes of measurement, data collection, treatment of samples, computation, transportation to research

TABLE 1.5 Simple hydrocarbons and the variations.

Number of C atoms	Alkane (single bond)	Alkene (double bond)	Alkyne (triple bond)	Cycloalkane	Alkadiene
1	Methane	—	—	—	—
2	Ethane	Ethylene (ethene)	Acetylene (ethyne)	—	—
3	Propane	Propylene (propene)	Propyne	Cyclopropane	Propadiene
4	Butane	Butylene (butene)	Butyne	Cyclobutane	Butadiene
5	Pentane	Pentene	Pentyne	Cyclopentane	Pentadiene
6	Hexane	Hexene	Hexyne	Cyclohexane	Hexadiene
7	Heptane	Heptene	Heptyne	Cycloheptane	Heptadiene
8	Octane	Octene	Octyne	Cyclooctane	Octadiene
9	Nonane	Nonene	Nonyne	Cyclononane	Nonadiene
10	Decane	Decene	Decyne	Cyclodecane	Decadiene
11	Undecane	Undecene	Undecyne	Cycloundecane	Undecadiene
12	Dodecane	Dodecene	Dodecyne	Cyclododecane	Dodecadiene

sites, sample collection, protection from hazardous materials, and communication. More and more new instruments and techniques are being developed through technology that makes it possible to advance various lines of scientific research.

However, technology does not just provide tools for science; it also may provide motivation and direction for theory and research. Scientists and engineers see patterns in phenomena to make the world as understandable as possible and being able to be manipulated. Technology also pushes scientists and engineers to show that theories fit the data and to show logical proof of abstract connections as well as demonstrable designs that work.

Technology affects the social system and culture with immediate implications for the success or failure of human enterprises and for personal benefit and harm. Technological decisions, whether in designing an irrigation system or a crude oil recovery project, inevitably involve social and personal values as well as scientific and engineering judgments. This leads to the issues regarding the supply of hydrocarbon derivatives (in the form of crude oil and natural gas) and the future of these valuable chemicals (Speight, 2014, 2017).

The rumors of the death of the hydrocarbon culture are greatly exaggerated (to paraphrase Mark Twain who observed, "The rumors of my death are greatly exaggerated"). The world is not about to run out of hydrocarbon derivatives, and perhaps it is not going to run out of crude oil or natural gas from unconventional sources any time soon. However, cheap crude oil will be difficult to obtain because the reserves that remain are not only difficult to recover but the crude oil is a low-grade and will be more difficult (costly) to refine to produce the desired hydrocarbon fuels (Speight, 2010, 2014).

As conventional oil becomes less important, it is important to invest in a different source of energy, one freeing us for the first time from our dependence on hydrocarbon derivatives (Speight, 2008). However, renewable energy technologies require further development but some do show advantages over hydrocarbon derivatives in terms of economic reliability, accessibility, and environmental benefits. With proper attention to environmental concerns, biomass-based energy generation is competitive in some cases relative to conventional hydrocarbon-based energy generation. By contrast, liquid-fuel production from grain and solar thermal power has a relatively low economic return on investment. But is does depend on the investment required to keep a fleet on alert offshore of various oil producing countries as well as the willingness of the population to pay an additional per gallon of gasoline or per gallon of fuel oil amount for a higher measure of energy independence.

Government intervention, in concert with ongoing private investment, will speed up the process of sorting the wheat from the chaff in the portfolio of feasible renewable energy technologies. It is time to think about possibilities other than the next cheapest hydrocarbon derivatives. If for no other reason than to protect the environment, all of the available technologies should be brought to bear on this task.

5. Hydrocarbons

A hydrocarbon is an organic compound consisting of carbon and hydrogen only. The inclusion of any atom other than carbon and hydrogen disqualified the compound from being considered as a hydrocarbon. The majority of hydrocarbon derivatives found naturally occur in crude oil (crude oil) and natural gas, where decomposed organic matter provides an abundance of many individual varieties of hydrocarbon derivatives. Hydrocarbon derivatives are the simplest organic compounds—they can be straight-chain, branched chain, or cyclic molecules (Fig. 1.4). Nevertheless, in spite of the variations in molecular structure of the various hydrocarbon derivatives, there are five specific families for hydrocarbon derivatives: (i) alkanes, (ii) alkenes, (iii) alkynes, (iv) cycloalkanes, and (v) aromatic hydrocarbon derivatives, also called *arenes*.

1. *Alkanes* (*paraffins*) are saturated hydrocarbon derivatives in which all of the four valence bonds of carbon are satisfied by hydrogen or by another carbon. Alkanes can have straight or branched chains, but without any ring structure.
2. *Alkenes* (*olefins*) are unsaturated hydrocarbon derivatives insofar as not all of the carbon valencies are satisfied by another atom and have a double bond

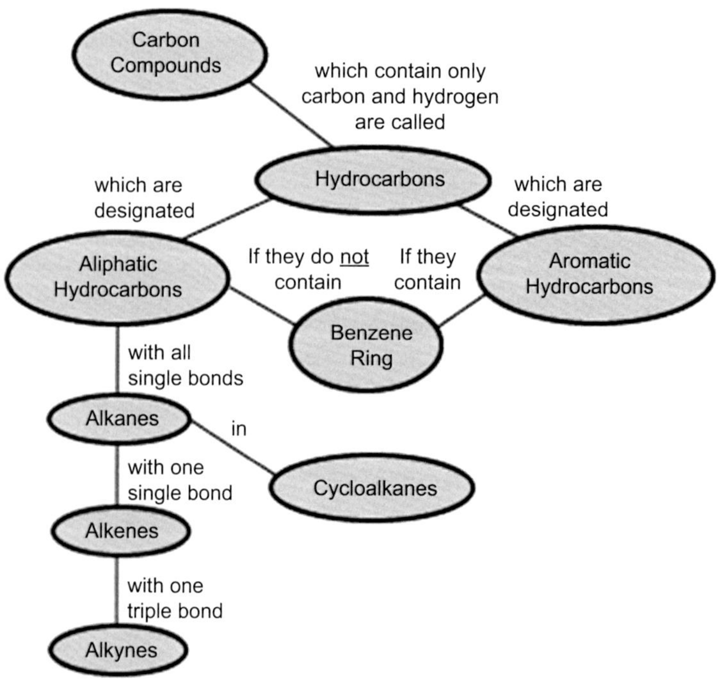

FIGURE 1.4 Types of hydrocarbons and their interrelationship.

(C=C) between carbon atoms. Alkenes have the general formula C_nH_{2n}, assuming there are no ring structures in the molecule. Alkenes may have more double than one double bond between carbon atoms, in which case the formula is reduced by two hydrogen atoms for each additional double bond. For example, an alkene with two double bonds in the molecule has the general formula C_nH_{2n}. Because of their reactivity and the time involved in crude oil maturation, alkenes do not usually occur in crude oil.

3. *Alkyne derivatives* (*acetylene derivatives*) are hydrocarbon derivatives which contain a triple bond (CC) and have the general formula C_nH_{2n-2}. Acetylene hydrocarbon derivatives are highly reactive and, as a consequence, are very rare in crude oil.
4. *Cycloalkane derivatives* (*naphthene derivatives*) are saturated hydrocarbon derivatives containing one or more rings, each of which may have one or more paraffinic side chains (more correctly known as *alicyclic hydrocarbon derivatives*). The general formula for a saturated hydrocarbon containing one ring is C_nH_{2n}.
5. *Aromatic hydrocarbon derivatives (arenes* or *arene derivatives*) are hydrocarbon derivatives containing one or more aromatic nuclei, such as benzene, naphthalene, and phenanthrene ring systems, which may be linked up with (substituted) naphthene rings and/or paraffinic side chains.

5.1 Bonding in hydrocarbons

Since carbon adopts the tetrahedral geometry when there are four σ bonds, only two bonds can occupy a plane simultaneously. The other two bonds are directed to the rear or to the front of the plane. In order to represent the tetrahedral geometry in two dimensions, solid wedges are used to represent bonds pointing out of the plane of the drawing toward the viewer, and dashed wedges are used to represent bonds pointing out of the plane or to the rear of the plane.

For example, in a representation of the methane molecule, the hydrogen connected by a solid wedge points to the front of the plane and the hydrogen connected by the dashed wedge points to the rear of the paper while the two hydrogen joined by solid single lines are in the plane (of the paper in this case):

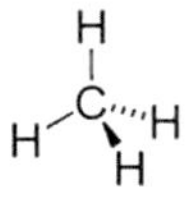

Fortunately, while there is the need to understand such stereochemistry (the existence of molecules in space), hydrocarbon derivatives can be represented in a shorthand notation called a *skeletal structure*.

In a skeletal structure, only the bonds between carbon atoms are represented. Individual carbon and hydrogen atoms are not drawn, and bonds to hydrogen are not drawn. In the case that the molecule contains just single bonds (sp^3 bonds), these bonds are drawn in a *zigzag* fashion. This is because in the tetrahedral geometry all bonds point as far away from each other as possible and the structure is not linear. For example:

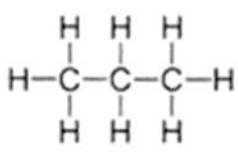

Structure of propane

Skeletal structure of propane

Only the bonds between carbons have been drawn, and these have been drawn in a *zigzag* manner and there is no evidence of hydrogen atoms in a skeletal structure. Since, in the absence of double or triple bonds, carbon makes four bonds total, the presence of hydrogens is implicit. Whenever an insufficient number of bonds to a carbon atom are specified in the structure, it is assumed that the rest of the bonds are made to hydrogens.

For example, if the carbon atom makes only one explicit bond, there are three hydrogens implicitly attached to it. If it makes two explicit bonds, there are two hydrogens implicitly attached, etc. Two lines are sufficient to represent three carbon atoms. It is the bonds only that are being drawn out, and it is understood that there are carbon atoms (with three hydrogens attached to each) at the terminal ends of the structure.

5.2 Nomenclature

The large number of organic compounds identified with each passing day, together with the fact that many of these compounds are isomers of other compounds, requires that a systematic nomenclature system be developed. Just as each distinct compound has a unique molecular structure which can be designated by a structural formula, each compound must be given a characteristic and unique name. As organic chemistry developed, many compounds were given trivial names, which are now commonly used and recognized.

The IUPAC (International Union of Pure and Applied Chemistry) systematic approach to nomenclature is a rational nomenclature system that should do at least two things. First, it should indicate how the carbon atoms of a given compound are bonded together in a characteristic lattice of chains and rings. Second, it should identify and locate any functional groups present in the compound. Since hydrogen is such a common component of organic compounds, its amount and locations can be assumed from the tetravalency of carbon, and need not be specified in most cases.

The IUPAC nomenclature system is a set of logical rules devised and used by organic chemists to circumvent problems caused by arbitrary nomenclature. Knowing these rules and given a structural formula, one should be able to write a unique name for every distinct compound. Likewise, given an IUPAC name, one should be able to write a structural formula. In general, an IUPAC name will have three essential features: (i) a root or base indicating a major chain or ring of carbon atoms found in the molecular structure, (ii) a suffix or other element(s) designating functional groups that may be present in the compound, and (iii) the names of substituent groups, other than hydrogen, that complete the molecular structure.

As defined by the IUPAC nomenclature of organic chemistry, the classifications for hydrocarbons are:

1. Saturated hydrocarbons are the simplest of the hydrocarbon species. They are composed entirely of single bonds and are saturated with hydrogen. The formula for acyclic saturated hydrocarbons (alkanes) is C_nH_{2n+2} (Table 1.5). The most general form of saturated hydrocarbons is $C_nH_{2n+2(1-r)}$, where r is the number of rings. Those with exactly one ring are the cycloalkane derivatives. Saturated hydrocarbons are the basis of fuels from crude oil and are found as either linear or branched species. Substitution reaction is their characteristic property (such as chlorination to form chloroform). Hydrocarbon derivatives with the same molecular formula but different structural formula are called structural isomers. The number of structural isomers increases phenomenally with the number of carbon atoms in the molecule (Table 1.6)
2. Unsaturated hydrocarbon derivatives have one or more double bonds (>C=C<, alkenes) or triple bonds (>C≡C<, alkynes) between carbon atoms. Those with one double bond have the formula C_nH_{2n} (assuming the absence of cyclic structures). Those containing one triple bond have the formula C_nH_{2n-2}.
3. Aromatic hydrocarbon derivatives (also known as arenes) are hydrocarbon derivatives that have at least one aromatic ring in the molecule—typically an aromatic ring placed so that it (the ring) dictates the reactivity and behavior of the molecule.

Hydrocarbon derivatives can be gases (such as methane, ethane, propane, and butane) liquids (such as, for example, pentane, hexane, and benzene), waxes or low melting solids or polymers (such as polyethylene, polypropylene, and polystyrene).

5.2.1 Alkanes

Alkanes are named using a prefix for the number of carbon atoms they contain, followed by the suffix *-ane* (Table 1.5). When one of the hydrogen atoms is replaced by another nonhydrogen atom or nonhydrocarbon group, the atom or group which replaces the hydrogen or carbon is called a *substituent*.

TABLE 1.6 Increase in the number of potential isomers with carbon number.

Number of carbon atoms	Number of potential isomers
1	1
2	1
3	1
4	2
5	3
6	5
7	9
8	18
9	35
10	75
15	4347
20	366,319
25	36,797,588
30	4,111,846,763
40	62,491,178,805,831

For example, when one of the hydrogen atoms in pentane is replaced by a methyl group, the resulting molecule must be named for identification:

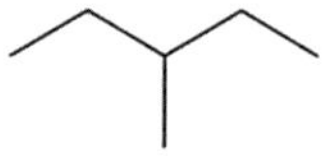

There are a set of rules to name such a molecule:

1. Identify the longest chain of carbon atoms. This is the alkane that serves as the root name for the molecule. In the example above, the root name is pentane.
2. Number the carbon atoms, starting at the end that gives the substituent the lowest number. In the example above, counting can commence from either end and arrive at 3 for the substituent.
3. The substituent is named as if it is an independent alkane but the suffix *-ane* is replaced with *yl*, which will serve as the prefix. In the example above, methane is the substituent, so it is called *methyl.*
4. The compound is named number-prefix-root name and the molecule is named 3-methyl pentane

If the alkane has more than one substituent, the rules above are followed, and the carbons on the longest chain are numbered to give the lowest number possible to one of the substituent. The substituents are then all named in the prefix (e.g., 2-ethyl, 3-methyl). If more than one substituent is attached to the same carbon atom, the number of that carbon atom is repeated to indicate the number of substituents and the prefixes di- (ii) or tri- (iii) are used. If there is more than one substituent on different carbon atoms, the prefixes are ordered alphabetically (e.g., ethene before methane). The prefixes di- and tri-are ignored when considering alphabetical order. Thus:

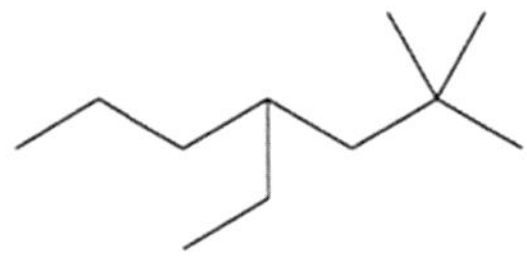

4-ethyl-2,2-dimethyl heptane

The longest carbon chain has seven carbon atoms, so the root name is heptane. Numbering from the right gives the lowest number to the first substituent. There are two methyl substituents at the second carbon atom, so the prefix 2,2-dimethyl is used. There is another substituent on the fourth carbon atom, so the prefix ethyl is used. Ethyl comes before methyl alphabetically, hence: 4-ethyl-2,2-dimethyl heptane.

5.2.2 Alkenes

Alkenes are named using the same general naming rules for alkanes, except that the suffix is *ene*. There are a few other small differences:

1. The main chain of carbon atoms must contain both carbons in the double bond. The main chain is numbered so that the double bond gets the smallest number.
2. Before the root name, the number of the carbon atom at which the double bond starts (the smaller number) is written.
3. If more than one double bond is present, prefixes such as di-, tri-, tetra- are used before the *ene*.

5.2.3 Alkynes

Alkynes are named using the same general procedure used for alkenes, replacing the suffix with *yne*. If a molecule contains both a double and a triple bond, the carbon chain is numbered so that the first multiple bond gets a lower number. If both bonds can be assigned the same number, the double bond takes precedence. The molecule is then named "n-ene-n-yne," with the double bond root name preceding the triple bond root name (e.g., 2-heptene-4-yne).

5.2.4 Cycloalkanes

Alkanes exist as linear and branched structures (above) and also as ring structures (cycloalkanes):

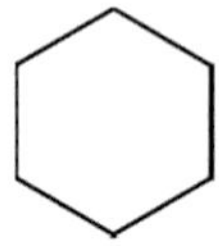

Skeletal structure of cyclohexane

Stable cycloalkanes cannot be formed with carbon chains of any length since carbon adopts the sp^3 tetrahedral geometry in which the angles between bonds are 109.5 degrees. For certain cycloalkanes, the angle between bonds must deviate from this ideal angle (*angle strain*, *bond strain*). In addition, some hydrogen atoms may come into closer proximity with each other than is desirable (become eclipsed) (*torsional strain*).

These destabilizing effects, which compromise *ring strain,* are evident in the lower molecular weight cycloalkanes such as cyclopropane and cyclobutane because the bond angles deviate substantially from 109.5 degrees and the hydrogen atoms tend to eclipse each other.

On the other hand, cyclopentane is a more stable molecule with a small amount of ring strain, while *cyclohexane* is able to adopt the perfect geometry for a cycloalkane in which all of the bond angles are the ideal 109.5 degrees and none of the hydrogen atoms are eclipsed—the molecule has no ring strain at all. Cycloalkanes larger than cyclohexane have ring strain and are not commonly encountered in organic chemistry.

Most of the time, cyclohexane adopts the fully staggered, ideal angle *chair conformation* in which the carbon-carbon bonds exist with the substituents in the staggered conformation and possess the ideal bond angle of 109.5 degrees.

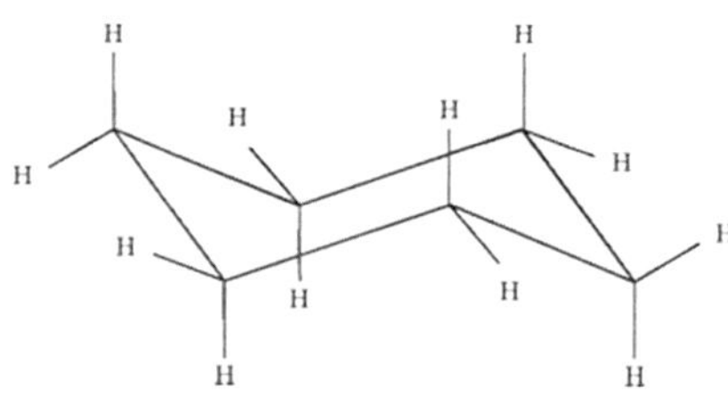

Cyclohexane (chair conformation)

In the chair conformation, the hydrogen atoms are labeled according to their location. Those hydrogens which exist above or below the plane of the molecule (shown with red bonds above) are called *axial* while those hydrogens which exist in the plane of the molecule are called *equatorial*.

Although the chair conformation is the most stable conformation that cyclohexane can adopt, there is enough thermal energy for it to also pass through less favorable conformations before returning to a different *conformation*. When it does so, the axial and equatorial substituents change places. The passage of cyclohexane from one chair conformation to another occurs when the axial substituents switch places with the equatorial substituents (a *ring flip*).

5.2.5 Aromatic hydrocarbons

The aromatic system is a *conjugated system* which contains a series of alternating single and double bonds in which there is a *p* orbital on each atom. Owing to resonance, in a conjugated system of alternating bonds, the double and single bonds are able to switch places, producing an overall more stable structure. Conjugated systems can also exist in cyclic molecules.

The classic example of an aromatic system involves a six-membered ring (benzene) and there are two possible chemical structures for a conjugated six-membered benzene ring:

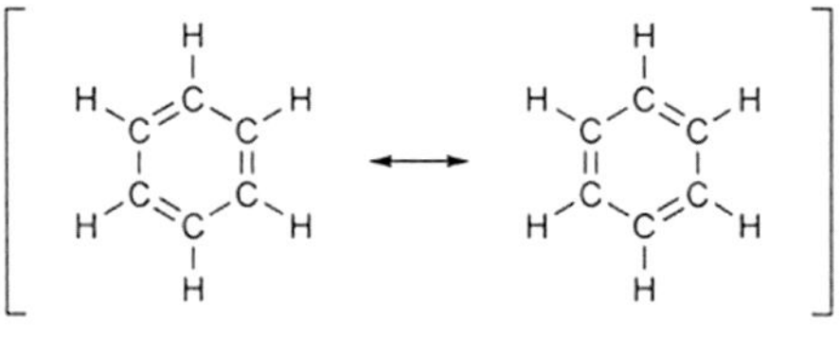

Resonance structures of benzene

In an aromatic system like benzene, each atom has a *p* orbital, the electrons of which are delocalized about the system. Benzene and other aromatic compounds exhibit chemistry very different from ordinary, nonaromatic hydrocarbon derivatives (*aliphatic hydrocarbon derivatives*).

Benzene and other aromatic compounds can have substituents. When benzene itself is a substituent, it is called a *phenyl group*. Benzene is typically drawn in such a way that the hybrid between the resonance structures is emphasized:

However, not every conjugated cyclic system is aromatic since not all are stabilized by resonance, mainly due to differences in filling molecular orbitals with electrons.

Benzene is obviously an unsaturated hydrocarbon because it has far less hydrogen than the equivalent saturated hydrocarbon: C_6H_{14}. But benzene is too stable to be an alkene or alkyne. Alkenes and alkynes rapidly add Br_2 to the C=C or CC bonds, whereas benzene only reacts with bromine in the presence of a catalyst: $FeBr_3$. Furthermore, when benzene reacts with Br_2 in the presence of $FeBr_3$, the product of this reaction is a compound in which a bromine atom has been *substituted* for a hydrogen atom, not *added* to the compound the way an alkene adds bromine.

$$C_6H_6 + Br_2 \rightarrow C_6H_5Br + HBr$$

Other compounds were eventually isolated from coal that had similar properties. Their formulas suggested the presence of multiple C=C bonds, but these compounds were not reactive enough to be alkenes.

The structure of benzene was a recurring problem throughout most of the 19th century. The first step toward solving this problem was taken by Friedrich August Kekulé in 1865. (The interest shown by Kekulé in the structure of organic compounds may have resulted from the fact that he first enrolled at the University of Giessen as a student of architecture.) One day, while dozing before a fire, Kekulé dreamed of long rows of atoms twisting in a snakelike motion until one of the snakes seized hold of its own tail. This dream led Kekulé to propose that benzene consists of a ring of six carbon atoms with alternating C−C single bonds and C=C double bonds. Because there are two ways in which these bonds can alternate, Kekulé proposed that benzene was a mixture of two compounds in equilibrium.

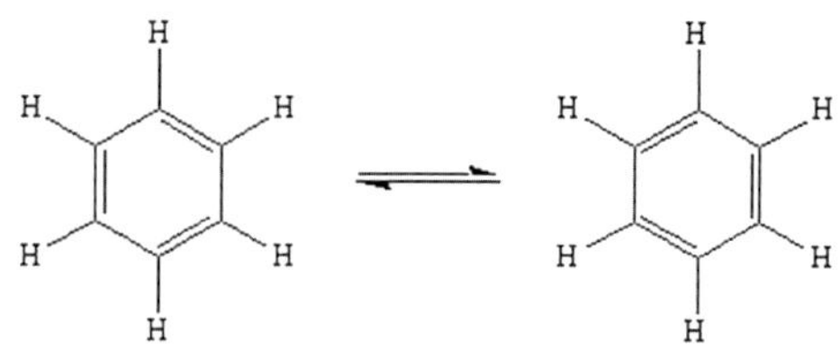

Kekule's explanation of the structure of benzene

The structure devised by Kekulé explained the formula of benzene, but it did not explain why benzene failed to behave like an alkene. The unusual stability of benzene wasn't understood until the development of the theory of resonance. This theory states that molecules for which two or more satisfactory Lewis structures can be drawn are an average, or hybrid, of these structures. Benzene, for example, is a resonance hybrid of the two Kekulé structures.

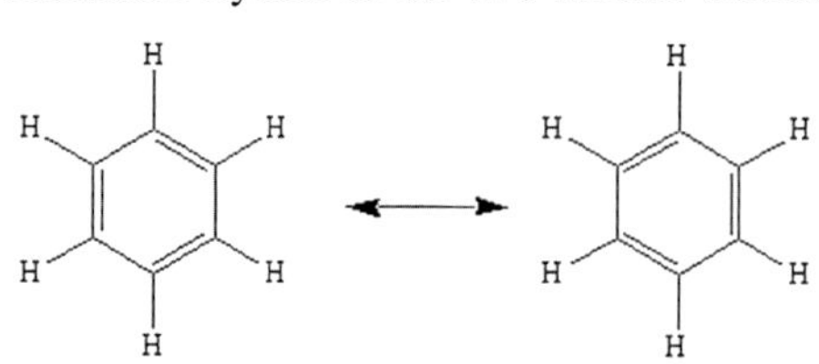

The resonance structure for benzene

The difference between the equilibrium and resonance descriptions of benzene is subtle, but important. In the equilibrium approach, a pair of arrows is used to describe a reversible reaction, in which the molecule on the left is converted into the one on the right, and vice versa. In the resonance approach, a double-headed arrow is used to suggest that a benzene molecule does not shift back and forth between two different structures; it is a hybrid mixture of these structures.

One way to probe the difference between the Kekulé idea of equilibrium between two structures and the resonance theory in which benzene is a hybrid mixture of these structures would be to study the lengths of the carbon-carbon bonds in benzene. If this idea was correct, we would expect to find a molecule in which the bonds alternate between relatively long carbon-carbon (C—C) single bonds (0.154 nm) and significantly shorter carbon-carbon (C=C) double bonds (0.133 nm). When benzene is cooled until it crystallizes, and the structure of the molecule is studied by X-ray diffraction, we find that the six carbon-carbon bonds in this molecule are the same length (0.1395 nm). The crystal structure of benzene is therefore more consistent with the resonance model of bonding in benzene than the original Kekulé structures.

The resonance theory does more than explain the structure of benzene —. It also explains why benzene is less reactive than an alkene. The resonance theory assumes that molecules that are hybrids of two or more Lewis structures are more stable than those that aren't. It is this extra stability that makes benzene and other aromatic derivatives less reactive than normal alkenes. To emphasize the difference between benzene and a simple alkene, many chemists replace the Kekulé structures for benzene and its derivatives with an aromatic ring in which the circle in the center of the ring indicates that the electrons in the ring are delocalized; they are free to move around the ring.

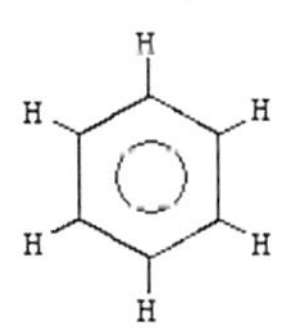

The significance of the circle in the center of this aromatic ring in which each of the carbon atoms is sp^2 hybridized. This leaves one electron in a $2p$ orbital on each of the six carbon atoms.

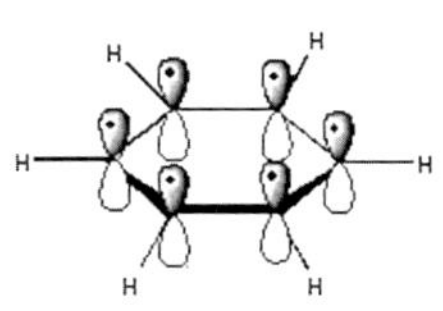

It is this delocalization of electrons around the aromatic ring that is conveyed by the circle that is often written inside the ring. It is also the delocalization of electrons that makes benzene less reactive than a simple alkene.

6. Isomers

Alkanes with more than three carbon atoms can be arranged in numerous different ways, forming different structural isomers. An isomer is like a chemical anagram, in which the atoms of a chemical compound are arranged or joined together in a different order. The simplest isomer of an alkane is the one in which the carbon atoms are arranged in a single chain with no branches. This isomer is sometimes called the *n-isomer* (n for "normal," although it is not necessarily the most common). However, the chain of carbon atoms may also be branched at one or more points. The number of possible isomers increases rapidly with the number of carbon atoms. For example:

C1: 0 isomers: methane
C2: 0 isomers: ethane
C3: 0 isomers: propane
C4: 2 isomers: n-butane, iso-butane
C5: 3 isomers: pentane, iso-pentane, neo-pentane
C6: 5 isomers: hexane
C12: 355 isomers
C32: 27,711,253,769 isomers
C60: 22,158,734,535,770,411,074,184 isomers, many of which are only on-paper isomers and do not exist.

Branched alkanes can be chiral: 3-methylhexane and its higher homologues are chiral due to their stereogenic center at carbon atom number 3. In addition to these isomers, the chain of carbon atoms may form one or more loops (cycloalkanes).

In the benzene system, there are three ways in which a pair of substituents can be placed on an aromatic ring. In the *ortho* (*o*) isomer, the substituents are in adjacent positions on the ring. In the *meta* (*m*) isomer, they are separated by one carbon atom. In the *para* (*p*) isomer, they are on opposite ends of the ring, as for example in the isomers of dimethyl benzene (xylene):

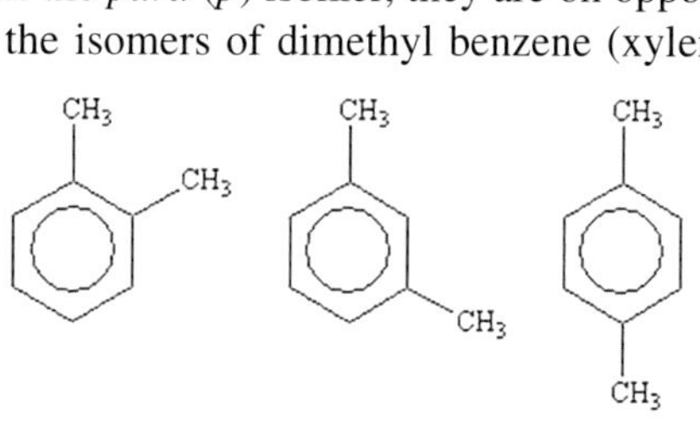

Ortho-xylene o-xylene | Meta-xylene m-xylene | Para Xylene p-xylene

7. Nonhydrocarbons

Atoms of other elements can be joined to the carbons in place of one or more hydrogens. Oxygen, nitrogen, and the halogens are the most common atoms that replace hydrogens. The resulting compound is called a *substituted* hydrocarbon. Sometimes a combination of two of these other elements will be found in place of hydrogens. These other elements give rise to what are called *functional groups*. The presence of different functional groups (Fig. 1.5, Fig. 1.6) causes the substituted hydrocarbon to be one of several classes of organic compounds and it is (in the absence of steric factors) the functional group that dictates the reactivity and behavior of the molecule.

While the subject of this text is the hydrocarbons, a general description of the nonhydrocarbons is worthy of a brief description here as an aid to understanding the structure and properties of the corresponding nonhydrocarbon derivatives and their relation to the various hydrocarbons.

7.1 Acids

Organic acids contain the carboxyl group (-COOH) and the presence of one or more of these groups, therefore, causes the compound to be acid in nature. Among the simplest examples are the formic acid (HCOOH) that occurs in ants, and acetic acid (CH_3COOH) that gives vinegar its sour taste.

Acids with two or more carboxyl groups are called *dicarboxylic*, *tricarboxylic*, signifying the presence of two or three acid groups. The simplest dicarboxylic example is oxalic acid (HOOC−COOH), which has two bonded carboxylic acid groups.

Carboxylic acids are the most common type of organic acids. When the carboxyl group is deprotonated, the conjugate base is resonance stabilized increasing its stability—this causes carboxylic acids to be more acidic than alcohols.

7.2 Alcohols

The most common of these functional groups is the hydroxyl (−OH) and an aliphatic hydrocarbon that has one hydroxyl group attached to a carbon is called an alcohol. The simplest alcohol is methyl alcohol, or methanol (CH_3OH).

In the more complicated molecules, the hydroxyl group can be attached to either end carbon (same compound in either case) or to the middle compound,

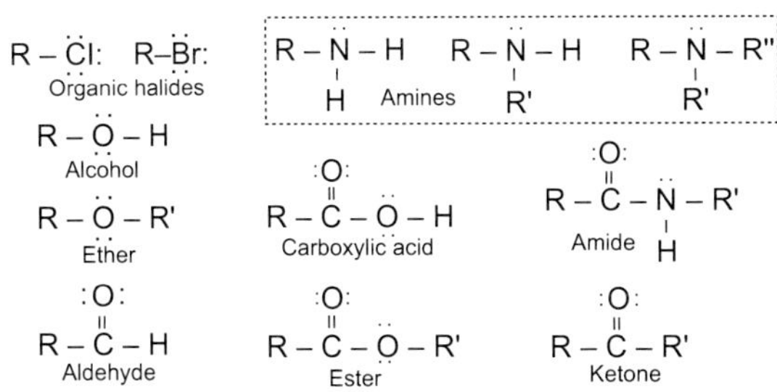

FIGURE 1.5 Various types of functional groups.

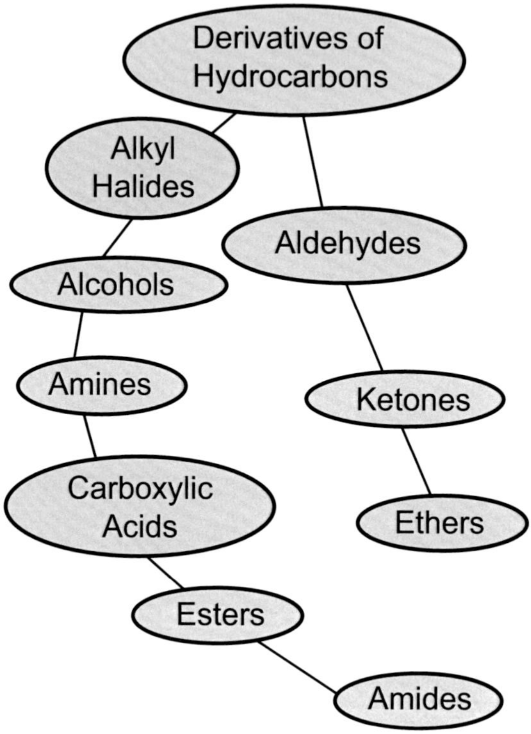

FIGURE 1.6 Different classes of organic compounds derived from hydrocarbons.

which produces a slightly different compound. The first case represents the compound n-propanol (normal propanol or 1-propanol, C_3H_7OH). The use of the number indicates the position to which the hydroxyl is attached. The second carbon placement of the hydroxyl group gives rise to the name 2-propanol, or, using an older system of naming compounds, iso-propanol, or iso-propyl alcohol. This is probably the type of rubbing alcohol in the medicine cabinet in your homes. More specific formulas are: $CH_3CH_2CH_2OH$ (1-propanol) and $CH_3CHOHCH_3$ (2-propanol).

7.3 Aldehydes

A new class of substituted hydrocarbon derivatives arises when an oxygen atom is double bonded to the carbon at the end of the chain. In this case there are two less hydrogen atoms, so instead of three end hydrogens, there is the C=O and only one hydrogen. The simplest aldehyde is formaldehyde (CH_2O) – the IUPAC name is methanal. These compounds have the generic formula H–R=O.

Aldehydes have properties that are diverse and which depend on the remainder of the molecule. Smaller aldehydes are more soluble in water, formaldehyde and acetaldehyde completely so. The volatile aldehydes have pungent odors. Aldehydes degrade in air via the process of *autoxidation*.

Both of the important aldehydes, formaldehyde and acetaldehyde, have complicated behavior because of their tendency to oligomerize or polymerize. They also tend to hydrate in the presence of water, forming the germinal diol. These properties are often not appreciated because the oligomers/polymers and the hydrates exist in equilibrium with the parent aldehyde.

7.4 Amides

An *amide* is usually an organic compound that contains a functional group consisting of an acyl group (R—C=O) linked to a nitrogen atom:

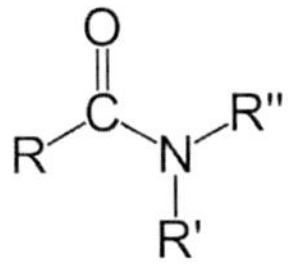

The simplest amides are derivatives of ammonia (NH_3) in which one hydrogen atom has been replaced by an acyl group. Closely related and even more numerous are amides derived from primary amines (R'NH_2) with the formula RC(O)NHR'. Amides are regarded as derivatives of carboxylic acids in which the hydroxyl group has been replaced by an amine or ammonia.

In the typical nomenclature, the term *amide* is added to the stem of the parent acid's name— the simplest amide derived from acetic acid is *acetamide* (CH_3CONH_2). When the amide is derived from a primary or secondary amine, the substituents on nitrogen are indicated first in the name. Thus the amide formed from dimethylamine and acetic acid is *N,N*-dimethylacetamide (CH_3CONMe_2, where Me = CH_3). Cyclic amides are called lactams and are necessarily secondary or tertiary amides.

Compared to amines, amides are very weak bases and do not have clearly defined acid-base properties in water. On the other hand, amides are much stronger bases than esters, aldehydes, and ketones.

7.5 Amines

The generic formula for amines is R—NH_2, where one hydrogen has been replaced by an amino group (-NH_2). The simplest amine is methyl amine, where the "R" group is methyl. This kind of amine is called a *primary amine*. There can also be *secondary amines* and *tertiary amines*, with the generic formulas R_2-NH and R_3-N, respectively, with a second, and a third hydrogen replaced with an "R" group. The "R" groups can all be the same, or they can be different.

Primary amines arise when one of three hydrogen atoms in ammonia (NH_3) is replaced by an alkyl group. Important primary alkyl amines include methylamine (CH_3NH_2) and ethanolamine (2-aminoethanol, $H_2NCH_2CH_2OH$).

Secondary amines have two alkyl substituents bound to nitrogen in addition to the single hydrogen atom. Important representatives include dimethylamine [$(CH_3)_2NH$] and methyl ethanolamine ($CH_3HNCH_2CH_2OH$). In tertiary amines, all three hydrogen atoms are replaced by organic substituents. Examples include trimethylamine [$(CH_3)_3N$], which has a distinctively *fishy* odor.

Primary amine	Secondary amine	Tertiary amine
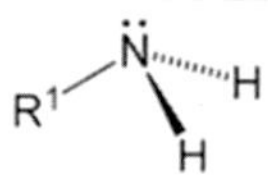	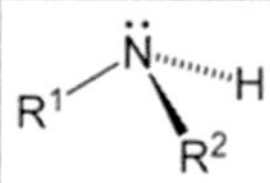	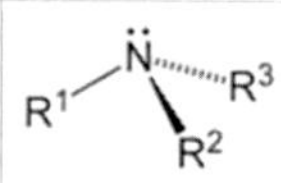

It is also possible to have four alkyl substituents on the nitrogen. These compounds [$R_4N^+X^-$] are not amines but are *quaternary ammonium compounds* and have a positively charged nitrogen center and a negatively charge ion (anion).

7.6 Esters

Esters have the generic formula, R–COO–R′, which is similar to that of the organic acid, but the H of the –COOH has been replaced by a hydrocarbon group. The ending of the name of an ester is *ate*, such as in ethyl acetate.

Esters are usually derived from an inorganic acid or organic acid in which at least one hydroxyl (OH) group is replaced by an alkoxy (-O-alkyl) group, and most commonly from carboxylic acids and alcohols.

Esters occur widely in nature—many naturally occurring fats and oils are the fatty acid esters of glycerol. Esters with low molecular weight are commonly used as fragrances and found in essential oils and pheromones. Nitrate esters, such as nitroglycerin, are known for their explosive properties, while polyesters are important plastics, with monomers linked by ester moieties.

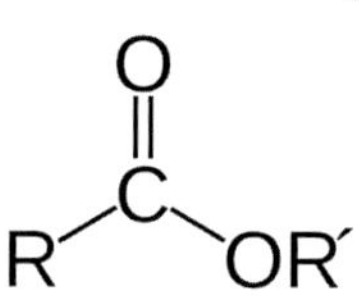

A carboxylic acid ester. R and R′ denote any alkyl or aryl (aromatic) group.

7.7 Ethers

An ether results when there is an oxygen atom between two carbons in the chain. The simplest is dimethyl ether (CH_3OCH_3), which has the same molecular formula as ethanol (C_2H_6O).

The names for simple ethers (i.e., those with none or few other functional groups) are a composite of the two substituents followed by *ether*, e.g., Methyl

ethyl ether ($CH_3OC_2H_5$), diphenyl ether ($C_6H_5OC_6H_5$). The rules of the IUPAC are often not followed for simple ethers. As for other organic compounds, very common ethers acquired names before rules for nomenclature were formalized. Diethyl ether is simply called *ether*, but was once called *sweet oil of vitriol*. Methyl phenyl ether is *anisole*, because it was originally found in aniseed. The aromatic ethers include furans. Acetals (α-alkoxy ethers R–CH(OR)OR) are another class of ethers with characteristic properties.

In the IUPAC system of nomenclature, which is rarely encountered, ethers are named using the general formula *alkoxy alkane*, for example, $CH_3CH_2OCH_3$ is *methoxy ethane*. If the ether is part of a more complex molecule, it is described as an alkoxy substituent, so $-OCH_3$ would be considered a *methoxy* group. The simpler alkyl radical is written in front, so $CH_3OCH_2CH_3$ would be given as *methoxy* (CH_3O) *ethane* (CH_2CH_3).

7.8 Halides

A halide, specifically an alkyl halide is another name for a halogen-substituted alkane. The carbon atom, which is bonded to the halogen atom, has sp^3 hybridized bonding orbitals and exhibits a tetrahedral shape. Due to electronegativity differences between the carbon and halogen atoms, the σ covalent bond between these atoms is polarized, with the carbon atom becoming slightly positive and the halogen atom partially negative.

Halogen atoms increase in size and decrease in electronegativity as we go down the family in the periodic table. Therefore, the bond length between carbon and halogen becomes longer and less polar as the halogen atom changes from fluorine to iodine.

Alkyl halides are named using the IUPAC rules for alkanes. Naming the alkyl group attached to the halogen and adding the inorganic halide name for the halogen atom creates common names.

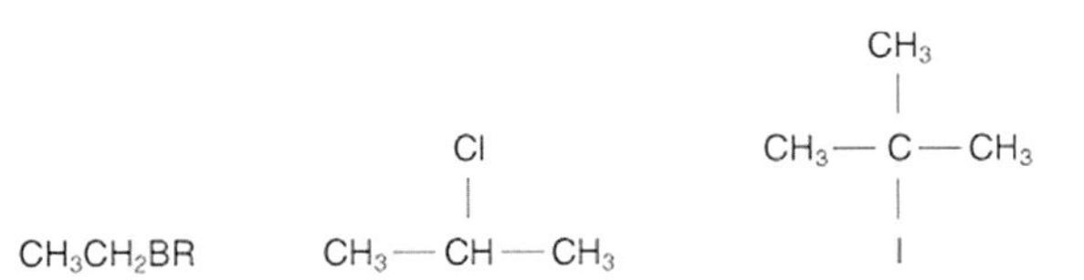

bromoethane (ethyl bromide) 2-chloropropane (isopropyl chloride) 2-iodo-2-methylpropane (tertiary-butyliodide)

7.9 Ketones

A different class of organic compounds results if the C=O occurs somewhere along the chain other than on the end carbon. The simplest ketone has three carbons and has the common name acetone. The more correct name is dimethyl ketone or propanone. The generic formula for ketones is RC = O(R′).

The carbonyl group (C=O) is polar as a consequence of the fact that the electronegativity of the oxygen center is greater than that for carbonyl carbon. Thus, ketones are nucleophilic at oxygen and electrophilic at carbon.

The carbonyl group interacts with water by hydrogen bonding and ketones are typically more soluble in water than the related methylene compounds. Ketones are hydrogen-bond acceptors. Ketones are not usually hydrogen-bond donors and cannot hydrogen-bond to themselves. Because of their inability to serve both as hydrogen-bond donors and acceptors, ketones tend not to *self-associate* and are more volatile than alcohols and carboxylic acids of comparable molecular weight. These factors relate to pervasiveness of ketones in perfumery and as solvents.

8. Properties of hydrocarbons

The properties of hydrocarbon derivatives are varied and depend upon the molecular structure and also on the three-dimensional structure. The individual properties are presented in more detail elsewhere in this book (Chapter 9) but it is appropriate to briefly mention here an introduction to the properties of hydrocarbon derivatives using the hydrocarbon derivatives isolated from natural gas as examples.

In contrast to many inorganic materials, organic compounds typically melt and many boil. In earlier times, the melting point (m.p.) and boiling point (b.p.) provided crucial information on the purity and identity of organic compounds. The melting and boiling points correlate with the polarity of the molecules and their molecular weight. Some organic compounds, especially symmetrical ones, sublime, that is, they evaporate without melting. Organic compounds are usually not very stable at temperatures above 300°C (570°F), although some exceptions exist.

Because of differences in molecular structure, the empirical formula remains different between hydrocarbon derivatives. In linear alkanes, alkenes, and alkynes, the amount of bonded hydrogen lessens in alkenes and alkynes due to the self-bonding of carbon preventing entire saturation of the hydrocarbon by the formation of double or triple bonds.

The composition of natural gas varies depending on the field, the formation, or the reservoir from which it is extracted and that are an artifact of its formation (Mokhatab et al., 2006; Speight, 2014). Because of this variability of composition, the properties of unrefined natural gas are also variable. Therefore, the properties and behavior of natural gas are best understood by investigating the properties and behavior of the constituents.

Thus, assuming that the natural gas has been cleaned (i.e., any constituents such as carbon dioxide and hydrogen sulfide) have been removed and the only constituents remaining are hydrocarbon derivatives, the properties and behavior of natural gas become a study of the properties and behavior of the relevant hydrocarbon derivatives (Speight, 2017).

The different hydrocarbon derivatives that form natural gas can be separated using their different physical properties as weight, boiling point, or vapor pressure (Chapter 4). Depending on its content of higher molecular weight hydrocarbon components, natural gas can be considered as rich (five or six gallons or more of recoverable hydrocarbon components per cubic feet) or lean (less than one gallon of recoverable hydrocarbon components per cubic feet).

In this section the common properties and behavior of hydrocarbon derivatives (separated from natural gas) up to and including *n*-octane (C_8H_{18}) are presented (Table 1.5).

8.1 Behavior

In general terms, mixtures of volatile hydrocarbons are now used in preference to the chlorofluorocarbons as a propellant for aerosol, due to the impact of the chlorofluorocarbons on the ozone.

Methane (CH_4) and ethane (C_2H_6) are gaseous at ambient temperatures and cannot be readily liquefied by pressure alone. However, propane (C_3H_8) is easily liquefied while butane (C_4H_{10}) is so easily liquefied that it provides a safe, volatile fuel for small pocket lighters. Pentane (C_5H_{12}) is a colorless liquid at room temperature, commonly used in chemistry and industry as a powerful nearly odorless solvent of waxes and high molecular weight organic compounds, including greases. Hexane (C_6H_{14}) is also a widely used nonpolar, nonaromatic solvent, as well as a significant fraction of common of naphtha and gasoline. The C_6 through C_{10} alkane derivatives, alkene derivatives, and isomeric cycloalkane derivatives are the top components of naphtha, gasoline, and kerosene and specialized industrial solvent mixtures. With the progressive addition of carbon units, the simple nonring structured hydrocarbon derivatives have higher viscosity, higher boiling point, and higher solidification temperature.

In terms of the gaseous hydrocarbon derivatives, an ideal gas is a gas in which all collisions between atoms or molecules are perfectly elastic and in which there are no intermolecular attractive forces. An ideal gas can be characterized by three variables: (i) absolute pressure (P), (ii) volume (V), and (iii) absolute temperature (T). The relationship between them is called the *ideal gas law*:

$$PV = nRT = NkT$$

n = number of moles, R = universal gas constant (= 8.3145 J/mol K), N = number of molecules, k = Boltzmann constant (= 1.38,066 x 10^{-23} J/K = 8.617,385 x 10^{-5} eV/K), k = R/N_A, N_A = Avogadro's number = 6.0221 x 10^{23}/mol.

The ideal gas law can arise from the pressure of gas molecules colliding with the walls of a container. And 1 mole of an ideal gas at standard temperature and pressure occupies 22.4 L. However, natural gas is a nonideal gas and does not obey the ideal gas law but obeys the modified gas law:

$$PV = nZRT$$

P is the pressure, V is the volume, T is the absolute temperature (degree Kelvin), Z is the compressibility, n is the number of kilo-moles of the gas, and R is the gas constant.

For example, if all other factors remained constant, when the volume of a certain mass of gas is reduced by 50%, the pressure would double and so on. As a gas, it would expand to fill any volume it is in. However the compressibility, Z, is the factor which differentiates natural gas from an ideal gas. For methane, Z is 1 at 1 bar but decreases to 0.85 at 100 atm, both at 25°C, that is, it compresses to a smaller volume than the proportional relationship.

8.2 Combustion and the heat of combustion

Hydrocarbons are currently the main source of the electric energy and heat sources (such as home heating) because of the energy produced by combustion. Often this energy is used directly as heat such as in home heaters, which use crude oil, crude oil products, or natural gas.

During combustion, hydrocarbon derivatives produce steam, carbon dioxide, and heat as, for example, during the combustion of methane:

$$CH_4 + 2O_2 \rightarrow 2H_2O + CO_2 + \text{energy}$$

In an inadequate supply of air, carbon monoxide and water vapor are formed:

$$2CH_4 + 3O_2 \rightarrow 2CO + 4H_2O$$

Another example is the combustion of propane:

$$C_3H_8 + 5O_2 \rightarrow 4H_2O + 3\ CO_2 + \text{energy}$$

And finally, for any liner alkane derivative containing *n* carbon atoms, the general equation for the exothermic reaction is as follows:

$$C_nH_{2n+2} + (3n+ {}^1\!/_2\ )O_2 \rightarrow (n + 1)H_2O + n\ CO_2 + \text{energy.}$$

The *heat of combustion* (*energy content*) of natural gas is the amount of energy that is obtained from the burning of a volume of natural gas, which is measured in British thermal units (Btu). The value of natural gas is calculated by its Btu content. One *Btu* is the quantity of heat required to raise the temperature of one pound of water by 1 degree Fahrenheit at atmospheric pressure. A cubic foot of natural gas has an energy content of approximately 1031

Btu, but the range of values is between 500 and 1500 Btu depending upon the composition of the gas (Speight, 2017).

Thus, the energy content of natural gas is variable because natural gas has variations in the amount and types of energy gases (methane, ethane, propane, butane) it contains: the more noncombustible the gases in the natural gas, the lower the energy (Btu). In addition, the volume mass of energy gases which are present in a natural gas accumulation also influence the Btu value of natural gas. The more carbon atoms in a hydrocarbon gas, the higher its Btu value. It is necessary to conduct the Btu analysis of natural gas at each stage of the supply chain. Gas chromatographic process analyzers are used in order to conduct fractional analysis of the natural gas streams, separating natural gas into identifiable components. The components and their concentrations are converted into a gross heating value in Btu-cubic foot. In the USA, at retail, natural gas is often sold in units of therms (th) (1 therm = 100,000 BTU. Wholesale transactions are generally done in decatherms (Dth) or in 1000 decatherms (MDth) or in million decatherms (MMDth). A million decatherm is roughly a billion cubic feet of natural gas.

The gross heats of combustion of crude oil and its products are given with fair accuracy by the equation:

$$Q = 12{,}400 - 2100d^2$$

Where d is the 60/60°F specific gravity. Deviation from the formula is generally less than 1%.

8.3 Density

Density is the mass of a substance contained in a unit volume (simply, density is mass divided by volume). In the SI system of units, the ratio of the density of a substance to the density of water at 15°C is known as the *specific gravity* (*relative density*). Various units of density, such as kg/m^3, $lb\text{-}mass/ft^3$, and g/cm^3, are commonly used. In addition, molar densities or the density divided by the molecular weight is often specified.

Density values (including those of natural gas hydrocarbon derivatives) are given at room temperature unless otherwise indicated by the superscript figure; for example, 2.487^{15} indicates a density of 2.487 g/cm for the substance at 15°C (59°F). A superscript 20 over a subscript 4 indicates a density at 20°C relative to that of water at 4°C (39°F). For gases the value of the density is given in grams per liter (g/L). Another term, *specific gravity*, is commonly used is relation to the properties of hydrocarbon derivatives. The specific gravity of a substance is a comparison of its density to that of water. Density as a physical property of matter is a measure of the relative *heaviness* of hydrocarbon derivatives and other chemicals at a constant volume, and each constituent of natural gas has a unique density associated with it.

For most chemical compounds (i.e., those that are solid or liquid), the density is measured relative to water (1.00). For gases, the density is more likely to be compared to the density of air (called the *vapor density* in which the density of air is given the number 1.00 but this is arbitrary and bears no relationship to the density of water). As a comparison, the density of liquefied natural gas (LNG) is approximately 0.41−0.5 kg/L, depending on temperature, pressure, and composition; in comparison the density of water is 1.0 kg/L.

However, the density of raw natural gas which is a mixture of several hydrocarbon and nonhydrocarbon components is not an accurate measure of the character of natural gas. The density of any gas compared to the density of air is the vapor density and is a very important characteristic of the constituents of natural gas and natural gas constituents. Put simply, if the constituents of natural gas are less dense (lighter) than air, they will dissipate into the atmosphere whereas if the constituents of natural gas are denser (heavier) than air, they will sink and be less likely to dissipate into the atmosphere. Of the hydrocarbon constituents of natural gas, methane is the only one that is less dense than air.

The statement is often made that natural gas is lighter than air. This statement often arises because of the continued insistence by engineers and scientists that the properties of a mixture are determined by the mathematical average of the properties of the individual constituents of the mixture. Such mathematical bravado and inconsistency of thought is detrimental to safety and needs to be qualified.

Relative to air, methane is less dense (Table 1.7) but the other hydrocarbon constituents of unrefined natural gas (i.e., ethane, propane, butane, etc.) are denser than air. Therefore, should a natural gas leak occur in field operations, especially where the natural gas contains constituents other than methane, only methane dissipates readily into the air whereas the other hydrocarbon constituents that are heavier than air do not readily dissipate into the atmosphere. This poses considerable risk if these constituents of natural gas accumulate or pool at ground level when it has been erroneously assumed that natural gas is lighter than air.

8.4 Substitution and addition reactions

Substitution reactions only occur in saturated hydrocarbons (single carbon−carbon bonds). In this reaction, an alkane reacts with a chlorine molecule. One of the chlorine atoms displaces a hydrogen atom which leads to the formation of hydrochloric acid (HCl, if gaseous and nonaqueous, the correct term is hydrogen chloride) as well as the hydrocarbon with one chlorine atom. If the reaction is not halted, the reaction will progress to the formation of carbon tetrachloride (CCl_4):

$$CH_4 + Cl_2 \rightarrow CH_3Cl + HCl$$

$$CH_3Cl + Cl_2 \rightarrow CH_2Cl_2 + HCl$$

$$CH_2Cl_2 + Cl_2 \rightarrow CHCl_3 + HCl$$

$$CHCl_3 + Cl_2 \rightarrow CCl_4 + HCl$$

TABLE 1.7 General properties of the constituents of natural gas up to n-octane (C_8H_{18}), including toluene, ethyl benzene, and xylene.

	Molecular weight	Specific gravity	Vapor density, air = 1	Boiling point, °C	Ignition temperature, °C	Flash point, °C
Methane	16	0.553	0.56	−160	537	−221
Ethane	30	0.572	1.04	−89	515	−135
Propane	44	0.504	1.50	−42	468	−104
Butane	58	0.601	2.11	−1	405	−60
Pentane	72	0.626	2.48	36	260	−40
Hexane	86	0.659	3.00	69	225	−23
Benzene	78	0.879	2.80	80	560	−11
Heptane	100	0.668	3.50	98	215	−4
Octane	114	0.707	3.90	126	220	13
Toluene	92	0.867	3.20	161	533	4
Ethyl benzene	106	0.867	3.70	136	432	15
Xylene	106	0.861	3.70	138	464	17

Data extracted from Speight, 2003.

On the other hand, addition reactions occur when there is unsaturation in the hydrocarbon molecule and involve alkene derivatives and alkyne derivatives. In this reaction a halogen molecule breaks the double or triple bond in the hydrocarbon and forms a bond as in the case of the addition of chlorine to ethylene:

$$CH_2{=}CH_2 + Cl_2 \rightarrow CH_2ClCH_2Cl$$

8.5 Volatility, flammability, and explosive properties

The *boiling point* (boiling temperature) of a substance is the temperature at which the vapor pressure of the substance is equal to atmospheric pressure.

At the boiling point, a substance changes its state from liquid to gas. A stricter definition of boiling point is the temperature at which the liquid and vapor (gas) phases of a substance can exist in equilibrium. When heat is applied to a liquid, the temperature of the liquid rises until the *vapor pressure* of the liquid equals the pressure of the surrounding atmosphere (gases). At this point there is no further rise in temperature, and the additional heat energy supplied is absorbed as *latent heat* of vaporization to transform the liquid into gas. This transformation occurs not only at the surface of the liquid (as in the case of *evaporation*) but also throughout the volume of the liquid, where bubbles of gas are formed. The boiling point of a liquid is lowered if the pressure of the surrounding atmosphere (gases) is decreased. On the other hand, if the pressure of the surrounding atmosphere (gases) is increased, the boiling point is raised. For this reason, it is customary when the boiling point of a substance is given to include the pressure at which it is observed, if that pressure is other than standard, i.e., 760 mm of mercury or 1 atm (STP, Standard Temperature and Pressure).

The boiling points of crude oil fractions are rarely, if ever, distinct temperatures. It is, in fact, more correct to refer to the boiling ranges of the various fractions; the same is true of natural gas. To determine these ranges, the material in question is tested in various methods of distillation, either at atmospheric pressure or at reduced pressure. Thus, the boiling points of the hydrocarbon constituents of natural gas increase with molecular weight and the initial boiling point of natural gas corresponds to the boiling point of the most volatile constituents (i.e., methane).

Purified natural gas is neither corrosive nor toxic, its ignition temperature is high, and it has a narrow flammability range, making it an apparently safe fossil fuel compared to other fuel sources. In addition, purified natural gas (i.e., methane) having a specific gravity (0.60) lower than that of air (1.00) rises if escaping and dissipates from the site of any leak. However, methane is highly flammable, burns easily, and almost completely. Therefore, natural gas can also be hazardous to life and property because of the tendency for explosion.

When natural gas is confined, such as within a house or in a coal mine, concentration of the gas can reach explosive mixtures that, if ignited, results in blasts that could destroy buildings.

The *flash point* of crude oil or a crude oil product, including natural gas, is the temperature to which the product must be heated under specified conditions to give off sufficient vapor to form a mixture with air that can be ignited momentarily by a specified flame (ASTM D56, ASTM D92, ASTM D93). As with other properties, the flash point is dependent on the composition of the gas and the presence of other hydrocarbon constituents.

The *fire point* is the temperature to which the gas must be heated under the prescribed conditions of the method to burn continuously when the mixture of vapor and air is ignited by a specified flame (ASTM D92).

From the viewpoint of safety, information about the flash point is of most significance at or slightly above the maximum temperatures (30°C −60°C, 86°F to 140°F) that may be encountered in storage, transportation, and use of liquid crude oil products, in either closed or open containers. In this temperature range the relative fire and explosion hazard can be estimated from the flash point. For products with flash point below 40°C (104°F) special precautions are necessary for safe handling. Flash points above 60°C (140°F) gradually lose their safety significance until they become indirect measures of some other quality. The flash point of a crude oil product is also used to detect contamination. A substantially lower flash point than expected for a product is a reliable indicator that a product has become contaminated with a more volatile product, such as gasoline. The flash point is also an aid in establishing the identity of a particular petroleum product.

A further aspect of volatility that receives considerable attention is the vapor pressure of petroleum and its constituent fractions. The *vapor pressure* is the force exerted on the walls of a closed container by the vaporized portion of a liquid. Conversely, it is the force that must be exerted on the liquid to prevent it from vaporizing further (ASTM D323). The vapor pressure increases with temperature for any given gasoline, liquefied crude oil gas, or other product. The temperature at which the vapor pressure of a liquid, either a pure compound of a mixture of many compounds, equals 1 atm (14.7 psi, absolute) is designated as the boiling point of the liquid.

The flammable range is expressed by the lower explosive limit (LEL) and the upper explosive limit (UEL). The lower explosive limit is the concentration of natural gas in the air below which the propagation of a flame will not occur on contact with an ignition source. The lower explosive limit for natural gas is 5% by volume in air and, in most cases the smell of gas would be detected well before combustion conditions are met. The upper explosive limit is the concentration of natural gas in the air above which the propagation of a flame will not occur on contact with an ignition source. The natural gas upper explosive limit is 15% by volume in air.

Explosions caused by natural gas leaks occur a few times each year. Frequently, the blast will be enough to significantly damage a building but leave it standing. Occasionally, the gas can collect in high enough quantities to cause a deadly explosion, disintegrating one or more buildings in the process.

In any form, a minute amount of odorant (odorizer) that has an obvious smell is added to the otherwise colorless and odorless gas, so that leaks can be detected before a fire or explosion occurs. Odorants are considered nontoxic in the extremely low concentrations occurring in natural gas delivered to the end user.

References

Ali, M., El Ali, B.M., Speight, J.G., 2005. Handbook of Industrial Chemistry: Organic Chemicals. McGraw-Hill, New York.

Ancheyta, J., Speight, J.G., 2007. Hydroprocessing of Heavy Oils and Residua. CRC Press, Taylor & Francis Group, Boca Raton, Florida.

Badger, W.L., Baker, E.M., 1941. Inorganic Chemical Technology, second ed. McGraw-Hill Book Company Inc., New York.

Hall, C.A.S., Lindenberger, D., Kummel, R., Kroeger, T., Eichhorn, W., 2001. The need to reintegrate the natural sciences with economics. BioScience 51, 663–673.

Hall, C.A.S., Tharakan, P.J., Hallock, J., Cleveland, C., Jefferson, M., 2003. Hydrocarbons and the evolution of human culture. Nature 426 (20), 318–322.

Henglein, F.A., 1969. Chemical Technology. Pergamon Press Ltd., London, United Kingdom.

Jess, A., Wasserscheid, P., 2013. Chemical Technology. Wiley-VCH, Weinheim, Germany.

Mokhatab, S., Poe, W.A., Speight, J.G., 2006. Handbook of Natural Gas Transmission and Processing. Elsevier, Amsterdam, Netherlands.

Speight, J.G., 2003. Perry's Standard Tables and Formulas for Chemical Engineers. McGraw-Hill, New York, 2003.

Speight, J.G., 2008. Synthetic Fuels Handbook: Properties, Processes, and Performance. McGraw-Hill, New York.

Speight, J.G., 2010. Oil: An Introduction to the Technology, Economics, and Politics of a Nonrenewable Resource. M.E. Sharpe, Armonk, New York.

Speight, J.G., 2014. The Chemistry and Technology of Petroleum, fifth ed. CRC Press, Taylor and Francis Group, Boca Raton, Florida.

Speight, J.G. (Ed.), 2017. Lange's Handbook of Chemistry, seventeenth ed. McGraw-Hill Education, New York.

Tharakan, P.J., Kroeger, T., Hall, C.A.S., 2001. Twenty-five years of industrial development: a study of resource use rates and macro-efficiency indicators for five Asian countries. Environmental Science and Policy 4, 319–332.

Further reading

ASTM, 2009. Annual Book of Standards. American Society for Testing and Materials, West Conshohocken, PA.

Speight, J.G., 2019. Natural Gas: A Basic Handbook, second ed. Gulf Publishing Company, Elsevier, Cambridge, MA.

Chapter 2

Sources of hydrocarbons

1. Introduction

By definition, a *hydrocarbon* is an organic chemical compound composed exclusively of hydrogen and carbon atoms (Chapter 1). Hydrocarbons, from which, one hydrogen atom or one carbon atom has been replaced by a non-hydrogen or a noncarbon atom is not a hydrocarbon. Collections of such nonhydrocarbon derivatives as occuring in the high-boiling crude oil residues, in coal, in shale oil, and in biomass are not hydrocarbons and, for the convenience of this text are called hydrocarbonaceous species. Hydrocarbons do occur in nature and are found in crude oil, natural gas, some types of biomass, and wax deposits whereas hydrocarbons that are produced from coal and from oil shale— the organic ingredient of oil shale is the carbonaceous material known as kerogen— and generally from nonhydrocarbon biomass are the products of thermal processes (Chapter 5, Chapter 6, Chapter 7). The combustion of hydrocarbon derivatives in the presence of sufficient oxygen produces carbon dioxide, water, and heat (Chapter 10), which is the reason why hydrocarbon derivatives are desirable as gaseous fuels, liquid fuels, and, on occasion, solid fuels.

Methane, the main component of natural gas, is the simplest hydrocarbon because of the way it is structured. The five classes of hydrocarbon derivatives are (i) alkanes, also called paraffins, (ii) alkenes, also called olefins, (iii) alkynes, also called acetylenes, (iv) cycloalkanes, also called naphthenes, and (v) arenes, also called aromatic hydrocarbon derivatives. The way the hydrogen and carbon atoms are arranged and the types of chemical bonds that connect them determine what product they create. Crude oil is a major source of hydrocarbon derivatives such as propane; butane and methane are gas hydrocarbon derivatives. Coal also contains hydrocarbon derivatives, but the geological process that produce solid coal press out most of the hydrogen and oxygen atoms, leaving a carbonaceous rich solid. Although hydrocarbon derivatives are technically just some of molecules contained within crude oil, natural gas, and coal; these three substances are often referred to as hydrocarbon derivatives in and of themselves.

Hydrocarbon derivatives form naturally from plant and animal remains that are compressed through temperature and pressure over millennia, deep within

Handbook of Industrial Hydrocarbon Processes. https://doi.org/10.1016/B978-0-12-809923-0.00002-3

the Earth, in porous rocks like sandstone, limestone, and shale. These types of rocks exist in large bodies of water, especially oceans, and the natural gas and crude oil gradually rise through the rock and closer to the surface of the water (but still thousands of feet deep) and form a reservoir. Different techniques are used to extract hydrocarbon derivatives depending on their type and the material they're contained in. For example, hydraulic fracturing is used to extract natural gas and crude oil from shale formations (and from other low-permeability formations) by cracking the rock and using pressurized liquid to force the gas up through a well to the surface of the Earth (Speight, 2016a). Similarly, tar sand formations (called oil sand formations in Canada) are unconventional deposits of bitumen mixed in with sand or partially formed sandstone, requiring the deposit to be mined (Speight, 2014a, 2016b). A conventional vertical or horizontal well is used when a productive formation is easily accessed.

On the other hand, hydrocarbon fuels (gas, liquid, and solid) are those combustible or energy-generating molecular species that can be harnessed to create mechanical energy. Most liquid fuels, in widespread use, are derived from fossil fuels.

Crude oil–based hydrocarbon fuels are well-established products that have served industry and consumers for more than 100 years. However, the time is running out when these fuel sources, once considered inexhaustible, are now being depleted at a rapid rate. In fact, there is little doubt that the supplies of crude oil are being depleted with each year that passes. However, in spite of all of the argument, it is not clear just how long it will take to reach the bottom of the well—but for the most part and based on current estimates of reserves, it should be assumed that the time frame for depletion to occur is within the next 50 years. Thus, as the amount of available crude oil decreases, the need for alternate technologies to produce hydrocarbon fuels that could potentially help prolong the *liquid fuels culture* and mitigate the forthcoming effects of the shortage of transportation fuels that has been suggested to occur under the Hubbert peak oil theory (Speight, 2010; Speight and Islam, 2016) has increased.

The Hubbert peak oil theory is based on the fundamental observation that the amount of oil under the ground is finite and proposes that for any given geographical area, from an individual oil field to the planet as a whole, the rate of crude oil production tends to follow a bell-shaped curve. The theory also proposed the method that can be used to calculate the point of maximum production in advance based on discovery rates, production rates, and cumulative production. Early in the curve (prepeak), the production rate increases due to the discovery rate and the addition of infrastructure. Late in the curve (postpeak), production declines due to resource depletion.

There is no doubt that crude oil is being consumed at a steady rate but whether or not the Hubbert peak oil theory will affect the consumption of oil is another issue (Speight and Islam, 2016). It is a theory that is based on reserve estimates and reserve consumption. No one will disagree that hydrocarbon resources (in the form of crude oil and natural gas) are finite resources and will run out at some future point in time. The issue is the timing of this event—

whether it is tomorrow, next week, next month, next year, or in 50 years is not certain.

Whatever the timing, the modern world is based on a hydrocarbon culture and this will continue using crude oil and natural gas as the sources of hydrocarbon derivatives for another 50 years or more. However, it is time for procrastination to cease and this will not help in getting beyond the depletion of crude oil and natural gas resources, and we must look to the future for other sources of hydrocarbon derivatives. To mitigate the influence of the peak production of crude oil and the subsequent depletion of supplies, unconventional (or non-crude oil-derived) fuels are becoming major issues in the consciousness of oil importing countries.

On the other hand, alternate hydrocarbon fuels (also called synthetic fuels or synfuels), such as gasoline and diesel from sources other than crude oil or the blend stock from natural gas liquids, are making headway into the fuel balance. For example, biodiesel (not a true hydrocarbon but a mixture of esters) from plant sources is usable in diesel engines, but has differences that include higher cetane rating (45 to 60 compared to 45 to 50 for crude oil—derived diesel), and it acts as a cleaning agent to get rid of dirt and deposits. As with alcohols and gasoline engines, taking advantage of the high cetane rating of biodiesel potentially overcomes the energy deficit compared to ordinary number 2 diesel. Furthermore, coal (coal-to-liquids), natural gas (gas-to-liquids), oil shale (shale-to-liquids), and biomass (biomass-to-liquids) have been touted, and used to some extent, as sources of hydrocarbon for decades. At this time, the potential for hydrocarbon fuels from various types of biomass is also seeing prominence. Shortages of the supply of crude oil and the wish for various measures of energy independence are a growing part of the national psyche of many countries.

However, the production of hydrocarbon fuels from sources other than crude oil has a checkered history. The on-again-off-again efforts that are the result of political maneuvering have seen to it that the race to secure self-sufficiency by the production of nonconventional fuels has never got much further than the starting gate! This is due in no small part to the price fluctuations of crude oil (i.e., gasoline) and the lack of foresight by various levels of government. It must be realized that for decades the price of crude oil, the main source of hydrocarbon derivatives, has always been maintained at a level that was sufficiently low to discourage the establishment of a higher-cost synthetic fuels industry. However, we are close to the time when the lack of preparedness for the production of nonconventional fuels may set any national government on its heels (Speight, 2011a).

The dynamics are now coming into place for the establishment of hydrocarbon production by way of a *synthetic fuels industry* and it is up to various levels of government not only to promote the establishment of such an industry but to lead the way recognizing that it is not only supply and demand but the available and variable technology. For example, the technology of the tar sand industry is not same as it was in the 1970s. The processes for recovery of the raw materials and the processing options have changed in an attempt to increase the efficiency of oil production. Various national events (for the United States) and international events (for other countries) have made it essential that we move ahead to develop fuels from nonconventional sources.

Voices are being raised for the establishment of an industry that produces and develops hydrocarbon fuels from nonconventional sources but there is still a long way to go. Incentives are still needed to develop such resources. There is a cone of silence in many government capitals that covers the cries to develop nonconventional fuel sources. Hopefully, the silence will end within the near future, before it is too late.

2. Natural sources

In the context of this text, a natural product is a chemical compound produced by a living organism that typically has a degree of pharmacological or biological activity and which can find use in pharmaceutical drug design. A natural product can be considered as such even if it can be prepared by total synthesis in the laboratory or in an industrial setting.

The natural sources of hydrocarbon derivatives include crude oil and natural gas which contain hydrocarbon derivatives in the natural state. For example, the isoprenoid hydrocarbons, such as pristane and phytane, which occur in crude oils and ancient sediments, are presumed to have been formed from chlorophyll by various reaction routes. Neither pristane nor phytane is a normal constituent of land plants (Brooks et al., 1969).

Briefly considering the phytol-phytane-pristane family, phytol is an abundant acyclic isoprenoid compound that is present in the biosphere, and its degradation products have been used as biogeochemical tracers in aquatic environments. Phytane is the isoprenoid alkane formed when phytol— a constituent of chlorophyll—loses the hydroxyl group (Rontani and Volkman, 2003). Pristane is a natural saturated terpenoid derivative (a member of the alkane family); when pristane is deficient of one carbon atom (with the accompanying two hydrogen atoms), the product is pristane (Chung et al., 1989).

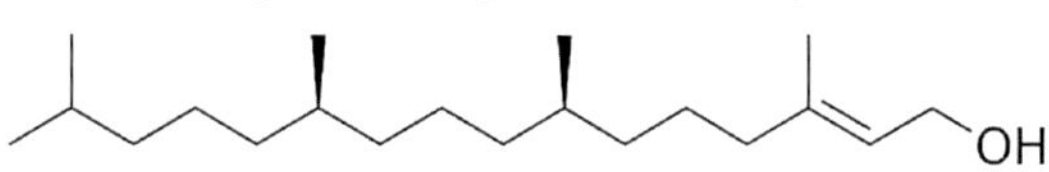

Phytol ($C_{20}H_{40}O$)

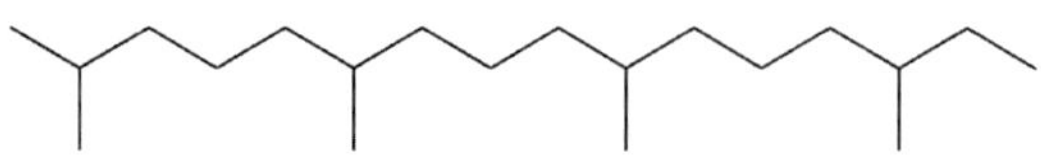

Phytane ($C_{20}H_{42}$)

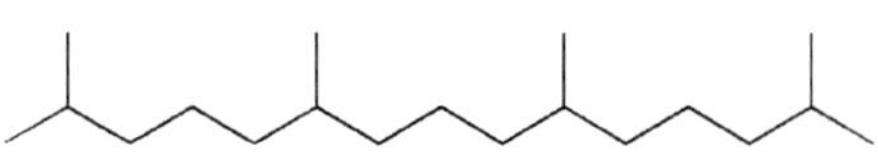

Pristane ($C_{19}H_{40}$)

There are other sources (often referred to as unconventional sources of hydrocarbon derivatives) insofar as hydrocarbon derivatives do occur to some extent but are generally produced from these sources by application of one or more thermal processes and these are: (i) extra heavy oil, (ii) tar sand bitumen, (iii) coal, (iv) oil shale, (v) wax, (vi) biomass, and (vii) municipal and industrial waste.

The fossil fuels represent the remains of animals and plants which died millions of years ago and the remains have become deposited and transformed by various maturation processes into organic sediments as a result of the burial within the crust of the Earth. However, for convenience, the sources of hydrocarbon derivatives as used in this text are divided into (i) the sources that contain naturally occurring hydrocarbon derivatives and (ii) the sources that do not contain naturally occurring hydrocarbon derivatives but from which hydrocarbon derivatives can be produced. Wax, the exception to these general categories, is composed of naturally occurring hydrocarbon derivatives but is not typically recognized as a source of hydrocarbon derivatives.

2.1 Crude oil

By definition, crude oil is a naturally occurring, unrefined liquid (which also may occur in gaseous and/or solid form) composed of hydrocarbon derivatives and other organic materials containing the so-called heteroatoms nitrogen, oxygen, sulfur, and metals such as iron, copper, nickel, and vanadium. Typically, crude oil can be refined to produce usable products such as gasoline, diesel fuel, fuel oils, lubricating oil, wax, and various forms of petrochemicals (Parkash, 2003; Gary et al., 2007; Speight, 2014a, 2017a; Hsu and Robinson, 2017, 2019a). Crude oil is found in a reservoir, which is a subsurface collection of hydrocarbon derivatives contained in porous or fractured rock formation. The hydrocarbon derivatives are trapped by impermeable underlying and overlying rock formations (Fig. 2.1) (Speight, 2014a). Natural gas also occurs with crude oil as a gas cap (*associated natural gas*) or it may occur on its own in a gas reservoir (*unassociated natural gas*). However, crude oil is a nonrenewable resource which cannot be replaced naturally at the rate that it is consumed, and it is, therefore, a limited resource with a current lifetime in the order of 50 years (Speight, 2011a, 2011b; Speight and Islam, 2016).

Crude oil is a naturally occurring mixture of hydrocarbon derivatives, generally in a liquid state, which may also include compounds of sulfur, nitrogen, oxygen, metals, and other elements (Speight, 2009; 2015). Thus, the common use of the term *crude oil* includes all gaseous, liquid and solid hydrocarbon derivatives. In the reservoir, the proportions of crude oil constituents that occur in the gaseous phase, the liquid phase, and the solid phase depend on subsurface temperature and pressure. However, under the temperature and pressure conditions prevalent on the surface (i.e., at the wellhead), the lower

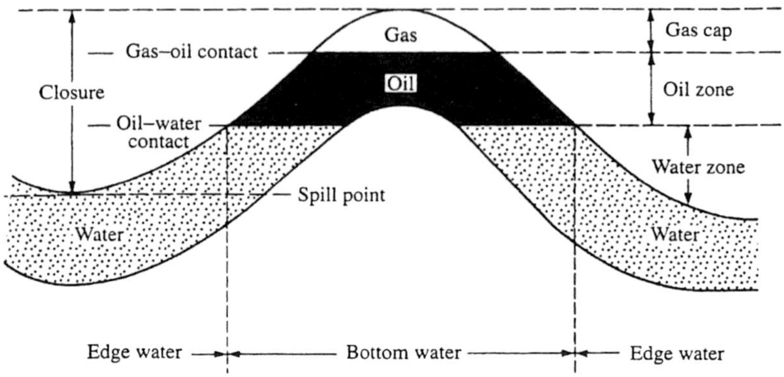

FIGURE 2.1 Schematic illustration of a syncline-anticline petroleum reservoir.

boiling hydrocarbon derivatives (methane, ethane, propane, and butane) surface emerge from the crude oil as gases while pentane and the higher boiling (higher molecular weight) hydrocarbon derivatives remain in the liquid phase. Higher molecular weight hydrocarbon derivatives can occur in the solid phase (i.e., wax derivatives) and remain as solids dissolved in the liquid phase.

In addition, the crude oils available to the modern refinery are quite different in composition and properties to those available to refiners 50 years (or more) ago (Speight, 2014a, 2017a). The current crude oils are somewhat more viscous insofar as they have higher proportions of nonvolatile (asphaltic) constituents than the previous refinery feedstock. In fact, changes in feedstock character, such as this tendency to heavier (higher boiling) materials (heavy oils), require adjustments to refinery operations to handle these heavier crude oils to reduce the amount of coke formed during processing and to balance the overall product slate (Speight, 2014a, 2017a).

Because of the need for a thorough understanding of the sources of hydrocarbon derivatives from the various members of the crude oil family, it is essential that the definitions and the terminology of crude oil science and technology be given prime consideration. While there is terminology that is recommended for crude oil and crude oil products, there is little in the way of standard terminology for heavy oil, extra heavy oil, and tar sand bitumen (Speight, 2014a). At best, the terminology is ill-defined and subject to changes from one governing body company to another. Particularly troublesome, and more confusing, are those terms that are applied to the more viscous materials, for example, the use of the terms *bitumen* and *asphalt*. An example of an irrelevant term is *black oil* which, besides the color of the oil, offers nothing in the way of explanation of the properties of the oil material and certainly adds nothing to any scientific and/or engineering understanding of the oil. Thus this term, i.e., *black oil,* is not used in this book.

The hydrocarbon derivatives in crude oil are predominantly alkane derivatives, cycloalkane derivatives, as well as various aromatic hydrocarbon derivatives, while the other organic compounds contain nitrogen, oxygen, sulfur, and trace amounts of metal derivatives such as iron, copper, nickel, and vanadium. The exact molecular composition of crude oil varies widely from formation to formation, but the proportion of chemical elements tends to vary only over fairly narrow limits (Speight, 2014a), thus:

Carbon: 83%−85% w/w
Hydrogen: 10%−14% w/w
Nitrogen: 0.1%−2% w/w
Oxygen: 0.05%−1.5% w/w
Sulfur: 0.05%−6% w/w
Metals: 100 to 5000 ppm w/w

Thus, the term *crude oil* covers a wide assortment of materials consisting of mixtures of hydrocarbon derivatives and other compounds containing variable amounts of sulfur, nitrogen, and oxygen, which may vary widely in properties. Metal-containing constituents, notably those compounds that contain vanadium and nickel, usually occur in the more viscous crude oils in amounts up to several thousand parts per million and can have serious consequences during processing of these feedstock (Parkash, 2003; Gary et al., 2007; Speight, 2014a, 2017a; Hsu and Robinson, 2017). Because crude oil is a mixture of widely varying constituents and proportions, its physical properties also vary widely and the color varies from near colorless to black.

The molecular boundaries of crude oil cover a wide range of boiling points and carbon numbers of hydrocarbon compounds and other compounds containing nitrogen, oxygen, and sulfur, as well as metal-containing (porphyrin) constituents (Speight, 2014a). However, the actual boundaries of such a *crude oil map* can only be arbitrarily defined in terms of boiling point and carbon number. In fact, crude oil is so diverse that materials from different sources exhibit different boundary limits, and for this reason alone it is not surprising that crude oil has been difficult to *map* in a precise manner.

Before passing on to heavy oil, there are three types of conventional crude oil that need to be addressed in order to compete this subsection on crude oil, and these are: (i) opportunity crude oil, (ii) high acid crude oil, and (iii) foamy oil.

2.1.1 Opportunity crude oil

There is also the need for a refinery to be configured to accommodate *opportunity crude oils* and/or *high acid crude oils,* which for many purposes are often included with high-boiling feedstocks (Speight, 2014a, 2014b; Yeung, 2014). *Opportunity crude oils* are either new crude oils with unknown or poorly understood properties relating to processing issues or are existing crude oils with well-known properties and processing concerns (Ohmes, 2014). Opportunity crude oils are often, but not always, heavy crude oils but in either case are more difficult to process due to high levels of solids (and other contaminants) produced with the oil, high levels of acidity, and high viscosity. These crude oils may also be incompatible with other oils in the refinery feedstock blend and cause excessive equipment fouling when processed either in a blend or separately (Speight, 2015). There is also the need for a refinery to be configured to accommodate *opportunity crude oils* and/or *high acid crude oils* which, for many purposes are often included with high-boiling feedstocks.

Opportunity crude oils, while offering initial pricing advantages, may have composition problems which can cause severe problems at the refinery, harming infrastructure, yield, and profitability. Before refining, there is the need for comprehensive evaluations of opportunity crudes, giving the potential buyer and seller the needed data to make informed decisions regarding fair pricing and the suitability of a particular opportunity crude oil for a refinery. This will assist the refiner to manage the ever-changing crude oil quality input to a refinery— including quality and quantity requirements and situations, crude oil variations, contractual specifications, and risks associated with such opportunity crudes.

2.1.2 High acid crude oil

High acid crude oils are crude oils that contain considerable proportions of naphthenic acids which, as commonly used in the crude oil industry, refer collectively to all of the organic acids present in the crude oil (Shalaby, 2005; Speight, 2014b). In many instances, the high acid crude oils are actually the heavier crude oils (Speight, 2014a, 2014b). The total acid matrix is therefore complex, and it is unlikely that a simple titration, such as the traditional methods for measurement of the total acid number, can give meaningful results to use in predictions of problems. An alternative way of defining the relative organic acid fraction of crude oils is therefore a real need in the oil industry, both upstream and downstream.

Typically, the end result of formation of low molecular weight acidic species is treated in the overheads in refineries. A combined approach to front end treating at crude inlet to heaters and preheat exchangers should be considered. It is commonly assumed that acidity in crude oils is related to carboxylic acid species, i.e., components containing a −COOH functional group. While it is clear that carboxylic acid functionality is an important feature (60% of the ions have two or more oxygen atoms), a major portion (40%) of the acid types are not carboxylic acids. In fact, naphthenic acids are a mixture of different compounds which may be polycyclic and may have unsaturated bonds, aromatic rings, and hydroxyl groups. Even the carboxylic acids are more diverse than expected, with approximately 85% containing more heteroatoms than the two oxygen atoms needed to account for the carboxylic acid groups. Examining the distribution of component types in the acid fraction reveals that there is a broad distribution of species.

Thermal decarboxylation can occur during the distillation process (during which the temperature of the crude oil in the distillation column can be as high as 400°C (750°F):

$$RCO_2H \rightarrow R{-}H + CO_2$$

However not all acidic species in crude oil are derivatives of carboxylic acids (−COOH) and some of the acidic species are resistant to high temperatures. For example, acidic species appear in the vacuum residue after having been subjected to the inlet temperatures of an atmospheric distillation tower and a vacuum distillation tower (Speight and Francisco, 1990). In addition, for the acid species that are volatile, naphthenic acids are most active at their boiling point and the most severe corrosion generally occurs on condensation from the vapor phase back to the liquid phase.

2.1.3 Foamy oil

Foamy oil is oil-continuous foam that contains dispersed gas bubbles produced at the well head from heavy oil reservoirs under solution gas drive. The nature of the gas dispersions in oil distinguishes foamy oil behavior from conventional heavy oil. The gas that comes out of solution in the reservoir does not coalesce into large gas bubbles nor into a continuous flowing gas phase. Instead it remains as small bubbles entrained in the crude oil, keeping the effective oil viscosity low while providing expansive energy that helps drive the oil toward the producing well. Foamy oil accounts for unusually high production in heavy oil reservoirs under solution gas drive (Sun et al., 2013). Thus, foamy oil is formed in solution gas drive reservoirs when gas is released from solution with a decline in reservoir pressure.

Thus, when, for example, heavy oil contains associated gas, foaming may occur at any point in the reservoir, wellbore, flowlines, or production equipment at the surface. Foaming occurs when gas reaches the bubble point and comes out of suspension as the pressure and temperature of the fluids change. Gas separates from less viscous crude oil much faster and more predictably than it does from heavy oil. The ability of the oil to foam depends on its viscosity; so higher viscosity oil is more likely to foam, and the foam it makes will be longer lasting.

2.1.4 Heavy oil

The name *heavy oil* can often be misleading as it has also been used in reference to (i) fuel oil that contains residuum left over from distillation, i.e., high-boiling fuel oil or residual fuel oil, (ii) coal tar creosote, or (iii) viscous crude oil. Thus for the purposes of this book the term is used to mean *viscous crude oil*. Extra heavy oil has been included in many of these categories. On the other hand, tar sand bitumen (often called simply *bitumen*) is often confused with manufactured asphalt (confusingly referred to as bitumen in many countries). To add further to this confusion, in some countries, tar sand bitumen is referred to as natural asphalt!

Heavy oil is a viscous type of crude oil that contains higher level of sulfur that conventional crude oil and occurs in similar locations to crude oil (Ancheyta and Speight, 2007). The nature of heavy oil is a problem for recovery operations and for refining—the viscosity of the oil may be too high thereby leading to difficulties in recovery and/or difficulties in refining the oil (Speight, 2016b, 2017a). However, tar sand bitumen, in terms of properties and behavior, are a world apart from conventional crude oil (Table 2.1, Table 2.2). Success with this material and with extra heavy oil depends as much on understanding the fluid (or nonfluid) properties of the material and the behavior of the fluids in the deposit in which they occur as it does on knowing the geology of the deposit (2014a). The reason is that the chemical and physical differences between heavy oil, extra heavy oil, and tar sand bitumen oil ultimately affect the viscosity and other relevant properties which, in turn, influence the individual aspects of recovery and refining operations.

Heavy oil is a *type* of crude oil that is different from conventional crude oil insofar as they are much more difficult to recover from the subsurface reservoir. These materials have a much higher viscosity (and lower API gravity) than conventional crude oil, and recovery of these crude oil types usually requires thermal stimulation of the reservoir. When crude oil occurs in a reservoir that allows the crude material to be recovered by pumping operations as a free-flowing dark to light colored liquid, it is often referred to as *conventional crude oil*. Very simply, heavy oil is viscous and does not flow easily. The common characteristic properties (relative to conventional crude oil) are high specific gravity, low hydrogen to carbon ratios, high carbon residues, and high contents of asphaltenes, heavy metal, sulfur, and nitrogen. Specialized recovery and refining processes are required to produce more useful fractions, such as naphtha, kerosene, and gas oil.

A very general definition of heavy oils has been, and remains, based on the API gravity or viscosity, and the definition is quite arbitrary although there have been attempts to rationalize the definition based upon viscosity, API gravity, and density (Fig. 2.2). For example, heavy oil is considered (arbitrarily) to be those crude oils that had gravity somewhat less than 20° API with the heavy oil falling into the API gravity range 10 to 15°. For example, Cold Lake heavy crude oil has an API gravity equal to 12° and extra heavy oil, such as tar sand bitumen, usually have an API gravity in the range 5°−10° (Athabasca bitumen = 8° API). Residua would vary depending upon the temperature at which distillation was terminated but usually vacuum residua are in the range 2°−8° API (Ancheyta and Speight, 2007; Speight, 2017a).

TABLE 2.1 Simplified differentiation between conventional crude oil, tight oil, heavy oil, extra heavy oil, and tar sand bitumen.

Conventional crude oil
Mobile in the reservoir
API gravity: >25°
High-permeability reservoir
Primary recovery
Secondary recovery
Tight oil
Similar properties to the properties of conventional crude oil
API gravity: >25°
Immobile in the reservoir
Low-permeability reservoir
Horizontal drilling into reservoir
Fracturing (typically multifracturing) to release fluids/gases
Medium crude oil
Similar properties to the properties of conventional crude oil
API gravity: 20–25°
High-permeability reservoir
Primary recovery
Secondary recovery
Heavy crude oil
More viscous than conventional crude oil
API gravity: 10–20°
Mobile in the reservoir
High-permeability reservoir
Secondary recovery
Tertiary recovery (enhanced oil recovery—EOR; e.g., steam stimulation)
Extra heavy oil
Fluid and/or mobile in the reservoir
Similar properties to the properties of tar sand bitumen
API gravity: <10°

Continued

TABLE 2.1 Simplified differentiation between conventional crude oil, tight oil, heavy oil, extra heavy oil, and tar sand bitumen.—cont'd

High-permeability reservoir
Secondary recovery
Tertiary recovery (enhanced oil recovery: EOR; such as steam stimulation)
Tar sand bitumen
Immobile (solid to near-solid) in the deposit
API gravity: <10°
High-permeability reservoir
Mining (often preceded by explosive fracturing)
Steam assisted gravity draining (SAGD)
Solvent methods (VAPEX)
Extreme heating methods
Innovative methods[a]

[a] *Innovative methods exclude tertiary recovery methods and methods such as SAGD and VAPEX and also includes variants or hybrids thereof.*

2.1.5 Crude oil production

The production of hydrocarbon derivatives from crude oil could be traced back to over 5000 years to the times when asphalt materials and oils were isolated from areas where natural seepage occurred (Abraham, 1945; Forbes, 1958; Hoiberg, 1960). Any treatment of the asphalt (such as hardening in the air prior to use) or of the oil (such as allowing for more volatile components to escape prior to use in lamps) may be considered to be refining under the general definition of refining.

An undeveloped reservoir may be under sufficient pressure to push hydrocarbon derivatives to surface. As the fluids are produced, the pressure will often decline, and production will falter. The reservoir may respond to the withdrawal of fluid in a way that tends to maintain the pressure. Artificial drive methods may be necessary and these are (i) solution gas drive, (ii) gas cap drive, (iii) water drive, (iv) water injection, and (v) gas injection methods.

Solution gas drive depends on the associated gas of the oil. The virgin reservoir may be entirely liquid, but will be expected to have gaseous hydrocarbon derivatives in solution due to the pressure. As the reservoir depletes, the pressure falls below the bubble point, and the gas comes out of solution to form a gas cap at the top. This gas cap pushes down on the liquid helping to maintain pressure. *Gas cap drive* occurs in reservoirs already having a gas cap (the pressure is already below bubble point) (Fig. 2.1), the gas cap expands

TABLE 2.2 Simplified use of pour point and reservoir/deposit temperature to differentiate between heavy oil, extra heavy oil, and tar sand bitumen.

Oil	Location	Temperature	Effect on oil
Heavy oil	Reservoir or deposit	Higher than oil pour point	Fluid and/or mobile
			Mobile
	Surface/ambient	Higher than oil pour point	Fluid and/or mobile
			Mobile
Extra heavy oil	Reservoir or deposit	Higher than oil pour point	Fluid and/or mobile
			Mobile
	Surface/ambient	Lower than oil pour point	Solid to near-solid
			Fluidity much reduced
			Immobile
Tar sand bitumen	Reservoir or deposit	Lower than oil pour point	Solid to near-solid
			Not fluid
			Immobile
	Surface/ambient	Lower than oil pour point	Solid to near-solid
			Not fluid
			Immobile

with the depletion of the reservoir, pushing down on the liquid sections applying extra pressure.

Water drive requires the presence of water in the reservoir, usually as a layer below the crude oil (Fig. 2.1). Water is compressible and as the hydrocarbon derivatives are depleted, the reduction in pressure in the reservoir causes the water to expand slightly. Although this expansion is minute, if the aquifer is large enough, this will translate into a large increase in volume, which will push up on the hydrocarbon derivatives, maintaining pressure. *Water and gas injection methods* are usually activated when the natural drives are insufficient, as they very often are, then the pressure can be artificially maintained by injecting water into the aquifer or gas into the gas cap.

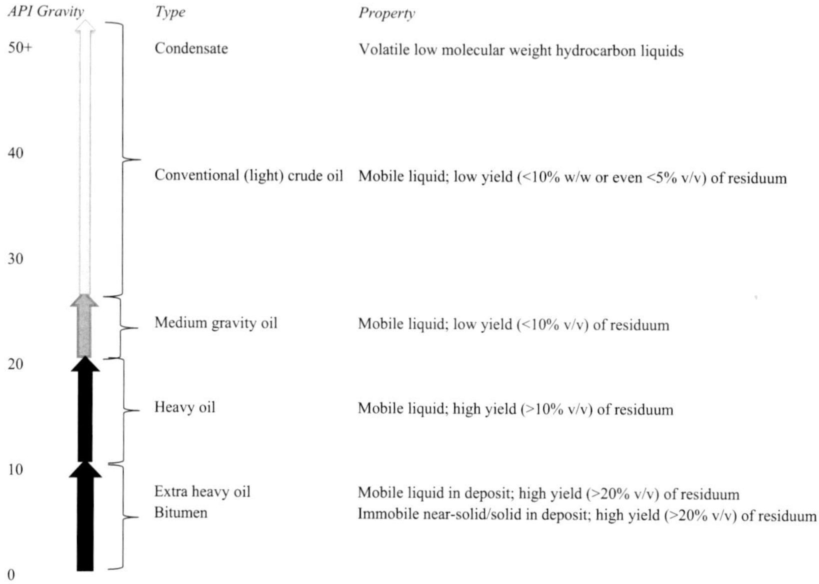

FIGURE 2.2 General description of various feedstock by API gravity.

Enhanced recovery methods are brought into play prior to the exhaustion of crude oil recovery by the above methods. The available enhanced recovery methods are variable and are often applied to heavy oil reservoirs and usually require thermal stimulation (such as steam) of the crude oil to move it to a production well (Speight, 2009).

2.1.6 Crude oil refining

Crude oil refining is the separation of recovered crude oil into fractions and the subsequent treating of these fractions to yield marketable products (Table 2.3) (Speight, 2014a, 2019a). In fact, refining is the means by which hydrocarbon derivatives are separated or produced from crude oil. A refinery is essentially a group of manufacturing plants which vary in number with the variety of products produced (Parkash, 2003; Gary et al., 2007; Speight, 2014a, 2017a; Hsu and Robinson, 2017). As the basic elements of crude oil, hydrogen and carbon form the main input into a refinery, combining into thousands of individual constituents and the economic recovery of these constituents varies with the individual crude oil according to its particular individual qualities, and the processing facilities of a particular refinery. In general, crude oil, once refined, yields three basic groupings of products that are produced when it is broken down into cuts or fractions (Speight, 2014a, 2017a). The complexity of crude oil is emphasized insofar as the actual proportions of low-boiling,

TABLE 2.3 Fractions obtained from crude petroleum by distillation.

	Boiling range[a]	
Fraction	**°C**	**°F**
Low-boiling naphtha	−1−150	30−300
Gasoline	−1−180	30−355
Heavy naphtha	150−205	300−400
Kerosene	205−260	400−500
Low-boiling gas oil	260−315	400−600
Heavy gas oil	315−425	600−800
Lubricating oil	>400	>750
Vacuum gas oil	425−600	800−1100
Residuum	>510	>950

[a]*For convenience, boiling ranges are converted to the nearest 5°.*

medium-boiling, and high-boiling fractions vary significantly from one crude oil to another.

Naphtha, a precursor to gasoline and solvents, is extracted from both the low-boiling and middle range of distillate cuts and is also used as a feedstock for the petrochemical industry. The middle distillates refer to hydrocarbon products from the middle boiling range of crude oil and include kerosene, diesel fuel, distillate fuel oil, and low-boiling gas oil. Waxy distillate and lower boiling lubricating oils are sometimes included in the middle distillates. The remainder of the crude oil includes the higher boiling lubricating oil fractions, gas oil, and residuum (the nonvolatile fraction of the crude oil). The residuum can also produce high-boiling lubricating oils and waxes but is more often used for asphalt production.

Refinery processes must be selected and products manufactured to give a balanced operation in which crude oil is converted into a variety of products in amounts that are in accord with the demand for each (Chapter 3). For example, the manufacture of hydrocarbon products from the lower-boiling portion of crude oil automatically produces a certain amount of higher-boiling hydrocarbon components. If the latter cannot be sold as, say, high-boiling fuel oil, these products will accumulate until refinery storage facilities are full. To prevent the occurrence of such a situation, the refinery must be flexible and be able to change operations as needed. This usually means more processes: thermal processes to change an excess of high-boiling fuel oil into more gasoline with coke as the residual product, or a vacuum distillation process to separate the heavy oil into lubricating oil stocks and asphalt.

The refining industry has been the subject of the four major forces that affect most industries and which have hastened the development of new crude oil refining processes: (i) the demand for hydrocarbon products such as gasoline, diesel, fuel oil, and jet fuel, (ii) feedstock supply, specifically the changing quality of crude oil and geopolitics between different countries and the emergence of alternate feed supplies such as bitumen from tar sand, natural gas, and coal, (iii) environmental regulations that include more stringent regulations in relation to sulfur in gasoline and diesel, and (iv) technology development such as new catalysts and processes to produce more hydrocarbon derivatives from the barrel of oil.

In the early days of the 20th century, refining processes were developed to extract kerosene for lamps. Any other products were considered to be unusable and were usually discarded. Thus, first refining processes were developed to purify, stabilize, and improve the quality of kerosene. However, the invention of the internal combustion engine led (at approximately the time of World War I) to a demand for gasoline for use in increasing quantities as motor fuel for cars and trucks. This demand on the lower boiling products increased, particularly when the market for aviation fuel developed. Thereafter, refining methods had to be constantly adapted and improved to meet the quality requirements and needs of car and aircraft engines.

Since then, the general trend throughout refining has been to produce more products from each barrel of crude oil and to process those products in different ways to meet the product specifications for use in modern engines. Overall, the demand for gasoline has rapidly expanded and demand has also developed for gas oils and fuels for domestic central heating, and fuel oil for power generation, as well as for low-boiling distillates and other inputs, derived from crude oil, for the petrochemical industries.

As the need for the lower boiling products developed, crude oil yielding the desired quantities of the lower boiling products became less available and refineries had to introduce conversion processes to produce greater quantities of lower boiling products from the higher boiling fractions. The means by which a refinery operates in terms of producing the relevant products depends not only on the nature of the crude oil feedstock but also on its configuration (i.e., the number of types of the processes that are employed to produce the desired product slate) and the refinery configuration is, therefore, influenced by the specific demands of a market. Therefore, refineries need to be constantly adapted and upgraded to remain viable and responsive to ever changing patterns of crude supply and product market demands. As a result, refineries have been introducing increasingly complex and expensive processes to gain higher yields of lower boiling products from the higher boiling fractions and residua. To convert crude oil into desired products in an economically feasible and environmentally acceptable manner. Refinery process for crude oil are generally divided into three categories: (i) separation processes, of which distillation is the prime example, (ii) conversion processes, of which coking and catalytic cracking are prime examples, and (iii) finishing processes, of which hydrotreating to remove sulfur is a prime example (Speight, 2014a, 2017a).

The simplest refinery configuration is the *topping refinery*, which is designed to prepare feedstocks for petrochemical manufacture or for production of industrial fuels in remote oil-production areas. The topping refinery consists of tankage, a distillation unit, recovery facilities for gases and low-boiling hydrocarbon derivatives, and the necessary utility systems (steam, power, and water-treatment plants). Topping refineries produce large quantities of unfinished oils and are highly dependent on local markets, but the addition of hydrotreating and reforming units to this basic configuration results in a more flexible *hydroskimming refinery*, which can also produce desulfurized distillate fuels and high-octane gasoline. These refineries may produce up to half of their output as residual fuel oil, and they face increasing market loss as the demand for low-sulfur (even no-sulfur) and high-sulfur fuel oil increases.

The most versatile refinery configuration today is known as the *conversion refinery*, which incorporates all the basic units found in both the topping and hydroskimming refineries, but it also features gas oil conversion plants such as catalytic cracking and hydrocracking units, olefin conversion plants such as alkylation or polymerization units, and, frequently, coking units for sharply reducing or eliminating the production of residual fuels. Modern conversion refineries may produce two-thirds of their output as unleaded gasoline, with the balance distributed between liquefied crude oil gas, jet fuel, diesel fuel, and a small quantity of coke. Many such refineries also incorporate solvent extraction processes for manufacturing lubricants and petrochemical units with which to recover propylene, benzene, toluene, and xylenes for further processing into polymers.

Finally, the yields and quality of refined crude oil products produced by the configuration of refineries may vary from refinery to refinery. Some refineries may be more oriented toward the production of gasoline (large reforming and/or catalytic cracking) whereas the configuration of other refineries may be more oriented toward the production of middle distillates such as jet fuel and gas oil.

The gas and gasoline fractions form the lower boiling products and are usually more valuable than the higher boiling fractions and provide hydrocarbon gas (liquefied petroleum gas, LPG) and hydrocarbon fractions such as naphtha, kerosene, aviation fuel, fuel oil, and feedstocks for the petrochemical industry (Speight, 2014a, 2017a; 2019a).

2.2 Natural gas

Natural gas (also called marsh gas and swamp gas in older texts) is a gaseous fossil fuel that is found in crude oil reservoirs (and called associated gas) and in natural gas reservoirs (and called nonassociated gas). While natural gas is commonly grouped in with other fossil fuels and sources of energy, there are many characteristics of natural gas that make it unique. The term natural gas is often extended to gases and liquids from the recently developed shale formations (Kundert and Mullen, 2009; Aguilera and Radetzki, 2014; Khosrokhavar et al., 2014; Speight, 2017b) as well as gas (biogas) produced from biological sources. However, for the purposes of this book, the petroliferous natural gas is placed under the category of *conventional gas* while petroliferous gas from tight formations and the nonpetroliferous gases (such as biogas and landfill gas) are placed under the term *nonconventional gas* (sometime called *unconventional gas*).

Thus, natural gas is a gaseous hydrocarbon-based fossil fuel which consists primarily of methane but contains significant quantities of ethane, propane, butane, and other hydrocarbon derivatives up to octane as well as carbon dioxide, nitrogen, helium, and hydrogen sulfide (Table 2.4). Natural gas is found with crude oil in crude oil reservoirs (*associated natural gas*) (Fig. 2.1), in natural gas reservoirs (*nonassociated natural gas*), and in coal beds (*coalbed methane*) (Speight, 2014a; 2019b).

TABLE 2.4 Composition of associated natural gas from a petroleum well.

Category	Component	Amount (% v/v)
Paraffins	Methane (CH_4)	70–98
	Ethane (C_2H_6)	1–10
	Propane (C_3H_8)	Trace-5
	Butane (C_4H_{10})	Trace-2
	Pentane (C_5H_{12})	Trace-l
	Hexane (C_6H_{14})	Trace-0.5
	Heptane and higher molecular weight (C_{7+})	Trace
Cycloparaffins	Cyclohexane (C_6H_{12})	Trace
Aromatics	Benzene (C_6H_6) + other aromatics	Trace
Nonhydrocarbons	Nitrogen (N_2)	Trace-15
	Carbon dioxide (CO_2)	Trace-1
	Hydrogen sulfide (H_2S)	Trace-1
	Helium (He)	Trace-5
	Other sulfur and nitrogen compounds	Trace
	Water (H_2O)	Trace-5

2.2.1 Composition

Natural gas, which is predominantly methane, occurs in underground reservoirs separately or in association with crude oil (Chapter 4) (Speight, 2014a, 2018). The principal types of hydrocarbon derivatives produced from natural gas are methane (CH_4) and varying amounts of higher molecular weight hydrocarbon derivatives from ethane (CH_3CH_3) to octane [$CH_3(CH_2)_6CH_3$]. Generally the higher molecular weight liquid hydrocarbon derivatives from pentane to octane and collective are referred to as *gas condensate*.

While natural gas is predominantly a mixture of combustible hydrocarbon derivatives (Table 2.4), many natural gases also contain nitrogen (N_2) as well as carbon dioxide (CO_2) and hydrogen sulfide (H_2S). Trace quantities of helium and other sulfur and nitrogen compounds may also be present. However, raw natural gas varies greatly in composition and the constituents can be several of a group of saturated hydrocarbon derivatives from methane to higher molecular weight hydrocarbon derivatives, especially natural gas that has been associated with crude oil in the reservoir, and nonhydrocarbon constituents (Table 2.4). The treatment required to prepare natural gas for distribution as an industrial or household fuel is specified in terms of the use and environmental regulations.

Briefly, natural gas contains hydrocarbon derivatives and nonhydrocarbon gases. Hydrocarbon gases are methane (CH_4), ethane (C_2H_6), propane (C_3H_8), butane (C_4H_{10}), pentane (C_5H_{12}), hexane (C_6H_{14}), heptane (C_7H_{16}), and sometimes trace amounts of octane (C_8H_{18}), and higher molecular weight hydrocarbon derivatives. For example:

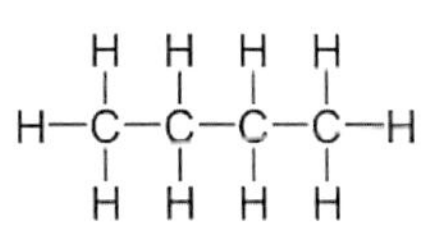

$CH_3CH_2CH_2CH_3$

n-Butane

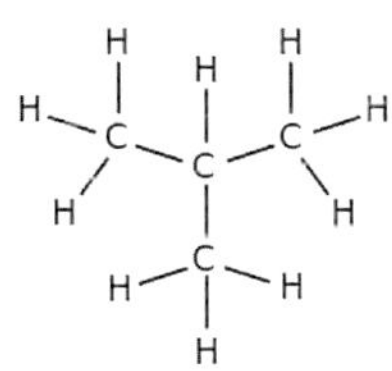

$(CH_3)_3CH$ or $(CH_3)_2CHCH_3$

iso-Butane

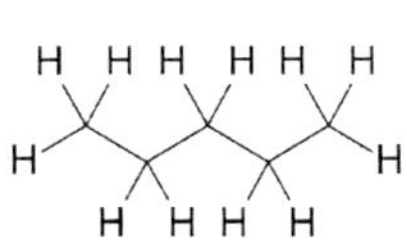

$CH_3CH_2CH_2CH_2CH_3$

n-Pentane

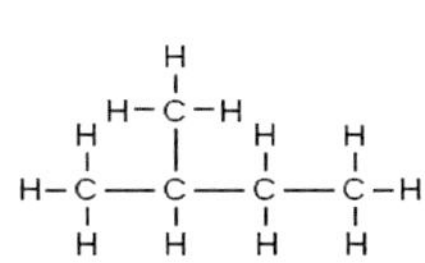

$(CH_3)_2CHCH_2CH_3$

iso-Pentane

As illustrated above, an iso-paraffin is an isomer having a methyl group branching from carbon number 2 of the main chain.

The higher-boiling hydrocarbon constituents than methane (CH_4) are often referred to as *natural gas liquids* (NGLs) and the natural gas may be referred to as *rich gas*. The constituents of natural gas liquids are hydrocarbon derivatives such as ethane (CH_3CH_3), propane ($CH_3CH_2CH_3$), butane ($CH_3CH_2CH_2CH_3$, as well as iso-butane), pentane derivatives ($CH_3CH_2CH_2CH_2CH_3$, as well as iso-pentane) and higher molecular weight hydrocarbon derivatives which have found wide use in the petrochemical industry (Chapter 12) (Speight, 2014a, 2019a). Some aromatic derivatives [BTX— benzene (C_6H_6), toluene ($C_6H_5CH_3$), and the xylene isomers (*o*-, *m*-, and *p*-$CH_3C_6H_4CH_3$)] can also be present, raising safety issues due to their toxicity. The nonhydrocarbon gas portion of the natural gas contains nitrogen (N_2), carbon dioxide (CO_2), helium (He), hydrogen sulfide (H_2S), water vapor (H_2O), and other sulfur compounds (such as carbonyl sulfide (COS) and mercaptans (e.g., methyl mercaptan, CH_3SH) and trace amounts of other gases.

Carbon dioxide and hydrogen sulfide are commonly referred to as *acid gases* since they form corrosive compounds in the presence of water. Nitrogen, helium, and carbon dioxide are also referred to as *diluents* since none of these burn, and thus they have no heating value. Mercury can also be present either as a metal in vapor phase or as an organometallic compound in liquid fractions. Concentration levels are generally very small, but even at very small concentration levels, mercury can be detrimental due its toxicity and its corrosive properties (reaction with aluminum alloys).

The higher molecular weight constituents (i.e., the C_{5+} product) are also commonly referred to as gas condensate or natural gasoline or sometimes, on occasion, as *casinghead gas* because of the tendency of these constituents to condense at the top of the well casing. When referring to natural gas liquids in the gas stream, the term *gallon per 1000 cubic feet* is used as a measure of high molecular weight hydrocarbon content. On the other hand, the composition of nonassociated gas (sometimes called well gas) is deficient in natural gas liquids. The gas is produced from geological formations that typically do not contain much, if any, hydrocarbon liquids. Furthermore, within the natural gas family, the composition of associated gas (a byproduct of oil production and the oil recovery process) is extremely variable, even within the gas from a crude oil reservoir (Speight, 2014a, 2019b). After the production fluids are brought to the surface, they are separated at a tank battery at or near the production lease into a hydrocarbon liquid stream (crude oil or condensate), a produced water stream (brine or salty water), and a gas stream.

Nonassociated natural gas is found in reservoirs containing no oil (dry wells). Associated gas, on the other hand, is present in contact with and/or dissolved in crude oil and is coproduced with it. The principal component of most natural gases is methane. Higher molecular weight paraffin hydrocarbon derivatives (C_2 to C_7, even to C_{10} in some cases) are usually present in smaller amounts with the natural gas mixture, and their ratios vary considerably from one gas field to another. Nonassociated gas normally contains a higher methane ratio than associated gas, while the latter contains a higher ratio of higher molecular weight hydrocarbon derivatives (Speight, 2014a, 2019b).

Crude oil–related gases (including associated natural gas) and refinery gases (process gases), as well as product gases produced from crude oil upgrading, are a category of saturated and unsaturated gaseous hydrocarbon derivatives, predominantly in the C_1 to C_6 carbon number range. Some gases may also contain inorganic compounds, such as hydrogen, nitrogen, hydrogen sulfide, carbon monoxide, and carbon dioxide. As such, crude oil and refinery gases (unless produced as a saleable product that must meet specifications prior to sale) are of often unknown or variable composition and toxic. The site-restricted crude oil and refinery gases (i.e., those not produced for sale) often serve as fuels consumed onsite, as intermediates for purification and recovery of various gaseous products, or as feedstock for isomerization and alkylation processes within a facility. Thus, natural gas is a combustible mixture of hydrocarbon gases that, in addition to methane, also includes ethane, propane, butane, and pentane. The composition of natural gas can vary widely before it is refined (Table 2.4, Table 2.5) (Mokhatab et al., 2006; Speight, 2009). In its

TABLE 2.5 General properties of unrefined natural gas (left-hand column) and refined natural gas (right-hand column).

Property	Unrefined gas	Refined gas
Carbon, % w/w	75	95
Hydrogen, % w/w	27	25
Oxygen, % w/w	0.4	0
Hydrogen-to-carbon atomic ratio	3.5 or less	4.0
Vapor density (air = 1.0) @ 15°C (59°F)	1.5	0.6
Methane, % v/v]	80	1000
Ethane, % v/v	<5	50
Propane	<2	2
Butane	<2	2
Nitrogen, % v/v	15	0
Carbon dioxide, % v/v	5	0
Sulfur, ppm w/w	5	0

purest form, the natural gas that is delivered to the consumer is almost pure methane.

2.2.2 Gas hydrates

Natural gas hydrates (*gas hydrates*) are crystalline solids in which a hydrocarbon, usually methane, is trapped in a lattice of ice. They occur in the pore spaces of sediments, and may form cements, nodes, or layers. Gas hydrates are found in naturally occurring deposits under ocean sediments or within continental sedimentary rock formations. The worldwide amounts of carbon bound in gas hydrates are conservatively estimated to total twice the amount of carbon to be found in all known fossil fuels on Earth.

Gas hydrates occur abundantly in nature, both in Arctic regions and in marine sediments. Gas hydrate is a crystalline solid consisting of gas molecules, usually methane, each surrounded by a cage of water molecules. It looks very much like ice. Methane hydrate is stable in ocean floor sediments at water depths greater than 300 m, and where it occurs, it is known to cement loose sediments in a surface layer several hundred meters thick.

Methane trapped in marine sediments as a hydrate represents such an immense hydrocarbon reservoir that it must be considered a dominant factor in estimating unconventional energy resources; the role of methane as a *greenhouse gas* also must be carefully assessed. Hydrates have major implications for energy resources and climate, but the natural controls on hydrates and their impacts on the environment are very poorly understood.

Extraction of methane from hydrates could provide an enormous energy and crude oil feedstock resource. Additionally, conventional gas resources appear to be trapped beneath methane hydrate layers in ocean sediments. The immense volumes of gas and the richness of the deposits may make methane hydrates a strong candidate for development as an energy resource. Because the gas is held in a crystal structure, gas molecules are more densely packed than in conventional or other unconventional gas traps. Gas-hydrate-cemented strata also act as seals for trapped free gas. These traps provide potential resources, but they can also represent hazards to drilling, and therefore must be well understood. Production of gas from hydrate-sealed traps may be an easy way to extract hydrate gas because the reduction of pressure caused by production can initiate a breakdown of hydrates and a recharging of the trap with gas.

Seafloor slopes of 5° and less should be stable on the Atlantic continental margin, yet many landslide scars are present. The depth of the top of these scars is near the top of the hydrate zone, and seismic profiles indicate less hydrate in the sediment beneath slide scars. Evidence available suggests a link between hydrate instability and occurrence of landslides on the continental margin. A likely mechanism for initiation of land sliding involves a breakdown of hydrates at the base of the hydrate layer. The effect would be a change from a semicemented zone to one that is gas-charged and has little strength, thus facilitating sliding. The cause of the breakdown might be a reduction in pressure on the hydrates due to a sea-level drop, such as occurred during glacial periods when ocean water became isolated on land in great ice sheets.

2.2.3 Gas processing

In addition to hydrogen sulfide and carbon dioxide, gas may contain other contaminants, such as mercaptans (also called *thiols*, R–SH) and carbonyl sulfide (COS). The presence of these impurities may eliminate some of the sweetening processes, since some processes remove large amounts of acid gas but not to a sufficiently low concentration. On the other hand, there are those processes that are not designed to remove (or are incapable of removing) large amounts of acid gases. However, these processes are also capable of removing the acid gas impurities to very low levels when the acid gases are there in low to medium concentrations in the gas.

Initially, natural gas receives a degree of cleaning at the wellhead (Speight, 2019b). The extent of the cleaning depends upon the specification that the gas must meet to enter the pipeline system. For example, natural gas from high-pressure wells is usually passed through field separators at the well to remove hydrocarbon condensate and water. Natural gasoline, butane, and propane are usually present in the gas, and gas processing plants are required for the recovery of these liquefiable constituents (Chapter 4).

Absorption is a process in which the absorbed gas is ultimately distributed throughout the absorbent (liquid). The process depends only on physical solubility and may include chemical reactions in the liquid phase (*chemisorption*). Common absorbing media used are water, aqueous amine solutions, caustic, sodium carbonate, and nonvolatile hydrocarbon oils, depending on the type of gas to be absorbed. Usually, the gas-liquid contactor designs which are employed are plate columns or packed beds.

Absorption is achieved by dissolution (a physical phenomenon) or by reaction (a chemical phenomenon). Chemical adsorption processes adsorb sulfur dioxide onto a carbon surface where it is oxidized (by oxygen in the flue gas) and absorbs moisture to give sulfuric acid impregnated into and on the adsorbent.

Adsorption differs from *absorption* in that it (adsorption) is a physical-chemical phenomenon in which the gas is concentrated on the surface of a solid or liquid to remove impurities. Usually, carbon is the adsorbing medium, which can be regenerated upon *desorption* (Speight, 2019b). The quantity of material adsorbed is proportional to the surface area of the solid and, consequently, adsorbents are usually granular solids with a large surface area per unit mass. Subsequently, the captured gas can be desorbed with hot air or steam either for recovery or for thermal destruction.

The number of steps and the type of process used to produce pipeline-quality natural gas most often depends upon the source and makeup of the wellhead production stream. In some cases, several of the steps may be integrated into one unit or operation, performed in a different order or at alternative locations or not required at all. In many instances pressure relief at the wellhead will cause a natural separation of gas from oil (using a conventional closed tank, where gravity separates the gas hydrocarbon derivatives from the higher boiling crude oil). In some cases, however, a multistage gas-oil separation process is needed to separate the gas stream from the crude oil. These *gas-oil separators* are commonly closed cylindrical shells, horizontally mounted with inlets at one end, an outlet at the top for removal of gas, and an outlet at the bottom for removal of oil. Separation is accomplished by alternately heating and cooling (by compression) the flow stream through multiple steps; some water and condensate, if present, will also be extracted as the process proceeds.

At some stage of the processing, the gas flow is directed to a unit that contains a series of filter tubes. As the velocity of the stream reduces in the unit, primary separation of remaining contaminants occurs due to gravity. Separation of smaller particles occurs as gas flows through the tubes, where they combine into larger particles which flow to the lower section of the unit. Further, as the gas stream continues through the series of tubes, a centrifugal force is generated which further removes any remaining water and small solid particulate matter.

Once purified, natural gas is *methane* which is colorless in its pure form and is a combustible mixture of hydrocarbon gases and while the major constituent is methane, ethane, propane, butane, and pentane are also present but the composition of natural gas varies widely.

In addition to the higher molecular weight hydrocarbon derivatives (often called *gas condensate*), methane can also be used to produce alternative liquid fuels (often referred to as gas-to-liquid, GTL). The term *alternative fuel* includes methanol, ethanol, and other alcohols, mixtures containing methanol, and other alcohols with gasoline or other fuels, biodiesel, fuels (other than alcohol) derived from biological materials, and any other fuel that is substantially not a crude oil product.

The production of hydrocarbon derivatives (either for fuel use or chemical use) from sources other than crude oil broadly covers liquid fuels that are produced from tar sand (oil sand) bitumen, coal, oil shale, and natural gas. Synthetic liquid fuels have characteristics approaching those of liquid fuels generated from crude oil but differ because the constituents of synthetic liquid fuels do not occur naturally in the source material used for their production (Speight, 2009). Thus, the creation of hydrocarbon to be used as fuels from sources other than natural crude oil broadly defines synthetic liquid fuels. For much of the 20th century, the synthetic fuels' emphasis was on liquid products derived from coal upgrading or by extraction or hydrogenation of organic matter in coke liquids, coal tars, tar sands, or bitumen deposits.

Projected shortages of crude oil make it clear that, for the remaining 21st century, alternative sources of liquid fuels are necessary. Such sources (for example, natural gas) are available but the exploitation technologies are in general not as mature as for crude oil. The feasibility of the upgrading of natural gas to valuable chemicals, especially liquid fuels, has been known for years. However, the high cost of the steam reforming and the partial oxidation processes, used for the conversion of natural gas to synthesis gas, has hampered the widespread exploitation of natural gas. Other sources include tar sand (also called oil sand or bituminous sand) (Speight, 1990, 2014a) and coal (Speight, 2013) that are also viable sources of liquid fuels.

The potential of natural gas, which typically has 85%–95% methane, has been recognized as a plentiful and clean alternative feedstock to crude oil. Currently, the rate of discovery of proven natural gas reserves is increasing faster than the rate of natural gas production. Many of the large natural gas deposits are located in areas where abundant crude oil resources lie such as in the Middle East. However, huge reserves of natural gas are also found in many other regions of the world, providing oil-deficient countries access to a plentiful energy source. The gas is frequently located in remote areas far from centers of consumption, and pipeline costs can account for as much as one-third of the total natural gas cost. Thus, tremendous strategic and economic incentives exist for conversion of gas to liquids, especially if this can be accomplished onsite or at a point close to the wellhead where shipping costs become a minor issue.

Gas hydrates are crystalline solids in which a hydrocarbon, usually methane, is trapped in a lattice of ice. They occur in the pore spaces of sediments, and may form cements, nodes, or layers. Gas hydrates are found in naturally occurring deposits under ocean sediments or within continental sedimentary rock formations. The worldwide amounts of carbon bound in gas hydrates are conservatively estimated to total twice the amount of carbon to be found in all known fossil fuels on Earth.

Methane trapped in marine sediments as a hydrate represents such an immense carbon reservoir that it must be considered a dominant factor in estimating unconventional energy resources; the role of methane as a *greenhouse gas* also must be carefully assessed. Hydrates store immense amounts of methane, with major implications for energy resources and climate, but the natural controls on hydrates and their impacts on the environment are very poorly understood.

3. Unconventional sources

Conventional and unconventional sources of hydrocarbon derivatives differ in their geologic locations and accessibility; conventional fuels are often found in discrete, easily accessible reservoirs, while unconventional fuels may be found within pore spaces throughout a wide geologic formation, requiring advanced extraction technologies. Typically, the conventional sources of hydrocarbons are crude oil and natural gas and this categorization is retained throughout this book. Thus the unconventional sources of hydrocarbon derivatives are: (i) extra heavy oil, (ii) tar sand bitumen, (iii) coal, (iv) oil shale, (v) wax, (vi) biomass, and (vii) municipal and industrial waste.

3.1 Extra heavy oil

The term *heavy oil* has also been arbitrarily (incorrectly) used to describe both the heavy oils that require thermal stimulation of recovery from the reservoir and the bitumen in bituminous sand (tar sand) formations from which the high-boiling bituminous material is recovered by a mining operation. *Extra heavy oil* is a nondescript term (related to viscosity) of little scientific meaning that is usually applied to tar sand bitumenlike material. An example is the extra heavy oil of the Zaca-Sisquoc extra heavy oil (sometime referred to as the Zaca-Sisquoc bitumen) that has an API gravity on the order of 4.0–6.0°. The reservoir has an average depth of 3500 ft., average thickness 1700 ft., average temperature 51°C −71°C (125°F −160°F), and sulfur in the range 6.8%–8% (Isaacs, 1992; USGS, 2006; Villarroel1 and Hernández, 2013). The deposit temperature is certainly equal to or above the pour point of the oil (Isaacs, 1992). This renders the oil capable of being pumped as a liquid from the deposit because of the high deposit temperature (which is higher than the pour point of the oil). The same rationale is applied to the extra heavy oil found in the Venezuelan Orinoco deposits.

The general difference is that extra heavy oil, which may have properties similar to tar sand bitumen in the laboratory but, unlike immobile tar sand bitumen in the deposit, has some degree of mobility in the reservoir or deposit (Delbianco and Montanari, 2009; Speight, 2014a). Extra heavy oil can flow at reservoir temperature and can be produced economically, without additional viscosity-reduction techniques, through variants of conventional processes such as long horizontal wells, or multilaterals. This is the case, for instance, in the Orinoco basin (Venezuela) or in offshore reservoirs of the coast of Brazil but, once outside of the influence of the high reservoir temperature, these oils are too viscous at surface to be transported through conventional pipelines and require heated pipelines for transportation. Alternatively, the oil must be partially upgraded or fully upgraded or diluted with a low-boiling hydrocarbon solvent (such as aromatic naphtha) to create a mix that is suitable for transportation (Speight, 2014a).

Thus, extra heavy oil is a material that occurs in the solid or near-solid state and generally has mobility under reservoir conditions. While this type of oil resembles tar sand bitumen and does not flow easily, extra heavy oil is generally recognized as having mobility in the reservoir compared to tar sand bitumen, which is typically incapable of mobility (free flow) under reservoir conditions. It is likely that the mobility of extra heavy oil is due to a high reservoir temperature (that is higher than the pour point of the extra heavy oil) or due to other factors is variable and subject to localized conditions in the reservoir.

3.2 Tar sand bitumen

Tar sand, also called *oil sand* (in Canada), or the more geologically correct term *bituminous sand* is commonly used to describe a sandstone reservoir that is impregnated with a heavy, viscous bituminous material. Tar sand is actually a mixture of sand, water, and bitumen but many of the tar sand deposits in countries other than Canada lack the water layer that is believed to facilitate the hot water recovery process. The heavy bituminous material has a high viscosity under reservoir conditions and cannot be retrieved through a well by conventional production techniques. The bitumen is another source of hydrocarbon fuels that is distinctly separate from conventional crude oil.

Geologically, the term *tar sand* is commonly used to describe a sandstone reservoir that is impregnated with bitumen, a naturally occurring material that is solid or near solid and is substantially immobile under reservoir conditions. The bitumen cannot be retrieved through a well by conventional production techniques, including currently used enhanced recovery techniques. In fact, tar sand is defined (FE-76-4) in the United States as (US Congress, 1976):

> *The several rock types that contain an extremely viscous hydrocarbon which is not recoverable in its natural state by conventional oil well production methods including currently used enhanced recovery techniques. The hydrocarbon-bearing rocks are variously known as bitumen-rocks oil, impregnated rocks, tar sands, and rock asphalt.*

In addition to this definition, there are several tests that must be carried out to determine whether or not, in the first instance, a resource is a tar sand deposit. Most of all, a core taken from a tar sand deposit, the bitumen isolated therefrom, are certainly not identifiable by the preliminary inspections (sight and touch) alone.

In the United States, the final determinant is whether or not the material contained therein can be recovered by primary, secondary, or tertiary (enhanced) recovery methods (US Congress 1976).

The relevant position of tar sand bitumen in nature is best illustrated by comparing its position relevant to crude oil and heavy oil. Thus, crude oil is referred to generically as a *fossil energy resource* and is further classified as a hydrocarbon resource and, for illustrative (or comparative) purposes in this report, coal and oil shale kerogen are also being included in this classification. However, the inclusion of coal and oil shale under the broad classification of hydrocarbon resources has required (incorrectly) that the term hydrocarbon be expanded to include these resources. It is essential to recognize that resources such a coal, oil shale kerogen, and tar sand bitumen contain large proportions of heteroatomic species. Heteroatomic species are those organic constituents that contain atoms other than carbon and hydrogen, e.g., nitrogen, oxygen, sulfur, and metals (nickel and vanadium).

Use of the term *organic sediments* is more correct and preferred. The inclusion of coal and oil shale kerogen in the category hydrocarbon resources is due to the fact that these two natural resources (coal and oil shale kerogen) will produce hydrocarbon derivatives by thermal decomposition (high-temperature processing). Therefore, if either coal and/or oil shale kerogen is to be included in the term hydrocarbon resources, it is more appropriate that they be classed as hydrocarbon-producing resources under the general classification of organic sediments (Speight, 2014a). Thus, tar and bitumen are different to crude oil and heavy oil not only from the method of recovery but also from the means by which hydrocarbon derivatives are produced (Table 2.1, Table 2.2).

Also, it is incorrect to refer to tar sand bitumen as *tar* or *pitch*. In many parts of the world *bitumen* is used as the name for *road asphalt*. Although the word tar is somewhat descriptive of the black bituminous material, it is best to avoid its use with respect to natural materials. More correctly, the name tar is usually applied to the high-boiling product remaining after the destructive distillation of coal or other organic matter. *Pitch* is the distillation residue of the various types of tar.

Physical methods of fractionation of tar sand bitumen can also produce the four generic fractions: saturates, aromatics, resins, and asphaltenes. However, for tar sand bitumen, the fractionation process shows that bitumen contains high proportions of *asphaltenes* and *resins*, even in amounts up to 50% w/w (or higher) of the bitumen with much lower proportions of saturates and aromatics than crude oil or heavy oil. In addition, the presence of ash-forming metallic constituents, including organometallic compounds such as vanadium and nickel, is also a distinguishing feature of bitumen.

Currently, the only commercial production of bitumen from tar sand deposits occurs in northeastern Alberta (Canada) where mining operations are currently used to recover the tar sand. After mining, the tar sands are transported to an extraction plant, where a hot water process separates the bitumen from sand, water, and minerals. The separation takes place in separation cells. Hot water is added to the sand, and the resulting slurry is piped to the extraction plant where it is agitated. The combination of hot water and agitation releases bitumen from the oil sand, and causes tiny air bubbles to attach to the bitumen droplets that float to the top of the separation vessel, where the bitumen can be skimmed off. Further processing removes residual water and solids. The bitumen is then transported and converted to *synthetic crude oil* by thermal processes into synthetic crude oil. Approximately two tons of tar sand are required to produce one barrel of oil.

Both mining and processing of tar sands involve a variety of environmental impacts, such as global warming and greenhouse gas emissions, disturbance of mined land, impacts on wildlife and air and water quality. The development of commercial tar sands industry in the United States would also have significant social and economic impacts on local communities. Of special concern is the relatively arid western United States where large amount of water is required for tar sands processing. Currently, tar sands extraction and processing require several barrels of water for each barrel of oil produced, though some of the water can be recycled.

To some observers, this proves the viability of the entire process while to others the energy requirements for the production of the synthetic crude oil make it marginally feasible for a significant percentage of world oil production to be extracted from tar sand (Speight, 2013a; Lee et al., 2014). Nevertheless synthetic crude oil is a product that has given Canada a measure of self-sufficiency (at a cost) that is currently moving toward 1,000,000 barrels of synthetic crude oil per day.

3.3 Coal

Coal is a fossil fuel formed in swamp ecosystems where plant remains were saved by water and mud from oxidation and biodegradation. Coal is a combustible black or brownish-black organic rock and is composed primarily of carbon along with assorted other elements, including hydrogen and oxygen. It is extracted from the ground by coal mining—either underground mining or open-pit mining (surface mining) (Speight, 2013b).

Coal included the following classifications: (i) lignite—also referred to as brown coal and is the lowest rank of coal and used almost exclusively as fuel for steam-electric power generation; (ii) subbituminous coal—the properties of which range from those of lignite to those of bituminous coal and are used primarily as fuel for steam-electric power generation; (iii) bituminous coal—a dense coal, usually black, sometimes dark brown, often with well-defined bands of bright and dull material, used primarily as fuel in steam-electric power generation, with substantial quantities also used for heat and power applications in manufacturing and to make coke; and (iv) anthracite—the highest rank coal which is a hard, glossy, black coal used primarily for residential and commercial space heating.

Despite reduced prominence, coal technology continues to be a viable option for the production of hydrocarbon derivatives in the future. World crude oil production is expected ultimately to level off and then decline and despite apparent surpluses of natural gas, production is expected to suffer a similar decline. Coal gasification to synthesis gas is utilized to synthesize liquid fuels in much the same manner as natural gas steam reforming technology. But the important aspect is to use the natural gas reserves when they are available and to maximize the use of these reserves by conversion of natural gas to liquid fuels.

The crude oil price has been sharply rising in the 21st century and there are indications that the high crude oil price is here to stay, rather than a temporary phenomenon. Even after considering the changes in various economic factors involving energy industries, production of transportation fuels or fuel oils via coal liquefaction is certainly an outstanding option for the sustainable future. Further, the products of coal liquefaction can be refined and formulated to possess the properties of conventional transportation fuels, as such requiring neither change in distribution nor life style changes for consumers.

There are inherent technological advantages with the conversion of coal to liquid products since *coal liquefaction* can produce clean liquid fuels that can be sold as transportation fuels such as gasoline and diesel. There are two principal routes by which liquid fuels can be produced from solid coal: (i) direct conversion to liquids and (ii) indirect conversion to liquids.

The direct liquefaction of coal by the Bergius process (liquefaction by hydrogenation) is also available. Several other direct liquefaction processes have been developed (Speight, 2013b). Another process to manufacture liquid hydrocarbon derivatives from coal is low temperature carbonization in which coal is heated at temperatures between 450°C and 700°C (840°F and 1290°F). These temperatures optimize the production of coal tars richer in lower-boiling hydrocarbon derivatives that are suitable for fuel production. The Bergius process has not been used outside Germany, where such processes were operated both during World War I and World War II. Several other direct liquefaction processes have been developed, among these being the SRC-I and SRC-II (Solvent Refined Coal) processes developed by (the now defunct) Gulf Oil and implemented as pilot plants in the United States in the 1960s and 1970s (Speight, 2013).

The direct hydrogenation of coal was explored by the NUS Corporation in 1976 and involved the thermal conversion of dried, pulverized coal mixed with roughly 1% by weight molybdenum catalyst. The process yielded a limited amount of propane and butane, a synthetic naphtha (the precursor to gasoline), small amounts of NH_3, and significant amounts of carbon dioxide.

Another process to manufacture liquid hydrocarbon derivatives from coal is low temperature carbonization (LTC (Karrick) process). Coal is converted to coke by heating at temperatures between 450°C and 700°C (840 and 1,290°F) compared to temperatures in the range 800°C to 1000°C (1470°F to 1830F) which are employed for the production of metallurgical coke. The lower temperatures optimize the production of coal tar that is richer in lower-boiling hydrocarbon derivatives than high-temperature coal tar. The coal tar is then further processed into hydrocarbon fuels.

Coal can also be converted into liquid fuels by indirect liquefaction which involves gasification of coal to mixtures of carbon monoxide and hydrogen (synthesis gas) followed by application of the Fischer-Tropsch process in which the synthesis gas is converted to hydrocarbon derivatives under catalytic conditions of temperature and pressure. The Fischer-Tropsch process for the *indirect* synthesis of liquid hydrocarbon derivatives was used in Germany for many years and is currently used by Sasol in South Africa. In the process, coal is gasified to produce synthesis gas (syngas; a balanced purified mixture of carbon monoxide and hydrogen) and the syngas condensed using Fischer-Tropsch catalysts to produce low-boiling hydrocarbon derivatives which are further processed into gasoline and diesel. Syngas can also be converted to methanol, which can be used as a fuel, fuel additive, or further processed into gasoline via the Mobil M-gas process.

In the process, methanol is first made from methane (natural gas) in a series of three reactions:

Steam reforming:	$CH_4 + H_2O \rightarrow CO + 3H_2$	$\Delta rH = +206$ kJ/mol
Water shift reaction:	$CO + H_2O \rightarrow CO_2 + H_2$	$\Delta rH = +206$ kJ/mol
Methanol synthesis:	$2H_2 + CO \rightarrow CH_3OH$	$\Delta rH = -92$ kJ/mol
Overall reaction:	$CO_2 + CO + 5\,H_2 \rightarrow 2\,CH_3OH + H_2O + \text{heat}$	

The methanol is then converted to gasoline by (i) a dehydration step to produce dimethyl ether:

$$2CH_3OH \rightarrow CH_3OCH_3 + H_2O$$

This is then further dehydrated over a zeolite catalyst, ZSM-5, to give a hydrocarbon mixture that has the same boiling range as gasoline.

Many of the methods for the production of hydrocarbon fuels from coal (as well as the conversion of coal to synthesis gas) release carbon dioxide (CO_2) in the conversion process, far more than is released in the production of liquid fuels from crude oil. If these methods were adopted to replace declining crude oil supplies, carbon dioxide emissions would be greatly increased on a global scale. Hence, carbon dioxide sequestration has been proposed to avoid releasing it into the atmosphere, though no pilot projects have confirmed the feasibility of this approach on a wide scale. Sequestration, however, may well add to the costs of synthetic fuels.

3.4 Oil shale

Oil shale is a fine-grained sedimentary rock containing relatively large amounts of organic matter (called *kerogen*) from which significant amounts of shale oil and combustible gas can be extracted by destructive distillation.

Oil shale, or the kerogen contained therein, does not have definite geological definition or specific chemical formula. Oil shales vary by the chemical construct, type of kerogen, age, and depositional history, including the organisms from which they were derived. Based upon environment of deposition, oil shales could be divided into three groups which are of terrestrial origin, lacustrine (lake) origin, and marine origin.

The term *oil shale* is a misnomer. The shale does not contain oil nor is it commonly shale. The organic material is chiefly kerogen, and the *shale* is usually a relatively hard rock, called *marl*. Properly processed, kerogen can be converted into a substance somewhat similar to crude oil. However, the kerogen in oil shale has not gone through the *oil window* by which crude oil is produced and to be converted into a liquid hydrocarbon product, it must be heated to a high temperature. By this process the organic material is converted into a liquid, which must be further processed to produce oil which is said to be better than the lowest grade of oil produced from conventional oil deposits, but of lower quality than the upper grades of conventional oil.

Oil shale occurs in many parts of the world ranging from deposits of little or no economic value to those that occupy thousands of square miles and contain many billions of barrels of potentially extractable shale oil. Total world resources of oil shale are conservatively estimated at 2.6 trillion (2.6×10^{12}) barrels of oil equivalent. With the continuing decline of crude oil supplies, accompanied by increasing costs of crude oil–based products, oil shale presents opportunities for supplying some of the fossil energy needs of the world in the years ahead.

The organic content of oil shale is much higher than those of normal and ordinary rocks, and typically range from 1% to 5% by mass (lean shale) to 15%–20% by mass (rich shale). This natural resource is widely scattered in the entire world, and occurrences are scientifically closely linked to the history and geological evolution of the Earth. Due to its abundance and wide distribution throughout the world, its utilization has a long history, both documented and undocumented. It is also obvious that these shales must have been relatively easy sources for domestic energy requirements for the ancient world. Mainly due to the ease of handling and transportation, solid fuels were more convenient in the earlier human history and the examples are plentiful, including wood and coal.

There are several advantages and merits associated with oil shale commercialization and exploitation. They are (a) worldwide abundance and distribution, (b) politically less sensitive fossil fuel resource, (c) source for high quality crude products, (d) source for aliphatic liquid fuels, and (e) surface mining or in-situ processing possibilities. The mostly aliphatic nature of the shale oil is very attractive from the environmental and processing standpoints, since aromatics in liquid fuel are generally viewed negatively due to the high potential for evaporative and fugitive emission that introduces a high level of volatile organic compounds (VOCs) into the atmosphere.

The quality of oil shale can be very simply represented by its oil content in the shale. To compare the oil contents as recoverable amounts of hydrocarbon from a wide variety of oil shale, a standardized method of oil content determination is needed. Fischer assay is most generally used for this purpose and it has definite merits based on its simplicity and use of a common apparatus (Fischer assay).

There are two conventional approaches to oil shale processing. In one, the shale is fractured in-situ and heated to obtain gases and liquids by wells. The second is by mining, transporting, and heating the shale to approximately 450°C (840°F), adding hydrogen to the resulting product, and disposing of and stabilizing the waste. Both processes use considerable amounts of water. The total energy and water requirements together with environmental and monetary costs (to produce shale oil in significant quantities) have so far made production uneconomic. During and following the oil crisis of the 1970s, major oil companies, working on some of the richest oil shale deposits in the world in western United States, spent several billion dollars in various unsuccessful attempts to commercially extract shale oil.

The amount of shale oil that can be recovered from a given deposit depends upon many factors. Some deposits or portions thereof, such as large areas of the Devonian black shale in eastern United States, may be too deeply buried to economically mine in the foreseeable future. Surface land uses may greatly restrict the availability of some oil shale deposits for development, especially those in the industrial Western countries. The bottom line in developing a large oil shale industry will be governed by the price of crude oil. When the price of shale oil is comparable to that of crude oil because of diminishing resources of crude, then shale oil may find a place in the world fossil energy mix.

In order to extract hydrocarbon derivatives, the oil shale is typically subjected to a thermal treatment, scientifically categorized as *destructive distillation*. A collective scientific term for hydrocarbon derivatives in oil shale is called *kerogen*, an ill-defined macromolecule which, when heated, undergoes both physical and chemical changes. Physical changes involve phase changes, softening, expansion, and oozing through pores, while chemical changes typically involve bond cleavages mostly on carbon−carbon bonds that result in smaller and simpler molecules. The chemical change is often termed as *pyrolysis* or *thermal decomposition*. The pyrolysis reaction is endothermic in nature, requiring heat, and produces lower-boiling molecules thereby increasing the pressure.

In addition to the kerogen pyrolysis reaction, carbonate decomposition reactions are also included here as principal chemical reactions, due to their abundant existence and also to their reaction temperature ranges that overlap the kerogen pyrolysis temperature range. Other mineral matters of oil shale that are worthy of note are alumina, nahcolite, and dawsonite. Some of the processes are designed to recover these mineral matters for economic benefit to the overall process. The pyrolysis reaction is quite active at a temperature above 400°C, where most of the commercial retorting processes are operated.

Most of the ex-situ processes utilize the spent (processed) shale as a char source to supply the process heat, thus accomplishing higher energy efficiency for the process. The typical temperature required to carry out such pyrolysis reaction is in the range of 450°C −520°C (840°F −960°F). In order to make the efficiency of oil extraction higher, oil shale rocks need to be ground to finer particle sizes, thus alleviating mass transfer resistance and at the same time facilitating smoother flow for cracked hydrocarbon derivatives to escape out of the rock matrix. Due to the poor porosity or permeability of oil shale rock, the rock matrix often goes through stress fracture during pyrolysis operation, typically noticed as crackling.

Major drawbacks of this type of process involve (a) "mining first" operation, which is costly, (b) transportation or conveying the mined shales to retorting facilities, (c) size reduction such as rubbelizing, grinding, or milling, and (d) returning the spent shale back to the environment. Often, the mass percentage of oil content of oil shale or the volume of recoverable oil from unit mass of oil shale is used as a measuring parameter. The latter is called Fischer Assay, which is a standardized laboratory test for determining the oil yield from oil shale to be expected from a conventional shale oil extraction. In the test method, a 100 gram sample of oil shale sample is crushed to <2.38 mm, then heated in a small aluminum retort to 500°C (930°F) at a rate of 12 °C/min (22°F/min), and held at that temperature for 40 minutes. The distilled vapors of oil, gas, and water are passed through a condenser and cooled with ice water into a graduated centrifuge tube. The oil yields achieved by other technologies are often reported as a percentage of the Fischer Assay oil yield. However, this value should not be considered as the maximum recoverable oil content for the shale or the oil content itself in the shale. As a result the standard test method [ASTM D3904. Test Method for Oil From Oil Shale (Resource Evaluation by the Fischer Assay Procedure)] has been withdrawn from the list of standard ASTM test methods and there is no replacement.

Several of the ex-situ retorting processes have been commercially tested on large scales and also proven effective for designed objectives. Some of the successfully demonstrated processes include: (a) Gas Combustion Retort process, (b) TOSCO (The Oil Shale Corporation) process, (c) Union Oil retorting process, (d) Lurgi-Ruhrgas process, (d) Superior Oil multi-mineral process, and (e) Petróleo Brasileiro (Petrobrás) process. In gas combustion retort, partial combustion of residual char provides the thermal energy for heating the shale via direct contact, thus achieving energy efficiency. TOSCO process uses heated ceramic balls to provide the thermal energy for heating the shale by direct contact, and also successfully implements multilevels of heat recovery and energy integration strategy. The Union Oil retorting process is unique and innovative with utilizing well-designed rock pumps and adopting a number of designs for heating shales in the retort. The Lurgi-Ruhrgas (LR) process produces hydrocarbon derivatives from oil shale by bringing raw shale in contact with hot fine-grained solid heat carrier, which can be just spent shale (ii). The Petrobrás process was operated for approximately 10 years in southern Brazil, treating over 3,500,000 tons of Irati (Permian age) oil shale to produce more than 1,500,000 barrels of shale oil and 20,000 tons of sulfur.

In-situ retorting of oil shale does not involve any mining operation, except starter holes and implementation digging. Therefore, in-situ retorting does not require any transportation of shale out of the oil shale field. In-situ retorting is often called subsurface retorting. The advantages of in-situ retorting processes include (a) no need for mining, (b) no need for oil shale transportation, and (c) cost and labor effectiveness. However, difficulties are in the domain of (a) process control and reliability, (b) environmental and ecological impact before and after the processing, (c) long-term ground water contamination, and (d) process efficiency. Sites for in-situ processing are put back to normal vegetated areas or to the original forms of the environment as close as possible, upon completion.

Oil shale can be ignited and burst into fire, if conditions are met. Depending upon the shale types and their hydrocarbon contents, the self-ignition temperature (SIT) of dry shale in the atmosphere varies widely from as low as 135°C–420°C (275–790°F). The finer the particle, the stronger is the possibility of catching fire spontaneously. However, it is generally too expensive to grind oil shale to fine meshes for processing. This threshold value is not generally set for all types of oil shale or processes; however, it is estimated to be approximately 1–3 mm as a minimum. Oozing oils from raw or spent shale can complicate the safety matters by exposing not only potentially hazardous air pollutants (HAPs) to the environment, but also highly combustible matters in contact with air. This can be especially true with spent shale transportation, if the residual hydrocarbon derivatives are not burnt off for heat recovery for the process. Reburial or disposal of spent shale potentially renders an ecological and environmental concern. Since spent shale is the shale that has gone through a thermal treatment process, it is more likely to become a source for leaching of minerals and organics that may be harmful to the ecological constituents and contaminating the ground waterway.

3.5 Wax

Naturally occurring wax, often referred to as *mineral wax*, occurs as a yellow to dark brown, solid substance that is composed largely of paraffins. Fusion points vary from 60°C (140°F) to as high as 95°C (203°F). They are usually found associated with considerable mineral matter, as a filling in veins and fissures or as an interstitial material in porous rocks. The similarity in character of these native products is substantiated by the fact that, with minor exceptions where local names have prevailed, the original term *ozokerite* (*ozocerite*) has served without notable ambiguity for mineral wax deposits.

Ozokerite (*ozocerite*), from the Greek meaning *odoriferous wax*, is a naturally occurring hydrocarbon material composed chiefly of solid paraffins and cycloparaffins (i.e., hydrocarbon derivatives). Ozocerite usually occurs as stringers and veins that fill rock fractures in tectonically disturbed areas. It is predominantly paraffinic material (containing up to 90% nonaromatic hydrocarbon derivatives) with a high content (40%–50%) of normal or slightly branched paraffins as well as cyclic paraffin derivatives. Ozocerite contains approximately 85% carbon, 14% hydrogen, and 0.3% each of sulfur and nitrogen and is, therefore, predominantly a mixture of pure hydrocarbon derivatives; any nonhydrocarbon constituents are in the minority.

Ozocerite deposits are believed to have originated in much the same way as mineral veins, the slow evaporation and oxidation of crude oil having resulted in the deposition of its dissolved paraffin hydrocarbon derivatives in the fissures and crevices previously occupied by the liquid. As found native, ozocerite varies from a very soft wax to a black mass as hard as gypsum.

Deposits of ozocerite occur in Scotland, Northumberland (England), and Wales, as well as from approximately 30 different countries, including the United States (Utah)— no systematic effort has been made to ascertain the quantity of ozokerite in Utah but the veins are usually several inches wide and may continue for several hundred feet. The main sources of commercial supply are in Galicia, Dzwiniacz, and Starunia, though the mineral is found at other points on both flanks of the Carpathians.

Pure ozocerite is generally odorless but may have a slight odor, which in keeping with it being a mixture of long chain hydrocarbon derivatives. Ozocerite is mainly a mixture of n-alkanes (C_{20} to C_{50}) that occasionally accompany deposits of crude oil, coal, or lignite. In most samples, n-alkanes near C_{30} were most abundant. Crude ozocerite is black; after refining, its color varies from yellow to white. It hardens on aging and the hardness varies according to its source and refinement. Ceresin is a white to yellow waxy mixture of paraffins and hydrocarbon derivatives obtained by purification of ozocerite.

The specific gravity of ozocerite ranges from 0.85 to 0.96, and the melting point falls in the range 60°C –95°C (140°F –200°F). The flash point is high, on the order of 205°C (400°F). Ceresin (ceresine, cerasin), a chemical relative of ozocerite, is lower melting 55°C –72°C (130°F –160°F). Both waxes are nontoxic and nonhazardous, thus permitting use in personal-care applications. Ozocerite is soluble in solvents that are commonly employed for dissolution of crude oil derivatives, e.g., toluene, benzene, carbon disulfide, chloroform, and ethyl ether. Ozocerite varies in color from light yellow to dark brown, and frequently appears green owing to dichroism. Chemically, ozocerite consists of a mixture of various hydrocarbon derivatives, containing 85%–87% w/w/ carbon and 13%–14% w/w/hydrogen. Ozocerite is stable under normal conditions of storage and handling. This material may burn but will not ignite readily and combustion can yield major amounts of oxides of carbon and minor amounts of oxides of sulfur and nitrogen.

If discarded as produced, ozocerite is not an RCRA listed or characteristic hazardous waste. Use of the wax, which results in chemical or physical change or contamination, may subject it to hazardous waste regulations. Wax has also been classified using techniques such as gas chromatography, Fourier transform infrared spectroscopy, proton magnetic resonance, urea adduction, and solid liquid chromatography. The multitechnique approach used for wax reflects the potential problems that can arise for classifying crude oil, especially when the complexity of crude oil vis-à-vis wax is considered.

Ozocerite is recovered by mining and the workings of an ozocerite mine may extend to a depth of 700 feet. In these mines there are the usual main shafts and galleries, the ozocerite being reached by levels driven along the strike of the deposit. The wax, as it reaches the surface, varies in purity, and, in new workings; especially, only hand-picking is needed to separate the pure material. In other cases much earthy matter is mixed with the material, and then the rock or shale having been eliminated by hand-picking, the *wax-stone* is boiled with water in large coppers, when the pure wax rises to the surface. This is again melted without water, and the impurities are skimmed off, the material being then run into slightly conical cylindrical molds, and thus made into blocks for the market.

The crude ozokerite is refined by treatment first with oil of vitriol, and subsequently with charcoal. The refined ozokerite, which usually has a melting point of 61°C −78°C (142°F −172°F), is largely used as an adulterant of beeswax, and is frequently colored artificially to resemble that product in appearance. On distillation in a current of superheated steam, ozokerite yields a candle-making material resembling the paraffin obtained from crude oil and shale oil but of higher melting point, and therefore of greater value if the candles made from it are to be used in hot climates. They are also used in the distillation of low-boiling oils and a product resembling Vaseline. The residue stills consists of a hard, black, waxy substance, which in a mixture with India rubber was employed under the name of *okonite* as an electrical insulator. From the residue a form of the material known as *heel-ball*, used to impart a polished surface to the heels and soles of boots, was also manufactured.

Mining of ozokerite fell off after 1940 due to competition from hydrocarbon derivatives manufactured from crude oil, but as it has a higher melting point than most crude oil waxes, it is still favored for some applications, such as electrical insulators and candles, or in extrasoft paper tissues.

3.6 Biomass

Biomass refers to a) energy crops grown specifically to be used as fuel, such as fast-growing trees or switch grass, b) agricultural residues and byproducts, such as straw, sugarcane fiber, and rice hulls, and c) residues from forestry, construction, and other wood-processing industries (NREL, 2003).

Biomass is material that is derived from plants (Wright et al., 2006) and there are many types of biomass resources currently used and potentially available. *Biomass* is a term used to describe any material of recent biological origin, including plant materials such as trees, grasses, agricultural crops, and even animal manure. Other biomass components, which are generally present in minor amounts, include triglycerides, sterols, alkaloids, resins, terpenes, terpenoids and waxes. This includes everything from *primary sources* of crops and residues harvested/collected directly from the land to *secondary sources* such as sawmill residuals, to *tertiary sources* of postconsumer residuals that often end up in landfills. A *fourth source*, although not usually categorized as such, includes the gases that result from anaerobic digestion of animal manures or organic materials in landfills (Wright et al., 2006).

The production of hydrocarbon derivatives from renewable plant-based feedstock utilizing state-of-the-art conversion technologies presents an opportunity to maintain competitive advantage and contribute to the attainment of national environmental targets. Bioprocessing routes have a number of compelling advantages over conventional petrochemicals production; however, it is only in the last decade that rapid progress in biotechnology has facilitated the commercialization of a number of plant-based chemical processes. It is widely recognized that further significant production of plant-based chemicals will only be economically viable in highly integrated and efficient production complexes producing a diverse range of chemical products. This biorefinery concept is analogous to conventional oil refineries and petrochemical complexes that have evolved over many years to maximize process synergies, energy integration, and feedstock utilization to drive down production costs.

Plants offer a unique and diverse feedstock for hydrocarbon derivatives. Plant biomass can be gasified to produce synthesis gas; a basic chemical feedstock and also a source of hydrogen for a future hydrogen economy. In addition, the specific components of plants such as (i) carbohydrates, (ii) vegetable oils, (iii) plant fiber, and complex organic molecules known as primary and secondary metabolites can be utilized to produce a range of valuable monomers, chemical intermediates, pharmaceuticals, and materials:

Carbohydrates (starch, cellulose, sugars): starch readily obtained from wheat and potato, while cellulose is obtained from wood pulp. The structures of these polysaccharides can be readily manipulated to produce a range of biodegradable polymers with properties similar to those of conventional plastics such as polystyrene foams and polyethylene film. In addition, these polysaccharides can be hydrolyzed, catalytically or enzymatically to produce sugars, a valuable fermentation feedstock for the production of ethanol, citric acid, lactic acid and dibasic acids such as succinic acid.

Vegetable oils are obtained from seed oil plants such as palm, sunflower, and soya. The predominant source of vegetable oils in many countries is rapeseed oil. Vegetable oils are a major feedstock for the oleochemicals industry (surfactants, dispersants, and personal care products) and are now successfully entering new markets such as diesel fuel, lubricants, polyurethane monomers, functional polymer additives, and solvents. (iii) Plant fibers including lignocellulosic fibers that are extracted from plants such as hemp and flax can replace cotton and polyester fibers in textile materials and glass fibers in insulation products. Specialties include plants can synthesize highly complex bioactive molecules often beyond the power of laboratories and a wide range of chemicals is currently extracted from plants for a wide range of markets from crude herbal remedies through to very high value pharmaceutical intermediates. These products represent a range of chemicals that can be used as such or converted to hydrocarbon derivatives.

Many different types of biomass can be grown for the express purpose of energy production and, also, for hydrocarbon production. The production of hydrocarbon derivatives depends to a large extent on the nature of the primary products and the technology available for conversion of these products to hydrocarbon derivatives. The biomass feedstock include crops specifically grown for bioenergy, and various agricultural residues, wood residues, and waste streams. Their costs and availability vary widely. Collection and transportation costs are often critical. Biorefining offers a key method to accessing the integrated production of chemicals, materials, and fuels. The biorefinery concept is analogous to that of an oil refinery.

In a manner similar to the crude oil refinery, a biorefinery would integrate a variety of conversion processes to produce multiple product streams such as motor fuels and other chemicals from biomass. In short, a biorefinery would combine the essential technologies to transform biological raw materials into a range of industrially useful intermediates. However, the type of biorefinery would have to be differentiated by the character of the feedstock. For example, the *crop biorefinery* would use raw materials such as cereals or maize and the *lignocellulose biorefinery* would use raw material with high cellulose content, such as straw, wood, and paper waste.

For example, a biorefinery using lignin as a feedstock would produce a range of valuable organic chemicals and liquid fuels that, at the present time, could supplement or even replace equivalent or identical products currently obtained from crude oil, coal, or gas. Thus, the biorefinery is analogous to an oil refinery in which crude oil is separated into a series of products, such as gasoline, heating oil, jet fuel, and petrochemicals.

By producing multiple products, a biorefinery can take advantage of the differences in biomass components and intermediates and maximize the value derived from the biomass feedstock. A biorefinery might, for example, produce one or several low-volume, but high-value, chemical products and a low-value, but high-volume liquid transportation fuel, while generating electricity and process heat for its own use and perhaps enough for sale of electricity. The high-value products enhance profitability, the high-volume fuel helps meet national energy needs, and the power production reduces costs and avoids greenhouse-gas emissions.

As a feedstock, biomass can be converted by thermal or biological routes to a wide range of useful forms of energy including process heat, steam, electricity, as well as liquid fuels, chemicals, and synthesis gas. As a raw material, biomass is a nearly universal feedstock due to its versatility, domestic availability, and renewable character. At the same time, it also has its limitations. For example, the energy density of biomass is low compared to that of coal, liquid crude oil, or crude oil–derived fuels. The heat content of biomass, on a dry basis (7000 to 9000 Btu/lb) is at best comparable with that of a low-rank coal or lignite, and substantially (50%–100%) lower than that of anthracite, most bituminous coals, and crude oil. Most biomass, as received, have a high burden of physically adsorbed moisture, up to 50% by weight. Thus, without substantial drying, the energy content of a biomass feed per unit mass is even less.

By analogy with crude oil every element of the plant feedstock will be utilized including the low value lignin components. However, the different compositional nature of the biomass feedstock, compared to crude oil, will require the application of a wider variety of processing tools in the biorefinery. Processing of the individual components will utilize conventional thermochemical operations and state-of-the-art bioprocessing techniques. The production of biofuels in the biorefinery complex will service existing high volume markets, providing economy-of-scale benefits and large volumes of byproduct streams at minimal cost for upgrading to valuable chemicals. A pertinent example of this is the glycerol byproduct produced in biodiesel plants. Glycerol has high functionality and is a potential platform chemical for conversion into a range of higher value chemicals. The high volume product streams in a biorefinery need not necessarily be a fuel but could also be a large volume chemical intermediate such as ethylene or lactic acid. Flash pyrolysis can be used to convert biomass into a type of fuel-type product (biooil). The process (*fast pyrolysis*, *flash pyrolysis*) occurs when solid fuels are heated at temperatures between 350°C and 500°C (660°F –930°F) for a very short period of time (<2 s). The biooils currently produced are suitable for use in boilers for electricity generation.

In another process, the feedstock is fed into a fluidized bed (at 450°C –500°C, 840°F–930°F) and the feedstock flashes and vaporizes. The resulting vapors pass into a cyclone where solid particles, char, are extracted. The gas from the cyclone enters a quench tower where they are quickly cooled by heat transfer using biooil already made in the process. The biooil condenses into a product receiver and any noncondensable gases are returned to the reactor to maintain process heating. The entire reaction from injection to quenching takes only 2 s.

More important, in terms of hydrocarbon production, biomass can be gasified to produce synthesis gas (syngas) composed primarily of hydrogen and carbon monoxide, also called *biosyngas* (Cobb, 2007). The production of high-quality syngas from biomass, which is later used as a feedstock for biomass-to-liquids (BTL) production, requires particular attention. This is due to the fact that the production of synthesis gas from biomass is indeed the novel component in the gas-to-liquid concept—obtaining syngas from fossil raw materials (natural gas and coal) is a relatively mature technology.

In principle, the larger the carbon and hydrogen content in raw materials employed in gas-to-liquids processing is, the easier and more efficient the carbon monoxide and hydrogen. Hence, the natural gas pathway is the most convenient one, since natural gas is gaseous and contains virtually carbon and hydrogen only. Solid raw materials (biomass, coal) involve more processing, because first they have to be gasified and the product gas should be cleaned up from other components such as nitrogen oxides (NO_X), sulfur oxides (SO_X), and particulate matter to the extent of getting as high as possible purity of syngas. Two basic types of biomass raw materials are distinguished, viz., woody material and herbaceous material. Currently woody material accounts for approximately 50% of total world bioenergy potential. Another 20% is strawlike feedstock, obtained as a byproduct from agriculture. The dedicated cultivation of strawlike energy crops could increase the herbaceous share up to 40% (Boerrigter and Van der Drift, 2004).

There are three main types of gasifiers—fixed bed, fluidized bed, and entrained flow. The air-blown direct gasifiers operated at atmospheric pressure and used in power generation—fixed bed updraft and downdraft and fluidized bed bubbling and circulating bed gasifiers—are not suitable for biomass-to-liquids production. In addition, downdraft fixed bed gasifiers face severe constraints in scaling and are fuel inflexible, being able to process only fuels with well-defined properties. Updraft fixed bed gasifiers have fewer restrictions in scaling but the produced gas contains a lot of tars and methane.

The production of hydrocarbon derivatives form biomass biofuels to supplement (even replace thought time) oil and natural gas is in active development, focusing on the use of cheap organic matter (usually cellulose, agricultural waste, and sewage waste) in the efficient production of liquid and gas biofuels which yield high net energy gain. The carbon in biofuels was recently extracted from atmospheric carbon dioxide by growing plants, so burning it does not result in a net increase of carbon dioxide in the atmosphere of the Earth. As a result, biofuels are seen by many as a way to reduce the amount of carbon dioxide released into the atmosphere by using them to replace nonrenewable sources of energy. Other observers are not supporting of this concept.

Thus, biomass-to-liquids (BTL) both produce synthetic fuels out of biomass in the so-called Fischer-Tropsch process. The synthetic biofuel containing oxygen is used as additive in high quality diesel and petrol. Furthermore, the diesel fraction produced from biomass is suitable for use in diesel engines. Biologically produced crude oil can be refined into kerosene, crude oil, diesel, and other fractions. Feedstock for such products are (i) vegetable oil, (ii) waste vegetable oil such as waste cooking oils and greases produced in quantity mostly by commercial kitchens, (iii) transesterification of animal fats and vegetable oil can yield biodiesel that is directly usable in crude oil–diesel engines, (iv) biologically derived crude oil— a misnomer— is produced together with biogas and carbonaceous solid via the thermal depolymerization of complex organic materials including non-oil-based materials (for example, waste products such as old tires, offal, wood, and plastic), (v) pyrolysis oil may be produced out of biomass and products such as wood waste using heat only in the flash pyrolysis process but the oil has to be treated before using in conventional fuel systems or internal combustion engines (water + pH).

Biomass currently supplies 14% of the energy needs of the world, but has the theoretical potential to supply 100%. Most present day production and use of biomass for energy is carried out in a very unsustainable manner with a great many negative environmental consequences. If biomass is to supply a greater proportion of the energy needs of the world (in the form of hydrocarbon derivatives), the challenge will be to produce biomass and to convert and use it without harming the natural environment. Technologies and processes exist today which, if used properly, make biomass-based fuels less harmful to the environment than fossil fuels. Applying these technologies and processes on a site-specific basis in order to minimize negative environmental impacts is a prerequisite for sustainable use of biomass energy in the future.

3.7 Municipal and industrial waste

Municipal solid waste (MSW), commonly known as trash or garbage in the United States, consists of items that are discarded by the public. The term *garbage* can also refer specifically to food waste. The current municipal solid waste (MSW) landfill sites are a large source of human-related methane emissions and, in some countries such as the United States, can account for up to a substantial portion of these emissions (Johnston et al., 2000; Pawlowska, 2014). At the same time, methane emissions from landfills represent a lost opportunity to capture and use a significant energy resource. Some landfills are also used for waste management purposes, such as the temporary storage, consolidation, and transfer, or processing of waste material (sorting, treatment, or recycling). In addition, landfill gas can be upgraded to natural gas and another advantage manifests itself by having a specific location for disposal that can be monitored, where waste can be processed to remove all recyclable materials before tipping.

Generally, waste can be classified in several ways and typical examples of such classifications are: (i) biodegradable waste, (ii) recyclable materials such as paper, cardboard, glass, metals, tires, and batteries. (iii) electronic waste, and (iv) discarded clothing. Waste material such as toxic waste and biomedical waste are regulated to be sent for disposal in specified landfills and through high temperature combustion, such as in a cement kiln. The collection of such wastes not only includes the gathering of the waste and recyclable materials, but also the transport of these materials to the location where the collection vehicle is emptied and the waste is discharged into the landfill. This location may be a material processing facility, a transfer station, or a landfill disposal site.

Municipal waste (sometime referred to as urban waste) includes municipal solid waste (MSW) generated by household and commercial activities and liquid waste or sewage. Municipal solid waste is produced and collected each year with the vast majority being disposed of in open fields. Most municipal solid waste is currently disposed of in landfill sites. However, the disposal of this waste is a growing problem worldwide. Much of the waste could be used for energy production though incineration and processes. It is also possible to use the methane produced in landfill sites for energy production.

On the other hand, industrial waste—which may be gaseous, liquid, or solid—is the waste produced by industrial activity which includes any material that is rendered useless during a manufacturing process such as that of factories, industries, mills, and mining operations. Types of industrial waste include dirt and gravel, masonry and concrete, scrap metal, oil, solvents, chemicals, scrap lumber, even vegetable matter from restaurants. Industrial waste may be hazardous or nonhazardous. Hazardous waste may be toxic, ignitable, corrosive, reactive, or radioactive. Industrial waste may pollute the air, the soil, or nearby water sources, eventually ending up in the sea. Also, industrial waste is often mixed into municipal waste, making accurate assessments difficult.

Urban wood and yard wastes are similar in nature to agricultural residues in many regards. A biomass facility will rarely need to purchase urban wood and yard wastes, and most likely can charge a tipping fee to accept the fuel. Many landfills are already sorting waste material by isolating wood waste. This waste could be diverted to a biomass project, and although the volume currently accepted at the landfills would not be enough on its own to fuel a biomass project, it could be an important supplemental fuel and could provide more value to the community in which the landfill resides through a biomass project than it currently does as daily landfill cover.

The biomass resource in municipal solid waste comprises the putrescible materials, paper, and plastic and averages 80% of the total municipal solid waste collected. Municipal solid waste can be converted into energy by direct combustion, or by natural anaerobic digestion in the engineered landfill. At the landfill sites the gas produced by the natural decomposition of municipal solid waste (approximately 50% methane and 50% carbon dioxide) is collected from the stored material and scrubbed and cleaned before feeding into internal combustion engines or gas turbines to generate heat and power. The organic fraction of municipal solid waste can be anaerobically stabilized in a high-rate digester to obtain biogas for electricity or steam generation.

3.8 Wood and wood wastes

Waste wood arises from various sectors including municipal arisings, retail, the wood and wood products sector, furniture, transport, and packaging. The majority of the residual wood waste is mixed, containing contaminants in the form of glue, resins in chipboard and medium density fiberboard, paint, and plastic coatings. Other wood waste includes wood derived from natural forests, natural woodlands and forestry plantations, namely fuelwood and charcoal from these sources. These wastes include sawdust and other residues from forestry and wood processing activities.

The size of the wood waste resource depends upon how much wood is harvested for lumber, pulp, and paper. Finally, fuelwood can be grown in plantations like a crop. Fast growing species such as poplar, willow, or eucalyptus can be harvested every few years. With short-rotation poplar coppices grown in three 7-year rotations, it is now possible to obtain 10–13 tons of dry matter per hectare annually on soil of average or good quality. Waste wood from the forest products industry such as bark, sawdust, and board ends are widely used for energy production. This industry, in many cases, is now a net exporter of electricity generated by the combustion of wastes.

Overall, wood wastes of all types make excellent biomass fuels and can be used in a wide variety of biomass technologies. Combustion of woody fuels to generate steam or electricity is a proven technology and is the most common biomass-to-energy process. Different types of woody fuels can typically be mixed together as a common fuel, although differing moisture content and chemical makeup can affect the overall conversion rate or efficiency of a biomass project.

There are at least six subgroups of woody fuels: (i) forestry residues which include in-forest woody debris and slash from logging and forest management activities, (ii) mill residues which include byproducts such as sawdust, hog fuel, and wood chips from lumber mills, plywood manufacturing, and other wood processing facilities, (iii) agricultural residues which includes byproducts of agricultural activities including crop wastes, waste from vineyard and orchard pruning, and rejected agricultural products, (iv) urban wood and yard wastes which include residential organics collected by municipal programs or recycling centers and construction wood wastes, (v) dedicated biomass crops which includes trees, corn, oilseed rape, and other crops grown as dedicated feedstocks for a biomass project, and (vi) chemical recovery fuels—some time known as black liquor—which includes woody residues recovered out of the chemicals used to separate fiber for the pulp and paper industry. Mill residues are a much more economically attractive fuel than forestry residues, since the in-forest collection and chipping are already included as part of the commercial mill operations. Biomass facilities collocated with and integral to the mill operation have the advantage of eliminating transportation altogether and thus truly achieve a no-cost fuel.

Softwood residues are generally in high demand as feedstocks for paper production, but hardwood timber residues have less demand and fewer competing uses. In the past, as much as 50% of the trees were left onsite at the time of harvest. Whole tree harvest systems for pulp chips recover a much larger fraction of the wood. Wood harvests for timber production often generate residues which may be left on the site or recovered for pulp production. Economics of wood recovery depend greatly on accessibility and local demand. Underutilized wood species include Southern red oak, poplar, and various small diameter hardwood species. Unharvested dead and diseased trees can comprise a major resource in some regions. When such timber accumulates in abundance, it comprises a fire hazard and must be removed. Such low grade wood generally has little value and is often removed by prescribed burns in order to reduce the risk of wildfires.

3.9 Agricultural residues

Agricultural residues are basically biomass materials that are byproducts of agriculture. This includes materials such as cotton stalks, wheat and rice straw, coconut shells, maize and jowar cobs, jute sticks, and rice husk. Many developing countries have a wide variety of agricultural residues in ample quantities. Large quantities of agricultural plant residues are produced annually worldwide and are vastly underutilized. The most common agricultural residue is the rice husk, which makes up 25% of rice by mass.

Agricultural residues include rice straw, wheat straw, rice husk, and corn stover, which are mostly left on the fields after harvests and used for fodder and landfill material or burnt in many places. Forestry residues consist of branches, leaves, bark, and other portions of wood. Lignocellulosic biomass in general consists of 35%−55% w/w cellulose, 25%−40% w/w hemicellulose, and 15%−25% w/w lignin with small percentage of extractives, protein, and mineral matter (which is manifested in the form of mineral ash after combustion). Most of the woody biomass, free from leaves and needles, has a mineral content less than 2% w/w whereas that of agricultural residues such as rice husk could have a mineral content as high as 20% w/w. When it comes to elemental compositions, a little higher carbon and hydrogen are present in wood residues compared with agricultural residues.

Corn stalks and wheat straws are the two agricultural residues produced in the largest quantities. However, many other residues such as potato and beet waste may be prevalent in some regions. In addition to quantity it is necessary to consider density and water content (which may restrict the feasibility of transportation) and seasonality which may restrict the ability of the conversion plant to operate on a year-round basis. Facilities designed to use seasonal crops will need adequate storage space and should also be flexible enough to accommodate alternative feedstocks such as wood residues or other wastes in order to operate year-around. Some agricultural residues need to be left in the field in order to increase tilth (the state of aggregation of soil and its condition for supporting plant growth and to reduce erosion) but some residues such as corncobs can be removed and converted without much difficultly.

Agricultural residues can provide a substantial amount of biomass fuel. Similar to the way mill residues provide a significant portion of the overall biomass consumption in areas that are copiously forested, agricultural residues from sugar cane harvesting and processing provide a significant portion of the total biomass consumption in other parts of the world. One significant issue with agricultural residues is the seasonal variation of the supply. Large residue volumes follow harvests, but residues throughout the rest of the year are minimal. Biomass facilities that depend significantly on agricultural residues must either be able to adjust output to follow the seasonal variation, or have the capacity to stockpile a significant amount of fuel.

Animal wastes include manures, renderings, and other wastes from livestock finishing operations. Although animal wastes contain energy, the primary motivation for biomass processing of animal wastes is mitigation of a disposal issue rather than generation of energy. This is especially true for animal manures. Animal manures are typically disposed of through land application to farmlands. Tightening regulations on nutrient management, surface and groundwater contamination, and odor control are beginning to force new manure management and disposal practices.

Dry animal manure, which is typically defined as having a moisture content less than 30% w/w, is produced by feedlots and livestock corrals, where the manure is collected and removed only once or twice a year. Manure that is scraped or flushed on a more frequent schedule can also be separated, stacked, and allowed to dry. Dry manure can be composted or can fuel a biomass-to-energy combustion project. Animal manure does have value to farmers as fertilizer, and a biomass-to-energy project would need to compete for the manure. However, the total volume of manure produced in many livestock operations exceeds the amount of fertilizer required for the farmlands and, in some areas/countries, nutrient management plans are beginning to limit the overfertilization of farmlands. Therefore, although there are competitive uses for the manure and low-cost disposal options at this time, manure disposal is going to become more costly over time, and the demand for alternative disposal options, including biomass-to-energy, will only increase.

Biomass technologies present attractive options for mitigating many of the environmental challenges of manure wastes. The most common biomass technologies for animal manures are combustion, anaerobic digestion, and composting. Moisture content of the manure and the amount of contaminants, such as bedding, determine which technology is most appropriate.

References

Abraham, H., 1945. Asphalts and Allied Substances, vol. I. Van Nostrand, New York.

Aguilera, R.F., Radetzki, M., 2014. The shale revolution: global gas and oil markets under transformation. Mineral Economics 26 (3), 75–84.

Ancheyta, J., Speight, J.G., 2007. Hydroprocessing of Heavy Oils and Residua. CRC-Taylor and Francis Group, Boca Raton, Florida.

Boerrigter, H., Van Der Drift, A., 2004. Biosyngas: Description of R&D Trajectory Necessary to Reach Large-Scale Implementation of Renewable Syngas from Biomass. Energy Research Center of the Netherlands, Petten, The Netherlands.

Brooks, J.D., Gould, K., Smith, J.W., 1969. Isoprenoid hydrocarbons in coal and petroleum. Nature 222, 257–259.

Chung, J.-G., Garrett, L.R., Byers, P.E., Cuchens, M.A., 1989. A survey of the amount of pristane in common fruits and vegetables. Journal of Food Composition and Analysis 2 (1), 22–27.

Cobb Jr., J.T., 2007. Production of synthesis gas by biomass gasification. In: Proceedings. Spring National Meeting AIChE. Houston, Texas, April 22-26, 2007.

Congress, U.S., 1976. Public Law FEA-76-4. Congress of the United States of America, Washington, DC.

Delbianco, A., Montanari, R., 2009. Encyclopedia of Hydrocarbons, Volume III/New Developments: Energy, Transport, Sustainability. Eni S.p.A., Rome, Italy.

Forbes, R.J., 1958. A History of Technology, V. Oxford University Press, Oxford, England.

Gary, J.H., Handwerk, G.E., Kaiser, M.J., 2007. Petroleum Refining. CRC Press, Taylor & Francis Group, Boca Raton, Florida.

Hoiberg, A.J., 1960. Bituminous Materials: Asphalts, Tars and Pitches, I & II. Interscience Publishers, New York.

Hsu, C.S., Robinson, P.R. (Eds.), 2017. Handbook of Petroleum Technology. Springer International Publishing AG, Cham, Switzerland.

Isaacs, C.M., 1992. Preliminary Petroleum Geology Background and Well Data for Oil Samples in the Cooperative Monterey Organic Geochemistry Study. Santa Maria and Santa Barbara-Ventura Basins, California. Open-File Report No. USGS 92-539-F. United States Geological Survey, Reston, Virginia.

Johnston, A.G., Edwards, J.S., Owen, J., Marshall, S.J., Young, S.D., 2000. Research into methane emissions from landfills. In: Proceedings. Waste 2000: Waste Management at the Dawn of the Third Millennium. Stratford-upon-Avon, United kingdom, pp. 13–20.

Khosrokhavar, R., Griffiths, S., Wolf, K.-H., 2014. Shale gas formations and their potential for carbon storage: opportunities and outlook. Environmental Processes 1 (4), 595–611.

Kundert, D., Mullen, M., 2009. Proper evaluation of shale gas reservoirs leads to a more effective hydraulic-fracture stimulation. Paper No. SPE 123586. In: Proceedings. SPE Rocky Mountain Petroleum Technology Conference, Denver, Colorado. April 14-16.

Lee, S., Speight, J.G., Loyalka, S., 2014. Handbook of Alternative Fuel Technologies, second ed. CRC Press, Taylor & Francis Group, Boca Raton, Florida.

Mokhatab, S., Poe, W.W., Speight, J.G., 2006. Handbook of Natural Gas Transmission and Processing. Elsevier, Amsterdam, The Netherlands, 2006.

NREL, 2003. Dollars from Sense. National Renewable Energy Laboratory, Golden, Colorado. http://www.nrel.gov/docs/legosti/fy97/20505.pdf.

Ohmes, R., 2014. Characterizing and Tracking Contaminants in Opportunity Crudes. Digital Refining. http://www.digitalrefining.com/article/1000893,Characterising_and_tracking_contaminants_in_opportunity_crudes_.html#.VJhFjV4AA.

Parkash, S., 2003. Refining Processes Handbook. Gulf Professional Publishing, Elsevier, Amsterdam, Netherlands.

Pawlowska, M., 2014. Mitigation of Landfill Gas Emissions. CRC Press, Taylor & Francis Group, Boca Raton, Florida.

Rontani, J.-F., Volkman, J.K., 2003. Phytol degradation products as biogeochemical tracers in aquatic environments. Organic Geochemistry 34 (1), 1–35.

Shalaby, H.M., 2005. Refining of Kuwait's heavy crude oil: materials challenges. In: Proceedings. Workshop on Corrosion and Protection of Metals. Arab School for Science and Technology. December 3-7, Kuwait.

Speight, J.G., 1990. Fuel Science and Technology Handbook. Marcel Dekker Inc., New York.

Speight, J.G., 2009. Enhanced Recovery Methods for Heavy Oil and Tar Sands. GPC Books, Gulf Publishing Company, Houston, Texas.

Speight, J.G., 2010. Oil: An Introduction to the Technology, Economics, and Politics of a Non-renewable Resource. M.E. Sharpe Armonk, New York.

Speight, J.G., 2011a. An Introduction to Petroleum Technology, Economics, and Politics. Scrivener Publishing, Beverly, Massachusetts.

Speight, J.G., 2011b. The Refinery of the Future. Gulf Professional Publishing, Elsevier, Oxford, United Kingdom.

Speight, J.G., 2013. The Chemistry and Technology of Coal, third ed. CRC Press, Taylor & Francis Group, Boca Raton, Florida.

Speight, J.G., 2013a. Oil Sand Production Processes. Gulf Professional Publishing Company, Elsevier, Oxford, United Kingdom.

Speight, J.G., 2013b. The Chemistry and Technology of Coal, third ed. CRC Press, Taylor & Francis Group, Boca Raton, Florida.

Speight, J.G., 2014a. The Chemistry and Technology of Petroleum, fifth ed. CRC Press, Taylor & Francis Group, Boca Raton, Florida.

Speight, J.G., 2014b. High Acid Crudes. Gulf Professional Publishing, Elsevier, Oxford, United Kingdom.

Speight, J.G., 2015. Handbook of Petroleum Product Analysis, second ed. John Wiley & Sons Inc., Hoboken, New Jersey.

Speight, J.G., 2016a. Handbook of Hydraulic Fracturing. John Wiley & Sons Inc., Hoboken, New Jersey.

Speight, J.G., 2016b. Introduction to Enhanced Recovery Methods for Heavy Oil and Tar Sands, second ed. Gulf Professional Publishing Company, Elsevier, Oxford, United Kingdom.

Speight, J.G., 2017a. Handbook of Petroleum Refining. CRC Press, Taylor & Francis Group, Boca Raton, Florida.

Speight, J.G., 2017b. Deep Shale Oil and Gas. Gulf Professional Publishing, Elsevier, Oxford, United Kingdom.

Speight, J.G., 2018. Handbook of Natural Gas Analysis. John Wiley & Sons Inc., Hoboken, New Jersey.

Speight, J.G., 2019a. Handbook of Petroleum Refining. CRC Press, Taylor & Francis Group, Boca Raton, Florida.

Speight, J.G., 2019b. Natural Gas: A Basic Handbook, second ed. GPC Books, Gulf Publishing Company, Houston, Texas.

Speight, J.G., Francisco, M.A., 1990. Studies in petroleum composition IV: changes in the nature of chemical constituents during crude oil distillation. Revue de l'Institut Français du Petrole 45, 733.

Speight, J.G., Islam, M.R., 2016. Peak Energy – Myth or Reality. Scrivener Publishing, Beverly, Massachusetts.

Sun, X., Zhang, Y., Li, X., Cui, G., Gu, J., 2013. A case Study on foamy oil characteristics of the Orinoco Belt, Venezuela. Advances in Petroleum Exploration and Development 5 (1), 37–41.

USGS, 2006. Natural Bitumen Resources of the Unbitted States. Fact Sheet 2006-3133. United States Geological Survey, Reston, Virginia.

Villarroel, T., Hernández, R., 2013. Technological Developments for Enhancing Extra Heavy Oil Productivity in Fields of the Faja Petrolifera del Orinoco (FPO), Venezuela. In: Proceedings. AAPG Annual Convention and Exhibition, Pittsburgh, Pennsylvania. May 19-22. American Association of Petroleum Geologists, Tulsa, Oklahoma.

Wright, L., Boundy, R., Perlack, R., Davis, S., Saulsbury, B., 2006. Biomass Energy Data Book: Edition 1. Office of Planning, Budget and Analysis. Energy Efficiency and Renewable Energy, United States Department of Energy, Oak Ridge, Tennessee. Contract No. DE-AC05-00OR22725. Oak Ridge National Laboratory.

Yeung, T.W., 2014. Evaluating Opportunity Crude Processing. Digital Refining. http://www.digitalrefining.com/article/1000644.

Further reading

ASTM, 2019. Annual Book of Standards. ASTM International, West Conshohocken, Pennsylvania.

Cobb, C., Goldwhite, H., 1995. Creations of Fire: Chemistry's Lively History from Alchemy to the Atomic Age. Plenum Press, New York.

McKetta, J.J. (Ed.), 1992. Petroleum Processing Handbook. Marcel Dekker Inc., New York.

Chapter 3

Hydrocarbons from crude oil

1. Introduction

The constant demand for hydrocarbon products such as liquid fuels is one of the major driving forces behind the crude oil industry. However, the other driving force is a major group of hydrocarbon products—petrochemicals—that are the basis of a major industry. There is a myriad of products that have evolved through the short life of the crude oil industry, either as bulk fractions or as single hydrocarbon products (Tables 1.1 and 1.2). And the complexities of product composition have matched the evolution of the products. In fact, it is the complexity of product composition that has served the industry well and, at the same time, had an adverse effect on product use. In addition, any nonhydrocarbon atoms such as sulfur, must be removed for product purity (Song and Ma, 2004; Ancheyta and Speight, 2007).

Crude oil products (in contrast to *petrochemicals*) are those hydrocarbon fractions that are derived from crude oil and have commercial value as a bulk product (Table 3.1). A major group of hydrocarbon products from crude oil (petrochemicals) are the basis of a major industry. They are, in the strictest sense different to crude oil products insofar as the petrochemicals are the basic building blocks of the chemical industry. Furthermore, the constant demand for hydrocarbon products such as liquid fuels is a major driving force behind the crude oil industry.

A *petrochemical* is a chemical product developed from crude oil that has become an essential part of the modern chemical industry (Parkash, 2003; Gary et al., 2007; Speight, 2014, 2017; Hsu and Robinson, 2017). The *chemical industry* is, in fact, the *chemical process industry* by which a variety of chemicals are manufactured. The chemical process industry is, in fact, subdivided into other categories that are: (i) the chemicals and allied products industries in which chemicals are manufactured from a variety of feedstock and may then be put to further use, (ii) the rubber and miscellaneous products industries which focus on the manufacture of rubber and plastic materials, and (iii) crude oil refining and related industries which, on the basis of prior chapters in this text, is now self-explanatory. Thus, the petrochemical industry falls under the subcategory of *crude oil and related industries*.

There is a myriad of other products that have evolved through the short life of the crude oil industry, either as single hydrocarbon derivatives or as

Handbook of Industrial Hydrocarbon Processes. https://doi.org/10.1016/B978-0-12-809923-0.00003-5

TABLE 3.1 The various distillation fractions of crude oil.

Product	Lower carbon number[a]	Upper carbon number[a]	Lower b.p., °C[a]	Upper b.p., °C[a]	Lower b.p., °F[a]	Upper b.p., °F[a]
Liquefied petroleum gas	C3	C4	−42	−1	−44	31
Naphtha	C5	C17	36	302	97	575
Kerosene	C8	C18	126	258	302	575
Low-boiling gas oil	C12	>C20	216	421	>345	>650
High-boiling gas oil	>C20		>345		>650	
Residuum	>C20		>345		>660	

[a]*The carbon number and boiling point are difficult to assess accurately because of variations in production parameters from refinery to refinery and are inserted for illustrative purposes only.*

hydrocarbon fractions (Table 3.2). And the complexities of product composition have matched the evolution of the products. In fact, it is the complexity of product composition that has served the industry well and, at the same time, had an adverse effect on product use. Product complexity has made the industry unique among industries. Indeed, current analytical techniques that are accepted as standard methods, for example, the aromatics content of fuels (ASTM D1319; ASTM D2425; ASTM D2549; ASTM D2786; ASTM D2789) as well as proton and carbon nuclear magnetic resonance methods, yield different information. Each method will yield the *% w/w aromatics* in the sample but the data must be evaluated within the context of the method.

The customary processing of crude oil does not usually involve the separation and handling of pure hydrocarbon derivatives (Fig. 3.1). Indeed, crude oil–derived products are always mixtures: occasionally simple but more often very complex. Thus, for the purposes of this chapter, such materials as the gross fractions of crude oil (such as gasoline, naphtha, kerosene, and gas oil) which are usually obtained by distillation and/or refining are classed as *crude oil products*; asphalt and other solid products (e.g., wax) are also included in this division. This type of classification separates this group of products from those obtained as crude oil chemicals (petrochemicals), for which the emphasis is on separation and purification of single chemical compounds, which are in fact starting materials for a host of other chemical products.

2. Gaseous products

Natural gas, which is predominantly methane (Chapter 4), occurs in underground reservoirs separately or in association with crude oil. The principal types of gaseous hydrocarbon derivatives are crude oil (distillation) gas, reformed natural gas, and reformed propane or liquefied petroleum gas (LPG) (Table 3.3) (Parkash, 2003; Gary et al., 2007; Speight, 2014, 2017; Hsu and Robinson, 2017).

Thus, the gaseous hydrocarbons that are produced from crude oil comprise mixtures that are predominantly natural gas and liquefied petroleum gas. The constituents of each type of gas may be similar but the variations of the amounts of these constituents can cover wide ranges. Each type of gas may be analyzed by similar methods although the presence of high boiling hydrocarbons and nonhydrocarbon species such as carbon dioxide and hydrogen sulfide may require slight modifications to the analytical test methods.

Natural gas (predominantly methane) denoted by the chemical structure CH_4 has the lowest boiling and is the least complex of all hydrocarbons. Natural gas from an underground reservoir contains hydrocarbons and nonhydrocarbon gases. Hydrocarbon gases are methane (CH_4), ethane (C_2H_6), propane (C_3H_8), butanes (C_4H_{10}), pentanes (C_5H_{12}), hexane (C_6H_{14}), heptane (C_7H_{16}), sometimes trace amounts of octane (C_8H_{18}), and higher molecular weight hydrocarbons. Some aromatics [BTX—benzene (C_6H_6), toluene ($C_6H_5CH_3$), and xylene ($CH_3C_6H_4CH_3$)] can also be present, raising safety

TABLE 3.2 Properties of hydrocarbon products from petroleum.

	Molecular weight	Specific gravity	Boiling point, °F	Ignition temperature, °F	Flash point, °F	Flammability limits in air, % v/v
Benzene	78.1	0.879	176.2	1040	12	1.35–6.65
n-Butane	58.1	0.601	31.1	761	−76	1.86–8.41
iso-Butane	58.1		10.9	864	−117	1.80–8.44
n-Butene	56.1	0.595	21.2	829	Gas	1.98–9.65
iso-Butene	56.1		19.6	869	Gas	1.8–9.0
Diesel fuel	170–198	0.875			100–130	
Ethane	30.1	0.572	−127.5	959	Gas	3.0–12.5
Ethylene	28.0		−154.7	914	Gas	2.8–28.6
Fuel oil no. 1		0.875	304–574	410	100–162	0.7–5.0
Fuel oil no. 2		0.920		494	126–204	
Fuel oil no. 4	198.0	0.959		505	142–240	
Fuel oil no. 5		0.960			156–336	
Fuel oil no. 6		0.960			150	

Gasoline	113.0	0.720	100–400	536	−45	1.4–7.6
n-Hexane	86.2	0.659	155.7	437	−7	1.25–7.0
n-Heptane	100.2	0.668	419.0	419	25	1.00–6.00
Kerosene	154.0	0.800	304–574	410	100–162	0.7–5.0
Methane	16.0	0.553	−258.7	900–1170	Gas	5.0–15.0
Naphthalene	128.2		424.4	959	174	0.90–5.90
Neohexane	86.2	0.649	121.5	797	−54	1.19–7.58
Neopentane	72.1		49.1	841	Gas	1.38–7.11
n-Octane	114.2	0.707	258.3	428	56	0.95–3.2
iso-Octane	114.2	0.702	243.9	837	10	0.79–5.94
n-Pentane	72.1	0.626	97.0	500	−40	1.40–7.80
iso-Pentane	72.1	0.621	82.2	788	−60	1.31–9.16
n-Pentene	70.1	0.641	86.0	569	-	1.65–7.70
Propane	44.1		−43.8	842	Gas	2.1–10.1
Propylene	42.1		−53.9	856	Gas	2.00–11.1
Toluene	92.1	0.867	321.1	992	40	1.27–6.75
Xylene	106.2	0.861	281.1	867	63	1.00–6.00

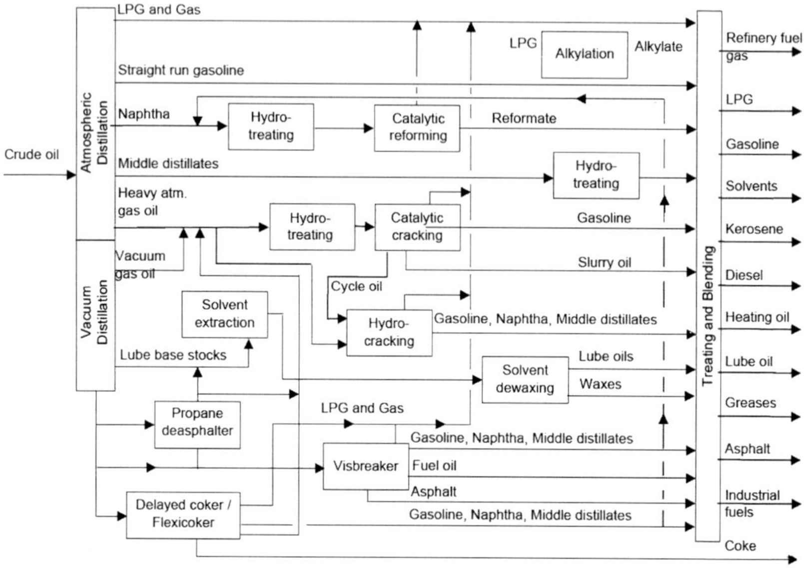

FIGURE 3.1 Schematic of a modern refinery illustrating the various hydrocarbon product streams from fuel gas to grease.

issues due to their toxicity. The nonhydrocarbon gas portion of the natural gas contains nitrogen (N_2), carbon dioxide (CO_2), helium (He), hydrogen sulfide (H_2S), water vapor (H_2O), and other sulfur compounds (such as carbonyl sulfide (COS) and mercaptans (e.g., methyl mercaptan, CH_3SH) and trace amounts of other gases. Carbon dioxide and hydrogen sulfide are commonly referred to as *acid gases* since they form corrosive compounds in the presence of water.

2.1 Manufacture

Liquefied petroleum gas (*LPG*) is the term applied to certain specific hydrocarbon derivatives and their mixtures, which exist in the gaseous state under atmospheric ambient conditions but can be converted to the liquid state under conditions of moderate pressure at ambient temperature. These are the low-boiling hydrocarbon derivatives fraction of the paraffin series, derived from refinery processes, crude oil stabilization plants, and natural gas processing plants comprising propane ($CH_3CH_2CH_3$), butane ($CH_3CH_2CH_2CH_3$), isobutane [$CH_3CH(CH_3)CH_3$], and to a lesser extent propylene ($CH_3CH{=}CH_2$) or butylene ($CH_3CH_2CH{=}CH_2$). The most common commercial products are propane, butane, or some mixture of the two (Table 3.4) and are generally extracted from natural gas or crude oil. Propylene and the butylene isomers result from cracking other hydrocarbon derivatives in a crude oil

TABLE 3.3 Illustration of the production of hydrocarbons from natural gas and crude oil.

Feedstock	Process	Product
Natural gas	Processing/refining	Methane
		Ethane
		Propane
		Butane
Crude oil	Distillation	Low-boiling hydrocarbons
		Methane
		Ethane
		Propane
		Butane
	Catalytic cracking	Ethylene
		Propylene
		Butylene isomers
		Higher-boiling olefins
	Catalytic reforming	Benzene
		Toluene
		Xylene isomers
	Coking	Ethylene
		Propylene
		Butylenes
		Higher-boiling olefins

refinery and are two important chemical feedstock. *Mixed gas* is a gas prepared by adding natural gas or liquefied petroleum gas to a manufactured gas, giving a product of better utility and higher heat content or Btu value.

2.2 Composition

The principal constituent of natural gas is methane (CH_4). Other constituents are paraffinic hydrocarbon derivatives such as ethane (CH_3CH_3), propane ($CH_3CH_2CH_3$), and the butanes [$CH_3CH_2CH_2CH_3$ and/or $(CH_3)_3CH$]. Many natural gases contain nitrogen (N_2) as well as carbon dioxide (CO_2) and

TABLE 3.4 Properties of propane and butane.

	Propane	Butane
Formula	C_3H_8	C_4H_{10}
Boiling point F.	−44°	32°
Specific gravity—gas (air = 1.00)	1.53	2.00
Specific gravity—liquid (water = 1.00)	0.51	0.58
lb/gallon—liquid @ 60°F.	4.24	4.81
BTU/gallon—gas @ 60°F.	91,690	102,032
BTU/lb. —gas	21,591	21,221
BTU/ft.^{3}—gas @ 60°F.	2516	3280
Flash point, F.	−156	−96
Ignition temperature in air, °F.	920–1020	900-1000
Maximum flame temperature in air, °F.	3595	3615
Octane number (Iso-octane = 100)	100+	92

hydrogen sulfide (H_2S). Trace quantities of argon, hydrogen, and helium may also be present. Generally, the hydrocarbon derivatives having a higher molecular weight than methane, carbon dioxide, and hydrogen sulfide are removed from natural gas prior to its use as a fuel. Gases produced in a refinery contain methane, ethane, ethylene, propylene, hydrogen, carbon monoxide, carbon dioxide, and nitrogen, with low concentrations of water vapor, oxygen, and other gases.

Unless produced specifically as a product (e.g., liquefied petroleum gas), the gaseous products of refinery operations are mixtures of various gases. Each gas is a byproduct of a refining process. Thus, the compositions of natural, manufactured, and mixed gases can vary so widely, no single set of specifications could cover all situations.

As already noted, the compositions of natural, manufactured, and mixed gases can vary so widely, no single set of specifications could cover all situations. The requirements are usually based on performances in burners and equipment, on minimum heat content, and on maximum sulfur content. Gas utilities in most states come under the supervision of state commissions or regulatory bodies and the utilities must provide a gas that is acceptable to all types of consumers and that will give satisfactory performance in all kinds of consuming equipment. However, there are specifications for liquefied petroleum gas (ASTM D1835) which depend upon the required volatility.

Since natural gas as delivered to pipelines has practically no odor, the addition of an odorant is required by most regulations in order that the presence of the gas can be detected readily in case of accidents and leaks. This odorization is provided by the addition of trace amounts of some organic sulfur compounds to the gas before it reaches the consumer. The standard requirement is that a user will be able to detect the presence of the gas by odor when the concentration reaches 1% of gas in air. Since the lower limit of flammability of natural gas is approximately 5%, this 1% requirement is essentially equivalent to one-fifth the lower limit of flammability. The combustion of these trace amounts of odorant does not create any serious problems of sulfur content or toxicity.

The different methods for gas analysis include absorption, distillation, combustion, mass spectroscopy, infrared spectroscopy, and gas chromatography (ASTM D2163; ASTM D2650; ASTM D4424). Absorption methods involve absorbing individual constituents one at a time in suitable solvents and recording of contraction in volume measured. Distillation methods depend on the separation of constituents by fractional distillation and measurement of the volumes distilled. In combustion methods, certain combustible elements are caused to burn to carbon dioxide and water, and the volume changes are used to calculate composition. Infrared spectroscopy is useful in particular applications. For the most accurate analyses, mass spectroscopy and gas chromatography are the preferred methods.

2.3 Properties and uses

Liquefied petroleum gas (LPG) is the term applied to certain specific hydrocarbons and their mixtures, which exist in the gaseous state under atmospheric ambient conditions but can be converted to the liquid state under conditions of moderate pressure at ambient temperature. Typically, fuel gas with four or less carbon atoms in the hydrogen-carbon combination have boiling points that are lower than room temperature and these fuels are gases at ambient temperature and pressure.

Liquefied petroleum gas is a hydrocarbon mixture containing propane ($CH_3CH_2CH_3$, boiling point: −42°C, −44°F), butane ($CH_3CH_2CH_2CH_3$, boiling point: 0°C, 32°F), and iso-butane [$CH_3CH(CH_3)CH_3$], boiling point: −11.7°C (10.9°F). The most common commercial fuel consists of propane and butane. In addition, liquefied petroleum gas is usually available in different grades (usually specified as: Commercial Propane, Commercial Butane, Commercial Propane-Butane (P-B) Mixtures, and Special Duty Propane). During the use of liquefied petroleum gas, the gas must vaporize completely and burn satisfactorily in the appliance without causing any corrosion or producing any deposits in the system.

Commercial propane consists predominantly of propane and/or propylene while commercial butane is mainly composed of butanes and/or butylenes. Both must be free from harmful amounts of toxic constituents and free from mechanically entrained water (that may be further limited by specifications). Commercial propane-butane mixtures are produced to meet particular requirements such as volatility, vapor pressure, specific gravity, hydrocarbon composition, sulfur and its compounds, corrosion of copper, residues, and water content. These mixtures are used as fuels in areas and at times where low ambient temperatures are less frequently encountered. Special duty propane is intended for use in spark-ignition engines and the specification includes a minimum motor octane number to ensure satisfactory antiknock performance. Propylene ($CH_3CH{=}CH_2$) has a significantly lower octane number than propane, so there is a limit to the amount of this component that can be tolerated in the mixture. Analysis by gas chromatography is also advocated. The presence of water in liquefied petroleum gas (or in natural gas) is undesirable since it can produce hydrates that will cause, for example, line blockage due to the formation of hydrates under conditions where the water *dew point* is attained. If the amount of water is above acceptable levels, the addition of a small methanol will counteract any such effect.

The specific gravity of product gases, including liquefied petroleum gas, may be determined conveniently by a number of methods and a variety of instruments (ASTM D1070; ASTM D4891). The *heat value* of gases is generally determined at constant pressure in a flow calorimeter in which the heat released by the combustion of a definite quantity of gas is absorbed by a measured quantity of water or air. A continuous recording calorimeter is available for measuring heat values of natural gases (ASTM D1826).

The lower and upper limits of *flammability* of organic compounds indicate the percentage of combustible gas in air below which and above which flame will not propagate. When flame is initiated in mixtures having compositions within these limits, it will propagate and therefore the mixtures are flammable. Knowledge of flammable limits and their use in establishing safe practices in handling gaseous fuels is important, e.g., when purging equipment used in gas service, in controlling factory or mine atmospheres, or in handling liquefied gases.

Many factors enter into the experimental determination of flammable limits of gas mixtures, including the diameter and length of the tube or vessel used for the test, the temperature and pressure of the gases, and the direction of flame propagation—upward or downward. For these and other reasons, great care must be used in the application of the data. In monitoring closed spaces where small amounts of gases enter the atmosphere, often the maximum concentration of the combustible gas is limited to one-fifth of the concentration of the gas at the lower limit of flammability of the gas-air mixture.

3. Naphtha

Naphtha (*petroleum naphtha*) is a generic term applied to refined, partly refined, or unrefined crude oil fuels and liquid fuels of natural gas which distill below 240°C (465°F) and is the volatile fraction of the crude oil, which is used as a solvent or as a precursor to gasoline and is produced by a variety of processes (Table 3.5) (Parkash, 2003; Gary et al., 2007; Speight, 2014, 2017; Hsu and Robinson, 2017). In fact, not less than 10% v/v of material should distil below 75°C (167°F); not less than 95% v/v of the material should distil below 240°C (465°F) under standard distillation conditions, although there are different grades of naphtha within this extensive boiling range that have different boiling ranges.

The term *petroleum solvent* describes the liquid hydrocarbon fractions obtained from crude oil and used in industrial processes and formulations (Parkash, 2003; Gary et al., 2007; Speight, 2014, 2017; Hsu and Robinson, 2017). These fractions are also referred to *naphtha* or as *industrial naphtha*. By definition the solvents obtained from the petrochemical industry (Chapter 12) such as alcohols, ethers, and the like are not included in this chapter. A refinery is capable of producing hydrocarbon derivatives of a high degree of purity and at the present time crude oil solvents are available covering a wide range of solvent properties including both volatile and high boiling qualities.

Naphtha (often referred to as *naft* in the older literature) is actually a generic term applied to refined, partly refined, or unrefined crude oil products. In the strictest sense of the term, not less than 10% of the material should distill below 175°C (345°F); not less than 95% of the material should distill below 240°C (465°F) under standardized distillation conditions (ASTM D86).

Naphtha has been available since the early days of the crude oil industry. Indeed, the infamous *Greek fire* documented as being used in warfare during the last three millennia is a crude oil derivative. It was produced either by distillation of crude oil isolated from a surface seepage or (more likely) by destructive distillation of the bituminous material obtained from bitumen

TABLE 3.5 Naphtha production.

Primary process	Primary product	Secondary process	Secondary product
Atmospheric distillation	Naphtha		Low-boiling naphtha
			High-boiling naphtha
	Gas oil	Catalytic cracking	Naphtha
	Gas oil	Hydrocracking	Naphtha
Vacuum distillation	Gas oil	Catalytic cracking	Naphtha
		Hydrocracking	Naphtha
	Residuum	Coking	Naphtha
		Hydrocracking	Naphtha

seepages (of which there are/were many known during the heyday of the civilizations of the Fertile Crescent. The bitumen obtained from the area of Hit (Tuttul) in Iraq (Mesopotamia) is an example of such an occurrence (Abraham, 1945; Forbes, 1958a, 1958b, 1959).

Other crude oil products boiling within the naphtha boiling range include (i) *industrial spirit* and *white spirit. Industrial spirit* comprises liquids distilling between 30°C and 200°C (−1°F−390°F), with a temperature difference between 5% volume and 90% volume distillation points, including losses, of not more than 60°C (140°F). There are several (up to eight) grades of industrial spirit, depending on the position of the cut in the distillation range defined above. On the other hand, *white spirit* is an industrial spirit with a flash point above 30°C (99°F) and has a distillation range from 135°C to 200°C (275°F−390°F).

3.1 Manufacture

In general, naphtha may be prepared by any one of several methods, which include (i) fractionation of straight-run, cracked, and reforming distillates, or even fractionation of unrefined crude oil; (ii) solvent extraction; (iii) hydrogenation of cracked distillates; (iv) polymerization of unsaturated compounds (olefins); and (v) alkylation processes (Table 3.5). In fact, the naphtha may be a combination of product streams from more than one of these processes.

The more common method of naphtha preparation is distillation. Depending on the design of the distillation unit, either one or two naphtha steams may be produced: (i) a single naphtha with an end point of approximately 205°C (400°F) and similar to straight-run naphtha, i.e., naphtha that is distilled from crude oil without thermal alteration, or (ii) this same fraction divided into a low-boiling naphtha and a high-boiling naphtha. The end point of the low-boiling naphtha is varied to suit the subsequent subdivision of the naphtha into narrower boiling fractions and may be of the order of 120°C (250°F). Before the naphtha is redistilled into a number of fractions with boiling ranges suitable for aliphatic solvents, the naphtha is usually treated to remove sulfur compounds, as well as aromatic hydrocarbon derivatives, which are present in sufficient quantity to cause an odor. Aliphatic solvents that are specially treated to remove aromatic hydrocarbon derivatives are known as *deodorized solvents*. *Odorless solvent* is the name given to high-boiling alkylate used as an aliphatic solvent, which is a byproduct in the manufacture of aviation alkylate.

Sulfur compounds are most commonly removed or converted to a harmless form by chemical treatment with lye, doctor solution, copper chloride, or similar treating agents. Hydrorefining processes are also often used in place of chemical treatment. Solvent naphtha is a solvent selected for its low sulfur content, and the usual treatment processes, if required, removes only sulfur compounds. Naphtha with a small aromatic content has a slight odor, but the aromatic constituents increase the solvent power of the naphtha and there is no need to remove aromatics unless an odor-free solvent is specified.

Naphtha that is either naturally sweet (sulfur-free and no odor), or has been treated until sweet, is subdivided into several fractions in efficient fractional distillation towers frequently called pipe stills, columns, and column steam stills. A typical arrangement consists of primary and secondary fractional distillation towers and a stripper. High-boiling naphtha, for example, is heated by a steam heater and passed into the primary tower, which is usually operated under vacuum. The vacuum permits vaporization of the naphtha at the temperatures obtainable from the steam heater.

The primary tower separates the naphtha into three parts:

1. A high boiling hydrocarbon fraction that is removed as a bottom product and sent to a cracking unit.
2. A side stream hydrocarbon product of narrow boiling range that, after passing through the stripper, may be suitable for the aliphatic solvent Varsol.
3. An overhead hydrocarbon product that is sent the secondary (vacuum) tower where the overhead product from the primary tower is divided into an overhead and a bottom product in the secondary tower, which operates under a partial vacuum with steam injected into the bottom of the tower to assist in the fractionation. The overhead and bottom products are finished aliphatic solvents, or if the feed to the primary tower is low-boiling naphtha instead of high-boiling naphtha, other aliphatic solvents of different boiling ranges are produced.

Superfractionation (Parkash, 2003; Gary et al., 2007; Speight, 2014, 2017; Hsu and Robinson, 2017) is a highly efficient fractionating tower used to separate ordinary crude oil products and isolate narrow-boiling hydrocarbon fractions. For example, to increase the yield of furnace fuel oil, high-boiling naphtha may be redistilled in a tower that is capable of making a better separation of the naphtha and the fuel oil components. The latter, obtained as a bottom product, is diverted to furnace fuel oil.

Fractional distillation as normally carried out in a refinery does not completely separate one crude oil fraction from another. One product overlaps another, depending on the efficiency of the fractionation, which in turn depends on the number of trays in the tower, the amount of reflux used, and the rate of distillation. Kerosene, for example, normally contains a small percentage of hydrocarbon derivatives that (according to their boiling points) belong in the naphtha fraction and a small percentage that should be in the gas oil fraction. Complete separation is not required for the ordinary uses of these materials, but certain materials, such as solvents for particular purposes (hexane, heptane, and aromatics), are required as essentially pure compounds. Since they occur in mixtures of hydrocarbon derivatives, they must be separated by distillation and with no overlap of one hydrocarbon with another. This requires highly efficient fractional distillation towers specially designed for the purpose and referred to as superfractionators. Several towers with 50–100 trays operated with a *high reflux ratio* may be required to separate a single compound with the necessary purity.

Azeotropic distillation (Parkash, 2003; Gary et al., 2007; Speight, 2014, 2017; Hsu and Robinson, 2017) is the use of a third component to separate two close-boiling components by means of the formation of an azeotropic mixture between one of the original components and the third component to increase the difference in the boiling points and facilitates separation by distillation.

All compounds have definite boiling temperatures, but a mixture of chemically dissimilar compounds sometimes causes one or both of the components to boil at a temperature other than that expected. For example, benzene boils at 80°C (176°F), but if it is mixed with hexane, it distills at 69°C (156°F). A mixture that boils at a temperature lower than the boiling point of either of the components is called an azeotropic mixture.

Two main types of azeotropes exist, i.e., the *homogeneous azeotrope*, where a single liquid phase is in equilibrium with a vapor phase; and the *heterogeneous azeotropes*, where the overall liquid composition which forms two liquid phases is identical to the vapor composition. Most methods of distilling azeotropes and low relative volatility mixtures rely on the addition of specially chosen chemicals to facilitate the separation. The five methods for separating azeotropic mixtures are:

1. *Extractive distillation* and *homogeneous azeotropic distillation:* where the liquid separating agent is completely miscible.
2. *Heterogeneous azeotropic distillation* or more commonly *azeotropic distillation:* where the liquid separating agent (the *entrainer*) forms one or more azeotropes with the other components in the mixture and causes two liquid phases to exist over a wide range of compositions. This immiscibility is the key to making the distillation sequence work.
3. *Distillation using ionic salts:* the salts dissociate in the liquid mixture and alter the relative volatilities sufficiently that the separation become possible.
4. *Pressure-swing distillation:* where a series of columns operating at different pressures are used to separate binary azeotropes which change appreciably in composition over a moderate pressure range or where a separating agent which forms a pressure-sensitive azeotrope is added to separate a pressure-insensitive azeotrope.
5. *Reactive distillation:* where the separating agent reacts preferentially and reversibly with one of the azeotropic constitutes. The reaction product is then distilled from the nonreacting components, and the reaction is reversed to recover the initial component.

In *simple distillation* (Parkash, 2003; Gary et al., 2007; Speight, 2014, 2017; Hsu and Robinson, 2017) a multicomponent liquid mixture is slowly boiled in a heated zone and the vapors are continuously removed as they form and, at any instant in time, the vapor is in equilibrium with the liquid remaining on the still. Because the vapor is always richer in the more volatile components than the liquid, the liquid composition changes continuously with time, becoming more and more concentrated in the least volatile species. A simple *distillation residue curve* (Parkash, 2003; Gary et al., 2007; Speight, 2014, 2017; Hsu and Robinson, 2017) is a means by which the composition of the liquid residue curves on the pot changes over time. A *residue curve map* is a collection of the liquid residue curves originating from different initial compositions. Residue curve maps contain the same information as phase diagrams, but represent this information in a way that is more useful for understanding how to synthesize a distillation sequence to separate a mixture.

All of the residue curves originate at the lowest boiling pure component in a region, move toward the intermediate boiling component, and end at the highest boiling pure component in the same region. The lowest temperature nodes are termed as *unstable nodes*, as all trajectories leave from them; the highest temperature points in the region are termed *stable nodes*, as all trajectories ultimately reach them. The point that the trajectories approach from one direction and end in a different direction (as always is the point of intermediate boiling component) are termed *saddle point*. Residue curves that divide the composition space into different distillation regions are called distillation boundaries.

Many different residue curve maps are possible when azeotropes are present. Ternary mixtures containing only one azeotrope may exhibit six possible residue curve maps that differ by the binary pair forming the azeotrope and by whether the azeotrope is minimum or maximum boiling. By identifying the limiting separation achievable by distillation, residue curve maps are also useful in synthesizing separation sequences combining distillation with other methods.

However, the separation of components of similar volatility may become economical if an *entrainer* can be found that effectively changes the relative volatility. It is also desirable that the entrainer be reasonably cheap, stable, nontoxic, and readily recoverable from the components. In practice it is probably this last criterion that severely limits the application of extractive and azeotropic distillation. The majority of successful processes, in fact, are those in which the entrainer and one of the components separate into two liquid phases on cooling if direct recovery by distillation is not feasible.

A further restriction in the selection of an azeotropic entrainer is that the boiling point of the entrainer be in the range 10°C −40°C (18°F −72°F) below that of the components. Thus, although the entrainer is more volatile than the components and distills off in the overhead product, it is present in a sufficiently high concentration in the rectification section of the column.

Extractive distillation (Parkash, 2003; Gary et al., 2007; Speight, 2014, 2017; Hsu and Robinson, 2017) is the use of a third component to separate two close-boiling components in which one of the original components in the mixture is extracted by the third component and retained in the liquid phase to facilitate separation by distillation.

Using acetone-water as an extractive solvent for butanes and butenes, butane is removed as overhead from the extractive distillation column with acetone-water charged at a point close to the top of the column. The bottoms product of butenes and the extractive solvent are fed to a second column where the butenes are removed as overhead. The acetone-water solvent from the base of this column is recycled to the first column. Extractive distillation may also be used for the continuous recovery of individual aromatics, such as benzene, toluene, or xylene(s), from the appropriate crude oil fractions. *Prefractionation* concentrates a single aromatic cut into a close-boiling cut, after which the aromatic concentrate is distilled with a solvent (usually phenol) for benzene or toluene recovery. Mixed cresylic acids (cresols and methyl phenol derivatives) are used as the solvent for xylene recovery.

Extractive distillation is successful because the solvent is specially chosen to interact differently with the components of the original mixture, thereby altering their relative volatilities. Because these interactions occur predominantly in the liquid phase, the solvent is continuously added near the top of the extractive distillation column so that an appreciable amount is present in the liquid phase on all of the trays below. The mixture to be separated is added through second feed point further down the column. In the extractive column, the component having the greater volatility, not necessarily the component having the lowest boiling point, is taken overhead as a relatively pure distillate. The other component leaves with the solvent from below the columns. The solvent is separated from the remaining components in a second distillation column and then recycled back to the first column.

Several methods, involving *solvent extraction* (Parkash, 2003; Gary et al., 2007; Speight, 2014, 2017; Hsu and Robinson, 2017) or *destructive hydrogenation* (*hydrocracking*) (Parkash, 2003; Gary et al., 2007; Speight, 2014, 2017; Hsu and Robinson, 2017) can accomplish the removal of aromatic hydrocarbon derivatives from naphtha. By this latter method, aromatic hydrocarbon constituents are converted into odorless, straight-chain paraffin hydrocarbon derivatives that are required in aliphatic solvents.

The *Edeleanu* process (Parkash, 2003; Gary et al., 2007; Speight, 2014, 2017; Hsu and Robinson, 2017) was originally developed to improve the burning characteristics of kerosene by extraction of the smoke-forming aromatic compounds. Thus, it is not surprising that its use has been extended to the improvement of other products as well as to the segregation of aromatic hydrocarbon derivatives for use as solvents. Naphtha fractions rich in aromatics may be treated by the Edeleanu process for the purpose of recovering the aromatics, or the product stream from a catalytic reformer unit— particularly when the unit is operated to produce maximum aromatics—may be Edeleanu treated to recover the aromatics. The other most widely used processes for this purpose are the extractive distillation process and the *Udex* processes. Processes such as the *Arosorb* process and cyclic adsorption are used to a lesser extent.

The *Udex* process (Parkash, 2003; Gary et al., 2007; Speight, 2014, 2017; Hsu and Robinson, 2017) is also employed to recover aromatic streams from reformate fractions. This process uses a mixture of water and diethylene glycol to extract aromatics. Unlike extractive distillation, an aromatic concentrate is not required and the solvent removes all the aromatics, which are separated from one another by subsequent fractional distillation.

The reformate is pumped into the base of an extractor tower. The feed rises in the tower countercurrent to the descending diethylene glycol-water solution, which extracts the aromatics from the feed. The nonaromatic portion of the feed leaves the top of the tower, and the aromatic-rich solvent leaves the bottom of the tower. Distillation in a solvent stripper separates the solvent from the aromatics, which are sulfuric acid and clay treated and then separated into individual aromatics by fractional distillation.

Silica gel (SiO_2) is an adsorbent for aromatics and has found use in extracting aromatics from refinery streams (Arosorb and cyclic adsorption processes) (Parkash, 2003; Gary et al., 2007; Speight, 2014, 2017; Hsu and Robinson, 2017). Silica gel is manufactured from amorphous silica that is extremely porous and has the property of selectively removing and holding certain chemical compounds from mixtures. For example, silica gel selectively removes aromatics from a crude oil fraction, and after the nonaromatic portion of the fraction is drained from the silica gel, the adsorbed aromatics are washed from the silica gel by a stripper (or desorbent). Depending on the kind of feedstock, xylene, kerosene, or pentane may be used as the desorbent.

However, silica gel can be poisoned by contaminants, and the feedstock must be treated to remove water as well as nitrogen, oxygen, and sulfur-containing compounds by passing the feedstock through beds of alumina and/or other materials that remove impurities. The treated feedstock then enters one of several silica gel cases (columns) where the aromatics are adsorbed. The time period required for adsorption depends on the nature of the feedstock; for example, reformate product streams have been known to require substantially less treatment time than kerosene fractions.

3.2 Composition

Naphtha contains varying amounts of paraffins, olefins, naphthene constituents, and aromatics and olefins in different proportions in addition to potential isomers of paraffin that exist in naphtha boiling range. As a result, naphtha is divided predominantly into two main types: (i) aliphatic naphtha and (ii) aromatic (naphtha). The two types differ in two ways: first, in the kind of hydrocarbons making up the solvent, and second, in the methods used for their manufacture. Aliphatic solvents are composed of paraffinic hydrocarbons and cycloparaffins (naphthenes), and may be obtained directly from crude oil by distillation. The second type of naphtha contains aromatics, usually alkyl-substituted benzene, and is very rarely, if at all, obtained from petroleum as straight-run product.

Thus, naphtha is divided into two main types, aliphatic and aromatic. The two types differ in two ways: first, in the kind of hydrocarbon derivatives making up the solvent, and second, in the methods used for their manufacture. Aliphatic solvents are composed of paraffinic hydrocarbon derivatives and cycloparaffins (naphthenes), and may be obtained directly from crude oil by distillation. The second type of naphtha contains aromatics, usually alkyl-substituted benzene, and is very rarely, if at all, obtained from crude oil as straight-run materials.

Stoddard solvent is a crude oil distillate widely used as a dry cleaning solvent and as a general cleaner and degreaser. It may also be used as paint thinner, as a solvent in some types of photocopier toners, in some types of printing inks, and in some adhesives. Stoddard solvent is considered to be a form of mineral spirits, white spirits, and naphtha but not all forms of mineral spirits, white spirits, and naphtha are considered to be Stoddard solvent. Stoddard solvent consists of linear alkanes (30%–50%), branched alkanes (20%–40%), cycloalkanes (30%–40%), and aromatic hydrocarbon derivatives (10%–20%). The typical hydrocarbon chain ranges from C_7 through C_{12} in length.

3.3 Properties and uses

Naphtha is required to have a low level of odor to meet the specifications for use (Pandey et al., 2004), which is related to the chemical composition—generally paraffin hydrocarbons possess the mildest odor, and the aromatic hydrocarbons have a much stronger odor. Naphtha containing higher proportions of aromatic constituents may be pale yellow—usually, naphtha is colorless (water white) and can be tested for the level of contaminants (ASTM D156).

Generally, naphtha is valuable as for solvents because of good dissolving power. The wide range of naphtha available from the ordinary paraffin straight-run to the highly aromatic types and the varying degree of volatility possible offer products suitable for many uses (Boenheim and Pearson, 1973; Hadley and Turner, 1973).

The main uses of naphtha fall into the general areas of (i) solvents (diluents) for paints, for example; (ii) drycleaning solvents; (iii) solvents for cutback asphalt; (iv) solvents in the rubber industry; and (v) solvents for industrial extraction processes. Naphtha is also used as a blend stock for automotive fuel, engine fuel, and jet-B (naphtha type). Broadly, naphtha is classified as *light naphtha* and *heavy naphtha*. Light naphtha is used as rubber solvent, lacquer diluent, while heavy naphtha finds its application as varnish solvent, dyer's naphtha, and cleaner's naphtha.

Turpentine, the older more conventional solvent for paints, has now been almost completely replaced with the discovery that the cheaper and more abundant crude oil naphtha is equally satisfactory. The differences in application are slight: naphtha causes a slightly greater decrease in viscosity when added to some paints than does turpentine, and depending on the boiling range, may also show difference in evaporation rate.

The boiling ranges of fractions that evaporate at rates permitting the deposition of good films have been fairly well established. Depending on conditions, products are employed as low-boiling as those boiling from 38°C to 150°C (100°F −300°F) and as high-boiling as those boiling between 150°C and 230°C (300°F and 450°F). The latter are used mainly in the manufacture of backed and forced-drying products.

The solvent power required for conventional paint diluents is low and can be reached by distillates from paraffinic crude oils, which are usually recognized as the poorest solvents in the crude oil naphtha group. In addition to solvent power and correct evaporation rate, a paint thinner should also be resistant to oxidation, i.e., the thinner should not develop bad color and odor during use. The thinner should be free of corrosive impurities and reactive materials, such as certain types of sulfur compounds, when employed with paints containing lead and similar metals. The requirements are best met by straight-run distillates from paraffinic crude oils that boil from 120°C to 205°C (250°F−400°F). The components of enamels, vanishes, nitrocellulose lacquers, and synthetic resin finishes are not as soluble in paraffinic naphtha as the materials in conventional paints, and hence naphthenic and aromatic naphtha are favored for such uses.

Naphtha is used in the rubber industry for dampening the play and tread stocks of automobile tires during manufacture to obtain better adhesion between the units of the tire. They are also consumed extensively in making rubber cements (adhesives) or are employed in the fabrication of rubberized cloth, hot-water bottles, bathing caps, gloves, overshoes, and toys. These cements are solutions of rubber and were formerly made with benzene, but crude oil naphtha is now preferred because of the less toxic character.

Crude oil hydrocarbon distillates are also added in amounts up to 25% and higher at various stages in the polymerization of butadiene-styrene to synthetic rubber. Those employed in oil extended rubber are of the aromatic type. These distillates are generally high boiling fractions and preferably contain no wax, boil from 425°C to 510°C (800°F−950°F), have characterization factors of 10.5−11.6, a viscosity index lower than 0, bromine numbers of 6−30, and API gravity of 3−24.

Naphtha is used for extraction on a fairly wide scale. They are supplied in extracting residual oil from castor beans, soybeans, cottonseed, and wheat germ and in the recovery of *grease* from mixed garbage and refuse. The solvent employed in these cases is a hexane cut, boiling from approximately 65°C to 120°C (150°F −250°F). When the oils recovered are of edible grade or intended for refined purposes, stable solvents completely free of residual odor and taste are necessary, and straight-run streams from low-sulfur, paraffinic crude oils are generally satisfactory.

Cutback asphalt is asphalt cement diluted with a crude oil distillate to make it suitable for direct application to road surfaces with little or no heating. Asphalt cement, in turn, is a combination of hard asphalt with a heavy distillate or with a viscous residuum of an asphaltic crude oil. The products are classified as rapid, medium, and slow curing, depending on the rate of evaporation of the solvent. A rapid-curing product may contain 40%−50% of material distilling up to 360°C (680°F); a slow-curing mixture may have only 25% of such material. Gasoline naphtha, kerosene, and light fuel oils boiling from 38°C to 330°C (100°F−30°F) are used in different products and for different purposes; the use may also dictate the nature of the asphaltic residuum that can be used for the asphalt.

4. Gasoline

Gasoline, also called *gas* (United States and Canada), or *petrol* (Great Britain) or *benzine* (Europe) is a mixture of volatile, flammable liquid hydrocarbon derivatives derived from crude oil and used as fuel for internal-combustion engines (Parkash, 2003; Gary et al., 2007; Speight, 2014, 2017; Hsu and Robinson, 2017). It is also used as a solvent for oils and fats. Originally a byproduct of the crude oil industry (kerosene being the principal product), gasoline became the preferred automobile fuel because of its high energy of combustion and capacity to mix readily with air in a carburetor.

There is no single process in a refinery that produces gasoline. In fact, gasoline is a mixture of hydrocarbon derivatives that usually boil below 180°C (355°F) or, at most, below 200°C (390°F). The hydrocarbon constituents in this boiling range are those that have 4 to 12 carbon atoms in their molecular structure and fall into three general types: paraffins (including the cycloparaffins and branched materials), olefins, and aromatics.

Gasoline is still in great demand as a major product from crude oil. The network of interstate highways that links towns and cities in the United States is dotted with frequent service centers where motorists can obtain refreshment not only for themselves but also for their vehicles.

4.1 Manufacture

Gasoline was at first produced from crude oil by distillation, simply separating the volatile, more valuable fractions of crude oil and was actually equivalent to the distillate now referred to as naphtha. Up to and during the first decade of the present century, the gasoline produced was that originally present in crude oil or that could be condensed from natural gas. However, it was soon discovered that if the higher-boiling portions of crude oil (such as the fraction that boiled higher than kerosene, e.g., gas oil) were heated to more severe temperatures, thermal degradation (or cracking) occurred to produce smaller molecules within the range suitable for gasoline. Therefore, gasoline that was not originally in the unrefined crude oil could be manufactured. Later processes, designed to raise the yield of gasoline from crude oil, decomposed higher molecular weight constituents into lower molecular weight products by processes known as *cracking*. And like typical gasoline, several processes produce the blending stocks for gasoline manufacture (Fig. 3.2).

Thermal cracking, employing heat and high pressures, was introduced in 1913 but was accompanied in refineries after 1937 by catalytic cracking, the application of catalysts that facilitates chemical reactions producing more gasoline. Thermal cracking processes (Table 3.6) for converting the high-boiling fraction of crude oil to lower-boiling products still play an important role in the modern refinery through upgradation of heavy residue and improving the economics of the refinery through the production of lower-boiling distillates and other valuable products such as hydrocarbon gases and petroleum coke. The predominant reactions of thermal cracking are: (i) cracking of side chains aromatic nuclei, (ii) dehydrogenation of naphthene derivatives to form aromatic derivatives, (iii) condensation of aliphatic derivatives to form aromatic products, (iv) condensation of aromatic derivatives to form higher molecular weight aromatic derivatives, and (v) dimerization or oligomerization. Other methods used to improve the quality of gasoline and

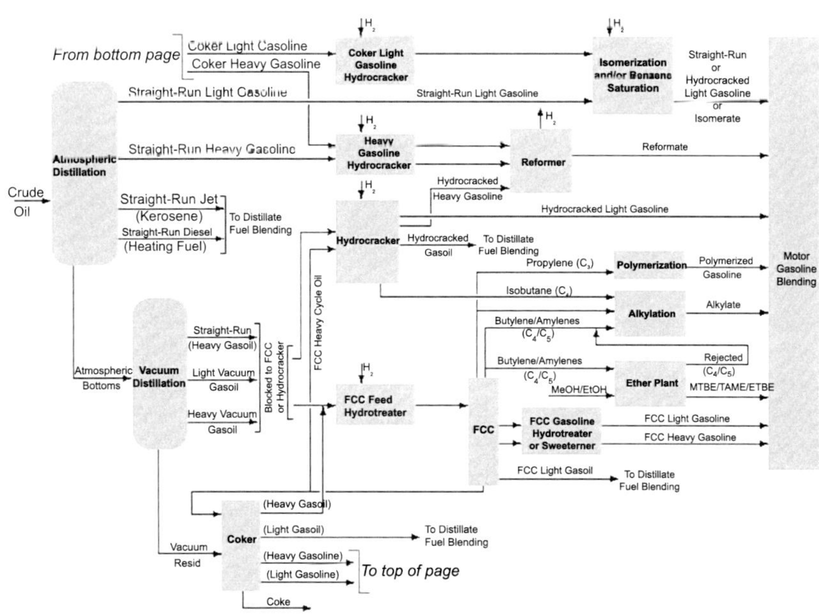

FIGURE 3.2 Refinery streams that are blended to produce gasoline.

TABLE 3.6 Thermal cracking processes and process conditions.

Process	Conditions
Visbreaking	Mild thermal cracking (low severity)
	Mild (470°C–500°C, 880°F to 930°F), 50 to 200 psi
	Low conversion (10% w/w)
	Residence time 1–3 min
	Heated coil or drum
Delayed coking	Operates in semi batch mode
	Moderate (480°C–515°C, 900°F to 960°F), 90 psi
	Soak drums (450°C–480°C, 845°F to 900°F)
	Coke form on walls of drum
	Coked until drum solid
	Coke (removed hydraulically) 20%–40% w/w on feed
Fluid coking	Server (510°C –520°C, 940°F to 970°F) heating At 10 psig
	Feedstock contacts refractory coke
	Bed fluidized with steam-even heating
	Higher yield of light ends ($<C_5$)
	Lower coke yield than delayed coking
Flexicoking	Continuous fluidized bed technology
	Converts high-boiling residue to lower-boiling products
	Eliminates the coke production by gasification o the coke
	Temperature 525°C (975°F)

increase its supply include *polymerization*, *alkylation*, *isomerization*, and *reforming*.

Polymerization is the conversion of gaseous olefins, such as propylene and butylene, into larger molecules in the gasoline range. A*lkylation* is a process combining an olefin and paraffin (such as *iso*-butane). *Isomerization* is the conversion of straight-chain hydrocarbon derivatives to branched-chain hydrocarbon derivatives. *Reforming* is the use of either heat or a catalyst to rearrange the molecular structure.

Aviation gasoline is form of motor gasoline that has been especially prepared for use for aviation piston engines. It has an octane number suited to the engine, a freezing point of −60°C (−76°F), and a distillation range usually within the limits of 30°C−180°C (86°F−356°F) compared to −1°C to 200°C (30°F−390°F) for automobile gasoline. The narrower boiling range ensures better distribution of the vaporized fuel through the more complicated induction systems of aircraft engines. Aircraft operate at altitudes at which the prevailing pressure is less than the pressure at the surface of the Earth (pressure at 17,500 feet is 7.5 psi compared to 14.7 psi at the surface of the Earth). Thus, the vapor pressure of aviation gasoline must be limited to reduce boiling in the tanks, fuel lines, and carburetors. Thus, the aviation gasoline does not usually contain the gaseous hydrocarbon derivatives (butanes) that give automobile gasoline the higher vapor pressures.

Aviation gasoline is strictly limited regarding hydrocarbon composition. The important properties of the hydrocarbon derivatives are the highest octane numbers economically possible, boiling points in the limited temperature range of aviation gasoline, maximum heat contents per pound (high proportion of combined hydrogen), and high chemical stability to withstand storage. Aviation gasoline are composed of paraffins and iso-paraffins (50%−60%), moderate amounts of naphthenes (20%−30%), small amounts of aromatics (10%), and usually no olefins, whereas motor gasoline may contain up to 30% olefins and up to 40% aromatics.

Under conditions of use in aircraft, olefins tend to form gum, cause pre-ignition, and have relatively poor antiknock characteristics under lean mixture (cruising) conditions; for these reasons olefins are detrimental to aviation gasoline. Aromatics have excellent antiknock characteristics under rich mixture (takeoff) conditions, but are much like the olefins under lean mixture conditions; hence the proportion of aromatics in aviation gasoline is limited. Some naphthenes with suitable boiling temperatures are excellent aviation gasoline components but are not segregated as such in refinery operations. They are usually natural components of the straight-run naphtha (aviation base stocks) used in blending aviation gasoline. The lower boiling paraffins (pentane and hexane), and both the high-boiling and low-boiling iso-paraffins (iso-pentane to iso-octane) are excellent aviation gasoline components. These hydrocarbon derivatives have high heat content per pound and are chemically stable, and the iso-paraffins have high octane numbers under both lean and rich mixture conditions.

The manufacture of aviation gasoline is thus dependent on the availability and selection of fractions containing suitable hydrocarbon derivatives. The lower boiling hydrocarbons are usually found in straight-run naphtha from certain crude oils. These fractions have high contents of iso-pentanes and iso-hexane and provide needed volatility, as well as high octane number components. Higher boiling iso-paraffins are provided by aviation alkylate, which consists mostly of branched octanes. Aromatics, such as benzene, toluene, and xylene, are obtained from catalytic reforming or a similar source.

To increase the proportion of higher boiling octane components, such as aviation alkylate and xylenes, the proportion of lower boiling components must also be increased to maintain the proper volatility. Iso-pentane and, to some extent, iso-hexane are the lower boiling components used. Iso-pentane and iso-hexane may be separated from selected naphtha by superfractionators or synthesized from the normal hydrocarbons by isomerization. In general, most aviation gasoline is made by blending a selected straight-run naphtha fraction (aviation base stock) with iso-pentane and aviation alkylate.

4.2 Composition

Gasoline is manufactured to meet specifications and regulations and not to achieve a specific distribution of hydrocarbon derivatives by class and size. However, chemical composition often defines properties. For example, volatility is defined by the individual hydrocarbon constituents and the lowest boiling constituent(s) defines the volatility as determined by certain test methods.

Automotive gasoline typically contains approximately almost 200 (if not several hundred) hydrocarbon compounds. The relative concentrations of the compounds vary considerably depending on the source of crude oil, refinery process, and product specifications. Typical hydrocarbon chain lengths range from C_4 through C_{12} with a general hydrocarbon distribution consisting of alkanes (4%−8%), alkenes (2%−5%), iso-alkanes 25%−40%, cycloalkanes (3%−7%), cycloalkenes (1%−4%), and aromatics (20%−50%). However, these proportions vary greatly.

The majority of the members of the paraffin, olefin, and aromatic series (of which there are approximately 500) boiling below 200°C (390°F) have been found in the gasoline fraction of crude oil. However, it appears that the distribution of the individual members of straight-run naphtha (i.e., naphtha distilled from crude oil without thermal alteration) is not even.

Highly branched paraffins, which are particularly valuable constituents of gasoline(s), are not usually the principal paraffinic constituents of straight-run gasoline. The more predominant paraffinic constituents are usually the normal (straight-chain) isomers, which may dominate the branched isomer(s) by a factor of 2 or more. This is presumed to indicate the tendency to produce long uninterrupted carbon chains during crude oil maturation rather than those in which branching occurs. However, this trend is somewhat different for the cyclic constituents of gasoline, i.e., cycloparaffins (naphthenes) and aromatics. In these cases, the preference appears to be for several short side chains rather than one long substituent.

Gasoline can vary widely in composition: even those with the same octane number may be quite different, not only in the physical makeup but also in the molecular structure of the constituents. The variation in the content of normal paraffins, branched paraffins, cyclopentanes, and cyclohexane derivatives involve characteristics of any one individual crude oil and may in some instances be used for crude oil identification. Furthermore, straight-run gasoline generally shows a decrease in paraffin content with an increase in molecular weight, but the cycloparaffins (naphthenes) and aromatics increase with increasing molecular weight. Indeed, the hydrocarbon type variation may also vary markedly from process to process.

The reduction of the lead content of gasoline and the introduction of reformulated gasoline has been very successful in reducing automobile emissions (Absi-Halabi et al., 1997). Further improvements in fuel quality have been proposed for the years 2000 and beyond. These projections are accompanied by a noticeable and measurable decrease in crude oil quality and the reformulated gasoline will help meet environmental regulations for emissions for liquid fuels.

4.3 Properties and uses

Despite the diversity of the processes within a modern crude oil refinery, no single hydrocarbon stream meets all the requirements of gasoline. Thus, the final step in gasoline manufacture is *blending* the various streams into a finished product (Fig. 3.2). It is not uncommon for the finished gasoline to be made up of six or more streams and several factors make this flexibility critical: (i) the requirements of the gasoline specification (ASTM D4814) and the regulatory requirements and (ii) performance specifications that are subject to local climatic conditions and regulations.

The early criterion for gasoline quality was Baumé (or API) gravity. For example, a 70°API gravity naphtha contained fewer, if any, of the higher-boiling naphtha constituents than a 60°API naphtha. Therefore, the 70°API naphtha was a higher quality and, hence, economically more valuable naphtha. However, apart from being used as a rough estimation of quality (not only for crude oil products but also for crude oil), specific gravity is no longer of any significance as a true indicator of naphtha quality.

Gasoline performance and hence quality of an automobile gasoline is determined by its resistance to knock, for example, *detonation* or *ping* during service. The antiknock quality of the fuel limits the power and economy that an engine using that fuel can produce: the higher the antiknock quality of the fuel, the more the power, and efficiency of the engine.

Octane numbers are obtained by the two test procedures, those obtained by the first method are called *motor octane numbers* (indicative of high-speed performance) (ASTM D2700 and ASTM D2723). Those obtained by the second method are called *research octane numbers* (indicative of normal road performance) (ASTM D2699 and ASTM D2722). Octane numbers quoted are usually, unless stated otherwise, research octane numbers. In the test methods used to determine the antiknock properties of gasoline, comparisons are made with blends of two pure hydrocarbons, *n*-heptane and iso-octane (2,2,4-trimethylpentane). Iso-octane has an octane number of 100 and is high in its resistance to knocking; *n*-heptane is quite low (with an octane number of 0) in its resistance to knocking.

Extensive studies of the octane numbers of individual hydrocarbons have brought to light some general rules. For example, normal paraffins have the least desirable knocking characteristics, and these become progressively worse as the molecular weight increases. Iso-paraffins have higher octane numbers than the corresponding normal isomers, and the octane number increases as the degree of branching of the chain is increased. Olefins have markedly higher octane numbers than the related paraffins; naphthenes are usually better than the corresponding normal paraffins but rarely have very high octane numbers; aromatics usually have quite high octane numbers.

Blends of *n*-heptane and iso-octane thus serve as a reference system for gasoline and provide a wide range of quality used as an antiknock scale. The exact blend, which matches identically the antiknock resistance of the fuel under test, is found, and the percentage of iso-octane in that blend is termed the *octane number* of the gasoline. For example, gasoline with a knocking ability which matches that of a blend of 90% iso-octane and 10% *n*-heptane has an octane number of 90. However, many pure hydrocarbons and even commercial gasoline have antiknock quality above an octane number of 100. In this range it is common practice to extend the reference values by the use of varying amounts of tetraethyllead in pure iso-octane.

With an accurate and reliable means of measuring octane numbers, it was possible to determine the cracking conditions—temperature, cracking time, and pressure—that caused increases in the antiknock characteristics of cracked naphtha. In general it was found that higher cracking temperatures and lower pressures produced higher octane naphtha, but unfortunately more gas, cracked residua, and coke were formed at the expense of the volume of cracked naphtha.

To produce higher-octane naphtha, cracking coil temperatures were pushed up to 510°C (950°F), and pressures dropped from 1000 to 350 psi. This was the limit of thermal cracking units, for at temperatures over 510°C (950°F) coke formed so rapidly in the cracking coil that the unit became inoperative after only a short time on-stream. Hence it was at this stage that the nature of the naphtha-producing process was reexamined, leading to the development of other processes, such as reforming, polymerization, and alkylation for the production of naphtha components having suitably high octane numbers.

It is worthy of note here that the continued decline in crude oil reserves and the issue of environmental protection have emerged as of extreme importance in the search for alternatives to crude oil. In this light, oxygenates, either neat or as additives to fuels, appear to be the principal alternative fuel candidates beyond the crude oil refinery.

5. Kerosene and related fuels

Kerosene (spelled *kerosine* in some countries), also called liquid paraffin or paraffin oil, is a flammable pale-yellow or colorless oily liquid with a characteristic odor (Parkash, 2003; Gary et al., 2007; Speight, 2014, 2017; Hsu and Robinson, 2017). It is obtained from crude oil and used for burning in lamps and domestic heaters or furnaces, as a fuel or fuel component for jet engines, and as a solvent for greases and insecticides.

Kerosene is intermediate in volatility between naphtha and gas/diesel oil. It is a medium oil distilling between 150°C and 300°C (300°F −570°F). Kerosene has a flash point of approximately 25°C (77°F) and is suitable for use as an illuminant when burned in a wide lamp. The term *kerosene* is also too often incorrectly applied to various fuel oils, but a fuel oil is actually any liquid or liquid crude oil product that produces heat when burned in a suitable container or that produces power when burned in an engine.

5.1 Manufacture

Kerosene was first manufactured in the 1850s from coal tar, hence the name coal oil is often applied to kerosene, but crude oil became the major source after 1859. From that time, the kerosene fraction is and has remained a distillation fraction of crude oil. However, the quantity and quality vary with the type of crude oil, and although some crude oils yield excellent kerosene quite simply, others produce kerosene that requires substantial refining. In the modern refinery, kerosene is largely produced by cracking the less volatile portion of crude oil at atmospheric pressure and elevated temperatures.

In the early days, the poorer quality kerosene was treated with large quantities of sulfuric acid to convert them to marketable products. However, this treatment resulted in high acid and kerosene losses, but the later development of the *Edeleanu* process overcame these problems (Parkash, 2003; Gary et al., 2007; Speight, 2014, 2017; Hsu and Robinson, 2017).

Kerosene is a very stable product, and additives are not required to improve the quality. Apart from the removal of excessive quantities of aromatics by the Edeleanu process, kerosene fractions may need only a lye wash or a doctor treatment if hydrogen sulfide is present to remove mercaptans.

5.2 Composition

Chemically, kerosene is a mixture of hydrocarbons; the chemical composition depends on its source, but it usually consists of approximately 10 different hydrocarbons, each containing from 10 to 16 carbon atoms per molecule; the constituents include *n*-dodecane (n-$C_{12}H_{26}$), alkyl benzenes, and naphthalene and its derivatives. Kerosene is less volatile than naphtha; it boils between approximately 140°C (285°F) and 320°C (610°F).

Kerosene, because of its use as burning oil, must be free of aromatic and unsaturated hydrocarbons, as well as free of the more obnoxious sulfur compounds. The desirable constituents of kerosene are saturated hydrocarbons, and it is for this reason that kerosene is manufactured as a straight-run fraction, not by a cracking process.

Although the kerosene constituents are predominantly saturated materials, there is evidence for the presence of substituted tetrahydronaphthalene. Dicycloparaffin derivatives also occur in substantial amounts in kerosene. Other hydrocarbons with both aromatic and cycloparaffin rings in the same molecule, such as substituted indan, also occur in kerosene. The predominant structure of the dinuclear aromatics appears to be that in which the aromatic rings are condensed, such as naphthalene whereas the *isolated* two-ring compounds, such as biphenyl, are only present in traces, if at all.

5.3 Properties and uses

Kerosene is by nature a fraction distilled from crude oil that has been used as a fuel oil from the beginning of the crude oil-refining industry. The combustion of kerosene as a fuel can be represented by the simple equation that is based on the combustion of dodecane ($C_{12}H_{26}$):

$$2C_{12}H_{26} + 37O_2 \rightarrow 24CO_2 + 26H_2O$$

As such, low proportions of aromatic and unsaturated hydrocarbons are desirable to maintain the lowest possible level of smoke during burning. Although some aromatics may occur within the boiling range assigned to kerosene, excessive amounts can be removed by extraction; that kerosene is not usually prepared from cracked products almost certainly excludes the presence of unsaturated hydrocarbons.

The essential properties of kerosene are flash point, fire point, distillation range, burning, sulfur content, color, and cloud point. In the case of the flash point (ASTM D56), the minimum flash temperature is generally placed above the prevailing ambient temperature; the fire point (ASTM D92) determines the fire hazard associated with its handling and use.

The boiling range (ASTM D86) is of less importance for kerosene than for naphtha, but it can be taken as an indication of the viscosity of the product, for which there is no requirement for kerosene. The ability of kerosene to burn steadily and cleanly over an extended period (ASTM D187) is an important property and gives some indication of the purity or composition of the product.

The significance of the total sulfur content of a fuel oil varies greatly with the type of oil and the use to which it is put. Sulfur content is of great importance when the oil to be burned produces sulfur oxides that contaminate the surroundings. The color of kerosene is of little significance, but a product darker than usual may have resulted from contamination or aging, and in fact a color darker than specified (ASTM Dl56) may be considered by some users as unsatisfactory. Finally, the cloud point of kerosene (ASTM D2500;ASTM D2500) gives an indication of the temperature at which the wick may become coated with wax particles, thus lowering the burning qualities of the oil.

Kerosene was the major refinery product before the onset of the *automobile age*, but now kerosene can be termed one of several secondary crude oil products after the primary refinery product— naphtha. Kerosene originated as a straight-run crude oil fraction that boiled between approximately 205°C and 260°C (400°F −500°F) (Walmsley, 1973). Some crude oils, for example, those from the crude oil fields where the crude oil is paraffinic in nature, contain kerosene fractions of very high quality, but other crude oils, such as those having an asphalt base, must be thoroughly refined to remove aromatics and sulfur compounds before a satisfactory kerosene fraction can be obtained.

Jet fuel comprises both gasoline-type and kerosene-type jet fuels meeting specifications for use in aviation turbine power units and is often referred to as *gasoline-type jet fuel* and *kerosene-type jet fuel*. The major jet fuels used by the military are JP-4, JP-5, JP-6, JP-7, and JP-8. Briefly, JP-4 is a wide-cut fuel developed for broad availability. JP-6 is a higher cut than JP-4 and is characterized by fewer impurities. JP-5 is specially blended kerosene, and JP-7 is high flash point special kerosene used in advanced supersonic aircraft. JP-8 is kerosene modeled on Jet A-l fuel (used in civilian aircraft). From what data are available, typical hydrocarbon chain lengths characterizing JP-4 range from C_4 to C_{16}. Aviation fuels consist primarily of straight and branched alkanes and cycloalkanes. Aromatic hydrocarbons are limited to 20%−25% of the total mixture because they produce smoke when burned. A maximum of 5% alkenes is specified for JP-4. The approximate distribution by chemical class is (i) straight chain alkanes, 32% v/v, (ii) branched alkanes, 31% v/v, (iii) cycloalkanes, 16% v/v, and (iv) aromatic hydrocarbons, 21% v/v.

Gasoline type jet fuel includes all low-boiling hydrocarbon oils for use in aviation turbine power units that distill between 100°C and 250°C (212°F −480°F). It is obtained by blending kerosene and gasoline or naphtha in such a way that the aromatic content does not exceed 25% in volume. Additives can be included to improve fuel stability and combustibility.

Kerosene type jet fuel is a medium distillate product that is used for aviation turbine power units. It has the same distillation characteristics and flash point as kerosene (between 150°C and 300°C, 300°F and 570°F, but not generally above 250°C, 480°F). In addition, it has particular specifications (such as freezing point) which are established by the International Air Transport Association (IATA).

6. Diesel fuel

Diesel fuel oil is essentially the same as furnace fuel oil, but the proportion of cracked gas oil is usually less since the high aromatic content of the cracked gas oil reduces the cetane value of the diesel fuel (Parkash, 2003; Gary et al., 2007; Speight, 2014, 2017; Hsu and Robinson, 2017). Diesel fuel was originally a straight-run product obtained from the distillation of crude oil. However, with the use of various cracking processes to produce diesel constituents, diesel fuels also may contain varying amounts of selected cracked distillates to increase the volume available for meeting the growing demand. Care is taken to select the cracked stocks in such a manner that specifications are met as simply as possible.

Under the broad definition of diesel fuel, many possible combinations of characteristics (such as volatility, ignition quality, viscosity, gravity, stability, and other properties) exist. To characterize diesel fuels and thereby establish a framework of definition and reference, various classifications are used in different countries. An example is ASTM D975 in the United States in which grades No. l-D and 2-D are distillate fuels, the types most commonly used in high speed engines of the mobile type, in medium speed stationary engines, and in railroad engines. Grade 4-D covers the class of more viscous distillates and, at times, blends of these distillates with residual fuel oils. No. 4-D fuels are applicable for use in low and medium speed engines employed in services involving sustained load and predominantly constant speed.

6.1 Manufacture

Diesel fuel derived from crude oil (also called *petrodiesel* or fossil diesel) is the most common type of diesel fuel and is produced from the fractional distillation of crude oil between 200°C and 350°C (390°F and 660°F) at atmospheric pressure, resulting in a mixture of hydrocarbon derivatives that typically contain between 9 and 25 carbon atoms per molecule (Parkash, 2003; Gary et al., 2007; Speight, 2014, 2017; Hsu and Robinson, 2017).

Under the broad definition of diesel fuel, many possible combinations of characteristics (such as volatility, ignition quality, viscosity, gravity, stability, and other properties) exist.

The diesel fuel produced in a modern refinery is (like gasoline) a blend of all the appropriate available streams, such as: (i) a straight-run product, (ii) the light cycle oil from a fluid catalytic cracking unit, and (iii) hydrocracked gas oil. The straight-run diesel may be acceptable as is, or may need minor upgrading for use in diesel fuel prepared for off-road use. The refiner must blend the available streams to meet all performance, regulatory, economic, and inventory requirements. Typically, the refiner really has limited control over the detailed composition of the final diesel blend which is determined primarily by the composition of the crude oil feedstock. While the chemical reactions that occur in the conversion processes involve compositional changes, they are not specific enough to allow for much tailoring of the products.

On the other hand, synthetic diesel fuel can be produced from any carbonaceous material, including biomass, biogas, natural gas, coal, and many others. The raw material is gasified into synthesis (a mixture of carbon monoxide and hydrogen), which after purification is converted by the Fischer-Tropsch process to a series of synthetic hydrocarbon derivatives (Chapter 8) (Chadeesingh, 2011; Luque and Speight, 2015). This type of diesel fuel is typically paraffinic in nature and has a zero sulfur content of and very low content of aromatic derivatives thereby reducing the emissions of environmentally unfriendly hydrocarbons, sulfur oxides, and particulate matter (PM).

For clarification, biodiesel is not a true hydrocarbon product—it consists of alkyl esters of long chain fatty acids and is a transportation fuel that has gradually grown in popularity over the past decade (Speight, 2011). By way of clarification, the transesterification reaction is the reaction where an organic group (R″) of an ester is exchanged with the organic group (R′) of an alcohol and the reaction is catalyzed by the addition of an acid catalyst or a base catalyst:

Common feedstock for the production of biodiesel include soybean oil, algal oil, higuereta oil, canola oil, vegetable oil, palm oil, and corn oil. The common alcohol is methyl alcohol (CH_3OH).

6.2 Composition

Diesel fuel is a member of the class of crude oil products known as middle distillates. As the name implies, these fuels are higher boiling than gasoline but lower boiling than gas oil. Middle distillates cover the boiling range from approximately 175°C to 375°C (350°F −700°F) and the carbon number range from about C_8 to C_{24}.

Diesel fuel derived from crude oil is composed of approximately saturated hydrocarbon derivatives (75% v/v, primarily paraffin hydrocarbons including *n*-paraffins, iso-paraffins, and cycloparaffins), and aromatic hydrocarbon derivatives (25% v/v, including alkyl benzenes and naphthalene derivatives). Diesel fuels predominantly contain a mixture of C_{10} to C_{19}. Jet fuels are based primarily on straight-run kerosene, as well as additives.

6.3 Properties and uses

The general appearance, or color, of diesel fuel is a useful indicator against contamination by residual (higher boiling) constituents, water or fine solid particles. Therefore, it is prudent to check by visual inspection that clear fuel is being delivered. Color, being part of the appearance of diesel fuel should also be determined since the color of crude oil fuels is used for manufacturing control purposes. In some cases the color may serve as an indication of the degree of refinement of the material. Typically the methods require a visual determination of color using colored glass disks or reference materials. Similarly, acceptance is important with regard to odor, and it is usually required that diesel fuel is reasonably free of contaminants, such as mercaptan derivatives (RSH, also called thiol derivatives), which impart unpleasant odors to the fuel.

The principal measure of diesel fuel quality is the cetane number which is a measure of the delay of ignition of a diesel fuel and is based upon the ignition characteristics of two hydrocarbons *n*-hexadecane (cetane) and 2,3,4,5,6,7,8-heptamethylnonane. Cetane has a short delay period during ignition and is assigned a cetane number of 100; heptamethylnonane has a long delay period and has been assigned a cetane number of 15. Just as the octane number is meaningful for automobile fuels, the cetane number is a means of determining the ignition quality of diesel fuels and is equivalent to the percentage by volume of cetane in the blend with heptamethylnonane, which matches the ignition quality of the test fuel (ASTM D613).

$$\text{Cetane Number} = \%\ \text{v/v}\ n\text{-cetane} + 0.15\%\ \text{v/v heptamethylnonane)}$$

One of the most widely used methods is based on the calculated Cetane Index formula. This formula represents a method for estimating the cetane number of distillate fuels from API gravity and mid-boiling point. The index value as computed from the formula is designated as a calculated cetane index (ASTM D976). A high cetane number indicates that the fuel ignites more readily when sprayed into hot compressed air.

7. Gas oil and fuel oil

Gas oil is a distillate fraction that is intermediate in boiling range and viscosity between kerosene and lubricating oil which is used as a fuel oil. More specifically, light gas oil (low-boiling gas oil) is typically the last fraction distilled from crude oil at atmospheric pressure (without thermal decomposition) while heavy gas oil (high-boiling gas oil) is the first fraction distilled from the atmospheric residuum of crude oil under reduced pressure (without thermal decomposition) (Table 3.1) (Parkash, 2003; Gary et al., 2007; Speight, 2014, 2017; Hsu and Robinson, 2017).

Fuel oil is classified in several ways but generally may be divided into two main types: *distillate fuel oil* and *residual fuel oil*. Distillate fuel oil is vaporized and condensed during a distillation process and thus have a definite boiling range and do not contain high-boiling constituents. A fuel oil that contains any amount of the residue from crude distillation of thermal cracking is a residual fuel oil. The terms *distillate fuel oil* and *residual fuel oil* are losing their significance, since fuel oil is now made for specific uses and may be either distillates or residuals or mixtures of the two. The terms *domestic fuel oil*, *diesel fuel oil*, and *heavy fuel oil* are more indicative of the uses of fuel oils. Domestic fuel oil is fuel oil that is used primarily in the home. This category of fuel oil includes kerosene, stove oil, and furnace fuel oil; they are distillate fuel oils.

7.1 Manufacture

The manufacture of fuel oil at one time largely involved using what was left after removing desired products from crude oil. Now fuel oil manufacture is a complex matter of selecting and blending various crude oil fractions to meet definite specifications, and the production of a homogeneous, stable fuel oil requires experience backed by laboratory control.

Heavy fuel oil comprises all residual fuel oils and the constituents range from distillable constituents to residual (nondistillable) constituents that must be heated to 260°C (500°F) or more before they can be used. The kinematic viscosity is above 10 cSt at 80°C (176°F). The flash point is always above 50°C (122°F) and the density is always higher than 0.900. In general, heavy fuel oil usually contains cracked residua, reduced crude, or cracking coil heavy product which is mixed (cut back) to a specified viscosity with cracked gas oils and fractionator bottoms. For some industrial purposes in which flames or flue gases contact the product (ceramics, glass, heat treating, and open hearth furnaces), fuel oils must be blended to contain minimum sulfur contents, and hence low-sulfur residues are preferable for these fuels.

Stove oil, like kerosene, is always a straight-run fraction from suitable crude oils, whereas other fuel oils are usually a blend of two or more fractions, one of which is usually cracked gas oil. The straight-run fractions available for blending into fuel oils are heavy naphtha, light and heavy gas oils, reduced crude, and pitch. Cracked fractions such as light and heavy gas oils from catalytic cracking, cracking coil tar, and fractionator bottoms from catalytic cracking, may also be used as blends to meet the specifications of the different fuel oils.

Since the boiling ranges, sulfur contents, and other properties of even the same fraction vary from crude oil to crude oil and with the way the crude oil is processed, it is difficult to specify which fractions are blended to produce specific fuel oils. In general, however, furnace fuel oil is a blend of straight-run gas oil and cracked gas oil to produce a product boiling in the 175°C –345°C (350°C –50°F) range.

7.2 Composition

Fuel oil No. 1 and fuel oil No. 2 are distillate fuels which consist of distilled process streams. Residual fuel oil, such as fuel oil No. 4, is composed of the residuum remaining after distillation or cracking, or blends of such residues with distillates. Diesel fuel is approximately similar to fuel oil used for heating (fuel oils No. 1, fuel oil No. 2, and fuel oil No. 4).

All fuel oils consist of complex mixtures of aliphatic and aromatic hydrocarbons, the relative amounts depending on the source and grade of the fuel oil. The aliphatic alkanes (paraffins) and cycloalkane constituents (naphthene constituents) are hydrogen saturated and compose as much as 90% w/w of the fuel oil. Aromatic constituents (e.g., benzene) and olefin constituents compose up to 20% v/v and 1% v/v, respectively, of the fuel oils. Fuel oil No. 1 (straight-run kerosene) is a distillate which consists primarily of hydrocarbons in the C_9 to C_{16} range while fuel oil No. 2 is a higher-boiling, usually blended, distillate with hydrocarbons in the C_{11} to C_{20} range.

In some countries, residual fuel oil is also known as *heavy fuel oil* which also consists primarily of the residue from distillation or cracking units in the refinery. Historically, fuel oils were based on residua from the atmospheric distillation column and were known as straight-run fuels. However, the increasing demand for transportation fuels such as gasoline, kerosene, and diesel fuel has led to an increased value for the atmospheric residue as a feedstock for vacuum distillation and for cracking processes. As a consequence, most heavy fuel oils are currently based on vacuum residua from thermal and catalytic cracking operations. These fuels differ in character from straight-run fuels in that the density and mean molecular weight are higher—the atomic hydrogen-carbon is lower indicating a higher degree of aromaticity. The density of some heavy fuel oils can be above that of water (>1.000) which has environmental implications in the event of a spillage into water systems.

Residual fuel oil (and/or heavy fuel oil) is typically more complex in composition and impurities than distillate fuel oil. Therefore, a specific composition cannot always be determined—the sulfur content in residual fuel oil has been reported to vary up to 5% w/w. Residual fuel oils are complex mixtures of high molecular weight compounds having a typical boiling range from 350°C to 650°C (660°F–1200°F). They consist of aromatic, aliphatic, and naphthenic hydrocarbons, typically having carbon numbers from C_{20} to C_{50}, together with asphaltene constituents and smaller amounts of heterocyclic compounds containing sulfur, nitrogen, and oxygen. They have chemical characteristics similar to liquid asphalt and hence, are considered to be stabilized suspensions of asphaltene constituents in an oily medium.

Residual fuel oil also contains organometallic compounds from their presence in the original crude oil— the most important of which are nickel and vanadium. The metals (especially vanadium) are of particularly major significance for fuels burned in both diesel engines and boilers because when combined with sodium (perhaps from brine contamination from the reservoir or remaining after the refinery dewatering/desalting process) and other metallic compounds in critical proportions can lead to the formation of high melting point ash which is corrosive to engine parts. Other elements that occur in heavy fuel oils include iron, potassium, aluminum, and silicon—the latter two metals are mainly derived from refinery catalyst fines.

7.3 Properties and uses

Diesel fuel oil is also a distillate fuel oil that distils between 180°C and 380°C (356°F–716°F). Several grades are available depending on the use: diesel oil for diesel compression ignition (cars, trucks, and marine engines) and low-boiling heating oil for industrial and commercial uses.

Heavy fuel oil comprises all residual fuel oils (including those obtained by blending). Heavy fuel oil constituents range from distillable constituents to residual (nondistillable) constituents that must be heated to 260°C (500°F) or more before they can be used. The kinematic viscosity is above 10 cSt at 80°C (176°F). The flash point is always above 50°C (122°F) and the density is always higher than 0.900. In general, heavy fuel oil usually contains cracked residua, reduced crude, or cracking coil high-boiling product which is mixed (cut back) to a specified viscosity with cracked gas oils and fractionator bottoms. For some industrial purposes in which flames or flue gases contact the product (ceramics, glass, heat treating, and open hearth furnaces) fuel oils must be blended to contain minimum sulfur contents, and hence low-sulfur residues are preferable for these fuels.

No. 1 fuel oil is a crude oil distillate that is one of the most widely used of the fuel oil types. It is used in atomizing burners that spray fuel into a combustion chamber where the tiny droplets burn while in suspension. It is also used as a carrier for pesticides, as a weed killer, as a mold release agent in the ceramic and pottery industry, and in the cleaning industry. It is found in asphalt coatings, enamels, paints, thinners, and varnishes. No. 1 fuel oil is a low-boiling crude oil distillate (straight-run kerosene) consisting primarily of hydrocarbons in the range C_9-C_{16}. Fuel oil #l is very similar in composition to diesel fuel; the primary difference is in the additives.

No. 2 fuel oil is a crude oil distillate that may be referred to as domestic or industrial. The domestic fuel oil is usually lower boiling and a straight-run product. It is used primarily for home heating. Industrial distillate is a cracked product or a blend of both. It is used in smelting furnaces, ceramic kilns, and packaged boilers. No. 2 fuel oil is characterized by hydrocarbon chain lengths in the $C_{11}-C_{20}$ range. The composition consists of aliphatic hydrocarbons (straight-chain alkanes and cycloalkanes) (64%), unsaturated hydrocarbons (alkenes) (1% to 2%), and aromatic hydrocarbons (including alkyl benzenes and 2-ring, 3-ring aromatics) 35%, but contains only low amounts of the polycyclic aromatic hydrocarbons (<5%1989b).

No. 6 fuel oil (also called *Bunker C oil* or *residual fuel oil*) is the residuum from crude oil after naphtha-gasoline, No. 1 fuel oil, and No. 2 fuel oil have been removed. No. 6 fuel oil can be blended directly to heavy fuel oil or made into asphalt. Residual fuel oil is more complex in composition and impurities than distillate fuels. Limited data are available on the composition of No. 6 fuel oil. Polycyclic aromatic hydrocarbons (including the alkylated derivatives) and metal-containing constituents are components of No. 6 fuel oil.

Stove oil, like kerosene, is always a straight-run fraction from suitable crude oils, whereas other fuel oils are usually a blend of two or more fractions, one of which is usually cracked gas oil. The straight-run fractions available for blending into fuel oils are high-boiling naphtha, low-boiling gas oil and high-boiling gas oils, reduced crude, and pitch. Cracked fractions such as low-boiling and high-boiling gas oils from catalytic cracking, cracking coil tar, and fractionator bottoms from catalytic cracking may also be used as blends to meet the specifications of the different fuel oils.

Since the boiling ranges, sulfur contents, and other properties of even the same fraction vary from crude oil to crude oil and with the way the crude oil is processed, it is difficult to specify which fractions are blended to produce specific fuel oils. In general, however, furnace fuel oil is a blend of straight-run gas oil and cracked gas oil to produce a product boiling in the 175°C−345°C (350°F−50°F) range.

The manufacture of fuel oils at one time largely involved using what was left after removing desired products from crude oil. Now fuel oil manufacture is a complex matter of selecting and blending various crude oil fractions to meet definite specifications, and the production of a homogeneous, stable fuel oil requires experience backed by laboratory control.

8. Lubricating oil

After kerosene the early crude oil refiners wanted paraffin wax for the manufacture of candles, and *lubricating oil* was, at first, byproduct of wax manufacture (Parkash, 2003; Gary et al., 2007; Speight, 2014, 2017; Hsu and Robinson, 2017). The preferred lubricants in the 1860s were lard oil, sperm oil, and tallow. The demand that existed for kerosene did not develop for crude oil−derived lubricating oils. In fact, oils were used to supplement the animal and vegetable oils used as lubricants. However, as the trend to heavier industry increased, the demand for mineral lubricating oils increased, and after the 1890s crude oil displaced animal and vegetable oils as the source of lubricants for most purposes.

Mineral oils are often used as lubricating oils but also have medicinal and food uses. A major type of hydraulic fluid is the mineral oil class of hydraulic fluids. The mineral-based oils are produced from the high-boiling crude oil distillates. Hydrocarbon numbers ranging from C_{15} to C_{50} occur in the various types of mineral oils, with the higher-boiling distillates having higher percentages of the higher carbon number compounds.

Crankcase oil (*motor oil*) may be either mineral-based or synthetic. The mineral-based oils are more widely used than the synthetic oils and may be used in automotive engines, railroad and truck diesel engines, marine equipment, jet and other aircraft engines, and most small 2- and 4-stroke engines. The mineral-based oils contain hundreds to thousands of hydrocarbon compounds, including a substantial fraction of nitrogen- and sulfur-containing compounds. The hydrocarbons are mainly mixtures of straight and branched chain hydrocarbons (alkanes), cycloalkanes, and aromatic hydrocarbons. Polynuclear aromatic hydrocarbons (and the alkyl derivatives) and metal-containing constituents are components of motor oils and crankcase oils, with the used oils typically having higher concentrations than the new unused oils. Typical carbon number chain lengths range from C_{15} to C_{50}.

8.1 Manufacture

Lubricating oil manufacture was well established by 1880, and the method depended on whether the crude oil was processed primarily for kerosene or for lubricating oils. Usually the crude oil was processed for kerosene, and primary distillation separated the crude into three fractions, naphtha, kerosene, and a residuum. To increase the production of kerosene the cracking distillation technique was used, and this converted a large part of the gas oils and lubricating oils into kerosene. The cracking reactions also produced coke products and asphaltlike materials, which gave the residuum a black color, and hence it was often referred to as *tar* (Parkash, 2003; Gary et al., 2007; Speight, 2014, 2017; Hsu and Robinson, 2017).

The production of lubricating oils is well established (Sequeira, 1992) and consists of four basic processes: (i) distillation to remove the lower boiling and lower molecular weight constituents of the feedstock, (ii) solvent refining, such as deasphalting, and/or hydrogen treatment to remove the nonhydrocarbon constituents and to improve the feedstock quality, (iii) dewaxing to remove the wax constituents and improve the low-temperature properties, (iv) and clay treatment or hydrogen treatment to prevent instability of the product.

Chemical, solvent, and hydrogen refining processes have been developed and are used to remove aromatics and other undesirable constituents, and to improve the viscosity index and quality of lube base stocks. Traditional chemical processes that use sulfuric acid and clay refining have been replaced by solvent extraction/refining and hydrotreating which are more effective, cost efficient, and generally more environmentally acceptable. Chemical refining is used most often for the reclamation of used lubricating oils or in combination with solvent or hydrogen refining processes for the manufacture of specialty lubricating oils and byproducts.

8.1.1 Chemical refining processes

Acid-alkali refining, also called *wet refining*, is a process where lubricating oils are contacted with sulfuric acid followed by neutralization with alkali. Oil and acid are mixed and an acid sludge is allowed to coagulate. The sludge is removed or the oil is decanted after settling, and more acid is added and the process repeated.

Acid-clay refining, also called *dry refining* is similar to acid-alkali refining with the exception that clay and a neutralizing agent are used for neutralization. This process is used for oils that form emulsions during neutralization. Neutralization with aqueous and alcoholic caustic, soda ash lime, and other neutralizing agents is used to remove organic acids from some feedstock. This process is conducted to reduce organic acid corrosion in downstream units or to improve the refining response and color stability of lube feedstock.

8.1.2 Hydroprocessing

Hydroprocessing, which has been generally replaced with solvent refining, consists of lube hydrocracking as an alternative to solvent extraction, and hydrorefining to prepare specialty products or to stabilize hydrocracked base stocks. Hydrocracking catalysts consist of mixtures of cobalt, nickel, molybdenum, and tungsten on an alumina or silica-alumina–based carrier. Hydrotreating catalysts are proprietary but usually consist of nickel-molybdenum on alumina. The hydrocracking catalysts are used to remove nitrogen, oxygen, and sulfur, and convert polynuclear aromatics and polynuclear naphthenes to mononuclear naphthenes, aromatics, and iso-paraffins which are typically desired in lube base stocks. Feedstock consist of unrefined distillates and deasphalted oils, solvent extracted distillates, cycle oils, hydrogen refined oils, and mixtures of these hydrocarbon fractions.

Lube hydrorefining processes are used to stabilize or improve the quality of lube base stocks from lube hydrocracking processes and for manufacture of specialty oils. Feedstocks are dependent on the nature of the crude source but generally consist of waxy or dewaxed-solvent extracted or hydrogen-refined paraffinic oils and refined or unrefined naphthenic and paraffinic oils from some selected crude oils.

8.1.3 Solvent refining processes

Feedstock from solvent refining processes consist of paraffinic and naphthenic distillates, deasphalted oils, hydrogen refined distillates, cycle oils, and dewaxed oils. The products are refined oils destined for further processing or finished lube base stocks. The byproducts are aromatic extracts which are used in the manufacture of rubber, carbon black, petrochemicals, catalytic cracking feedstock, fuel oil, or asphalt. The major solvents in use are N-methyl-2-pyrrolidone (NMP) and furfural, with phenol and liquid sulfur dioxide used to a lesser extent.

The solvents are typically recovered in a series of flash towers. Steam or inert gas strippers are used to remove traces of solvent, and a solvent purification system is used to remove water and other impurities from the recovered solvent.

Lube feedstocks typically contain increased wax content resulting from deasphalting and refining processes. These waxes are normally solid at ambient temperatures and must be removed to manufacture lube oil products with the necessary low temperature properties.

Catalytic dewaxing and solvent dewaxing (the most prevalent) are processes currently in use. Older technologies include cold settling, pressure filtration, and centrifuge dewaxing.

8.1.4 Catalytic dewaxing

Because solvent dewaxing is relatively expensive for the production of low pour point oils, various catalytic dewaxing (selective hydrocracking) processes have been developed for the manufacture of lube oil base stocks. The basic process consists of a reactor containing a proprietary dewaxing catalyst followed by a second reactor containing a hydrogen finishing catalyst to saturate olefins created by the dewaxing reaction and to improve stability, color, and demulsibility of the finished lube oil.

8.1.5 Solvent dewaxing

Solvent dewaxing consists of the following steps: crystallization, filtration, and solvent recovery. In the crystallization step, the feedstock is diluted with the solvent and chilled, solidifying the wax components. The filtration step removes the wax from the solution of dewaxed oil and solvent. Solvent recovery removes the solvent from the wax cake and filtrate for recycling by flash distillation and stripping. The major processes in use today are the ketone dewaxing processes. Other processes that are used to a lesser degree include the Di/Me process and the propane dewaxing process. The most widely used ketone processes are the Texaco solvent dewaxing process and the Exxon Dilchill Process. Both processes consist of diluting the waxy feedstock with solvent while chilling at a controlled rate to produce a slurry. The slurry is filtered using rotary vacuum filters and the wax cake is washed with cold solvent. The filtrate is used to chill the feedstock and solvent mixture. The primary wax cake is diluted with additional solvent and filtered again to reduce the oil content in the wax. The solvent is recovered from the dewaxed oil and wax cake by flash vaporization and recycled back into the process.

The Texaco solvent dewaxing process (also called the MEK process) uses a mixture of MEK and toluene as the dewaxing solvent, and sometimes uses mixtures of other ketones and aromatic solvents. The Exxon Dilchill dewaxing process uses a direct cold solvent dilution-chilling process in a special crystallizer in place of the scraped surface exchangers used in the Texaco process. The Di/Me dewaxing process uses a mixture of dichloroethane and methylene dichloride as the dewaxing solvent. The propane dewaxing process is essentially the same as the ketone process except for the following: propane is used as the dewaxing solvent and higher pressure equipment is required, and chilling is done in evaporative chillers by vaporizing a portion of the dewaxing solvent. Although this process generates a better product and does not require crystallizers, the temperature differentia between the dewaxed oil and the filtration temperature is higher than for the ketone processes (higher energy costs), and dewaxing aids are required to get good filtration rates.

8.1.6 Finishing processes

Hydrogen finishing processes have largely replaced acid and clay finishing processes. The hydrogen finishing processes are mild hydrogenation processes used to improve the color, odor, thermal, oxidative stability, and demulsibility of lube base stocks.

The process consists of fixed bed catalytic reactors that typically use a nickel-molybdenum catalyst to neutralize, desulfurize, and denitrify lube base stocks. These processes do not saturate aromatics or break carbon-carbon bonds as in other hydrogen finishing processes. Sulfuric acid treating is still used by some refiners for the manufacture of specialty oils and the reclamation of used oils. This process is typically conducted in batch or continuous processes similar to the chemical refining processes with the exception that the amount of acid used is much lower than that used in acid refining.

Clay contacting involves mixing the oil with fine bleaching clay at elevated temperature followed by separation of the oil and clay. This process improves color and chemical, thermal, and color stability of the lube base stock, and is often combined with acid finishing. Clay percolation is a static bed absorption process used to purify, decolorize, and finish lube stocks and waxes. It is still used in the manufacture of refrigeration oils, transformer oils, turbine oils, white oils, and waxes.

8.1.7 Older processes

Because of cracking distillation in the primary distillation and the high temperatures used in the still, the paraffin distillate contained dark-colored, sludge-forming asphaltic materials. These undesirable materials were removed by treatment with sulfuric acid followed by lye washing. Then, to separate the wax from the acid-treated paraffin distillate, the latter was chilled and filtered. The chilled, semisolid paraffin distillate was then squeezed in canvas bags in a knuckle or rack press (similar to a cider press) so that the oil would filter through the canvas, leaving the wax crystals in the bag. Later developments saw chilled paraffin distillate filtered in hydraulically operated plate and frame presses, and the use of these continued almost to the present time.

The oil from the press was known as *pressed distillate*, which was subdivided into three fractions by redistillation. Two overhead fractions of increasing viscosity, the heavier with a Society of Automotive Engineers (SAE) viscosity of approximately 10, were called *paraffin oils*. The residue in the still (viscosity equivalent to a low-viscosity SAE 30) was known as *red oil*. All three fractions were again acid- and lye-treated and then washed with water. The treated oils were pumped into shallow pans in the bleacher house, where air blown through the oil and exposure to the sun through the glass roof of the bleacher house or pan removed cloudiness or made the oils bright. Further treatment of the paraffin oil produced pale oil; thus if the paraffin oil was filtered through bone charcoal, fuller's earth, clay, or similar absorptive material, the color was changed from a deep yellow to a pale yellow. The filtered paraffin oil was called pale oil to differentiate it from the nonfiltered paraffin oil, which was considered of lower quality.

The wax separated from paraffin distillate by cold pressing contained approximately 50% oil and was known as *slack wax*. The slack wax was melted and cast into cakes, which were again pressed in a hot or hard press. This squeezed more oil from the wax, which was known as *scale wax*. By a process known as *sweating*, the scale wax was subdivided into several paraffin waxes with different melting points.

In contrast, crude oil processed primarily as a source of lubricating oil was handled differently from crude oils processed primarily for kerosene. The primary distillation removed naphtha and kerosene fractions, but without using temperatures high enough to cause cracking. The yield of kerosene was thus much lower, but the absence of cracking reactions increased the yield of lubricating oil fractions. Furthermore, the residuum was distilled using steam, which eliminated the need for high distillation temperatures, and cracking reactions were thus prevented. Thus, various overhead fractions suitable for lubricating oils and known as neutral oils were obtained; many of these were sufficiently low-boiling that they did not contain wax and did not need dewaxing; the more viscous oils could be dewaxed by cold pressing.

If the wax in the residual oil could not be removed by *cold pressing,* it was removed by *cold settling*. This involved admixture of the residual oil with a large volume of naphtha, which was then allowed to stand for as long as necessary in a tank exposed to low temperature, usually climatic cold (winter). This caused the waxy components to congeal and settle to the bottom of the tank. In the spring the supernatant naphtha-oil mixture was pumped to a steam still, where the naphtha was removed as an overhead stream; the bottom product was known as steam-refined stock. If the *steam-refined stock* (*bright stock*) was filtered through charcoal or a similar filter material, the improvement in color caused the oil to be known as bright stock. Mixtures of steam-refined stock with the much lower-boiling paraffin, pale, red, and neutral oils produced oils of any desired viscosity. The wax material that settled to the bottom of the cold settling tank was crude petrolatum. This was removed from the tank, heated, and filtered through a vessel containing clay, which changed its red color to brown or yellow. Further treatment with sulfuric acid produced white grades of *petrolatum*.

The development of vacuum distillation led to a major improvement in both paraffinic and naphthenic (low cold-test) oils. By vacuum distillation the more viscous paraffinic oils (even oils suitable for bright stocks) could be distilled overhead and could be separated completely from residual asphaltic components. Vacuum distillation provided the means of separating more suitable lubricating oil fractions with predetermined viscosity ranges and removed the limit on the maximum viscosity that might be obtained in the distillate oil.

However, although vacuum distillation effectively prevented residual asphaltic material from contaminating lubricating oils, it did not remove other undesirable components. The naphthenic oils, for example, contained components (naphthenic acids) that caused the oil to form emulsions with water. In particular, naphthenic oils contained components that caused oil to thicken excessively when cold and become very thin when hot. The degree to which the viscosity of oil is affected by temperature is measured on a scale that originally ranged from 0 to 100 and is called the viscosity index. An oil that changes the least in viscosity when the temperature is changed has a high viscosity index. Naphthenic oils have viscosity indices of 35 or less, compared to 70 or more for paraffinic oils.

8.2 Composition

Lubricating oil is distinguished from other fractions of crude oil by their usually high (>400°C, >750°F) boiling point, as well as their high viscosity. Materials suitable for the production of lubricating oils are comprised principally of hydrocarbons containing from 25 to 35 or even 40 carbon atoms per molecule, whereas residual stocks may contain hydrocarbons with 50 or more (up to 80 or so) carbon atoms per molecule. The composition of lubricating oil may be substantially different from the lubricant fraction from which it was derived, since wax (normal paraffins) is removed by distillation or refining by solvent extraction and adsorption preferentially removes nonhydrocarbon constituents as well as polynuclear aromatic compounds and the multiring cycloparaffins.

Normal paraffins up to C_{36} have been isolated from crude oil, but it is difficult to isolate any hydrocarbon from the lubricant fraction of crude oil. Various methods have been used in the analysis of products in the lubricating oil range, but the most successful procedure involves a technique based on the correlation of simple physical properties, such as refractive index, density, and molecular weight or viscosity, refractive index, and density. Results are obtained in the form of carbon distribution and the methods may also be applied to oils that have not been subjected to extensive fractionation. Although they are relatively rapid methods of analysis, the lack of information concerning the arrangement of the structural groups within the component molecules is a major disadvantage.

Nevertheless, there are general indications that the lubricant fraction contains a greater proportion of normal and branched paraffins than the lower boiling portions of crude oil. For the polycycloparaffin derivatives, a good proportion of the rings appear to be in condensed structures, and both cyclopentyl and cyclohexyl nuclei are present. The methylene groups appear principally in unsubstituted chains at least four carbon atoms in length, but the cycloparaffin rings are highly substituted with relatively short side chains.

Mono-, di-, and trinuclear aromatic compounds appear to be the main constituents of the aromatic portion, but material with more aromatic nuclei per molecule may also be present. For the dinuclear aromatics, most of the material consists of naphthalene types. For the trinuclear aromatics, the phenanthrene type of structure predominates over the anthracene type. There are also indications that the greater part of the aromatic compounds occurs as mixed aromatic-cycloparaffin compounds.

8.3 Properties and uses

Lubricating oil may be divided into many categories according to the types of service they are intended to perform. However, there are two main groups: (i) oils used in intermittent service, such as motor and aviation oils, and (ii) oils designed for continuous service, such as turbine oils. Lubricating oil is distinguished from other fractions of crude oil by a high (>400°C, >750°F) boiling point, as well as a high viscosity and, in fact, lubricating oil is identified by viscosity.

This classification is based on the SAE (Society of Automotive Engineers) J 300 specification. The single grade oils (such as the SAE 20 oil) correspond to a single class and have to be selected according to engine manufacturer specifications, operating conditions, and climatic conditions. At −20°C (68°F), multigrade lubricating oil such as SAE 10W-30 possesses the viscosity of a 10W oil and at 100°C (212°F) the multigrade oil possesses the viscosity of SAE 30 oil.

Oils used in intermittent service must show the least possible change in viscosity with temperature; that is, their viscosity indices must be high. These oils must be changed at frequent intervals to remove the foreign matter collected during service. The stability of such oils is therefore of less importance than the stability of oils used in continuous service for prolonged periods without renewal.

Oils used in continuous service must be extremely stable, but their viscosity indices may be low because the engines operate at fairly constant temperature without frequent shutdown.

9. Wax

Crude oil−derived wax (paraffin wax or petroleum wax) is a soft colorless solid, derived from crude oil which consists of a mixture of hydrocarbon derivative molecules containing between 20 and 40 carbon atoms. The wax is solid at room temperature and begins to melt above approximately 37°C (99°F) with a boiling point of > 370°C (700°F) (Parkash, 2003; Gary et al., 2007; Speight, 2014, 2017; Hsu and Robinson, 2017). The crude oil−derived wax is of two general types: (i) *paraffin wax* in crude oil distillates and (ii) *microcrystalline wax* in crude oil residua. The melting point of wax is not directly related to the boiling point because waxes contain hydrocarbon derivatives of different chemical nature. Nevertheless, waxes are graded according to their melting point and oil content.

9.1 Manufacture

The feedstock for paraffin is slack wax which is a mixture of oil and wax, a byproduct from the refining of lubricating oil. The first step in making paraffin wax is to remove the oil (deoiling or dewaxing) from the slack wax which is wax from a solvent dewaxing operation and the processes employed for the production of waxes are aimed at deoiling the slack wax (petroleum wax concentrate).

The oil is separated by crystallization. Most commonly, the slack wax is heated, mixed with one or more solvents such as a ketone derivative, and then cooled. As it cools, wax crystallizes out of the solution, leaving only oil. This mixture is filtered into two streams: solid (wax plus some solvent) and liquid (oil and solvent). After the solvent is recovered by distillation, the resulting products are called "product wax" (or "press wax") and "foots oil." The lower the percentage of oil in the wax, the more refined it is considered (semirefined vs. fully refined).[18] The product wax may be further processed to remove colors and odors. The wax may finally be blended together to give certain desired properties such as melt point and penetration. Paraffin wax is sold in either liquid or solid form.

Wax sweating was originally used in Scotland to separate wax fractions with various melting points from the wax obtained from shale oils. Wax sweating is still used to some extent but is being replaced by the more convenient recrystallization process. In wax sweating, a cake of slack wax is slowly warmed to a temperature at which the oil in the wax and the lower melting waxes become fluid and drip (or sweat) from the bottom of the cake, leaving a residue of higher melting wax. However, wax sweating can be carried out only when the residual wax consists of large crystals that have spaces between them, through which the oil and lower melting waxes can percolate; it is therefore limited to wax obtained from low-boiling paraffin distillate.

The amount of oil separated by sweating is now much smaller than it used to be owing to the development of highly efficient solvent dewaxing techniques. In fact, wax sweating is now more concerned with the separation of slack wax into fractions with different melting points. A wax sweater consists of a series of approximately nine shallow pans arranged one above the other in a sweater house or oven, and each pan is divided horizontally by a wire screen. The pan is filled to the level of the screen with cold water. Molten wax is then introduced and allowed to solidify, and the water is then drained from the pan leaving the wax cake supported on the screen.

A single sweater oven may contain more than 600 barrels of wax, and steam coils arranged on the walls of the oven slowly heat the wax cakes, allowing oil and the lower melting waxes to sweat from the cakes and drip into the pans. The first liquid removed from the pans is called *foots oil*, which melts at 38°C (100°F) or lower, followed by *interfoots* oil, which melts in the range 38°C–44°C (100°F–112°F). Crude scale wax next drips from the wax cake and consists of wax fractions with melting points over 44°C (112°F).

When oil removal was an important function of sweating, the sweating operation was continued until the residual wax cake on the screen was free of oil. When the melting point of the wax on the screen has increased to the required level allowing the oven to cool, this terminates sweating. The wax on the screen is a sweated wax with the melting point of a commercial grade of paraffin wax, which after a finished treatment becomes refined paraffinic wax. The crude scale wax obtained in the sweating operation may be recovered as such or treated to improve the color, in which case it is white crude scale wax. The crude scale wax and interfoots, however, are the sources of more waxes with lower melting points. The crude scale wax and interfoots are resweated several times to yield sweated waxes, which are treated to produce a series of refined paraffin waxes with melting points ranging from approximately 50°C to 65°C (125°F–150°F). Sweated waxes generally contain small amounts of unsaturated aromatic and sulfur compounds, which are the source of unwanted color, odor, and taste that reduce the ability of the wax to resist oxidation; the commonly used method of removing these impurities is clay treatment of the molten wax.

Wax recrystallization, like wax sweating, separates slack wax into fractions, but instead of using the differences in melting points, it makes use of the different solubility of the wax fractions in a solvent, such as the ketone used in the dewaxing process. When a mixture of ketone and slack wax is heated, the slack wax usually dissolves completely, and if the solution is cooled slowly, a temperature is reached at which a crop of wax crystals is formed. These crystals will all be of the same melting point, and if they are removed by filtration, a wax fraction with a specific melting point is obtained. If the clear filtrate is further cooled, a second crop of wax crystals with a lower melting point is obtained. Thus by alternate cooling and filtration the slack wax can be subdivided into a large number of wax fractions, each with different melting points.

This method of producing wax fractions is much faster and more convenient than sweating and results in a much more complete separation of the various fractions. Furthermore, recrystallization can also be applied to the microcrystalline waxes obtained from intermediate and high-boiling paraffin distillates, which cannot be sweated. Indeed, the microcrystalline waxes have higher melting points and differ in their properties from the paraffin waxes obtained from low-boiling paraffin distillates; thus wax recrystallization makes new kinds of waxes available.

9.2 Composition

Paraffin wax is a solid crystalline mixture of straight-chain (normal) hydrocarbons ranging from C_{20} to C_{30} and possibly higher, that is, CH_3 $(CH_2)_n$ CH_3 where $n \geq 18$. It is distinguished by its solid state at ordinary temperatures (25°C, 77°F) and low viscosity (35–45 SUS at 99°C, 210°F) when melted. However, in contrast to petroleum wax, petrolatum (*petroleum jelly*), although solid at ordinary temperatures, does in fact contain both solid and liquid hydrocarbons. It is essentially a low-melting, ductile, microcrystalline wax.

Although many natural waxes contain esters, paraffin waxes are hydrocarbon derivatives, mixtures of alkane derivatives usually in a homologous series of chain lengths. These materials represent a significant fraction of crude oil and are refined by vacuum distillation. Paraffin waxes are mixtures of saturated *n*- and iso-alkanes, naphthenes, and alkyl-substituted and naphthene-substituted aromatic compounds. A typical alkane paraffin wax's chemical composition comprises hydrocarbons with the general formula C_nH_{2n+2}. The degree of branching has an important influence on the properties.

9.3 Properties and uses

The melting point of paraffin wax (ASTM D87) has both direct and indirect significance in most wax utilization. All wax grades are commercially indicated in a range of melting temperatures rather than at a single value, and a range of 1°C (2°F) usually indicates a good degree of refinement. Other common physical properties that help to illustrate the degree of refinement of the wax are color (ASTM D156), oil content (ASTM D721), API gravity (ASTM D287), flash point (ASTM D92), and viscosity (ASTM D88 and ASTM D445), although the last three properties are not usually given by the producer unless specifically requested.

Petroleum waxes (and petrolatum) find many uses in pharmaceuticals, cosmetics, paper manufacturing, candle making, electrical goods, rubber compounding, textiles, and many others. For additional information, more specific texts on petroleum waxes should be consulted.

References

Abraham, H., 1945. Asphalt and Allied Substances, fifth ed., vol. I. Van Nostrand Inc., New York, p. 1.

Absi-Halabi, M., Stanislaus, A., Qabazard, H., 1997. Hydrocarbon Processing 76 (2), 45.

Ancheyta, J., Speight, J.G., 2007. Hydroprocessing of Heavy Oils and Residua. CRC Press, Taylor & Francis Group, Boca Raton, Florida.

ASTM D56, 2019. Standard Test Method for Flash Point by Tag Closed Cup Tester. Annual Book of Standards. ASTM International, West Conshohocken, Pennsylvania.

ASTM D86, 2019. Standard Test Method for Distillation of Petroleum Products and Liquid Fuels at Atmospheric Pressure. Annual Book of Standards. ASTM International, West Conshohocken, Pennsylvania.

ASTM D87, 2019. Standard Test Method for Melting Point of Petroleum Wax (Cooling Curve). Annual Book of Standards. ASTM International, West Conshohocken, Pennsylvania.

ASTM D88, 2019. Standard Test Method for Saybolt Viscosity. Annual Book of Standards. ASTM International, West Conshohocken, Pennsylvania.

ASTM D92, 2019. Standard Test Method for Flash and Fire Points by Cleveland Open Cup Tester. Annual Book of Standards. ASTM International, West Conshohocken, Pennsylvania.

ASTM D156, 2019. Standard Test Method for Saybolt Color of Petroleum Products (Saybolt Chromometer Method). Annual Book of Standards. ASTM International, West Conshohocken, Pennsylvania.

ASTM D187, 2019. Standard Test Method for Burning Quality of Kerosene. Annual Book of Standards. ASTM International, West Conshohocken, Pennsylvania.

ASTM D287, 2019. Standard Test Method for API Gravity of Crude Petroleum and Petroleum Products (Hydrometer Method). Annual Book of Standards. ASTM International, West Conshohocken, Pennsylvania.

ASTM D445, 2019. Standard Test Method for Kinematic Viscosity of Transparent and Opaque Liquids (And Calculation of Dynamic Viscosity). Annual Book of Standards. ASTM International, West Conshohocken, Pennsylvania.

ASTM D613, 2019. Standard Test Method for Cetane Number of Diesel Fuel Oil. Annual Book of Standards. ASTM International, West Conshohocken, Pennsylvania.

ASTM D721, 2019. Standard Test Method for Oil Content of Petroleum Waxes. Annual Book of Standards. ASTM International, West Conshohocken, Pennsylvania.

ASTM D975, 2019. Standard Specification for Diesel Fuel Oils. Annual Book of Standards. ASTM International, West Conshohocken, Pennsylvania.

ASTM D976, 2019. Standard Test Method for Calculated Cetane Index of Distillate Fuels. Annual Book of Standards. ASTM International, West Conshohocken, Pennsylvania.

ASTM D1070, 2019. Standard Test Methods for Relative Density of Gaseous Fuels. Annual Book of Standards. ASTM International, West Conshohocken, Pennsylvania.

ASTM D1319, 2019. Standard Test Method for Hydrocarbon Types in Liquid Petroleum Products by Fluorescent Indicator Adsorption. Annual Book of Standards. ASTM International, West Conshohocken, Pennsylvania.

ASTM D1826, 2019. Standard Test Method for Calorific (Heating) Value of Gases in Natural Gas Range by Continuous Recording Calorimeter. Annual Book of Standards. ASTM International, West Conshohocken, Pennsylvania.

ASTM D1835, 2019. Standard Specification for Liquefied Petroleum (LP) Gases. Annual Book of Standards. ASTM International, West Conshohocken, Pennsylvania.

ASTM D2163, 2019. Standard Test Method for Determination of Hydrocarbons in Liquefied Petroleum (LP) Gases and Propane/Propene Mixtures by Gas Chromatography. Annual Book of Standards. ASTM International, West Conshohocken, Pennsylvania.

ASTM D2425, 2019. Standard Test Method for Hydrocarbon Types in Middle Distillates by Mass Spectrometry. Annual Book of Standards. ASTM International, West Conshohocken, Pennsylvania.

ASTM D2500, 2019. Standard Test Method for Cloud Point of Petroleum Products and Liquid Fuels. Annual Book of Standards. ASTM International, West Conshohocken, Pennsylvania.

ASTM D2549, 2019. Standard Test Method for Separation of Representative Aromatics and Nonaromatics Fractions of High-Boiling Oils by Elution Chromatography. Annual Book of Standards. ASTM International, West Conshohocken, Pennsylvania.

ASTM D2650, 2019. Standard Test Method for Chemical Composition of Gases by Mass Spectrometry. Annual Book of Standards. ASTM International, West Conshohocken, Pennsylvania.

ASTM D2699, 2019. Standard Test Method for Research Octane Number of Spark-Ignition Engine Fuel. Annual Book of Standards. ASTM International, West Conshohocken, Pennsylvania.

ASTM D2700, 2019. Standard Test Method for Motor Octane Number of Spark-Ignition Engine Fuel. Annual Book of Standards. ASTM International, West Conshohocken, Pennsylvania.

ASTM D2786, 2019. Standard Test Method for Hydrocarbon Types Analysis of Gas-Oil Saturates Fractions by High Ionizing Voltage Mass Spectrometry. Annual Book of Standards. ASTM International, West Conshohocken, Pennsylvania.

ASTM D2789, 2019. Standard Test Method for Hydrocarbon Types in Low Olefinic Gasoline by Mass Spectrometry. Annual Book of Standards. ASTM International, West Conshohocken, Pennsylvania.

ASTM D4424, 2019. Standard Test Method for Butylene Analysis by Gas Chromatography. Annual Book of Standards. ASTM International, West Conshohocken, Pennsylvania.

ASTM D4814, 2019. Standard Specification for Automotive Spark-Ignition Engine Fuel. Annual Book of Standards. ASTM International, West Conshohocken, Pennsylvania.

ASTM D4891, 2019. Standard Test Method for Heating Value of Gases in Natural Gas and Flare Gases Range by Stoichiometric Combustion. Annual Book of Standards. ASTM International, West Conshohocken, Pennsylvania.

Boenheim, A.F., Pearson, A.J., 1973. In: Hobson, G.D., Pohl, W. (Eds.), Modern Petroleum Technology. Applied Science Publishers Inc., Barking, Essex, England (Chapter 19).

Chadeesingh, R., 2011. The Fischer-Tropsch process. In: Speight, J.G. (Ed.), The Biofuels Handbook, The Royal Society of Chemistry, pp. 476–517. London, United Kingdom. Part 3, (Chapter 5).

Forbes, R.J., 1958a. A History of Technology, vol. V. Oxford University Press, Oxford, England, p. 102.

Forbes, R.J., 1958b. Studies in Early Petroleum Chemistry. E. J. Brill, Leiden, The Netherlands.

Forbes, R.J., 1959. More Studies in Early Petroleum Chemistry. In: Brill, E.J. (Ed.) (Leiden, The Netherlands).

Gary, J.G., Handwerk, G.E., Kaiser, M.J., 2007. Petroleum Refining: Technology and Economics, fifth ed. CRC Press, Taylor & Francis Group, Boca Raton, Florida.

Hadley, D.J., Turner, L., 1973. In: Hobson, G.D., Pohl, W. (Eds.), Modern Petroleum Technology. Applied Science Publishers Inc., Barking, Essex, England (Chapter 12).

Hsu, C.S., Robinson, P.R. (Eds.), 2017. Handbook of Petroleum Technology. Springer International Publishing AG, Cham, Switzerland.

Luque, R., Speight, J.G. (Eds.), 2015. Gasification for Synthetic Fuel Production: Fundamentals, Processes, and Applications. Woodhead Publishing, Elsevier, Cambridge, United Kingdom.

Pandey, S.C., Ralli, D.K., Saxena, A.K., Alamkhan, W.K., 2004. Physicochemical characterization and application of naphtha. Journal of Scientific and Industrial Research 63 (3), 276–282.

Parkash, S., 2003. Refining Processes Handbook. Gulf Professional Publishing, Elsevier, Amsterdam, Netherlands.

Sequeira Jr., A., 1992. In: McKetta, J.J. (Ed.), Petroleum Processing Handbook. Marcel Dekker Inc., New York, p. 634.

Song, C., Ma, X., 2004. Ultra-deep desulfurization of liquid hydrocarbon fuels: chemistry and process. International Journal of Green Energy 1 (2), 167–191.

Speight, J.G. (Ed.), 2011. The Biofuels Handbook. Royal Society of Chemistry, London, United Kingdom.

Speight, J.G., 2014. The Chemistry and Technology of Petroleum, fifth ed. CRC Press, Taylor & Francis Group, Boca Raton, Florida.

Speight, J.G., 2017. Handbook of Petroleum Refining. CRC Press, Taylor & Francis Group, Boca Raton, Florida.

Walmsley, A.G., 1973. In: Hobson, G.D., Pohl, W. (Eds.), Modern Petroleum Technology. Applied Science Publishers Inc., Barking, Essex, England (Chapter 17).

Further reading

Hobson, G.D., Pohl, W., 1973. Modern Petroleum Technology. Applied Science Publishers, Barking, Essex, England.

Chapter 4

Hydrocarbons from natural gas and natural gas hydrates

1. Introduction

Natural gas, which is predominantly methane, occurs in underground reservoirs separately or in association with crude oil (Chapter 2, Chapter 3) (Parkash, 2003; Gary et al., 2007; Speight, 2011, 2014, 2019a; Hsu and Robinson, 2017; Speight, 2017). The principal types of hydrocarbons produced from natural gas are methane (CH_4) and varying amounts of higher molecular weight hydrocarbons from ethane (CH_3CH_3) to octane [$CH_3(CH_2)_6CH_3$]. Generally the higher molecular weight liquid hydrocarbons from pentane to octane are collectively referred to as *gas condensate*. While natural gas is predominantly a mixture of combustible hydrocarbons (Table 4.1), many natural gases also contain nitrogen (N_2) as well as carbon dioxide (CO_2) and hydrogen sulfide (H_2S). Trace quantities of argon, hydrogen, and helium may also be present (Table 4.1).

In the 1800s, natural gas was usually produced as a byproduct of crude oil production, since the lower molecular weight crude oil-soluble hydrocarbons come out of solution as pressure reduction occurred from the reservoir to the surface. However, the market for natural gas was limited, most cities finding it preferable to use gas from coal for lighting and heating. Unwanted natural gas was usually burned off at the well site. Often, unwanted gas (or *stranded gas* without a market) is pumped back into the reservoir through an *injection well* for disposal or repressurizing the formation to encourage additional production of crude oil.

The gas from coal (town gas) is a mixture of methane and other gases, mainly carbon monoxide, which can be used in a similar way to natural gas. Although coal gasification is not usually economic at current gas prices, the depletion of crude oil and gas reserves, and related infrastructure considerations, allows coal to be a viable future option for gas production and (via the Fischer-Tropsch process) a plentiful source of hydrocarbons.

The majority of the town gas plants in the late nineteenth century and early twentieth century were coke ovens in which heated bituminous coal (contained in air-tight chambers) produced the coke and gas as a byproduct. The gas

Handbook of Industrial Hydrocarbon Processes. https://doi.org/10.1016/B978-0-12-809923-0.00004-7

TABLE 4.1 Constituents of natural gas from a crude oil well.

Paraffins	Methane (CH_4)	70–98
	Ethane (C_2H_6)	1–10
	Propane (C_3H_8)	Trace-5
	Butane (C_4H_{10})	Trace-2
	Pentane (C_5H_{12})	Trace-l
	Hexane (C_6H_{14})	Trace-0.5
	Heptane and higher molecular weight (C_{7+})	Trace
Cycloparaffins	Cyclohexane (C_6H_{12})	Trace
Aromatics	Benzene (C_6H_6) + other aromatics	Trace
Nonhydrocarbons	Nitrogen (N_2)	Trace-15
	Carbon dioxide (CO_2)	Trace-1
	Hydrogen sulfide (H_2S)	Trace-1
	Helium (He)	Trace-5
	Other sulfur and nitrogen compounds	Trace
	Water (H_2O)	Trace-5

driven off from the coal was collected and distributed through town-wide networks of pipes to residences and other buildings where it was used for cooking and lighting purposes. By the time gas heating came into widespread use in the last half of the 20th century, natural gas was being used to supplant gas from coal. The coal tar that collected in the bottoms of the coke ovens was often used for roofing and other waterproofing purposes, and mainly as a source of chemical from which further yields of individual hydrocarbons (such as benzene, toluene, the xylenes, and aromatic naphtha) could be produced.

As the 20th century evolved, the market for natural gas expanded and in addition to the standard uses of natural gas (e.g., use of gas as a fuel) gas-to-liquids technology as a means of producing a range of hydrocarbons from the naphtha-gasoline-range hydrocarbons to kerosene-diesel-range hydrocarbons (Chapter 8).

Currently, raw natural gas is recovered from three types of wells: oil wells, gas wells, and condensate wells. Natural gas that comes from oil wells is typically termed *associated gas*. This gas can exist separate from oil in the formation (free gas), or dissolved in the crude oil (dissolved gas). Natural gas from gas and condensate wells, in which there is little or no crude oil, is termed *non-associated gas*. Gas wells typically produce raw natural gas by itself, while condensate wells produce free natural gas along with a semiliquid hydrocarbon condensate. Whatever the source of the natural gas, once separated from crude oil (if present) it commonly exists in mixtures with other hydrocarbons; principally ethane, propane, butane, and pentanes. In addition, raw natural gas contains water vapor, hydrogen sulfide (H_2S), carbon dioxide, helium, nitrogen, and other compounds. In fact, associated hydrocarbons, known as *natural gas liquids* (NGLs) can be very valuable byproducts of natural gas processing. Natural gas liquids include ethane, propane, butane, iso-butane, and natural gasoline that are sold separately and have a variety of different uses; including enhancing oil recovery in oil wells, providing raw materials for oil refineries or petrochemical plants, and as sources of energy.

Future sources of methane include landfill gas, biogas, and methane hydrates (also called natural gas hydrates). Landfill gas is a type of biogas, but biogas usually refers to gas produced from organic material that has not been mixed with other waste. Biogas, especially landfill gas, is already used in some areas, but their use could be greatly expanded (Speight, 2016, 2019b).

2. Gas processing

While natural gas is predominantly a mixture of combustible hydrocarbon derivatives (Table 4.1), many natural gases also contain nitrogen (N_2) as well as carbon dioxide (CO_2) and hydrogen sulfide (H_2S). Trace quantities of helium and other sulfur and nitrogen compounds may also be present. However, raw natural gas varies greatly in composition and the constituents can be several of a group of saturated hydrocarbon derivatives from methane to higher molecular weight hydrocarbon derivatives, especially natural gas that has been associated with crude oil in the reservoir, and nonhydrocarbon constituents (Table 4.1). The treatment required to prepare natural gas for distribution as an industrial or household fuel is specified in terms of the use and environmental regulations.

Briefly, natural gas contains hydrocarbon derivatives and nonhydrocarbon gases. Hydrocarbon gases are methane (CH_4), ethane (C_2H_6), propane (C_3H_8), butanes (C_4H_{10}), pentanes (C_5H_{12}), hexane (C_6H_{14}), heptane (C_7H_{16}), and sometimes trace amounts of octane (C_8H_{18}), and higher molecular weight hydrocarbon derivatives. For example:

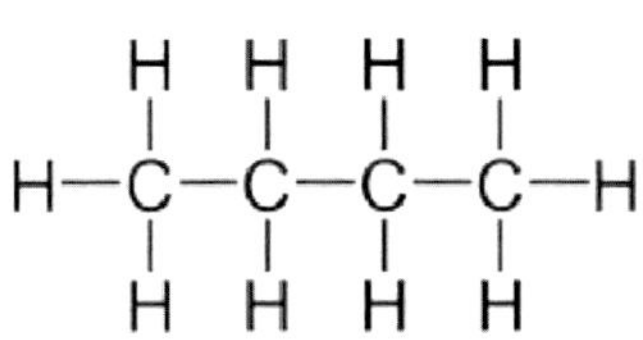

$CH_3CH_2CH_2CH_3$
n-Butane

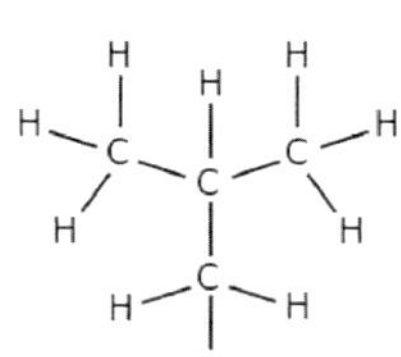

$(CH_3)_3CH$ or $(CH_3)_2CHCH_3$
iso-Butane

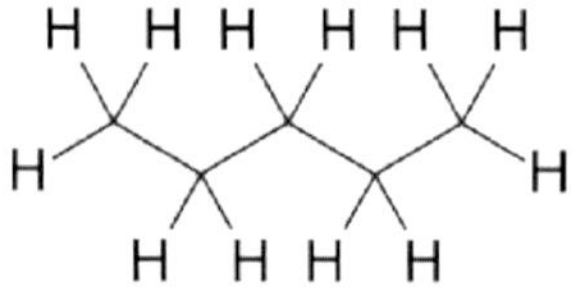

$CH_3CH_2CH_2CH_2CH_3$
n-Pentane

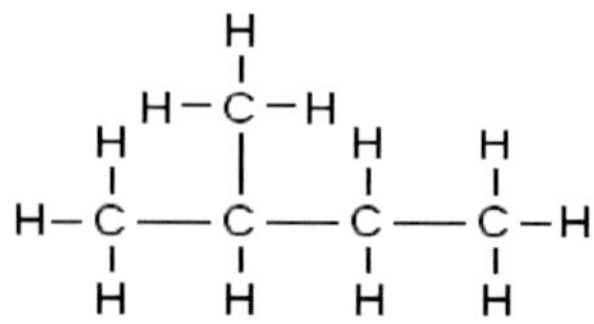

$(CH_3)_2CHCH_2CH_3$
iso-Pentane

As illustrated above, an iso-paraffin is an isomer having a methyl group branching from carbon number 2 of the main chain.

The higher-boiling hydrocarbon constituents than methane (CH_4) are often referred to as *natural gas liquids* (NGLs) and the natural gas may be referred to as *rich gas*. The constituents of natural gas liquids are hydrocarbon derivatives such as ethane (CH_3CH_3), propane ($CH_3CH_2CH_3$), butane ($CH_3CH_2CH_2CH_3$, as well as iso-butane), pentane derivatives ($CH_3CH_2CH_2CH_2CH_3$, as well as iso-pentane), and higher molecular weight hydrocarbon derivatives which have found wide use in the petrochemical industry (Chapter 12). Some aromatic derivatives [BTX—benzene (C_6H_6), toluene ($C_6H_5CH_3$), and the xylene isomers (*o*-, *m*-, and *p*-$CH_3C_6H_4CH_3$)] can also be present, raising safety issues due to their toxicity. The nonhydrocarbon gas portion of the natural gas contains nitrogen (N_2), carbon dioxide (CO_2), helium (He), hydrogen sulfide (H_2S), water vapor (H_2O), and other sulfur compounds (such as carbonyl sulfide (COS) and mercaptans (e.g., methyl mercaptan, CH_3SH)) and trace amounts of other gases. In addition, the composition of a gas stream from a source or at a location can also vary over time which can cause difficulties in resolving the data from the application of standard test methods (Klimstra, 1978; Liss and Thrasher, 1992).

Carbon dioxide and hydrogen sulfide are commonly referred to as *acid gases* since they form corrosive compounds in the presence of water. Nitrogen, helium, and carbon dioxide are also referred to as *diluents* since none of these burn, and thus they have no heating value. Mercury can also be present either as a metal in vapor phase or as an organometallic compound in liquid fractions. Concentration levels are generally very small, but even at very small concentration levels, mercury can be detrimental due its toxicity and its corrosive properties (reaction with aluminum alloys).

The higher molecular weight constituents (i.e., the C_{5+} product) are also commonly referred to as gas condensate or natural gasoline or sometimes, on occasion, as *casinghead gas* because of the tendency of these constituents to condense at the top of the well casing. When referring to natural gas liquids in the gas stream, the term gallon per 1000 cubic feet is used as a measure of high molecular weight hydrocarbon content. On the other hand, the composition of nonassociated gas (sometimes called well gas) is deficient in natural gas liquids. The gas is produced from geological formations that typically do not contain much, if any, hydrocarbon liquids. Furthermore, within the natural gas family, the composition of associated gas (a byproduct of oil production and the oil recovery process) is extremely variable, even within the gas from a petroleum reservoir (Speight, 2014; 2018). After the production fluids are brought to the surface, they are separated at a tank battery at or near the production lease into a hydrocarbon liquid stream (crude oil or condensate), a produced water stream (brine or salty water), and a gas stream.

The gaseous mixtures considered in this volume are mixtures of various constituents that may or may not vary over narrow limits. The defining characteristics of the various gas streams in the context of this book are that the gases (i) exist in a gaseous state at room temperature, (ii) may contain hydrocarbon constituents with one–four carbons, i.e., methane, ethane, propane, and butane isomers, (iii) may contain diluents and inert gases, and (iv) may contain contaminants in the form of nonhydrocarbon constituents. Each constituent of the gas influences the properties.

Typically, these gases fall into the general category of fuel gases and each gas is any one of several fuels that, at standard conditions of temperature and pressure, are gaseous. Before sale of the gas to the consumer actions, it is essential to give consideration of the variability of the composition of gas streams before and after processing (Table 2.2) and the properties of the individual constituents and their effects on gas behavior, even when considering the hydrocarbon constituents only. If not, the properties of the gas may be unstable and the ability of the gas to be used for the desired purpose will be seriously affected.

Before natural gas can be used as a fuel or for petrochemical feedstock, it must undergo processing (refining) to remove almost all materials other than methane. The byproducts of gas processing include ethane, propane, butanes, pentanes, and higher molecular weight hydrocarbons as well as hydrogen sulfide, thiols (mercaptans), carbon dioxide, water vapor, and sometime helium and nitrogen. Gas processing (*gas refining*) (Mokhatab et al., 2006; Speight, 2019a) consists of separating all of the various hydrocarbons and fluids from the pure natural gas (Fig. 4.1).

Major transportation pipelines usually impose restrictions on the make-up of the natural gas that is allowed into the pipeline. That means that before the natural gas can be transported it must be purified. While ethane, propane, butane, and pentanes must be removed from natural gas, this does not mean that they are all waste products. Gas processing is necessary to ensure that the natural gas intended for use is as clean and pure as possible, making it the clean burning and environmentally sound energy choice. Thus, natural gas, as

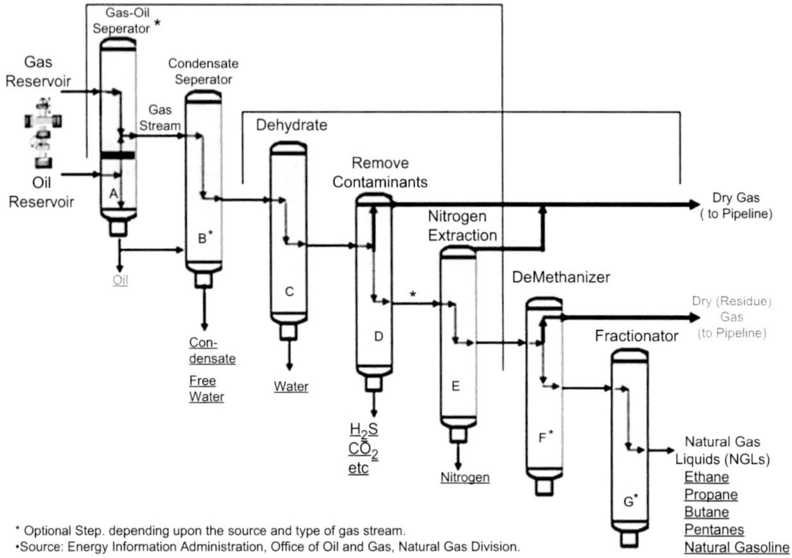

FIGURE 4.1 Gas processing.

it is used by consumers, is much different from the natural gas that is brought from underground up to the wellhead. Although the processing of natural gas is in many respects less complicated than the processing and refining of crude oil, it is equally as necessary before its use by end users. The natural gas used by consumers is composed almost entirely of methane. However, natural gas found at the wellhead, although still composed primarily of methane, is by no means as pure. Raw natural gas comes from three types of wells: (i) oil wells, (ii) gas wells, and (iii) condensate wells.

Natural gas that comes from oil wells is typically termed *associated gas*. This gas can exist separate from oil in the formation (free gas), or dissolved in the crude oil (dissolved gas). Natural gas from gas and condensate wells, in which there is little or no crude oil, is termed *nonassociated gas*. Gas wells typically produce raw natural gas by itself, while condensate wells produce free natural gas along with a liquid hydrocarbon condensate. Whatever the source of the natural gas, once separated from crude oil (if present) it commonly exists in mixtures with other hydrocarbons; principally ethane, propane, butane, and pentanes. In addition, raw natural gas contains water vapor, hydrogen sulfide (H_2S), carbon dioxide, helium, nitrogen, and other compounds.

Although the processing of natural gas is in many respects less complicated than the processing and refining of crude oil, it is equally as necessary before its use by end users.

2.1 Water removal

Water is a common impurity in gas streams, and removal of water is necessary to prevent condensation of the water and the formation of ice or gas hydrates ($C_nH_{2n+2}.xH_2O$). Water in the liquid phase causes corrosion or erosion problems in pipelines and equipment, particularly when carbon dioxide and hydrogen sulfide are present in the gas. The simplest method of water removal (refrigeration or cryogenic separation) is to cool the gas to a temperature at least equal to or (preferentially) below the dew point (Fig. 4.2). Absorption occurs when the water vapor is taken out by a dehydrating agent. Adsorption occurs when the water vapor is condensed and collected on the surface. In a majority of cases, cooling alone is insufficient and, for the most part, impractical for use in field operations. Other, more convenient, water removal options use (i) *hygroscopic* liquids (e.g., diethylene glycol or triethylene glycol) and (ii) solid adsorbents or desiccants (e.g., alumina, silica gel, and molecular sieves). Ethylene glycol can be directly injected into the gas stream in refrigeration plants.

An example of absorption dehydration is known as *glycol dehydration* and diethylene glycol, the principal agent in this process, has a chemical affinity for water and removes water from the gas stream. In this process, a liquid desiccant dehydrator serves to absorb water vapor from the gas stream. Essentially, glycol dehydration involves using a glycol solution, usually either diethylene glycol (DEG) or triethylene glycol (TEG), which is brought into contact with the wet gas stream in a *contactor*. The glycol solution will absorb water from the wet gas and, once absorbed, the glycol particles become heavier and sink to the bottom of the contactor where they are removed. The natural gas, having been stripped of most of its water content, is then transported out of the dehydrator. The glycol solution, bearing all of the water

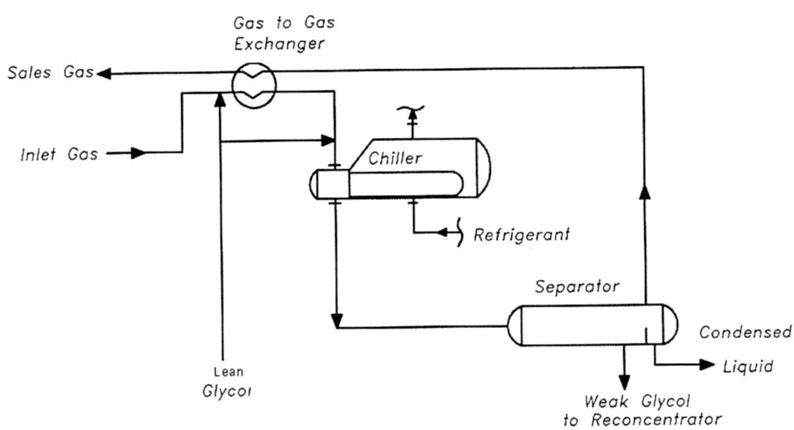

FIGURE 4.2 The glycol refrigeration process (Geist, 1985).

stripped from the natural gas, is put through a specialized boiler designed to vaporize only the water out of the solution. The boiling point differential between water (100°C, 212°F) and glycol (204°C, 400°F) makes it relatively easy to remove water from the glycol solution, allowing it to be reused in the dehydration process.

As well as absorbing water from the wet gas stream, the glycol solution occasionally carries with it small amounts of methane and other compounds found in the wet gas. In the past, this methane was simply vented out of the boiler. In addition to losing a portion of the natural gas that was extracted, this venting contributes to air pollution and the greenhouse effect. In order to decrease the amount of methane and other compounds that are lost, flash tank separator-condensers work to remove these compounds before the glycol solution reaches the boiler. Essentially, a flash tank separator consists of a device that reduces the pressure of the glycol solution stream, allowing the methane and other hydrocarbons to vaporize (*flash*). The glycol solution then travels to the boiler, which may also be fitted with air or water cooled condensers, which serve to capture any remaining organic compounds that may remain in the glycol solution. The regeneration (stripping) of the glycol is limited by temperature: diethylene glycol and triethylene glycol decompose at or before their respective boiling points. Such techniques as stripping of hot triethylene glycol with dry gas (e.g., heavy hydrocarbon vapors, the *Drizo process*) or vacuum distillation are recommended. In practice, absorption systems recover 90%−99% by volume of methane that would otherwise be flared into the atmosphere (Emam, 2015).

Solid adsorbent dehydration (*solid-desiccant dehydration*) is the primary form of dehydrating natural gas using adsorption, and usually consists of two or more adsorption towers, which are filled with a solid desiccant. Typical desiccants include activated alumina or a granular silica gel material. Wet natural gas is passed through these towers, from top to bottom. As the wet gas passes around the particles of desiccant material, water is retained on the surface of these desiccant particles. Passing through the entire desiccant bed, almost all of the water is adsorbed onto the desiccant material, leaving the dry gas to exit the bottom of the tower.

Silica gel (SiO_2) and alumina (Al_2O_3) have good capacities for water adsorption (up to 8% by weight). Bauxite (crude alumina, Al_2O_3) adsorbs up to 6% by weight water, and molecular sieves adsorb up to 15% by weight water. Silica is usually selected for dehydration of sour gas because of its high tolerance to hydrogen sulfide and to protect molecular sieve beds from plugging by sulfur. Alumina *guard beds* (which serve as protectors by the act of attrition and may be referred to as an *attrition catalyst*) (Ancheyta and Speight, 2007) may be placed ahead of the molecular sieves to remove the sulfur compounds. Downflow reactors are commonly used for adsorption processes, with an upward flow regeneration of the adsorbent and cooling in the same direction as adsorption.

Membrane separation processes are very versatile and are designed to process a wide range of feedstocks and offer a simple solution for removal and recovery of higher boiling hydrocarbons (natural gas liquids) from natural gas (Foglietta, 2004). The separation process is based on high-flux membranes that selectively permeate higher boiling hydrocarbons (compared to methane) and are recovered as a liquid after recompression and condensation. The residue stream from the membrane is partially depleted of higher boiling hydrocarbons, and is then sent to sales gas stream. Gas permeation membranes are usually made with vitreous polymers that exhibit good selectivity but, to be effective, the membrane must be very permeable with respect to the separation process.

2.2 Fractionation

Natural gas is considered *dry* when it is almost pure methane, having had most of the other commonly associated higher molecular weight hydrocarbons removed. When other hydrocarbons are present, the natural gas is *wet*. The higher molecular weight hydrocarbons start with ethane up to measurable amount of octane. These hydrocarbons are commonly referred to as *natural gas liquids* (NGLs). In a well that produces only natural gas (and not crude oil), any natural gas liquids are usually referred to as *gas condensate* which is removed from the gas stream at the wellhead. In most instances, natural gas liquids have a higher value as separate products, and it is thus economical to remove them from the gas stream.

The removal of natural gas liquids usually takes place in a relatively centralized processing plant, and uses techniques similar to those used to dehydrate natural gas. Recovery of the liquid hydrocarbons can be justified either because it is necessary to make the gas salable or because economics dictates this course of action. The justification for building a liquid recovery (or a liquid removal) plant depends on the price differential between the enriched gas (containing the higher molecular weight hydrocarbons) and lean gas with the added value of the extracted liquid.

There are two basic steps to the treatment of natural gas liquids in the natural gas stream. First, the liquids must be extracted from the natural gas. Second, these natural gas liquids must be separated themselves, down to their base components. These two processes account for approximately 90% of the total production of natural gas liquids.

Fractionation processes are very similar to those processes classed as *liquids removal* processes but often appear to be more specific in terms of the objectives: hence the need to place the fractionation processes into a separate category. The fractionation processes are those processes that are used (i) to remove the more significant product stream first, or (ii) to remove any unwanted light ends from the heavier liquid products.

In the general practice of natural gas processing, the first unit is a deethanizer (which separates ethane from the hydrocarbon stream) followed by a depropanizer (which separates propane from the hydrocarbon stream) then by a debutanizer (which separates the butanes from the pentanes and higher molecular weight hydrocarbons) and, finally, a butane fractionator (which separates the butane constituents into *n*-butane and iso-butane). Thus each column can operate at a successively lower pressure, thereby allowing the different gas streams to flow from column to column by virtue of the pressure gradient, without necessarily the use of pumps.

The purification of hydrocarbon gases by any of these processes is an important part of refinery operations, especially in regard to the production of liquefied petroleum gas (LPG). This is actually a mixture of propane and butane, which is an important domestic fuel, as well as an intermediate material in the manufacture of petrochemicals (Speight, 2019a). The presence of ethane in liquefied petroleum gas must be avoided because of the inability of this lighter hydrocarbon to liquefy under pressure at ambient temperatures and its tendency to register abnormally high pressures in the liquefied petroleum gas containers. On the other hand, the presence of pentane in liquefied petroleum gas must also be avoided, since this particular hydrocarbon (a liquid at ambient temperatures and pressures) may separate into a liquid state in the gas lines.

There are two principle techniques for removing hydrocarbons other than methane from natural gas: (i) the absorption method and (ii) the cryogenic expander process.

2.2.1 Absorption process

The absorption method of extraction is very similar to using absorption for dehydration. The main difference is that, in the absorption of natural gas liquids, absorbing oil is used as opposed to glycol. This absorbing oil has an affinity for natural gas liquids in much the same manner as glycol has an affinity for water. Before the oil has picked up any natural gas liquids, it is termed *lean* absorption oil.

The *oil absorption process* involves the countercurrent contact of the lean (or stripped) oil with the incoming wet gas with the temperature and pressure conditions programmed to maximize the dissolution of the liquefiable components in the oil. The *rich* absorption oil (sometimes referred to as *fat* oil), containing natural gas liquids, exits the absorption tower through the bottom. It is now a mixture of absorption oil, propane, butanes, pentanes, and other higher boiling hydrocarbons. The rich oil is fed into lean oil stills, where the mixture is heated to a temperature above the boiling point of the natural gas liquids but below that of the oil. This process allows for the recovery of around 75% by volume of the butanes, and 85%–90% by volume of the pentanes and higher boiling constituents from the natural gas stream.

The basic absorption process above can be modified to improve its effectiveness, or to target the extraction of specific natural gas liquids. For example, in the refrigerated oil absorption method, where the lean oil is cooled through refrigeration, propane recovery can be upwards of 90% by volume and approximately 40% by volume of the ethane can be extracted from the natural gas stream. Extraction of the other, higher boiling natural gas liquids can be close to 100% by volume using this process.

The AET process (Fig. 4.3) for recovery of liquefied petroleum gas utilizes noncryogenic absorption to recover ethane, propane, and higher boiling constituents from natural gas streams. The absorbed gases in the rich solvent from the bottom of the absorber column are fractionated in the solvent regenerator column which separates gases (as an overhead fraction) and lean solvent (as a bottoms fraction). After heat recuperation, the lean solvent is presaturated with absorber overhead gases. The chilled solvent flows in the top of the absorber column. The separated gases are sent to storage. Depending upon the economics of ethane recovery, the operation of the plant can be switched online from ethane plus recovery to propane plus recovery without affecting the propane recovery levels. The AET liquefied petroleum gas plant uses lower boiling lean oils. For most applications there are no solvent make-up requirements.

2.2.2 Cryogenic process

In the cryogenic process, a turboexpander is used to produce the necessary refrigeration and very low temperatures and high recovery of light components, such as ethane and propane, can be attained. The natural gas is first

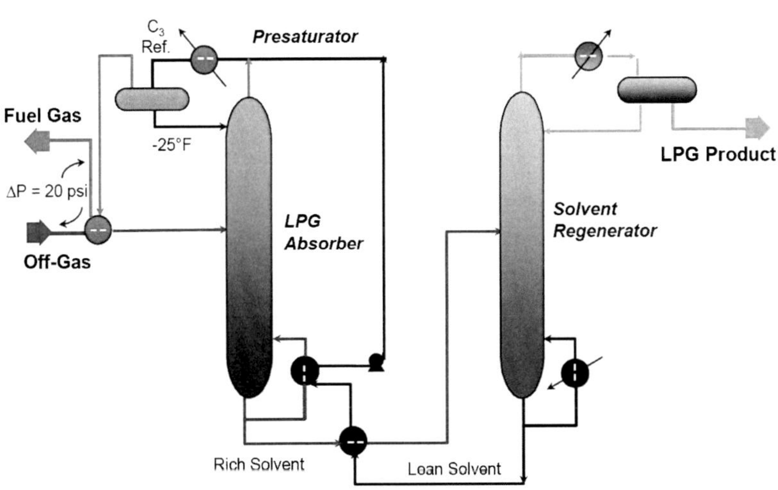

FIGURE 4.3 The AET process.

dehydrated using a molecular sieve followed by cooling. The separated liquid containing the higher molecular weight hydrocarbon fractions is then demethanized (i.e., the methane is removed), and the cold gases are expanded through a turbine that produces the cooling that is necessary for the process.

The expander outlet is a two-phase stream that is fed to the top of the demethanizer column. This serves as a separator in which: (i) the liquid is used as the column reflux and the separator vapors combined with vapors stripped in the demethanizer are exchanged with the feed gas, and (ii) the heated gas, which is partially recompressed by the expander compressor, is further recompressed to the desired distribution pressure in a separate compressor. This process allows for the recovery of approximately 90%–95% by volume of the ethane originally in the gas stream. In addition, the expansion turbine is able to convert some of the energy released when the natural gas stream is expanded into recompressing the gaseous methane effluent, thus saving energy costs associated with extracting ethane.

The extraction of natural gas liquids from the natural gas stream produces both cleaner, purer natural gas, as well as the valuable hydrocarbons that are the natural gas liquids themselves.

2.2.3 Fractionation of natural gas liquids

Natural gas liquids (lease condensate, natural gasoline, NGL) are components of natural gas that are liquid at surface in gas or oil field facilities or in gas processing plants. The composition of the natural gas liquids is dependent upon the type of natural gas and the composition of the natural gas. Natural gas liquids are the nonmethane constituents such as ethane, propane, butane, pentanes, and higher molecular weight hydrocarbon constituents which can be separated as liquids during gas processing (Mokhatab et al., 2006; Speight, 2014; 2019a). While NGLs are gaseous at underground pressure, the molecules condense at atmospheric pressure and turn into liquids. The composition of natural gas can vary by geographic region, the geological age of the deposit, the depth of the gas, and many other factors. Natural gas that contains a lot of NGLs and condensates is referred to as *wet gas*, while gas that is primarily methane, with little to no liquids in it when extracted, is referred to as *dry gas*.

The higher molecular weight constituents of natural gas (i.e., the C_{5+} product) are commonly referred to as gas condensate or natural gasoline. Rich gas will have a high heating value and a high hydrocarbon dew point. When referring to natural gas liquids in the gas stream, the term gallon per 1000 cubic feet is used as a measure of high molecular weight hydrocarbon content. On the other hand, the composition of nonassociated gas (sometimes called well gas) is deficient in natural gas liquids. The gas is produced from geological formations that typically do not contain much, if any, hydrocarbon liquids.

Generally, the hydrocarbon derivatives having a higher molecular weight than methane as well as any acid gases (carbon dioxide and hydrogen sulfide) are removed from natural gas prior to use of the gas as a fuel. However, since the composition of natural gas is never constant, there are standard test methods by which the composition and properties of natural gas can be determined and, thus, prepared for use. It is not the intent to cover the standard test methods in any detail in this chapter since descriptions of the test methods are available elsewhere (Speight, 2015; ATM, 2018; Speight, 2018).

Also, by way of a further reminder, natural-gas condensate (also called *condensate*, or *gas condensate*, or *natural gasoline*) is a low-density mixture of hydrocarbon liquids that are present as gaseous components in the raw natural gas produced from many natural gas fields ((Mokhatab et al., 2006; Speight, 2014; 2019a). Some gas constituents within the raw (unprocessed) natural gas will condense to a liquid state if the temperature is reduced to below the hydrocarbon dew point temperature at a set pressure. There are many condensate sources, and each has its own unique gas condensate composition.

Natural gas condensate (condensate, gas condensate, natural gasoline) is a low-density mixture of hydrocarbon liquids that are present as gaseous components in the raw natural gas produced from many natural gas fields. Gas condensate condenses out of the raw natural gas if the temperature is reduced to below the hydrocarbon dew point temperature of the raw gas. The composition of the gas condensate liquids is dependent upon the type of natural gas and the composition of the natural gas. Similarities exist between the composition of natural gas liquids and gas condensate—to the point that the two names are often (sometimes erroneously) used interchangeably. On a strictly comparative basis, the constituents of gas condensate represent the higher boiling constituents of natural gas liquids.

The fraction known as pentanes-plus is a mixture of pentane isomers and higher molecular weight constituents (C_{5+}) that is a liquid at ambient temperature and pressure, and consists mostly of pentanes and higher molecular weight (higher carbon number) hydrocarbon derivatives. Pentanes plus includes, but is not limited to, normal pentane, iso-pentane, hexanes-plus (natural gasoline), and condensate.

To separate the condensate from a natural gas feedstock from a gas well or a group of wells, the stream is cooled to lower the gas temperature to below the hydrocarbon dew point at the feedstock pressure and that condenses a good part of the gas condensate hydrocarbon derivatives (Mokhatab et al., 2006; Speight, 2014; 2019a). The feedstock mixture of gas, liquid condensate, and water is then routed to a high pressure separator vessel where the water and the raw natural gas are separated and removed. The raw natural gas from the high pressure separator is sent to the main gas compressor.

The gas condensate from the high pressure separator flows through a throttling control valve to a low pressure separator. The reduction in pressure across the control valve causes the condensate to undergo a partial vaporization referred to as a flash vaporization. The raw natural gas from the low pressure separator is sent to a booster compressor which raises the gas pressure and sends it through a cooler and on to the main gas compressor. The main gas compressor raises the pressure of the gases from the high and low pressure separators to whatever pressure is required for the pipeline transportation of the gas to the raw natural gas processing plant. The main gas compressor discharge pressure will depend upon the distance to the raw natural gas processing plant and it may require that a multistage compressor be used.

At the raw natural gas processing plant, the gas will be dehydrated and acid gases and other impurities will be removed from the gas. Then the ethane, propane, butanes, and pentanes plus higher molecular weight hydrocarbon derivatives (referred to as C5+) will also be removed and recovered as byproducts. The water removed from both the high and low pressure separators will probably need to be processed to remove hydrogen sulfide before the water can be disposed of or reused in some fashion.

Natural gas liquids can be classified according to their vapor pressures as low (condensate), intermediate (natural gasoline), and high (liquefied petroleum gas) vapor pressure. Natural gas liquids include propane, butane, pentane, hexane, and heptane, but not methane and not always ethane, since these hydrocarbon derivatives need refrigeration to be liquefied. A more general definition of natural gas liquids includes the nonmethane hydrocarbon derivatives from natural gas that are separated from the gas as liquids through the process of absorption, condensation, adsorption, or other methods in gas processing or cycling plants. Generally, under this definition, such liquids consist of ethane, propane, butane, and higher molecular weight hydrocarbon derivatives. For further use, the hydrocarbon derivatives are fractionated using a system which, after deethanization of the natural gas liquids, produces propane, butanes, and naphtha (C_5^+) (Mokhatab et al., 2006; Speight, 2019a).

After separation of the natural gas liquids from the natural gas stream, the hydrocarbons must be fractionated into their base components to be useful. The entire fractionation process is broken down into steps, starting with the removal of the lower boiling hydrocarbons from the stream. The fractionation process involves the use of fractionation towers (columns) to separate and remove various hydrocarbons. They towers can be controlled to produce pure-vapor-phase products from the overhead by optimizing the inlet feed flow rate, reflux flow rate, reboiler temperature, reflux temperature, and column pressure.

The particular fractionators are used in the following order: (i) deethanizer that separates the ethane from the stream of natural gas liquids, (ii) depropanizer that separates the propane from the deethanized stream, (iii) debutanizer that separates the butanes, leaving the pentanes and higher boiling hydrocarbons (naphtha) in the stream (Fig. 4.4), (iv) the butane splitter or deisobutanizer that separates the iso-butane and *n*-butane.

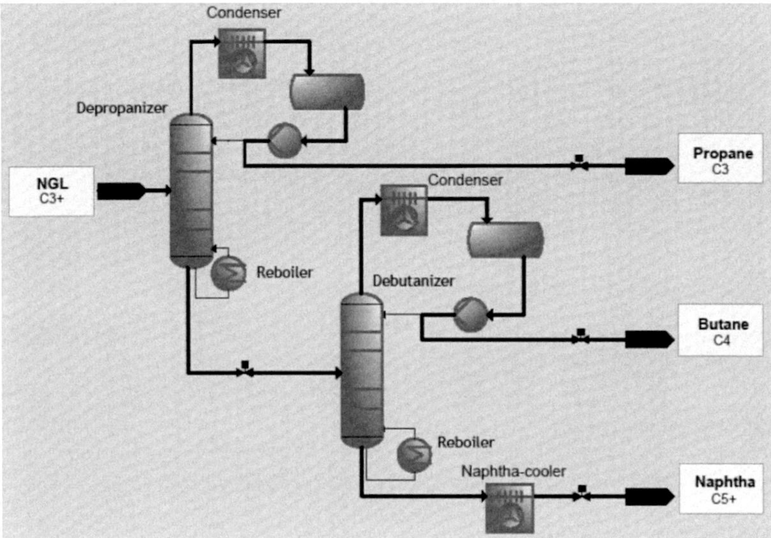

FIGURE 4.4 Fractionation of natural gas Liquids.

2.3 Acid gas removal

In addition to water and higher molecular weight hydrocarbons, one of the most important parts of gas processing involves the removal of hydrogen sulfide and carbon dioxide. Natural gas from some wells contains significant amounts of hydrogen sulfide and carbon dioxide and is usually referred to as *sour gas*. Sour gas is undesirable because the sulfur compounds it contains can be extremely harmful, even lethal, to breathe and the gas can also be extremely corrosive. The process for removing hydrogen sulfide from sour gas is commonly referred to as *sweetening* the gas.

The primary process for sweetening sour natural gas is quite similar to the processes of glycol dehydration and removal of natural gas liquids by absorption. In this case, however, amine (*olamine*) solutions are used to remove the hydrogen sulfide (the *amine process*) (Fig. 4.5). The sour gas is run through a tower, which contains the olamine solution. There are two principle amine solutions used, monoethanolamine (MEA) and diethanolamine (DEA) and either of these compounds, in liquid form, will absorb sulfur compounds from hydrocarbon streams. Other olamines are also used (Table 4.2). The effluent gas is virtually free of sulfur compounds, and thus loses its sour gas status. Like the process for the extraction of natural gas liquids and glycol dehydration, the amine solution used can be regenerated for reuse.

Although most sour gas sweetening involves the amine absorption process, it is also possible to use solid desiccants like iron oxide (iron sponge) to remove hydrogen sulfide and carbon dioxide.

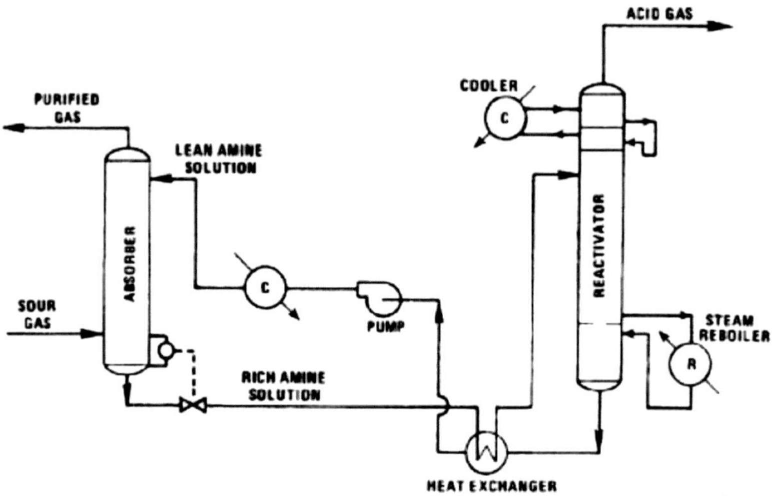

FIGURE 4.5 The amine (olamine) process.

The most well-known hydrogen sulfide removal process is based on the reaction of hydrogen sulfide with iron oxide (often also called the *iron sponge process* or the *dry box method*) in which the gas is passed through a bed of wood chips impregnated with iron oxide (Duckworth and Geddes, 1965; Anerousis and Whitman, 1984; Zapffe, 1963).

In the process (Fig. 4.6) the sour gas is passed down through the bed. In the case where continuous regeneration is to be utilized, a small concentration of air is added to the sour gas before it is processed. This air serves to continuously regenerate the iron oxide, which has reacted with hydrogen sulfide, which serves to extend the on-stream life of a given tower but probably serves to decrease the total amount of sulfur that a given weight of bed will remove.

The use of iron sponge process for sweetening sour gas is based on adsorption of the acid gases on the surface of the solid sweetening agent followed by chemical reaction of ferric oxide (Fe_2O_3) with hydrogen sulfide:

$$2Fe_2O_3 + 6H_2S \rightarrow 2Fe_2S_3 + 6H_2O$$

The reaction requires the presence of slightly alkaline water and a temperature below 43°C (110°F) and bed alkalinity (pH +8−10) should be checked regularly, usually on a daily basis. The pH level is be maintained through the injection of caustic soda with the water. If the gas does not contain sufficient water vapor, water may need to be injected into the inlet gas stream.

TABLE 4.2 Olamines used for removal of acid gases from hydrocarbon streams.

Olamine	Formula	Derived name	Molecular weight	Specific gravity	Melting point, °C	Boiling point, °C	Flash point, °C	Relative capacity, %
Ethanolamine (monoethanolamine)	$HOC_2H_4NH_2$	MEA	61.08	1.01	10	170	85	100
Diethanolamine	$(HOC_2H_4)_2NH$	DEA	105.14	1.097	27	217	169	58
Triethanolamine	$(HOC_2H_4)_3NH$	TEA	148.19	1.124	18	335[d]	185	41
Diglycolamine (hydroxyethanolamine)	$H(OC_2H_4)_2NH_2$	DGA	105.14	1.057	−11	223	127	58
Diisopropanolamine	$(HOC_3H_6)_2NH$	DIPA	133.19	0.99	42	248	127	46
Methyldiethanolamine	$(HOC_2H_4)_2NCH_3$	MDEA	119.17	1.03	−21	247	127	51

[d] *with decomposition*

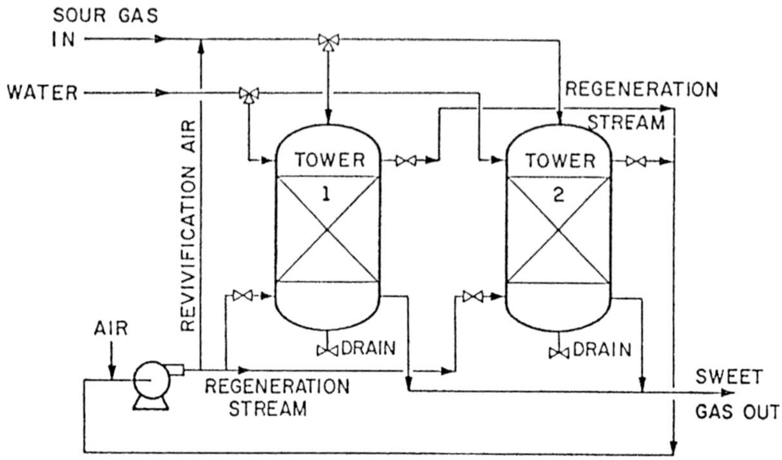

FIGURE 4.6 The iron oxide process.

The ferric sulfide produced by the reaction of hydrogen sulfide with ferric oxide can be oxidized with air to produce sulfur and regenerate the ferric oxide:

$$2Fe_2S_3 + 3O_2 \rightarrow 2Fe_2O_3 + 6S$$

$$S_2 + 2O_2 \rightarrow 2SO_2$$

The regeneration step is exothermic and air must be introduced slowly so the heat of reaction can be dissipated. If air is introduced quickly the heat of reaction may ignite the bed. Some of the elemental sulfur produced in the regeneration step remains in the bed. After several cycles this sulfur will form a cake over the ferric oxide, decreasing the reactivity of the bed. Typically, after 10 cycles the bed must be removed and a new bed introduced into the vessel.

The iron oxide process is one of several metal oxide–based processes that scavenge hydrogen sulfide and organic sulfur compounds (mercaptans) from gas streams through reactions with the solid-based chemical adsorbent (Kohl and Riesenfeld, 1985). They are typically nonregenerable, although some are partially regenerable, losing activity upon each regeneration cycle.

In the zinc oxide process, the zinc oxide media particles are extruded cylinders 3–4 mm in diameter and 4–8 mm in length (Kohl and Nielsen, 1997) and react readily with the hydrogen sulfide:

$$ZnO + H_2S \rightarrow ZnS + H_2O$$

At increased temperatures (205°C–370°C, 400°F–700°F), zinc oxide has a rapid reaction rate, therefore providing a short mass transfer zone, resulting in a short length of unused bed and improved efficiency.

Removal of larger amounts of hydrogen sulfide from gas streams requires a continuous process, such as the *Ferrox* process or the Stretford process. The *Ferrox process* is based on the same chemistry as the iron oxide process except that it is fluid and continuous. The *Stretford* process employs a solution containing vanadium salts and anthraquinone disulfonic acid (Maddox, 1974). Most hydrogen sulfide removal processes return the hydrogen sulfide unchanged, but if the quantity involved does not justify installation of a sulfur recovery plant (usually a Claus plant; Fig. 4.7) it is necessary to select a process that directly produces elemental sulfur.

The processes using ethanolamine and potassium phosphate are now widely used. The ethanolamine process, known as the *Girbotol* process, removes acid gases (hydrogen sulfide and carbon dioxide) from liquid hydrocarbons as well as from natural and from refinery gases. The Girbotol process uses an aqueous solution of ethanolamine ($H_2NCH_2CH_2OH$) that reacts with hydrogen sulfide at low temperatures and releases hydrogen sulfide at high temperatures. The ethanolamine solution fills a tower called an absorber through which the sour gas is bubbled. Purified gas leaves the top of the tower, and the ethanolamine solution leaves the bottom of the tower with the absorbed acid gases. The ethanolamine solution enters a reactivator tower where heat drives the acid gases from the solution. Ethanolamine solution, restored to its original condition, leaves the bottom of the reactivator tower to go to the top of the absorber tower, and acid gases are released from the top of the reactivator. When only carbon dioxide is to be removed in large quantities or when only partial removal is necessary, a hot carbonate solution or one of the physical solvents is the most economical selection.

The process using potassium phosphate is known as phosphate desulfurization, and it is used in the same way as the Girbotol process to remove acid

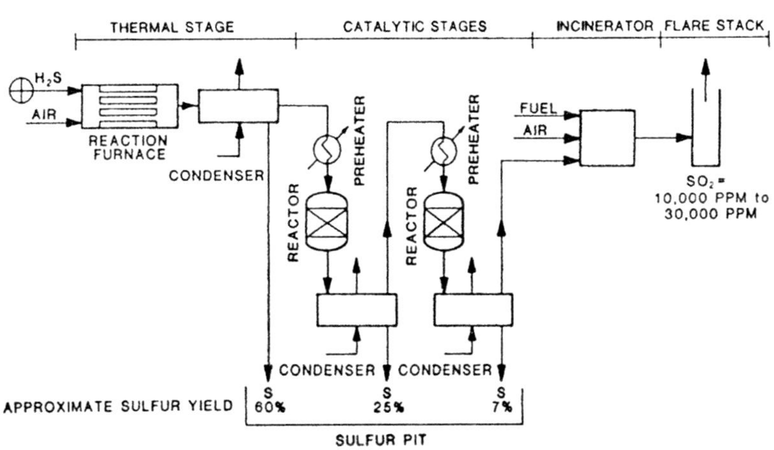

FIGURE 4.7 The Claus process (Maddox, 1974).

gases from liquid hydrocarbons as well as from gas streams. The treatment solution is a water solution of potassium phosphate (K_3PO_4), which is circulated through an absorber tower and a reactivator tower in much the same way as the ethanolamine is circulated in the Girbotol process; the solution is regenerated thermally.

Moisture may be removed from hydrocarbon gases at the same time as hydrogen sulfide is removed. Moisture removal is necessary to prevent harm to anhydrous catalysts and to prevent the formation of hydrocarbon hydrates (e.g., $C_3H_8.18H_2O$) at low temperatures. A widely used dehydration and desulfurization process is the glycolamine process, in which the treatment solution is a mixture of ethanolamine and a large amount of glycol. The mixture is circulated through an absorber and a reactivator in the same way as ethanolamine is circulated in the Girbotol process. The glycol absorbs moisture from the hydrocarbon gas passing up the absorber; the ethanolamine absorbs hydrogen sulfide and carbon dioxide. The treated gas leaves the top of the absorber; the spent ethanolamine-glycol mixture enters the reactivator tower, where heat drives off the absorbed acid gases and water.

Other processes include the *Alkazid process* for removal of hydrogen sulfide and carbon dioxide using concentrated aqueous solutions of amino acids. The hot potassium carbonate process decreases the acid content of natural and refinery gas from as much as 50% to as low as 0.5% and operates in a unit similar to that used for amine treating. The *Giammarco-Vetrocoke* process is used for hydrogen sulfide and/or carbon dioxide removal. In the hydrogen sulfide removal section, the reagent consists of sodium or potassium carbonates containing a mixture of arsenite and arsenate salts. The carbon dioxide removal section utilizes hot aqueous alkali carbonate solution activated by arsenic trioxide or selenous acid or tellurous acid.

Molecular sieves are highly selective for the removal of hydrogen sulfide (as well as other sulfur compounds) from gas streams and over continuously high absorption efficiency. They are also an effective means of water removal and thus offer a process for the simultaneous dehydration and desulfurization of gas. Gas that has excessively high water content may require upstream dehydration, however (Rushton and Hays, 1961).

The *molecular sieve process* is similar to the iron oxide process. Regeneration of the bed is achieved by passing heated clean gas over the bed. As the temperature of the bed increases, it releases the adsorbed hydrogen sulfide into the regeneration gas stream. The sour effluent regeneration gas is sent to a flare stack, and up to 2% of the gas heated can be lost in the regeneration process (Rushton and Hays, 1961). A portion of the natural gas may also be lost by the adsorption of hydrocarbon components by the sieve.

In this process, unsaturated hydrocarbon components, such as olefins and aromatics, tend to be strongly adsorbed by the molecular sieve (Mokhatab et al., 2006). Molecular sieves are susceptible to poisoning by such chemicals as glycols and require thorough gas cleaning methods before the adsorption step. Alternatively, the sieve can be offered some degree of protection by the use of *guard beds* in which a less expensive catalyst is placed in the gas stream before contact of the gas with the sieve, thereby protecting the catalyst from poisoning. This concept is analogous to the use of guard beds or attrition catalysts in the crude oil industry (Ancheyta and Speight, 2007).

3. Natural gas hydrates

The concept of natural gas production from *methane hydrate* (also called *gas hydrate*, *methane clathrate*, *natural has hydrate*, *methane ice*, *hydromethane*, *methane ice*, *fire ice*) is relatively new but does offer the potential to recover hitherto unknown reserves of methane that can be expected to extend the availability of natural gas (Giavarini et al., 2003, 2005; Giavarini and Maccioni, 2004; Makogon et al., 2007; Makogon, 2010; Wang and Economides, 2012; Yang and Qin, 2012). In terms of gas availability from this resource, 1 liter of solid methane hydrate can contain up to 168 L of methane gas).

Gas hydrates were first obtained by Joseph Priestley in 1778 in the laboratory by bubbling sulfur dioxide (SO_2) through cold water (0°C, 32°F) at atmospheric pressure and low room temperature. However, when describing the crystals that he produced, Priestley did not name them as *hydrates*. In 1811, similar crystals of aqueous chlorine were named hydrate of gas by Humphrey Davy. In both cases, the gas hydrates were not hydrocarbon hydrates, but they were gas hydrates.

Since their discovery in the early 19th century, gas hydrates have gone from being merely a laboratory curiosity to a serious problem for the natural gas industry to potentially becoming the largest source of methane. The emerging gas hydrate technologies have the potential not only to provide a huge source of methane, but may also one day be a means for natural gas storage and transportation and for various separations. However, in order to shift to these processes from the conceptual stage to becoming commercially feasible, it is still necessary to further enhance current understandings of hydrate science and engineering.

Natural gas hydrates are an unconventional source of energy and occur abundantly in nature, both in Arctic regions and in marine sediments (Bishnoi and Clarke, 2006). The formation of gas hydrate occurs when water and natural gas are present at low temperature and high pressure. Such conditions often exist in oil and gas wells and pipelines. Gas hydrates offer a source of energy as well as a source of hydrocarbons for the future (Fig. 4.8).

Gas hydrates are an icelike material which is made up methane molecules contained in a cage of water molecules and held together by hydrogen bonds. This material occurs in large underground deposits found beneath the ocean floor on continental margins and in places north of the Arctic Circle such as Siberia. It is estimated that gas hydrate deposits contain twice as much carbon as all other fossil fuels on Earth. This source, if proven feasible for

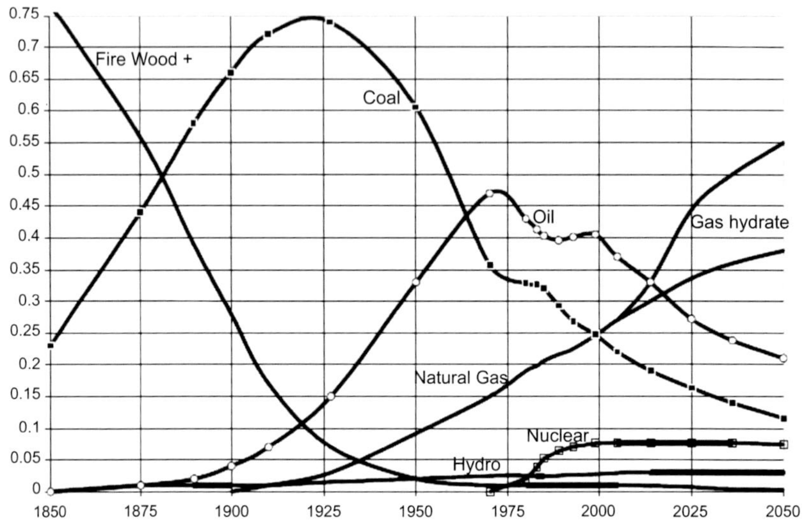

FIGURE 4.8 Fractional distribution of the various sources of energy and hydrocarbons since 1850.

recovery, could be a future energy as well as chemical source for petrochemicals.

Due to its physical nature (a solid material only under high pressure and low temperature), it cannot be processed by conventional methods used for natural gas and crude oils. One approach is by dissociating this cluster into methane and water by injecting a warmer fluid such as sea water. Another approach is by drilling into the deposit. This reduces the pressure and frees methane from water. However, the environmental effects of such drilling must still be evaluated.

The methane in gas hydrates is predominantly generated by bacterial degradation of organic matter in low oxygen environments. Organic matter in the uppermost few centimeters of sediments is first attacked by aerobic bacteria, generating carbon dioxide, which escapes from the sediments into the water column. In this region of aerobic bacterial activity sulfate derivatives (-SO_4) are reduced to sulfide derivatives (-S). If the sedimentation rate is low (<1 cm per 1000 years), the organic carbon content is low (<1% w/w), and oxygen is abundant, and the aerobic bacteria use up all the organic matter in the sediments. However, when sedimentation rate is high, and the organic carbon content of the sediment is high, the pore waters in the sediments are anoxic at depths of less than one foot or so and methane is produced by anaerobic bacteria.

The two major conditions that promote hydrate formation are thus: (i) high gas pressure and low gas temperature and (ii) the gas at or below its water dew point with free water present (Sloan, 1998b, 2006; Collett et al., 2009, 2015). The hydrates are believed to form by migration of gas from depth along geological faults, followed by precipitation, or crystallization, on contact of the rising gas stream with cold sea water. At high pressures methane hydrates remain stable at temperatures up to 18°C (64°F) and the typical methane hydrate contains one molecule of methane for every six molecules of water that forms the ice cage. However, the methane (hydrocarbon)-water ratio is dependent on the number of methane molecules that fit into the cage structure of the water lattice.

Chemically, gas hydrates are nonstoichiometric compounds formed by a lattice of hydrogen bonded molecules (host) which engage low molecular weight gases or volatile liquids (guest) with specific properties that differentiate them from ice (Bishnoi and Clarke, 2006). No actual chemical bond exists between guest and host molecules. Hydrate formation is favored by low temperature and high pressure (Makogon, 1997; Sloan, 1998a, 2006; Lorenson and Collett, 2000; Carrol, 2003; Seo et al., 2009). Most methane hydrate deposits also contain small amounts of other hydrocarbon hydrates; these include ethane hydrate and propane hydrate. In fact, gas hydrates of current interest are composed of water and the following molecules: methane, ethane, propane, isobutane, normal butane, nitrogen, carbon dioxide, and hydrogen sulfide. However, other nonpolar components such as argon (Ar) and ethyl cyclohexane ($C_6H_{11}C_2H_5$) can also form hydrates. Typically, gas hydrates form at temperatures on the order of 0°C (32°F) and elevated pressures (Sloan, 1998a, 2006).

3.1 Deposits

There are two basic forms of gas hydrate deposit: (i) primary and (ii) secondary. A primary deposit is the one which did not melt after its formation. Primary deposits are usually found in deep water, where temperatures do not change rapidly over time. Primary deposits are formed by the gases dissolved in the reservoir water and are located in the near seafloor sediments, which are characterized by high porosity, low temperature, and low rock strength. Frequently, a primary gas hydrate deposit does not have good barriers or seals. The hydrate begins to form in the pore space and eventually plugs the migration paths which traps more hydrate.

The hydrate can also act as cement holding the rock together. After the decomposition of hydrate, the porous media may revert back to a permeable unconsolidated state. For a primary gas hydrate deposit, the gas can be found over large areas that do not depend on the presence of structures. Free oil or gas may be present in the case of primary gas hydrate deposit.

Secondary gas hydrate deposits are usually located in the Arctic onshore. They are associated with natural gas reservoirs, located under the impermeable cap rocks in structural or stratigraphic traps. Upon temperature decrease in the formation (lower than the equilibrium temperature for the existing gas of this composition) hydrates may form. The temperature of rock layers on the continents is cyclic during the geologic time. During these cycles, the gas hydrates in the rocks will form and melt repeatedly. Often, there is free gas or oil under the hydrate layers. The most promising regions to look for commercial deposits of gas hydrate are the deep-water shelves, continental slopes, and continental abyssal trenches, with the depths of water ranging from 2500 to 7500 ft. However, the most promising resources of gas hydrate are concentrated in only 9%−12% of the ocean floor.

The process of hydrate formation is a heterogeneous process having similarities with crystallization processes (Bishnoi and Clarke, 2006). The difference in the two processes is that in the hydrate formation the solute (hydrate former) is supplied from another fluid phase (gas or liquid) to the aqueous liquid phase where it combines with water and crystallizes as solid hydrate. Also, the process is generally conducted at high pressures.

It is of interest to avoid hydrate formation or modify its flow characteristics to circumvent the problem of plugging natural gas pipelines or process equipment, leading to explosions. Activities to exploit the huge natural deposits of gas hydrates as an energy resource must be carefully planned and controlled. Second-guessing the behavior of gas hydrates will only lead to problems.

3.2 Composition

Gas hydrates are a carrier substance for natural gas and are generally considered to be an unconventional gas resource. The hydrates are widely distributed in oceans and polar regions and the gas contained in gas hydrates may be at least 10 times more than the global conventional gas reserves and twice the total carbon content in all world resources of crude oil, natural gas, and coal. Under standard conditions, one unit of gas hydrate could contain approximately 180 units of natural gas (Zou, 2017).

The composition of natural gas hydrates is determined by the composition of gas and water, and the pressure and temperature which existed at the time of their formation (Ruffine et al., 2009). Over geologic time, there will be changes in the thermodynamic conditions and the vertical and lateral migration of gas and water; therefore, the composition of hydrate can change both due to the absorption of free gas and the recrystallization of already-formed hydrate. In very general terms, the gas of gas hydrates is composed mainly of C1 to C8 hydrocarbons, carbon dioxide, and hydrogen sulfide (Yang et al., 2014). Thus, the pretreatment process for the gas produced from gas hydrates can follow the existing gas treatment process.

In the hydrate structure, methane is trapped within a cagelike crystal structure composed of water molecules in a structure that resembles packed snow or ice (Lorenson and Collett, 2000). The hydrate usually consists of methane with small ad mixtures of heavier components. However, in a number of cases the hydrate contains a significant volume of higher molecular weight hydrocarbon derivatives (Table 2.4) (Taylor, 2002). The presence of higher molecular weight hydrocarbon, other than methane, in the hydrates may be an indicator of the presence of petroleum reservoirs in the formation below the gas hydrate deposit.

Under the appropriate pressure, gas hydrates can exist at temperatures significantly above the freezing point of water, but the stability of the hydrate derivatives depends on pressure and gas composition and is also sensitive to temperature changes (Stern et al., 2000; Stoll and Bryan, 1979; Collett, 2001, 2010; Belosludov et al., 2007). For example, methane plus water at 600 psia forms hydrate at 5°C (41°F), while at the same pressure, methane with 1% v/v propane forms a gas hydrate at 9.4°C (49°F). Hydrate stability can also be influence by other factors, such as salinity.

Methane hydrates are restricted to the shallow lithosphere (i.e., at depths less than 6000 feet below the surface). The necessary conditions for the formation of hydrates are found only either in polar continental sedimentary rocks where surface temperatures are less than 0°C (32°F) or in oceanic sediment at water depths greater than 1000 feet where the bottom water temperature is on the order of 2°C (35°F).

Caution is advised when drawing generalities about the formation and the stability of gas hydrates. Methane hydrates are also formed during natural gas production operations, when liquid water is condensed in the presence of methane at high pressure. Higher molecular weight hydrocarbon derivatives such as ethane and propane can also form hydrates, although larger molecules (butane hydrocarbon derivatives and pentane hydrocarbon derivatives) cannot fit into the water cage structure and, therefore, tend to destabilize the formation of hydrates (Belosludov et al., 2007). However, for this text, the emphasis is focused on methane hydrates.

3.3 Other types of gases

The composition and properties of any gas stream depends on the characterization and properties of the hydrocarbon derivatives that make up the stream and calculation of the properties of a mixture depends on the properties of its constituents. However, calculation of the property of a mixture based on an average calculation neglects any interactions between the constituents. This makes the issue of modeling properties a difficult one because of the frequent omission of any chemical or physical interactions between the gas stream constituents.

The defining characteristics of the various gas streams in the context of this book are that the gases (i) exist in a gaseous state at room temperature, (ii) may contain hydrocarbon constituents with one−four carbons, i.e., methane,

ethane, propane, and butane isomers, (iii) may contain diluents and inert gases, and (iv) may contain contaminants in the form of nonhydrocarbon constituents. Each constituent of the gas influences the properties. Thus:

Hydrocarbons	Provide the calorific value of natural gas when it is burned
Diluents/inert gases	Typical gases are carbon dioxide, nitrogen, helium, and argon
Contaminants	Present in low concentrations; will affect processing operations

Many hydrocarbon gases do contain C_5 and C_6 hydrocarbon derivatives and apart from gas streams produced as process byproducts in a refinery, the C_{5+} constituents are typically found at lower concentrations (% v/v) in gases than the C_1 to C_4 constituents. There are also a few category members that may contain C_7 and even C_8 hydrocarbon derivatives, although such streams would necessarily be at elevated temperature and/or pressure to maintain the heptane (C_7H_{16}) and octane (C_8H_{18}) constituents in the gaseous state. Hydrocarbon compounds such as pentane (C_5H_{12}), hexane (C_6H_{14}), heptane (C_7H_{16}), and octane (C_8H_{18}) derivatives are typically found predominantly in naphtha derived from crude oil.

Typically, natural gas produced from shale reservoirs and other tight reservoirs has been classed under the general title *unconventional gas*. The production process requires stimulation by horizontal drilling coupled with hydraulic fracturing because of the pack of permeability in the gas-bearing formation. The boundary between conventional gas and unconventional gas resources is not well defined, because they result from a continuum of geologic conditions. Coal seam gas, more frequently called coal bed methane, is frequently referred to as unconventional gas. Tight shale gas and gas hydrates are also placed into the category of *unconventional gas*.

In addition to gas hydrate derivatives, there are several types of unconventional gas resources that arise from different sources and/or are currently produced by methods other than those used for conventional gas production and require processing before sale to the consumer. In this section, it would be remiss not to mention prominent gases produced from biomass and waste materials, viz., biogas and landfill gas. Both types of gas contain methane and carbon dioxide as well as various other constituents and are often amenable to gas processing methods that are applied to natural gas (Mokhatab et al., 2006; Speight, 2014; 2019a). The other types of gases are listed alphabetically rather than on the basis of current importance and are: (i) biogas, (ii) coalbed methane, (iii) coal gas, (iv) flue gas, (iv) gas in geopressurized zones, (v) gas in tight formations, (vii) landfill gas, (viii) refinery gas, and (ix) shale gas (Mokhatab et al., 2006; Speight, 2011, 2013, 2014).

Based on the cores taken while drilling in gas hydrate deposits, the hydrate usually consists of methane with small ad mixtures of heavier components. However, in a number of cases the hydrate contains a significant volume of higher molecular weight hydrocarbons (Table 4.3). The presence of higher molecular

TABLE 4.3 Composition of gas produced from various gas hydrates (Taylor, 2002).

	Gas composition, mol %							
Gas hydrate deposit	CH_4	C_2H_6	C_3H_8	iC_4H_{10}	nC_4H_{10}	C_5+	CO_2	N_2
Haakon Mosby mud volcano	99.5	0.1	0.1	0.1	0.1	0.1		
Nankai trough, Japan	99.3						0.63	
Bush Hill White	72.1	11.5	13.1	2.4	1	0		
Bush Hill Yellow	73.5	11.5	11.6	2	1	0.3	0.1	
Green Canyon White	66.5	8.9	15.8	7.2	1.4	0.2		
Green Canyon Yellow	69.5	8.6	15.2	5.4	1.2	0		
Bush Hill	29.7	15.3	36.6	9.7	4	4.8		
Messoyakha, Russia	98.7	0.03					0.5	0.77
Mallik, Canada	99.7	0.03	0.27					
Nankai trough −1, Japan	94.3	2.6	0.57	0.09	0.8		0.24	1.4
Blake Ridge, USA	99.9	0.02						0.08

weight hydrocarbon, other than methane, in the hydrates may be an indicator of the presence of crude oil reservoirs in the formation below the gas hydrate deposit.

3.4 Properties

Gas hydrate is a crystalline solid consisting of gas molecules, usually methane, each surrounded by a cage of water molecules. Thus it is *similar to ice*, except that the crystalline structure is stabilized by the guest gas molecule within the cage of water molecules.

Gas hydrates are gas concentrators. The decomposition of one unit volume of methane hydrate at a pressure of 1 atmosphere produces approximately 160 unit volumes of gas.

Many gases have molecular sizes suitable to form hydrate, including such naturally occurring gases as carbon dioxide, hydrogen sulfide, and several low-carbon-number hydrocarbons, but most marine gas hydrates that have been analyzed are methane hydrates. Methane hydrate is stable in ocean floor sediments at water depths greater than 1000 ft., and where it occurs, it is known to cement loose sediments in a surface layer several hundred to one thousand or more feet thick.

The morphology of gas hydrate crystals depends on water and gas compositions, pressure, temperature, and phase state of water (liquid, vapor, or solid) and gas. More than 10,000 different forms of crystals were studied. Nuclei of hydrate crystals usually start to form at a gas-water interface and grow to a total coverage of the interface. Following that, crystals grow in free gas phase or in water. There are three basic morphologic forms of hydrate crystals: massive, whiskery, and gellike.

Natural gas hydrates are metastable minerals, where the formation and dissociation depend on the pressure and temperature, composition of gas, salinity of the reservoir water, and the characteristics of the porous medium in which they were formed. Hydrate crystals in reservoir rocks can be dispersed in the pore space without the destruction of pores; however, in some cases, the rock is affected. Hydrates can be in the form of small nodules (from 5 to 12 cm in size), in the form of small lenses, or in the form of layers that can be several feet thick.

Gas hydrate is a mineral of the clathrate hydrate group. Hydrates have six different forms: (i) molecular sieves, characterized by interconnected trough cavities and/or passages, (ii) channel complexes when hydrate forming molecules form a crystalline lattice with tubular cavities, (iii) layered complexes forming clathrates with interlaced molecular layers, (iv) complexes which form large molecules having concavities or niches in which an inclusion molecule resides, (v) linear polymeric complexes formed by clathrate molecules, having a tubelike shape, (vi) clathrates which form in cases when inclusion molecules fill in the closed cavities close in shape to a sphere. Hydrates of gases and volatile liquids are related to the latter type of clathrates.

Gas hydrates may formally be referred to as chemical compounds because they have a fixed composition at a certain pressure and temperature. However, hydrates are the compounds of a molecular type. They form as a result of the Van der Waals attraction forces between the molecules. Covalent bond is absent in the gas hydrates because during their formation there is no pairing of valence electrons and no spatial redistribution of electron cloud density.

In many respects, the acoustic, strength, thermal, and rheological properties of gas hydrates are similar to those of ice. However, there are a few properties, including the dielectric constant and the thermal conductivity, which differ significantly from that of ice. The vast majority of data available on the physical properties of gas hydrates is only for methane hydrates (Davidson, 1983).

Gas hydrates are stable at the temperatures and pressures that occur in ocean floor sediments at water depths greater than approximately 1500 feet, and at these pressures they are stable at temperatures above those for ice stability. Gas hydrates also are stable in association with permafrost in the polar regions, both in offshore and onshore sediments. Gas hydrates bind immense amounts of methane in sea floor sediments. Hydrate is a gas concentrator; the breakdown of a unit volume of methane hydrate at a pressure of one atmosphere produces approximately 160 unit volumes of gas. The worldwide amount of methane in gas hydrates is considered to contain at least 1×10^4 gigatons of carbon in a very conservative estimate—this is *approximately twice the amount of carbon held in all fossil fuels on Earth.*

3.5 Development

To eventually produce natural gas economically from gas hydrate deposits, it is important to determine not only the potential gas-in-place, but also what amount can be extracted economically (Makogon et al., 2005, 2007; Chong et al., 2017). The effectiveness of the extraction is determined by the geological and thermodynamic conditions, and by the concentration of gas hydrate in the deposit. To produce the free gas, the hydrate must be first changed from a solid to a fluid.

Thus, it is necessary to use much of the energy contained in the gas hydrate deposit for heating the rock layers near the gas hydrate deposit. Preliminary estimates show that the coefficient of extraction of the gas hydrate can be as high as 50%–70%. However, from total world potential resource it has been estimated that the coefficient of extraction should average from 17% to 20%.

Hydrate development (hydrate crystal decomposition) like the hydrate growth is a deterministic process (Bishnoi and Clarke, 2006). The process is a heterogeneous process where liquid water and gas are released as the solid shrinks due to its decomposition. To eventually produce natural gas economically from gas hydrates, it is important to determine not only the potential gas-in-place, but also what amount can be extracted economically.

The effectiveness of the extraction is determined by the geological and thermodynamic conditions, and by the concentration of gas hydrate in the deposit. To produce the free gas, the hydrate must be first changed from a solid to a fluid.

Methods of gas recovery from hydrates generally deal with dissociating or melting in-situ gas hydrates by heating the reservoir beyond the temperature of hydrate formation, or decreasing the reservoir pressure below hydrate equilibrium. Computer models have been developed to evaluate natural gas production from hydrates by both heating and depressurization. Depressurization is considered to be the most economically promising method for the production of natural gas from gas hydrates. Also, thermal stimulation, inhibitor injection, or a combination of these methods has been considered as possible means of producing gas hydrate. Another method for gas hydrate production involves the injection of carbon dioxide in which the carbon dioxide is exchanged for the methane in the hydrate. The concept involves injecting carbon dioxide gas, which is then allowed to equilibrate with methane hydrate along the three-phase equilibrium boundary. Because of the difference in chemical affinity for carbon dioxide versus methane in the hydrate structure, the mole fraction of methane would be reduced to approximately 0.48 in the hydrate and rise to a value of 0.7 in the gas phase at equilibrium.

For offshore conditions, effective production of gas from gas hydrate decomposition in the majority of the cases may occur when hydrate saturation of porous media exceeds 30%–40%. However, each geologic region will have to be studied in detail to establish the minimal hydrate saturation that is required. To change gas hydrate decomposition to natural gas, it is necessary to (i) decrease reservoir pressure to lower than equilibrium one, (ii) increase the temperature to higher than equilibrium one, (iii) inject active reagents, which facilitate the decomposition of hydrate, and (iv) use some new technology. The easiest method is to lower the reservoir pressure in gas hydrate decomposition.

For offshore conditions, with the depths of water ranging from one halt mile to one mile or more, effective production of gas from gas hydrate deposits in the majority of the cases may occur when hydrate saturation of porous media exceeds 30%–40%. However, each geologic region will have to be studied in detail to establish the minimal hydrate saturation that is required. To change a gas hydrate deposit to natural gas and other hydrocarbons, it is necessary to (i) decrease reservoir pressure to lower than equilibrium one, (ii) increase the temperature to higher than equilibrium one, (iii) inject active reagents, which facilitate the decomposition of hydrate, and (iv) use some new technology. The easiest method is to lower the reservoir pressure in gas hydrate deposit. Clearly, this method is only feasible when free gas is found below the gas hydrate deposit.

Obviously, the energy concentrated in natural gas hydrates can serve as an unconventional energy source very important to sustain the growing energy needs for several decades. Natural gas hydrates are more evenly distributed on the planet than sources of hydrocarbons. The production of gas from gas hydrate deposits will be accessible to many countries. Many existing technologies can be used to find and develop gas hydrate deposits. The economic and ecological aspects of producing a gas hydrate deposit must be evaluated. Both the economics and the ecological aspects depend upon the developing technologies.

4. Hydrocarbon products

4.1 Methane

Methane (CH_4) (Fig. 4.9) commonly (often incorrectly) known as natural gas is the lowest molecular weight alkane hydrocarbon (Table 4.4). It is colorless and naturally odorless, and burns efficiently without many byproducts. http://www.pmel.org/NaturalGas-Glossary.htm - TOP#TOP. Methane is the simplest alkane, and the principal component of natural gas (usually 70%–90% v/v). In addition, there is a large, but unknown, amount of methane in gas hydrates (methane clathrates) in the ocean floors and significant amounts of methane are produced anaerobically by methanogenesis. Other sources include mud volcanoes, such as those that occur regularly in Trinidad, which are connected with deep geological faults, landfills, and livestock (primarily ruminants) from enteric fermentation.

At room temperature, methane is a gas less dense than air; it melts at −183°C (−297°F) and boils at −161°C (−258°F). Methane is a colorless, odorless gas—the smell characteristic of commercial natural gas is an artificial safety measure caused by the addition of an odorant, such as methanethiol (CH_3SH) or ethanethiol (C_2H_5SH) or *tertiary*-butyl mercaptan [$(CH_3)_3CSH$], to detect leaks. Methane is also an asphyxiant, especially if the oxygen

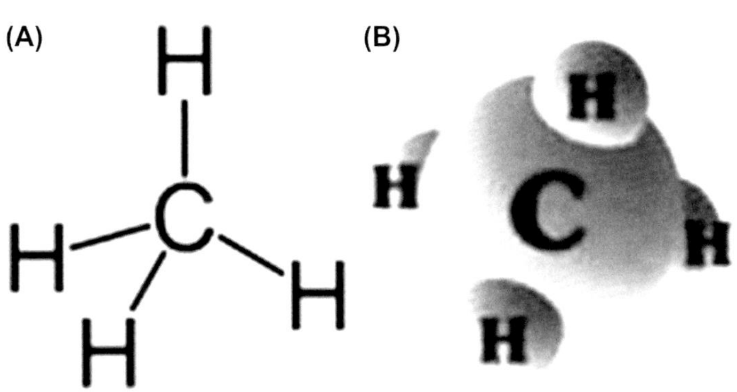

FIGURE 4.9 Simplified representation of methane as (A) a two-dimensional formula and (B) a three-dimensional formula.

TABLE 4.4 General properties of the lower molecular weight alkane hydrocarbons (up to and including *n*-octane).

Hydrocarbon	Formula	Melting point (C)	Boiling point (°C)	Density (20°C)	Physical state (20°C)
Methane	CH_4	−182	−164	0.668 g/L	Gas
Ethane	C_2H_6	−183	−89	1.265 g/L	Gas
Propane	C_3H_8	−190	−42	1.867 g/L	Gas
Butane	C_4H_{10}	−138	−1	2.493 g/L	Gas
Pentane	C_5H_{12}	−130	36	0.626 g/mL	Liquid
Hexane	C_6H_{14}	−95	69	0.659 g/mL	Liquid
Octane	C_8H_{18}	−57	125	0.703 g/mL	Liquid

concentration is reduced to below 20% v/v 19.5% by displacement in an enclosed space.

The concentrations at which flammable or explosive mixtures form are much lower than the concentration at which asphyxiation risk is significant. When structures are built on or near landfill sites, methane off-gas can penetrate the interior of the buildings and expose the occupants to significant levels of methane. Some buildings have specially engineered recovery systems below their basements to actively capture such fugitive off-gas and vent it away from the building.

Methane is a relatively potent greenhouse gas and compared with carbon dioxide, it has a high global warming potential of 72 (calculated over a period of 20 years) or 25 (for a time period of 100 years). Methane in the atmosphere is eventually oxidized, producing carbon dioxide and water. As a result, methane in the atmosphere has a half-life of 7 years.

The major reactions of methane are: (i) combustion, (ii) steam reforming to synthesis gas, and (iii) halogenation. Generally, the reactions of methane are difficult to control. For example, the partial oxidation of methane (CH_4) to methanol (CH_3OH) is difficult to achieve and the reaction typically progresses all the way to carbon dioxide (CO_2) and water (H_2O).

The combustion of methane is believed to form formaldehyde (HCHO or H_2CO) as an intermediate which then gives a formyl radical (HCO) leading to the formation of carbon monoxide (CO):

$$CH_4 + O_2 \rightarrow CO + H_2 + H_2O$$

The hydrogen oxidizes to water, releasing heat:

$$2H_2 + O_2 \rightarrow 2H_2O$$

Finally, the carbon monoxide oxidizes to carbon dioxide, releasing more heat:

$$2CO + O_2 \rightarrow 2CO_2$$

Thus:

$$CH_4(g) + 2O_2(g) \rightarrow CO_2(g) + 2H_2O(l) + 891\ \text{kJ/mol}$$

The strength of the carbon-hydrogen covalent bond in methane is among the strongest in all hydrocarbons, and thus its use as a chemical feedstock is limited. Despite the high activation barrier for breaking the carbon hydrogen bond, methane is still the principal starting material for manufacture of hydrogen in steam reforming.

Steam-methane reforming is a method of producing hydrogen or other useful products from natural gas or other fossil fuels, after gasification. This is achieved in a processing device called a reformer which reacts steam at high temperature with the fossil fuel.

Steam reforming of natural gas or synthesis gas (steam methane reforming, SMR) is the most common method of producing commercial bulk hydrogen as well as the hydrogen used in the industrial synthesis of ammonia. At high temperatures (700°C–1100°C) and in the presence of a metal-based catalyst (nickel), steam reacts with methane to yield carbon monoxide and hydrogen:

$$CH_4 + H_2O \rightarrow CO + 3H_2$$

Additional hydrogen can be recovered by a lower-temperature gas-shift reaction with the carbon monoxide produced:

$$CO + H_2O \rightarrow CO_2 + H_2$$

The first reaction is strongly endothermic (consumes heat), the second reaction is mildly exothermic (produces heat).

The steam methane reformer is widely used in industry to make hydrogen. There is also interest in the development of much smaller units based on similar technology to produce hydrogen as a feedstock for fuel cells. Small-scale steam reforming units to supply fuel cells are currently the subject of research and development, typically involving the reforming of methanol or natural gas but other fuels are also being considered such as propane, gasoline, diesel fuel, and ethanol.

Methane reacts with all of the halogens given appropriate reaction conditions, temperature, and pressure:

$$CH_4 + X_2 \rightarrow CH_{3X} + HX$$

X is a halogen: fluorine (F), chlorine (Cl), bromine (Br), or iodine (I). When X is chlorine, the mechanism has the following form:

1. Homolytic scission:

$$Cl_2 + \text{uv energy} \rightarrow 2Cl\bullet \text{ (free radicals)}$$

The needed energy can also arise from heat energy. Then:

2. Radical exchange:

$$CH_4 + Cl\bullet \rightarrow CH_3\bullet\cdot + HCl$$

$$CH_3\bullet\cdot + Cl_2 \rightarrow CH_3Cl + Cl\bullet$$

3. Radical termination:

$$2\ Cl\bullet\cdot \rightarrow Cl_2$$

$$CH_3\bullet + Cl\bullet \rightarrow CH_3Cl$$

$$2CH_3\bullet \rightarrow CH_3CH_3 + 347\ kJ$$

If methane and the halogen (represented as X) are used in equimolar quantities, CH_2X_2, CHX_3, and even CX_4 are formed. Using an excess of methane reduces the production of CH_2X_2, CHX_3, CX_4, and thus more CH_3X is formed.

4.2 Ethane and higher homologs

Ethane (C_2H_6) is a two-carbon alkane (Table 4.5) that, at standard temperature and pressure, is a colorless, odorless gas. Ethane is isolated on an industrial

TABLE 4.5 The BTEX family of aromatic hydrocarbon derivatives.

Compound	Description
Benzene	C_6H_6; a clear, colorless, and flammable liquid.
	Found in ambient air as a result of burning fuels, such as crude oil, coal, and wood.
Toluene	$C_6H_5CH_3$; also known as methylbenzene.
	A colorless liquid, with a strong, solvent-like smell. Toluene is inexpensive and simple to produce, and is widely used in industrial processes as a solvent.
Ethylbenzene	$C_6H_5CH_2CH_3$; a colorless liquid.
	Widely used in industrial processes for the manufacture of styrene ($C_6H_5CH=CH_2$), which is then used for polystyrene manufacture.
Xylene isomers	$C_6H_4(CH3)_2$; the term used to describe the three isomers of dimethyl benzene—*m*-xylene, *p*-xylene, and *o*-xylene.
	Usually concentrations of each are added together as total xylenes.
	Xylene is refined from crude oil, and is a clear liquid.

scale from natural gas and as a byproduct of crude oil refining. Its chief use is as petrochemical feedstock for ethylene production, usually by pyrolysis:

$$CH_3CH_3 \rightarrow CH_2{=}CH_2 + H_2$$

After methane, ethane is the second-largest component of natural gas. Natural gas from different gas fields varies in ethane content from less than 1% to more than 6% v/v. Prior to the 1960s, ethane and larger molecules were typically not separated from the methane component of natural gas, but simply burnt along with the methane as a fuel. Currently, ethane is an important petrochemical feedstock, and it is separated from the other components of natural gas in most gas processing plants (Fig. 4.10). Ethane can also be separated from crude oil gas, a mixture of gaseous hydrocarbons that arises as a byproduct of crude oil refining.

Ethane is most efficiently separated from methane by liquefying it at cryogenic temperatures. Various refrigeration strategies exist: the most economical process presently in wide use employs turboexpansion, and can recover over 90% of the ethane in natural gas. In this process, chilled gas expands through a turbine; as it expands, its temperature drops to approximately −100°C. At this low temperature, gaseous methane can be separated from the liquefied ethane and heavier hydrocarbons by distillation. Further distillation then separates ethane from the propane and heavier hydrocarbons.

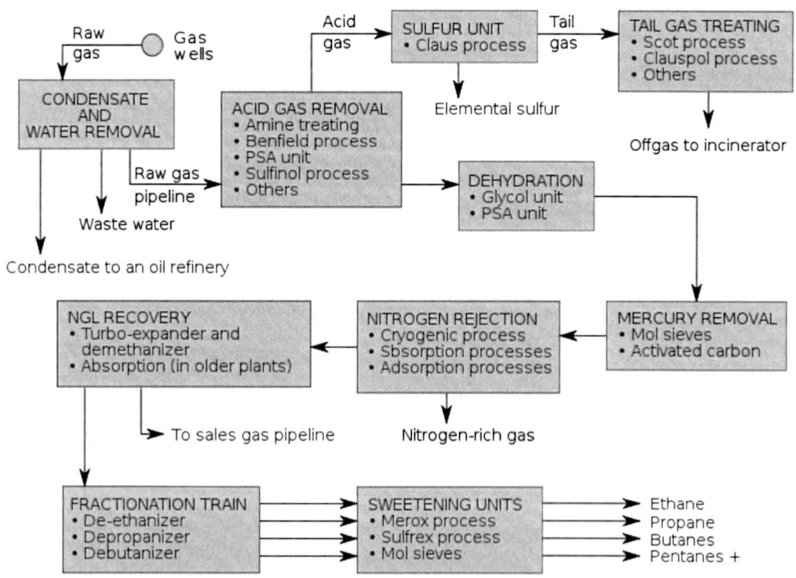

FIGURE 4.10 Gas processing.

4.3 Natural gas liquids

Natural gas liquids (lease condensate, natural gasoline, NGL) are components of natural gas that are liquid at surface in gas or oil field facilities or in gas processing plants. The composition of the natural gas liquids is dependent upon the type of natural gas and the composition of the natural gas. The constituents of natural gas liquids (NGLs) are hydrocarbon derivatives that are from the same family of molecules as natural gas and crude oil and are composed exclusively of carbon and hydrogen. Ethane, propane, butane, isobutane, and pentane (Table 4.4) are all natural gas liquids and have many uses, such as: (i) input for petrochemical plants, (ii) combusted for space heat and cooking, and (iii) as a blend stock into vehicle fuel.

The chemical composition of these hydrocarbons is similar, yet their applications vary widely. Ethane occupies the largest share of the natural gas liquids field production and is used almost exclusively to produce ethylene, which is then turned into plastics (Speight, 2019c). Much of the propane, by contrast, is burned for heating, although a substantial amount is used as petrochemical feedstock. A blend of propane and butane, sometimes referred to as autogas, is a popular fuel in some countries (such as various countries of Europe, Turkey, and Australia). Natural gasoline (the C%+ fraction) can be blended into various kinds of fuel for combustion engines. For further use, the hydrocarbons are fractionated using a system (Fig. 4.4), which, after deethanization of the natural gas liquids, produces propane, butanes, and naphtha (C_5^+).

Natural gas liquids can be classified according to their vapor pressures as low (condensate), intermediate (natural gasoline), and high (liquefied petroleum gas) vapor pressure. Natural gas liquids include propane, butane, pentane, hexane, and heptane, but not methane and not always ethane, since these hydrocarbons need refrigeration to be liquefied. A more general definition of natural gas liquids includes the nonmethane hydrocarbons from natural gas that are separated from the gas as liquids through the process of absorption, condensation, adsorption, or other methods in gas processing or cycling plants. Generally, under this definition, such liquids consist of ethane, propane butane, and higher molecular weight hydrocarbons.

4.4 Gas condensate

Natural gas condensate (condensate, gas condensate, natural gasoline) is a low-density mixture of hydrocarbon liquids that are present as gaseous components in the raw natural gas produced from many natural gas fields. Gas condensate condenses out of the raw natural gas if the temperature is reduced to below the hydrocarbon dew point temperature of the raw gas.

There are many sources of condensate sources, and each source has its own unique gas condensate composition. In general, gas condensate has a specific gravity in the range 0.5–0.8, and is composed of hydrocarbons such as propane, butane, pentane, hexane, and higher homologs (Table 4.4). Natural gas compounds with more carbon atoms (such as pentane, or blends of butane, pentane, and other hydrocarbons with additional carbon atoms) exist as liquids at ambient temperatures. Additionally, condensate may contain additional impurities such as(i) hydrogen sulfide, (ii) thiols derivatives, also called mercaptan derivatives, which are denoted by RSH where R is an organic group such as methyl, CH_3-, ethyl, C_2H_5-, propyl, C_3H_7-, and butyl, C_4H_9-, (iii) carbon dioxide, (iv) straight chain alkanes, also called linear alkanes or *n*-alkanes, having 2 to 12 carbon atoms per molecule, (v) cyclohexane and other cycloparaffin derivatives, also called naphthene derivative, and (vi) aromatic derivatives such as benzene, toluene, ethylbenzene, and the xylene isomers (Table 4.5).

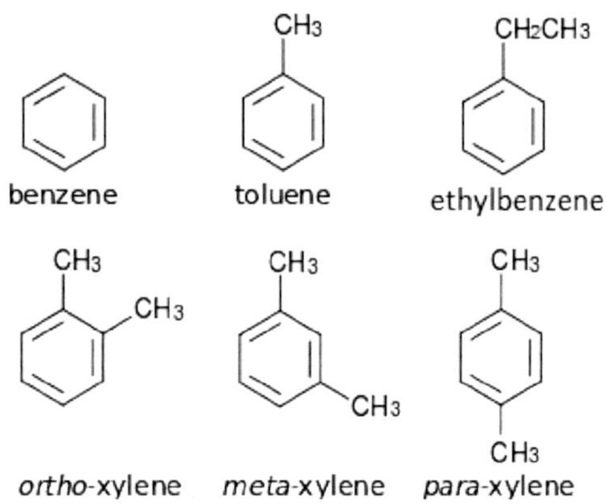

The composition of the gas condensate liquids is dependent upon the type of natural gas and the composition of the natural gas. Similarities exist between the composition of natural gas liquids and gas condensate—to the point that the two names are often (sometimes erroneously) used interchangeably. On a strictly comparative basis, the constituents of gas condensate represent the higher boiling constituents of natural gas liquids. Pentanes plus (C_{5+}) is a mixture of hydrocarbons that is a liquid at ambient temperature and pressure, and consists mostly of pentanes and higher molecular weight (higher carbon number) hydrocarbons. Pentanes plus includes, but is not limited to, *n*-pentane, iso-pentane, hexanes-plus (natural gasoline), and condensate.

To separate the condensate from a natural gas feedstock from a gas well or a group of wells (Fig. 4.11), the stream is cooled to lower the gas temperature to below the hydrocarbon dew point at the feedstock pressure and that condenses a good part of the gas condensate hydrocarbons. The feedstock mixture of gas, liquid condensate, and water is then routed to a high pressure separator

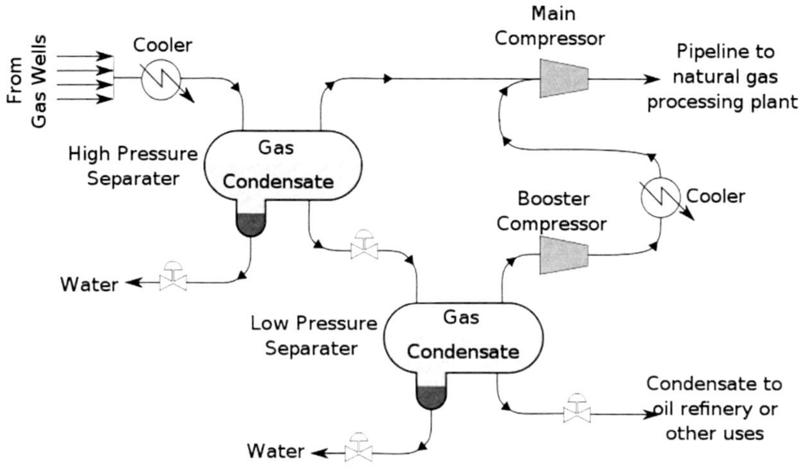

FIGURE 4.11 Gas condensate separation.

vessel where the water and the raw natural gas are separated and removed. The raw natural gas from the high pressure separator is sent to the main gas compressor.

The gas condensate from the high pressure separator flows through a throttling control valve to a low pressure separator. The reduction in pressure across the control valve causes the condensate to undergo a partial vaporization referred to as a flash vaporization. The raw natural gas from the low pressure separator is sent to a booster compressor which raises the gas pressure and sends it through a cooler and on to the main gas compressor. The main gas compressor raises the pressure of the gases from the high and low pressure separators to whatever pressure is required for the pipeline transportation of the gas to the raw natural gas processing plant. The main gas compressor discharge pressure will depend upon the distance to the raw natural gas processing plant and it may require that a multistage compressor be used.

At the raw natural gas processing plant, the gas will be dehydrated and acid gases and other impurities will be removed from the gas. Then the ethane, propane, butanes, and pentanes plus higher molecular weight hydrocarbons (referred to as C_{5+}) will also be removed and recovered as byproducts. The water removed from both the high and low pressure separators will probably need to be processed to remove hydrogen sulfide before the water can be disposed of or reused in some fashion.

4.5 Synthesis gas

Although not a hydrocarbon, synthesis gas is a source of industrial hydrocarbons. It is, therefore, worthy of inclusion at this point. *Synthesis gas* is a mixture of carbon monoxide (CO) and hydrogen (H_2) that is the beginning of a wide range of chemicals (Table 4.6).

The production of synthesis gas, i.e., mixtures of carbon monoxide and hydrogen, has been known for several centuries. But it is only with the commercialization of the Fischer-Tropsch reaction that the importance of synthesis gas has been realized. The thermal cracking (pyrolysis) of crude oil or fractions thereof was an important method for producing gas in the years following its use for increasing the heat content of water gas. Many water-gas sets operations converted into oil-gasification units; some have been used for base-load city gas supply but most find use for peak-load situations in the winter. In addition to the gases obtained by distillation of crude oil, further gaseous products are produced during the processing of naphtha and middle distillate to produce gasoline. Hydrodesulfurization processes involving treatment of naphtha, distillates, and residual fuels and from the coking or similar thermal treatment of vacuum gas oils and residual fuel oils also produce gaseous products.

The chemistry of the oil-to-gas conversion has been established for several decades and can be described in general terms although the primary and secondary reactions can be truly complex. The composition of the gases produced from a wide variety of feedstocks depends not only on the severity of cracking but often to an equal or lesser extent on the feedstock type. In general terms, gas heating values are on the order of 950−1350 Btu/ft^3 (30−50 MJ/

TABLE 4.6 Examples of chemicals from synthesis gas.

Starting material	Reaction type	Product
Synthesis gas (carbon monoxide + hydrogen)	Oxo reaction	Oxo products
	Shift reaction	Hydrogen
	Shift reaction	Methyl alcohol
	Shift reaction	Ammonia
	Shift reaction and methanation	Substitute natural gas
	Organic synthesis	Hydroquinone
	Homologation	Ethyl alcohol
	Carbonylation	Acetic acid
	Fischer-Tropsch	Ethylene
	Fischer-Tropsch	Paraffins
	Glycol synthesis	Ethylene glycol

m^3). A second group of refining operations which contribute to gas production are the catalytic cracking processes, such as fluid-bed catalytic cracking, and other variants, in which heavy gas oils are converted into gas, naphtha, fuel oil, and coke.

The catalysts will promote steam-reforming reactions that lead to a product gas containing more hydrogen and carbon monoxide and fewer unsaturated hydrocarbon products than the gas product from a noncatalytic process. The resulting gas is more suitable for use as a medium heat-value gas than the rich gas produced by straight thermal cracking. The catalyst also influences the reactions rates in the thermal cracking reactions, which can lead to higher gas yields and lower tar and carbon yields.

Almost all crude oil fractions can be converted into gaseous fuels, although conversion processes for the heavier fractions require more elaborate technology to achieve the necessary purity and uniformity of the manufactured gas stream. In addition, the thermal yield from the gasification of heavier feedstocks is invariably lower than that of gasifying light naphtha or liquefied petroleum gas since, in addition to the production of synthesis gas components (hydrogen and carbon monoxide) and various gaseous hydrocarbons, heavy feedstocks also yield some tar and coke.

Synthesis gas can be produced from heavy oil by partially oxidizing the oil:

$$[2CH]_{\text{crude oil}} + O_2 \rightarrow 2CO + H_2$$

The initial partial oxidation step consists of the reaction of the feedstock with a quantity of oxygen insufficient to burn it completely, making a mixture consisting of carbon monoxide, carbon dioxide, hydrogen, and steam. Success in partially oxidizing heavy feedstocks depends mainly on details of the burner design. The ratio of hydrogen to carbon monoxide in the product gas is a function of reaction temperature and stoichiometry and can be adjusted, if desired, by varying the ratio of carrier steam to oil fed to the unit.

Reactor temperatures vary from 1095°C to 1490°C (2000°F to 2700°F), while pressures can vary from approximately atmospheric pressure to approximately 2000 psi. The process has the capability of producing high-purity hydrogen although the extent of the purification procedure depends upon the use to which the hydrogen is to be put. For example, carbon dioxide can be removed by scrubbing with various alkaline reagents, while carbon monoxide can be removed by washing with liquid nitrogen or, if nitrogen is undesirable in the product, the carbon monoxide should be removed by washing with copper-amine solutions.

The synthesis gas generation process is a noncatalytic process for producing synthesis gas (principally hydrogen and carbon monoxide) for the ultimate production of high-purity hydrogen from gaseous or liquid hydrocarbons. In this process, a controlled mixture of preheated feedstock and oxygen is fed to the top of the generator where carbon monoxide and hydrogen emerge as the products. Soot, produced in this part of the operation, is removed in a water scrubber from the product gas stream and is then extracted from the resulting carbon-water slurry with naphtha and transferred to a fuel oil fraction. The oil-soot mixture is burned in a boiler or recycled to the generator to extinction to eliminate carbon production as part of the process. The soot-free synthesis gas is then charged to a shift converter where the carbon monoxide reacts with steam to form additional hydrogen and carbon dioxide at the stoichiometric rate of 1 mol of hydrogen for every mole of carbon monoxide charged to the converter.

This particular partial oxidation technique has also been applied to a whole range of liquid feedstocks for hydrogen production. There is now serious consideration being given to hydrogen production by the partial oxidation of solid feedstocks such as crude oil coke (from both delayed and fluid-bed reactors), lignite, and coal, as well as crude oil residua. Although these reactions may be represented conveniently using simple equations of this type, the reactions can be complex and result in carbon deposition on parts of the equipment thereby requiring careful inspection of the reactor.

Synthesis gas forms the basis of what has become to be known as Fischer-Tropsch chemistry and which leads to a whole range of hydrocarbon products (Chapter 8) (Speight, 2019c). Synthesis gas consists primarily of carbon monoxide and hydrogen with (before purification) small amounts of carbon dioxide. Synthesis gas is combustible and often used as a fuel source or as an intermediate for the production of other chemicals. Synthesis gas for use as a fuel is most often produced by gasification of the carbonaceous feedstock or municipal waste mainly by the following paths:

$$C + O_2 \rightarrow CO_2$$

$$CO_2 + C \rightarrow 2CO$$

$$C + H_2O \rightarrow CO + H_2$$

When used as an intermediate in the large-scale, industrial synthesis of hydrogen and ammonia, it is also produced from natural gas (via the steam reforming reaction):

$$CH_4 + H_2O \rightarrow CO + 3H_2$$

The synthesis gas produced in large waste-to-energy gasification facilities is used as fuel to generate electricity.

Although the focus of this section is the production of hydrocarbon derivatives from synthesis gas, it is worthy of note that all or part of the clean synthesis gas can also be used (i) as chemical *building blocks* to produce a broad range of chemicals using processes well established in the chemical and petrochemical industry, (ii) as a fuel producer for highly efficient fuel cells (which run off the hydrogen made in a gasifier) or perhaps in the future, hydrogen turbines and fuel cell-turbine hybrid systems, and (iii) as a source of hydrogen that can be separated from the gas stream and used as a fuel or as a feedstock for refineries (which use the hydrogen to upgrade petroleum products) (Parkash, 2003; Gary et al., 2007; Speight, 2011, 2014, 2017; Hsu and Robinson, 2017). However, the decreasing availability and increased price of petroleum has renewed the worldwide interest in the production of liquid hydrocarbon derivatives from carbon monoxide and hydrogen using metal catalysts, also known as Fischer-Tropsch synthesis or Fischer-Tropsch process.

Gasification to produce synthesis gas can proceed from just about any organic material, including biomass and plastic waste. The resulting synthesis gas burns cleanly into water vapor and carbon dioxide. Alternatively, synthesis gas may be converted efficiently to methane via the Sabatier reaction, or to a diesellike synthetic fuel via the Fischer-Tropsch process. Inorganic components of the feedstock, such as metals and minerals, are trapped in an inert and environmentally safe form as char, which may have use as a fertilizer.

In principle, synthesis gas can be produced from any hydrocarbon feedstock. These include natural gas, naphtha, residual oil, petroleum coke, coal, and biomass. The lowest cost routes for synthesis gas production, however, are based on natural gas. The cheapest option is remote or stranded reserves. Current economic considerations dictate that the production of liquid fuels from synthesis gas translates into using natural gas as the hydrocarbon source. Nevertheless, the synthesis gas production operation in a gas-to-liquids plant amounts to greater than half of the capital cost of the plant. The choice of technology for synthesis gas production also depends on the scale of the synthesis operation. Synthesis gas production from solid fuels can require an even greater capital investment with the addition of feedstock handling and more complex synthesis gas purification operations. The greatest impact on improving gas-to-liquids plant economics is to decrease capital costs associated with synthesis gas production and improve thermal efficiency through better heat integration and utilization. Improved thermal efficiency can be obtained by combining the gas-to-liquids plant with a power generation plant to take advantage of the availability of low-pressure steam.

The manufacture of gas mixtures of carbon monoxide and hydrogen has been an important part of chemical technology for about a century. Originally, such mixtures were obtained by the reaction of steam with incandescent coke and were known as *water gas*. Used first as a fuel, water gas soon attracted attention as a source of hydrogen and carbon monoxide for the production of chemicals, at which time it gradually became known as synthesis gas. Eventually, steam reforming processes, in which steam is reacted with natural gas (methane) or petroleum naphtha over a nickel catalyst, found wide application for the production of synthesis gas.

A modified version of steam reforming known as autothermal reforming, which is a combination of partial oxidation near the reactor inlet and conventional steam reforming further along the reactor, improves the overall reactor efficiency and increases the flexibility of the process. Partial oxidation processes using oxygen instead of steam also found wide application for synthesis gas manufacture, with the special feature that they could utilize low-value feedstocks such as heavy petroleum residua. In recent years, catalytic partial oxidation employing very short reaction times (milliseconds) at high temperatures (850°C −1000°C; 1560°F to 1830°F) is providing still another approach to synthesis gas manufacture. Nearly complete conversion of methane, with close to 100% selectivity to hydrogen and carbon monoxide, can be obtained with a rhenium monolith under well-controlled conditions. Experiments on the catalytic partial oxidation of *n*-hexane conducted with added steam give much higher yields of hydrogen than can be obtained in experiments without steam, a result of much interest in obtaining hydrogen-rich streams for fuel cell applications.

The route of coal to synthetic automotive fuels, as practiced by SASOL is technically proven and products with favorable environmental characteristics are produced. As is the case in essentially all the carbonaceous feedstock conversion processes where air or oxygen is used for the utilization or partial conversion of the energy in the carbonaceous feedstock, the carbon dioxide burden is a drawback as compared to crude oil.

The uses of synthesis gas include use as a chemical feedstock and in gas-to-liquid processes which use Fischer-Tropsch chemistry to make liquid fuels as feedstock for chemical synthesis, as well as being used in the production of fuel additives, including diethyl ether and methyl t-butyl ether (MTBE), acetic acid and its anhydride, synthesis gas could also make an important contribution to chemical synthesis through conversion to methanol. There is also the option in which stranded natural gas is converted to synthesis gas production followed by conversion to liquid fuels.

The hydroformylation synthesis (also known as the oxo synthesis or the oxo process) is an industrial process for the production of aldehyde derivatives from alkene derivatives. This chemical reaction entails the addition of a formyl group (CHO) and a hydrogen atom to a carbon-carbon double bond. A key consideration of hydroformylation is the production of normal isomers or the production of iso-isomers in the product(s). For example, the hydroformylation of propylene can yield two isomeric products, butyraldehyde or iso-butyraldehyde:

$$H_2 + CO + CH_3CH{=}CH_2 \rightarrow CH_3CH_2CH_2CHO \text{ (}n\text{-butyraldehyde)}$$

$$H_2 + CO + CH_3CH{=}CH_2 \rightarrow (CH_3)_2CHCHO \text{ (iso-butyraldehyde)}$$

These isomers reflect the regio-chemistry (the preference of one direction of chemical bond formation or chemical bond scission over all other possible directions) of the insertion of the alkene into the M−H bond. Generally, both products are not equally desirable.

As an example of the hydroformylation process, the Exxon process, also Kuhlmann-oxo process or the PCUK-oxo process, is used for the hydroformylation of C_6 to C_{12} olefin derivatives using cobalt-based catalysts. In order to recover the catalyst, an aqueous sodium hydroxide solution or sodium carbonate is added to the organic phase. By extraction with olefin and neutralization by addition of sulfuric acid solution under carbon monoxide pressure the metal-carbonyl hydride can be recovered. The recovered hydride is stripped out with synthesis gas, absorbed by the olefin, and returned to the reactor. The Exxon process, similar to the BASF process, is carried out at a temperature on the order of 160°C −180°C (320°F −355°F) and a pressure suitable to the reactants and products.

The Fischer-Tropsch reaction can be described as the synthesis of hydrocarbon derivatives via the hydrogenation of carbon monoxide using transition metal catalysts. The major catalysts used industrially are Fe and Co, but can also be Ru and Ni. From a mechanism perspective the reactions can be regarded as a carbon chain building process where methylene (CH_2) groups are attached sequentially in a carbon chain (Table 10.3). Thus:

$$nCO + [n + m/2]H_2 \rightarrow C_nH_m + nH_2O$$

For example:

$$CO + 2H_2 \rightarrow \text{-}[CH_2]\text{-} + H_2O$$

A common and salient feature of the reactions is the exothermicity of the reactions. As a general rule of thumb, the reactions which produce water and carbon dioxide as a product tend to be more exothermic on account of the very high heat of formation of these species. Some of the reactions proposed are as follows:

$$CO_2 + 3H2 \rightarrow \text{-}[CH_2]\text{-} + 2H_2O \quad \Delta H = -125\ \text{kJ/mol}$$

$$CO + 2H_2 \rightarrow \text{-}[CH_2] + H_2O \quad \Delta H = -165\ \text{kJ/mol}$$

$$2CO + H_2 \rightarrow \text{-}[CH_2]\text{-} + CO_2 \quad \Delta H = -204\ \text{kJ/mol}$$

$$3CO\ H_2 \rightarrow \text{-}[CH_2]\text{-} + 2CO_2 \quad \Delta H = -244\ \text{kJ/mol}$$

There is also the water-gas-shift reaction:

$$CO + H_2O \rightarrow H_2 + CO_2 \quad \Delta H = -39\ \text{kJ/mol}$$

Due to the very high exothermic nature of the Fischer-Tropsch reactions, as illustrated in the reactions above, an important issue is, not surprisingly, the need to avoid an increase in temperature. The need for cooling is thus of critical importance in order to: (i) maintain stable reaction conditions, (ii) avoid the tendency to produce lighter hydrocarbon derivatives, and (iii) prevent catalyst sintering and hence reduction in activity. Since the total heat of reaction is in the order of approximately 25% of the heat of combustion of the synthesis gas (i.e., reactants if the Fischer-Tropsch process), a theoretical limit on the maximum efficiency of the Fischer-Tropsch process is imposed (Rauch, 2001).

Two main chemical characteristics of Fischer-Tropsch synthesis are the unavoidable production of a wide range of hydrocarbon products (olefin derivatives, paraffin derivatives, and oxygenated products) and the liberation of a large amount of heat from the highly exothermic synthesis reactions. Product distributions are influenced by temperature, feed gas composition (hydrogen/carbon monoxide), pressure, catalyst type, and catalyst composition. Fischer-Tropsch products are produced in four main steps: synthesis gas generation, gas purification, Fischer-Tropsch synthesis, and product upgrading. Depending on the types and quantities of Fischer-Tropsch products desired, either low (200°C–240°C; 390°F–465°F) or high temperature (300°C–350°C; 570°F to 660°F) synthesis is used with either an iron (Fe) or cobalt catalyst (Co).

The required gas mixture of carbon monoxide and hydrogen (synthesis gas) is created through a reaction of coke or the carbonaceous feedstock with water steam and oxygen, at temperatures over 900°C. In the past, town gas and gas for lamps were a carbon monoxide-hydrogen mixture, made by gasifying coke in gas works. In the 1970s it was replaced with imported natural gas (methane). Gasification of carbonaceous feedstocks and Fischer-Tropsch hydrocarbon synthesis together bring about a two-stage sequence of reactions which allows the production of liquid fuels like gasoline and diesel out of the solid, combustible, carbonaceous feedstock.

The Fischer-Tropsch synthesis is, in principle, a carbon chain building process, where methylene groups are attached to the carbon chain. The actual reactions that occur have been, and remain, a matter of controversy, as it has been the last century since 1930s.

$$(2n+1)H_2 + nCO \rightarrow C_nH_{(2n+2)} + nH_2O$$

Even though the overall Fischer-Tropsch process is described by the above relatively simple equation, the initial reactants in the above reaction (i.e., carbon monoxide and H_2) can be produced by other reactions such as the partial combustion of a hydrocarbon derivative:

$$2C_nH_{2n+2} + nO_2 \rightarrow 2(n+1)H_2 + 2nCO$$

For example (when n = 1), methane (in the case of gas-to-liquids applications) the reaction is represented by:

$$2CH_4 + O_2 \rightarrow 4H_2 + 2CO$$

Or by the gasification of any carbonaceous source, such as biomass, the reaction is:

$$C + H_2O \rightarrow H_2 + CO$$

The energy needed for this endothermic reaction is usually provided by the combustion with air or oxygen (an exothermic reaction):

$$2C + O_2 \rightarrow 2CO$$

Such reactions are highly exothermic, and to avoid an increase in temperature, which results in lighter hydrocarbon derivatives, it is important to have sufficient cooling, to secure stable reaction conditions. The total heat of reaction amounts to approximately 25% of the heat of combustion of the synthesis gas, and lays thereby a theoretical limit on the maximal efficiency of the Fischer-Tropsch process.

In practice, the Fischer-Tropsch reaction is generally carried out at temperatures in the range 200°C–350°C (390°F–660°F) and at pressures of 75–4000 psi; the hydrogen/carbon monoxide ratio is usually at ca. 2.2:1 or 2.5:1. Since up to three volumes of hydrogen may be required to achieve the next stage of the liquids production, the synthesis gas must then be converted by means of the water-gas shift reaction to the desired level of hydrogen after which the gaseous mix is purified (acid gas removal, etc.) and converted to a wide variety of hydrocarbon derivatives.

$$CO + H_2O \rightarrow CO_2 + H_2$$

$$CO + (2n+1)H_2 \rightarrow C_nH_{2n+2} + H_2O$$

These reactions result primarily in low- and medium-boiling aliphatic compounds; present commercial objectives are focused on the conditions that result in the production of *n*-hydrocarbon derivatives as well as olefin derivatives and oxygenated materials.

The reaction is dependent of a catalyst, mostly an iron or cobalt catalyst where the reaction takes place. There is either a low or high temperature process (low temperature Fischer-Tropsch, high temperature Fischer-Tropsch), with temperatures ranging between 200°C and 240°C (390°F–465°F) for low temperature Fischer-Tropsch and 300°C–350°C (570°F–660°F) for the high temperature Fischer-Tropsch process. The high temperature Fischer-Tropsch process uses an iron catalyst, and the low temperature Fischer-Tropsch either an iron or a cobalt catalyst. The different catalysts include also nickel-based and ruthenium-based catalysts, which also have enough activity for commercial use in the process. But the availability of ruthenium is limited and the nickel-based catalyst has high activity but produces too much methane, and additionally the performance at high pressure is poor, due to production of volatile carbonyls. This leaves only cobalt and iron as practical catalysts, and this study will only consider these two. Iron is cheap, but cobalt has the advantage of higher activity and longer life, though it is on a metal basis 1000 times more expensive than iron catalyst.

References

Ancheyta, J., Speight, J.G., 2007. Hydroprocessing of Heavy Oils and Residua. CRC Press, Taylor & Francis Group, Boca Raton, Florida.

Anerousis, J.P., Whitman, S.K., 1984. An updated examination of gas sweetening by the iron sponge process. Paper No. SPE 13280. In: SPE Annual Technical Conference and Exhibition, Houston, Texas. September.

Belosludov, V.R., Subbotin, O.S., Krupskii, D.S., Belosludov, R.V., Kawazoe, Y., Kudoh, J., 2007. Physical and chemical properties of gas hydrates: theoretical aspects of energy storage application. Materials Transactions 48 (4), 704–710.

Bishnoi, P.R., Clarke, M.A., 2006. Encyclopedia of Chemical Processing. CRC Press, Taylor & Francis Group, Boca Raton, Florida, pp. 1849–1863.

Carrol, J.J., 2003. Natural Gas Hydrates. Gulf Professional Publishing, Burlington, Vermont.

Chong, R., Yin, Z., Zhao, J., Ling, P., 2017. Recovering natural gas from gas hydrates using horizontal wellbore. Energy Procedia 143, 780-78.

Collett, T.S., 2001. Natural-gas hydrates; resource of the twenty-first century? Journal of the American Association of Petroleum Geologists 74, 85–108.

Collett, T.S., Johnson, A.H., Knapp, C.C., Boswell, R., 2009. Natural gas hydrates: a review. AAPG Memoir No. 89. In: Collett, T.S., Johnson, A.H., Knapp, C.C., Boswell, R. (Eds.), Natural Gas Hydrates – Energy Resource Potential and Associated Geologic Hazards. American Association of Petroleum Geologists, Tulsa, Oklahoma, pp. 146–219.

Collett, T.S., 2010. Physical properties of gas hydrates: a review. Journal of Thermodynamics 2010. https://doi.org/10.1155/2010/271291. Article ID 271291. https://www.hindawi.com/journals/jther/2010/271291/.

Collett, T.S., Bahk, J.J., Baker, R., Boswell, R., Divins, D., Frye, M., Goldberg, D., Husebø, J., Koh, C., Malone, M., Morell, M., Myers, G., Shipp, C., Torres, M., 2015. Methane hydrates in nature – current knowledge and challenges. Journal of Chemical & Engineering Data 60 (2), 319–329.

Davidson, D., 1983. Gas hydrates as clathrate ices. In: Cox, J. (Ed.), Natural Gas Hydrates – Properties, Occurrence, and Recovery. Butterworth, Woburn, Massachusetts.

Duckworth, G.L., Geddes, J.H., 1965. Natural gas desulfurization by the iron sponge process. Oil & Gas Journal 63 (37), 94–96.

Emam, E.A., 2015. Gas flaring in industry: an overview. Petroleum & Coal 57 (5), 532–555. https://www.researchgate.net/publication/303027684_GAS_flaring_in_industry_An_overview.

Foglietta, J.H., 2004. Dew point turboexpander process: a solution for high pressure fields. Proceedings. IAPG Gas Conditioning Conference, Neuquen, Argentina. October 18.

Gary, J.G., Handwerk, G.E., Kaiser, M.J., 2007. Petroleum Refining: Technology and Economics, fifth ed. CRC Press, Taylor & Francis Group, Boca Raton, Florida.

Geist, J.M., 1985. Refrigeration cycles for the future. Oil & Gas Journal 83 (5), 56–60.

Giavarini, C., Maccioni, F., Santarelli, M.L., 2003. Formation kinetics of propane hydrate. Industrial & Engineering Chemistry Research 42, 1517–1521.

Giavarini, C., Maccioni, F., 2004. Self-preservation at low pressure of methane hydrates with various gas contents. Industrial & Engineering Chemistry Research 43, 6616–6621.

Giavarini, C., Maccioni, F., Santarelli, M.L., 2005. Characterization of gas hydrates by modulated differential scanning calorimetry. Petroleum Science and Technology 23, 327–335.

Hsu, C.S., Robinson, P.R. (Eds.), 2017. Handbook of Petroleum Technology. Springer International Publishing AG, Cham, Switzerland.

Kohl, A.L., Nielsen, R.B., 1997. Gas Purification. Gulf Publishing Company, Houston, Texas.

Klimstra, J., 1978. Interchangeability of gaseous fuels – the importance of the Wobbe-index. Report No. SAE 861578. In: Society of Automotive Engineers, SAE International, Warrendale , Pennsylvania.

Kohl, A.L., Riesenfeld, F.C., 1985. Gas Purification, fourth ed. Gulf Publishing Company, Houston, Texas.

Liss, W.E., Thrasher, W.R., 1992. Variability of Natural Gas Composition in Select Major Metropolitan Areas of the U.S. Report No. GRI-92/0123. Gas Research Institute, Chicago, Illinois.

Lorenson, T.D., Collett, T.S., 2000. Gas content and composition of gas hydrate from sediments of the southeastern North American continental margin. In: Paull, C.K., Matsumoto, R., Wallace, P.J., Dillon, W.P. (Eds.), Proceedings of the Ocean Drilling Program, Scientific Results, vol. 164, pp. 37–46.

Maddox, R.N., 1974. Gas and Liquid Sweetening, second ed. Campbell Publishing Co., Norman, Oklahoma.

Makogon, Y.F., 1997. Hydrates of Hydrocarbons. PennWell Books, Tulsa, Oklahoma.

Makogon, Y.F., Holditch, S.A., Makogon, T.Y., 2005. Russian field illustrates gas-hydrate production. Oil & Gas Journal 103, 43–47.

Makogon, Y.F., Holditch, S.A., Makogon, T.Y., 2007. Natural gas hydrates – a potential energy source for the 21st century. Journal of Petroleum Science and Engineering 56, 14–31.

Makogon, Y.F., 2010. Natural gas hydrates – a promising source of energy. Journal of Natural Gas Science and Engineering 2 (1), 49–59.

Mokhatab, S., Poe, W.A., Speight, J.G., 2006. Handbook of Natural Gas Transmission and Processing. Elsevier, Amsterdam, The Netherlands.

Parkash, S., 2003. Refining Processes Handbook. Gulf Professional Publishing, Elsevier, Amsterdam, Netherlands.

Ruffine, L., Donval, J.P., Charlou, J.L., 2009. Gas hydrates from the sea of Marmara: variability in composition and origin of the hydrate-forming gases. In: Proceedings. Fall Meeting. American Geophysical Union, Washington DC.

Rushton, D.W., Hayes, W., 1961. Oil & Gas Journal 59 (38), 102.

Seo, Y., Kang, S.P., Jang, W., 2009. Structure and composition analysis of natural gas hydrates: ^{13}C NMR spectroscopic and gas uptake measurements of mixed gas hydrates. Journal of Physical Chemistry 113 (35), 9641–9649.

Sloan Jr., E.D., 1998a. Gas hydrates: review of physical/chemical properties. Energy & Fuels 12 (2), 191–196.

Sloan Jr., E.D., 1998b. Clathrate Hydrates of Natural Gases, second ed. Marcel Dekker Inc., New York.

Sloan Jr., E.D., 2006. Clathrate Hydrates of Natural Gases, third ed. Marcel Dekker Inc., New York.

Speight, J.G., 2011. The Refinery of the Future. Gulf Professional Publishing, Elsevier, Oxford, United Kingdom.

Speight, J.G., 2014. The Chemistry and Technology of Petroleum, fifth ed. CRC Press, Taylor and Francis Group, Boca Raton, Florida.

Speight, J.G., 2015. Handbook of Petroleum Product Analysis, second ed. John Wiley & Sons Inc., Hoboken, New Jersey.

Speight, J.G., 2016. Production of syngas, synfuel, bio-oils, and biogas from coal, biomass, and opportunity fuels. In: Oakey, J. (Ed.), Fuel Flexible Energy Generation: Solid, Liquid, and Gaseous Fuels. Woodhead Publishing, Woodhead Publishing, Elsevier, Waltham, Massachusetts (Chapter 6).

Speight, J.G., 2018. Handbook of Natural Gas Analysis. John Wiley & Sons Inc., Hoboken, New Jersey.

Speight, J.G., 2019a. Natural Gas: A Basic Handbook, second ed. Gulf Publishing Company, Elsevier, Cambridge, Massachusetts.

Speight, J.G., 2019b. Biogas: Production and Properties. Nova Science publishers, New York.

Speight, J.G., 2019c. Handbook of Petrochemical Processes. CRC Press, Taylor and Francis Group, Boca Raton, Florida.
Stern, L., Kirby, S., Durham, W., Circone, S., Waite, W.F., 2000. Laboratory synthesis of pure methane hydrate suitable for measurement of physical properties and decomposition behavior. In: Max, M.D. (Ed.), Proceedings. Natural Gas Hydrate in Oceanic and Permafrost Environments. Kluwer Academic Publishers, Dordrecht, Netherlands, pp. 323–348.
Stoll, R.G., Bryan, G.M., 1979. Physical properties of sediments containing gas hydrates. Journal of Geophysical Research 84 (B4), 1629–1634.
Taylor, C., 2002. Formation Studies of Methane Hydrates with Surfactants. 2nd International Workshop On Methane Hydrates October, Washington, DC.
Wang, X., Economides, M.J., 2012. Natural gas hydrates as an energy source – revisited 2012. In: Proceedings. SPE International Petroleum Technology Conference 2012, vol. 1. Society of Petroleum Engineers, Richardson, Texas, pp. 176–186.
Yang, X., Qin, M., 2012. Natural gas hydrate as potential energy resources in the future. Advanced Materials Research 462, 221–224.
Yang, S., Xu, D., Xiu, L., Duan, C., Pang, B., 2014. Composition characteristics of gas hydrate produced gas and pretreatment research. Energy and Power Engineering 6, 481–486.
Zapffe, F., 1963. Iron sponge process removes mercaptans. Oil & Gas Journal 61 (33), 103–104.
Zou, C., 2017. Natural gas hydrate. In: Unconventional Petroleum Geology, second ed. Elsevier, Amsterdam, Netherland, pp. 337–354 (Chapter 10).

Further reading

Nisha, G., Harini, C.G., 2012. Enhance recovery of methane from gas hydrate reservoirs with CO_2 sequestration. Paper No. SPE-162521-MS. In: Proceedings. Abu Dhabi International Petroleum Conference and Exhibition, Abu Dhabi, UAE. November 11-14. Society of Petroleum Engineers, Richardson, Texas.

Chapter 5

Hydrocarbons from coal

1. Introduction

Coal (the term is used generically throughout the book to include all types of coal) is a black or brownish-black *organic sedimentary rock* of biochemical with variable proportions of hydrogen, nitrogen, oxygen, and sulfur. More specifically, coal is a readily combustible black or brownish-black organic sedimentary rock normally which occurs in rock strata as *layers or veins* (*coal beds*, *coal seams*).

Coal has also been considered to be a *metamorphic rock*, which is the result of heat and pressure on organic sediments such as peat. However, the discussion in favor of coal is as a sedimentary rock because most sedimentary rocks undergo some heat and pressure and the association of coal with typical sedimentary rocks and the mode of formation of coal usually keep low grade coal in the sedimentary classification system. Anthracite, on the other hand, undergoes more heat and pressure and is associated with low grade *metamorphic rocks* such as slate and quartzite. Subducted coal may become graphite in igneous rocks or even the carbonate rich rocks (carbonatites).

The degree of metamorphosis results in differing coal types, each of which has different quality. However, peat is not actually a rock but no longer just organic matter and is a major source of energy for many nonindustrialized countries. The unconsolidated plant matter is lacking the metamorphic changes found in coal. Thus, coal is classified into four main types, depending on the amount of carbon, oxygen, and hydrogen present. The higher the carbon content, the more energy the coal contains. However, coal contains very few natural-occurring hydrocarbon derivatives and is composed primarily of carbon along with variable quantities of other elements, such as nitrogen, oxygen, and sulfur. But coal is one of the several natural products which—although not containing hydrocarbon derivatives—can be converted to hydrocarbon derivatives (Speight, 2013).

Coal is also a fossil fuel formed in swamp ecosystems where plant remains were saved by water and mud from oxidation and biodegradation. Coal is a combustible organic sedimentary rock (composed primarily of carbon, hydrogen, and oxygen) formed from ancient vegetation and consolidated between other rock strata to form coal seams. Coal is composed primarily of

Handbook of Industrial Hydrocarbon Processes. https://doi.org/10.1016/B978-0-12-809923-0.00005-9

carbon along with assorted other elements, such as hydrogen, nitrogen, oxygen, and sulfur. Of particular importance is the carbon content of the coal, which is part of the basis for the modern classification system of coal and also serves as the backbone of produced hydrocarbon derivatives. Thus, whereas the carbon content of the coals of the world varies over a wide range (approximately 75.0%−95.0% w/w), crude oil, on the other hand, does not exhibit such a wide variation in carbon content; all of the crude oil, heavy oil, and bitumen (natural asphalt) that occur throughout the world fall into the range of 82.0%−88.0% w/w carbon.

There is no evidence that coal was of great importance in Britain, and certainly not in the United States, before the 1000 AD. Mineral coal came to be referred to as sea coal because it was found on the beaches of north-eastern England having fallen from the exposed coal seams on cliffs above the shore or washed out of underwater coal seam outcrops. By the 13th century, when underground mining from shafts or adits was developed and the onset of the Industrial Revolution led to the expansive mining industry of 19th-century and 20th-century England.

2. Occurrence and reserves

Deposits of coal, sandstone, shale, and limestone are often found together in sequences hundreds of feet thick. Coal is found as successive layers, or seams, sandwiched between strata of sandstone and shale. Deposits of coal, sandstone, shale, and limestone are often found together in sequences hundreds of feet thick. This period is recognized in the United States as the Mississippian and Pennsylvanian time periods due to the significant sequences of these rocks found in those states (i.e., Mississippi and Pennsylvania) (Table 5.1). Other notable coal bearing ages are the Cretaceous, Triassic, and Jurassic periods. The more recently aged rocks are not as productive for some reason, but lignite and peat are common in younger deposits but generally, the older the deposit, the better the grade (higher rank) of coal.

Prior to commencement of mining operations, technical and economic feasibility are evaluated based on: (i) regional geologic conditions, (ii) overburden characteristics, (iii) coal seam continuity, (iv) coal seam thickness, (v) coal seam structure, (vi) coal seam depth, (vii) coal quality, (viii) strength of materials above and below the seam for roof and floor conditions, (ix) topography—especially altitude and slope, (x) surface drainage patterns, as well as capital investment requirements.

Coal is extracted from the ground by mining, either underground mining or open-pit mining (surface mining). The choice of mining method depends primarily on depth of burial, density of the overburden, and thickness of the coal seam (in addition to the items noted in the previous paragraph). Seams relatively close to the surface, at depths less than approximately 180 ft, are usually surface mined. Coal that occurs at depths of 180−300 ft is usually

TABLE 5.1 The geological timescale.

Era	Period	Epoch	Duration (years $\times 10^6$)	Years ago (years $\times 10^6$)
Cenozoic	Quaternary	Holocene	0.01 to the present	
		Pleistocene	2	0.01
	Tertiary	Pliocene	11	2
		Miocene	12	13
		Oligocene	11	25
		Eocene	22	36
		Paleocene	71	58
Mesozoic	Cretaceous		71	65
	Jurassic		54	136
	Triassic		35	190
Paleozoic	Permian		55	225
	Carboniferous		65	280
	Devonian		60	345
	Silurian		20	405
	Ordovician		75	425
	Cambrian		100	500
Precambrian			3380	600

deep mined but, in some cases, surface mining techniques can be used. After mining, coal extracted from both surface and underground mines require washing in a coal preparation plant.

At the current rates of recovery and consumption, the world global coal reserves have been variously estimated to have a reserves/production ratio of at least 155 years. However, as with all estimates of resource longevity, coal longevity is subject to the assumed rate of consumption remaining at the current rate of consumption and, moreover, to technological developments that dictate the rate at which the coal can be mined. And, moreover, coal is a fossil fuel and an *unclean* energy source that will only add to global warming. In fact, the next time electricity is advertised as a clean energy source, consider the means by which the majority of electricity is produced—almost 50% of the electricity generated in the United States is from coal.

In spite of improvements in mining methods and hazardous gas monitoring, the risks of rock falls, explosions, and unhealthy air quality is still a safety issue. While statistical analyses might show a decrease in the rate of injuries and deaths in mines, there is still *a clear and ever-present danger* to the miners. Mining remains a dangerous occupation. In addition to the risks to miners, coal mining can result in a number of adverse effects on the environment. Surface mining of coal completely eliminates existing vegetation, destroys the genetic soil profile, displaces or destroys wildlife and habitat, degrades air quality, alters current land uses, and to some extent permanently changes the general topography of the area mined—the movement, storage, and redistribution of soil, the community of microorganisms, and nutrient cycling processes can be disrupted. This results in a scarred landscape with little immediate value to the flora and fauna and certainly no immediate scenic value. Rehabilitation or reclamation mitigates some of these concerns and is required by law in many countries but reclamation takes time.

Mine tailing dumps produce acid mine drainage which can seep into waterways and aquifers, with consequences on ecological and human health. If underground mine tunnels collapse, this typically causes subsidence of land surfaces. During actual mining operations, methane (*firedamp*) may be released into the air. Firedamp is explosive at concentrations between 5% v/v and 15% v/v, with most violence at around 10%, and still causes loss of life in coal mines. Release of methane to the atmosphere causes an increase in greenhouse gas concentration in the air.

3. Formation and types

Coal is not deposited into the Earth as coal but begins as layers of plant matter that have accumulated at the bottom of a body of water after which maturation (coal formation) commences. For the process to continue the plant matter must be protected from biodegradation and oxidation, usually by mud or acidic water; such protection was available in the wide shallow seas of the

Carboniferous period (Table 5.1). This trapped atmospheric carbon in the ground in peat bogs that eventually were covered over and deeply buried by sediments under which the organic deposits metamorphosed into coal. Over time, the chemical and physical properties of the organic material were changed by geological action to create a solid material.

3.1 Coal formation

The precursors to coal were plant remains (containing carbon, hydrogen, and oxygen) that were deposited in the Carboniferous period, between 345 and 280 million years ago. As the plant remains became submerged under water, decomposition occurred in which oxygen and hydrogen were lost from the remains to leave a deposit with a high percentage of carbon. With the passage of time, layers of inorganic material such as sand and mud settled from the water and covered the deposits. The pressure of these overlying layers, as well as movements of the crust of the Earth acted to compress and harden the deposits, thus producing coal from the vegetal matter.

The plant material (vegetal matter) is composed mainly of carbon, hydrogen, oxygen, nitrogen, sulfur, and some inorganic mineral elements. When this material decays under water, in the absence of oxygen, the carbon content increases. The initial product of this decomposition process is known as peat. The transformation of peat to lignite is the result of pressure exerted by sedimentary materials that accumulate over the peat deposits. Even greater pressures and heat from movements of the crust of the Earth (as occurs during mountain building), and occasionally from igneous intrusion, cause the transformation of lignite to bituminous and anthracite coal.

Coal is believed to have been formed under anoxic conditions (i.e., conditions under which the level of oxygen has been totally decreased). Such conditions exist in peat bogs or wetlands but for coal formation the peat bogs and wetlands existed millions of years ago. For example, coal has been reported in rocks as old as Proterozoic (possibly two billion years old) and as young as Pliocene (2 million years old) but the great majority of the coal in the world was laid down during the Carboniferous period, a 60-million-year stretch (approximately 350 to 290 million years ago (Table 5.1) when sea level was high and forests of tall ferns and cycads grew in gigantic tropical swamps.

Coal formation began during the Carboniferous period (known as the *first coal age*), which spanned 360 million to 290 million years before present. The Carboniferous period is divided into two parts. The Lower Carboniferous, also called the Mississippian, began approximately 360 million years ago and ended 310 million years ago. The Upper Carboniferous, or Pennsylvanian, extended from approximately 310 to 290 million years ago, the beginning of the Permian period. Coal formation continued throughout the Permian, Triassic, Jurassic, Cretaceous, and Tertiary periods (known collectively as the *second coal age*), which spanned 290 million to 1.6 million years before present. During that time, coal formation started from thick layers of dead plants that piled up in ancient swamps. The dead-plant layers were buried deeper and deeper under even thicker layers of sand and mud. During burial, pressure and heat changed the plant material into coal. Coal from the Western United States usually formed during or after the era of the dinosaurs (approximately 50−100 million years ago).

In geological terms, coal is often defined as banded, streaky, black mineral which is readily combustible and contains more than 50% (by weight) or 70% (by volume) of carbonaceous material. However, to be considered as coal, the black rock must substantially be a deposit of compacted, decayed plant material. The principal geological periods into which Earth's history has been divided (Table 5.1) have been established on the basis of the evolution of life forms recorded by fossils found in the successive layers of rocks deposited through geological time. There are also major breaks in the continuity of depositional environments that are recognized in the rock sequences.

Most sedimentary rocks were deposited as particles in water and the mud, silt, or sand deposited by streams ultimately formed shale (bedded mudstone), clay-stone (nonbedded mudstone), siltstone, or sandstone. Early in geological time (although not yet generally recognized in the very earliest sedimentary rocks) animals and simple marine plants developed. Primitive algal remains, fungal remains, and wormlike burrows and trails have been recognized in pre-Cambrian sediments, but by earliest Cambrian time a relatively diverse assemblage of small marine animals had evolved. Many of the marine animals and plants extracted calcium carbonate from the seawater. When their remains accumulated on the sea floor, their shells and fine calcium carbonate muds formed deposits commonly from a few feet up to hundreds of feet thick that are now limestone and dolomite.

At various times in the geological past, large areas of the sea were apparently cut off from the rest, and evaporation first made the water too salty to support life and then resulted in deposits of salt and/or gypsum and anhydride. Many other deposits are also recognized and include wind-blown deposits and deposits in river valleys and lakes. However, somewhat unique in character among the sedimentary rocks are the vast accumulations of plant material that formed at many different times to produce coal seams.

The major areas of coal distribution are principally in the Northern Hemisphere; with the exception of Australia, the southern continents are relatively deficient in coal deposits. This relatively uneven distribution is the result of the deposition and maturation of the plant at different times in the geological past in predominantly tropical latitudes, and the subsequent drift of the continents to their present-day positions. The oldest coals of any economic significance date from the Middle Carboniferous period—the earliest geological strata in which coal has been identified are of Devonian age but they are generally of little economic significance. With the exception of parts of the Triassic period, major coal deposits have been forming somewhere in the world throughout the last 320 million years. Sedimentary sequences of the last two-to-three million years do not contain coal—there has been insufficient time for them to develop from plant debris. However, not all regions have coal reserves. Generally, regions where post-Devonian sediments were not deposited or have been completely eroded have no coal whatsoever, such as Scandinavia and much of Africa. In addition, rocks with the potential for coal-bearing, once deposited, may have been eroded away and some regions which do have rocks with the potential for coal-bearing may be buried deeply under younger sedimentary sequences and cannot be worked economically.

Individual coal seams vary in thickness from a fraction of an inch up to many feet and may be as much as 300 feet (91.5 m) in some areas. Most of the bituminous coal seams are less than 20 feet (6.1 m) thick, and most mining has been in seams from 3 to 10 feet (0.9−3.5 m) thick. The total sequence of sedimentary rocks may reach several thousand feet (but generally much less) and the coal-bearing seams that have been mapped throughout the world commonly occur at depths less than 5000 feet. Most coal seams in the United States lie at depths up to 3000 feet whereas most of the mining operations have been carried out at depths ranging up to 1000 feet.

As a result of the formation processes, coal contains organic (carbon-containing) compounds transformed from ancient plant material (Speight, 2013). The original plant material was composed of cellulose, the reinforcing material in plant cell walls; lignin, the substance that cements plant cells together; tannins, a class of compounds in leaves and stems; and other organic compounds, such as fats and waxes. In addition to carbon, these organic compounds contain hydrogen, oxygen, nitrogen, and sulfur. After a plant dies and begins to decay on a swamp bottom, hydrogen and oxygen (and smaller amounts of other elements) gradually dissociate from the plant matter, increasing its relative carbon content.

As a consequence of the sequence of events that lead to the formation of coal, inorganic constituents also occur in coal. The mineral matter includes minerals such as pyrite (FeS_2) and the related mineral marcasite (FeS_2), as well as other minerals formed from metals that accumulated in the living tissues of the ancient plants. Quartz, clay, and other minerals are also added to coal deposits by wind and groundwater. The mineral matter lowers the fixed carbon content of coal, leading to a decrease in the heating value of coal.

3.2 Coal types

Coal occurs in different forms or types. Variations in the nature of the source material and local or regional variations in the coalification processes cause the vegetal matter to evolve differently. Thus, various classification systems exist to define the different types of coal. Thus, as geological processes increase their effect over time, the coal precursors are transformed over time into:

1. Lignite— also referred to as *brown coal*, is the lowest rank of coal and used almost exclusively as fuel for steam-electric power generation. Lignite is the least mature of the coal types and provides the least yield of energy; it is often crumbly, relatively moist, and powdery. It is the lowest rank of coal, with a heating value of 4000 to 8300 Btu per pound. Most lignite mined in the United States comes from Texas. Lignite is mainly used to produce electricity.
2. Sub-bituminous coal is poorly indurated and brownish in color, but more like bituminous coal than lignite. It typically contains less heating value (8300–13,000 Btu per pound) and more moisture than bituminous coal. The properties range from those of lignite to those of bituminous coal and are used primarily as fuel for steam-electric power generation.
3. Bituminous coal is a dense coal, usually black, sometimes dark brown, often with well-defined bands of bright and dull material. This type of coal was formed by added heat and pressure on lignite and is the black, soft, slick rock and the most common coal used around the world. Made of many tiny layers, bituminous coal looks smooth and sometimes shiny. It is the most abundant type of coal found in the United States and has two to three times the heating value of lignite. Bituminous coal contains 11,000 to 15,500 Btu per pound. Bituminous coal is used to generate electricity and is an important fuel for the steel and iron industries.
4. Anthracite—the highest rank; a harder, glossy, black coal used primarily for residential and commercial space heating. *Anthracite* is usually considered to be the highest grade of coal and is actually considered to be metamorphic. Compared to other coal types, anthracite is much harder, has a glassy luster, and is denser and blacker with few impurities. It is largely used for heating domestically as it burns with little smoke. It is deep black and looks almost metallic due to its glossy surface. Like bituminous coal, anthracite coal is a big energy producer, containing nearly 15,000 Btu per pound.

With increasing coalification (i.e., progressing from lignite to subbituminous coal to bituminous coal to anthracite), the moisture content decreases while the carbon content and the energy content both increase.

Finally, *steam coal*, which is not a specific rank of coal, is a grade of coal that falls between bituminous coal and anthracite, once widely used as a fuel for steam locomotives. In this specialized use it is sometimes known as *sea-coal* in the United States. Small steam coal (*dry small steam nuts*, DSSN) was used as a fuel for domestic water heating. In addition, the material known as *jet* is the gem variety of coal. Jet is generally derived from anthracite and lacks a crystalline structure, so it is considered to be a mineraloid. Mineraloids are often mistaken for minerals and are sometimes classified as minerals, but lack the necessary crystalline structure to be truly classified as a mineral. Jet is, in to being one of the products of an organic process, which remains removed from full mineral status.

Coal classification systems are based on the degree to which coals have undergone coalification. Such varying degrees of coalification are generally called coal *ranks* (or *classes*). The *rank* of a coal indicates the progressive changes in carbon, volatile matter, and probably ash and sulfur that take place as coalification progresses from the lower-rank lignite through the higher ranks of sub-bituminous, high-volatile bituminous, low-volatile bituminous, and anthracite. The rank of a coal should not be confused with its grade.

In terms of *coal grade*, the grade of a coal establishes its economic value for a specific end use. The grade of coal refers to the amount of mineral matter that is present in the coal and is a measure of coal quality. Sulfur content, ash fusion temperature (i.e., the temperature at which measurement the ash melts and fuses), and quantity of trace elements in coal are also used to grade coal. Although formal classification systems have not been developed around grade of coal, grade is important to the coal user.

A high rank (e.g., anthracite) represents coal from a deposit that has undergone the greatest degree of metamorphosis and contains very little mineral matter, ash, and moisture. On the other hand, any rank of coal, when cleaned of impurities through coal preparation will be of a higher grade. In the United States and some other countries, a commonly used system of classification is based on fixed carbon content, volatile matter content, and calorific value (Speight, 2013). In addition to the major ranks (lignite, sub-bituminous, bituminous, and anthracite), each rank may be subdivided into coal groups, such as *high-volatile bituminous coal*. Other designations, such as *coking coal* and *steam coal*, have been applied to coals, but they tend to differ from country to country.

The term *coal type* is also employed to distinguish between banded coals and nonbanded coals. Banded coals contain varying amounts of vitrinite and opaque material. They include bright coal, which contains more than 80% vitrinite, and splint coal, which contains more than 30% opaque matter. The nonbanded varieties include boghead coal, which has a high percentage of algal remains, and cannel coal with a high percentage of spores. The usage of all the above terms is quite subjective.

By analogy to the term *mineral*, which is applied to inorganic material, the term *maceral* is used to describe organic constituents present in coals. Three major maceral groups are generally recognized: *vitrinite*, *exinite*, and *inertinite*. The vitrinite group is the most abundant and is derived primarily from cell walls and woody tissues. Several varieties are recognized, e.g., telinite (the brighter parts of vitrinite that make up cell walls) and collinite (clear vitrinite that occupies the spaces between cell walls).

Coal analysis may be presented in the form of *proximate* and *ultimate* analyses, whose analytic conditions are prescribed by organizations such as the

ASTM International. A typical proximate analysis includes the moisture content, ash yield (that can be converted to mineral matter content), volatile matter content, and fixed carbon content. It is important to know the moisture content and the mineral matter content (manifested in the analysis as ash) of a coal because they do not contribute to the heating value of a coal. In most cases ash becomes an undesirable residue and a source of pollution, but for some purpose (e.g., use as a chemical feedstock or for liquefaction) the presence of mineral matter may be desirable. Most of the heat value of a coal comes from its volatile matter, excluding moisture, and fixed carbon content. For most coals, it is necessary to measure the actual amount of heat released upon combustion, which is expressed in British thermal units (BTU) per pound. Fixed carbon is the material, other than ash, that does not vaporize when heated in the absence of air. It is determined by subtracting the weight percent sum of the moisture, ash, and volatile matter—in weight percent from 100%.

Ultimate analyses are used to determine the carbon, hydrogen, sulfur, nitrogen, ash, oxygen, and moisture contents of a coal. For specific applications, other chemical analyses may be employed. These may involve, for example, identifying the forms of sulfur present; sulfur may occur in the form of sulfide minerals (pyrite and marcasite, FeS_2), sulfate minerals (gypsum, Na_2SO_4), or organically bound sulfur. In other cases the analyses may involve determining the trace elements present (e.g., mercury, chlorine), which may influence the suitability of a coal for a particular purpose or help to establish methods for reducing environmental pollution.

3.3 Coalbed methane

As a conclusion to this section some mention must be made of coalbed methane—methane that occurs within the coal seam as part of the coal structure. The methane that occurs in coal seams is also a valuable commodity as well as a potential danger to the miners and mining operations. In fact, it has been suggested (Levine, 1991a, 1991b) with sufficient justification that the materials comprising a coalbed fall broadly into the following two categories: (i) volatile low-molecular-weight materials (components) that can be liberated from the coal by pressure reduction, mild heating, or solvent extraction and (ii) materials that will remain in the solid state after the separation of volatile components.

Methane adsorbed to the surface of coal is a very old issue, mainly because this explosive gas has made underground coal mines dangerous both from the risk of explosion and the possibility of an oxygen-poor atmosphere that would not support life. The main concern of miners with coal bed methane (coalbed methane) has been how to get rid of it (Simpson et al., 2003). In the past several decades it had become clear that the coalbed methane is a significant economic resource and capturing coalbed methane for sale is often profitable even for exploiting coal seams that cannot be economically developed.

Coal bed methane (coalbed methane, CBM) is predominantly methane extracted from coal beds (coal seams) and has become an important source of energy in United States, Canada, and other countries. The gas is adsorbed into the solid matrix of the coal and is often referred to as *sweet gas* because of its lack of hydrogen sulfide. Unlike natural gas from conventional reservoirs (Mokhatab et al., 2006; Speight, 2019), coalbed methane contains very little higher molecular weight hydrocarbon derivatives such as propane or butane or condensate. It often contains up to a few percent carbon dioxide. Some coal seams contain little methane, with the predominant coal seam gas being carbon dioxide.

Techniques to detect and deal with coalbed methane in mines have ranged from (i) the classic canary in a cage to detect an oxygen-poor atmosphere, (ii) ventilation fans to force the replacement of a methane-rich environment with outside air, and (iii) drilling coalbed methane wells in front of the coal face to try to degas the coal prior to exposing the mine to the coalbed methane. Each method has met with a degree of success but none can prevent coalbed methane from escaping from the coal seam and entering the air in an underground mine.

With the unique method of gas storage of coalbed methane, the preponderance of the gas is available only to very low coal-face pressures. The coal-face pressure is set by a combination of flowing wellhead pressure and the hydrostatic head exerted by standing liquid within the wellbore. Effective removal of higher molecular weight hydrocarbon derivatives and other constituents (*liquids removal* or *deliquefaction*) techniques can reduce or remove the backpressure caused by accumulated liquid.

4. Mining and preparation

Early coal mining (i.e., the extraction of coal from the seam) was small-scale, the coal lying either on the surface, or very close to it. Typical methods for extraction included drift mining and bell pits. In Britain, some of the earliest drift mines date from the medieval period. As well as drift mines, small-scale shaft mining was used. This took the form of bell pit mining, the extraction working outward from a central shaft, or a technique called room and pillar mining in which rooms of coal were extracted with pillars left to support the roofs.

Deep shaft mining started to develop in England in the late 18th century, although rapid expansion occurred throughout the 19th and early 20th century. The counties of Durham and Northumberland were the leading coal producers, and they were the sites of the first deep pits. Before 1800 a great deal of coal was left in places as support pillars and, as a result in the deep pits (300−1000 ft deep) of these two northern counties only approximately 40% w/w of the coal could be extracted. The use of wood props to support the roof was an innovation first introduced approximately in 1800. The critical factor was circulation of air and control of explosive gases.

In the current context, coal mining depends on the following criteria: (i) seam thickness, (ii) the overburden thickness, (iii) the ease of removal of the overburden (surface mining), (iv) the ease with which a shaft can be sunk to reach the coal seam (underground mining), (v) the amount of coal extracted relative to the amount that cannot be removed, and (vi) the market demand for the coal.

There are two predominant types of mining methods that are employed for coal recovery. The first group consists of surface mining methods, in which the strata (overburden) overlying the coal seam are first removed after which the coal is extracted from the exposed seam. Underground mining currently accounts for recovery of approximately 60% of the world recovery of coal.

4.1 Surface mining

Surface mining is the application of coal removal methods to reserves that are too shallow to be developed by other mining methods. The surface mining methods are: (i) open pit mining, (ii) drift mining, (iii) slope mining, (iv) contour mining, and (v) auger mining.

The characteristic that distinguishes *open pit mining* is the thickness of the coal seam insofar as it is virtually impossible to backfill the immediate mined out area with the original overburden when extremely thick seams of coal are involved. Thus, the coal is removed either by taking the entire seam down to the seam basement (i.e., floor of the mine) or by benching (the staged mining of the coal seam). Frequent use is made of a *drift mine* in which a horizontal seam of coal outcrops to the surface in the side of a hill or mountain, and the opening into the mine can be made directly into the coal seam. This type of mine is generally the easiest and most economical to open because excavation through rock is not necessary.

Another surface mine, the *slope mine,* is one in which an inclined opening is used to trap the coal seam (or seams). A slope mine may follow the coal seam if the seam is inclined and outcrops to the surface, or the slope may be driven through rock strata overlying the coal to reach a seam that is below drainage. Coal transportation from a slope mine can be by conveyor or by track haulage (using a trolley locomotive if the grade is not severe) or by pulling mine cars up the slope using an electric hoist and steel rope if the grade is steep. The most common practice is to use a belt conveyor where grades do not exceed 18 degrees. On the other hand *contour mining* prevails in mountainous and hilly terrain taking its name from the method in which the equipment follows the contours of the Earth.

Auger mining is frequently employed in open pit mines where the thickness of the overburden at the high-wall section of the mine is too great for further economic mining. This, however, should not detract from the overall concept and utility of auger mining as it is also applicable to underground operations. As the coal is discharged from the auger spiral, it is collected for transportation to the coal preparation plant or to the market. Additional auger lengths are added as the cutting head of the auger penetrates further under the high wall into the coal. Penetration continues until the cutting head drifts into the top or bottom, as determined by the cuttings returned, into a previous hole, or until the maximum torque or the auger is reached.

4.2 Underground mining

The second is underground (or deep) mining, in which the coal is extracted from a seam without removal of the overlying strata, by means of a *shaft mine* that enters the Earth by a vertical opening from the surface and descends to the coal seam. In the mine, the coal is extracted from the seam by conventional mining, or by *continuous mining*, or by *longwall mining*, or by *shortwall mining*, or by *room and pillar mining*.

Conventional mining (also called *cyclic* mining) involves a sequence of operations in the order (i) supporting the roof, (ii) cutting, (iii) drilling, (iv) blasting, (v) coal removal, and (vi) loading. After the roof above the seam has been made safe by timbering or by roof bolting, one or more slots (a few inches wide and extending for several feet into the coal) are cut along the length of the coal face by a large, mobile cutting machine. The cut, or slot, provides a free face and facilitates the breaking up of the coal, which is usually blasted from the seam by explosives. These explosives (*permissible explosives*) produce an almost flame-free explosion and markedly reduce the amount of noxious fumes relative to the more conventional explosives. The coal may then be transported by rubber-tired electric vehicles (shuttle cars) or by chain (or belt) conveyor systems.

Continuous mining involved the use of a single machine (*continuous miner*) that breaks the coal mechanically and loads it for transport. Roof support is then installed, ventilation is advanced, and the coal face is ready for the next cycle. The method of secondary transportation is located immediately behind the continuous miner and required installation of mobile belt conveyors.

The *longwall mining* system involves the use of a mechanical self-advancing roof in which large blocks of coal are completely extracted in a continuous operation. Hydraulic or self-advancing jacks (*chocks*) support the roof at the immediate face as the coal is removed. As the face advances, the strata are allowed to collapse behind the support units. Coal recovery is near that attainable with the conventional or continuous systems as well as efficient mining under extremely deep cover or overburden or when the roof is weak.

The *shortwall mining* system is a combination of the continuous mining and longwall mining concepts and offers good recovery of the in-place coal with a marked decrease in the costs for roof support.

Room and pillar mining is a means of developing a coal face and, at the same time, retaining supports for the roof. Thus, by means of this technique, rooms are developed from large tunnels driven into the solid coal with the intervening pillars of coal supporting the roof. The percentage of coal recovered from a seam depends on the number and size of protective pillars of coal thought necessary to support the roof safely and of the percentage of pillar recovery.

4.3 Coal preparation

As-mined coal (*run-of-mine coal*) contains a mixture of different size fractions, sometimes together with unwanted impurities such as rock and dirt. Thus, another sequence of events is necessary to make coal of a consistent quality and salable. Such events are called *coal cleaning*. Effective preparation of coal prior to combustion improves the homogeneity of coal supplied, reduces transport costs, improves the utilization efficiency, produces less ash for disposal at the power plant, and may reduce the emissions of oxides of sulfur.

Coal cleaning (*coal preparation, coal beneficiation*) is the stage in coal production when the run-of-mine coal is processed into a range of clean, graded, and uniform coal products suitable for the commercial market. In some cases, the run-of-mine coal is of such quality that it meets the user specification without the need for beneficiation, in which case the coal would merely be crushed and screened to deliver the specified product.

A number of physical separation technologies are used in the washing and beneficiation of coals. After the raw run-of-mine coal is crushed, it is separated into various size fractions for optimum treatment. Larger material (10–150 mm lumps) is usually treated using *dense medium separation*—the coal is separated from other impurities by being floated across a tank containing a liquid of suitable specific gravity, usually a suspension of finely ground magnetite. The coal being less dense floats and is separated off, while heavier rock and other impurities sink and are removed as waste. Any magnetite mixed with the coal is separated using water sprays, and is then recovered, using magnetic drums, and recycled.

The smaller size fractions are treated in a variety of ways—usually based on gravity differentials. In the *froth flotation* method, coal particles are removed in a froth produced by blowing air into a water bath containing chemical reagents. The bubbles attract the coal but not the waste and are skimmed off to recover the coal fines. After treatment, the various size fractions are screened and dewatered or dried, and then recombined before going through final sampling and quality control procedures.

5. Properties

Chemically, coal is a hydrogen-deficient hydrocarbon with an atomic hydrogen-to-carbon ratio near 0.8, as compared to crude oil hydrocarbon derivatives, which have an atomic hydrogen-to-carbon ratio approximately equal to 2, and methane (CH_4) that has an atomic carbon-to-hydrogen ratio equal to 4. For this reason, any process used to convert coal to hydrocarbon fuels must add hydrogen. The chemical composition of the coal is defined in terms of its proximate and ultimate (elemental) analysis (Speight, 2013). The parameters of proximate analysis are moisture, volatile matter, mineral matter (as combustion ash), and fixed carbon. Elemental or ultimate analysis encompasses the quantitative determination of carbon, hydrogen, nitrogen, oxygen, and sulfur within the coal.

Knowledge of coal properties is an important aspect of coal characterization and has been used as a means of determining the suitability of coal for commercial use for decades, perhaps even centuries. In fact, the molecular characterization of coal (Speight, 2013) is seen to be of little or no importance, by some consumers. Therefore the properties outlined in this chapter must always be borne in mind when consideration is being given to the suitability of coal for a particular use, in light of various environmental regulations and how the data might be used to predict the suitability of coal for use in an environmentally clean and safe manner.

The properties discussed in this section are those properties that are relevant to the handling and decomposition of coal to produce hydrocarbon products.

5.1 Physical properties

The physical properties and the behavior of coal play an important part in dictating the methods by which coal should be handled and utilized. This section considers those properties such as, for example, density and hardness, which are definitely physical in nature, in contrast to the properties (e.g., spectroscopic properties) which arise by virtue of the molecular structural types that occur within the coal (Speight, 2013). At first consideration, there may appear to be little, if any, relationship between the physical and chemical behavior of coal but in fact the converse is very true. For example, the pore size of coal (which is truly a physical property) is a major factor in determining the chemical reactivity of coal. And chemical effects which result in the swelling and caking of coal(s) have a substantial effect on the means by which coal should be "handled" either prior to or during a coal conversion operation.

5.1.1 Density (specific gravity)

For porous solids, such as coal, there are three different density measurements, true density, particle density, and apparent density. The *true density* is usually determined by displacement of a fluid, but because of the porous nature of coal and also because of physicochemical interactions, the observed density data vary with the particular fluids employed (Agrawal, 1959). The *apparent density* of coal is determined by immersing a weighed sample of coal in a liquid followed by accurate measurement of the liquid that is displaced. For this procedure, the liquid should (i) wet the surface of the coal, (ii) not absorb strongly to the coal surface, (iii) not cause swelling, and (iv) penetrate the pores of the coal.

The *in-place density* of coal is the means by which coal in the seam can be expressed as tons per acre per foot of seam thickness and/or tons per square mile per foot of seam thickness. The bulk density of the solid is the mass of the solid particles per unit of volume they occupy. Major factors influencing the bulk density of coal are: moisture content, particle surface properties, particle shape, particle size distribution, and particle density (Speight, 2013).

The density of coal shows a notable variation with rank for carbon content (Speight, 2013) and, in addition, the methanol density is generally higher than the helium density because of the contraction of adsorbed helium in the coal pores as well as by virtue of interactions between the coal and the methanol which result in a combined volume that is notably less than the sum of the separate volumes. Similar behavior has been observed for the water density of coals having 80%−84% w/w carbon. Coal with more than 85% w/w carbon usually exhibits a greater degree of hydrophobic character than the lower rank coals with the additional note that the water density may be substantially lower than the helium density; for the coals with 80%−84% w/w carbon, there is generally little, if any, difference between the helium density and the water density.

An additional noteworthy trend is the tendency for the density of coal to exhibit a minimum value at approximately 85% carbon. For example, a 50%−55% carbon coal will have a density of approximately 1.5 g/cm and this will decrease to, say, 1.3 g/cm for an 85% carbon coal followed by an increase in density to ca. 1.8 g/cm for a 97% carbon coal. On a comparative note, the density of graphite (2.25 g/cm) also falls into this trend.

The *bulk density* is not an intrinsic property of coal and varies depending on how the coal is handled. Bulk density is the mass of many particles of coal divided by the total volume occupied by the particles—the total volume includes particle volume, interparticle void volume, and internal pore volume. This allows the density of coal to be expressed in terms of the cubic foot weight of crushed coal (ASTM D291), which varies with particle size of the coal and packing in a container. In addition, the bulk density of coal varies with rank: (i) anthracite: 50−58 lb/ft^3 which is 800−929 kg/m^3, (ii) bituminous coal: 42−57 lb/ft^3 which is 673−913 kg/m^3, and (iii) lignite: 40−54 lb/ft^3 which is 641−865 kg/m^3. The bulk density of coal decreases with increasing moisture content until a minimum is reached (at approximately 6%−8% moisture) but then increases with moisture content. With the use of aqueous solutions of wetting additives, the coal bulk density can be increased by up to 15%.

Due to the hydrophobic nature of higher rank coals, their bulk density rapidly decreases with increasing moisture, as the water stays on the surface of the coal in between the particles, increasing the volume of the bulk solid (decreasing bulk density). Further increase in moisture leads to the minimum bulk density attained at about 6%–8% moisture content. Beyond this level of moisture, more water penetrates the spaces in between the particles and forms a water layer around them allowing for aggregation which leads to tighter packing configuration. As a result particles are packed into smaller volume which further leads to an increase in bulk density of the sample. Furthermore, small additions of chemical reagents reducing surface tension of water resulted in an increased bulk density. These experiments showed that the bulk density could be increased by 13%–15% with the use of such additives.

The aggregation of fines, leading to tight packing of particles, affects the bulk density of coal samples. Wettability appears to be a controlling factor in the aggregation of fine coal particles. As a result, different patterns of bulk density can be anticipated with moisture increase for hydrophilic coal samples and also for hydrophobic coal samples (Holuszko and Laskowski, 2010).

5.1.2 Porosity and surface area

Coal is a porous material and, thus, the porosity and surface area of coal have a large influence on coal behavior during mining, preparation, and utilization. As already noted with respect to coal density the porosity of coal decreases with carbon content and has a minimum at approximately 89% w/w carbon followed by a marked increase in porosity. The nature of the porosity also appears to vary with carbon content (rank); for example, the macropores are usually predominant in the lower carbon (rank) coals while higher carbon (rank) coals contain predominantly micropores. Thus, pore volume can be calculated from the relationship:

$$V_p = 1/\rho_{Hg} - 1\ \rho_{He}$$

In this equation, ρ_{Hg} is the mercury density and ρ_{He} is the helium density, which decreases with carbon content. In addition, the surface area of coal varies over the range 10–200 m^2/g and also tends to decrease with the carbon content of the coal.

Porosity and surface area are two very important properties with respect to the gasification of coal since the reactivity of coal increases as the porosity and surface area of the coal increases. Thus, the gasification rate of coal is greater for the lower rank coals than for the higher rank coals. The porosity of coal is calculated from the relationship:

$$\rho = 100\rho_{Hg}(1/\rho_{Hg} - 1\ \rho_{He})$$

By determining the apparent density of coal in fluids of different, but known, dimensions, it is possible to calculate the pore size (pore volume) distribution. The open pore volume (V), i.e., the pore volume accessible to a particular fluid, can be calculated from the relationship:

$$V = 1\rho_{Hg} - 1\rho_a$$

ρ_a is the apparent density in the fluid.

The size distribution of the pores within a coal can be determined by immersing the coal in mercury and progressively increasing the pressure. Surface tension effects prevent the mercury from entering the pores having a diameter smaller than a given value d for any particular pressure p such that

$$P = 4\sigma \cdot \cos \theta / d$$

σ is the surface tension and θ is the angle of contact.

By recording the amount of mercury entering the coal for small increments of pressure, it is possible to build up a picture of the distribution of sizes of pores. However, the total pore volume accounted for by this method is substantially less than that derived from the helium density, thereby giving rise to the concept that coal contains two pore systems: (i) a macropore system accessible to mercury under pressure and (ii) a micropore system that is inaccessible to mercury but accessible to helium. By using liquids of various molecular sizes it is possible to investigate the distribution of micropore sizes. However, the precise role or function of these micropores as part of the structural model of coal is not fully understood, although it has been suggested that coal may behave in some respects like a molecular sieve.

5.1.3 Reflectance

Coal may be examined in visible light by either transmission or reflectance. The former is a measure of light absorbance at various wavelengths and may be determined for thin sections of coal or finely divided coal pressed into a potassium bromide disk or solutions of coal extracts in solvents such as pyridine or films of coal that have been deposited by evaporation of a dispersing liquid.

Coal reflectance (ASTM D2798) is very useful because it indicates several important properties of coal. Among these are the content of several macerals (ASTM D2799) and carbonization temperature. Coal reflectance is determined by the relative degree to which a beam of polarized light is reflected from a polished coal surface. The coal is crushed to pass a number 20 screen (850 micron screen) with minimal fines production and the particles are formed into a briquette held together with a binder. One side of the briquette is polished using successively finer abrasives until a smooth surface that is scratch-free, smear-free, and char-free is obtained. A metallurgical, or opaque-ore, microscope is employed to determine the reflectance with vertical illumination using polarized light.

Coal reflectance is important in aiding the determination of the maceral composition of coal which, in turn, is helpful for the prediction of behavior in processing. For example, the basic use for which determination of vitrinite reflectance is applied is the measurement of the degree of metamorphism or rank of coals and the organic components of sediments. It is possible to relate the reflectance of a coal to parameters such as carbon content, volatile matter, and calorific value.

5.2 Mechanical properties

In contrast to the proximate analysis, ultimate (elemental) analysis and certain of the physical properties, the mechanical properties of coal have been little used. But these properties are of importance and should be of consideration in predicting coal behavior during mining, handling, and preparation. For example, the mechanical properties of coal are of value as a means of predicting the strength of coal and its behavior in mines when the strength of coal pillars and stability of coal faces are extremely important factors. The mechanical properties of coal are also of value in areas such as coal winning (for the design and operation of cutting machinery), comminution (design and/or selection of mills), storage (flow properties, failure under shear), handling (shatter and abrasion during transport), as well as in many other aspects of coal technology.

5.2.1 Hardness

Considerable information is available on the properties of coal that relate to its hardness, such as friability and grindability, but little is known of the hardness of coal as an intrinsic property. The scratch hardness of coal can be determined by measuring the load on a pyramidal steel point required to make a scratch 100 microns in width on the polished surface of a specimen. The scratch hardness of anthracite is approximately 6 times that of a soft coal whereas pyrite is almost 20 times as hard as soft coal. Similar data were noted for anthracite and cannel coal but durain, the reputedly hard component of coal, was found to be only slightly harder than vitrain and cannel coal.

Although the resistance of coal to abrasion may have little apparent commercial significance, the abrasiveness of coal is, on the other hand, a factor of considerable importance. Thus, the wear of grinding elements due to the abrasive action of coal results in maintenance charges that constitute one of the major items in the cost of grinding coal for use as pulverized fuel. Moreover, as coals vary widely in abrasiveness, this factor must be considered when coals are selected for pulverized fuel plants. A standardized, simple laboratory method of evaluating the abrasiveness of coal would assist, like the standard grindability test now available, in the selection of coals suitable for use in pulverized form. Actually the abrasiveness of coal may be determined more by the nature of its associated impurities than by the nature of the coal substance. For example, pyrite is 20 times harder than coal, and the individual grains of sandstone, another common impurity (in coal), also are hard and abrasive.

5.2.2 Friability

One measure of the strength of coal is its ability to withstand degradation in size on handling. The tendency toward breakage on handling (friability) depends on toughness, elasticity, and fracture characteristics as well as on strength, but despite this fact the friability test is the measure of coal strength used most frequently.

Friability is of interest primarily because friable coals yield smaller proportions of the coarse sizes which may (depending on use) be more desirable and there may also be an increased amount of surface in the friable coals. This surface allows more rapid oxidation; hence conditions are more favorable for spontaneous ignition (Speight, 2013), loss in coking quality in coking coals, and other changes that accompany oxidation. These economic aspects of the friability of coal have provided the incentive toward development of laboratory friability tests.

The tumbler test for measuring coal friability (ASTM D441) employs a cylindrical porcelain jar mill (7.25 in., 18.4 cm in size) fitted with three lifters that assist in tumbling the coal. A 1000-g sample of coal sized between 1.5- and 1.05-in. (38.1 and 26.7 mm) square-hole screens is tumbled in the mill (without grinding medium) for 1 h at 40 rpm. The coal is then removed and screened on square-hole sieves with openings of 1.05 inch (26.7 mm), 0.742 in. (18.8 mm), 0.525 in. (13.3 mm), 0.371 in. (9.42 mm), 0.0369 in. (1.19 mm), and 0.0117 in. (0.297 mm). Friability is reported as the percentage reduction in the average particle size during the test. For example, if the average particle size of the tumbled coal was 75% that of the original sample, the friability would be 25%.

A drop shatter test has also been described for determining the friability of coal (ASTM D440) which is similar to the standard method used as a shatter test for coke (ASTM D3038). In this method, a 50-lb sample of coal (2–3 in., 4.5–7.6 cm) is dropped twice from a drop-bottom box onto a steel plate 6 ft below the box. The materials shattered by the two drops are then screened over round-hole screens with 3.0 in. (76.2 mm), 2.0 in. (50.8 mm), 1.5 in. (38.1 mm), 1.0 in. (25.4 mm). 0.75 in. (19.05 mm), and 0.5 in. (12.7 mm) openings and the average particle size is determined. The average size of the material, expressed as a percentage of the size of the original sample, is termed the sizeability, and its complement, the percentage of reduction in average particle size, is termed the percentage friability. Provision is made for testing sizes other than that stipulated for the standard test to permit comparison of different sizes of the same coal.

The relationship between the friability of coal and its rank has a bearing on its tendency to undergo spontaneous heating and ignition. The friable, low-volatile coals, because of their high rank, do not oxidize readily despite the excessive fines and the attendant increased surface they produce on handling. Coals of somewhat lower rank, which oxidize more readily, usually are relatively nonfriable, hence they resist degradation in size with its accompanying increase in the amount of surface exposed to oxidation. But above all, the primary factor in coal stockpile instability is unquestionably oxidation by atmospheric oxygen while the role of any secondary factors such as friability is to exacerbate the primary oxidation effect.

5.2.3 Grindability

The grindability of coal (i.e., the ease with which coal may be ground fine enough for use as pulverized fuel) is a composite physical property embracing other specific properties such as hardness, strength, tenacity, and fracture. Several methods of estimating relative grindability utilize a porcelain jar mill in which each coal may be ground for, say, 400 revolutions and the amount of new surface is estimated from screen analyses of the feed and of the ground product. Coals are then rated in grindability by comparing the amount of new surface found in the test with that obtained for a standard coal.

The test for grindability (ASTM D409) utilizes a ball-and-ring-type mill in which a 50 g sample of closely sized coal is ground for 60 revolutions after which the ground product is screened through a 200-mesh sieve and the grindability index is calculated from the amount of undersize produced using a calibration chart. The results are converted into the equivalent Hardgrove grindability index. High grindability index (GI) numbers indicate easy-to-grind coals. There is a rough relationship between volatile content and grindability in the low-, medium-, and high-volatile bituminous rank coals. Among these the low-volatile coals exhibit the highest values of the grindability index, often over 100. The grindability indexes of the high-volatile coals range from the mid-50s to the high-30s. There is also a correlation of sorts between friability and grindability. Soft, easily fractured coals generally exhibit relatively high values of the grindability index.

5.3 Thermal properties

The thermal properties of coal are important in determining the applicability of coal to a variety of conversion processes. For example, the heat content (also called the heating value or calorific value) is often considered to be the most important thermal property. However, there are other thermal properties that are of importance insofar as they are required for the design of equipment that is to be employed for the utilization (conversion, thermal treatment) of coal in processes such as combustion, carbonization, gasification, and liquefaction. Plastic and agglutinating properties as well as phenomena such as the agglomerating index give indications of how coal will behave in a reactor during a variety of thermal processes.

5.3.1 Calorific value

The calorific or heating value of a coal is a direct indication of the energy content and therefore is probably the most important property for determining the usefulness of coal. It is the amount of energy that a given quantity of coal will produce when burned. Heating value is used in determining the rank of coals. It is also used to determine the maximum theoretical fuel energy available for the production of steam. It is used to determine the quantity of fuel that must be handled, pulverized, and fired in the boiler.

The calorific value Q of coal can be determined experimentally using a calorimeter or Q can be calculated using the Dulong formula, when the oxygen content is less than 10%:

$$Q = 337C + 1442(H - O/8) + 93S$$

C is the mass percent of carbon, H is the mass percent of hydrogen, O is the mass percent of oxygen, and S is the mass percent of sulfur in the coal. With these constants, Q is given in kilojoules per kilogram (1 kJ per kilogram = 2.326 Btu/lb).

The calorific value is reported as gross calorific value, with a correction made if net calorific value is of interest (ASTM D121; ASTM D2015; ASTM D3286; ASTM D5865). For solid fuels such as coal, the gross heat of combustion is the heat produced by the combustion of a unit quantity of the coal in a bomb calorimeter with oxygen and under a specified set of conditions. The unit is calories per gram, which may be converted to the alternate units (1.0 kcal/kg = 1.8 Btu/lb = 4.187 kJ/kg).

The experimental conditions require an initial oxygen pressure of 300–600 psi and a final temperature in the range 20°C–35°C (68°F–95°F) with the products in the form of ash, water, carbon dioxide, sulfur dioxide, and nitrogen. Thus, once the gross calorific value has been determined, the net calorific value (i.e., the net heat of combustion) is calculated from the gross calorific value (at 20°C, 68°F) by deducting 1030 Btu/lb (2.4 × 103 kJ/kg) to allow for the heat of vaporization of the water. The deduction is not actually equal to the heat of vaporization of water (1055 Btu/lb; 2.45 × 103 kJ/kg) because the calculation is to reduce the data from a gross value at constant volume to a net value at constant pressure. Thus, the differences between the gross calorific value (GCV) and the net calorific value (NCV) are given by:

$$\text{NCV (Btu/lb)} = \text{GCV} - (1030 \times \text{total hydrogen} \times 9)/100$$

There are also reports of the use of differential thermal analysis for the determination of the calorific value of coal.

5.3.2 Heat capacity

The heat capacity of a material is the heat required to raise the temperature of one unit weight of a substance by 1° and the ratio of the heat capacity of one substance to the heat capacity of water at 15°C (60°F) in the specific heat. The heat capacity of coal can be measured by standard calorimetric methods for mixtures (e.g., ASTM C351). The units for heat capacity are Btu per pound per degree Fahrenheit (Btu/lb/°F) or calories per gram per degree centigrade (cal/g/°C), but the specific heat is the ratio of two heat capacities and is therefore dimensionless. The heat capacity of water is 1.0 Btu/lb/°F (= 4.2 × 103 J/kg/°K) and, thus, the heat capacity of any material will always be numerically equal to the specific heat. Consequently, there has been the tendency to use the terms heat capacity and specific heat synonymously. The specific heat of coal usually (i) increases with its moisture content, (ii) decreases with carbon content, and (iii) increases with volatile matter content, with ash content exerting a somewhat lesser influence (Speight, 2013).

The values for the specific heats of various coals fall into the general range 0.25–0.37 but, as with other physical data, comparisons should only be made on an equal (moisture content, ash content, etc.) basis. Estimates of the specific heat of coal have also been made on the assumption that the molecular heat of a solid material is equal to the sum of the atomic heats of the constituents (Kopp's law); the atomic heat so derived is divided by the atomic weight to give the (approximate) specific heat. Thus, from the data for various coals, it has been possible to derive a formula which indicates the relationship between the specific heat and the elemental analysis of coal (mmf basis):

$$C_p = 0.189C + 0.874H + 0.491N + 0.3600 + 0.215S$$

C, H, N, O, and S are the respective amounts (% w/w) of the elements in the coal.

5.4 Thermal conductivity

Thermal conductivity is the rate of transfer of heat by conduction through a unit area across a unit thickness for a unit difference in temperature:

$$Q = kA(t_2 - t_1)/d$$

Q = heat, expressed as kcal/sec cm °C or as Btu/ft h °F (1 Btu/ft h °F = 1.7 J/s m °K), A = area, $t_2 - t_1$ = temperature differential for the distance (d), and k = thermal conductivity (Carslaw and Jaeger, 1959).

However, the banding and bedding planes in coal (Speight, 2013) can complicate the matter to such an extent that it is difficult, if not almost impossible, to determine a single value for the thermal conductivity of a particular coal. Nevertheless, it has been possible to draw certain conclusions from the data available. Thus, monolithic coal is considered to be a medium conductor of heat with the thermal conductivity of anthracite being on the order of 5 to 9×10^{-4} kcal/s cm °C while the thermal conductivity of monolithic bituminous coal falls in the range 4 to 7×10^{-4} kcal/s/cm/°C. For example, the thermal conductivity of pulverized coal is lower than that of the corresponding monolithic coal. For example, the thermal conductivity of pulverized bituminous coal falls in the range $2.5–3.5 \times 10^{-4}$ kcal/s cm °C.

The thermal conductivity of coal generally increases with an increase in the apparent density of the coal as well as with volatile matter yield, ash yield, and temperature. In addition, the thermal conductivity of the coal parallel to the bedding plane appears to be higher than the thermal conductivity perpendicular to the bedding plane.

5.4.1 Plastic and agglutinating properties

Plasticity refers to the melting and bonding behavior of the coal and: (i) is an indication of the initial softening, chemical reaction, gas liberation, and resolidification process within the coke oven, (ii) is an important requirement in the coke blend and is required for end product coke strength, and (iii) the fluidity of the plastic stage is a major factor in determining what proportion of coal is used in a blend.

All coals undergo chemical changes when heated, but there are certain types of coal which also exhibit physical changes when subjected to the influence of heat. These particular types of coals are generally known as "caking" coals whereas the remaining coals are referred to as *noncaking coals*. The caking coals pass through a series of physical changes during the heating process insofar as they soften, melt, fuse, swell, and solidify within a specific temperature range. This temperature has been called the "plastic range" of coal and thus the physical changes which occur within this range have been termed the *plastic properties* (*plasticity*) of coal. The caking tendency of coals increases with the volatile matter content of the coal and reaches a maximum in the range 25%−35% w/w volatile matter but then tends to decrease. In addition, the caking tendency of coal is generally high in the 81%−92% w/w carbon coals (with a maximum at 89% carbon); the caking tendency of coal also increases with hydrogen content but decreases with oxygen content and with mineral matter content. When noncaking (nonplastic) coals are heated, the residue is pulverent and noncoherent. On the other hand, the caking coals produce residues that are coherent and have varying degrees of friability and swelling. In the plastic range, caking coal particles tend to form agglomerates (cakes) and may even adhere to surfaces of process equipment, thereby giving rise to reactor plugging problems.

Thus, the plastic properties of coal are an important means of projecting and predicting how coal will behave under various process conditions as well as assisting in the selection of process equipment. For example, the plasticity of coal is a beneficial property in terms of the production of metallurgical coke but may have an adverse effect on the suitability of coal for conversion to gases and liquids (Speight, 2013). As an example, the Lurgi gasifier is unable to adequately gasify caking coals unless the caking properties are first nullified by a prior oxidative treatment.

The plastic behavior of coal is of practical importance for semiquantitative evaluation of metallurgical coal and coal blends used in the production of coke for the steel industry. When bituminous coals are heated in the absence of air through the range 300°C −550°C (570°F −1020°F), volatile materials are released and the solid coal particles soften (to become a plasticlike mass) which swells and eventually resolidifies.

The Gieseler test is a standard test method which attempts to measure the actual extent of the plasticity of fluidity attained. The Giesler test is used to characterize coals with regard to thermo plasticity and is an important method used for coal blending for commercial coke manufacture. The maximum fluidity determined by the Gieseler is very sensitive to weathering (oxidation) of the coal. Briefly, the Giesler plastometer (ASTM Dl812, ASTM D2639) is a vertical instrument consisting of a sample holder, a stirrer with four small rabble arms attached at its lower end with the means of (i) applying a torque to the stirrer, (ii) heating the sample that includes provision for controlling temperature and rate of temperature rise, and (iii) measuring the rate of turning of the stirrer.

Data usually obtained with the Giesler plastometer are (i) the softening temperature (the temperature at which stirrer movement is equal to 0.5 dial divisions per minute) which may be characterized by other rates but if so the rate must be reported, (ii) the maximum fluid temperature (the temperature at which stirrer movement reaches maximum rate in terms of dial divisions per minute), (iii) the solidification temperature (the temperature at which stirrer movement stops), and (iv) the maximum fluidity (the maximum rate of stirrer movement in dial divisions per minute).

In terms of the elemental composition of coal, there is a relative hydrogen deficiency, but there are theories that admit to the presence of hydrogen-rich liquid (and mobile) hydrocarbon derivatives that are enclosed within the coal matrix and which are often (erroneously) called bitumen and which should not be confused with the bitumen from tar sand deposits (Chapter 2) that occur in various deposits throughout the world. The application of heat results in the liberation of these hydrocarbon liquids and forms other hydrocarbons (*thermobitumen*) by scission of hydrocarbon fragments from the coal structure, and the overall effect is the formation of a high-carbon coke and a hydrocarbon tar, the latter being responsible for the fluidity of the mass. With increased heating, the tar partly volatilizes and partly reacts to form nonfluid material ultimately leading to the coke residue.

An additional property of coal that is worthy of mention at this time is the softening point, which is generally defined as the temperature at which the particles of coal begin to melt and become rounded. The softening point indicates the onset of the plasticity stage and is (as should be anticipated) a function of the volatile matter content of coal. For example, coal with 15% volatile matter will have a softening point of the order of 440°C (825°F) which will decrease to a limiting value of ca. 340°C (645°F) for coal with 30% w/w volatile matter.

5.4.2 Agglomerating index and free swelling index

The *agglomerating index* is a grading index based on the nature of the residue from an 1 g sample of coal when heated at 950°C (1740°F) in the volatile matter determination (ASTM D3175). The agglomerating index has been adopted as a requisite physical property to differentiate semianthracite from low-volatile C bituminous coal and also high-volatile C bituminous coal from sub-bituminous coal.

The *free swelling index* (FSI) of coal is a measure of the increase in volume of a coal when it is heated (without restriction) under prescribed conditions (ASTM D720). The nature of the volume increase is associated with the plastic properties of coal (Loison et al., 1963) and, as might be anticipated, coals which do not exhibit plastic properties when heated do not, therefore, exhibit free swelling. Although this relationship between free swelling and plastic properties may be quite complex, it is presumed that when the coal is in a plastic (or semifluid) condition the gas bubbles formed as a part of the thermal decomposition process within the fluid material cause the swelling phenomenon which, in turn, is influenced by the thickness of the bubble walls, the fluidity of the coal, and the interfacial tension between the fluid material and the solid particles that are presumed to be present under the test conditions.

Other effects which can influence the free swelling index of coal include the weathering (oxidation) of the coal. Hence, it is advisable to test coal as soon as possible after collection and preparation. There is also evidence that the size of the sample can influence the outcome of the free swelling test; an excess of fine (−100 mesh) coal in a sample has reputedly been responsible for excessive swelling to the extent that the FSI numbers can be up to two numbers higher than is the true case.

6. Hydrocarbon products

Coal can be converted to liquid hydrocarbon derivatives (liquefaction) by either direct processes or by indirect processes (i.e., by using the gaseous products obtained by breaking down the chemical structure of coal) to produce liquid products (Table 5.2). Four general methods are used for liquefaction: (i) pyrolysis and hydrocarbonization (coal is heated in the absence of air or in a stream of hydrogen, respectively), (ii) solvent extraction (coal hydrocarbon derivatives are selectively dissolved and hydrogen is added to produce the desired liquids), (iii) catalytic liquefaction (hydrogenation takes place in the presence of a catalyst), and (iv) indirect liquefaction (carbon monoxide and hydrogen are combined in the presence of a catalyst).

Producing hydrocarbon fuels such as gasoline and diesel fuel from coal can be done through converting coal to syngas, a combination of carbon monoxide, hydrogen, carbon dioxide, and methane. The syngas is reacted through the Fischer-Tropsch synthesis to produce hydrocarbon derivatives that can be

TABLE 5.2 Examples of chemicals produced from the thermal decomposition of coal.

	Initial products	Secondary products	Tertiary products
Coal	Gas	Fuel gas	
		Sulfur	
	Light oil	Naphtha	Cyclopentadiene
			Indene
		Benzol	Benzene
		Toluols	Toluene
		Xylols	*o*-xylene
			m-xylene
			p-xylene
	Coal tar	Pitch (nondistillable)	
		Tar (distillable)	Fluorene
			Diphenylene oxide
			Acenaphthene
			Methyl-naphthalene derivatives
		Creosote (extractable)	
		tar acids (extractable)	Phenol
			o-cresol
			m-cresol
			p-cresol
			3,5-Xylenol
		tar bases (extractable)	Pyridine
			Alpha-picoline
			Beta-picoline
			Gamma-picoline
			2,6-Lutidine
			Quinoline
		Naphthalene	
		Crude anthracene	Anthracene
			Phenanthrene
			Carbazole
	Coke		Carbon

refined into liquid fuels. By increasing the quantity of high-quality fuels from coal while reducing the costs, research into this process could help ease the dependence on ever-more expensive but depleting stocks of crude oil. Furthermore, by improving the catalysts used in directly converting coal into liquid hydrocarbon derivatives, without the generation of the intermediate syngas, less power could be required to produce a product suitable for upgrading in existing crude oil refineries. Such an approach could reduce energy requirements and improve yields of desired products.

Coal, which can be viewed as the critical factor in the growth of the industrial age in the 1700s and the organic chemicals industry in the mid-1800s, may be ready to once again attain a key role in the global chemical industry. Certainly coal lost its key role to low-priced oil and gas in the middle of the 20th century, but it may be poised for a comeback now that conventional oil and gas production is being strained by the rate of global economic growth and the rate of depletion for many larger reserves. For many regions around the world, coal now appears to offer a realistic and available chemicals starting point when compared to the alternatives of importing LNG, LPG, naphtha, or crude oil.

Coal chemicals are obtained during the processing of metallurgical coke from coal. The aromatic compounds that are obtained as byproducts during such processing are used as intermediates during the process of synthesis of some solvents, dyes, drugs, and antiseptics. Most of the byproducts of coal chemicals are used as fuel. So far, many chemical compounds have been identified and isolated from coal tar, which is a byproduct of coal chemicals.

Coal chemicals are typically mixtures of: (i) methane, (ii) carbon monoxide, (iii) hydrogen, (iv) small amounts of higher molecular weight hydrocarbon derivatives, (v) ammonia, and (vi) hydrogen sulfide. Some aromatic hydrocarbon derivatives, such as toluene and xylene, are now largely produced from crude oil refineries byproducts. Furthermore, coal chemicals byproducts like benzene, naphthalene, anthracene, phenanthrene, pyridines, and quinolines are not obtained from crude oil refineries. In addition, a substantial quantity of phenol, cresols, and xylenols are still obtained as byproducts of coal chemicals.

Coal gasification is a well-proven technology (Speight, 2013; Luque and Speight, 2015) that has had many applications ranging from the earliest uses of coal gas for heating and lighting in urban areas ("town gas"), progressing to the production of synthetic fuels, such as liquid hydrocarbon derivatives and synthetic natural gas (SNG) chemicals, and most recently to large-scale IGCC (integrated gasification combined cycle) power generation. Coal can also be gasified as a cofeedstock with other carbonaceous feedstocks (Speight, 2013, 2014b; Luque and Speight, 2015).

By definition, the petrochemical industry is based on feedstocks derived from natural gas or crude oil. However, before 1940, many of these same organic chemicals were frequently referred to as "coal chemicals." In the

United States, oil and gas have, for the last 60 years or more, been abundant, leading to a situation where the preponderance of organic chemicals has been manufactured from these feedstocks. In other countries (e.g., South Africa, India, and possibly most importantly, China) coal has been an important feedstock during recent years.

Essentially all of the important first-stage organic petrochemicals were made from coal during the period of approximately 1900–1930. The coke oven industry provided as byproducts ammonia, ammonium sulfate, benzene, toluene, and phenols. In the fuels and fuel-chemicals sector, success was achieved in producing straight chain hydrocarbon derivatives, alcohols, and other organic chemicals from synthesis gas, as exemplified by the work of Bergius, Fischer and Tropsch, and others.

With the current tightness in North American natural gas supplies and high cost incremental supplies placing a floor under the price of natural gas, the prospects for coal and/or coal gasification as a source of petrochemicals and power are becoming a much more realistic alternative. As crude oil and natural gas supplies decrease relative to demand, prices are expected to continue rising, making coal a more economic and competitive feedstock. But, as crude and natural gas prices have continued to rise, coal prices have remained relatively flat.

As crude oil and natural gas continue to rise in price, alternate production of petrochemicals from coal becomes more of a cost reality. Regions with large coal reserves, such as China, are now reexamining the potential for coal to chemicals. For example, the Shenhua Group and Dow Chemical recently announced that they have agreed to evaluate the feasibility of coal-to-olefins projects in China. It can be expected that similar initiatives will follow elsewhere. In the United States, there is the need for greater use of coal to help relieve U.S. energy demand and there is also the need to develop technology for use of coal as a feedstock for synthetic crude oil. As the technology for coal gasification for power and fuels advances, the competitiveness of coal to chemicals in the United States will logically follow.

Though coal tar, a product of coke ovens, will continue to be a source of certain chemicals (e.g., anthracene to carbon black, naphthalene to phthalic anhydride, among others), the gasification of coal is considered to be a major potential "on-purpose" source of commodity petrochemicals and hydrocarbon derivatives. Another "on-purpose" route for coal to chemicals is via the production of acetylene, which can be used to produce a variety of chemicals, including vinyl chloride monomer and polyvinyl chloride. The economics for this production becomes attractive as the crude/coal price spread increases, and this route may prove important in areas with large coal reserves.

The attractiveness of the use of coal for chemicals production is primarily dependent upon three key factors: (i) the price and availability of alternate feedstocks, i.e., gas and crude oil, (ii) advances in gasification technology, and (iii) advances in environmental protection technology. The fundamentals of

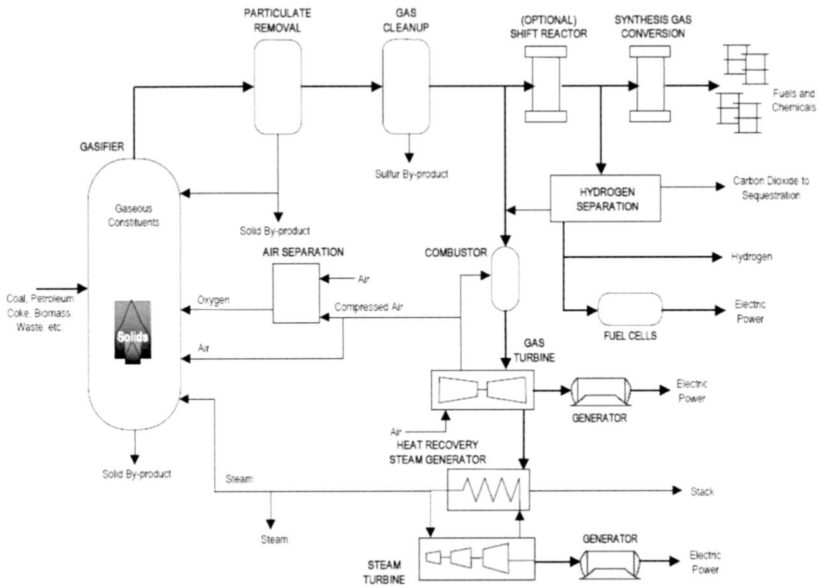

FIGURE 5.1 Gasification-based energy conversion. *Source: "Major Environmental Aspects of Gasification-Based Power Generation Technologies", SAIC Report for DOE, December 2002.*

these three factors have changed dramatically in the favor of coal since the mid-1900s when coal lost its role as the basic feedstock for organic chemicals. Chemicals can be produced from the three principal products of coal gasification; synthesis gas and hydrogen, as well as carbon monoxide (Fig. 5.1).

Various gasification and environmental cleanup technologies convert coal (or other carbon-based feedstocks), oxidant, and water to syngas for further conversion into marketable products: electricity, fuels, chemicals, steam, hydrogen, and others. There are multiple reasons for the choice of gasification as the means to utilize coal, compared to direct use as a combustion fuel, including environmental reasons, current and improving cost competitiveness, and feedstock flexibility (gasifiers can operate on a wide variety of feedstocks). Given the economies of scale envisioned for a coal-to-chemicals facility, it is likely that the technology will typically best suit large volume commodity chemicals. However, smaller-scale chemicals and specialty chemicals can also certainly be produced under the correct economic situations.

An especially intriguing coal-to-chemical application is acetylene and acetylene-based chemicals. Though not a product of coal gasification, this is another example of "on-purpose" use of coal for the production of major petrochemicals. Though not as dependent upon advances in technology, such as for gasification, the use of coal to produce acetylene and downstream derivatives gains attractiveness as the price difference between coal and crude oil widens.

6.1 Gasification processes

Gasification processes are segregated according to the bed types, which differ in their ability to accept (and use) caking coals and are generally divided into four categories based on reactor (bed) configuration: (i) fixed-bed, (ii) fluidized bed, (iii) entrained bed, and (iv) molten salt.

In a fixed-bed process the coal is supported by a grate and combustion gases (such as steam, air, and oxygen) pass through the supported coal whereupon the hot produced gases exit from the top of the reactor. Heat is supplied internally or from an outside source, but caking coals cannot be used in an unmodified fixed-bed reactor.

The fluidized bed system uses finely sized coal particles and the bed exhibits liquidlike characteristics when a gas flows upward through the bed. Gas flowing through the coal produces turbulent lifting and separation of particles and the result is an expanded bed having greater coal surface area to promote the chemical reaction, but such systems have a limited ability to handle caking coals.

An entrained-bed system uses finely sized coal particles blown into the gas steam prior to entry into the reactor and combustion occurs with the coal particles suspended in the gas phase; the entrained system is suitable for both caking and noncaking coals.

The molten salt system employs a bath of molten salt to convert coal (Speight, 2013).

6.1.1 Fixed-bed processes

As an example of a fixed-bed process, the Lurgi process was developed in Germany before World War II and is a process that is adequately suited for large-scale commercial production of synthetic natural gas. The older Lurgi process is a dry ash gasification process which differs significantly from the more recently developed slagging process (Baughman, 1978; Massey, 1979). The dry ash Lurgi gasifier is a pressurized vertical kiln which accepts crushed (1/4 × 1−3/4 in.; 6 × 44 mm) noncaking coal and reacts the moving bed of coal with steam and either air or oxygen. The coal is gasified at 350−450 psi and devolatilization takes place in the temperature range 615°C−760°C (1140°F to 1400°F); residence time in the reactor is approximately 1 h. Hydrogen is supplied by injected steam and the necessary heat is supplied by the combustion of a portion of the product char. The revolving grate, located at the bottom of the gasifier supports the bed of coal, removes the ash, and allows steam and oxygen (or air) to be introduced.

The Lurgi product gas has high methane content relative to the products from nonpressurized gasifiers. With oxygen injection, the gas has a heat content of approximately 450 Btu/ft^3 (16.8 MJ/m^3). The crude gas which leaves the gasifier contains tar, oil, phenols, ammonia, coal fines, and ash particles. The steam is first quenched to remove the tar and oil and, prior to methanation, part of the gas passes through a shift converter and is then washed to remove naphtha and unsaturated hydrocarbon derivatives; a subsequent step removes the acid gases. The gas is then methanated to produce a high heat-content pipeline quality product.

The Wellman Galusha process— another fixed-bed process— has been in commercial use for more than 50 years (Howard-Smith and Werner, 1976). These are two types of gasifiers, the standard type and the agitated type and the rated capacity of an agitated unit may be 25% or more higher than that of a standard gasifier of the same size. In addition, an agitated gasifier is capable of treating volatile caking bituminous coals.

The gasifier is water-jacketed and, therefore, the inner wall of the vessel does not require a refractory lining. Agitated units include a varying speed revolving horizontal arm which also spirals vertically below the surface of the coal bed to minimize channeling and to provide a uniform bed for gasification. A rotating grate is located at the bottom of the gasifier to remove the ash from the bed uniformly. Steam and oxygen are injected at the bottom of the bed through tuyeres. Crushed coal is fed to the gasifier through a lock hopper and vertical feed pipes. The fuel valves are operated so as to maintain a relatively constant flow of coal to the gasifier to assist in maintaining the stability of the bed and, therefore, the quality of the product gas.

6.1.2 Entrained-bed processes

The Koppers-Totzek Process (Baughman, 1978) is perhaps the best known of the entrained-solids processes and operates at atmospheric pressure. The reactor is a relatively small, cylindrical, refractory-lined vessel into which coal, oxygen, and steam are charged. The reactor typically operates at an exit temperature of approximately 1480°C (2700°F) and the pressure is maintained slightly above atmospheric pressure.

Gases and vaporized hydrocarbon derivatives produced by the coal at medium temperatures immediately pass through a zone of very high temperature in which they decompose so rapidly that coal particles in the plastic stage do not agglomerate, and thus any type of coal can be gasified irrespective of caking tendencies, ash content, or ash fusion temperature. The gas product contains no ammonia, tars, phenols, or condensables and can be upgraded to synthesis gas by reacting all or part of the carbon monoxide content with steam to produce additional hydrogen plus carbon dioxide.

6.1.3 Molten salt processes

Molten salt processes feature the use of a molten bath (>1550°C; >2820°F) into which coal, steam, and oxygen are injected. The coal devolatilizes with some thermal cracking of the volatile constituents. The product gas, which leaves the gasifier, is cooled, compressed, and fed to a shift converter where a portion of the carbon monoxide is reacted with steam to attain a carbon monoxide to hydrogen ratio of 1:3. The carbon dioxide so produced is removed and the gas is again cooled and enters a methanator where carbon monoxide and hydrogen react to form methane.

6.1.4 Underground gasification

The aim of underground (or in situ) gasification of coal is to convert the coal into combustible gases by combustion of a coal seam in the presence of air, oxygen, or oxygen and steam. Thus, seams that were considered to be inaccessible, unworkable, or uneconomical to mine could be put to use. In addition, strip mining and the accompanying environmental impacts, the problems of spoil banks, acid mine drainage, and the problems associated with use of high-ash coal are minimized or even eliminated.

The principles of underground gasification are very similar to those involved in the above-ground gasification of coal. The concept involves the drilling and subsequent linking of two boreholes so that gas will pass between the two. Combustion is then initiated at the bottom of one borehole (injection well) and is maintained by the continuous injection of air. In the initial reaction zone (combustion zone), carbon dioxide is generated by the reaction of oxygen (air) with the coal:

$$[C]_{coal} + O_2 \rightarrow CO_2$$

The carbon dioxide reacts with coal (partially devolatilized) further along the seam (reduction zone) to produce carbon monoxide:

$$[C]_{coal} + CO_2 \rightarrow 2CO$$

In addition, at the high temperatures that can frequently occur, moisture injected with oxygen or even moisture inherent in the seam may also react with the coal to produce carbon monoxide and hydrogen:

$$[C]_{coal} + H_2O \rightarrow CO + H_2$$

The gas product varies in character and composition but usually falls into the low-heat (low Btu) category ranging from 125 to 175 Btu/ft^3.

6.1.5 Gasifiers

The gasification of coal can be used to produce synthesis gas (syngas), a mixture of carbon monoxide (CO) and hydrogen (H_2) gas, which can be converted into hydrocarbon fuels such as gasoline and diesel through the Fischer-Tropsch process (Chapter 8). Alternatively, the hydrogen obtained from gasification can be used for upgrading fossil fuels to hydrocarbon fuels. In the process, the coal is mixed with oxygen and steam while are also being heated and pressurized. During the reaction, the coal is oxidized to carbon monoxide (CO) while also releasing hydrogen (H_2) gas. This process has been conducted in surface facilities and underground:

$$(\text{Coal}) + O_2 + H_2O \rightarrow H_2 + CO$$

If it is desired to produce gasoline, the synthesis gas is collected at this stage and routed into a Fischer-Tropsch reactor.

If hydrogen is the desired end-product, however, the synthesis gas is fed to a water-gas shift reactor where more hydrogen is produced:

$$CO + H_2O \rightarrow CO_2 + H_2$$

In the past, coal was converted to *coal gas* (*town gas*), which was piped to customers to burn for illumination, heating, and cooking. At present, natural gas is preferred over coal gas. Four types of gasifiers are currently available for commercial use: (i) countercurrent fixed-bed technology, or (ii) cocurrent fixed-bed technology, or (iii) fluid bed technology, or (iv) entrained flow technology.

The countercurrent fixed-bed (up draft) gasifier consists of a fixed bed of carbonaceous fuel (e.g., coal or biomass) through which the "gasification agent" (steam, oxygen, and/or air) flows in countercurrent configuration. The ash is either removed dry or as a slag. The slagging gasifiers require a higher ratio of steam and oxygen to carbon in order to reach temperatures higher than the ash fusion temperature. The nature of the gasifier means that the fuel must have high mechanical strength and must be noncaking so that it will form a permeable bed, although recent developments have reduced these restrictions to some extent. The throughput for this type of gasifier is relatively low. Thermal efficiency is high as the gas exit temperatures are relatively low. However, this means that tar and methane production is significant at typical operation temperatures, so product gas must be extensively cleaned before use or recycled to the reactor.

The cocurrent fixed-bed (down draft) gasifier is similar to the countercurrent type, but the gasification agent gas flows in cocurrent configuration with the fuel (downwards, hence the name down draft gasifier). Heat needs to be added to the upper part of the bed, either by combusting small amounts of the fuel or from external heat sources. The produced gas leaves the gasifier at a high temperature, and most of this heat is often transferred to the gasification agent added in the top of the bed, resulting in energy efficiency on level with the countercurrent type. Since all tars must pass through a hot bed of char in this configuration, tar levels are much lower than the countercurrent type.

In the fluid bed gasifier, the fuel is fluidized in oxygen (or air) and steam. The ash is removed dry or as high-density agglomerates that defluidize the bed. The temperatures are relatively low in dry ash gasifiers, so the fuel must be highly reactive; low-grade coals are particularly suitable. The agglomerating gasifiers have slightly higher temperatures, and are suitable for higher rank coals. Fuel throughput is higher than for the fixed bed, but not as high as for the entrained flow gasifier. The conversion efficiency is rather low, so recycle or subsequent combustion of solids is necessary to increase conversion. Fluidized bed gasifiers are most useful for fuels that form highly corrosive ash that would damage the walls of slagging gasifiers. Biomasses generally contain high levels of such ashes.

In the entrained flow gasifier a dry pulverized solid, an atomized liquid fuel, or a fuel slurry is gasified with oxygen (much less frequent: air) in cocurrent flow. The gasification reactions take place in a dense cloud of very fine particles. Most coals are suitable for this type of gasifier because of the high operating temperatures and because the coal particles are well separated from one another. The high temperatures and pressures also mean that a higher throughput can be achieved. However thermal efficiency is somewhat lower as the gas must be cooled before it can be cleaned with existing technology. The high temperatures also mean that tar and methane are not present in the product gas; however, the oxygen requirement is higher than for the other types of gasifiers. All entrained flow gasifiers remove the major part of the ash as a slag as the operating temperature is well above the ash fusion temperature. A smaller fraction of the ash is produced either as a very fine, dry fly ash or as black colored fly ash slurry. Some fuels, in particular certain types of biomasses, can form slag that is corrosive for ceramic inner walls that serve to protect the gasifier outer wall.

However, some entrained bed type of gasifiers do not possess a ceramic inner wall but have an inner water or steam cooled wall covered with partially solidified slag. These types of gasifiers do not suffer from corrosive slag. Some fuels have ashes with very high ash fusion temperatures. In this case mostly limestone is mixed with the fuel prior to gasification. Addition of a little limestone will usually suffice for lowering the fusion temperatures. The fuel particles must be much smaller than for other types of gasifiers. This means the fuel must be pulverized, which requires somewhat more energy than for the other types of gasifiers. By far the most energy consumption related to entrained bed gasification is not the milling of the fuel but the production of oxygen used for the gasification.

6.2 Liquefaction processes

Coal liquefaction is the process (or collection of various processes) used to convert coal, a solid fuel, into a substitute for liquid fuels such as gasoline and diesel fuel. Coal liquefaction has historically been used in countries without a secure supply of petroleum, such as Germany (during World War II) and South Africa (since the early 1970s). The technology used in coal liquefaction is quite old, and was first implemented during the 19th century to provide gas for indoor lighting. Coal liquefaction may be used in the future to produce oil for transportation and heating, in case crude oil supplies are ever disrupted.

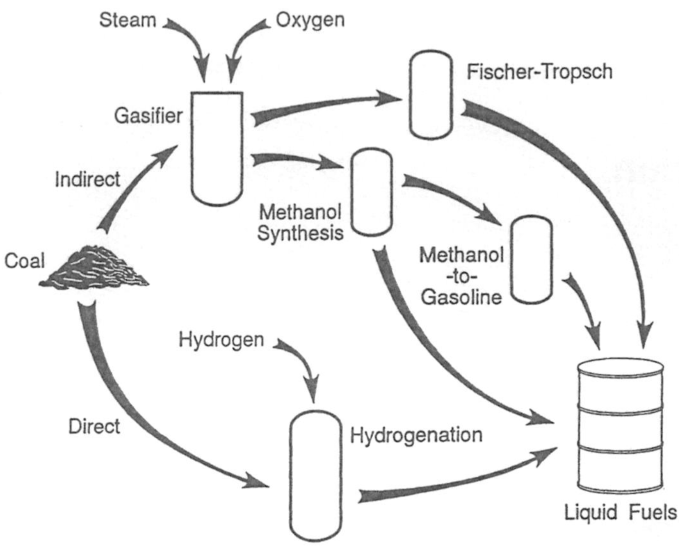

FIGURE 5.2 Routes for the production of hydrocarbon products from coal.

The production of hydrocarbon fuels from coal is not new and has received considerable attention since the concept does represent alternate pathways to liquid fuels (Fig. 5.2) (Speight, 2013). In fact, the concept is often cited as a viable option for alleviating projected shortages of liquid fuels as well as offering some measure of energy independence for those countries with vast resources of coal who are also net importers of crude oil.

The thermal decomposition of coal to a mix of solid, liquid, and gaseous products is usually achieved by the use of temperatures up to 1500°C (2730°F) (Whitehurst et al., 1980). But, coal carbonization is not a process which has been designed for the production of liquids as the major products. The chemistry of coal liquefaction is also extremely complex, not so much from the model compound perspective but more from the interactions that can occur between the constituents of the coal liquids. Even though many schemes for the chemical sequences, which ultimately result in the production of liquids from coal, have been formulated, the exact chemistry involved is still largely speculative, largely because the interactions of the constituents with each other are generally ignored. Indeed, the so-called structure of coal itself is still only speculative.

Hydrogen can represent a major cost item of the liquefaction process and, accordingly, several process options have been designed to limit (or control) the hydrogen consumption or even to increase the hydrogen/carbon atomic ratio without the need for added gas phase hydrogen (Whitehurst et al., 1980; Speight, 2013). Thus, at best, the chemistry of coal liquefaction is only speculative. Furthermore, various structures have been postulated for the structure of coal (albeit with varying degrees of uncertainty) but the representation of coal as any one of these structures is extremely difficult and, hence, projecting a thermal decomposition route and the accompanying chemistry is even more precarious.

The majority of the coal liquefaction processes involve the addition of a coal-derived solvent prior to heating the coal to the desired process temperature. This is, essentially, a means of facilitating the transfer of the coal to a high-pressure region (usually the reactor) and also to diminish the sticking that might occur by virtue of the plastic properties of the coal (Whitehurst et al., 1980). The process options for coal liquefaction can generally be divided into four categories: (i) pyrolysis, (ii) solvent extraction, (iii) catalytic liquefaction, and (iv) indirect liquefaction.

6.2.1 Pyrolysis processes

The first category of coal liquefaction processes, pyrolysis processes, involves heating coal to temperatures in excess of 400°C (750°F), which results in the conversion of the coal to gases, liquids, and char. The char is hydrogen deficient thereby enabling intermolecular or intramolecular hydrogen transfer processes to be operative, resulting in relatively hydrogen-rich gases and liquids. Unfortunately, the char produced often amounts to more than 45% by weight of the feed coal and, therefore, such processes have often been considered to be uneconomical or inefficient use of the carbon in the coal.

In the presence of hydrogen (*hydrocarbonization*) the composition and relative amounts of the products formed may vary from the process without hydrogen but the yields are still very much dependent upon the process parameters such as heating rate, pressure, coal type (and product), residence time, coal particle size, and reactor configuration. The operating pressures for pyrolysis processes are usually less than 100 psi— more often between 5 and 25 psi but the hydrocarbonization process requires hydrogen pressures of the order of 300 to 1000 psi. In both categories of process, the operating temperature can be as high as 600°C (1110°F).

There are three types of pyrolysis reactors that are of interest: (L) a mechanically agitated reactor, (ii) an entrained-flow reactor, and (iii) a fluidized bed reactor.

The agitated reactor may be quite complex but the entrained-flow reactor has the advantage of either downflow or upflow operation and can provide short residence times. In addition, the coal can be heated rapidly, leading to higher yields of liquid (and gaseous) products that may well exceed the volatile matter content of the coal as determined by the appropriate test. The short residence time also allows a high throughput of coal and the potential for small reactors. Fluidized reactors are reported to have been successful for processing noncaking coals but are not usually recommended for caking coals.

6.2.2 Solvent extraction processes

Solvent extraction processes are those processes in which coal is mixed with a solvent (*donor* solvent) that is capable of providing atomic or molecular hydrogen to the system at temperatures up to 500°C (930°F) and pressures up to 5000 psi. High-temperature solvent extraction processes of coal have been developed in three different process configurations: (i) extraction in the absence of hydrogen but using a recycle solvent that has been hydrogenated in a separate process stage; (ii) extraction in the presence of hydrogen with a recycle solvent that has not been previously hydrogenated; and (iii) extraction in the presence of hydrogen using a hydrogenated recycle solvent. In each of these concepts, the distillates of process-derived liquids have been used successfully as the recycle solvent which is recovered continuously in the process.

The overall result is an increase (relative to pyrolysis processes) in the amount of coal that is converted to lower molecular weight, i.e., soluble, products. More severe conditions are more effective for sulfur and nitrogen removal to produce a lower boiling liquid product that is more amenable to downstream processing. A more novel aspect of the solvent extraction process type is the use of tar sand bitumen and/or heavy oil as process solvents (Moschopedis et al., 1980, 1982).

6.2.3 Catalytic liquefaction processes

The final category of direct liquefaction process employs the concept of catalytic liquefaction in which a suitable catalyst is used to add hydrogen to the coal. These processes usually require a liquid medium with the catalyst dispersed throughout or may even employ a fixed-bed reactor. On the other hand, the catalyst may also be dispersed within the coal whereupon the combined coal-catalyst system can be injected into the reactor.

Many processes of this type have the advantage of eliminating the need for a hydrogen donor solvent (and the subsequent hydrogenation of the spent solvent) but there is still the need for an adequate supply of hydrogen. The nature of the process also virtually guarantees that the catalyst will be deactivated by the mineral matter in the coal as well as by coke laydown during the process. Furthermore, in order to achieve the direct hydrogenation of the coal, the catalyst and the coal must be in intimate contact, but if this is not the case, process inefficiency is the general rule.

6.2.4 Indirect liquefaction processes

The other category of coal liquefaction processes invokes the concept of the indirect liquefaction of coal. In these processes, the coal is not converted directly into liquid products but involves a two-stage conversion operation in which coal is first converted (by reaction with steam and oxygen) to produce a gaseous mixture that is composed primarily of carbon monoxide and hydrogen (syngas; synthesis gas). The gas stream is subsequently purified (to remove sulfur, nitrogen, and any particulate matter) after which it is catalytically converted to a mixture of liquid hydrocarbon products.

The synthesis of hydrocarbon derivatives from carbon monoxide and hydrogen (synthesis gas) (the Fischer-Tropsch synthesis) is a procedure for the indirect liquefaction of coal.

Thus, coal is converted to gaseous products at temperatures in excess of 800°C (1470°F), and at moderate pressures, to produce synthesis gas:

$$[C]_{coal} + H_2O \rightarrow CO + H_2$$

The gasification may be attained by means of any one of several processes or even by gasification of coal in place (underground, or in situ, gasification of coal, Section 5.5).

In practice, the Fischer-Tropsch reaction is carried out at temperatures of 200°C–350°C (390°F–660°F) and at pressures of 75 to 4000 psi. The hydrogen/carbon monoxide ratio is usually 2.2:1 or 2.5:1. Since up to three volumes of hydrogen may be required to achieve the next stage of the liquids production, the synthesis gas must then be converted by means of the water-gas shift reaction) to the desired level of hydrogen:

$$CO + H_2O \rightarrow CO_2 + H_2$$

After this, the gaseous mix is purified and converted to a wide variety of hydrocarbon derivatives:

$$nCO + (2n + 1)H_2 \rightarrow C_nH_{2n+2} + nH_2O$$

These reactions result primarily in low- and medium-boiling aliphatic compounds suitable for gasoline and diesel fuel.

6.2.5 Reactors

Several types of reactors are available for use in liquefaction processes and any particular type of reactor can exhibit a marked influence on process performance.

The simplest type of reactor is the noncatalytic reactor which consists, essentially, of a vessel (or even an open tube) through which the reactants pass. The reactants are usually in the fluid state but may often contain solids such as would be the case for coal slurry. This particular type of reactor is usually employed for coal liquefaction in the presence of a solvent. The second type of noncatalytic reactor is the continuous-flow, stirred-tank reactor, which has the notable feature of encouraging complete mixing of all of the ingredients, and if there is added catalyst (suspended in the fluid phase) the reactor may be referred to as a slurry reactor.

The fixed-bed catalytic reactor contains a bed of catalyst particles through which the reacting fluid flows; the catalysis of the desired reactions occurs as the fluid flows through the reactor. The liquid may pass through the reactor in a downward flow or in an upward flow but the problems that tend to accompany the latter operation (especially with regard to the heavier, less conventional feedstocks) must be recognized. In the downward-flowing mode, the reactor may often be referred to as a trickle-bed reactor.

Another type of reactor is the fluidized bed reactor, in which the powdered catalyst particles are suspended in a stream of upflowing liquid or gas. A form of this type of reactor is the ebullating-bed reactor. The features of these two types of reactor are the efficient mixing of the solid particles (the catalyst) and the fluid (the reactant) that occurs throughout the whole reactor. The final type of reactor to be described is the entrained-flow reactor in which the solid particles travel with the reacting fluid through the reactor. Such a reactor has also been described as a dilute or lean-phase fluidized-bed with pneumatic transport of solids.

6.3 Gaseous hydrocarbon products

The gasification of coal or a derivative (i.e., char or coke produced from coal) is the conversion of coal (by any one of a variety of processes) to produce gaseous products that are combustible. In fact, coal gasification has considerable potential for producing hydrocarbon derivatives as well as a host of other chemicals (Fig. 5.3). With the rapid increase in the use of coal from the 15th century onwards, it is not surprising the concept of using coal to produce a flammable gas, especially the use of the water and hot coal, became

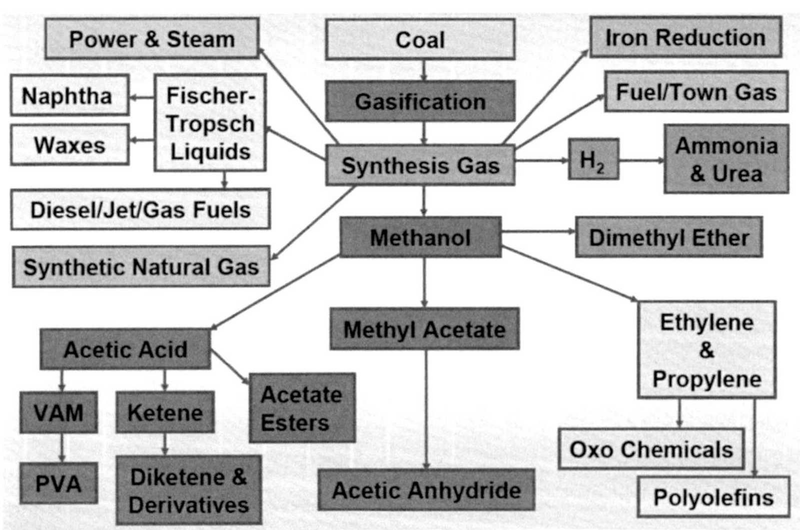

FIGURE 5.3 Potential for coal gasification. *Source: Lynn Schloesser, L., 2006. Gasification Incentives. Workshop on Gasification Technologies. June 28–29. Ramkota, Bismarck, North Dakota.*

common-place. In fact, the production of gas from coal has been a vastly expanding area of coal technology, leading to numerous research and development programs. As a result, the characteristics of rank, mineral matter, particle size, and reaction conditions are all recognized as having a bearing on the outcome of the process; not only in terms of gas yields but also on gas properties (Massey, 1974). The products from the gasification of coal may be of low, medium, or high heat content (high-Btu) as dictated by the process as well as by the ultimate use for the gas (Fryer and Speight, 1976; Cavagnaro, 1980; Probstein and Hicks, 1990; Lahaye and Ehrburger, 1991).

The mounting interest in gasification technology reflects a convergence of two changes in the electricity generation marketplace: (i) the maturity of gasification technology, and (ii) the extremely low emissions from integrated gasification combined cycle (IGCC) plants, especially air emissions, and the potential for lower cost control of greenhouse gases than other coal-based systems. Fluctuations in the costs associated with natural gas–based power, which is viewed as a major competitor to coal-based power, can also play a role. Gasification permits the utilization of coal resources to their fullest potential. Thus, power developers would be well advised to consider gasification as a means of converting coal to gas.

Coal gasification involves the thermal decomposition of coal and the reaction of the coal carbon and other pyrolysis products with oxygen, water, and hydrocarbon gases such as methane. The formation of the products is typically considered as the culmination of two subprocesses: (i) primary gasification and (ii) secondary gasification.

Primary gasification involves thermal decomposition of the raw coal via various chemical processes and many schemes involve pressures ranging from atmospheric to 1000 psi. Air or oxygen may be admitted to support combustion to provide the necessary heat. The product is usually a low heat content (low-Btu) gas ranging from a carbon monoxide/hydrogen mixture to mixtures containing varying amounts of carbon monoxide, carbon dioxide, hydrogen, water, methane, hydrogen sulfide, nitrogen, and typical products of thermal decomposition such as tar (themselves being complex mixtures), hydrocarbon oils, and phenols. A solid char product may also be produced, and may represent the bulk of the weight of the original coal. This type of coal being processed determines (to a large extent) the amount of char produced and the analysis of the gas product.

Secondary gasification usually involves the gasification of char from the primary gasifier. This is usually done by reacting the hot char with water vapor to produce carbon monoxide and hydrogen:

$$[C]_{char} + H_2O \rightarrow CO + H_2$$

The presence of oxygen, hydrogen, water vapor, carbon oxides, and other compounds in the reaction atmosphere during pyrolysis may either support or inhibit numerous reactions with coal and with the products evolved. The distribution of weight and chemical composition of the products are also influenced by the prevailing conditions (i.e., temperature, heating rate, pressure, residence time, and any other relevant parameters) and, last but not least, the nature of the feedstock.

If air is used as a means of combustion, the product gas will have a heat content of 150–300 Btu/ft^3 (5.6–11.2 MJ/m^3) (depending on process design characteristics) and will contain undesirable constituents such as carbon dioxide, hydrogen sulfide, and nitrogen. The use of pure oxygen, although expensive, results in a product gas having a heat content of 300–400 Btu/ft^3 (11.2–14.9 MJ/m^3) with carbon dioxide and hydrogen sulfide as byproducts (both of which can be removed from low or medium heat-content, low- or medium-Btu gas by any of several available processes).

If a high heat-content (high-Btu) gas (900–1000 Btu/ft^3; 33.5–37.3 MJ/m^3) is required, efforts must be made to increase the methane content of the gas. The reactions which generate methane are all exothermic and have negative values but the reaction rates are relatively slow and catalysts may, therefore, be necessary to accelerate the reactions to acceptable commercial rates. Indeed, the overall reactivity of coal and char may be subject to catalytic effects. It is also possible that the mineral constituents of coal and char may modify the reactivity by a direct catalytic effect (Cusumano et al., 1978; Davidson, 1983).

While there has been some discussion of the influence of physical process parameters and the effect of coal type on coal conversion, a note is warranted here regarding the influence of these various parameters on the gasification of coal. Most notable effects are those due to coal character, and often to the maceral content. In regard to the maceral content, differences have been noted between the different maceral groups with inertinite being the most reactive. In more general terms of the character of the coal, gasification technologies generally require some initial processing of the coal feedstock with the type and degree of pretreatment a function of the process and/or the type of coal. For example, the Lurgi process will accept lump coal (1 in., 25 mm, to 28 mesh), but it must be noncaking coal with the fines removed. The caking, agglomerating coals tend to form a plastic mass in the bottom of a gasifier and subsequently plug up the system thereby markedly reducing process efficiency. Thus, some attempt to reduce caking tendencies is necessary and can involve preliminary partial oxidation of the coal thereby destroying the caking properties.

Depending on the type of coal being processed and the analysis of the gas product desired, pressure also plays a role in product definition. In fact, some (or all) of the following processing steps will be required: (i) pretreatment of the coal (if caking is a problem); (ii) primary gasification of the coal; (iii) secondary gasification of the carbonaceous residue from the primary gasifier; (iv) removal of carbon dioxide, hydrogen sulfide, and other acid gases; (v) shift conversion for adjustment of the carbon monoxide/hydrogen mole ratio to the desired ratio; and (vi) catalytic methanation of the carbon monoxide/hydrogen mixture to form methane. If high heat-content (high-Btu) gas is desired, all of these processing steps are required since coal gasifiers do not yield methane in the concentrations required.

Typically, the products of coal gasification are varied insofar as the gas composition varies with the system employed. It is emphasized that the gas product must be first freed from any pollutants such as particulate matter and sulfur compounds before further use, particularly when the intended use is a water gas shift or methanation (Cusumano et al., 1978; Probstein and Hicks, 1990).

6.3.1 Low-Btu gas

Low-Btu gas (low heat-content gas) is also the usual product of in situ gasification of coal which is used essentially as a method for obtaining energy from coal without the necessity of mining the coal, especially if the coal cannot be mined or if mining is uneconomical. During the production of coal gas by oxidation with air, the oxygen is not separated from the air and, as a result, the gas product invariably has a low heat-content (150–300 Btu/ft^3). Several important chemical reactions, and a host of side reactions, are involved in the manufacture of low heat-content gas under the high temperature conditions employed. Low heat-content gas contains several components, four of which are always major components present at levels of at least several percent; a fifth component, methane, is marginally a major component.

The nitrogen content of low heat-content gas ranges from somewhat less than 33% v/v to slightly more than 50% v/v and cannot be removed by any reasonable means; the presence of nitrogen at these levels makes the product gas *low heat-content* by definition. The nitrogen also strongly limits the applicability of the gas to chemical synthesis. Two other noncombustible components (water, H_2O, and carbon dioxide, CO) further lower the heating value of the gas; water can be removed by condensation and carbon dioxide by relatively straightforward chemical means.

The two major combustible components are hydrogen and carbon monoxide; the H_2/CO ratio varies from approximately 2:3 to approximately 3:2. Methane may also make an appreciable contribution to the heat content of the gas. Of the minor components hydrogen sulfide is the most significant and the amount produced is, in fact, proportional to the sulfur content of the feed coal. Any hydrogen sulfide present must be removed by one, or more, of several procedures (Speight, 2014a, 2019).

Low heat-content gas is of interest to industry as a fuel gas or even, on occasion, as a raw material from which ammonia, methanol, and other compounds may be synthesized.

6.3.2 Medium-Btu gas

Medium-Btu gas (medium heat-content gas) has a heating value in the range 300–550 Btu/ft^3 and the composition is much like that of low heat-content gas, except that there is virtually no nitrogen. The primary combustible gases in medium heat-content gas are hydrogen and carbon monoxide (Kasem, 1979). Medium heat-content gas is considerably more versatile than low heat-content gas; like low heat-content gas, medium heat-content gas may be used directly as a fuel to raise steam, or used through a combined power cycle to drive a gas turbine, with the hot exhaust gases employed to raise steam, but medium heat-content gas is especially amenable to synthesize methane (by methanation), higher hydrocarbon derivatives (by Fischer-Tropsch synthesis), methanol, and a variety of synthetic chemicals. The reactions used to produce medium heat-content gas are the same as those employed for low heat-content gas synthesis, the major difference being the application of a nitrogen barrier (such as the use of pure oxygen) to keep diluent nitrogen out of the system.

In medium heat-content gas, the H_2/CO ratio varies from 2:3°C to 3:1 and the increased heating value correlates with higher methane and hydrogen contents as well as with lower carbon dioxide contents. Furthermore, the very nature of the gasification process used to produce the medium heat-content gas has a marked effect upon the ease of subsequent processing.

For example, the CO_2-acceptor product is quite amenable to use for methane production because it has (i) the desired H_2/CO ratio just exceeding 3:1, (ii) an initially high methane content, and (iii) relatively low water and carbon dioxide contents. Other gases may require appreciable shift reaction and removal of large quantities of water and carbon dioxide prior to methanation.

6.3.3 High-Btu gas

High-Btu gas (High heat-content gas) is essentially pure methane and often referred to as *synthetic natural gas* (SNG) (Kasem, 1979). However, to qualify as synthetic natural gas, a product must contain at least 95% methane; the energy content of synthetic natural gas is 980–1080 Btu/ft^3.

The commonly accepted approach to the synthesis of high heat-content gas is the catalytic reaction of hydrogen and carbon monoxide:

$$3H_2 + CO \rightarrow CH_4 + H_2O$$

To avoid catalyst poisoning, the feed gases for this reaction must be quite pure and, therefore, impurities in the product are rare. The large quantities of water produced are removed by condensation and recirculated as very pure water through the gasification system. The hydrogen is usually present in slight excess to ensure that the toxic carbon monoxide is reacted; this small quantity of hydrogen will lower the heat content to a small degree.

The carbon monoxide/hydrogen reaction is somewhat inefficient as a means of producing methane because the reaction liberates large quantities of heat. In addition, the methanation catalyst is troublesome and prone to poisoning by sulfur compounds and the decomposition of metals can destroy the catalyst. Thus, hydrogasification may be employed to minimize the need for methanation:

$$[C]_{coal} + 2H_2 \rightarrow CH_4$$

The product of hydrogasification is far from pure methane and additional methanation is required after hydrogen sulfide and other impurities are removed.

The gaseous product from a gasifier generally contains large amounts of carbon monoxide and hydrogen, plus lesser amounts of hydrocarbon gases. Carbon monoxide and hydrogen (if they are present in the mole ratio of 1:3) can be reacted in the presence of a catalyst to produce methane (Cusumano et al., 1978). However, some adjustment to the ideal (1:3) is usually required and, to accomplish this, all or part of the stream is treated according to the waste gas shift (*shift conversion*) reaction. This involves reacting carbon monoxide with steam to produce carbon dioxide and hydrogen whereby the desired 1:3 mol ratio of carbon monoxide to hydrogen may be obtained:

$$CO + H_2O \rightarrow CO_2 + H_2$$

Several exothermic reactions may occur simultaneously within a methanation unit. A variety of metals have been used as catalysts for the methanation reaction; the most common, and to some extent the most effective methanation catalysts, appear to be nickel and ruthenium, with nickel being the most widely used (Seglin, 1975; Cusumano et al., 1978; Watson, 1980). The synthesis gas must be desulfurized before the methanation step since sulfur compounds will rapidly deactivate (poison) the catalysts (Cusumano et al., 1978). A problem may arise when the concentration of carbon monoxide is excessive in the stream to be methanated since large amounts of heat must be removed from the system to prevent high temperatures and deactivation of the catalyst by sintering as well as the deposition of carbon (Cusumano et al., 1978). To eliminate this problem, temperatures should be maintained below 400°C (750°F).

Not all high-Btu gasification technologies depend entirely on catalytic methanation and, in fact, a number of gasification processes use hydrogasification, that is, the direct addition of hydrogen to coal under pressure to form methane:

$$[C]_{coal} + H_2 \rightarrow CH_4$$

The hydrogen-rich gas for hydrogasification can be manufactured from steam by using the char that leaves the hydrogasifier. Appreciable quantities of methane are formed directly in the primary gasifier and the heat released by methane formation is at a sufficiently high temperature to be used in the steam-carbon reaction to produce hydrogen so that less oxygen is used to produce heat for the steam-carbon reaction. Hence, less heat is lost in the low-temperature methanation step, thereby leading to higher overall process efficiency.

6.4 Liquid hydrocarbon products

An early process for the production of hydrocarbon fuels from coal involved the Bergius process. In the process, lignite or sub-bituminous coal is finely ground and mixed with high-boiling oil recycled from the process. Catalyst is typically added to the mixture and the mixture is pumped into a reactor. The reaction occurs at between 400°C and 500°C and under hydrogen pressure and produces gas, naphtha, middle distillate oil, as well as high-boiling oil:

$$nC_{coal} + (n+1)H_2 \rightarrow C_nH_{(2n+2)}$$

A number of catalysts have been developed over the years, including catalysts containing tungsten, molybdenum, tin, or nickel.

The different fractions can be sent to a refinery for further processing to yield synthetic fuel or a fuel blending stock of the desired quality. It has been reported that as much as 97% of the coal carbon can be converted to synthetic fuel but this very much depends on the coal type, the reactor configuration, and the process parameters. However, with reference to the refining of the liquids from coal liquefaction processes, current concepts for refining these products have relied, for the most part, on already-existing process units in crude oil refineries, although it must be recognized that the acidity of the coal liquids (i.e., phenol content) and the potential incompatibility of these liquids with conventional crude oil and crude oil products (including heavy oil and heavy oil products) may pose new issues within the refinery system (Speight, 2013, 2014a). In terms of the compatibility of liquid products derived from coal and crude oil product, the indirect liquefaction of coal and the production of liquids by the Fischer-Tropsch process represents the most attractive option and does not threaten to bring on incompatibility problems as can occur when phenols are present in the coal liquids.

More recently other processes have been developed for the conversion of coal to liquid fuels. The Fischer-Tropsch process of indirect synthesis of liquid hydrocarbon derivatives is today used by Sasol in South Africa. In the process, coal is to be gasified to make synthesis gas (syngas) (a purified mixture of carbon monoxide and hydrogen) and the syngas condensed using Fischer-Tropsch catalysts to make low-boiling hydrocarbon derivatives which are further processed into gasoline and diesel. Syngas can also be converted to methanol, which can be used as a fuel, fuel additive, or further processed into gasoline via the Mobil M-gas process.

Coal can also be converted into hydrocarbon fuels such as gasoline and/or diesel by several different processes. In the direct liquefaction processes, the coal is either hydrogenated at high temperature in the presence of hydrogen or sent through a carbonization process. Hydrogenation processes are the older

Bergius process (above), the SRC-I and SRC-II (Solvent Refined Coal) processes and the NUS Corporation hydrogenation process (Speight, 2013). In the low-temperature carbonization process, coal is heated at temperatures between 360°C and 750°C (680°F and 1380°F). These temperatures optimize the production of coal tars richer in lower boiling hydrocarbon derivatives than coal tar produced at higher temperatures. The coal tar is then further processed into hydrocarbon fuels. Alternatively, coal can be converted into a gas first, and then into a liquid, by using the Fischer-Tropsch process (Speight, 2013).

In spite of the interest in coal liquefaction processes that emerged during the 1970s and the 1980s, crude oil prices always remained sufficiently low to ensure that the initiation of a synthetic fuels industry based on noncrude oil sources would not become a commercial reality. Up to 1950, *benzene* was obtained almost exclusively from the products of coal carbonization—either scrubbed from the gas as low-boiling oil (light oil) or distilled from the tar stream. By 1940, production had risen from the depression lows to around 150 million gallons per year. During the fifties it reached a peak of almost 200 million gallons per year and has dropped significantly since. In 1950, crude oil benzene was included in the production statistics for the first time at 10 million gallons. Most of the benzene produced has been used as intermediate in the manufacture of chemicals that have only come to significance since the time of World War II. Styrene, cyclohexane, and phenol account for almost three-fourths of the benzene consumption. Since 1950, the specific addition of benzene to gasoline has been negligible in terms of the other uses.

As the demands of the World War I led to the production of *toluene* from byproduct ovens, so the greater demands of World War II led to the first significant production from crude oil. During the whole history of coke oven operation in the United States, the production of toluene from coal did not reach 50 million gallons per year. During the war, the production of toluene from crude oil in only 5 years rose from nothing to over 160 million gallons per year. At the end of the war, it dropped to less than 10 million gallons, and then started a climb that has not yet slowed down.

Much of the toluene produced is used for the hydrodealkylation to benzene; therefore a significant amount of benzene from crude oil is via toluene. Motor gasoline, solvents, and aviation gasoline are other major uses, and it is probably in these markets that most of the toluene from coal is used. *Xylene derivatives* from coal have not been of great importance in the past. During the fifties, production rose to above 10 million gallons per year for 7 years, after which it dropped. The synthesis of *phenol* (not a hydrocarbon but a chemical of interest in this context) was established as a commercial practice many decades ago, and by 1940 the synthetic production already amounted to three or four times the amount recovered from coke-oven operations.

Coke-oven operations have been the primary or exclusive source of *naphthalene* through substantially all of the period under consideration. However, some naphthalene was made from crude oil by hydrodealkylation in

1961, and by 1964 this accounted for over 40% of the total production. It has been estimated that the maximum amount available from coal tar would be approximately 650 million pounds per year. The total 1964 production (including crude oil–derived naphthalene) was 740 million pounds. Obviously, future increases in naphthalene supply will necessarily be of crude oil origin.

Of the *tar bases*, pyridine until the middle fifties was available only from coal tar, as were some of the homologs. The production of synthetic pyridine, the picoline derivatives, and other derivatives has made for a more stable market. and may in the future lead to the development of more widespread uses.

6.5 Solid hydrocarbon products

The most common solid product, coke, is a solid carbonaceous residue derived from low-ash, low-sulfur bituminous coal from which the volatile constituents are driven off by baking in an oven without oxygen at temperatures as high as 1000°C (1830°F) so that the fixed carbon and residual ash are fused together.

Coke is produced from coal by driving off (through the agency of heat) the volatile constituents of the coal using an airless furnace or oven at temperatures as high as 2000°C (3630°F). However, the coke does contain mineral constituents—the carbonization process is a concentration process in which all of the nonvolatile constituents (impurities) collect in the coke.

The volatile matter produced in the carbonization process is, in the current context, the more valuable product since it can be further refined to produce hydrocarbon derivatives. Thus different types of coal are proportionally blended to reach acceptable levels of volatility before the coking process begins. The coke is not a hydrocarbon but a carbonaceous mass that may be used as a fuel or to produce hydrocarbon derivatives through the gasification and treatment of the gases by the Fischer-Tropsch process.

References

Agrawal, P.L., 1959. Proceedings. Symposium on the Nature of Coal. Central Fuel Research Institute, Jealgora, India, p. 121.

ASTM C351, 2019. Test Method for Mean Specific Heat of Thermal Insulation. Annual Book of ASTM Standards. Section 04.06. ASTM International, West Conshohocken, Pennsylvania.

ASTM D121, 2019. Terminology of Coal and Coke. Annual Book of ASTM Standards. Section 05.05. ASTM International, West Conshohocken, Pennsylvania.

ASTM D440, 2019. Method for Drop Shatter Test for Coal. Annual Book of ASTM Standards. Section 05.05. ASTM International, West Conshohocken, Pennsylvania.

ASTM D441, 2019. Method for Tumbler Test for Coal. Annual Book of ASTM Standards. Section 05.05. ASTM International, West Conshohocken, Pennsylvania.

ASTM D720, 2019. Test Method for Free-Swelling Index of Coal. Annual Book of ASTM Standards. Section 05.05. ASTM International, West Conshohocken, Pennsylvania.

ASTM D2015, 2019. Test Method for Gross Calorific Value of Coal and Coke by the Adiabatic Bomb Calorimeter. Annual Book of ASTM Standards. Section 05.05. ASTM International, West Conshohocken, Pennsylvania.
ASTM D2639, 2019. Test Method for Plastic Properties of Coal by the Constant-Torque Gieseler Plastometer. Annual Book of ASTM Standards. Section 05.05. ASTM International, West Conshohocken, Pennsylvania.
ASTM D2798, 2019. Method for Microscopical Determination of the Reflectance of the Organic Components in a Polished Specimen of Coal. Annual Book of ASTM Standards. Section 05.05. ASTM International, West Conshohocken, Pennsylvania.
ASTM D2799, 2019. Method for Microscopical Determination of Volume Percent of Physical Components of Coal. Annual Book of ASTM Standards. Section 05.05. ASTM International, West Conshohocken, Pennsylvania.
ASTM D3038, 2019. Method for Drop Shatter for Coke. Annual Book of ASTM Standards. Section 05.05. ASTM International, West Conshohocken, Pennsylvania.
ASTM D3175, 2019. Test Method for Volatile Matter in the Analysis Sample of Coal and Coke. Annual Book of ASTM Standards. Section 05.05. ASTM International, West Conshohocken, Pennsylvania.
ASTM D3286, 2019. Test Method for Gross Calorific Value of Coal and Coke by the Isoperibol Bomb Calorimeter. Annual Book of ASTM Standards. Section 05.05. ASTM International, West Conshohocken, Pennsylvania.
Baughman, G.L., 1978. Synthetic Fuels Data Handbook. Cameron Engineers, Denver, Colorado.
Carslaw, H.S., Jaeger, J.C., 1959. Conduction of Heat in Solids, second ed. Oxford University Press, Oxford, p. 189.
Cavagnaro, D.M., 1980. Coal Gasification Technology. National Technical Information Service, Springfield, Virginia.
Cusumano, J.A., Dalla Betta, R.A., Levy, R.B., 1978. Catalysis in Coal Conversion. Academic Press Inc., New York.
Davidson, R.M., 1983. Mineral Effects in Coal Conversion. Report No. ICTIS/TR22. International Energy Agency, London.
Fryer, J.F., Speight, J.G., 1976. Coal Gasification: Selected Abstract and Titles. Information Series No. 74. Alberta Research Council, Edmonton, Canada.
Holuszko, M.E., Laskowski, J.S., 2010. Proceedings. International Coal Preparation Conference. International Coal Preparation Congress (ICPC). Lexington, Kentucky. April 25-29.
Howard-Smith, I., Werner, G.J., 1976. Coal Conversion Technology. Noyes Data Corp., Park Ridge, New Jersey, p. 71.
Kasem, A., 1979. Three Clean Fuels from Coal: Technology and Economics. Marcel Dekker Inc., New York.
Lahaye, J., Ehrburger, P. (Eds.), 1991. Fundamental Issues in Control of Carbon Gasification Reactivity. Kluwer Academic Publishers, Dordrecht, Netherlands.
Levine, J.R., 1991a. New methods for assessing gas resources in thin-bedded, high-ash coals. In: Proceedings. 1991 Coalbed Methane Symposium, Tuscaloosa, Alabama. Paper 9125, pp. 115−125.
Levine, J.R., 1991b. The impact of oil formed during coalification on generation and storage of natural gas in coalbed reservoir systems. In: Proceedings. 1991 Coalbed Methane Symposium, Tuscaloosa, Alabama. Paper 9126, pp. 307−315.
Loison, R., Peytavy, A., Boyer, A.F., Grillot, R., 1963. In: Lowry, H.H. (Ed.), A Chemistry of Coal Utilization. Supplementary Volume. John Wiley & Sons Inc., New York (Chapter 4).

Luque, R., Speight, J.G. (Eds.), 2015. Gasification for Synthetic Fuel Production: Fundamentals, Processes, and Applications. Woodhead Publishing, Elsevier, Cambridge, United Kingdom.
Massey, L.G. (Ed.), 1974. Coal Gasification. Advances in Chemistry Series No. 131. American Chemical Society, Washington, D.C.
Massey, L.G., 1979. In: Wen, C.Y., Lee, E.S. (Eds.), Coal Conversion Technology. Addison-Wesley Publishers Inc., Reading, Massachusetts, p. 313.
Mokhatab, S., Poe, W.A., Speight, J.G., 2006. Handbook of Natural Gas Transmission and Processing. Elsevier, Amsterdam, Netherlands.
Moschopedis, S.E., Hawkins, R.W., Fryer, J.F., Speight, J.G., 1980. The use of heavy oils (and derivatives) to process coal. Fuel 59, 647.
Moschopedis, S.E., Hawkins, R.W., Speight, J.G., 1982. Effects of process parameters on the liquefaction of coal using heavy oils and bitumen. Fuel Processing Technology 5, 213.
Probstein, R.F., Hicks, R.E., 1990. Synthetic Fuels. pH Press, Cambridge, Massachusetts (Chapter 4).
Seglin, L. (Ed.), 1975. Methanation of Synthesis Gas. Advances in Chemistry Series No. 146. American Chemical Society, Washington, D.C.
Simpson, D.A., Lea, J.F., Cox, J.C., 2003. Coal bed methane production. Paper SPE 80900. In: Proceedings. Production and Operations Symposium. Society of Petroleum Engineers, Richardson, Texas. March.
Speight, J.G., 2013. The Chemistry and Technology of Coal, third ed. CRC Press, Taylor and Francis Group, Boca Raton, Florida.
Speight, J.G., 2014a. The Chemistry and Technology of Petroleum, fifth ed. CRC Press, Taylor and Francis Group, Boca Raton, Florida.
Speight, J.G., 2014b. Gasification of Unconventional Feedstocks. Gulf Professional Publishing Company, Elsevier, Oxford, United Kingdom.
Speight, J.G., 2019. Natural Gas: A Basic Handbook, second ed. Gulf Publishing Company, Elsevier, Cambridge, Massachusetts.
Watson, G.H., 1980. Methanation Catalysts. Report ICTIS/TR09. International Energy Agency, London.
Whitehurst, D.D., Mitchell, T.O., Farcasiu, M., 1980. Coal Liquefaction: The Chemistry and Technology of Thermal Processes. Academic Press Inc., New York.

Further reading

Berthelot, M., 1869. Bull. Soc. Chim. France 11, 278.

Chapter 6

Hydrocarbons from oil shale

1. Introduction

The term *oil shale* is a misnomer since the kerogen-containing mineral is neither shale or oil and must be heated to approximately at 600°C (1110°F) to yield oil—often referred to as shale oil or pyrolysis oil. Nevertheless, oil shale comprises an enormous and largely untapped hydrocarbon resource. As readily accessible crude oil sources dwindle, utilization of the oil shale resource to meet world needs for hydrocarbon derivatives and hydrocarbon fuel will become both necessary and economically attractive.

Of late, the term *tight oil* has been introduced into the crude oil lexicon. The term tight oil refers to crude oil (primarily light sweet crude oil) and natural gas, respectively, which are contained in formations such as shale or tight sandstone, where the low permeability of the formation makes it difficult for producers to extract the crude oil or natural gas except by unconventional techniques such as horizontal drilling and hydraulic fracturing. The term *unconventional oil* is an umbrella terms for crude oil that is produced by methods that do not meet the criteria for conventional production. Thus, the term tight oil refers to crude oil trapped in organic-rich rocks dominated by shale as well as sandstone or limestone formations that exhibit very low permeability and such formations may also contain condensate. Given the low permeability of these reservoirs, the crude oil must be developed via special drilling and production techniques including fracture stimulation (hydraulic fracturing) in order to be produced commercially (Speight, 2016, 2017).

The term *light tight oil* is also used to describe oil from shale reservoirs and tight reservoirs because the crude oil produced from these formations is light crude oil. The term *light crude oil* refers to low-density petroleum that flows freely at room temperature and these light oils have a higher proportion of light hydrocarbon fractions resulting in higher API gravities (between 37 degrees and 42 degrees) (Speight, 2014). However, the crude oil contained in shale reservoirs and in tight reservoirs will not flow to the wellbore without assistance from advanced drilling (such as horizontal drilling) and fracturing (hydraulic fracturing) techniques. There has been a tendency to refer to this oil as *shale oil*. This terminology is incorrect insofar as it is confusing and the use of such terminology should be discouraged as illogical since shale oil has (for

Handbook of Industrial Hydrocarbon Processes. https://doi.org/10.1016/B978-0-12-809923-0.00006-0

decades, even centuries) been the name given to the distillate produced from oil shale by thermal decomposition (Scouten, 1990; Speight, 2012, 2014, 2016). There has been the recent (and logical) suggestion that shale oil can be referred to as *kerogen oil* (IEA, 2013). However in keeping with the usage of the terms and in the current context of this book, the term shale oil is used to describe oil that is generated from kerogen by thermal processes.

Also, in the context of this book, oil shale is a fine-grained sedimentary rock containing relatively large amounts of organic matter (kerogen) from which significant amounts of shale oil and combustible gas can be extracted by destructive distillation. Included in most definitions of *oil shale*, either stated or implied, is the potential for the profitable extraction of *shale oil* and combustible gas or for burning as a fuel. Oil shale differs from coal whereby the organic matter in coal has a lower atomic hydrogen/carbon atomic ratio and the organic matter/mineral matter ratio of coal is usually greater than 4.75/5.

Oil shale has been used since ancient times and, like coal, it can be used directly as a fuel. The role of oil shale in the production of energy and hydrocarbon derivatives is largely unknown (except for paper estimates) because the contribution to energy and hydrocarbon production is minimal compared to crude oil, natural gas, and coal. However, declining crude oil supplies is adding to speculation whether oil shale represents an important energy and hydrocarbon source for the increasing demands in the decades commencing in the middle of the current century. To date, the potential of the oil shale resources of the world have barely been touched, largely due to economics and environmental issues.

Oil shale is a complex and intimate mixture of organic and inorganic materials that vary widely in composition and properties. In general terms, oil shale is a fine-grained sedimentary rock that is rich inorganic matter and yields oil when heated. Some oil shale is genuine shale but others have been misclassified and are actually siltstones, impure limestone, or even impure coal. Oil shale does not contain oil and only produces oil when it is heated to approximately 500°C (approximately 930°F), when some of the organic material is transformed into a distillate similar to crude oil.

As already stated, there is no precise scientific definition of *oil shale* and the current definition is based on economics. However, Just like the term *oil sand* (tar sand in the United States), the term *oil shale* is a misnomer since the mineral does not contain oil nor is it always *shale*. The organic material is chiefly *kerogen* and the *shale* is usually a relatively hard rock, called marl. Properly processed, kerogen can be converted into a substance somewhat similar to crude oil which is often better than the lowest grade of oil produced from conventional oil reservoirs but of lower quality than conventional low-density light crude oil. *Shale oil*, sometimes termed *retort oil*, is the liquid oil condensed from the effluent in oil shale retorting and typically contains appreciable amounts of water and solids, as well as having an irrepressible tendency to form sediments. However, shale oils are sufficiently different from crude oil that processing shale oil presents some unusual problems.

Generally, oil shale is a mixture of carbonaceous molecules dispersed in an inorganic (mineral) matrix. It is called shale because it is found in a layered structure typical of sedimentary rocks, but the mineral composition can vary from true aluminosilicate shale to carbonate minerals. Thus, *oil shale* is a compact, laminated rock of sedimentary origin that yields over 33% of ash and containing insoluble organic matter that yields oil when distilled. *Kerogen* is the name given to the naturally occurring insoluble organic matter found in shale deposits. *Shale oil* is the synthetic fuel produced by the thermal decomposition of *kerogen* at high temperature (>500°C, >930°F). Shale oil is referred to as *synthetic crude oil* after hydrotreating. Oil shale is sedimentary marlstone rock that is embedded with rich concentrations of organic material known as kerogen. The oil shale deposits in the western United States contain approximately 15% organic material, by weight.

Oil (hydrocarbon) production potential from oil shale is measured by a laboratory pyrolysis method called Fischer Assay (Speight, 2013) and is reported in barrels (42 gal) per ton. Rich zones can yield more than 40 gallons per ton, while most shale falls in the range of 10–25 gallons per ton. Oil shale yields that are higher than 25 gal/ton are generally viewed as the most economically attractive, and hence, the most favorable for initial development.

Retorting is the process of heating oil shale in order to recover the organic material, predominantly as a liquid. To achieve economically attractive recovery of product, temperatures of 400°C −600°C (750°F −1100°F) are required. A retort is simply a vessel in which the oil shale is heated from which the product gases and vapors can escape to a collector. Retorting essentially involves destructive distillation (*pyrolysis*) of oil shale in the absence of oxygen. Pyrolysis (temperatures above 900°F) thermally breaks down (*cracks*) the kerogen to release the hydrocarbon derivatives and then cracks the hydrocarbon derivatives into lower-weight hydrocarbon molecules. Conventional refining uses a similar thermal cracking process, termed *coking*, to break down high-molecular weight residuum.

By heating oil shale to high temperatures, kerogen can be converted to a liquid that, once upgraded, can be refined into a variety of hydrocarbon fuels, gases, and high value chemical and mineral byproducts. The United States has vast known oil shale resources that could translate into as much as 2.6 trillion barrels (2.6×10^{12} bbls) of *oil-in-place*. Oil shale deposits concentrated in the Green River Formation in the states of Colorado, Wyoming, and Utah account for nearly three-quarters of this potential.

2. History

The use of oil shale can be traced back to ancient times. By the 17th century, oil shales were being exploited in several countries. One of the interesting oil shales is the Swedish alum shale of Cambrian and Ordovician age that is noted for its alum content and high concentrations of metals including uranium and vanadium.

As early as 1637, the alum shales were roasted over wood fires to extract potassium aluminum sulfate, a salt used in tanning leather and for fixing colors in fabrics. Late in the 1800s, the alum shales were retorted on a small scale for hydrocarbon derivatives. Production continued through World War II but ceased in 1966 because of the availability of cheaper supplies of crude oil.

An oil shale deposit at Autun, France, was exploited commercially as early as 1839. The Scottish oil shale industry began before 1859—the year that Colonel Drake drilled his pioneer well at Titusville, Pennsylvania. As many as 20 beds of oil shale were mined at different times. Mining continued during the 1800s and by 1881 oil shale production had reached one million metric tons per year. With the exception of the World War II years, between one and four million metric tons of oil shale were mined yearly in Scotland from 1881 to 1955 when production began to decline, then ceased in 1962. Canada produced some shale oil from deposits in New Brunswick and Ontario in the mid-1800s.

Common products made from oil shale from these early operations were hydrocarbon fractions, such as kerosene and lamp oil, paraffin, fuel oil, and lubricating oil. Hydrocarbon oil distilled from shale was first burnt for horticultural purposes in the 19th century, but it was not until the 1900s that larger investigations were made and the Office of Naval Petroleum and Oil Shale Reserves was established in 1912. The reserves were seen as a possible emergency source of fuel for the military, particularly the United States Navy, which had, at the beginning of the 20th century, converted its ships from coal to fuel oil, and the economy was transformed by gasoline-fueled automobiles and diesel fueled trucks and trains; concerns have been raised related to the assurance of an adequate supply of liquid fuels at affordable prices to meet the growing needs of the nation and its consumers.

The abundance of oil shale resources in the United States was initially eyed as a major source for hydrocarbon derivatives and hydrocarbon fuels. Numerous commercial entities sought to develop oil shale resources. The Mineral Leasing Act of 1920 made crude oil and oil shale resources on Federal lands available for development under the terms of federal mineral leases. Soon, however, discoveries of more economically producible and refinable liquid crude oil in commercial quantities caused interest in oil shale to decline.

Interest resumed after World War II, when military fuel demand and domestic fuel rationing and rising fuel prices made the economic and strategic importance of the oil shale resource more apparent. After the war, the booming postwar economy drove demand for fuels ever higher. Public and private research and development efforts were commenced, including the 1946 United States Bureau of Mines Anvil Point, Colorado oil shale demonstration project. Significant investments were made to define and develop the resource and to develop commercially viable technologies and processes to mine, produce, retort, and upgrade oil shale into viable refinery feedstocks and byproducts.

Once again, however, major crude oil discoveries in the lower-48 United States, offshore, and in Alaska, as well as other parts of the world reduced the foreseeable need for shale oil and interest and associated activities again diminished. Lower-48 United States crude oil reserves peaked in 1959 and lower-48 production peaked in 1970.

By 1970, oil discoveries were slowing, demand was rising, and crude oil imports, largely from Middle Eastern states, were rising to meet demand. Global oil prices, while still relatively low, were also rising reflecting the changing market conditions. Ongoing oil shale research and testing projects were reenergized and new projects were envisioned by numerous energy companies seeking alternative fuel feedstocks. These efforts were significantly amplified by the impacts of the 1973 Arab Oil Embargo which demonstrated the vulnerability of the United States to disruptions in oil supply, and were underscored by a new supply disruption associated with the 1979 Iranian Revolution.

By 1982, however, technology advances and new discoveries of offshore oil resources in the North Sea and elsewhere provided new and diverse sources for oil imports into the United States, and dampened global energy prices. Global political shifts promised to open previously restricted provinces to oil and gas exploration, and led economists and other experts to predict a long future of relatively low and stable oil prices. Despite significant investments by energy companies, numerous variations and advances in mining, restoration, retorting, and in-situ processes, the costs of oil shale production relative to foreseeable oil prices, made continuation of most commercial efforts impractical. During this time, numerous projects that were initiated and then terminated, primarily due to economic infeasibility relative to expected world oil prices or project design issues.

Several projects failed for technical and design reasons. Federal research and development, leasing, and other activities were significantly curtailed, and most commercial projects were abandoned. The collapse of world oil prices in 1984 seemed to seal the fate of oil shale as a serious player in the energy strategy of the United States, as well as in many other oil-importing countries.

Despite the huge resources, oil shale is an underutilized energy resource. In fact, one of the issues that arise when dealing with fuels from oil shale is the start-stop-start episodic nature of the various projects. The projects have varied in time and economic investment and viability. Nevertheless, oil shale has, though, a definite potential for meeting energy demand in an environmentally acceptable manner (Bartis et al., 2005; Andrews, 2006).

3. Origin

Oil shale represents a large and mostly untapped hydrocarbon resource. Like tar sand (*oil sand* in Canada) and coal, oil shale is considered unconventional because oil cannot be produced directly from the resource by sinking a well into the production zone well and pumping out the oil. Oil has to be produced thermally from the shale. The organic material contained in the shale is called *kerogen*, a solid material intimately bound within the mineral matrix (Baughman, 1978; Allred, 1982; Scouten, 1990; Speight, 2013, 2014).

Oil shale is distributed widely throughout the world with known deposits in every continent. Oil shale ranging from Cambrian to Tertiary in age occurs in many parts of the world (Scouten, 1990; Lee, 1991). Deposits range from small occurrences of little or no economic value to those of enormous size that occupy thousands of square miles and contain many billions of barrels of potentially extractable shale oil. However, crude oil is cheaper to produce than shale oil because of the additional costs of mining and extracting the energy from oil shale. Because of these higher costs, only a few deposits of oil shale are currently being exploited in China, Brazil, and Estonia. However, with the continuing decline of crude oil supplies, accompanied by increasing costs of crude oil—based products, oil shale presents opportunities for supplying some of the fossil energy needs of the world in the future (Bartis et al., 2005; Andrews, 2006).

Oil shale is not generally regarded as true shale by geologists nor does it contain appreciable quantities of free oil (Scouten, 1990). The fracture resistance of all oil shales varies with the organic content of the individual lamina and fractures preferentially initiate and propagate along the leaner horizontal laminas of the depositional bed. Oil shale was deposited in a wide variety of environments including freshwater to saline ponds and lakes, epicontinental marine basins, and related subtidal shelves as well as shallow ponds or lakes associated with coal-forming peat in limnic and coastal swamp depositional environments. This give rise to a variety of different oil shale types (Table 6.1) (Hutton, 1987, 1991) and it is not surprising, therefore, that oil shales exhibit a wide range in organic and mineral composition (Scouten, 1990; Mason, 2006; Ots, 2007; Wang et al., 2009). Most oil shales contain organic matter derived from varied types of marine and lacustrine algae, with some debris of land plants, depending upon the depositional environment and sediment sources.

During the creation of oil shale, source rocks are buried by natural geological processes and, over geologic time, converts the organic materials to solid (kerogen), liquids, and gases. The latter two products can migrate through cracks and pores in the rocks until it reaches the surface or is trapped by a tight overhead formation. The result is an oil and/or gas reservoir. The material that cannot migrate (kerogen) remains in the rock and gives rise to oil shale.

Oil shale precursors were deposited in a wide variety of environments including freshwater to saline ponds and lakes, epicontinental marine basins, and related subtidal shelves. They were also deposited in shallow ponds or lakes associated with coal-forming peat in limnic and coastal swamp depositional environments. It is not surprising, therefore, that oil shales exhibit a wide range in organic and mineral composition. Most oil shales contain

TABLE 6.1 General classification of oil shale.

Sedimentary rocks	Classification
Nonorganic	
Organic rich	
	Humic coal
	Bitumen-containing
	Tar sand (oil sand)
	Oil shale
	Terrestrial
	Cannel coal
	Lacustrine
	Lamosite
	Torbanite
	Marine
	Kukersite
	Marinite
	Tasmanite

organic matter derived from varied types of marine and lacustrine algae, with some debris of land plants, depending upon the depositional environment and sediment sources.

Organic matter in oil shale is a complex moisture and is derived from carbon-containing remains of algae, spores, pollen, plant cuticle and corky fragments of herbaceous and woody plants, plant resins, pant waxes, and other cellular remains of lacustrine, marine, and land plants (Scouten, 1990; Dyni, 2003, 2006). These materials are composed chiefly of carbon, hydrogen, oxygen, nitrogen, and sulfur. Generally, the organic matter is unstructured and is best described as amorphous (*bituminite*)— the origin of which has not been conclusively identified but is theorized to be a mixture of degraded algal or bacterial remains. Other carbon-containing materials such as phosphate and carbonate minerals may also be present which, although of organic origin, are excluded from the definition of organic matter in oil shale and are considered to be part of the mineral matrix of the oil shale.

Oil shale has often been called *high-mineral coal* but nothing can be further from reality. Maturation pathways for coal and kerogen are different and, in fact, the precursors of the organic matter in oil shale and coal also differ (Tissot and Welte, 1978; Durand, 1980; Scouten, 1990; Hunt, 1996; Speight, 2013). Furthermore, the origin of some of the organic matter in oil shale is obscure because of the lack of recognizable biologic structures that would help identify the precursor organisms, unlike the recognizable biological structures in coal (Speight, 2013). Such materials may be of (1) bacterial origin, (2) the product of bacterial degradation of algae, (3) other organic matter, or (4) all of the above.

Furthermore, oil shale does not undergo the maturation process that is often conveniently represented for crude oil and/or coal but produces the material that has come to be known as *kerogen* (Scouten, 1990). In fact, there are indications that kerogen may be a byproduct of the maturation process. The kerogen residue that remains in oil shale is formed during maturation and is then rejected from the organic matrix because of its insolubility and relative low reactivity under the maturation conditions (Speight, 2014). Furthermore, the fact that kerogen, under the high-temperature pyrolysis conditions imposed upon it in the laboratory, forms hydrocarbon distillates (albeit with relatively high amounts of nitrogen) does not guarantee that the kerogen of oil shale is a precursor to crude oil.

The thermal maturity of oil shale refers to the degree to which the organic matter has been altered by geothermal heating. If oil shale is heated to a high enough temperature—the actual historical temperature to which the shale has been heated is not known with any degree of accuracy and is typically speculative—as may be the case if the oil shale were deeply buried, the organic matter *may* thermally decompose to form liquids and gas. Under such circumstances, there is much unfounded speculation (other than high-temperature laboratory experiments) that oil shale sediments can be the source rocks for crude oil and natural gas.

Moreover, the fact that the high-temperature thermal decomposition of kerogen (in the laboratory) gives crude oil-like material is no guarantee that kerogen is or ever was a precursor to crude oil. The implied role of kerogen in crude oil formation is essentially that implied but without conclusive experimental foundation. However, caution is advised in choosing the correct definition of kerogen since there is the distinct possibility that kerogen, far from being a precursor to crude oil, is one of the byproducts of the crude oil generation and maturation processes and may not be a direct precursor to crude oil.

Crude oil precursors and crude oil itself is indeed subject to elevated temperatures (the *geothermal gradient*) in the subterranean formations due to the *geothermal gradient*. Although the geothermal gradient varies from place to place, it is generally on the order of 15°F/1000 ft or 120°C/1000 ft, i.e., 0.015°C per foot of depth or 0.012°C per foot of depth (25–30°C/km). This leaves serious questions related to whether or not the material has been subjected to temperatures in excess of 250°C (>480°F) (Speight, 2014).

Such experimental work is interesting insofar as it shows similar molecular moieties in kerogen and crude oil (thereby confirming similar origins for kerogen and crude oil). However, the absence of geologic time in the laboratory is not a reason to increase the temperature and it must be remembered that application of high temperatures (>250°C, <480°F) to a reaction not only increases the rate of reaction (thereby making up for the lack of geologic time) but can also change the *nature* and the *chemistry* of a reaction. In such a case, the geochemistry is altered. Furthermore, introduction of a pseudoactivation energy in which the activation energy of the kerogen conversion reactions are reduced leave much to be desired because of the assumption required to develop this pseudoactivation energy equation(s). In fact, not only will the oil window (the oil-producing phase) vary from kerogen-type to kerogen-type, it is not valid to use a fixed set of kinetic parameters within each of these groups.

It is claimed that the degree of thermal maturity of an oil shale can be determined in the laboratory by any one of several methods. One method is to observe the changes in color of the organic matter in samples collected from varied depths—assuming that the organic matter is subjected to geothermal heating (the temperature being a function of depth), the color of the organic matter might be expected to change from a lighter color (at relatively shallow depths) to a darker color (at relatively shallow depths). Then another unknown—the issue of shifting of the sedimentary strata—comes into play.

Suffice it to state that the role played by kerogen in the crude oil maturation process is not fully understood (Tissot and Welte, 1978; Durand, 1980; Hunt, 1996; Scouten, 1990; Speight, 2014). What obviously needs to be addressed more fully in terms of kerogen participation in crude oil generation is the potential to produce crude oil constituents from kerogen by low-temperature processes rather than by processes that involve the use of temperatures in excess of 250°C (>480°F) (Burnham and McConaghy, 2006; Speight, 2014).

If such geochemical studies are to be pursued, a thorough investigation is needed to determine the potential for such high temperatures being present during the main phase, or even various phases, of crude oil generation on order to give stronger indications that kerogen is a precursor to crude oil (Speight, 2014). Finally, much of the work performed on oil shale has referenced the old shale from the Green River formation in the western United States. Thus, unless otherwise stated, the shale referenced in the following text is the Green River shale.

Oil shale does not undergo that natural maturation process but produces the material that has come to be known as kerogen (Scouten, 1990). In fact, there are indications that kerogen, being different to crude oil, may be a byproduct of the maturation process. The kerogen residue that remains in oil shale is formed during maturation and is then rejected from the organic matrix because of its insolubility and relative unreactivity under the maturation conditions (Speight, 2014). Furthermore, the fact that kerogen, under the conditions imposed upon it in the laboratory by high-temperature pyrolysis, forms hydrocarbon products does not guarantee that the kerogen of oil shale is a precursor to crude oil.

Oil shale ranging from Cambrian to Tertiary in age occurs in many parts of the world. Deposits range from small occurrences of little or no economic value to those of enormous size that occupy thousands of square miles and contain many billions of barrels of potentially extractable shale oil. Total world resources of oil shale are conservatively estimated at 2.6 trillion barrels.

4. Occurrence

Oil shale ranges from Cambrian to Tertiary in age occur in many parts of the world. Deposits range from small occurrences of little or no economic value to those of enormous size that occupy thousands of square miles and contain many billions of barrels of potentially extractable shale oil. Total world resources of oil shale are conservatively estimated at 2.6 trillion barrels (2.6×10^{12} barrels) but can vary by one or more orders of magnitude above and below this figure depending upon the method of estimation and whether or not the deposits have been fully investigated. However, with the continuing decline of crude oil supplies, accompanied by increasing costs of crude oil–based products, oil shale presents opportunities for supplying some of the fossil energy needs of the world in the years ahead.

Oil shale is sedimentary marlstone rock that is embedded with rich concentrations of organic material known as kerogen. The western oil shale of the United States contains approximately 15% organic material, by weight. By heating oil shale to high temperatures, kerogen can be released and converted to a liquid that, once upgraded, can be refined into a variety of liquid fuels, gases, and high value chemical and mineral byproducts.

Oil shale represents a large and mostly untapped source of hydrocarbon fuels. Like oil sands, it is an unconventional or alternate fuel source and it does not contain oil. Oil is produced by thermal decomposition of the kerogen, which is intimately bound within the shale matrix and is not readily extractable.

Oil shale represents a large and mostly untapped hydrocarbon resource. Like tar sand (*oil sand* in Canada), oil shale is considered unconventional because oil cannot be produced directly from the resource by sinking a well and pumping. Oil has to be produced thermally from the shale. The organic material contained in the shale is called *kerogen*, a solid material intimately bound within the mineral matrix.

Oil shale occurs in nearly 100 major deposits in 27 countries worldwide. It is generally shallower (<3000 feet) than the deeper and warmer geologic zones required to form oil. Worldwide, the oil shale resource base is believed to contain approximately 2.6 trillion barrels (2.6×10^{12} barrels), of which the vast majority, or approximately two trillion barrels (including eastern and western shale), is located within the United States.

With an estimated resource of 2.1 billion barrels of shale oil, the United States has larger resources than any other country. The most economically attractive deposits, containing an estimated 1.5 trillion barrels (richness of >10 gal/ton) are found in the Green River Formation of Colorado (Piceance Creek Basin), Utah (Uinta Basin), and Wyoming (Green River and Washakie Basins). Eastern oil shale underlies 850,000 acres of land in Kentucky, Ohio, and Indiana. About 16 billion barrels, at a minimum grade of 25 gallons/ton, are located in the Kentucky Knobs region in the Sunbury shale and the New Albany/Ohio shale. Due to differences in kerogen type (compared to western shale) eastern oil shale requires different processing. Potential oil yields from eastern shale could someday approach yields from western shale, with processing technology advances (Johnson et al., 2004). However, in spite of all of the numbers and projections, it is difficult to gather production data (given either in shale oil or oil shale in weight or in volume) and few graphs have been issued. There are large discrepancies between percentages in reserve and in production because of the assumptions of *estimates* of the total resource and recoverable reserves. Thus, use of the data requires serious review.

When considering oil shale quality for liquid transportation feedstocks, it is most useful to assess the yield of oil that results from a shale sample in a laboratory retort. This is the most common type of analysis currently used to evaluate an oil shale resource. The method commonly used in the United States is called the "modified Fischer assay," first developed in Germany, then adapted by the U.S. Bureau of Mines for analyzing oil shale of the Green River Formation in the western United States. The method was subsequently standardized as the American Society for Testing and Materials Method D3904. Some laboratories have further modified the Fischer assay method to better evaluate different types of oil shale and different methods of oil shale processing.

5. Oil shale types

Mixed with a variety of sediments over a lengthy geological time period, shale forms a tough, dense rock ranging in color from light tan to black. Based on its apparent colors, shale may be referred to as *black shale* or *brown shale*. Oil shale has also been given various names in different regions. For example, the Ute Indians, on observing outcroppings burst into flames after being hit by lightning, referred to it as *the rock that burns*. Thus it is not surprising that definitions of the types of oil shale can be varied and confusing. It is necessary to qualify the source of the definition and the type of shale that fits within that particular definition.

For example, one definition is based on the mineral content of which three categories can recognized namely: (1) carbonate-rich oil shales, which contain a high proportion of carbonate minerals (such as calcite and dolomite) and which usually have the organic-rich layers sandwiched between carbonate-rich layers—these shales are hard formations that are resistant to weathering and are difficult to process using mining (ex-situ), (2) siliceous oil shales which are usually dark brown or black shales and are deficient in carbonate minerals but plentiful in siliceous minerals (such as quartz, feldspar, clay, chert, and opal)—these shales are not as hard and weather-resistant as the carbonate shales and may be better suited for extraction via mining (ex-situ) methods, and (3) cannel oil shales, which are typically usually dark brown or black shales and consist of organic matter that completely encloses other mineral grains—these shales are suitable for extraction via mining (ex-situ).

However, mineral content aside, it is more common to define oil shale on the basis of origin and formation of the shale as well as the character of the organic content of the shale. More specifically, the nomenclature is related to whether or not the shale is of (1) *terrestrial* origin, (2) *marine* origin, or (3) *lacustrine* origin (Hutton, 1987, 1991). This classification reflects differences in the composition of the organic matter and of the distillable products that can be produced from the shale. This classification also reflects the relationship between the organic matter found in sediment and the environment in which the organic precursors were deposited.

5.1 Terrestrial oil shale

The precursors to terrestrial oil shale (sometimes referred to as *cannel coal*) were deposited in stagnant, oxygen-depleted waters on land (such as coal-forming swamps and bogs). Cannel coal is brown to black oil shale composed of resins, spores, waxes, and cutinaceous and corky materials derived from terrestrial vascular plants together with varied amounts of vitrinite and inertinite. Cannel coals originate in oxygen-deficient ponds or shallow lakes in peat-forming swamps and bogs. This type of shale is usually rich in oil-generating lipid-rich organic matter derived from plant resins, pollen, spores, plant waxes and the corky tissues of vascular plants. The individual deposits usually are small in size, but they can be of a very high grade.

The latter also holds for lacustrine oil shales. This group of oil shales was deposited in freshwater, brackish, or saline lakes. The size of the organic-rich deposits can be small, or they can occur over tens of thousands of square miles as is the case for the Green River Formation in Colorado, Utah, and Wyoming. The main oil-generating organic compounds found in these deposits are derived from algae and/or bacteria. In addition, variable amounts of higher plant remains can be present as well.

5.2 Lacustrine oil shale

Lacustrine oil shales (lake-bottom-deposited shales) include lipid-rich organic matter derived from algae that lived in freshwater, brackish, or saline lakes. The lacustrine oil shales of the Green River formation which were discussed above are among the most extensively studied of sediments. However, their strongly basic depositional environment is certainly unusual, if not unique.

Therefore, it is useful to discuss the characteristics of the organic material in other lacustrine shales.

Lacustrine sequences from the Permian oil shales of Autun (France) and the Devonian bituminous flagstones of Caithness (Scotland) exhibited several series of biomarkers that were prominent in extracts from these shales: hopanes, steranes, and carotenoids. Algal remains were abundant in both shales. Blue-green algae, similar to those that contributed largely to the Green River oil shale kerogen, were found in the Devonian shale, for which a stratified lake environment similar to Green River has been proposed (Donovan and Scott, 1980).

In contrast, *Botryococcus* remains were found in the Permian Autun shale and are presumed to be the major source of organic matter, except for one sample. No *Botryococcus* remains were found in this sample and the oil produced by its retorting was nearly devoid of the straight-chain alkanes and 1-alkenes which are prominent in oils from *Botryococcus-derived* shales. Evidently, some as yet unidentified algae contributed to the organic matter in this stratum. Biodegradation cannot be ruled *out* but seems unlikely due to the lack of prominent iso- and ante-iso-alkanes. Straight-chain alkanes and 1-alkenes were also prominent in gas chromatograms of the retorted oils from the Devonian shale. However, in this case a pronounced hump, which usually indicates polycyclic derivatives, was also prominent. Both extracts and oil from the Devonian shale were found to be rich in steranes and tricyclic compounds. Diterpenoid derivatives and triterpenoid derivatives have been suggested as precursors for the di- and tricyclic compounds found in many oil shales. Rock-Eval pyrolysis results indicate that these shales have high hydrogen indices; the kerogen are all type I or type II, with one of the Devonian samples being clearly type I.

Lamosite is pale, grayish-brown and dark gray to black oil shale in which the chief organic constituent is lamalginite derived from lacustrine planktonic algae. Other minor components include vitrinite, inertinite, telalginite, and bitumen. The Green River oil-shale deposits in western United States and a number of the Tertiary lacustrine deposits in eastern Queensland, Australia, are Lamosite shale deposits. Other major lacustrine oil shale deposits include the Triassic shales of the Stanleyville Basin in Zaire and the Albert shales of New Brunswick, Canada (Mississippian).

Torbanite, named after Torbane Hill in Scotland, is a black oil shale whose organic matter is composed mainly of telalginite found in fresh-to brackish-water lakes. The deposits are commonly small, but can be extremely high grade.

5.3 Marine oil shale

Marine oil shales (marine-bottom-deposited shales) are composed of lipid-rich organic matter derived from marine algae, acritarchs (unicellular organisms of questionable origin), and marine dinoflagellates (Debyser and Deroo, 1969). Information related to the nature of organic matter in marine environments has resulted from studies of recent deposits and the contemporary oceans (Bader et al., 1960; Bordovskiy, 1965). Only a small part of primary production in the oceans reaches the bottom. Of an estimated annual production of 9×10^{19} tons of dry matter, it has also been estimated that approximately 2% reaches the floor in shallows and only approximately 0.02% in the open sea. The major part of marine primary production is consumed by predators; most of the rest by microbes. The principal marine microbial scavengers are bacteria that live free in the water or are attached to organic particles. In ocean water, organics occur in solution, in colloidal suspension, and as particulate matter comprising bodies and body fragments of living and dead organisms. Except in regions of a seaweed or plankton "bloom," the dissolved organics usually predominate. As a result, marine bacteria are most abundant only in the very upper part of the water column and in the organic detritus at the very bottom. Even in the oceans, the adsorption of organics onto inorganic detritus, such as the silica parts of diatoms, plays an important part in sedimentation.

After the organic sediment reaches the bottom, reworking begins. Bottom-dwelling (benthic) organisms feed on both the sediment and the dissolved organics and, in turn, are fed on by predators (e.g., crustaceans). In this sphere, the benthic bacteria are largely responsible for the decomposition of organics and the synthesis of new organics through enzymatic transformations. Approximately 60%–70% of the sedimentary organic carbon is typically liberated as carbon dioxide during this reworking, while most of the rest is converted into new compounds, resulting in an extremely complex mixture.

The various organic compound classes in oil shales include carbohydrates, lignins, humates and humic acids, lipid-derived waxes and the saturated and polyene acids in algal lipids which can serve as precursors of these waxes, and biological pigments and their derivatives (e.g., carotenoids, porphyrins). Only the latter three were judged to have sufficient inertness to be major contributors to oil shale kerogen (Cane, 1976).

The black marine shales formed in shallow seas have been extensively studied, as they occur in many places. These shales were deposited on broad, nearly flat sea bottoms, and therefore usually occur in thin deposits (10–50 m thick), but may extend over thousands of square miles. The Irati shale (Permian) in Brazil extends over more than 1000 miles from north to south. The Jurassic marine shales of Western Europe, Silurian shales of North Africa, and the Cambrian shales of northern Siberia and northern Europe are other examples of this kind of marine oil shale (Tissot and Welte, 1978).

Marinite is a gray to dark gray to black oil shale of marine origin in which the chief organic components are lamalginite and bituminite derived chiefly from marine phytoplankton. Marinite may also contain small amounts of bitumen, telalginite, and vitrinite. Marinites are deposited typically in an epeiric sea (a sea extending inland from a continental margin) such as on broad shallow marine shelves or inland seas where wave action is restricted and currents are minimal. The Devonian–Mississippian oil shales of eastern

United States are typical marinites. Such deposits are generally widespread covering hundreds to thousands of square kilometers, but they are relatively thin, often less than 300 feet.

Tasmanite, named from oil-shale deposits in Tasmania, is a brown to black oil shale. The organic matter consists of telalginite derived chiefly from unicellular algae of marine origin and lesser amounts of vitrinite, lamalginite, and inertinite.

Kukersite, which takes its name from Kukruse Manor near the town of Kohtla-Järve, Estonia, is a light brown marine oil shale. Its principal organic component is telalginite derived from green algae. Kukersite is the main type of oil shale in Estonia and western Russia. The organic matter of *kukersite* is considered to be entirely of marine origin, and consists almost entirely of accumulations of discrete bodies, telalginite derived from a colonial microorganism termed *Gloeocapsomorpha prisca.* As compared with other rocks containing telalginite, kukersite has low atomic H/C (1.48) and high atomic O/C (0.14) ratios and generally plots as Type II kerogen on the van Krevelen diagram (Cook and Sherwood, 1991).

6. Composition and properties

It is a fact the term *oil shale* describes an organic-rich rock from which little carbonaceous material can be removed by extraction (with common crude oil-based solvents) but which produces variable quantities of distillate (*shale oil*) when raised to temperatures in excess of 350°C (660°F). Thus, oil shale is assessed by the ability of the mineral to produce shale oil in terms of gallons per ton (g/t) by means of a test method (Fischer assay) in which the oil shale is heated to 500°C (930°F).

6.1 General properties

Oil shale is typically a fine-grained sedimentary rock containing relatively large amounts of organic matter (*kerogen*) from which significant amounts of shale oil and combustible gas can be extracted by thermal deposition with ensuing distillation from the reaction zone. However, oil shale does not contain any oil—this must be produced by a process in which the kerogen is thermally decomposed (cracked) to produce the liquid product (shale oil). Thus any estimates of shale oil reserves can only be based on speculative estimates from application of the Fischer assay test method to (often) nonrepresentative samples taken from an oil shale deposit and the assay data (in terms of oil yield in gallons per ton) must not to be taken as *proven reserves*.

Kerogen that has not thermally matured beyond the diagenesis (low-temperature) stage is typically due to the relatively shallow depth of burial. The Green River oil shale of Colorado has matured to the stage that heterocyclic constituents have formed and predominate, with up to 10% normal paraffins and isoparaffins that boil in the range that includes natural naphtha and gasoline constituents. The relatively high hydrogen/carbon ratio (1.6) is a significant factor in terms of yielding high quality fuels. However, the relatively high nitrogen content (1%–3% w/w) is a major issue in terms of producing stable fuels (crude oil typically contains less than 0.5% nitrogen), as well as producing environmentally detrimental nitrogen oxides during combustion.

In the United States there are two principal oil shale types, the shale from the Green River Formation in Colorado, Utah, and Wyoming, and the Devonian-Mississippian black shale of the East and Midwest (Table 6.2) (Baughman, 1978). The Green River shale is considerably richer, occurs in thicker seams, and has received the most attention for synthetic fuel. The mineral matter (shale) consists of fine-grained silicate and carbonate minerals. The ratio of kerogen-to-shale for commercial grades of oil shale is typically in the range 0.75:5 to 1.5:5—as a comparison, for coal the organic matter-to-mineral matter ratio is usually greater than 4.75:5 (Speight, 2013).

The common property of these two types of oil shale is the presence of the ill-defined kerogen. The chemical composition of the kerogen has been the subject of many studies (Scouten, 1990) but whether or not the data are indicative of the true nature of the kerogen is extremely speculative. Based on solubility/insolubility in various solvents (Koel et al., 2001) it is, however, a reasonable premise (remembering that regional and local variations in the flora that were the precursors to kerogen) led to differences in kerogen composition and properties from different shale samples—similar to the variance in quality, composition, and properties of crude oil from different reservoirs (Speight, 2014).

The organic matter that is derived from the varied types of marine and lacustrine algae, with some debris of land plants, is largely dependent on the

TABLE 6.2 Composition (% w/w) of the organic matter in the Mahogany Zone and New Albany Shale (Baughman, 1978).

Component %w/w	Green river mahogany zone	New albany
Carbon	80.5	82.0
Hydrogen	10.3	7.4
Nitrogen	2.4	2.3
Sulfur	1.0	2.0
Oxygen	5.8	6.3
Total	100.0	100.0
H/C atomic ratio	1.54	1.08

depositional environment and sediment sources. Bacterial processes were probably important during the deposition and early diagenesis of most oil shale deposits—these processes could produce significant quantities of biogenic methane, carbon dioxide, hydrogen sulfide, and ammonia. These gases in turn could react with dissolved ions in the sediment waters to form authigenic minerals (minerals generated where they were found or observed) such as calcite ($CaCO_3$), dolomite ($CaCO_3.MgCO_3$), pyrite (FeS_2), and even such rare authigenic minerals as buddingtonite (ammonium feldspar—$NH_4.Al.Si_3O_8.0.5H_2O$).

The organic matter in oil shale is composed chiefly of carbon, hydrogen, and oxygen with lesser amounts of sulfur and nitrogen. Because of the high molecular weight (best estimates are on the order of several thousand) and molecular complexity, oil shale kerogen is almost totally insoluble in crude oil–based solvents and conventional organic solvents (such as carbon disulfide) (Tissot and Welte, 1978; Durand, 1980; Scouten, 1990; Hunt, 1996; Speight, 2014). A portion of the organic matter in oil shale is soluble and is (incorrectly and confusingly) termed *bitumen*. The bitumen, which is soluble, is dispersed throughout the kerogen network, although even in finely crushed shale much of the bitumen may be inaccessible to the solvent. As a result, only a small fraction of the hydrocarbonaceous material in oil shale can be removed by conventional solvent-extraction techniques.

Briefly, the term *bitumen* is more correct when applied to the organic content of tar sand (oil sand) deposits, although the name also applied in Europe and other areas to road asphalt (Speight, 2014). Using this name in reference to the soluble portion of the organic constituents of oil shale is more for convenience than scientific correctness. Small amounts of oil shale bitumen that are soluble in organic solvents are present in some oil shale. Because of its insolubility, the organic matter must be retorted at temperatures on the order of 500°C (930°F) to decompose it into shale oil and gas. After thermal decomposition of the organic matter, some carbon (in the form of a carbonaceous deposit) remains with the shale residue after retorting but can be burned to obtain additional energy.

The organic matter of *kukersite* is considered to be entirely of marine origin, and consists almost entirely of accumulations of discrete bodies, telalginite derived from a colonial microorganism termed *Gloeocapsomorpha prisca*. As compared with other rocks containing telalginite, kukersite has a low atomic hydrogen-to-carbon ratio ($H/C = (1.48)$ and high atomic oxygen-to-carbon ratio ($O/C = 0.14$) and generally falls into the Type II kerogen on the van Krevelen diagram (Fig. 6.1) (Cook and Sherwood, 1991; Speight, 2014).

Major components of this kerogen are phenolic moieties with linear alkyl sidechains. In spite of the predominance of phenolic moieties kukersite appears as a highly aliphatic type II/I kerogen due to the presence of associated long, linear alkyl chains (Derenne et al., 1989). The formation of kukersite

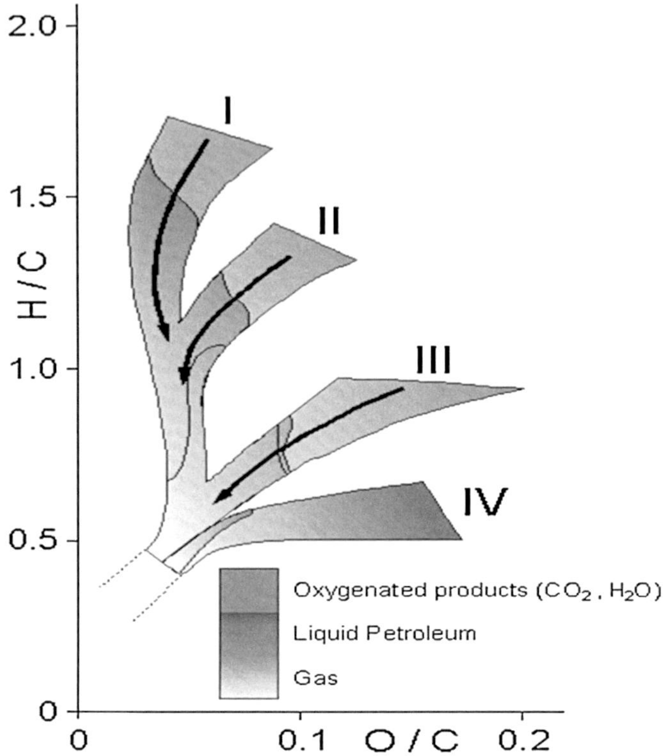

FIGURE 6.1 The Van Krevelen diagram showing the different types of kerogen.

kerogen is believed to have occurred via the selective preservation pathway and the phenolic moieties correspond to important basic structures of the resistant macromolecular material (Derenne et al., 1994). Different extraction methods give bitumen yield from kukersite on the level of 1%–3% w/w.

The yield of oil and gas under slow retorting conditions is not the same as under Fischer assay. Gas compositions reported for slow, modest pressure retorting indicate that energy content of the gas could be as much as 70% greater than for Fischer assay (Burnham and Singleton, 1983). This increase has at least three sources of uncertainty: (1) possible leaks in their gas collection system at the slowest heating rate at elevated pressure, (2) difficulty in recovering low-boiling hydrocarbon derivatives dissolved in the oil at elevated pressure, and (3) the likelihood that oil cracking at higher geological pressures in the liquid phase is less than in the self-purging reactor, which requires volatilization for expulsion. Nevertheless, it is likely that the gas yields will be higher for methane due to oil coking reactions, which was the main reason for the 70% increase, so it is likely that slow retorting would generate gases with good heat content (Burnham, 2003).

Finally, the gross heating value of oil shales on a dry-weight basis ranges from approximately 500 to 4000 kilocalories per kilogram (kcal/kg) of rock. The high-grade kukersite oil shale of Estonia, which fuels several electric power plants, has a heating value of approximately 2000 to 2200 kcal/kg. By comparison, the heating value of lignite ranges from 3500 to 4600 kcal/kg on a dry-mineral-free basis (Speight, 2013).

6.2 Oil shale grade

The grade of oil shale has been determined by many different methods with the results expressed in a variety of units (Scouten, 1990; Dyni, 2003, 2006). For example, the heating value is useful for determining the quality of an oil shale that is burned directly in a power plant to produce electricity. Although the heating value of a given oil shale is a useful and fundamental property of the rock, it does not provide information on the amounts of shale oil or combustible gas that would be yielded by retorting (destructive distillation). Alternatively, the grade of oil shale can be determined by measuring the yield of distillable oil produced from a shale sample in a laboratory retort (Scouten, 1990). This is perhaps the most common type of analysis that has been, and still is, used to evaluate an oil-shale resource—however, the end result of the evaluation depends upon the source of the sample and whether or not the sample is representative of the deposit.

The method commonly used in the United States is the *modified Fischer assay* test method (ASTM D3904). Some laboratories have further modified the Fischer assay method to better evaluate different types of oil shale and different methods of oil-shale processing. The standard Fischer assay test method (ASTM D3904, now withdrawn but still used in many laboratories) consists of heating a 100 gram sample crushed to −8 mesh (2.38 mm) screen in a small aluminum retort to 500°C (930°F) at a rate of 12°C (21.6°F) per minute and held at that temperature for 40 minutes. The distilled vapors of oil, gas, and water are passed through a condenser cooled with ice water into a graduated centrifuge tube. The oil and water are then separated by centrifuging. The quantities reported are the weight percentages of shale oil (and its specific gravity), water, shale residue, and (by difference) gas plus losses.

The Fischer assay method does not measure the total energy content of an oil shale sample because the gases, which include methane, ethane, propane, butane, hydrogen, hydrogen sulfide, and carbon dioxide can have significant energy content, but are not individually specified (Allix et al., 2011). Also, some retort methods, especially those that heat at a different rate or for different times, or that crush the rock more finely, may produce more oil than that produced by the Fischer assay method. Therefore, the method can only be used as a reference point and, at best, the data from the Fischer assay test method can only be employed to approximate the energy potential of an oil shale deposit.

Other retorting methods, such as the Tosco II process, are known to yield in excess of 100% of the yield reported by Fischer assay. In fact, some methods of retorting can increase oil yields of some oil shales by as much as three to four times the yield obtained by the Fischer assay method (Scouten, 1990; Dyni, 2003, 2006). Another method for characterizing organic richness of oil shale is a pyrolysis test developed by the Institut Français du Pétrole for analyzing source rocks (Allix et al., 2011). The Rock-Eval test heats a 50 mg to 100 mg sample through several temperature stages to determine the amounts of hydrocarbon and carbon dioxide generated. The results can be interpreted for kerogen type and potential for oil and gas generation. The method is faster than the Fischer assay and requires less sample material Kalkreuth and Macauley, 1987).

6.3 Mineral components

Oil shale has often been termed as (incorrectly and for various illogical reasons) high-mineral coal. Nothing could be further from the truth than this misleading terminology. Coal and oil shale are fraught with considerable differences (Speight, 2013) and such terminology should be frowned upon. Furthermore, the precursors of the organic matter in oil shale and coal also differ. Much of the organic matter in oil shale is of algal origin, but may also include remains of vascular land plants that more commonly compose much of the organic matter in coal (Scouten, 1990; Dyni, 2003, 2006; Speight, 2013). In addition, the lack of recognizable biologic structures in oil shale that would help identify the precursor organisms in oil shale makes it difficult to identify the origin of the organic matter.

In terms of mineral and elemental content, oil shale differs from coal in several distinct ways. Oil shale typically contains much larger amounts of inert mineral matter (60%–90%) than coal, which has been defined as containing less than 40% mineral matter (Speight, 2013). The organic matter of oil shale, which is the source of liquid and gaseous hydrocarbon derivatives, typically has a higher hydrogen and lower oxygen content than that of lignite and bituminous coal.

The mineral component of some oil shale deposits is composed of carbonates including calcite ($CaCO_3$), dolomite ($CaCO_3.MgCO_3$), siderite ($FeCO_3$), nahcolite ($NaHCO_3$), dawsonite [$NaAl(OH)_2CO_3$], with lesser amounts of aluminosilicates—such as alum [$KAl(SO_4)_2.12H_2O$]—and sulfur, ammonium sulfate, vanadium, zinc, copper, and uranium, which add byproduct value (Beard et al., 1974). For other deposits, the reverse is true—silicates including quartz (SiO_2), feldspar [$xAl(Al.Si)_3O_8$, where x can be sodium (Na), and/or calcium (Ca), and/or potassium (K)], and clay minerals are dominant and carbonates are a minor component. Many oil-shale deposits contain small, but ubiquitous, amounts of sulfides including pyrite (FeS_2) and marcasite (FeS_2, but which is physically and crystallographically distinct from pyrite), indicating that the sediments probably accumulated in dysaerobic (a depositional environment with 0.1–1.0 mL of dissolved oxygen per liter of water) to anoxic waters that prevented the destruction of the organic matter by burrowing organisms and oxidation.

Green River oil shale contains abundant carbonate minerals including dolomite, nahcolite, and dawsonite. The latter two minerals have potential byproduct value for their soda ash and alumina content, respectively. The oil shale deposits of the eastern United States are low in carbonate content but contain notable quantities of metals, including uranium, vanadium, molybdenum, and others which could add significant byproduct value to these deposits.

There is the potential for low emissions due to the inherent presence of carbonate minerals. Calcium carbonate present in oil shale ash binds sulfur dioxide and it is not necessary to add limestone for desulfurization:

$$CaCO_3 \rightarrow CaO + CO_2$$

$$2CaO + SO_2 + O_2 \rightarrow CaSO_4$$

Illite (a layered aluminosilicate $[K,H_3O)\ (Al,Mg,Fe)_2(Si,Al)_4O_{10}(OH)_2,\ (H_2O)]$ is ever-present in Green River oil shale— it is generally associated with other clay minerals but frequently occurs as the only clay mineral found in the oil shale (Tank, 1972). Smectite (a group of clay minerals that includes montmorillonite, which tends to swell when exposed to water) is present in all three members of the Green River Formation, but its presence frequently shows an inverse relationship to both analcime (a white, gray, or colorless tectosilicate mineral which consists of hydrated sodium aluminum silicate, $NaAlSi_2O_6.H_2O$) and loughlinite (a silicate of magnesium, $Na_2Mg_3\text{-}Si_6O_{16}.8(H_2O)$. Chlorite (a group of mostly monoclinic but also triclinic or orthorhombic micaceous phyllosilicate minerals) occurs only in the silty and sandy beds of the Tipton Shale Member. The distribution of random mixed-layer structures and amorphous material is irregular. Several independent lines of evidence favor an in situ origin for many of the clay minerals. Apparently the geochemical conditions favoring the accumulation of the oil shale also favored in situ generation of illite.

Finally, precious metals and uranium are contained in good amounts in oil shale of the eastern United States. It may not be in the near future to recover these mineral resources, since a commercially favorable recovery process has not yet been developed. However, there are many patents on recovery of alumina from Dawsonite-bearing beds $[NaAl(CO_3)\ (OH)_2]$ by leaching, precipitation, and calcination.

6.4 Thermal decomposition

Compared to coal, oil shale kerogen is relatively hydrogen-rich and can, therefore, be subjected to thermal conversion leading to higher yields of distillable oil and gas. This is in keeping with volatile products from fossil fuels being related to the hydrogen content of the fossil fuel (Scouten, 1990; Speight, 2013, 2014).

High-yield oil shale sustains combustion hence the older Native American name *the rock that burns,* but in the absence of air (oxygen) three carbonaceous end products result when oil shale is thermally decomposed. Distillable oil is produced as are noncombustible gases and a carbonaceous (high-carbon) deposit remains on the rock (the surface or in the pores) as char- or cokelike residue. The relative proportions of oil, gas, and char vary with the pyrolysis temperature and to some extent with the organic content of the raw shale. All three products are contaminated with nonhydrocarbon compounds and the amounts of the contaminants also vary with the pyrolysis temperature (Bozak and Garcia, 1976; Scouten, 1990).

At temperatures on the order of 500°C to 520°C (930°F −970°F), oil shale produces shale oil while the mineral matter of the oil shale is not decomposed. The yield and quality of the products depend on a number of factors, whose impact has been identified and quantified for some of the deposits, notably the US Green River deposits and the Estonian deposits (Brendow, 2003, 2009). A major factor is that oil shale ranges widely in organic content and oil yield. Commercial grades of oil shale, as determined by the yield of shale oil, range from approximately 25 to 50 gallons per ton of rock (typically using the Fischer assay method).

The correlation of the shale oil yield with the chemical and physical properties of oil shale and/or kerogen have been based on many different kinds of measurements, ranging from simple, qualitative tests that can be performed in the field to more complicated measurements in the laboratory. One simple aspect of the thermal decomposition of oil shale kerogen is the relationships of the organic hydrogen and nitrogen contents, and Fischer assay oil yields. Stoichiometry suggests that kerogen with a higher organic hydrogen-to-carbon atomic ratio can yield more oil per weight of carbon than kerogen that is relatively hydrogen-poor (Scouten, 1990). However, the hydrogen-to-carbon atomic ratio is not the only important factor. South African kerogen with an atomic hydrogen-to-carbon ratio of 1.35 has a lower oil yield than Brazilian kerogen with an atomic hydrogen-to-carbon ratio of 1.57. In general, the oil shale containing kerogen that is converted efficiently to oil contains relatively low levels of nitrogen (Scouten, 1990). Furthermore, variation of product distribution with time in the reaction zone can cause a change in product distribution (Fig. 6.2) (Hubbard and Robinson, 1950).

During retorting, kerogen decomposes into three organic fractions: (1) shale oil, (2) gas, and (3) carbonaceous residue. Oil shale decomposition begins at relatively low retort temperatures (300°C, 572°F) but proceeds more rapidly and more completely at higher temperatures (Scouten, 1990). The highest rate of kerogen decomposition occurs at retort temperatures of 480°C −520°C (895°F −970°F). In general, the yield of shale oil decreases, the yield

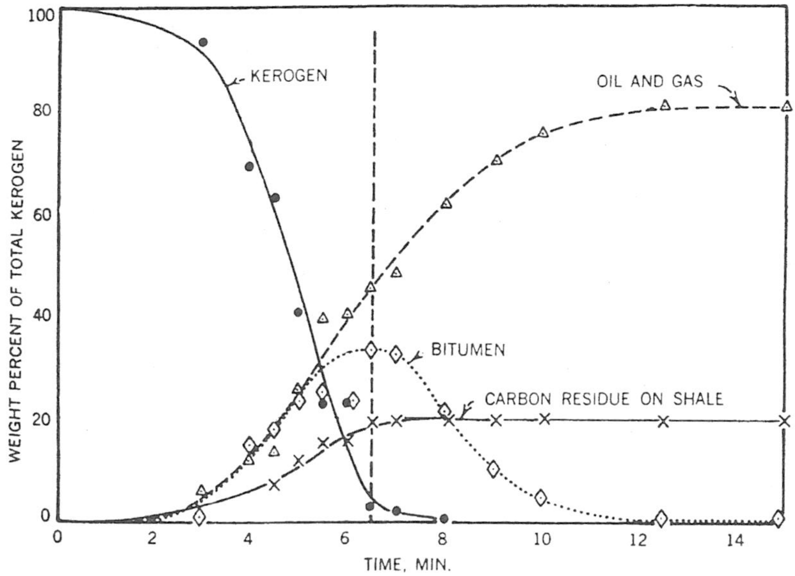

FIGURE 6.2 Variation of product yield with time (Hubbard and Robinson, 1950).

of gas increases, and the aromaticity of the oil increases with increasing decomposition temperature (Dinneen, 1976; Scouten, 1990).

However, there is an upper limit on optimal retorting temperature as the mineral content of the shale may decompose if the temperature is too high. For example, the predominant mineral component of Estonian kukersite shales is calcium carbonate, a compound that dissociates at high temperatures (600°C −750°C, 1112°F to 1380°F for dolomite, and 600°C −900°C, 1110°F to 1650°F for calcite). Thus carbon must be anticipated as a product of the oil shale decomposition process, which will dilute the off-gases (adding to emissions issues) produced from the retorting process. The gases and vapors leaving the retort are cooled to condense the reaction products, including oils and water.

The active devolatilization of oil shale begins at approximately 350°C −400°C (660°F −750°F), with the peak rate of oil evolution at approximately 425°C (800°F), and with devolatilization essentially complete in the range of 470°C −500°C (890°F −930°F) (Hubbard and Robinson, 1950; Shih and Sohn, 1980). At temperatures near 500°C (930°F), the mineral matter, consisting mainly of calcium/magnesium and calcium carbonates, begins to decompose yielding carbon dioxide as the principal product. The properties of crude shale oil are dependent on the retorting temperature, but more importantly on the temperature-time history because of the secondary reactions accompanying the evolution of the liquid and gaseous products. The produced shale oil is dark brown, odoriferous, and tending to waxy oil.

The processes for producing oil from oil shale involve heating (retorting) the shale to convert the organic kerogen to a raw shale oil (Janka and Dennison, 1979; Rattien and Eaton, 1976; Burnham and McConaghy, 2006). Conversion of kerogen to oil without the agency of heat has not yet been proven commercially, although there are schemes for accomplishing such a task but, in spite of claims to the contrary, these have not moved into the viable commercial or even demonstration stage.

However, since there are issues to consider when using the Fischer assay to determine the potential yields of oil from shale, there are other issues to consider relating to the rate of heating (Dyni et al., 1989; Allix et al., 2011). The reactions that convert kerogen to oil and gas are understood generally, but not in precise molecular detail and can only be represented by simple equations. The amount and composition of generated hydrocarbon derivatives depend on the heating conditions: the rate of temperature increase, the duration of exposure to heat, and the composition of gases present as the kerogen breaks down.

Generally, surface-based retorts heat the shale rapidly. The time scale for retorting is directly related to the particle size of the shale, which is why the rock is crushed before being heated in surface retorts. Pyrolysis of particles on the millimeter scale can be accomplished in minutes at 500°C (930°F) while pyrolysis of particles tens of centimeters in size takes much longer. In situ processes heat the shale more slowly. However, slow heating has advantages—the quality of the oil increases substantially. Coking and cracking reactions in the subsurface tend to leave the higher molecular weight (higher boiling) less desirable components in the ground. As a result, compared with surface processing, an in situ process can produce lower boiling products with fewer contaminants.

From the standpoint of shale oil as a substitute for crude oil products, its composition is of great importance. Paraffin-type shale oil is similar to paraffin crude oil. However, the composition of the kukersite shale oil of Estonia is more complicated and very specific—it contains abundant oxygen compounds, particularly phenols, which can be extracted from oil. The oil cannot serve directly as raw oil for high-quality engine fuel, but is well used as heating fuel. It has some specific properties as lower viscosity and pour point, and relatively low sulfur content, making it suitable for other uses such as marine fuel.

Contrary to other oil shales, obtaining high oil yields of distillable oil from kukersite needs specific conditions of processing. It can be explained by the fact that on thermal processing of kukersite, its elevated moisture percentage and the predominance of calcium carbonate in its mineral part result in high values of specific heat consumption in the process (Yefimov and Purre, 1993). Also shale is rich in organic matter and must pass the temperatures of thermobitumen formation and coking at a relatively high speed to avoid caking and secondary pyrolysis of oil.

One of the characteristics of kukersite causing considerable difficulties in its commercial-scale processing is the conversion to a bitumenlike on slow heating—the transition to the plastic state within the temperature range 350°C −400°C, 660°F to 750°F. The maximum yield of thermobitumen is produced at 390°C−395°C (735°F−745°F) and it constitutes 55%−57% w/w of the organic products. At these temperatures carbon content of solid residue (remaining after extraction with mixture of ethanol-benzene) is of the minimum value. However, as the heating continues to 510°C−520°C (950°F−970°F) the carbon content of the residue increases two- to threefold. As a result most of the carbonaceous residue in semicoke is of secondary origin formed at the pyrolysis of unstable components like oxygen-containing compounds (Yefimov and Purre, 1993)

The thermal characterization of Australian oil shale involved separation of the unique components of oil shale, the kerogen (organic component) and the clay minerals (inorganic components), using chemical and physical techniques (Berkovich et al., 2000). The heat capacity and enthalpy changes for the kerogen and clay minerals were measured using nonisothermal modulated differential scanning calorimetry from 25°C to 500°C (77°F−930°F). Heat capacity data was obtained over a temperature range spanning several hundred degrees in a single experiment. Heat capacity was also estimated by incorporating thermogravimetric data during regions where thermal reactions involving mass loss occurred. Enthalpy data for dehydration and pyrolysis of kerogen were also determined (Scouten, 1990; Berkovich et al., 2000).

The mineralogy of Green River Formation is radically changed when the raw oil shale is subjected to the extreme temperatures of processing (Milton, 1977; Smith et al., 1978; Essington et al., 1987). Mineral reactions from high temperature oil shale retorting can be summarized by two general steps, (1) decomposition of raw minerals and (2) crystallization from the melt. Complete decomposition of carbonate minerals and silicate minerals forms a pyrometamorphic melt containing the principle ions: Ca^{2+}, Na^{+}, Mg^{2+}, Fe^{2+}, Fe^{3+}, K^{+}, Si^{4+}, Al^{3+}, and O^{2+} (Park et al., 1979; Mason, 2006). Trace elements are abundant in the Green River Formation and are undoubtedly present in the melt, but low abundances are believed to make their contribution to the crystallization of new minerals negligible, although some partitioning has been recorded (Shendrikar and Faudel, 1978).

Silicate mineral products of high temperature oil shale processing fall into several general types; olivine group, melilite group, ortho- and clinopyroxenes, amphibole, feldspar group, quartz, and clay minerals. Amorphous silica (glass) is also common product in oil shale that has been processed at high temperatures then cooled rapidly. Although variation within the mineral groups can be in part due to minor differences in the composition of the raw oil shale, the final mineral suite appears to vary very little when examining material from different processes and localities (Mason, 2006). However, some oil-shale deposits contain minerals and metals that add byproduct dawsonite [$NaAl(OH)_2CO_3$], sulfur, ammonium sulfate, vanadium, zinc, copper, and uranium.

6.5 Porosity

The porosity (void fraction) is a measure of the void spaces in a material such as reservoir rock, and is a fraction of the volume of void space over the total volume, between and is expressed as a fractional number between 0 and 1, or as a percentage between 0 and 100. The porosity of porous material can be measured in a number of different ways, depending on what specific pores are looked at and how the void volumes are measured. They include: (1) interparticle porosity, (2) intraparticle porosity, (3) internal porosity, (4) porosity by liquid penetration, (5) porosity by saturation, (6) porosity by liquid absorption, (7) superficial porosity, (8) total open porosity, (9) bed porosity—the bed void fraction, and (10) packing porosity.

The porosity of the mineral matrix of oil shale cannot be determined by the methods used in determining porosity of crude oil reservoir rocks, because the organic matter in the shale exists in the form of solid and is essentially insoluble. However, inorganic particles contain some micropore structure, approximately 2.36%−2.66% v/v and although the mineral particles have an appreciable surface area, 4.24−4.73 m^2/g for oil shale capable of producing 29 to 75 gallons per ton in the Fischer assay, the measurement of porosity may be limited to the characteristics of the external surface rather than to the actual pore structure.

In the process of the production of shale oil from oil shale, both chemical and physical properties of oil shale play important roles. The low porosity, low permeability, and high mechanical strength of oil shale rock matrix make the extraction process less efficient by making the mass transport of reactants and products much harder as well as the process efficiency (Scouten, 1990; Eseme et al., 2007). Furthermore, the changes in properties as a function of temperature and pressure present implications of the evolution of these properties for in situ exploitation and basin modeling. While the mechanical properties at room temperature are well known, the existing data suggest a positive correlation between oil shale grade (organic matter content) and Poisson ratio, whereas tensile and compressive strength as well as modulus of elasticity show negative correlations. These properties are strongly affected by temperature—an increase in temperature results in loss of strength and decrease in Young's modulus (Scouten, 1990). Strength follows a logarithmic decrease with increasing temperature, depending on grade. Creep is much enhanced by elevated temperature. Extrapolation of laboratory data to nature suggests that tensile fracturing may occur more easily during crude oil generation, and creep is more prominent in oil shales than in other rocks at this depth in the crust (Eseme et al., 2007).

6.6 Permeability

Permeability is the ability, or measurement of the ability, of a rock to transmit fluids, typically measured in Darcies or millidarcies. Permeability is part of the proportionality constant in Darcy's law which relates the flow rate of the fluid and the fluid viscosity to a pressure gradient applied to the porous media. The permeability of raw oil shale is essentially zero, because the pores are filled with a nondisplaceable organic material. In general, oil shale constitutes a highly impervious system. Thus, one of the major challenges of any *in situ* retorting project is in the creation of a suitable degree of permeability in the formation. This is why an appropriate rubbelization technique is essential in the success of an in situ pyrolysis project.

Of practical interest is the dependency of porosity or permeability on temperature and organic contents. Upon heating to 510°C (950°F), an obvious increase in oil shale porosity is noticed. These porosities, which vary from 3% to 6% v/v of the initial bulk oil shale volume, represented essentially the volumes occupied by the organic matter before the retorting treatment. Therefore, the oil shale porosity increases as the extent of pyrolysis reaction proceeds.

In oil shale that produces a low yield of oil by the Fischer assay method (lean oil shale), structural breakdown of the cores is insignificant and the porosities are those of intact porous structures. However, in the high Fischer assay oil shales, i.e., rich oil shales, this is not the case because structural breakdown and mechanical disintegration due to retorting treatment become extensive and the mineral matrices no longer remain intact. Thermal decomposition of the mineral carbonates, such as magnesium carbonate ($MgCO_3$) and calcium carbonate ($CaCO_3$), actively occurring approximately at 380°C −900°C (715°F −1650°F) also results in an increase in porosity.

The increase in porosity from low to high Fischer assay oil shales varies from 2.82% to 50% (Table 6.3). These increased porosities constitute essentially the combined spaces represented by the loss of the organic matter and the decomposition of the mineral carbonates. Crackling of particles is also due to the devolatilization of organic matter that increases the internal vapor pressure of large nonpermeable pores to an extent the mechanical strength of the particle can no longer contain. Liberation of carbon dioxide from mineral carbonate decomposition also contributes to the pressure build-up in the oil shale pores.

6.7 Compressive strength

Raw oil shale has high compressive strengths both perpendicular and parallel to the bedding plane. After heating, the inorganic matrices of low-yield Fischer assay oil shale retain high compressive strength in both perpendicular and parallel planes. This indicates that a high degree of inorganic cementation

TABLE 6.3 Porosity and permeability of raw and treated oil shale (Chilingarian and Yen, 1978).

	Porosity			Permeability	
Fischer assay	Raw	Heated to 815°C	Plane	Raw	Heated to 815°C
1.0[a]	9.0[b]	11.9	A[c]		0.36[d]
			B		0.56
6.5	5.5	12.5	A		0.21
			B		0.65
13.5	0.5	16.4	A		4.53
			B		8.02
20.0	<0.03	25.0	A		
			B		
40.0	<0.03	50.0	A		
			B		

[a]*Fischer Assay in gal/ton.*
[b]*Numbers in percentages of the initial bulk volume. Porosity was taken as an isotropic property, i.e., property that is independent of measurement direction.*
[c]*Plane A is perpendicular to the bedding plane; plane B is parallel to the bedding plane.*
[d]*Units in millidarcy.*

exists between the mineral particles comprising each lamina and between adjacent laminae. With an increase in organic matter of oil shale the compressive strength of the respective organic-free mineral matrices decreases, and it becomes very low in those rich oil shales.

6.8 Thermal conductivity

Measurements of thermal conductivity of oil shale show that blocks of oil shale are anisotropic about the bedding plane and thermal conductivity as a function of temperature, oil shale assay and direction of heat flow, parallel to the bedding plane (parallel to the surface of the Earth for a flat oil shale bed), was slightly higher than the thermal conductivity perpendicular to the bedding plane. As layers of material were laid to form the oil shale bed over a long period of geological years, the resulting continuous strata have slightly higher resistance to heat flow perpendicular to the strata than parallel to the strata (Table 6.4).

The thermal conductivity of retorted and burnt shale is lower than those of the raw shales from which they are obtained (Table 6.4). This is attributable to

TABLE 6.4 Comparison between thermal conductivity values for Green River oil shale.

Temperature range (°C)	Fischer assay, gal/ton	Plane	Thermal conductivity (J/m-sec-°C)
38–593	7.2–47.9	–	0.69–1.56 (raw shales) 0.26–1.38 (retorted shales) 0.16–1.21 (burned shale)
25–420	7.7–57.5	A Average	0.92–1.92 1.00–1.82 (burned shale)
38–205	10.3–45.3	A	0.30–0.47
		B	0.22–0.28
20–380	5.5–62.3	A	1.00–1.42 (raw shale)
		B	0.25–1.75 (raw shale)

Note: A: parallel to the bedding plane.B: perpendicular to the bedding plane.Average - average of both directions.

the fact that the mineral matter is a better conductor of heat than the organic matter and, on the other hand, the organic matter is still a far better conductor than the voids created by its removal. While the first of the above hypotheses is well justified when one considers the contribution of the lattice conductivity to the overall value, the effect of the amorphous carbon formed from the decomposition of the organic matter could also be of importance in explaining the difference in thermal conductivity values for retorted shales and the corresponding burnt samples. The role of voids in determining the magnitude of the effective thermal conductivity is likely to be significant only for samples with high organic contents.

The thermal conductivity of oil shale is, in general, only weakly dependent on the temperature. However, extreme caution needs to be exercised in the interpretation of results at temperatures close to the decomposition temperature of the shale organic matter. This is due to the fact that the kerogen decomposition reaction (or, pyrolysis reaction) is endothermic in nature and as such the temperature transients can be confounded between the true rate of heat conduction and the rate of heat of reaction.

7. Kerogen

Kerogen is the naturally occurring, solid, insoluble organic matter that occurs in shale and can yield oil upon heating. Typically, kerogen has a high molecular weight and coexists with a lower molecular weight soluble organic fraction, usually referred to as *bitumen*, which should not be confused with tar sand bitumen (Chapter 2). Kerogen also yields oil when the shale containing kerogen is heated to temperatures sufficient to cause destructive distillation. Kerogen is a mixture of organic material, rather than a specific chemical, and cannot be given a chemical formula. Indeed the chemical composition of kerogen can vary substantially from sample to sample.

Kerogen has an implied role in the formation of crude oil and the term *kerogen* has also been used generally to indicate that the material is a precursor to crude oil (Tissot and Welte, 1978; Durand, 1980; Pelet and Durand, 1984; Hunt, 1996). However, caution is advised in choosing the correct definition since there is the distinct possibility that kerogen, far from being a precursor to crude oil, is one of the byproducts of the crude oil generation and maturation processes and may not be a direct precursor to crude oil.

In very general terms, the hydrogen content of kerogen falls between that of crude oil and that of coal, but this varies considerably with the source so that a range of values is found. This has been suggested as reflecting an overlap between terrestrial and aquatic origin. In fact, a high lipid content, consistent with the occurrence of aquatic plants in the source material, appears to be diminished in kerogen by lignin of terrestrial origin (Scouten, 1990). In fact, kerogen is best represented as a macromolecule that contains considerable amounts of carbon and hydrogen. Furthermore, it is the macromolecular and heteroatom nature of kerogen with up to 400 heteroatoms (nitrogen plus oxygen plus sulfur) for every 1000 carbon atoms occurring as an integral part of the macromolecule that classifies kerogen as a naturally occurring heteroatomic material.

In addition to being classified as a naturally occurring heteroatomic material, kerogen can be subclassified into three different types (I, II, and III). These types of kerogen originate because of the different kinds of debris deposited in the sediment and also because of the conditions that prevail in that sediment over geological time. As initially deposited in a *recent sediment*, each type of debris may have a characteristic range of composition that can depend upon local conditions, such as the types of flora and fauna that contribute to the debris. As the sediment is buried deeper and/or hotter and for a longer time the organic material in the sediment undergoes *maturation* to give oil, gas, or a mixture of the two (Tissot and Welte, 1978; Hunt, 1996).

Type I kerogen is rich in lipid-derived aliphatic chains and has a relatively low content of polynuclear aromatic systems and heteroatomic systems. The initial atomic H/C ratio is high (1.5 or more), and the atomic O/C ratio is generally low (0.1 or less). This type of kerogen is generally of lacustrine origin. Organic sources for the type I kerogen include the lipid-rich products of *algal blooms* and the finely divided and extensively reworked lipid-rich biomass deposited in stable stratified lakes.

Type II kerogen is characteristic of the marine oil shales. The organic matter in this type of kerogen is usually derived from a mixture of zooplankton, phytoplankton, and bacterial remains that were deposited in a reducing environment. Atomic H/C ratios are generally lower than for type I kerogen, but the O/C atomic ratios are generally higher for type II kerogen than for type I kerogen. Organic sulfur levels are also generally higher in the type II kerogen. The oil-generating potential of type II kerogen is generally lower than those of the type I kerogen (i.e., less of the organic material is liberated as oil upon heating a type II kerogen at the same level of maturation).

Type III kerogen is characteristic of coals and coaly shales. Easily identified fossilized plants and plant fragments are common, indicating that this type of kerogen is derived from woody terrestrial material. These materials have relatively low atomic H/C ratios (usually <1.0) and relatively high atomic O/C ratios (>0.2). Aromatic and heteroaromatic contents are high, and ether units (especially of the diaryl ethers) are important, as might be anticipated for a lignin-derived material. Oil-generating potentials are low, but gas-generating potentials are high.

The need to gather the very large mass of information related to kerogen structure into a compact form useful for guiding research and development has led to the development of models for kerogen structure. However, no one model can depict the molecular structure of kerogen. In fact, the kerogen models represent attempts, based on the available data, to depict a collection of skeletal fragments and functional groups as a three-dimensional network in a reasonable manner. Some efforts succeed, many efforts fail.

Retorting is the cracking process used in shale oil refining, and first breaks down the kerogen to release hydrocarbon derivatives, and then further cracks the hydrocarbon derivatives into lower weight products. Retorting can occur in an aboveground retort (after mining the oil shale) or may be conducted in situ. In situ processes require that the oil shale be heated, to release the hydrocarbon gases and prior to extraction from the ground.

The distillates from oil shale (kerogen) retorting typically favor the production of middle-distillates (diesel and kerosene), and have higher concentrations of nitrogen than crude oil distillates of the same boiling range. To produce lower boiling distillates (such as gasoline), additional processing, such as hydrocracking, is required to break down the higher boiling diesel and kerogen fractions. Also, the nitrogen must be removed through some hydrotreating process, comparable to hydrogen desulfurization to remove sulfur from crude oil, such as hydrodenitrogenation.

The thermal decomposition of kerogen occurs readily at moderate temperatures to produce a variety of products:

$$\text{Kerogen} \rightarrow \text{Hydrocarbon derivatives} + \text{Heteroatom compounds}$$

The precise mode of cracking (stepwise of successive) is still arguable but, in the current context, the production of hydrocarbon derivatives is by the thermal decomposition of kerogen. The occurrence of primary, secondary, and even tertiary reactions in the system must be considered and the role played by free radicals or minerals (as catalysts) in juxtaposition to the kerogen needs also to be resolved.

8. Hydrocarbon products

Interest in the oil shales of the western USA as a strategic reserve increased after the oil embargo of 1973, when the price of oil doubled, but was found to be commercially unviable in the 1980s. If oil shale should be considered as raw material for shale oil it must contain enough organic matter to yield more energy than it requires processing the rock. The organic content needs to be 8 to 10 weight percent (i.e., yielding approximately 12−15 gallons per ton), before it can be considered a source for hydrocarbon derivatives (synthetic fuel). Oil shale has been a difficult commodity to exploit economically. Since the early 1900s, many attempts have been made to wrest shale oil from the Green River deposits of the United States, but with little success. The higher costs of mining oil shale, the lack of a viable technology to economically recover oil from the shale and the cost of environmentally acceptable disposal of waste rock have been limiting factors in developing an oil shale industry.

Shale oil (*retort oil*) contains a large variety of hydrocarbon derivative compounds including paraffin derivatives, cycloparaffin derivatives, olefin derivatives, and aromatic derivatives, as well as heteroatom compounds (i.e., nonhydrocarbon derivatives). Furthermore, crude shale oil typically contains appreciable amounts of water and solids, as well as has an irrepressible tendency to form sediments. As a result, it must be upgraded to a synthetic crude oil (syncrude) before being suitable for pipelining or substitution for crude oil as a refinery feedstock.

It is difficult to generalize shale oil processing. Not only do the shale oil properties vary, refineries vary widely. For example, there are approximately 300 fluid catalytic cracking (FCC) units in free world refineries and these use more than 260 different cracking catalysts. Therefore, several of the reported large-scale studies have been selected to illustrate the major features of shale oil upgrading and refining. These studies have generally used one of the three approaches: (1) thermal conversion, such as visbreaking or coking followed by hydrotreating, (2) hydrotreating followed by fluid catalytic cracking, and (3) hydrotreating followed by hydrocracking.

However, the amount of hydrocarbon derivatives and hydrocarbon products that can be recovered from a given oil shale deposit depends upon many factors. Geothermal heating, for example, may have degraded a deposit, so that the amount of recoverable hydrocarbon derivatives may be significantly reduced. Some deposits may also be buried too deep to be mined economically in the foreseeable future. Also, surface land uses may greatly restrict the availability of some oil shale deposits, especially those in the industrial Western countries.

Assuming a deposit can produce hydrocarbon derivatives, there are processes for producing hydrocarbon oil from oil shale which involve heating (retorting) the shale to convert the organic kerogen to raw shale oil (Burnham and McConaghy, 2006). Conversion of kerogen to hydrocarbon derivatives without the agency of heat has not yet been proven commercially, although there are schemes for accomplishing such a task but, in spite of claims to the contrary, these have not moved into the viable commercial or even demonstration stage.

There are two basic oil shale retorting approaches for the production of shale oil and, therefore, hydrocarbon derivatives: (1) mining followed by retorting at the surface and (2) in situ retorting, i.e., heating the shale in place underground (Allred, 1982).

8.1 Mining and retorting

With the exception of in situ processes, oil shale must be mined before it can be converted to shale oil. Depending on the depth and other characteristics of the target oil shale deposits, either surface mining or underground mining methods may be used.

Open-pit mining has been the preferred method whenever the depth of the target resource is favorable to access through overburden removal. In general, open-pit mining is viable for resources where the over burden is less than 150 feet in thickness and where the ratio of overburden thickness to deposit thickness ratio is less, 1/1. Removing the ore may require blasting if the resource rock is consolidated. In other cases, exposed shale seams can be bulldozed. The physical properties of the ore, the volume of operations, and project economics determine the choice of method and operation.

When the depth of the overburden is too great, underground mining processes are required. Underground mining necessitates a vertical, horizontal, or directional access to the kerogen-bearing formation. Consequently, a strong *roof formation* must exist to prevent collapse or cave-ins, ventilation must be provided, and emergency egress must also be planned. Room and pillar mining has been the preferred underground mining option in the Green River formations. Technology currently allows for cuts up to 27 m in height to be made in the Green River formation, where ore-bearing zones can be hundreds of meters thick. Mechanical *continuous miners* have been selectively tested in this environment, as well.

Surface retorting involves transporting mined oil shale to the retort facility, retorting and recovering the raw kerogen oil, upgrading the raw oil to marketable products and disposing of the *spent* shale (Fig. 6.3). Retorting

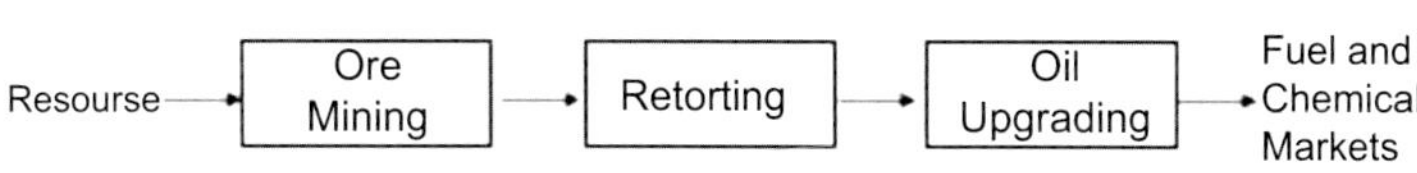

FIGURE 6.3 Process steps for mining and surface retorting (Bartis et al., 2005).

processes require mining more than a ton of shale to produce one barrel of oil. The mined shale is crushed to provide a desirable particle size, injected into a heated reactor (*retort*), where the temperature is increased to approximately 450°C (850°F). At this temperature, the kerogen decomposes to a mixture of liquid and gas. One way the various retorting processes differ is in how the heat is provided to the shale by hot gas, by a solid heat carrier, or by conduction through a heated wall.

Advances in mining technology continue in other mineral exploitation industries, including the coal industry. Open-pit mining is a well-established technology in coal mining, tar sand mining, and hard rock mining. Furthermore, room and pillar and underground mining have previously been proven at commercial scale for oil shale in the western United States. Costs for room and pillar mining will be higher than for surface mining, but these costs may be partially offset by having access to richer ore.

The fundamental issue with all oil shale technologies is the need to provide large amounts of heat energy to decompose the kerogen to liquid and gas products. More than one ton of shale must be heated to temperatures in the range 425°C to 525°C (850°F to 1000°F) for each barrel of oil generated, and the heat supplied must be of relatively high quality to reach retorting temperature. Once the reaction is complete, recovering sensible heat from the hot rock is very desirable for optimum process economics. This leads to three areas where new technology could improve the economics of oil recovery: (i) recovering heat from the spent shale, (ii) disposal of spent shale, especially if the shale is discharged at temperatures where the char can catch fire in the air, and (iii) concurrent generation of large volumes of carbon dioxide when the minerals contain limestone, as they do in Colorado and Utah.

Heat recovery from hot solids is generally not very efficient. The major exception to this generalization is in the field of fluidized bed technologies, where many of the lessons of fluids behavior can be applied. To apply fluidized bed technologies to oil shale would require grinding the shale to sizes less than approximately 1 mm, an energy-intensive task that would result in an expensive disposal problem. However, such fine particles might be used in a lower temperature process for sequestering CO_2, with the costs of grinding now spread over to the solution of this problem.

Disposal of spent shale is also a problem that must be solved in economic fashion for the largescale development of oil shale to proceed. Retorted shale contains carbon as a kind of char, representing more than half of the original carbon values in the shale. The char is potentially pyrophoric and can burn if dumped into the open air while hot. The heating process results in a solid that occupies more volume than the fresh shale because of the problems of packing random particles. A shale oil industry producing 100,000 barrels per day, related to the minimum for a world-scale operation, would process more than 100,000 tons of shale (density approximately 3 g/cc) and result in more than

35 m^3 of spent shale; this is equivalent to a block more than 100 feet on a side (assuming some effort at packing to conserve volume).

The Unocal process relied on direct contact between hot gases passing downward through a rising bed of crushed shale. This required that the retorting shale be pumped upward against gravity. Retorted shale reaching the top of the retort spilled over the sides and was cooled as it left the vessel. Oil formed in the process trickled down through the bed of shale, exchanged its heat with fresh shale rising in the roughly conical retort, and was drawn from the bottom. Unocal produced 4.5 million barrels from 1980 until 1991 from oil shale averaging 34 gallons per ton. The major problem that had to be overcome was formation of fine solids by decrepitation of the shale during retorting; the fines created problems in controlling solid flow in the retort and cooling shafts.

The Tosco (The Oil Shale Company) process used a rotating kiln that was reminiscent of a cement kiln in which heat was transferred to the shale by ceramic balls heated in an exterior burner. Retorted shale was separated from the balls using a coarse screen and the balls were recovered for recycle. Emerging vapors were cooled to condense product oil. The system was tested at the large pilot scale, but construction of a commercial retort was halted in 1982. One problem with the system was slow destruction of the ceramic balls by contact with the abrasive shale particles.

The Alberta Taciuk Processor (ATP), which was originally developed for oil recovery from tar sand, has been deployed in Australia. The unit involves a double-walled rotating kiln, with hot gas passing along the outer wall of the rotating retort, transferring heat through the wall to the retorting shale inside. Rotating seals are needed to contain all the components within the retort while excluding air.

8.2 In situ technologies

In situ processes introduce heat to the kerogen while it is still embedded in its natural geological formation. There are two general in situ approaches; true in situ in which there is minimal or no disturbance of the ore bed, and modified in situ, in which the bed is given a rubblelike texture, either through direct blasting with surface uplift or after partial mining to create void space. Recent technology advances are expected to improve the viability of oil shale technology, leading to commercialization.

In situ processes can be technically feasible where permeability of the rock exists or can be created through fracturing. The target deposit is fractured, air is injected, the deposit is ignited to heat the formation, and resulting shale oil is moved through the natural or man-made fractures to production wells that transport it to the surface (Fig. 6.4). However, difficulties in controlling the flame front and the flow of pyrolyzed oil can limit the ultimate oil recovery,

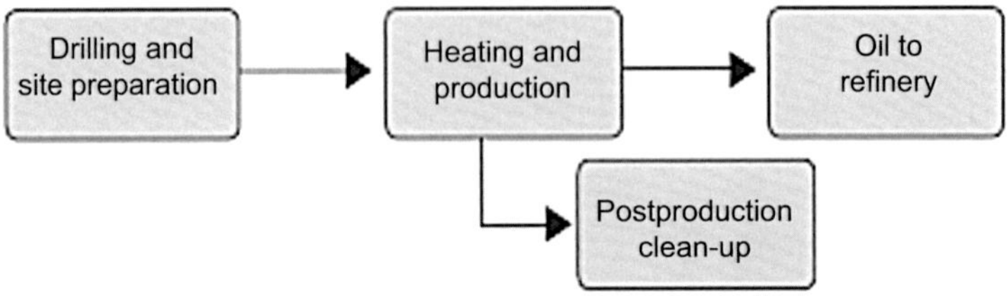

FIGURE 6.4 Process steps for thermal in situ conversion. *Source: Bartis, J.T., LaTourrette, T., Dixon, L., 2005. Oil Shale Development in the United States: Prospects and Policy Issues. Prepared for the National Energy Technology of the United States Department of Energy. Rand Corporation, Santa Monica, California.*

leaving portions of the deposit unheated and portions of the pyrolyzed oil unrecovered.

Thus, in situ processes avoid the need to mine the shale but require that heat be supplied underground and that product be recovered from a relatively nonporous bed. As such, the in situ processes tend to operate slowly, behavior that the Shell ICP process exploits by heating the resource to approximately 345°C (650°F) over a period of 3−4 years. This produces high yields of liquids with minimal secondary reactions.

In situ processes avoid the spent shale disposal problems because the spent shale remains where it is created but, on the other hand, the spent shale will contain uncollected liquids that can leach into ground water, and vapors produced during retorting can potentially escape to the aquifer (Karanikas et al., 2005). Modified in situ processes attempt to improve performance by exposing more of the target deposit to the heat source and by improving the flow of gases and liquid fluids through the rock formation, and increasing the volumes and quality of the oil produced. Modified in situ involves mining beneath the target oil shale deposit prior to heating. It also requires drilling and fracturing the target deposit above the mined area to create void space of 20% −25%. This void space is needed to allow heated air, produced gases, and pyrolyzed shale oil to flow toward production wells. The shale is heated by igniting the top of the target deposit. Condensed shale oil that is pyrolyzed ahead of the flame is recovered from beneath the heated zone and pumped to the surface.

Modified in situ processes attempt to improve performance by exposing more of the target deposit to the heat source and by improving the flow of gases and liquid fluids through the rock formation, and increasing the volumes and quality of the oil produced. Modified in situ involves mining beneath the target oil shale deposit prior to heating. It also requires drilling and fracturing the target deposit above the mined area to create void space of 20%−25%. This void space is needed to allow heated air, produced gases, and pyrolyzed shale oil to flow toward production wells. The shale is heated by igniting the top of the target deposit. Condensed shale oil that is pyrolyzed ahead of the flame is recovered from beneath the heated zone and pumped to the surface.

The Occidental vertical modified in situ process was developed specifically for the deep, thick shale beds of the Green River Formation. Approximately 20% of the shale in the retort area is mined; the balance is then carefully blasted using the mined out volume to permit expansion and uniform distribution of void space throughout the retort (Petzrick, 1995).

In this process, some of the shale was removed from the ground and explosively shattered the remainder to form a packed bed reactor within the mountain. Drifts (horizontal tunnels into the mountain) provided access to the top and bottom of the retort. The top of the bed was heated with burners to initiate combustion and a slight vacuum pulled on from the bottom of the bed to draw air into the burning zone and withdraw gaseous products. Heat from the combustion retorted the shale below, and the fire spread to the char left behind. Key to success was formation of shattered shale of relatively uniform particle size in the retort, at reasonable cost for explosives.

If the oil shale contains a high proportion of dolomite (a mixture of calcium carbonate and magnesium carbonate; e.g., Colorado oil shale) the limestone decomposes at the customary retorting temperatures to release large volumes of carbon dioxide. This consumes energy and leads to the additional problem of sequestering the carbon dioxide to meet global climate change concerns.

8.3 Refining shale oil

Crude shale oil, sometimes termed *retort oil*, is the organic (predominantly hydrocarbon) liquid oil condensed from the effluent in oil shale retorting. However, crude shale oil typically contains appreciable amounts of water and solids, as well as having an irrepressible tendency to form sediments. As a result, it must be upgraded to a synthetic crude oil (syncrude) before being suitable for pipelining or substitution for crude oil as a refinery feedstock. However, shale oil is sufficiently different from crude oils that processing shale oil presents some unusual problems.

Shale oil, especially shale oil from Green River oil shale has particularly high nitrogen content (typically on the order of 1.7%−2.2% w/w vs. 0.2% −0.3% w/w for typical crude oil). In many other shale oils (including those from shale deposits in the Eastern United States) nitrogen contents are lower than in the Green River shale oil, but still higher than those typical of crude oil. Because retorted shale oils are produced by a thermal cracking process, olefin and diolefin contents are high. In addition to olefin derivatives and diolefin derivatives, Green River shale oil contains appreciable amounts of aromatic derivatives, polar aromatic derivatives and pentane-insolubles (asphaltenes) (Tables 6.5 and 6.6). The concentration of polar aromatic derivatives and pentane-insoluble constituents in the higher-boiling fractions of shale oil parallels the nitrogen concentration in these fractions.

TABLE 6.5 Approximate elemental analysis of shale oil.

Element	% w/w
Carbon	84
Hydrogen	12
Nitrogen	<1
Oxygen	<1
Sulfur	<2
Metals	<0.1

TABLE 6.6 Major compound types in shale oil.

Saturates	Heteroatom systems
paraffins	Benzothiophene derivatives
cycloparaffins	Dibenzothiophene derivatives
Olefins	Phenol derivatives
Aromatics	Carbazole derivatives
Benzene derivatives	Pyridine derivatives
indan Derivatives	Quinoline derivatives
tetralin Derivatives	Nitrile derivatives
naphthalene Derivatives	Ketone derivatives
biphenyl Derivatives	Pyrrole derivatives
phenanthrene Derivatives	
chrysene Derivatives	

The oxygen content of shale oil is higher than those typically found in crude oil, but lower than the oxygen content of crude coal liquids. Crude shale oil also contains appreciable amounts of soluble arsenic, iron, and nickel that cannot be removed by filtration. However, it is the presence of the olefin derivatives and diolefin derivatives, in conjunction with high nitrogen content, which gives crude shale oils their characteristic instability and potential for sediment formation and poses difficulty in refining (Table 6.7). The sulfur content of shale oil varies widely, but is generally lower than those of high-sulfur crude oil and tar sand bitumen.

TABLE 6.7 Challenges for oil shale processing.

Particulates	–	Plugging on processing
		Product quality
Arsenic content		Toxicity
		Catalyst poison
High pour point		Oil not pipeline quality
Nitrogen content		Catalyst poison
		Contributes to instability
		Toxicity
Diolefin	–	Contributes to instability
		Plugging on processing

Upgrading, or partial refining, to improve the properties of a crude shale oil may be carried out using different options. Shale oils are rich in high-molecular weight, waxy paraffinic material. Thermal cracking lowers molecular weight, but yields straight-chain products of low octane number. Fluid catalytic cracking not only lowers molecular weight, but also causes isomerization to produce branched products with higher octane numbers. As a result, the naphtha produced by catalytic cracking is a more desirable feedstock for hydrotreating to make gasoline blend stock, than is the naphtha from thermal cracking or coking of shale oil.

Thermal conversions, coking, and visbreaking, are conceptually simple, noncatalytic methods for lowering the high pour point and viscosity of raw shale oils, in order to make the oil more suitable for hydrotreating that is needed to remove nitrogen and sulfur. Coking also separates suspended solids.

Visbreaking is a mild thermal treatment that lowers viscosity and pour point, to make the shale oil suitable for transportation by pipeline. Visbreaking also causes arsenic to separate but does little to reduce the contents of nitrogen, sulfur or olefin derivatives. In visbreaking, the oil is heated to approximately 500°C (930°F) for a short time (seconds to minutes), during which some product is cracked to gas, along with the desired cracking.

Delayed coking followed by hydrotreating was used in the upgrading of 8505 bbl of crude Paraho shale oil at the Gary Western Refinery in 1975. Coking was followed by severe hydrotreating, to produce experimental quantities of military jet and diesel fuels in the hydrogenation pilot plant at US Bureau of Mines facility in Bruceton, Pennsylvania. Delayed coking was also used to process approximately 3400 bbl of Occidental MIS crude shale oil at the Chevron refinery (Salt Lake City) but in this case the shale oil (13%–19% w/w) was coprocessed with the crude oil residuum.

In the Gary-Western test approximately 20% of the shale oil feed was converted to low-value gas and nearly 30% was converted to coke. Thus, the yield of high-value hydrocarbon transportation fuels amounted to only one-half the shale oil fed to the coker. Moreover, because of its high impurity content the shale-derived coke was not suitable for making carbon electrodes and could only be used for fuel. Nevertheless, the Gary-Western refining tests did demonstrate that shale oil can be processed into hydrocarbon fuels, using conventional refining technology with suitable adjustments of operating parameters.

For the production of hydrocarbon derivatives, hydrotreating is more flexible and less destructive than coking as a way to remove nitrogen, sulfur, oxygen, arsenic, and metals. One approach is to distill the crude shale oil, then to hydrotreat the fractions. However, catalysts activity will decline significantly if the shale oil is not first purified to some extent by removal of metals and nitrogen. Nevertheless, hydrotreating is the option of choice to produce a stable product.

In terms of refining and catalyst activity, the nitrogen content of shale oil is a disadvantage. But, in terms of the use of shale oil residua as a modifier for asphalt, where nitrogen species can enhance binding with the inorganic aggregate, the nitrogen content is beneficial. If not removed, the arsenic and iron in shale oil would poison and foul the supported catalysts used in hydrotreating.

Blending shale oil products with corresponding crude oil products, using shale oil fractions obtained from a very mildly hydrogen treated shale oil, yields kerosene and diesel fuel of satisfactory properties. Hydroprocessing shale oil products, either alone or in a blend with the corresponding crude oil fractions, is therefore necessary. The severity of the hydroprocessing has to be adjusted according to the particular properties of the feed and the required level of the stability of the product.

Gasoline from shale oil usually contains a high percentage of aromatic and naphthenic compounds that are not affected by the various treatment processes. The olefin content, although reduced in most cases by refining processes, will still remain significant. It is assumed that diolefin derivatives and the higher unsaturated constituents will be removed from the gasoline product by appropriate treatment processes. The same should be true, although to a lesser extent, for nitrogen- and sulfur-containing constituents.

The sulfur content of raw shale oil gasoline may be rather high due to the high sulfur content of the shale oil itself and the frequently even distribution of the sulfur compounds in the various shale oil fractions. Not only the concentration but also the type of the sulfur compounds is of importance when studying their effect on the gum formation tendency of the gasoline containing them.

Sulfides (R-S-R), disulfides (R-S-S-R), and mercaptans (R–SH) are, among the other sulfur compounds, the major contributors to the gum formation in gasoline. Sweetening processes for converting mercaptans to disulfides should therefore not be used for shale oil gasoline; sulfur extraction processes are preferred. Catalytic hydrodesulfurization processes are not a good solution for the removal of sulfur constituents from gasoline when high proportions of unsaturated constituents are present. A significant amount of the hydrogen would be used for hydrogenation of the unsaturated components. However, when hydrogenation of the unsaturated hydrocarbon derivatives is desirable, catalytic hydrogenation processes would be effective.

Gasoline derived from shale oil contains varying amounts of oxygen compounds. The presence of oxygen in a product, in which free radicals form easily, is a cause for concern. Free hydroxyl radicals are generated and the polymerization chain reaction is quickly brought to its propagation stage. Unless effective means are provided for the termination of the polymerization process, the propagation stage may well lead to an uncontrollable generation of oxygen bearing free radicals leading to gum and other polymeric products.

Diesel fuel derived from oil shale is also subject to the degree of unsaturation, the effect of diolefin derivatives, the effect of aromatic derivatives, the effect of nitrogen compounds, and the effect of sulfur compounds. Jet fuel produced from shale oil would have to be subjected to suitable refining treatments and special processes. The resulting product must be identical in its properties to corresponding products obtained from conventional crude oil. This can be achieved by subjecting the shale oil product to severe catalytic hydrogenation process with a subsequent addition of additives to ensure resistance to oxidation.

If antioxidants are used for a temporary reduction of shale oil instability, they should be injected into the shale oil (or its products) as soon as possible after production of the shale oil. The antioxidant types and their concentrations should be determined for each particular case separately. The antioxidants combine with the free radicals or supply available hydrogen atoms to mitigate the progress of the propagation and branching processes. When added to the freshly produced unstable product, the antioxidants may be able to fulfill this purpose. However, when added after some delay, i.e., after the propagation and the branching processes have advanced beyond controllable limits, the antioxidants would not be able to prevent formation of degradation products.

Exposure to oxygen is a major factor contributing to degradation product formation in shale oils. Peroxy radicals, which are readily formed when untreated shale oils or their products are exposed to oxygen, lead to rapid gum formation rate. Once oxygen is eliminated from such a system, the polymerization chain reaction tends to arrive to its termination stage. The termination stage of this polymerization chain reaction can take place by one of several ways, as for example, exhaustion of the reactive monomers or a combination of two free radicals. Chain reaction termination can be so affected by radical combination or disproportionation. In all cases free radicals have to be eliminated from the system. The chain termination can be also induced by certain constituents present naturally or added artificially in the form of antioxidants.

Thus, shale oil is different to conventional crude oils, and several refining technologies have been developed to deal with this. The primary problems identified in the past were arsenic, nitrogen, and the waxy nature of the crude. Nitrogen and wax problems were solved using hydroprocessing approaches, essentially classical hydrocracking and the production of high-quality lube stocks, which require that waxy materials be removed or isomerized. However, the arsenic problem remains.

In general, oil-shale distillates have a much higher concentration of high boiling-point compounds that would favor production of middle-distillates (such as diesel and jet fuels) rather than naphtha. Oil-shale distillates also had a higher content of olefin derivatives, oxygen, and nitrogen than crude oil, as well as higher pour points and viscosities. Above-ground retorting processes tend to yield lower API gravity oil than the in situ processes (a 25 degrees API gravity was the highest produced). Additional processing equivalent to hydrocracking would be required to convert oil-shale distillates to a lower-boiling hydrocarbon product (naphtha or gasoline). Removal of sulfur and nitrogen would, however, require hydrotreating.

By comparison, a typical 35 degrees API-gravity crude oil may be composed of up to 50% of gasoline and middle-distillate range hydrocarbon derivatives. West Texas Intermediate crude benchmark crude for trade in the commodity futures market) has 0.3% by weight sulfur, and Alaska North Slope crude has 1.1% by weight sulfur. The New York Mercantile Exchange (NYMEX) specifications for light sweet crude oil limits sulfur content to 0.42% or less (ASTM D4294) and an API gravity between 37 and 42° (ASTM D287).

A conventional refinery distills crude oil into various fractions, according to boiling point range, before further processing. In order of their increasing boiling range and density, the distilled fractions are fuel gases, low-boiling and high-boiling straight-run naphtha (90°F−380°F), kerosene (380°F−520°F), gas-oil (520°F−1050°F), and residuum (1050°F +) (Speight, 2014). Crude oil may contain 10%−40% gasoline, and early refineries directly distilled a straight-run gasoline (low-boiling naphtha) of low-octane rating. A hypothetical refinery may "crack" a barrel of crude oil into two-thirds gasoline and one-third distillate fuel (kerosene, jet, and diesel), depending on (i) the configuration of the refinery, (ii) the slate of crude oils refined, and (iii) the seasonal product demands of the market.

Just as natural clay catalysts help transform kerogen to crude oil through catagenesis, metallic catalysts help transform complex hydrocarbon derivatives to lower-boiling products in the refinery. The *catalytic-cracking* process developed during the World War II era enabled refineries to produce high-octane gasoline needed for the war effort. *Hydrocracking*, which entered commercial operation in 1958, improved on catalytic-cracking by adding hydrogen to convert residuum into high-quality motor gasoline and naphtha-based jet fuel. Many refineries rely heavily on *hydroprocessing* to convert low-value gas oils residuum to high-value transportation fuel demanded by the market. Middle-distillate range fuels (diesel and jet) can be blended from a variety of refinery processing streams. To blend jet fuel, refineries use desulfurized straight-run kerosene, kerosene boiling range hydrocarbon derivatives from a hydrocracking unit, and low-boiling coker gas-oil (cracked residuum). Diesel fuel can be blended from naphtha, kerosene, and low-boiling cracked-oils from coker and fluid catalytic cracking units. From the standard 42-gallon barrel of crude oil, United States refineries may actually produce more than 44 gallons of refined products through the catalytic reaction with hydrogen.

Oil derived from shale has been referred to as a synthetic crude oil and thus closely associated with synthetic fuel production. However, the process of retorting shale oil bears more similarities to conventional refining than to synthetic fuel processes. For the purpose of this report, the term *oil-shale distillate* is used to refer to middle-distillate range hydrocarbon derivatives produced by retorting oil shale. Two basic retorting processes were developed early on— aboveground retorting and underground, or in situ, retorting (*q.v.*). The retort is typically a large cylindrical vessel, and early retorts were based on rotary kiln ovens used in cement manufacturing. In situ technology involves mining an underground chamber that functions as a retort. A number of design concepts were tested from the 1960s through the 1980s.

Retorting essentially involves destructive distillation (*pyrolysis*) of oil shale in the absence of oxygen. Pyrolysis (temperatures above 900°F, 1650°F) thermally breaks down (*cracks*) the kerogen to release the hydrocarbon derivatives and then cracks the hydrocarbon derivatives into lower-weight hydrocarbon molecules. Conventional refining uses a similar thermal cracking process, termed *coking*, to break down high-molecular weight residuum.

As the demand for low boiling hydrocarbon fractions constantly increases, there is much interest in developing economical methods for recovering liquid hydrocarbon derivatives from oil shale on a commercial scale. However, the recovered hydrocarbon derivatives from oil shale are not yet economically competitive against the crude oil produced. Furthermore, the value of hydrocarbon derivatives recovered from oil shale is diminished because of the presence of undesirable contaminants. The major contaminants are sulfurous, nitrogenous, and metallic (and organometallic) compounds, which cause detrimental effects to various catalysts used in the subsequent refining processes. These contaminants are also undesirable because of their disagreeable odor, corrosive characteristics, and combustion products that further cause environmental problems.

Accordingly, there is great interest in developing more efficient methods for converting the higher-boiling hydrocarbon fractions obtained in a form of shale oil into lower-boiling (lower-molecular-weight) hydrocarbon derivatives. The conventional processes include catalytic cracking, thermal cracking, and coking. It is known that higher-boiling hydrocarbon fractions and refractory materials can be converted to lower-boiling materials by hydrocracking. These processes are most commonly used on liquefied coals or high-boiling residual or distillate oils for the production of substantial yields of low-boiling saturated products, and to some extent on intermediates that are used as domestic fuels, and still higher-boiling products that are used as lubricants. These destructive hydrogenation or hydrocracking processes may be operated on a strictly thermal basis or in the presence of a catalyst. Thermodynamically speaking, larger hydrocarbon molecules are broken into lower-boiling species when subjected to heat.

The H/C atomic ratio of such molecules is lower than that of saturated hydrocarbon derivatives, and abundantly supplied hydrogen improves this ratio by saturating reactions, thus producing liquid species. These two steps may occur simultaneously. However, the application of hydrocracking process has been hampered by the presence of certain contaminants in such hydrocarbon derivatives. The presence of sulfur- and nitrogen-containing compounds along with organometallic compounds in crude shale oils and various refined crude oil products has long been considered undesirable. Desulfurization and denitrification processes have been developed for this purpose.

The thermal cracking process is directed toward the recovery of gaseous olefin derivatives as the primarily desired cracked product, in preference to gasoline range liquids. By this process, it is claimed that at least 15%–20% of the feed shale oil is converted to ethylene, which is the most common gaseous product. Most of the feed shale oil is converted to other gaseous and liquid products. Other important gaseous products are propylene, 1,3-butadiene, ethane, and butanes. Hydrogen is also recovered as a valuable nonhydrocarbon gaseous product. Liquid products can comprise 40%–50% w/w or more of the total product. Recovered liquid products include benzene, toluene, xylene, gasoline-boiling-range liquids, and low-boiling and high-boiling oils. Coke is a solid product of the process and is produced by polymerization of unsaturated materials. Coke is typically formed in an oxygen-deficient environment via dehydrogenation and aromatization. Most of the formed coke is removed from the process as a deposit on the entrained inert heat carrier solids. The thermal cracking reactor does not require a gaseous hydrogen feed. In the reactor, entrained solids flow concurrently through the thermal riser at an average riser temperature of 700°C to 1400°C (1290°F to 2,550°F). The preferred high L-to-D ratio is in the range of a high 4:1 to 40:1, or 5:1 to 20:1 preferably.

The moving-bed hydroprocessing reactor is used to produce crude oil from oil shale or tar sands containing large amounts of highly abrasive particulate matter, such as rock dust and ash. The hydroprocessing takes place in a dual-function moving bed reactor, which simultaneously removes particulate matter by the filter action of the catalyst bed. The effluent from the moving bed reactor is then separated and further hydroprocessed in fixed bed reactors with fresh hydrogen added to the higher-boiling hydrocarbon fraction to promote desulfurization.

A preferred way of treating the shale oil involves using a moving bed reactor followed by a fractionation step to divide the wide-boiling-range crude oil produced from the shale oil into two separate fractions. The lower-boiling fraction is hydrotreated for the removal of residual metals, sulfur, and nitrogen, whereas the higher-boiling fraction is cracked in a second fixed bed reactor normally operated under high-severity conditions.

The fluidized bed hydroretort process eliminates the retorting stage of conventional shale upgrading, by directly subjecting crushed oil shale to a hydroretorting treatment in an upflow, fluidized bed reactor such as that used for the hydrocracking of heavy crude oil residues. This process is a *single stage retorting and upgrading* process. Therefore, the process involves: (1) crushing oil shale, (2) mixing the crushed oil shale with a hydrocarbon liquid to provide a pumpable slurry, (3) introducing the slurry along with a hydrogen-containing gas into an upflow, fluidized bed reactor at a superficial fluid velocity sufficient to move the mixture upwardly through the reactor, (4) hydroretorting the oil shale, (5) removing the reaction mixture from the reactor, and (6) separating the reactor effluent into several components.

The mineral carbonate decomposition is minimized, as the process operating temperature is lower than that used in retorting. Therefore, the gaseous product of this process has a greater heating value than that of other conventional methods. In addition, owing to the exothermic nature of the hydroretorting reactions, less energy input is required per barrel of product obtained. Furthermore, there is practically no upper or lower limit on the grade of oil shale that can be treated.

Hydrocracking is a cracking process in which higher molecular-weight hydrocarbon derivatives pyrolyze to lower-molecular-weight paraffins and olefin derivatives in the presence of hydrogen. The hydrogen saturates the olefin derivatives formed during the cracking process. Hydrocracking is used to process low-value stocks with high heavy metal content. It is also suitable for highly aromatic feeds that cannot be processed easily by conventional catalytic cracking. Shale oils are not highly aromatic, whereas coal liquids are very highly aromatic.

Middle-distillate (often called *mid-distillate*) hydrocracking is carried out with a noble metal catalyst. The typical reactor temperature is 480°C (895°F), and the average pressure is approximately 130–140 atm. The most common form of hydrocracking is carried out as a two-stage operation. The first stage is to remove nitrogen compounds and high-boiling aromatic derivatives from the raw crude, whereas the second stage is to carry out selective hydrocracking reactions on the cleaner oil from the first stage. Both stages are processed catalytically. Once the hydrocracking stages are over, the products go to a distillation section that consists of a hydrogen sulfide stripper and a recycle splitter. Commercial hydrocracking processes include Gulf HDS, H-Oil, IFP Hydrocracking, Isocracking, LC-Fining, Microcat-RC (also known as M-Coke), Mild Hydrocracking, Mild Resid Hydrocracking (MRH), Residfining, Unicracking, and Veba Combi-Cracking (VCC) (Speight, 2014, 2017).

Arsenic removed from the oil by hydrotreating remains on the catalyst, generating a material that is a carcinogen, an acute poison, and a chronic poison. The catalyst must be removed and replaced when its capacity to hold arsenic is reached. Unocal found that its disposal options were limited.

References

Allix, P., Burnham, A., Fowler, T., Herron, M., Kleinberg, R., Symington, B., 2011. Coaxing oil from shale. Oilfield Review 2010, 5015.

Allred, V.D. (Ed.), 1982. Oil Shale Processing Technology. Center for Professional Advancement, East Brunswick, New Jersey.

Andrews, A., 2006. Oil Shale: History, Incentives, and Policy. Specialist, Industrial Engineering and Infrastructure Policy Resources, Science, and Industry Division. Congressional Research Service, the Library of Congress, Washington, DC.

Bader, R.G., Hood, D.H., Smith, J.B., 1960. Recovery of dissolved organic matter in sea-water and organic sorption by particulate Material. Geochimica et Cosmochimica Acta 19, 236–243.

Bartis, J.T., LaTourrette, T., Dixon, L., 2005. Oil Shale Development in the United States: Prospects and Policy Issues. Prepared for the National Energy Technology of the United States Department of Energy. Rand Corporation, Santa Monica, California.

Baughman, G.L., 1978. Synthetic Fuels Data Handbook, second ed. Cameron Engineers, Inc., Denver, Colorado.

Beard, T.M., Tait, D.B., Smith, J.W., 1974. Nahcolite and dawsonite resources in the Green river formation, Piceance Creek Basin, Colorado. In: Guidebook to the Energy Resources of the Piceance Creek Basin, 25th Field Conference. Rocky Mountain Association of Geologists, Denver, Colorado, pp. 101–109.

Berkovich, A.J., John, H., Levy, J.H., Schmidt, S.J., Young, B.R., 2000. Heat capacities and enthalpies for some Australian oil shales from non-isothermal modulated DSC. Thermochimica Acta 357–358, 41–45.

Bordovskiy, O.K., 1965. Accumulation and transformation of organic substances in marine sediments. Marine Geology 3, 3–114.

Bozak, R.E., Garcia Jr., M., 1976. Chemistry in the oil shales. Journal of Chemical Education 53 (3), 154–155.

Brendow, K., 2003. Global oil shale issues and perspectives. Oil Shale 20 (1), 81–92.

Brendow, K., 2009. Oil shale – a local asset under global constraint. Oil Shale 26 (3), 357–372.

Burnham, A.K., Singleton, M.F., 1983. High-pressure pyrolysis of Green river oil shale. In: Miknis, F.P. (Ed.), Chemistry and Geochemistry of Oil Shale. Symp. Series No. 230, pp. 335–351. Washington, DC.

Burnham, A.K., McConaghy, J.R., 2006. Comparison of the acceptability of various oil shale processes. In: Proceedings. AICHE 2006 Spring National Meeting, Orlando, FL, March 23, 2006 through March 27.

Burnham, A.K., 2003. Slow Radio-Frequency Processing of Large Oil Shale Volumes to Produce Petroleum-like Shale Oil. Report No. UCRL-ID-155045. Lawrence Livermore National Laboratory, US Department of Energy, Livermore, California.

Cane, R.F., 1976. The origin and formation of oil shale. In: Yen, T.F., Chilingarian, G.V. (Eds.), Oil Shale. Elsevier, Amsterdam, Netherlands.

Chilingarian, G.V., Yen, T.F., 1978. Bitumens, Asphalts, and Tar Sands. Elsevier, Amsterdam, Netherlands (Chapter 1).

Cook, A.C., Sherwood, N.R., 1991. Classification of oil shales, coals and other organic-rich rocks. Organic Geochemistry 17 (2), 211–222.

Debyser, J., Deroo, G., 1969. Observations on the genesis of petroleum. Revue de l'Institut Francais du Petrole 24 (1), 21–48.

Derenne, S., Largeau, C., Casadevall, E., Sinninghie Damste, J.S., Tegelaar, E.W., deLeeuw, J.W., 1989. Characterization of Estonian kukersite by spectroscopy and pyrolysis: evidence for abundant alkyl phenolic moieties in an Ordovician, marine, type II/I kerogen. Organic Geochemistry 16 (4–6), 873–888.

Derenne, S., Largeau, C., Landais, P., Rochdi, A., 1994. Spectroscopic features of *Gloeocapsomorpha prisca* colonies and of interstitial matrix in kukersite as revealed by transmission micro-FT-IR: location of phenolic moieties. Fuel 73 (4), 626–628.

Dinneen, G.U., 1976. Retorting technology of oil shale. In: Yen, T.F., Chilingar, G.V. (Eds.), Oil Shale. Elsevier Science Publishing Company, Amsterdam, Netherlands, pp. 181–198.

Donovan, R.N., Scott, J., 1980. Lacustrine cycles, fish ecology and stratigraphic zonation in the middle devonian of caithness. The Journal of Geology 16, 35–50.

Durand, B., 1980. Kerogen: Insoluble Organic Matter from Sedimentary Rocks. Editions Technip, Paris, France.

Dyni, J.R., Anders, D.E., Rex, R.C., 1989. Comparison of hydroretorting, fischer assay, and rock-eval analyses of some world oil shales. In: Proceedings 1989 Eastern Oil Shale Symposium. Institute for Mining and Minerals Research, University of Kentucky, Lexington, Kentucky, pp. 270–286.

Dyni, J.R., 2003. Geology and resources of some world oil-shale deposits. Oil Shale 20 (3), 193–252.

Dyni, J.R., 2006. Geology and Resources of Some World Oil Shale Deposits. Report of Investigations 2005-5295. United States Geological Survey, Reston, Virginia.

Eseme, E., Urai, J.L., Krooss, B.M., Littke, R., 2007. Review of mechanical properties of oil shales: implications for exploitation and basin modelling. Oil Shale 24 (2), 159–174.

Essington, M.E., Spackman, L.K., Harbour, J.D., Hartman, K.D., 1987. Physical and Chemical Characteristics of Retorted and Combusted Western Reference Oil Shale. Report No. DOE/MC/11076-2453. United States Department of Energy, Washington, DC.

Hubbard, A.B., Robinson, W.E., 1950. A Thermal Decomposition Study of Colorado Oil Shale. Report of Investigations No. 4744. United States Bureau of Mines, Washington, DC.

Hunt, J.M., 1996. Petroleum Geochemistry and Geology, second ed. W.H. Freeman, San Francisco.

Hutton, A.C., 1987. Petrographic classification of oil shales. International Journal of Coal Geology 8, 203–231.

Hutton, A.C., 1991. Classification, organic petrography and geochemistry of oil shale. In: Proceedings. 1990 Eastern Oil Shale Symposium. Institute for Mining and Minerals Research, University of Kentucky, Lexington, Kentucky, pp. 163–172.

IEA, 2013. Resources to Reserves 2013: Oil, Gas and Coal Technologies for the Energy Markets of the Future. OECD Publishing. International Energy Agency, Paris, France.

Janka, J.C., Dennison, J.M., 1979. In: Devonian Oil Shale in Symposium Papers: Synthetic Fuels from Oil Shale, Atlanta, Georgia. December 3–6, pp. 21–116.

Johnson, H.R., Crawford, P.M., Bunger, J.W., March 2004. Strategic Significance of America's Oil Shale Resource. In: Volume II Oil Shale Resources. Technology and Economics. Office of Deputy Assistant Secretary for Petroleum Reserves. Office of Naval Petroleum and Oil Shale Reserves United States Department of Energy, Washington, D.C.

Kalkreuth, W.D., Macauley, G., 1987. Organic petrology and geochemical (Rock-Eval) studies on oil shales and coals from the pictou and antigonish areas, Nova scotia, Canada. Canadian Petroleum Geology Bulletin 35, 263–295.

Karanikas, J.M., de Rouffignac, E.P., Vinegar, H.J., Houston, T.X., Wellington, S., April 12, 2005. In Situ Thermal Processing of an Oil Shale Formation while Inhibiting Coking. United States Patent 6,877,555.

Koel, M., Ljovin, S., Hollis, K., Rubin, J., 2001. Using Neoteric solvents in oil shale studies. Pure and Applied Chemistry 73 (1), 153–159.

Mason, G.M., 2006. Fractional differentiation of silicate minerals during oil shale processing: a tool for prediction of retort temperatures. In: Proceedings. 26th Oil Shale Symposium. Colorado School of Mines, Golden Colorado. October 16–19.

Milton, C., 1977. Mineralogy of the Green river formation. The Mineralogical Record 8, 368–379.

Ots, A., 2007. Estonian oil shale properties and utilization in power plants. Energetika 53 (2), 8–18.

Park, W.C., Linderamanis, A.E., Robb, G.A., 1979. Mineral changes during oil shale retorting. In Situ 3 (4), 353–381.

Pelet, R., Durand, B., 1984. In: Perakis, L., Fraissard, J.P. (Eds.), Magnetic Resonance: Introduction, Advanced Topics, and Applications to Fossil Energy. D. Reidel, Norwell, Massachusetts.

Petzrick, P.A., 1995. Oil Shale and Tar Sand: Encyclopedia of Applied Physics, vol. 12. VCH Publishers Inc., Berlin, Germany, pp. 77–99.

Rattien, S., Eaton, D., 1976. Oil shale: the prospects and problems of an emerging energy industry. In: Hollander, J.M., Simmons, M.K. (Eds.), Annual Review of Energy, vol. 1, pp. 183–212.

Scouten, C., 1990. Oil shale. In: Speight, J.G. (Ed.), Fuel Science and Technology Handbook. Marcel Dekker Inc., New York.

Shendrikar, A.D., Faudel, G.B., 1978. Distribution of trace metals during oil shale retorting. Environmetnal Science and Technology 12 (3), 332–334.

Shih, S.M., Sohn, H.Y., 1980. Non-isothermal determination of the intrinsic kinetics of oil generation from oil shale. Industrial and Engineering Chemistry Process Design and Development 19, 420–426.

Smith, J.W., Robb, W.A., Young, N.B., 1978. High temperature mineral reactions of oil shale minerals and their benefit to oil shale processing in place. In: Proceedings. 11th Oil Shale Symposium. Colorado School of Miners, Golden, Colorado, pp. 100–112.

Speight, J.G., 2012. Shale Oil Production Processes. Gulf Professional Publishing, Elsevier, Oxford, United Kingdom.

Speight, J.G., 2013. The Chemistry and Technology of Coal, third ed. CRC Press, Taylor and Francis Group, Boca Raton, Florida.

Speight, J.G., 2014. The Chemistry and Technology of Petroleum, fifth ed. CRC Press, Taylor and Francis Group, Boca Raton, Florida.

Speight, J.G., 2016. Handbook of Hydraulic Fracturing. John Wiley & Sons Inc., Hoboken, New Jersey.

Speight, J.G., 2017. Deep Shale Oil and Gas. Gulf Professional Publishing Company, Elsevier, Oxford, United Kingdom.

Tank, R.W., 1972. Clay minerals of the Green river formation (eocene) of Wyoming. Clay Minerals 9, 297.

Tissot, B., Welte, D.H., 1978. Petroleum Formation and Occurrence. Springer-Verlag, New York.

Wang, D.-M., Xu, Y.-M., He, D.-M., Guan, J., Zhang, O.-M., 2009. Investigation of mineral composition of oil shale. Asia-pac. Journal of Chemical Engineering 4, 691–697.

Yefimov, V., Purre, T., 1993. Characteristics of kukersite oil shale, some regularities and features of its retorting. Oil Shale 10 (4), 313–319.

Further reading

ASTM, 2009. Annual Book of Standards. American Society for Testing and Materials, West Conshohocken, Pennsylvania.

Lee, S., 1996. Alternative Fuels. Taylor & Francis Publishers, Washington, DC.

Chapter 7

Hydrocarbons from biomass

1. Introduction

Biomass is the detritus or remains of living and recently dead biological material which can be used as fuel or for industrial production. Biomass also refers to (1) energy crops grown specifically to be used as fuel, such as fast-growing trees or switch grass, (ii) agricultural residues and byproducts, such as straw, sugarcane fiber, and rice hulls, and (iii) residues from forestry, construction, and other wood-processing industries. Biomass is a renewable energy source unlike other resources such as crude oil, natural gas, tar sand, coal, and oil shale which can be depleted with time and may have up to 50 years of use at current rates of depletion of these resources (Speight and Islam, 2016). Agricultural products specifically grown for biofuel production include crops such as corn, soybeans, rapeseed, wheat, sugar beet, sugar cane, palm oil, and jatropha oil, as well as wood.

Biofuel is derived from biomass and has the potential to produce fuels that are hydrocarbons by nature or are more environmentally benign than crude oil−based fuels. In addition, ethanol, a crop-based fuel alcohol, adds oxygen to gasoline thereby helping to improve vehicle performance and reduce air pollution. Biodiesel, an alternative or additive to crude oil diesel, is a nontoxic, renewable resource created from soybean or other oil crops (Metzger, 2006; Speight, 2011c).

Unlike other forms of renewable energy, biofuels do not reduce the amount of greenhouse gases in the atmosphere. The combustion of biofuels produces carbon dioxide and other greenhouse gases. The carbon in biofuels is often taken to have been recently extracted from atmospheric carbon dioxide during photosynthesis reactions that occur within plants as they grow. The potential for biofuels to be considered to be carbon neutral depends upon the carbon that is emitted being reused by further plant growth. Clearly however, cutting down trees in forests that have grown for hundreds or thousands of years for use as a biofuel without the replacement of this biomass would not have a carbon neutral effect. The production of biofuels to replace oil and natural gas as sources of hydrocarbon derivatives and hydrocarbon fuels is in active development, focusing on the use of cheap organic matter (usually cellulose, agricultural and sewage waste) in the efficient production of liquid and gas

Handbook of Industrial Hydrocarbon Processes. https://doi.org/10.1016/B978-0-12-809923-0.00007-2

biofuels which yield high net energy. One advantage of biofuel over most other fuel types is that it is biodegradable, and so relatively harmless to the environment if spilled.

The supply of crude oil, the basic feedstock for refineries and for the petrochemicals industry, is finite and its dominant position will become unsustainable as supply/demand issues erode its economic advantage over other alternative feedstocks. This situation will be mitigated to some extent by the exploitation of more technically challenging fossil resources and the introduction of new technologies for fuels and chemicals production from natural gas and coal.

However, the use of fossil resources at current rates will have serious and irreversible consequences for the global climate. Consequently, there is a renewed interest in the utilization of plant-based matter as a raw material feedstock for the chemicals industry. Plants accumulate carbon from the atmosphere via photosynthesis and the widespread utilization of these materials as basic inputs into the generation of power, fuels, and chemicals is a viable route to reduce greenhouse gas emissions. .

Thus, the crude oil and petrochemicals industries are coming under increasing pressure not only to compete effectively with global competitors utilizing more advantaged hydrocarbon feedstocks but also to ensure that its processes and products comply with increasingly stringent environmental legislation. The production of chemicals from renewable plant-based feedstocks utilizing state-of-the-art conversion technologies presents an opportunity to maintain competitive advantage and contribute to the attainment of national environmental targets (Metzger, 2006). Bioprocessing routes have a number of compelling advantages over conventional petrochemicals production; however, it is only in the last decade that rapid progress in biotechnology has facilitated the commercialization of a number of plant-based chemical processes. It is widely recognized that further significant production of plant-based chemicals will only be economically viable in highly integrated and efficient production complexes producing a diverse range of chemical products. This biorefinery concept is analogous to conventional oil refineries and petrochemical complexes that have evolved over many years to maximize process synergies, energy integration, and feedstock utilization to drive down production costs.

Reducing national dependence of any country on imported crude oil is of critical importance for long-term security and continued economic growth. Supplementing crude oil consumption with renewable biomass resources is a first step toward this goal. The realignment of the chemical industry from one of petrochemical refining to a biorefinery concept is, given time, feasible and has become a national goal of many oil-importing countries. However, clearly defined goals are necessary for increasing the use of biomass-derived feedstocks in industrial chemical production and it is important to keep the goal in perspective. In this context, the increased use of biofuels should be viewed as one of a range of possible measures for achieving self-sufficiency in energy, rather than a panacea (Crocker and Crofcheck, 2006).

However, for many staple food crops, a potentially large economic resource is effectively being thrown away. For example, the straw associated with the wheat crop in often ploughed back into the soil, even though only a small proportion is needed to maintain the level of organic matter. Thus, a huge renewable resource is not being usefully exploited since wheat straw contains a range of potentially useful chemicals. These include: (i) cellulose and related compounds which can be used for the production of paper and/or bioethanol, (ii) silica compounds which can be used as filter materials such as those necessary for water purification, and (iii) long-chain lipids which can be used in cosmetics or for other specialty chemicals (Figs. 7.1 and 7.2).

Biomass is material that is derived from plants (Wright et al., 2006) and there are many types of biomass resources currently used and potentially available. *Biomass* is a term used to describe any material of recent biological origin, including plant materials such as trees, grasses, agricultural crops, and even animal manure. Other biomass components, which are generally present in minor amounts, include triglycerides, sterols, alkaloids, resins, terpenes, terpenoids and waxes. This includes everything from primary sources of crops and residues harvested/collected directly from the land to secondary sources such as sawmill residuals, to tertiary sources of postconsumer residuals that often end up in landfills. A fourth source, although not usually categorized as such, includes the gases that result from anaerobic digestion of animal manures or organic materials in landfills (Wright et al., 2006).

Direct biofuels are biofuels that can be used in existing gasoline or diesel engines. Because engine technology changes all the time, direct biofuel can be hard to define; a fuel that works well in one unmodified engine may not work in another. In general, newer engines are more sensitive to fuel than older engines, but new engines are also likely to be designed with some amount of biofuel in mind. Straight vegetable oil can be used in many older diesel engines, but only in the warmest climates. Usually it is turned into biodiesel instead. No engine manufacturer explicitly allows any use of vegetable oil in their engines.

Biodiesel can be a direct biofuel. In some countries manufacturers cover many of their diesel engines under warranty for 100% biodiesel use. Many people have run thousands of miles on biodiesel without problem, and many studies have been made on 100% biodiesel.

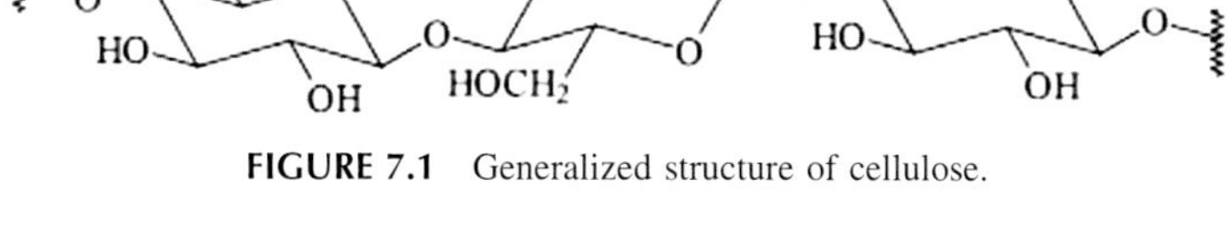

FIGURE 7.1 Generalized structure of cellulose.

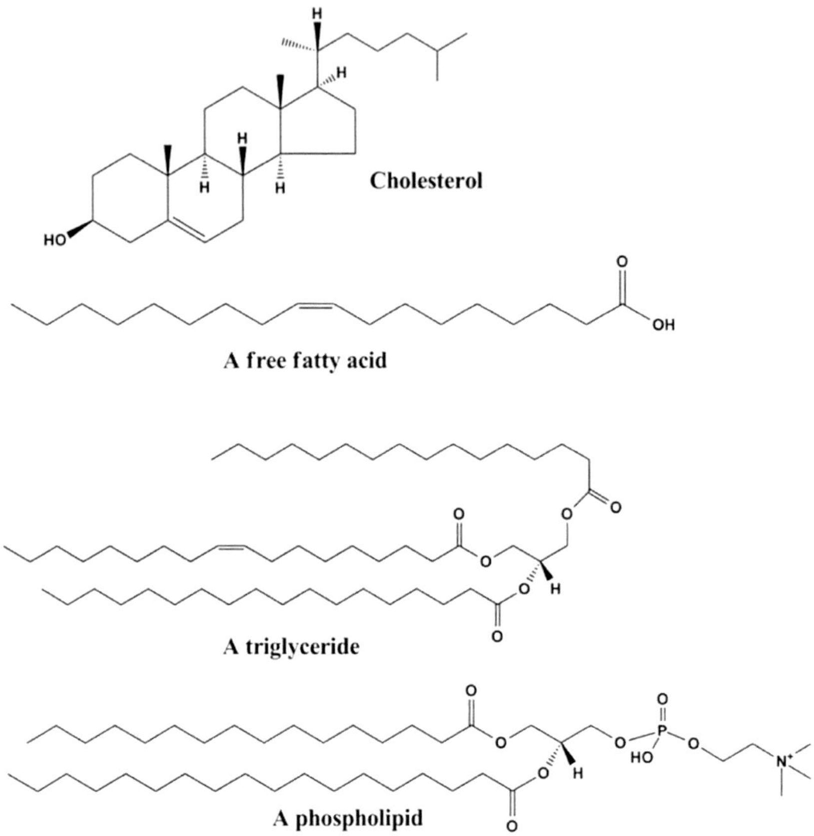

FIGURE 7.2 Chemical structures of some common lipid derivatives.

Butanol is often claimed as a direct replacement for gasoline. It is not in widespread production at this time, and engine manufacturers have not made statements related to the use of butanol. While on paper (and a few lab tests) it appears that butanol has sufficiently similar characteristics with gasoline such that it should work without problem in any gasoline engine, no widespread experience exists.

Ethanol is the most common biofuel, and over the years many engines have been designed to run on it. Many of these could not run on regular gasoline, so it is debatable whether ethanol is a replacement in them. In the late 1990s, engines started appearing that by design can use either fuel. Ethanol is a direct replacement in these engines, but it is debatable if these engines are unmodified, or factory modified for ethanol.

In reality, small amounts of biofuel are often blended with traditional fuels. The biofuel portion of these fuels is a direct replacement for the fuel they offset, but the total offset is small. For biodiesel, 5% or 20% are commonly approved by various engine manufacturers.

Plants offer a unique and diverse feedstock for chemicals. Plant biomass can be gasified to produce synthesis gas; a basic chemical feedstock and also a source of hydrogen for a future hydrogen economy. In addition, the specific components of plants such as carbohydrates, vegetable oils, plant fiber, and complex organic molecules known as primary and secondary metabolites can be utilized to produce a range of valuable monomers, chemical intermediates, pharmaceuticals, and materials.

Carbohydrates (starch, cellulose, sugars) is starch that is readily obtained from wheat and potato, while cellulose is obtained from wood pulp. The structures of these polysaccharides can be readily manipulated to produce a range of biodegradable polymers with properties similar to those of conventional plastics such as polystyrene foams and polyethylene film. In addition, these polysaccharides can be hydrolyzed, catalytically or enzymatically to produce sugars, a valuable fermentation feedstock for the production of ethanol, citric acid, lactic acid, and dibasic acids such as succinic acid.

Vegetable oils are obtained from seed oil plants such as palm oil, sunflower oil, and soy oil. The predominant source of vegetable oils in many countries is rapeseed oil. Vegetable oils are a major feedstock for the oleochemicals industry (surfactants, dispersants, and personal care products) and are now successfully entering new markets such as diesel fuel, lubricants, polyurethane monomers, functional polymer additives, and solvents.

Plant fibers, which are the lignocellulosic fibers extracted from plants such as hemp and flax can replace cotton and polyester fibers in textile materials and glass fibers in insulation products. Other product include specialty chemical and hydrocarbon derivatives since plants can synthesize highly complex bioactive molecules often beyond the power of laboratories and a wide range of chemicals is currently extracted from plants for a wide range of markets from crude herbal remedies to very high value pharmaceutical intermediates.

More generally, biomass feedstocks are recognized by the specific plant content of the feedstock or the manner in which the feedstocks are produced. For example, *primary biomass feedstocks* are thus primary biomass that is harvested or collected from the field or forest where it is grown. Examples of primary biomass feedstocks currently being used for bioenergy include grains and oilseed crops used for transportation fuel production, plus some crop residues (such as orchard trimmings and nut hulls) and some residues from logging and forest operations that are currently used for heat and power production. In the future it is anticipated that a larger proportion of the residues inherently generated from food crop harvesting, as well as a larger proportion of the residues generated from ongoing logging and forest operations, will be used for bioenergy (Smith, 2006). Additionally, as the bioenergy industry develops, both woody and herbaceous perennial crops will be planted and harvested specifically for bioenergy and product end-uses.

Secondary biomass feedstocks differ from primary biomass feedstocks in that the secondary feedstocks are a byproduct of processing of the primary feedstocks. By *processing* it is meant that there is substantial physical or chemical breakdown of the primary biomass and production of byproducts; *processors* may be factories or animals. Field processes such as harvesting, bundling, chipping, or pressing do not cause a biomass resource that was produced by photosynthesis (e.g., tree tops and limbs) to be classified as secondary biomass.

Specific examples of secondary biomass includes sawdust from sawmills, black liquor (which is a byproduct of paper making), and cheese whey (which is a byproduct of cheese-making processes). Manures from concentrated animal feeding operations are collectable secondary biomass resources. Vegetable oils used for biodiesel that are derived directly from the processing of oilseeds for various uses are also a secondary biomass resource.

Tertiary biomass feedstock includes postconsumer residues and wastes, such as fats, greases, oils, construction and demolition wood debris, other waste wood from the urban environments, as well as packaging wastes, municipal solid wastes, and landfill gases. A category *other wood waste from the urban environment* includes trimmings from urban trees, which technically fits the definition of primary biomass. However, because this material is normally handled as a waste stream along with other postconsumer wastes from urban environments (and included in those statistics), it makes most sense to consider it to be part of the tertiary biomass stream.

2. Biomass feedstocks

More generally, biomass feedstocks are recognized or classified by the specific plant content of the feedstock or the manner in which the feedstocks are produced. For example, *primary biomass feedstocks* are thus primary biomass that is harvested or collected from the field or forest where it is grown. Examples of primary biomass feedstocks currently being used for bioenergy include grains and oilseed crops used for transportation fuel production, plus some crop residues (such as orchard trimmings and nut hulls) and some residues from logging and forest operations that are currently used for heat and power production.

Secondary biomass feedstocks differ from *primary biomass feedstocks* in that the secondary feedstocks are a byproduct of processing of the primary feedstocks. By *processing* it is meant that there is substantial physical or chemical breakdown of the primary biomass and production of byproducts; *processors* may be factories or animals. Field processes such as harvesting, bundling, chipping, or pressing do not cause a biomass resource that was produced by photosynthesis (e.g., tree tops and limbs) to be classified as secondary biomass.

Tertiary biomass feedstocks often include *fats and greases*, which are byproducts of the reduction of animal biomass into component parts, since most fats and greases, and some oils, are not available for bioenergy use until after they become a postconsumer waste stream. Vegetable oils derived from processing of plant components and used directly for bioenergy (e.g., soybean oil used in biodiesel) would be a secondary biomass resource, though amounts being used for bioenergy are most likely to be tracked together with fats, greases, and waste oils.

The production of fuels and chemicals from renewable plant-based feedstocks utilizing state-of-the-art conversion technologies presents an opportunity to maintain competitive advantage and contribute to the attainment of national environmental targets. Bioprocessing routes have a number of compelling advantages over conventional petrochemicals production; however, it is only in the last decade that rapid progress in biotechnology has facilitated the commercialization of a number of plant-based chemical processes.

Plants offer a unique and diverse feedstock for chemicals and the production of *biofuels* from biomass requires some knowledge of the chemistry of biomass, the chemistry of the individual constituents of biomass, and the chemical means by which the biomass can be converted to fuel. It is widely recognized that further significant production of plant-based chemicals will only be economically viable in highly integrated and efficient production complexes producing a diverse range of chemical products. This biorefinery concept is analogous to conventional oil refineries and petrochemical complexes that have evolved over many years to maximize process synergies, energy integration, and feedstock utilization to drive down production costs.

In addition, the specific components of plants such as carbohydrates, vegetable oils, plant fiber, and complex organic molecules known as primary and secondary metabolites can be utilized to produce a range of valuable monomers, chemical intermediates, pharmaceuticals, and materials.

2.1 Carbohydrates

Plants capture solar energy as fixed carbon during which carbon dioxide is converted to water and sugars $(CH_2O)_x$:

$$CO_2 + H_2O \rightarrow (CH_2O)_x + O_2.$$

The sugars produced are stored in three types of polymeric macromolecules: (i) starch, (ii) cellulose, and (iii) hemicellulose.

In general sugar polymers such as cellulose and starch can be readily broken down to their constituent monomers by hydrolysis, preparatory to conversion to ethanol or other chemicals (Vasudevan et al., 2005). In contrast, lignin is an unknown complex structure containing aromatic groups that is totally hypothetical and is less readily degraded than starch or cellulose.

Although lignocellulose is one of the cheapest and most abundant forms of biomass, it is difficult to convert this relatively unreactive material into sugars. Among other factors, the walls of lignocellulose are composed of lignin, which must be broken down in order to render the cellulose and hemicellulose accessible to acid hydrolysis. For this reason, many efforts focused on ethanol production from biomass are based almost entirely on the fermentation of sugars derived from the starch in corn grain.

2.2 Vegetable oils

In many cases, it has been advocated that vegetable oil, and similar feedstocks, be used as feedstocks for a catalytic cracking unit. The properties of the product(s) can be controlled by controlling the process variables including the cracking temperature as well as the type of catalyst used. The production of biodiesel by direct esterification of fatty acids with short chain alcohols occurs in one step only whereby acidic catalysts can be used to speed up the reaction (Demirbaş, 2006).

2.3 Plant fibers

Lignocellulosic fibers extracted from plants such as hemp and flax can replace cotton and polyester fibers in textile materials and glass fibers in insulation products. Lignin is a complex chemical that is most commonly derived from wood and is an integral part of the cell wall of plants. The chemical structure of lignin is unknown and, at best, can only be represented by hypothetical formulas.

Lignin (Latin: *lignum— wood*) is one of most abundant organic chemicals on earth after cellulose and chitin. By way of clarification, chitin $[(C_8H_{13}O_5N)_n]$ is a long-chain polymeric polysaccharide of β-glucose that forms a hard, semitransparent material found throughout the natural world. Chitin is the main component of the cell walls of fungi and is also a major component of the exoskeletons of arthropods, such as the crustaceans (e.g., crab, lobster, and shrimp), and insects (e.g., ants, beetles, and butterflies), and the beaks of cephalopods (e.g., squids and octopuses).

Lignin makes up approximately one-quarter to one-third of the dry mass of wood and is generally considered to be a large, cross-linked hydrophobic, aromatic macromolecule with a molecular mass that is estimated to be in excess of 10,000. Lignin fills the spaces in the cell wall between cellulose, hemicellulose, and pectin components and is covalently linked (bonded) to hemicellulose. Lignin also forms covalent bonds with polysaccharides which enables cross-linking to different plant polysaccharides. Lignin confers mechanical strength to the cell wall (stabilizing the mature cell wall) and therefore the entire plant.

Lignin is a complex chemical compound that is most commonly derived from wood and is an integral part of the cell walls of plants, especially in tracheids, xylem fibers, and sclereids—small bundles of tissue in plants that form durable layers. Lignin was recognized as the carbon rich *encrusting material* which embedded cellulose in the wood (Sarkanen and Ludwig, 1971). Lignin fills the spaces in the cell wall between cellulose, hemicellulose, and pectin components and is covalently linked to hemicellulose. Lignin also forms covalent bonds to polysaccharides and thereby cross-links different plant polysaccharides (Erikkson and Lindgren, 1977; Karhunen et al., 1995). It confers mechanical strength to the cell wall (stabilizing the mature cell wall) and therefore the entire plant.

Lignin makes up approximately one-quarter to one-third of the dry mass of wood and is generally considered to be a large, cross-linked hydrophobic, aromatic macromolecule with molecular mass that is estimated to be in excess of 10,000. Degradation studies indicate that the molecule consists of various types of substructures which appear to repeat in a random manner.

Lignin is one of most abundant organic compounds on Earth after cellulose and chitin— chitin $(C_8H_{13}O_5N)_n$ is a long-chain polymeric polysaccharide of beta-glucose that forms a hard, semitransparent material found throughout the natural world. Chitin is the main component of the cell walls of fungi and is also a major component of the exoskeletons of arthropods, such as the crustaceans (e.g., crab, lobster, and shrimp), and the insects (e.g., ants, beetles, and butterflies), and of the beaks of cephalopods (e.g., squids, and octopuses).

Lignin has been speculatively described as a random, three-dimensional network polymer comprised of variously linked phenylpropane units (Sjöström, 1993) but the true chemical structure of lignin remains unknown and, at best, can only be represented by hypothetical formulas (Fig. 7.3). However, lignin is the second most abundant biological material on the planet, exceeded only by cellulose and hemicellulose, and comprises 15%–25% w/w of the dry weight of woody plants.

This macromolecule plays a vital role in providing mechanical support to bind plant fibers together. Lignin also decreases the permeation of water through the cell walls of the xylem, thereby playing an intricate role in the transport of water and nutrients. Finally, lignin plays an important function in the natural defense of a plant against degradation by impeding penetration of destructive enzymes through the cell wall (Sarkanen and Ludwig, 1971; Sjöström, 1993). Although lignin is necessary to trees, it is undesirable in most chemical papermaking fibers and is removed by pulping and bleaching processes.

Plant lignins can be broadly divided into three classes: softwood (gymnosperm), hardwood (angiosperm), and grass or annual plant (graminaceous) lignin (Pearl, 1967). Three different phenylpropane units, or monolignols, are responsible for lignin biosynthesis (Freudenberg and Neish, 1968). Guaiacyl lignin is composed principally of coniferyl alcohol units, while guaiacyl-

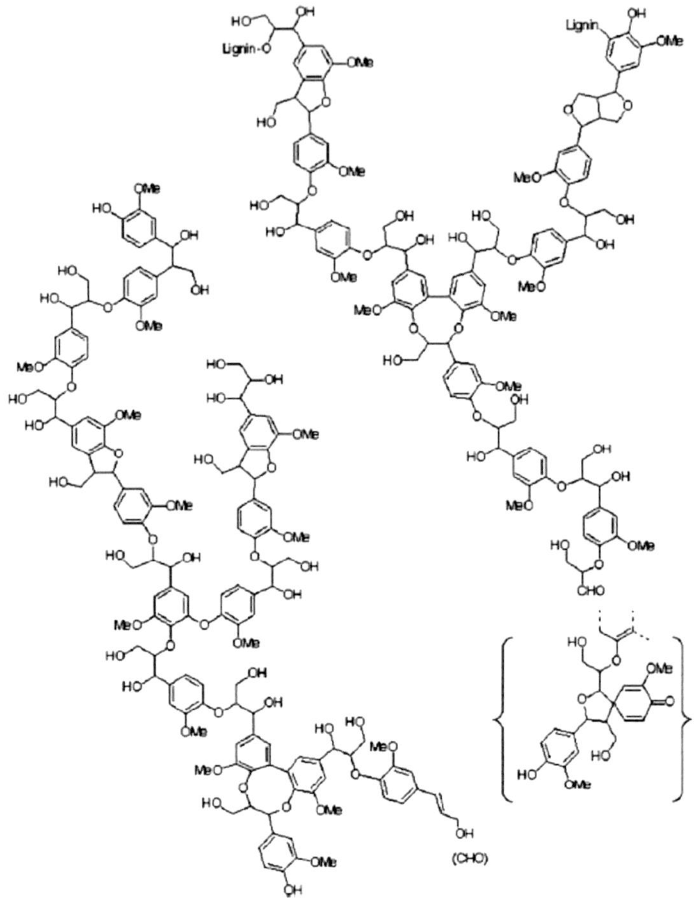

FIGURE 7.3 Hypothetical structural model for softwood lignin used here *only* to illustrate the potential complexity of the lignin molecule.

syringyl lignin contains monomeric units from coniferyl and sinapyl alcohol. In general, guaiacyl lignin is found in softwoods while guaiacyl-syringyl lignin is present in hardwoods. Graminaceous lignin is composed mainly of *p*-coumaryl alcohol units.

While the structure of native lignin remains unclear, the dominant structures in lignin have been elucidated as the methods for identification of the degradation products and for the synthesis of model compounds have improved. The results from these numerous studies have yielded what is believed to be an accurate representation of the structure of lignin. Examples of the elucidated structural features of lignin include the dominant linkages between the phenylpropane units and their abundance, as well as the abundance and frequency of some functional groups. Linkages between the phenylpropane units and the various functional groups on these units give lignin a unique and very complex structure.

The lignin macromolecule (Fig. 7.2) also contains a variety of phenylpropane functional groups that have an impact on its reactivity. In addition, lignin contains methoxyl groups, phenolic hydroxyl groups, and few terminal aldehyde groups. Only a small proportion of the phenolic hydroxyl groups are free, since most are occupied in linkages to neighboring phenylpropane linkages. Carbonyl and alcoholic hydroxyl groups are incorporated into the lignin structure during enzymatic dehydrogenation.

2.4 Waste

It would be remiss not to mention another potential feedstock for the production of chemicals—waste material that is not included under the general category of biomass (John and Singh, 2011). Nonbiomass waste is a byproduct of life and civilization; it is the material that remains after a useful component has been consumed. From an economic perspective, waste is a material involved in life or technology whose value is less than the cost of its utilization. From a regulatory viewpoint, waste is anything discarded or that can no longer be used for its original purpose.

Waste is the general term; though the other terms are used loosely as synonyms, they have more specific meanings The term *solid waste* includes not only solid materials but also liquid and gases. *Domestic waste* (also known as rubbish, garbage, trash, or junk) is unwanted or undesired material. Rubbish or trash are mixed household waste including paper and packaging; food waste or garbage (North America) is kitchen and table waste; and junk or scrap is metallic or industrial material.

The thermal pyrolysis of plastic wastes produces a broad distribution of hydrocarbon derivatives, from methane to waxy products. This process takes place at high temperatures. The gaseous compounds generated can be burned out to provide the process heat requirements, but the overall yield of valuable gasoline range hydrocarbon derivatives is poor, so that the pyrolysis process as a means for feedstock recycling of the plastic waste stream is rarely practiced on an industrial scale at present (Predel and Kaminsky, 2000; Kaminsky and Zoriqueta, 2007). In contrast, thermal cracking at low temperatures is usually aimed at the production of waxy oil fractions, which may be used in industrial units for steam cracking and in fluid catalytic cracking units (Aguado et al., 2002). An alternative to improve the yield of naphtha from the pyrolysis of plastic waste is to introduce suitable catalysts. High conversion and interesting product distribution is obtained when plastics are cracked over zeolites (Hernandez et al., 2007). Moreover the catalytic cracking of polymers has proven itself to be a very versatile process, since a variety of products can be obtained depending on parameters such as (i) the catalyst, (ii) the polymer feedstocks, (iii) the reactor type, and (iv) the process parameters, such as temperature, pressure, and residence time of the feedstock in the hot zone as well as (v) product removal from the hot zone (Aguado and Serrano. 1999; Demirbaş, 2004; Scheirs and Kaminsky, 2006; Marcilla et al., 2008; Al-Salem et al., 2009; Sarker et al., 2012).

In addition, urban waste (domestic and industrial) has considerable promise as a feedstock for gasification because it contains relatively more lignin, which biological processes cannot convert. Such waste is abundant in most countries and can be harnessed for production of fuels and petrochemical intermediates. Knowing the potential of the waste for gasification and subsequent fuel production is essential for reducing pressure on traditional energy sources.

Also, discarded tires can be reduced in size by grinding, chipping, pelletizing, and passed through a classifier to remove the steel belting after which the chips are pyrolyzed for 1 hour at a temperature of 300–500°C (570–930°F) and then heated for 2 hours in a closed retort to yield gas, distillable, and char. Discarded tires can also be shredded to 25 mm and ground to 24 mesh as a feed-preparation step for occidental flash pyrolysis that involves flash pyrolysis and product collection. The pyrolytic reaction occurs without the introduction of hydrogen or using a catalyst. This yields a gaseous stream that is passed to a quench tower from which fuel oil and gas (recycled to char fluidized and pyrolysis reactor as a supplemental fuel) and carbon black (35% w/w) is produced. In the Nippon Zeon process crushed tire chips undergo fluidized thermal cracking (fluidized bed, 400–600°C, 750–1110°F) which yields a gaseous stream that is passed to a quench tower from which gas and distillable oil is produced. All of the end products produced could be used directly as a supplemental fuel source at the plant or sent offsite for petrochemical manufacture.

3. Hydrocarbons from biomass

Plants as a source of hydrocarbon and rubber have been investigated periodically for many years. However, during the last few decades the need for additional sources has resurfaced since the world production of natural rubber is expected to be insufficient for the demand. Thus, biomass is a renewable energy source, unlike the fossil fuel resources (crude oil, coal, and natural gas) and can be subdivided into a variety of categories (Table 7.1). The energy of the sun is captured through the process of photosynthesis in growing plants. One advantage of biofuel in comparison to most other fuel types is it is biodegradable, and thus relatively harmless to the environment if spilled.

3.1 Isoprenoid hydrocarbons

Hydrocarbon derivatives in plants, such as natural rubber (polyisoprene) have chemical structures similar to many hydrocarbon derivatives derived from crude

TABLE 7.1 General categories of biomass[a].

Forest residues
Tree branches, tops of trunks, stumps, leaves.
Industrial waste
Citrus peels, sugarcane bagasse, olive husks, milling residues.
Energy crops
Switch grass, miscanthus, bamboo, sweet sorghum, tall fescue, wheatgrass.
Landfill gas and biogas
Bacteria and fungi consume dead plants and animals, causing them to rot or decay. A fungus on a rotting log is converting cellulose to sugars to feed itself. Although this process is slowed in a landfill, methane gas is still produced as the waste decays. Methane gas is colorless and odorless and can cause fires or explosions if it seeps into nearby homes and is ignited. Landfills can be set up to collect the methane gas, purify it, and use it as fuel. Methane can also be produced using energy from agricultural and human wastes. Biogas digesters are airtight containers or pits lined with steel or bricks. Waste placed into the containers is fermented without oxygen to produce a methane-rich gas.
Solid waste
Burning trash turns waste into a usable form of energy. Garbage is not all biomass; perhaps half of its energy content comes from plastics, which are made from petroleum and natural gas. Power plants that burn garbage for energy are called waste-to-energy plants.
Wood and agricultural products
Wood in the form of logs, chips, bark, and sawdust accounts for about 44% of biomass energy. Wood and wood waste are used to generate electricity. Much of the electricity is used by the industries making the waste. Paper mills and saw mills use much of their waste products to generate steam and electricity for their use.

[a]*Listed alphabetically rather than by any preference.*

oil. Natural rubber is the most common hydrocarbon polymer found in green plants. Low molecular weight natural rubber would be of interest as a plastic additive (processing aid) to rubber mixes, for making cements (adhesives), and if economically feasible as hydrocarbon feedstocks. Such materials, when fractured, will produce hydrocarbon derivatives of lower molecular weight which can be used as alternative energy sources for fuel and/or chemical raw materials that are used in the manufacture of a large number of products.

Polymeric isoprenoid hydrocarbon derivatives have also been identified. Rubber is undoubtedly the best known and most widely used compound of this kind. It occurs as a colloidal suspension called latex in a number of plants, ranging from the dandelion to the rubber tree (*Hevea brasiliensis*). Rubber is a polyene, and exhibits all the expected reactions of the C=C function. Bromine, hydrogen chloride, and hydrogen all add with a stoichiometry of 1 molar equivalent per isoprene unit. Ozonolysis of rubber generates a mixture of levulinic acid ($CH_3COCH_2CH_2CO_2H$) and the corresponding aldehyde. Pyrolysis of rubber produces the diene isoprene along with other products.

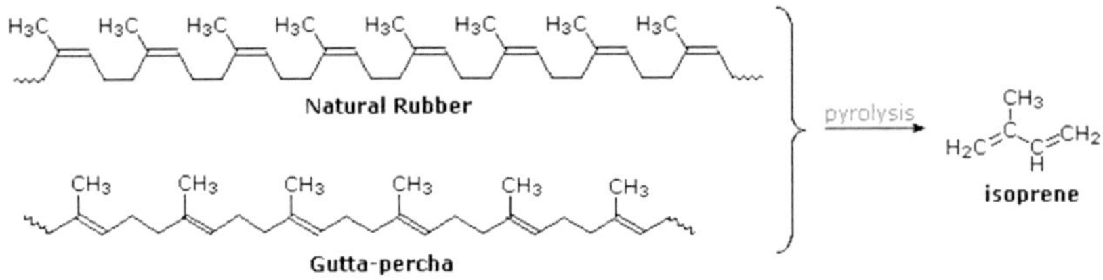

The double bonds in rubber all have a Z-configuration, which causes this macromolecule to adopt a kinked or coiled conformation. This is reflected in the physical properties of rubber. Despite its high molecular weight (approximately one million), crude latex rubber is a soft, sticky, elastic substance. Chemical modification of this material is normal for commercial applications. Gutta-percha (structure above) is a naturally occurring isomer of rubber. Here the hydrocarbon chains adopt a uniform zigzag or rodlike conformation, which produces a more rigid and tough substance. Uses of gutta-percha include electrical insulation and the covering of golf balls.

3.2 Waxes

In contrast to ozocerite, the waxes isolated from plants are esters of fatty acids with long-chain monohydric alcohols (one hydroxyl group). Natural waxes are often mixtures of such esters, and may also contain hydrocarbon derivatives. The formulas for three well-known waxes are given below:

Spermaceti	Beeswax	Carnauba wax
$CH_3(CH_2)_{14}CO_2\ (CH_2)_{15}$ CH_3	$CH_3(CH_2)_{24}CO_2(CH_2)_{29}$ CH_3	$CH_3(CH_2)_{30}CO_2(CH_2)_{33}$ CH_3

These and other similar waxes are widely distributed in nature. The leaves and fruits of many plants have waxy coatings, which may protect them from dehydration and small predators. The feathers of birds and the fur of some animals have similar coatings which serve as a water repellent.

As an example, Carnauba wax (Brazil wax, palm wax) is valued for its toughness and water resistance and is a wax of the leaves of the palm, *Copernicia prunifera*, a plant native to and grown only in northeastern Brazil. It comes in the form of hard yellow-brown flakes and is obtained from the leaves of the carnauba palm by collecting them, beating them to loosen the wax, then refining and bleaching the wax.

Carnauba wax contains mainly esters of fatty acids (80%–85%), fatty alcohols (10%–16%), acids (3%–6%), and hydrocarbon derivatives (1%–3%). Specific for carnauba wax is the content of esterified fatty diols (approximately 20%), hydroxylated fatty acids (approximately 6%), and cinnamic acid (approximately 10%).

3.3 Essential oils

Complex hydrocarbon derivatives and their derivatives are found throughout nature. Natural rubber, for example, is a hydrocarbon that contains long chains of alternating C=C double bonds and C–C single bonds.

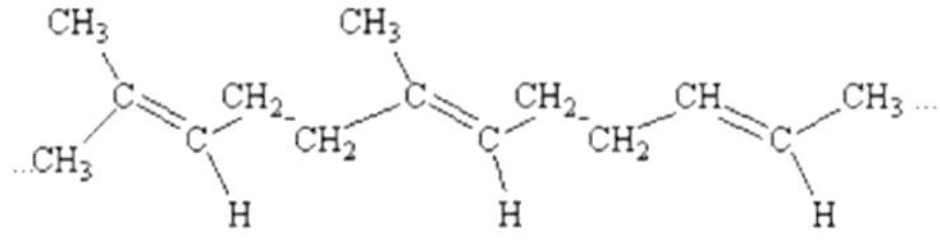

Writing the structure of complex hydrocarbon derivatives can be simplified by using a line notation in which a carbon atom is assumed to be present wherever a pair of lines intersects and enough hydrogen atoms are present to satisfy the tendency of carbon to form a total of four bonds.

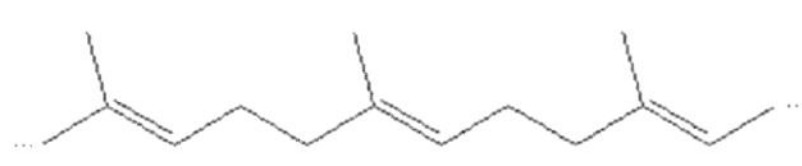

There are a variety of techniques for isolating both pleasant and foul-smelling compounds known as *essential oils* from natural sources, particularly from plants. These compounds are not "essential," in the sense of being vital to life. They were given that name because they give off a distinct "essence," or smell.

The essential oils are used in perfumes and medicines. Some of these compounds can be isolated by gently heating, or steam distilling, the crushed flowers of plants. Others can be extracted into nonpolar solvents, or absorbed onto grease-coated cloths in which the plants are wrapped. Many of these essential oils belong to classes of compounds known as *terpenes* and *terpenoids*. The terpenes are hydrocarbon derivatives that usually contain one or more C=C double bonds. The terpenoids are oxygen-containing analogs of the terpenes.

3.4 Terpenes

Terpenes are a common, yet unique group of hydrocarbon molecules that share structures based on multiple condensations of five-carbon (isoprene) building

TABLE 7.2 An example of the classification of terpene derivatives.

Terpene type	Description
Hemiterpenes	5 carbon atoms or 1 isoprene unit
Monoterpenes	10 carbon atoms or 2 isoprene units
Sesquiterpenes	15 carbon atoms or 3 isoprene units
Diterpenes	20 carbon atoms or 4 isoprene units
Sesterpenes	25 carbon atoms or 5 isoprene units
Triterpenes	30 carbon atoms or 6 isoprene units
Triterpenes may be further grouped into the following subclasses: Triterpenes (ursolic acid, lupeol)	
Steroids	
Saponins	
Sterolins	
Cardiac glycosides (Digitalis)	

blocks (Table 7.2). Organisms synthesize terpenes ranging in complexity and biological activity. Simple terpenes are volatile, evaporating quickly, and are considered the essential oils that imbue plant unique odors. These odors may attract or repel other organisms as needed for survival. More complex terpenes consisting of several isoprene units may be precursors to bioactive molecules like cholesterol, steroid hormones, or waxy substances that act as protective coverings. Notable plants with potent terpenes include: (i) neem (*Azadirachta indica*), (ii) menthol (*Plectranthus* sp *Menthol Plant*), (iii) common foxglove (*Digitalis purpurea*), and varuna (*Crataeva nurvala*).

Compounds classified as terpenes constitute what is arguably the largest and most diverse class of natural products. A majority of these compounds are found only in plants, but some of the larger and more complex terpenes (e.g., squalene and lanosterol) occur in animals. Terpenes incorporating most of the common functional groups are known, so this does not provide a useful means of classification. Instead, the number and structural organization of carbons is a definitive characteristic.

Terpenes may be considered to be made up of isoprene (more accurately isopentane) units, an empirical feature known as the *isoprene rule*. Because of this, terpenes usually have 5n carbon atoms (n is an integer), and are subdivided (Table 7.3). Isoprene itself, a C_5H_8 gaseous hydrocarbon, is emitted by the leaves of various plants as a natural byproduct of plant metabolism. Next to methane it is the most common volatile organic compound found in the atmosphere. The

TABLE 7.3 Classification of terpenes.

Classification	Isoprene units	Carbon atoms
Monoterpenes	2	C_{10}
Sesquiterpenes	3	C_{15}
Diterpenes	4	C_{20}
Sesterpenes	5	C_{25}
Triterpenes	6	C_{30}

isopentane units in most of these terpenes are easy to discern, and are defined by the shaded areas. In the case of the mono-terpene camphor, the units overlap to such a degree it is easier to distinguish them by coloring the carbon chains. This is also done for alpha-pinene. In the case of triterpene lanosterol we see an interesting deviation from the isoprene rule. This 30 carbon compound is clearly a terpene, and four of the six isopentane units can be identified.

Examples of terpenes include -pinene and -pinene, the primary components of turpentine that give rise to its characteristic odor.

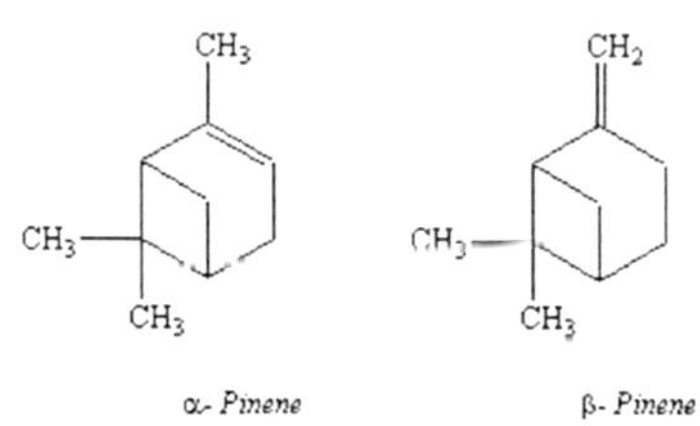

Camphor and menthol are examples of terpenoids.

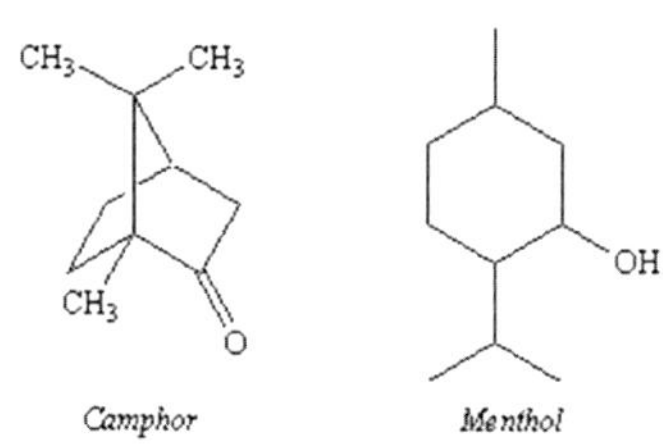

Both of these compounds have a fragrant, penetrating odor and taste cool. Camphor is used as a moth repellent. Menthol is a mild anesthetic that is added to some brands of cigarettes. The terpenoids also include compounds such as geranial and neral, a pair of cis/trans stereoisomers that can be found in lemon oil. Geranial has a strong lemon odor. Neral tastes sweeter, but has a less intense odor.

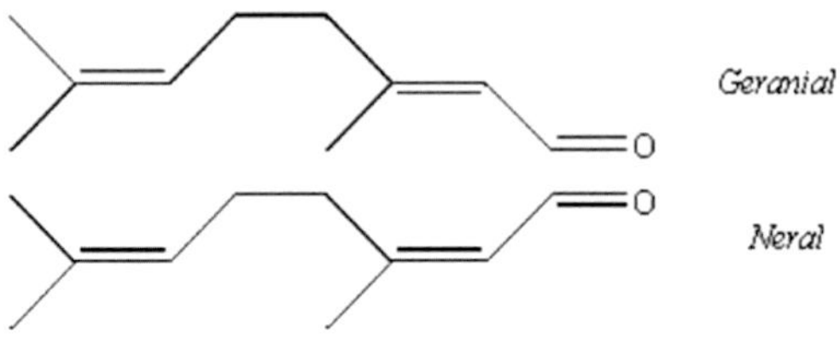

Although the terpenes and terpenoids discussed so far have very different structures, they have one important property in common: they all contain 10 carbon atoms, neither more nor less. Each of these compounds can be traced back to a reaction in which a pair of five-carbon molecules is fused. Thus, it is not surprising that sesquiterpenes (15 carbon atoms), diterpenes (20 carbon atoms), sesterpenes (25 carbon atoms), and triterpenes (30 carbon atoms) exist. Important examples of these compounds include vitamin A and the b-carotene that gives carrots their characteristic color.

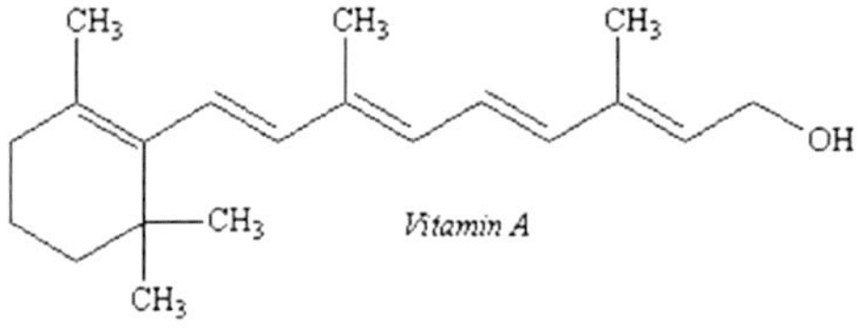

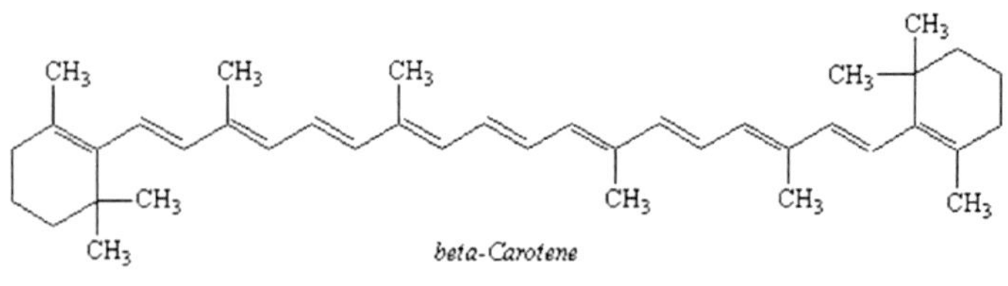

3.5 Steroids

By definition, *steroids* are compounds that have the basic structure formed by fusing three six-membered rings and a five-membered ring. The most important property of this molecule is the fact that, with the exception of the −OH group on the lower-left-hand corner of the molecule.

Steroids are not terpenes or terpenoids in the literal sense because they don't contain the characteristic number of carbon atoms. Consider cholesterol, for example, which is one of the most important steroids.

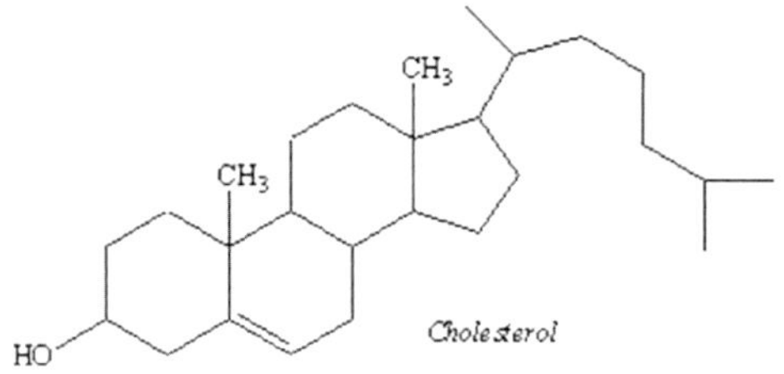

Analysis of this structure suggests formula $C_{27}H_{46}O$, which doesn't fit the pattern expected of a terpenoid. The biosynthetic precursor of this molecule, however, is a 30-carbon triterpene that is converted into cholesterol by a series of enzyme-catalyzed reactions.

The important class of lipids called **steroids** is actually metabolic derivatives of terpenes, but it is customarily treated as a separate group. Steroids may be recognized by their tetracyclic skeleton, consisting of three fused six-membered and one five-membered ring, as shown in the diagram to the right. The four rings are designated A, B, C & D as noted, and the peculiar numbering of the ring carbon atoms (shown in red) is the result of an earlier mis-assignment of the structure. The substituents designated by R are often alkyl groups, but may also have functionality. The R group at the A:B ring fusion is most commonly methyl or hydrogen, that at the C:D fusion is usually methyl. The substituent at C-17 varies considerably, and is usually larger than methyl if it is not a functional group. The most common locations of functional groups are C-3, C-4, C-7, C-11, C-12 & C-17. Ring A is sometimes aromatic. Since a number of tetracyclic triterpenes also have this tetracyclic structure, it cannot be considered a unique identifier.

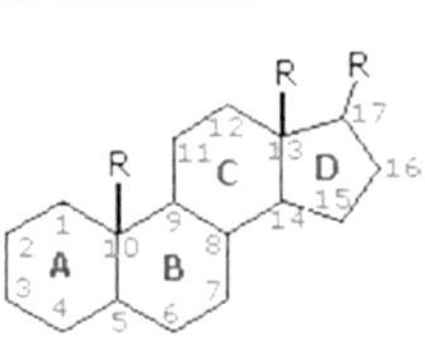

The Steroid
Carbon Skeleton

Steroids are widely distributed in animals, where they are associated with a number of physiological processes. Examples of some important steroids are shown in the following diagram.

Norethindrone is a synthetic steroid, all the other examples occur naturally. A common strategy in pharmaceutical chemistry is to take a natural compound, having certain desired biological properties together with undesired side effects, and to modify its structure to enhance the desired characteristics and diminish the undesired.

The generic steroid structure drawn above has seven chiral stereocenters (carbons 5, 8, 9, 10, 13, 14, and 17), which means that it may have as many as 128 stereoisomers. With the exception of C-5, natural steroids generally have a single common configuration. This is shown in the last of the toggled displays, along with the preferred conformations of the rings.

It is useful to recognize that it is incorrect to label products such as peanut butter as cholesterol-free. That is like saying that the Sahara desert is rain-free. Peanut butter is made from peanuts and cholesterol isn't a characteristic ingredient in plants; it is synthesized by animals, particularly mammals. It is also useful to note that placing someone on a cholesterol-free diet won't reduce their cholesterol level to zero. Even on a low-cholesterol diet, the individual will synthesize approximately 0.80 g of cholesterol per day. The key question is: Is there excess of cholesterol in the blood stream? If there is, a diet that reduces the intake of cholesterol might be important.

Cholesterol is the biosynthetic precursor for the synthesis of all of the major classes of *hormones*, the chemical messengers that coordinate the activity of different cells in a multicellular organism. The steroid hormones include the progestogens, estrogens, and androgens.

Progesterone is an example of a progestogen. This hormone plays a vital role in pregnancy. After ovulation, the corpus luteum secretes progesterone, which prepares the lining of the uterus for implantation of the fertilized ovum. Progesterone is then released by the placenta throughout pregnancy to suppress ovulation. Progesterone was therefore the model on which the first oral contraceptives were built. Progesterone itself is not a good oral contraceptive because this hormone is degraded in the digestive system. It therefore requires massive doses of progesterone to prevent pregnancy when this hormone is taken orally.

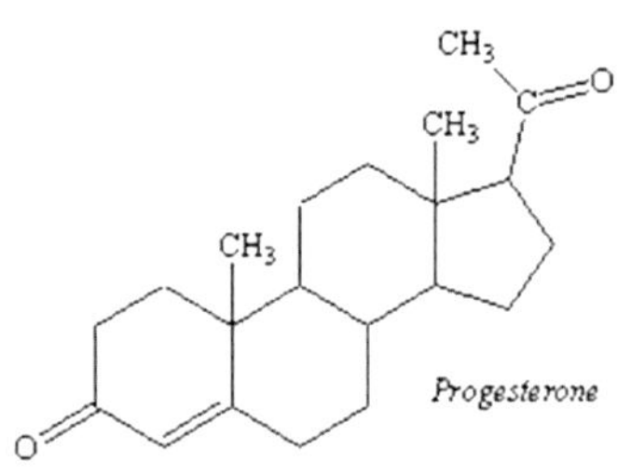

Progesterone

The estrogen hormones, such as estrone and estradiol serve three functions. First, they are responsible for the development of the secondary sex characteristics that appear at the onset of puberty in women. Second, they participate in both the ovarian and estrus cycles, and are therefore another model for the design of oral contraceptives. Third, they stimulate the development of the mammary glands during pregnancy.

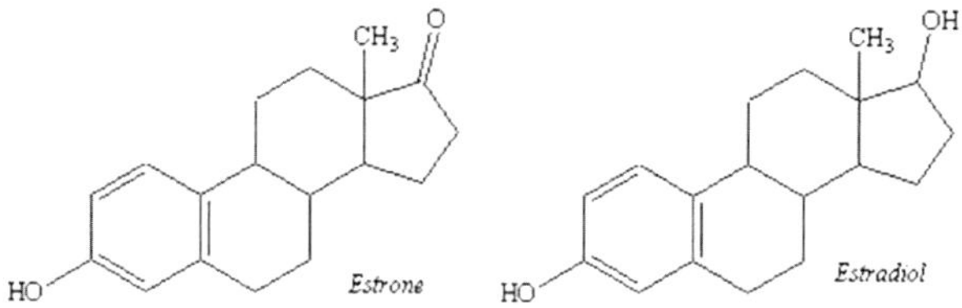

The androgen hormones, such as androsterone and testosterone play an equivalent role in men, where they are responsible for the secondary sex characteristics that appear at puberty. Testosterone is the true male sex hormone; androsterone is a metabolized form of this steroid that is excreted in the urine.

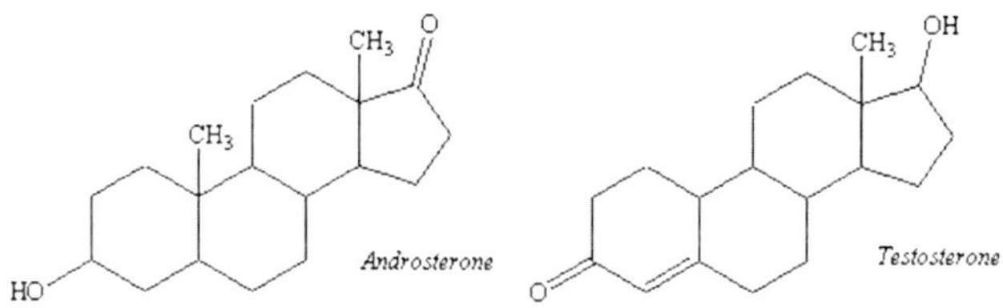

4. Production of Hydrocarbons by conversion

In addition to the synthesis and production of hydrocarbon derivatives by woody plants and herbaceous plants, biomass can also be converted into hydrocarbon fuels. Biomass can be converted into commercial fuels, suitable to substitute for fossil fuels. These can be used for transportation, heating, electricity generation, or anything else fossil fuels are used for. The conversion is accomplished through the use of several distinct processes which include both *thermal conversion* and *biochemical conversion* to produce gaseous, liquid, and solid fuels which have high energy content, are easily transportable, and are therefore suitable for use as commercial fuels.

4.1 Hydrocarbons from wood

The conversion of lignin and lignocellulosic material to hydrocarbon derivatives is difficult. Nevertheless there are three prominent pathways: (i) via methanol, (ii) via ethanol, and (iii) via gasification to synthesis gas.

4.2 Hydrocarbons via methanol and ethanol

Liquid fuels that could be suitable for use in transportation vehicles have been made from wood for a long time. Methanol was commonly called wood alcohol, and this term is still used. Cellulose which is the largest wood component could be dissolved in concentrated acid solutions and converted to sugar, a precursor for making ethanol. A dilute sulfuric acid hydrolysis process was used to make ethanol during World War I and wood hydrolysis received considerable attention in Europe during the period between the World Wars I and II. Wood hydrolysis plants continue to operate in Russia.

However, methanol and ethanol are not the only transportation fuels that might be made from wood. A number of possibilities exist for producing alternatives. The most promising biomass fuels, and closest to being competitive in current markets without subsidy are (i) ethanol, (ii) methanol, (iii) ethyl-t-butyl ether, and (iv) methyl-t-butyl ether. Other candidates include isopropyl alcohol, sec-butyl alcohol, t-butyl alcohol, mixed alcohols, and t-amyl methyl ether.

During the energy crisis of the 1970s and 1980s, alternatives to fuels derived from crude oil became necessary. Up to that time, only two processes of fuel synthesis had any commercial significance. The first was the Bergius process that used oil-coal slurry and an iron catalyst to produce synthetic crude oil. The second was the Fischer–Tropsch process, which produced hydrocarbon derivatives from coal. Both of these processes produced hydrocarbon derivatives with poor selectivity and quality. This problem was overcome by the Mobil methanol-to-gasoline (MTG) process. The Mobil process of methanol conversion over a highly selective zeolite catalyst makes possible the synthesis of a high quality, high octane gasoline.

The conversion of methanol-to-hydrocarbon derivatives (MTHC) on acidic zeolite catalysts is considered to be one of the most promising routes for producing hydrocarbon derivatives boiling in the gasoline range and chemicals (Jayamurthy and Vasudevan, 1996). With the increasing consumption demands for low-boiling olefins, the methanol-to-olefin (MTO) process, one close relative of methanol to hydrocarbon derivative process becomes more significant.

It has been well established that the first step of the methanol-to-olefin process is the dehydration of methanol to form the equilibrium mixture among methanol, dimethyl ether, and water. Subsequently, this equilibrium mixture converts to low-boiling olefins, which can further react to form paraffins, aromatics, naphthenes, and higher olefins by hydrogen transfer, alkylation, and polycondensation.

On addition, under steady-state conditions of the methanol-to-olefin process, the formation of large organic compounds acting as coke trapped in the cages of acidic zeolite catalysts is the most important reason of catalyst deactivation in industrial processes.

Methanol is made from synthesis gas (a mixture of CO and H_2), which is formed by steam reforming of natural gas or gasification of coal. First the methane is converted to carbon monoxide and hydrogen, and this is then reacted on $Cu/ZnO/Al_2O_3$ or similar catalysts to produce methanol. Natural gas is first converted via synthesis gas into methanol. The next step is the dehydration of methanol to dimethyl ether (DME) and water. The resulting equilibrium mixture of methanol, dimethyl ether, and water on an acidic solid catalyst is converted to primary low-boiling olefins, such as ethylene and propylene. In the last step, the low-boiling olefins react with paraffins, higher olefins, aromatics, and naphthenes by hydrogen transfer, alkylation, and polycondensation. In the past decades, most of the work concerning the conversion of methanol to hydrocarbon derivatives has been done on acidic zeolite catalysts, which have become an efficient means to selectively produce desired components while minimizing the production of undesired byproducts. In general, the structure of zeolites can be considered as a three-dimensional network of tetrahedra connecting four valence or three valence metal ions such as Si or Al, each having four oxygen atoms as neighbors. And vice versa, each oxygen atom has two metal ions as nearest neighbors.

While there has been at least 20 distinct mechanisms proposed for the methanol-to-olefin process, there is a consensus that the formation of low-boiling olefins is dominated by a *hydrocarbon-pool* route in which methanol is directly added onto these reactive organic compounds, while low-boiling olefins are formed via an elimination from these compounds. However, the first C—C bond formation and the detailed chemistry of the methanol-to-olefin process still remains a matter of debate.

4.2.1 Hydrocarbons from ethanol

The search for new energy sources has also initiated investigations of hydrocarbon production from ethanol.

Ethanol is a volatile, colorless liquid that has a strong characteristic odor. It burns with a smokeless blue flame that is not always visible in normal light. The physical properties of ethanol stem primarily from the presence of its hydroxyl group and the shortness of its carbon chain. The hydroxy group (-OH) of ethanol is able to participate in hydrogen bonding, rendering it more viscous and less volatile than less polar organic compounds of similar molecular weight.

The most obvious route to hydrocarbon derivatives from ethanol is dehydration, i.e., the removal of the elements of water to produce ethylene. Strong acid desiccants cause the dehydration of ethanol to form ethylene, although under certain conditions diethyl ether is also a product:

$$CH_3CH_2OH \rightarrow CH_2{=}CH_2 + H_2O$$

$$2CH_3CH_2OH \rightarrow CH_3CH_2OCH_2CH_3 + H_2O$$

The production of ethylene by this route involves an endothermic reaction. Also the reaction is reversible with the equilibrium being favored by higher temperatures and hindered by higher pressures and water vapor in the feed.

Once produced by whatever means, ethylene can be polymerized to polyethylene:

$$nCH_2 = CH_2 \rightarrow (CH_2CH_2)_n$$

Or it can be used in the crude oil industry to produce alkylate, itself a hydrocarbon with a high octane number for gasoline enhancement:

$$(CH_3)_3CH + CH_2{=}CH_2 \rightarrow (CH_3)_3CCH_2CH_3$$

The major difficulty encountered in the processing of ethanol directly to higher molecular weight hydrocarbon derivatives in the manner similar to the production of hydrocarbon derivatives from methanol is that the conversion generally stops at the production of ethylene, and if any higher molecular weight hydrocarbon derivatives happen to form, the yield is poor and not reproducible. The catalyst life time in the treatment of ethanol is also very short compared with the treatment of methanol.

For example, when the so-called protonated ZSM-5 zeolite catalyst was used under conditions set forth in the previously mentioned patents, ethanol was predominantly converted to ethylene. With the so-called acid-processed zeolite, ZSM-5H, ethanol was converted to a spectrum of higher hydrocarbon derivatives similar to that from methanol conversion, but the catalyst lifetime was considerably shortened to a time span of less than 5 hours, as compared to several tens of hours in the conversion of methanol.

Iron incorporation into ZSM-5 zeolites by different methods has led to a variety of chemical applications. Thus, hydrocarbon production from ethanol was evaluated using a [Fe,Al]ZSM-5 zeolite which was synthesized without nitrogenated templates, using ethanol and crystallization seeds and partially substituting iron for aluminum in the reaction mixture. Maximum production of liquid hydrocarbon derivatives was achieved with the zeolite with 0.5% iron. The procedure for obtaining the acid form of the zeolites, involving ammonium exchange and calcinations, has changed the iron species, probably with extraction from the structure, migration, and agglomeration (Machado et al., 2006).

However, one biofuel is beginning to gain a great deal of research (and investor) interest: algae. There are a number of strains of algae which, when allowed to react, produce a remarkably pure grade of composite hydrocarbon derivatives, from ethanol all the way up to octane and higher chains. Most oil and natural gas that currently exists in the world came not from decaying trees but rather as algae in shallow oceans and seas absorbed sunlight, photosynthesized various sugar energies, then died and drifted to the sea floors. Deprived of the oxygen-free radicals that would have decomposed them on land, the algae formed thick layers, hundreds or even thousands of feet deep, with the bottom-most layers becoming increasingly compressed by the weight of sludge and water on top of them.

Most of this natural process occurred over the course of millions of years during the Cretaceous and Jurassic eras, between 110 and 90 million years ago, and again during the late Triassic and early Cenozoic era, approximately 70 to 55 million years ago, when high global temperatures created inland seas that in turn slowly dried out as temperatures (and consequently sea levels) dropped. Similar activity occurred earlier as well, creating the necessary preconditions for coal to form.

One approach to hydrocarbon derivatives from algae is to grow algae in dedicated ponds for the purpose. This is probably the least costly approach, at least initially, but it suffers both from the danger of contamination (as other algae strains or chemical contaminants may end in the ponds as well) and the fact that it is necessary to have reasonably large bodies of water to grow the algae that aren't used for other purposes. The second approach involves the use of long plastic tubes filled with algae and medium, which can be exposed to sunlight or artificial light as appropriate to cause the algae to grow. A variation of this approach (and one that shows great promise) is to actually grow the algae in the dark, but to provide a medium high in sugar, which the algae then converts into high energy hydrocarbon chains. One consequence is that a much denser medium as algae on the interior of the tubes in sunlight-based systems are less likely to get the critical energy that they need, though this also comes at the cost of using sugar to feed the process.

Typically, a tube of algae can be grown in 10 days. Processing the algae then involves extracting and filtering out the algae (and resterilizing the growth environment) and reacting the algae down over the course of several days in what are called *bioreactors*. The resulting liquid tends to be rich in a number of different oil compounds, with the specific composition depending very much upon the algae strain itself. Recently, bioengineering of algae strains has made it possible to select for different compounds— one variety of algae produces gasoline grade fuel, a second, jet fuel (JP4, JP5, and JP8 fuels), while still others produce oils that can be used for lubrication or even food production, as such oils can be used for creating both saturated and unsaturated fatty acids.

This same process, though primarily via the sugar medium, is also used with a similar set of one-cell organisms—yeast, which does not photosynthesize but rather consumes simple sugars to build complex hydrocarbon derivatives, but otherwise both algae and yeast production can be adapted to create fuel grade products. Algae has somewhat of an advantage here, as algae can grow very quickly compared to yeast products, but yeast-to-oil production may prove to be more efficacious in terms of urban settings—portable bioreactors (that work better with yeastcan be created) which have a somewhat shorter overall production lifecycle.

4.3 Hydrocarbon from nonwoody plants

Plants have always been a rich source of chemicals, many of which are useful drugs and others have been the basis for synthetic drugs. In fact, plants provide a large bank of rich, complex, and highly varied structures which are unlikely to be synthesized in laboratories. Furthermore, evolution has already carried out a screening process itself whereby plants are more likely to survive if they contain potent compounds which deter animals or insects from eating them.

There has been the suggestion that certain plants rich in hydrocarbonlike materials might be cultivated for renewable photosynthetic products (Calvin, 1980). Indeed, there are certain species of flowering plants belonging to different families which convert a substantial amount of photosynthetic products into latex. The latex of such plants contains liquid hydrocarbon derivatives of high molecular weight (approximately 10,000). These hydrocarbon derivatives can be converted into high grade transportation fuel (i.e., such as fuels from crude oil). Therefore, hydrocarbon producing plants are often called *petroleum plants* or *petroplants* and their crop as *petrocrop*. Natural gas is also one of the products obtained from hydrocarbon derivatives. Thus, petroleum plants can be an alternative source for obtaining petroleum to be used in diesel engines. Normally, some of the latex-producing plants of families Euphorbiaceae, Apocynaceae, Asclepiadaceae, Sapotaceae, Moraceae, Dipterocarpaceae, etc. are petroplants. Similarly, sunflower (family Compositae), *Hardwickia pinnata* (family Leguminosae) are also petroplants. Some algae also produce hydrocarbon derivatives.

Euphorbia: Different species of Euphorbia of the family Euphorbiaceae serve as petroplants. The latex of *Euphorbia lathyrus* contains a fairly high percentage of terpenoids, which can be converted into high-grade transportation fuel. Similarly the carbohydrates (hexoses) from such plants can be used for ethanol formation.

Sugarcane and sugar beet (*Saccharum officinarum*, family: Gramineae) is the main source of raw material for sugar industry. The wastes from sugar industry include bagasse, molasses, and press mud. After extracting the cane juice for sugar production, the cellulosic fibrous residue that remains is called *bagasse*. It is used as the raw material (biomass) and processed variously for the production of fuel, alcohols, single cell protein, as well as in paper mills. Molasses is an important byproduct of sugar mills and contains 50%–55% fermentable sugars. One ton of molasses can produce approximately 280 L of ethanol. Molasses is used for the production of animal feed, liquid fuel, and alcoholic beverages. *Sugar beet* (*Beta vulgaris*, family: Chenopodiaceae) is yet another plant which contains a high percentage of sugars stored in fleshy storage roots. It is also an important source for production of sugar as well as ethanol.

In fact, plants offer a unique and diverse feedstock for chemicals. Plant biomass can be gasified to produce synthesis gas; a basic chemical feedstock and also a source of hydrogen for a future hydrogen economy. In addition, the specific components of plants such as carbohydrates, vegetable oils, plant fiber, and complex organic molecules known as primary and secondary metabolites can be utilized to produce a range of valuable monomers, chemical intermediates, pharmaceuticals, and materials: (i) carbohydrates, vegetable oils, (iii) plants fibers, and (iv) specialty molecules.

Specialty molecules such as highly complex bioactive molecules often beyond the power of laboratories and a wide range of chemicals are currently extracted from plants for a wide range of markets from crude herbal remedies through to very high-value pharmaceutical intermediates.

4.4 Hydrocarbons by anaerobic digestion

Most types of biomass can be used as feedstocks for gas production as long as they contain carbohydrates, proteins, fats, cellulose, and hemicelluloses as main components. Woody wastes are the exception, because they are largely unaffected by digestion, as most anaerobes are unable to degrade lignin. Xylophaous anaerobes (lignin consumers) or high temperature pretreatment, such as pyrolysis, can be used to break lignin down. Anaerobic digesters can also be fed with specially grown energy crops, such as silage, for dedicated gas production. A codigestion or cofermentation plant is typically an agricultural anaerobic digester that accepts two or more input materials for simultaneous digestion. However, when the feedstock is woody or contains high proportions of lignin, digestion becomes difficult—to obtain as efficient digestion, these feedstocks are combined in proportions that allow the digestion process to proceed. Predigestion and finely chopping will be helpful in the case of some materials—for example, animal wastes are predigested and plant wastes do not need predigestion but excessive amounts of plant material in the feedstock may cause digester inefficiency. The composition and yield of the gaseous products (referred to as gas) depends on (i) the feedstock type, (ii) the digestion system, and (iii) the retention time. The theoretical gas yield varies with the content of carbohydrate derivatives, protein derivatives, and fat derivatives. Only strong lignified organic substances, such as wood, are not suitable due to the slow rate of the anaerobic decomposition of lignin (Speight, 2019).

The gas composition and yield from the individual feedstocks vary considerably dependent on their origin, content of organic substance, and feedstock composition, although in the current context, the predominant component is a hydrocarbon (methane) (Table 7.4). Fat derivatives provide the highest gas yield but require a long retention time due to their poor bioavailability. Carbohydrate derivatives and protein derivatives have much higher rates of conversion than fats but the yield of the gas is lower. All feedstocks should be free of pathogens and other organisms; otherwise,

TABLE 7.4 Composition of various gas samples from anaerobic digestion of the waste.

Constituents	Household waste	Wastewater treatment plant sludge	Agricultural waste
Methane, % v/v	50–60	60–75	60–75
Carbon dioxide, % v/v	38–34	33–19	33–19
Nitrogen, % v/	5–0	1–0	1–0
Oxygen, % v/v	0–1	<0.5	<0.5
Water, % v/v	6	6	6
Hydrogen sulfide, mg/m^3	100–900	1000–4000	3000 - 10,000
Ammonia, mg/m^3	-	-	50–100
Aromatics mg/m^3	0–200	-	-
Properties:			
Density[a]	0.93		0.85
Wobbe index[b]	6.9		8.1

[a]*Natural gas: 0.57.*
[b]*Natural gas: 14.9.*

pasteurization at 70°C (158°F) or sterilization at 130°C (266°F) is necessary prior fermentation.

Semicontinuously feeding of high amounts of silage (fermented high-moisture stored fodder which can be fed to cattle and sheep or used as a feedstock for anaerobic digesters) results in a sudden change of the gas quality because carbon dioxide is stripped out due to the local reduction of the pH in the fermenter. A pretreatment by mechanical, thermal, chemical, or enzymatic processes can be applied to increase the rate of degradation of feedstocks. The decomposition process is faster with decreasing particle size but does not necessarily increase the methane yield (Mshandete et al., 2006) and as a result, feedstock crushing is usually directly connected to the feeding system by application of an extruder or by ultrasonic treatment of a side stream of the fermenter (Kim et al., 2003).

Treatment by a combination of a thermal pressure hydrolysis reaction (230°C, 445°F, 300 to 450 psi) results in degradation of organic polymers by hydrolysis into short chain, biologically good available compounds which increase the gas yield while the retention time in the digester can be reduced drastically. The addition of hydrolytic enzymes can improve the decomposition of structural polysaccharides resulting in an increased gas yield of up to 20%. The addition of enzymes reduces the viscosity of the feedstock mixture in the digester significantly and avoids the formation of floating layers. But the effect of enzymes can be strongly reduced if the protease enzymes of anaerobic microorganisms degrade the added enzymes. Protease enzymes are those enzymes that assist proteolysis—protein catabolism (breakdown of protein molecules into simpler ones) by hydrolysis of peptide bonds.

Large quantities of lignocellulosic waste are collected from agricultural, municipal, and other activities. Generally, the composition of lignocellulose is highly variable when it is selected from different sources—the composition is dependent on various conditions such as the origin and seasonal factors. Cellulose is a linear polymer and is linked by several b-1,4 glycoside bonds.

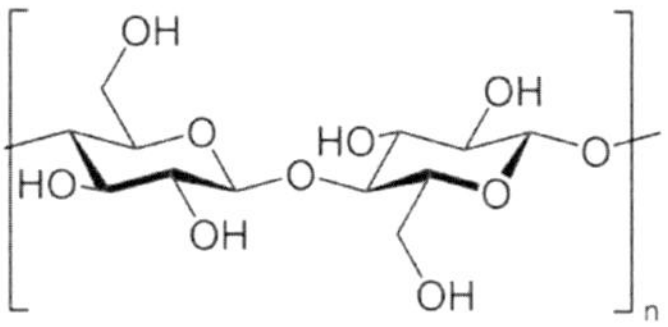

The structure contains parts with a crystalline structure and parts with an amorphous arrangement. The crystalline structure is based on hydrogen linkages which give higher toughness and solidity to the molecule. Moreover, crystalline cellulose can be converted to nonorganized structure-based cellulose by applying a temperature of 320°C (610°F) and a pressure of 3500 psi (Deguchi et al., 2006). The dominant compound in the hemicellulose is xylan (up to 90%) which is a group of hemicellulose derivatives that represents an abundant naturally occurring biopolymer but which can vary according to the origin of the feedstock.

Another natural polymer, lignin, is a heteropolymer of the cellular wall and appears in nature. The structure is complex and largely unknown although hypothetical structures have been proposed (Fig. 7.3). These structures contain covalent bonds and consist of three phenylpropane-based units (p-coumaryl, coniferyl, and sinapyl alcohol units) that are held together by linkages. However, the structural compactness of lignin provides resistance in microbial attack and the nonwater solubility of lignin makes for difficult degradation and the characteristics of lignin, such as composition and structure, that can positively affect the hydrolysis process in order to increase the gas production efficiency and pretreatment is necessary.

The gases formed are the waste products of the respiration of these decomposer microorganisms and the composition of the gases depends on the substance that is being decomposed. If the material consists of mainly carbohydrates, such as glucose and other simple sugars and high-molecular

TABLE 7.5 General properties of methane.

Chemical formula	CH_4
Molar mass	16.04 g/mol
Appearance	Colorless gas
Odor	Odorless[a]
Density	0.656 g/L (gas, 25°C, 1 atm/14.7 psi) 0.716 g/L (gas, 0°C/32°F, 1 atm/14.7 psi)
Melting point	−182.5°C; −296.4°F
Boiling point	−162°C; −260°F
Solubility in water	22.7 mg/L
Solubility	Soluble in ethanol, diethyl ether, benzene, toluene, methanol, acetone
Molecular shape	Tetrahedron
Flash point	−188°C (−306.4°F)
Autoignition temperature	537°C (999°F)
Explosive limits	4.4%−17% v/v in air

[a]*Natural gas which is predominantly methane has an odor because of the addition of odiferous compounds (such as t-butyl thiol, also known as t-butyl mercaptan, that has a skunklike odor) that is added by the seller so that inherently dangerous gas leaks can be detected immediately.*

compounds (polymers) such as cellulose and hemicellulose, the methane production is low. However, if the fat content is high, the methane production is likewise high.

Methane—a colorless and odorless gas with a boiling point of −162°C (−260°F) and burns with a blue flame—is the major combustible constituents of gas. Methane is also the main constituent (77%−90%) of natural gas. Chemically, methane belongs to the alkane series of hydrocarbon derivatives and is the simplest possible member of this series (C_nH_{2n+2}). At normal temperature and pressure, methane has a density of on the order of 0.66−0.72 g/L in the gas phase or 0.42 g/L in the liquid phase (Table 7.5). Due to carbon dioxide being of somewhat higher density (Table 7.6), the gas has a slightly higher density than methane and is on the order of 1.15 g/L. If the gas is mixed with 10%−20% v/v air, there is the high probability of an explosion—explosive air is the name often applied to such a mixture.

TABLE 7.6 General properties of carbon dioxide.

Chemical formula	CO_2
Molecular weight	44.01 g/mol
Appearance	Colorless gas
Odor	Low concentrations: barely detectable High concentrations: sharp; acidic
Density	1.84 g/L (NTP[a]); 1.96 g/L (STP[a])
Melting point	−56.6°C; −69.8°F
Solubility in water	1.45 g/L at 25°C (77°F)

[a]*NTP: 0°C (32°F) and 1 atm (14.696 psi); STP: 15.6°C (60°F) and 1 atm (14.696 psi).*

4.5 Hydrocarbons via synthesis gas

Wood can be used to make both liquid and gaseous fuels. When wood is heated in the absence of air, or with a reduced air supply it is possible to produce a liquid fuel which can be used in a similar way to conventional oil fuels. It can be used to run internal combustion engines in vehicles or generators. The gas produced from wood is a mixture of hydrogen and carbon monoxide which is similar to the coal gas which was made before the arrival of natural gas from the North Sea. This wood gas can be used in internal combustion engines or in gas turbines which can be used to power generators. Although the liquid fuels are rarely produced from wood at present, wood gas is important in other countries for producing electricity in more remote areas.

Thus, gasification technology is an attractive route for the production of fuel gases from biomass. By gasification, solid biomass is converted into a combustible gas mixture normally called *producer gas* consisting primarily of hydrogen (H_2) and carbon monoxide (CO), with lesser amounts of carbon dioxide (CO_2), water (H_2O), methane (CH_4), and higher hydrocarbon derivatives (C_xH_y), as well as nitrogen (N_2) and particulates.

Synthesis gas (syngas) is the name given to a gas mixture that contains varying amounts of carbon monoxide and hydrogen generated by the gasification of a carbon containing fuel to a gaseous product with a heating value. Examples include (i) steam reforming of natural gas or liquid hydrocarbon derivatives to produce hydrogen, (ii) the gasification of coal, and (iii) in some types of waste-to-energy gasification facilities. The name comes from their use as intermediates in creating synthetic natural gas (SNG) and for producing ammonia or methanol. Synthesis gas is also used as an intermediate in producing synthetic crude oil for use as a fuel or lubricant via the Fischer-Tropsch process.

Gasification to produce synthesis gas can proceed from almost any organic material, including biomass and plastic waste. The resulting synthesis gas burns cleanly into water vapor and carbon dioxide. Alternatively, synthesis gas may be converted efficiently to methane via the Sabatier reaction, or to a dieselike synthetic fuel via the Fischer-Tropsch process. Inorganic components of the feedstock, such as metals and minerals, are trapped in an inert and environmentally safe form as char, which may have use as a fertilizer.

The gasification is carried out at elevated temperatures, 500−1500°C (930−2730°F), and at atmospheric or elevated pressures. The process involves conversion of biomass, which is carried out in absence of air or with less air than the stoichiometric requirement of air for complete combustion. Partial combustion produces carbon monoxide as well as hydrogen which are both combustible gases. Solid biomass fuels, which are usually inconvenient and have low efficiency of utilization, can be converted into gaseous fuel. The energy in producer gas is 70%−80% percent of the energy originally stored in the biomass. The producer gas can serve in different ways: it can be burned directly to produce heat or used as a fuel for gas engines and gas turbines to generate electricity; in addition, it can also be used as a feedstock (syngas) in the production of chemicals, e.g., methanol. The diversified applications of the producer gas make the gasification technology very attractive.

Synthesis gas consists primarily of carbon monoxide and hydrogen with the occasional lesser amounts of carbon dioxide and has less than half the energy density of natural gas. Synthesis gas is combustible and often used as a fuel source or as an intermediate for the production of other chemicals. Synthesis gas for use as a fuel is most often produced by gasification of coal or municipal waste mainly by the following paths:

$$C + O_2 \rightarrow CO_2$$

$$CO_2 + C \rightarrow 2CO$$

$$C + H_2O \rightarrow CO + H_2$$

The synthesis gas generation process is a noncatalytic process for producing synthesis gas (principally hydrogen and carbon monoxide) for the ultimate production of high-purity hydrogen from gaseous or liquid hydrocarbon derivatives. In the process, a controlled mixture of preheated feedstock and oxygen is fed to the top of the generator where carbon monoxide and hydrogen emerge as the products. Soot, produced in this part of the operation, is removed in a water scrubber from the product gas stream and is then extracted from the resulting carbon-water slurry with naphtha and transferred to a fuel oil fraction. The oil-soot mixture is burned in a boiler or recycled to the generator to extinction to eliminate carbon production as part of the process.

The composition of the produced gases varies widely with the properties of the biomass, the gasifying agent, and the process conditions. Depending on the nature of the raw solid feedstock and the process conditions, the char formed from pyrolysis contains 20%−60% of the energy input. Therefore the gasification of char is an important step for the complete conversion of the solid biomass into gaseous products and for an efficient utilization of the energy in the biomass.

A variety of biomass gasifiers has been developed and can be grouped into four major classes: (i) fixed-bed updraft or countercurrent gasifier, (20 fixed-bed downdraft or cocurrent gasifier, (iii) bubbling fluidized-bed gasifier, and (iv) circulating fluidized bed gasifier. Differentiation is based on the means of supporting the biomass in the reactor vessel, the direction of flow of both the biomass and oxidant, and the way heat is supplied to the reactor. The processes occurring in any gasifier include drying, pyrolysis, reduction, and oxidation. The unique feature of the updraft gasifier is the sequential occurrence of these processes: they are separated spatially and therefore temporally. The soot-free synthesis gas is then charged to a shift converter where the carbon monoxide reacts with steam to form additional hydrogen and carbon dioxide at the stoichiometric rate of 1 mol of hydrogen for every mole of carbon monoxide charged to the converter.

The reactor temperatures vary from 1095 to 1490°C (2000−2,700°F), while pressures can vary from approximately atmospheric pressure to approximately 2000 psi. The process has the capability of producing high-purity hydrogen although the extent of the purification procedure depends upon the use to which the hydrogen is to be put. For example, carbon dioxide can be removed by scrubbing with various alkaline reagents, while carbon monoxide can be removed by washing with liquid nitrogen or, if nitrogen is undesirable in the product, the carbon monoxide should be removed by washing with copper-amine solutions.

This particular partial oxidation technique can be applied to a whole range of liquid feedstocks for hydrogen production. There is now serious consideration being given to hydrogen production by the partial oxidation of solid feedstocks such as crude oil coke (from both delayed and fluid-bed reactors), lignite, and coal, as well as crude oil residua.

The Fischer-Tropsch synthesis is, in principle, a carbon chain building process, where methylene groups are attached to the carbon chain. The actual reactions that occur have been, and remain, a matter of controversy, as it has been in the last century since 1930s.

$$(2n+1)H_2 + nCO \rightarrow C_nH_{(2n+2)} + nH_2O$$

Even though the overall Fischer-Tropsch process is described by the following chemical equation:

$$(2n+1)H_2 + nCO \rightarrow C_nH_{(2n+2)} + nH_2O$$

The initial reactants in the above reaction (i.e., CO and H_2) can be produced by other reactions such as the partial combustion of a hydrocarbon:

$$C_nH_{(2n+2)} + {}^{1}\!/_{2}\, nO_2 \rightarrow (n+1)H_2 + nCO$$

For example (when n = 1), methane (in the case of gas to liquids applications):

$$2CH_4 + O_2 \rightarrow 4H_2 + 2CO$$

Or by the gasification of any carbonaceous source, such as biomass:

$$C + H_2O \rightarrow H_2 + CO$$

The energy needed for this endothermic reaction is usually provided by the exothermic combustion with air or oxygen:

$$2C + O_2 \rightarrow 2CO$$

The reaction is dependent of a catalyst, mostly an iron or cobalt catalyst where the reaction takes place. There is either a low or high temperature process (LTFT, HTFT), with temperatures ranging between 200 and 240°C (390–465°F) for LTFT and 300–350°C (570–660°F) for HTFT. The HTFT uses an iron catalyst, and the LTFT either an iron or a cobalt catalyst. The different catalysts include also nickel-based and ruthenium-based catalysts, which also have enough activity for commercial use in the process. But the availability of ruthenium is limited and the nickel-based catalyst has high activity and produces methane. Iron is cheap, but cobalt has the advantage of higher activity and longer life, though it is on a metal basis 1000 times more expensive than iron catalyst.

4.6 Biorefining

A crude oil refinery is a series of integrated unit processes by which crude oil can be converted to a slate of useful (salable) products. A crude oil refinery, as currently configured, is unsuitable for processing raw, or even partially processed, biomass. A typical refinery might be suitable for processing products such as gas, liquid, or solids products from biomass processing. These products from biomass might be acceptable as a single feedstock to a specific unit or, more likely, as a feedstock to be blended with refinery streams to be coprocessed in various refinery units.

Thus, a biorefinery might, in the early stages of development, be a series of unit processes which covert biomass to a primary product that requires further processing to become the final saleable product. The analogy is in the processing of bitumen from tar sand which is fist processed to a synthetic crude oil (primary processing) and then sent to a refinery for conversion to saleable fuel products (Speight, 2014, 2017). Analogous, in many cases, to the thermal decomposition of crude oil constituents, in the flash pyrolysis (high-temperature cracking and short residence time), the products are ethylene, benzene, toluene, and the xylene isomers as well as carbon monoxide and carbon dioxide. The type of biomass (for example, wood) used influences the product distribution (Steinberg et al., 1992). Theoretically, the flash pyrolysis process can use a wide range of biomass sources. The process has much in common with the naphtha cracking process.

At this point in the context of flash pyrolysis it is worthy of note that plastic waste (while not a biomass material) can also be treated by flash pyrolysis to produce starting materials for petrochemical manufacture. In the process, the mixed plastic waste is heated in an oxygen-free atmosphere. At a temperature of several hundred degrees, the constituents of the waste decompose to yield a mixture of gases, liquids, and solids. The composition of the product depends on temperature and pressure—the higher the temperature, the more gaseous products are formed. An important fraction of this gaseous product is ethylene if plastics are used as feedstock.

Biorefining, in which biomass is converted to a variety of chemical products, is not new if activities such as production of vegetable oils, beer, and wine requiring pretreatment are considered. Many of these activities are known to have been in practice for millennia. Biomass can be converted into commercial fuels, suitable to substitute for fossil fuels. These can be used for transportation, heating, electricity generation, or anything else fossil fuels are used for. The conversion is accomplished through the use of several distinct processes which include both biochemical conversion and thermal conversion to produce gaseous, liquid, and solid fuels which have high energy contents, are easily transportable, and are therefore suitable for use as commercial fuels.

Biorefining offers a method to accessing the integrated production of chemicals, materials, and fuels. Although the concept of a biorefinery concept is analogous to that of an oil refinery, the differences in the various biomass feedstocks require a divergence in the methods used to convert the feedstocks to fuels and chemicals (Speight, 2014, 2017). Thus, a biorefinery, like a crude oil refinery, may need to be a facility that integrates biomass conversion processes and equipment to produce fuels, power, and chemicals from biomass. In a manner similar to the crude oil refinery, a biorefinery would integrate a variety of conversion processes to produce multiple product streams such as motor fuels and other chemicals from biomass such as the inclusion of gasification processes and fermentation processes to name only two possible processes options.

In short, a biorefinery should combine the essential technologies to transform biological raw materials into a range of industrially useful intermediates. However, the type of biorefinery would have to be differentiated by the character of the feedstock. For example, the *crop biorefinery* would use raw materials such as cereals or maize and the *lignocellulose biorefinery* would use raw material with high cellulose content, such as straw, wood, and paper waste.

As a crude oil refinery uses crude oil as the major input and processes it into many different products, a biorefinery would use feedstocks such as lignocellulosic biomass as the major input and process it into many different products. Currently wet-mill corn processing and pulp and paper mills can be categorized as biorefineries since they produce multiple products from biomass. Research is currently being conducted to foster new industries to convert biomass into a wide range of products, including ones that would otherwise be made from petrochemicals. The idea is for biorefineries to produce both high-volume liquid fuels and high-value chemicals or products in order to address national energy needs while enhancing operation economics.

However, the different compositional nature of the biomass feedstock, compared to crude oil, will require the application of a wider variety of processing tools in the biorefinery. Processing the individual components will utilize conventional thermochemical operations and state-of-the-art bioprocessing techniques. Although a number of new bioprocesses have been commercialized, it is clear that economic and technical barriers still exist before the full potential of this area can be realized. The biorefinery concept could significantly reduce production costs of plant-based chemicals and facilitate their substitution into existing markets. This concept is analogous to that of a modern oil refinery in that the biorefinery is a highly integrated complex that will efficiently separate biomass raw materials into individual components and convert these into marketable products such as energy, fuels, and chemicals. By analogy with crude oil; every element of the plant feedstock will be utilized including the low-value lignin components.

One aspect of designing a refinery for any feedstocks is the composition of the feedstocks. For example, a heavy oil refinery would differ somewhat from a conventional refinery and a refinery for tar sand bitumen would be significantly different to both (Speight, 2014, 2017). Furthermore, the composition of biomass is variable (Speight, 2011c) which is reflected in the range of heat value (heat content, calorific value) of biomass, which is somewhat lesser than for coal and much lower than the heat value for crude oil, generally falling in the range 6000 to 8500 Btu/lb (Speight, 2014). Moisture content is probably the most important determinant of heating value. Air-dried biomass typically has approximately 15%–20% moisture, whereas the moisture content for oven-dried biomass is around 0%. Moisture content is also an important characteristic of coals, varying in the range of 2%–30%. However, the bulk density (and hence energy density) of most biomass feedstocks is generally low, even after densification, approximately 10% and 40% of the bulk density of most fossil fuels.

A key requirement for the biorefinery is the ability of the refinery to develop process technology that can economically access and convert the five- and six-membered ring sugars present in the cellulose and hemicellulose fractions of the lignocellulosic feedstock. Although engineering technology exists to effectively separate the sugar containing fractions from the lignocellulose, the enzyme technology to economically convert the five-ring sugars to useful products requires further development.

Plants are very effective chemical minifactories or refineries insofar as they produce chemicals by specific pathways. The chemicals they produce are usually essential manufacture (called metabolites) include sugars and amino acids that are essential for the growth of the plant, as well as more complex compounds. Unlike crude oil—derived petrochemicals where most chemicals are built from the bottom up, biofeedstocks already have some valuable products to skim off the top before being broken down and used to build new molecules.

As a feedstock, biomass can be converted by thermal or biological routes to a wide range of useful forms of energy including process heat, steam, electricity, as well as liquid fuels, chemicals, and synthesis gas. As a raw material, biomass is a nearly universal feedstock due to its versatility, domestic availability, and renewable character. At the same time, it also has its limitations. For example, the energy density of biomass is low compared to that of coal, liquid crude oil, or crude oil—derived fuels. The heat content of biomass, on a dry basis (7000 to 9000 Btu/lb) is at best comparable with that of a low-rank coal or lignite, and substantially (50%—100%) lower than that of anthracite, most bituminous coals, and crude oil. Most biomass, as received, has a high burden of physically adsorbed moisture, up to 50% by weight. Thus, without substantial drying, the energy content of a biomass feed per unit mass is even less. These inherent characteristics and limitations of biomass feedstocks have focused the development of efficient methods of chemically transforming and upgrading biomass feedstocks in a refinery.

The sugar-base involves breakdown of biomass into raw component sugars using chemical and biological means. The raw fuels may then be upgraded to produce fuels and chemicals that are interchangeable with existing commodities such as transportation fuels, oils, and hydrogen.

Although a number of new bioprocesses have been commercialized, it is clear that economic and technical barriers still exist before the full potential of this area can be realized. One concept gaining considerable momentum is the biorefinery which could significantly reduce production costs of plant-based chemicals and facilitate their substitution into existing markets. This concept is analogous to that of a modern oil refinery in that the biorefinery is a highly integrated complex that will efficiently separate biomass raw materials into individual components and convert these into marketable products such as energy, fuels, and chemicals.

By analogy with crude oil, every element of the plant feedstock will be utilized including the low-value lignin components. However, the different compositional nature of the biomass feedstock, compared to crude oil, will require the application of a wider variety of processing tools in the biorefinery. Processing of the individual components will utilize conventional thermochemical operations and state-of-the-art bioprocessing techniques. The production of biofuels in the biorefinery complex will service existing high volume markets, providing economy-of-scale benefits and large volumes of byproduct streams at minimal cost for upgrading to valuable chemicals. A pertinent example of this is the production of glycerol (glycerin) as a byproduct in biodiesel plants.

Glycerol has high functionality and is a potential platform chemical for conversion to a range of higher value chemicals. The high volume product streams in a biorefinery need not necessarily be a fuel but could also be a large volume chemical intermediate such as ethylene or lactic acid. In addition, a variety of methods and techniques can be employed to obtain different product portfolios of bulk chemicals, fuels, and materials. Biotechnology-based conversion processes can be used to ferment the biomass carbohydrate content into sugars that can then be further processed. As one example, the fermentation path to lactic acid shows promise as a route to biodegradable plastics. An alternative is to employ thermochemical conversion processes which use pyrolysis or gasification of biomass to produce a hydrogen-rich synthesis gas which can be used in a wide range of chemical processes.

A key requirement for delivery of the biorefinery is the ability of the refinery to develop and use process technology that can economically access and convert the five- and six-membered ring sugars present in the cellulose and hemicellulose fractions of the lignocellulosic feedstock. Although engineering technology exists to effectively separate the sugar containing fractions from the lignocellulose, the enzyme technology to economically convert the five-ring sugars to useful products requires further development.

The construction of both large biofuel and renewable chemical production facilities coupled with the pace at which bioscience is being both developed and applied demonstrates that the utilization of nonfood crops will become more significant in the near term. The biorefinery concept provides a means to significantly reduce production costs such that a substantial substitution of petrochemicals by renewable chemicals becomes possible. However, significant technical challenges remain before the biorefinery concept can be realized.

If the biorefinery is truly analogous to an oil refinery in which crude oil is separated into a series of products, such as gasoline, heating oil, jet fuel, and petrochemicals, the biorefinery can take advantage of the differences in biomass components and intermediates and maximize the value derived from the biomass feedstock. A biorefinery might, for example, produce one or several low-volume, but high-value, chemical products and a low-value, but high-volume liquid transportation fuel, while generating electricity and process heat for its own use and perhaps enough for sale of electricity. The high-value products enhance profitability, the high-volume fuel helps meet national energy needs, and the power production reduces costs and avoids greenhouse-gas emissions.

The basic types of processes used to generate chemicals from biomass as might be incorporated into a biorefinery are: (i) pyrolysis, (ii) gasification, (iii) anaerobic digestion, and (iv) fermentation.

4.6.1 Pyrolysis

Pyrolysis is a medium temperature method which produces gas, oil, and char from crops which can then be further processed into useful fuels or feedstock (Boateng et al., 2007). Pyrolysis is the direct thermochemical conversion process that includes pyrolysis, liquefaction, and solvolysis (Kavalov and Peteves, 2005).

Wood and many other similar types of biomass which contain lignin and cellulose (such as agricultural wastes, cotton gin waste, wood wastes, and peanut hulls) can be converted through a thermochemical process, such as pyrolysis, into solid, liquid, or gaseous fuels. Pyrolysis, used to produce charcoal since the dawn of civilization, is still the most common thermochemical conversion of biomass to commercial fuel.

During pyrolysis, biomass is heated in the absence of air and breaks down into a complex mixture of liquids, gases, and a residual char. If wood is used as the feedstock, the residual char is what is commonly known as charcoal. With more modern technologies, pyrolysis can be carried out under a variety of conditions to capture all the components, and to maximize the output of the desired product be it char, liquid, or gas. Pyrolysis is often considered to be the gasification of biomass in the absence of oxygen. However, the chemistry of each process may differ significantly. In general, biomass does not gasify as easily as coal, and it produces other hydrocarbon compounds in the gas mixture exiting the gasifier; this is especially true when no oxygen is used. As a result, typically an extra step must be taken to reform these hydrocarbon derivatives with a catalyst to yield a clean syngas mixture of hydrogen, carbon monoxide, and carbon dioxide.

Fast pyrolysis is a thermal decomposition process that occurs at moderate temperatures with a high heat transfer rate to the biomass particles and a short hot vapor residence time in the reaction zone. Several reactor configurations have been shown to assure this condition and to achieve yields of liquid product as high as 75% based on the starting dry biomass weight. They include bubbling fluid beds, circulating and transported beds, cyclonic reactors, and ablative reactors.

Fast pyrolysis of biomass produces a liquid product, pyrolysis oil, or bio-oil that can be readily stored and transported. Pyrolysis oil is a renewable liquid fuel and can also be used for production of chemicals. Fast pyrolysis has now achieved commercial success for production of chemicals and is being actively developed for producing liquid fuels. Pyrolysis oil has been successfully tested in engines, turbines, and boilers, and been upgraded to high quality hydrocarbon fuels. In the 1990s several fast pyrolysis technologies reached near-commercial status and the yields and properties of the generated liquid product, bio-oil, depend on the feedstock, the process type and conditions, and the product collection efficiency.

Direct hydrothermal liquefaction involves converting biomass to an oily liquid by contacting the biomass with water at elevated temperatures (300°C −350°C, 570°F to 660°F) with sufficient pressure to maintain the water primarily in the liquid phase for residence times up to 30 minutes. Alkali may be added to promote organic conversion. The primary product is an organic liquid with reduced oxygen content (approximately 10%) and the primary byproduct is water containing soluble organic compounds.

The importance of the provisions for the supply of feedstocks as crops and other biomass are often underestimated, since it is assumed that the supplies are inexhaustible. While this may be true over the long term, short-term supply of feedstocks can be as much as risk as any venture.

4.6.2 Gasification

Alternatively, biomass can be converted into fuels and chemicals indirectly (by gasification to syngas followed by catalytic conversion to liquid fuels) (Molino et al., 2016). Biomass gasification is a mature technology pathway that uses a controlled process involving heat, steam, and oxygen to convert biomass to hydrogen and other products, without combustion, and represents an efficient process for the production of chemicals and hydrogen.

Gasification is a process that converts organic carbonaceous feedstocks into carbon monoxide, carbon dioxide, and hydrogen by reacting with the feedstock at high temperatures (>700°C, 1290°F), without combustion, with a controlled amount of oxygen and/or steam. The resulting gas mixture (*synthesis gas*, *syngas*, or *producer gas*) is itself a fuel. The power derived from carbonaceous feedstocks and gasification followed by the combustion of the product gas(es) is considered to be a source of renewable energy if the gaseous products are from a source (e.g., biomass) other than a fossil fuel. The carbon monoxide can then be reacted with water (steam) to form carbon dioxide and more hydrogen via a water-gas shift reaction. Adsorber or special membranes can separate the hydrogen from this gas stream. The simplified reaction is:

$$C_6H_{12}O_6 + O_2 + H_2O \rightarrow CO + CO_2 + H_2 + \text{other species}$$

$$CO + H_2O \rightarrow CO_2 + H_2 \text{ (water gas shift reaction)}$$

This reaction scheme uses glucose as a surrogate for cellulose, but it must be recognized that biomass has highly variable composition and complexity with cellulose as one major component.

Coal has, for many decades, been the primary feedstock for gasification units—coal can also be gasified in situ (in the underground seam) (Speight, 2013a) but that is not the subject of this chapter and is not discussed further. However, with the concern on the issue of environmental pollutants and the potential shortage of coal in some areas there is a move to feedstocks other than coal for gasification processes. Gasification permits the utilization of various feedstocks (coal, biomass, crude oil residues, and other carbonaceous wastes) to their fullest potential.

The advantage of the gasification process when a carbonaceous feedstock (a feedstock containing carbon) or hydrocarbonaceous feedstock (a feedstock containing carbon and hydrogen) is employed is that the product of focus—synthesis gas—is potentially more useful as an energy source and results in an overall cleaner process. The production of synthesis gas is a more efficient production of an energy source than, say, the direct combustion of the original feedstock because synthesis gas can be converted via the Fischer-Tropsch process into a range of synthesis liquid fuels suitable for use as blend stock in gasoline-fueled engines or diesel-fueled engines (Chapter 8) (Chadeesingh, 2011; Luque and Speight, 2015).

Biomass includes a wide range of materials that produce a variety of products which are dependent upon the feedstock (Balat, 2011; Demirbaş, 2011; Ramroop Singh, 2011; Speight, 2011a). For example, typical biomass wastes include wood material (bark, chips, scraps, and saw dust), pulp and paper industry residues, agricultural residues, organic municipal material, sewage, manure, and food processing byproducts. Agricultural residues such as straws, nut shells, fruit shells, fruit seeds, plant stalks and stover, green leaves, and molasses are potential renewable energy resources. Many developing countries have a wide variety of agricultural residues in ample quantities. Large quantities of agricultural plant residues are produced annually worldwide and are vastly underutilized. Agricultural residues, when used as fuel, through direct combustion, only a small percentage of their potential energy is available, due to inefficient burners used. Current disposal methods for these agricultural residues have caused widespread environmental concerns. For example, disposal of rice and wheat straw by open-field burning causes air pollution. In addition, the widely varying heat content of the different types of biomass varies widely and must be taken into consideration when designing any conversion process (Jenkins and Ebeling, 1985).

Raw materials that can be used to produce biomass fuels are widely available and arise from a large number of different sources and in numerous forms. Biomass can also be used to produce electricity—either blended with traditional feedstocks, such as coal, or by itself. However, each of the biomass materials can be used to produce fuel, but not all forms are suitable for all the different types of energy conversion technologies such as biomass gasification (Rajvanshi, 1986; Brigwater, 2003; Dasappa et al., 2004; Speight, 2011a; Basu, 2013). The main basic sources of biomass material are: (i) wood, including bark, logs, sawdust, wood chips, wood pellets, and briquettes, (ii) high-yield energy crops, such as wheat, that are grown specifically for energy applications, (iii) agricultural crop and animal residues, like straw or slurry, (iv) food waste, both domestic and commercial, and (v) industrial waste, such as wood pulp or paper pulp. For processing, a simple form of biomass such as untreated and unfinished wood may be cut into a number of physical forms, including pellets and wood chips, for use in biomass boilers and stoves.

Thermal conversion processes use heat as the dominant mechanism to convert biomass into another chemical form. The basic alternatives of combustion, torrefaction, pyrolysis, and gasification are separated principally by the extent to which the chemical reactions involved are allowed to proceed (mainly controlled by the availability of oxygen and conversion temperature) (Speight, 2011a).

Many forms of biomass contain a high percentage of moisture (along with carbohydrates and sugars) and mineral constituents—both of which can influence the viability of a gasification process (Luque and Speight, 2015)—the presence of high levels of moisture in the biomass reduces the temperature inside the gasifier, which then reduces the efficiency of the gasifier. Therefore, many biomass gasification technologies require that the biomass be dried to reduce the moisture content prior to feeding into the gasifier. In addition, biomass can come in a range of sizes. In many biomass gasification systems, the biomass must be processed to a uniform size or shape to feed into the gasifier at a consistent rate and to ensure that as much of the biomass is gasified as possible.

Biomass, such as wood pellets, yard and crop wastes, and the so-called *energy crops* such as switch grass and waste from pulp and paper mills can be used to produce ethanol and synthetic diesel fuel. The biomass is first gasified to produce the synthetic gas (synthesis gas), and then converted via catalytic processes to these downstream products. Furthermore, most biomass gasification systems use air instead of oxygen for the gasification reactions (which is typically used in large-scale industrial and power gasification plants). Gasifiers that use oxygen require an air separation unit to provide the gaseous/liquid oxygen; this is usually not cost-effective at the smaller scales used in biomass gasification plants. Air-blown gasifiers use the oxygen in the air for the gasification reactions.

In general, biomass gasification plants are much smaller than the typical coal or crude oil coke gasification plants used in the power, chemical, fertilizer, and refining industries—the sustainability of the fuel supply is often brought into question. As such, a biomass gasification plant is less expensive to construct and has a smaller environmental footprint. For example, while a large industrial gasification plant may take up 150 acres of land and process 2500 to 15,000 tons per day of feedstock (such as coal or crude oil coke), the smaller biomass plants typically process 25–200 tons of feedstock per day and take up less than 10 acres.

Biomass gasification has been the focus of research in recent years to estimate efficiency and performance of the gasification process using various types of biomass such as sugarcane residue (Gabra et al., 2001), rice hulls (Boateng et al., 1992), pine sawdust (Lv et al., 2004), almond shells (Rapagnà and Latif, 1997; Rapagnà et al., 2000), wheat straw (Ergudenler and Ghaly, 1993), food waste (Ko et al., 2001), and wood biomass (Pakdel and Roy, 1991; Bhattacharaya et al., 1999; Chen et al., 1992; Hanaoka et al., 2005). Recently, cogasification of various biomass and coal mixtures has attracted a great deal of interest from the scientific community. Feedstock combinations including Japanese cedar wood and coal (Kumabe et al., 2007), coal and saw dust, coal and pine chips (Pan et al., 2000), coal and silver birch wood (Collot et al., 1999), and coal and birch wood (Brage et al., 2000) have been reported in gasification practice. Cogasification of coal and biomass has some synergy— the process not only produces a low carbon footprint on the environment, but also improves the H_2/CO ratio in the produced gas which is required for liquid fuel synthesis (Sjöström et al., 1999; Kumabe et al., 2007). In addition, the inorganic matter present in biomass catalyzes the gasification of coal. However, cogasification processes require custom fittings and optimized processes for the coal and region-specific wood residues.

While cogasification of coal and biomass is advantageous from a chemical viewpoint, some practical problems are present on upstream, gasification, and downstream processes. On the upstream side, the particle size of the coal and biomass is required to be uniform for optimum gasification. In addition, moisture content and pretreatment (torrefaction) are very important during upstream processing.

While upstream processing is influential from a material handling point of view, the choice of gasifier operation parameters (temperature, gasifying agent, and catalysts) dictate the product gas composition and quality. Biomass decomposition occurs at a lower temperature than coal and therefore different reactors compatible to the feedstock mixture are required (Speight, 2011c, 2013a, 2013b; Brar et al., 2012). Furthermore, feedstock and gasifier type along with operating parameters not only decide product gas composition but also dictate the amount of impurities to be handled downstream.

Downstream processes need to be modified if coal is cogasified with biomass. Heavy metals and impurities such as sulfur and mercury present in coal can make synthesis gas difficult to use and unhealthy for the environment. Alkali present in biomass can also cause corrosion problems like high temperatures in downstream pipes. An alternative option to downstream gas cleaning would be to process coal to remove mercury and sulfur prior to feeding into the gasifier.

However, first and foremost, coal and biomass require drying and size reduction before they can be fed into a gasifier. Size reduction is needed to obtain appropriate particle sizes; however, drying is required to achieve moisture content suitable for gasification operations. In addition, biomass densification may be conducted to prepare pellets and improve density and material flow in the feeder areas.

It is recommended that biomass moisture content should be less than 15% w/w prior to gasification. High moisture content reduces the temperature achieved in the gasification zone, thus resulting in incomplete gasification.

Forest residues or wood has a fiber saturation point at 30%–31% moisture content (dry basis) (Brar et al., 2012). Compressive and shear strength of the wood increases with decreased moisture content below the fiber saturation point. In such a situation, water is removed from the cell wall leading to shrinkage. The long-chain molecules constituents of the cell wall move closer to each other and bind more tightly. A high level of moisture, usually injected in form of steam in the gasification zone, favors formation of a water-gas shift reaction that increases hydrogen concentration in the resulting gas.

The torrefaction process is a thermal treatment of biomass in the absence of oxygen, usually at 250–300°C (480–570°F) to drive off moisture, decompose hemicellulose completely, and partially decompose cellulose (Speight, 2011a). Torrefied biomass has reactive and unstable cellulose molecules with broken hydrogen bonds and not only retains 79%–95% of feedstock energy but also produces a more reactive feedstock with lower atomic hydrogen-carbon and oxygen-carbon ratios to those of the original biomass. Torrefaction results in higher yields of hydrogen and carbon monoxide in the gasification process.

Most small- to medium-sized biomass/waste gasifiers are air blown, and operate at atmospheric pressure and temperatures in the range 800 to 100°C (1470–2190°F). They face very different challenges compared to large gasification plants—the use of small-scale air separation plant should oxygen gasification be preferred. Pressurized operation, which eases gas cleaning, may not be practical.

Biomass fuel producers, coal producers and, to a lesser extent, waste companies are giving serious consideration to supplying cogasification power plants and realize the benefits of cogasification with alternate fuels (Speight, 2011a, 2013a, 2013b; Lee and Shah, 2013). The benefits of a co-gasification technology involving coal and biomass include the use of a reliable coal supply with gate-fee waste and biomass which allows the economies of scale from a larger plant to be supplied just with waste and biomass. In addition, the technology offers a future option of hydrogen production and fuel development in refineries. In fact, oil refineries and petrochemical plants are opportunities for gasifiers when the hydrogen is particularly valuable (Speight, 2011b, 2014).

In addition, while biomass may seem to some observers to be the answer to the global climate change issue, the advantages and disadvantages must be considered carefully. For example, the advantages are (i) biomass is a theoretically inexhaustible fuel source, (ii) when direct conversion of combustion of plant mass—such as fermentation and pyrolysis—is not used to generate energy, there is minimal environmental impact, (iii) alcohols and other fuels produced by biomass are efficient, viable, and relatively clean-burning, and (iv) biomass is available on a world-wide basis.

On the other hand, the disadvantages include (i) the highly variable heat content of different biomass feedstocks, (ii) the high water content that can affect the process energy balance, and (iii) there is a potential net loss of energy when a biomass plant is operated on a small scale— an account of the energy used to grow and harvest the biomass must be included in the energy balance.

In the primary processing stage of gasification, the volatile components of biomass are subjected to pyrolysis or combustion in the absence of oxygen, at temperatures ranging from 450 to 600°C and, depending on the residence time in the chamber, a variety of products can be achieved. If pyrolysis is rapid (short residence time), gaseous products, condensable liquids, and char are produced and overall yield of bio-oil can, under ideal conditions, make up 60%–75% of the original fuel mass. The oil produced can be used as a biofuel or as a feedstock for value-added chemical products (Garcia et al., 2000). If the pyrolysis is slow (long residence time), the products are more likely to be gaseous and consist of carbon monoxide, hydrogen, methane, carbon dioxide, and water as well as volatile tar.

Slow pyrolysis, like fast pyrolysis, leaves behind a solid residue of char (or charcoal) which comprise approximately 10%–25% by weight of the original feedstock. The char can be used as a fuel source to drive the pyrolysis process (Cetin et al., 2005). If the pyrolysis is carried out at the higher temperature range (550–600°C), the gaseous products consist of carbon monoxide, hydrogen, methane, volatile tar, carbon dioxide, and water (Cetin et al., 2005). Any char produced can be used as a fuel source to drive the pyrolysis process or can be gasified to produce synthesis gas (Cetin et al., 2005), so-called because of the presence of carbon monoxide and hydrogen in the product stream.

After the production of syngas, a number of pathways may be followed to create biofuels. Proven catalytic processes for syngas conversion to fuels and chemicals exist using syngas produced commercially from natural gas and coal. These proven technologies can be applied to biomass-derived syngas.

4.7 Biochemical conversion

Biochemical conversion of biomass is completed through alcoholic fermentation to produce liquid fuels and "anaerobic" digestion or fermentation, resulting in gas. Alcoholic fermentation of crops such as sugarcane and maize (corn) to produce ethanol for use in internal combustion engines has been practiced for years with the greatest production occurring in Brazil and the United States, where ethanol has been blended with gasoline for use in automobiles. With engine modifications, automobiles can operate on ethanol alone.

Anaerobic digestion of biomass has been practiced for almost a century, and is very popular in many developing countries such as China and India. The organic fraction of almost any form of biomass, including sewage sludge, animal wastes, and industrial effluents, can be broken down through anaerobic digestion into methane and carbon dioxide. This "gas" is a reasonably clean burning fuel which can be captured and put to many different end uses such as cooking, heating, or electrical generation.

Wood and many other similar types of biomass which contain lignin and cellulose (agricultural wastes, cotton gin waste, wood wastes, peanut hulls, etc.) can be converted through thermochemical processes into solid, liquid, or gaseous fuels. Pyrolysis, used to produce charcoal since the dawn of civilization, is still the most common thermochemical conversion of biomass to commercial fuel.

During pyrolysis, biomass is heated in the absence of air and breaks down into a complex mixture of liquids, gases, and a residual char. If wood is used as the feedstock, the residual char is what is commonly known as charcoal. With more modern technologies, pyrolysis can be carried out under a variety of conditions to capture all the components, and to maximize the output of the desired product be it char, liquid, or gas.

There is a consensus among scientists that biomass fuels used in a sustainable manner result in no net increase in atmospheric carbon dioxide (CO_2). Some would even go as far as to declare that sustainable use of biomass will result in a net decrease in atmospheric carbon dioxide. This is based on the assumption that all the carbon dioxide given off by the use of biomass fuels was recently taken in from the atmosphere by photosynthesis. Increased substitution of fossil fuels with biomass-based fuels would therefore help reduce the potential for global warming, caused by increased atmospheric concentrations of carbon dioxide.

Unfortunately, it may not be as simple as has been assumed above. Currently, biomass is being used all over the world in a very unsustainable manner, and the long-term effects of biomass energy plantations have not been proven. Also, the natural humus and dead organic matter in the forest soils are a large reservoir of carbon. Conversion of natural ecosystems to managed energy plantations could result in a release of carbon from the soil as a result of the accelerated decay of organic matter.

An ever increasing number of people are faced with hunger and starvation. It has been argued that the use of land to grow fuel crops will increase this problem. Hunger in developing countries, however, is more complex than just a lack of agricultural land. Many countries in the world, such as the United States, have food surpluses. Much fertile agricultural land is also used to grow tobacco, flowers, food for domestic pets, and other "luxury" items, rather than staple foods. Similarly, a significant proportion of agricultural land is used to grow feed for animals to support the highly wasteful, meat centered diet of the industrialized world. By feeding grain to livestock we end up with only approximately 10% of the caloric content of the grain. When considered as a necessity, it does not seem to be so unreasonable to use some fertile land to grow fuel. Marginal land and underutilized agricultural land can also be used to grow biomass for fuel.

Acid rain, which can damage to lakes and forests, is a byproduct of the combustion of fossil fuels, particularly coal and oil. The high sulfur content of these fuels together with hot combustion temperatures result in the formation of sulfur dioxide (SO_2) and nitrous oxides (NO_x), when they are burned to provide energy. The replacement of fossil fuels with biomass can reduce the potential for acid rain. Biomass generally contains less than 0.1% sulfur by weight compared to low sulfur coal with 0.5%–4% sulfur. Lower combustion temperatures and pollution control devices such as wet scrubbers and electrostatic precipitators can also keep emissions of NO_x to a minimum when biomass is burned to produce energy.

The final major environmental impact of biomass energy may be that of loss of biodiversity. Transforming natural ecosystems into energy plantations with a very small number of crops, as few as one, can drastically reduce the biodiversity of a region. Such "monocultures" lack the balance achieved by a diverse ecosystem, and are susceptible to widespread damage by pests or disease.

References

Aguado, J., Serrano, D., 1999. Feedstock Recycling of Plastic Wastes. Royal Society of Chemistry, Cambridge, United Kingdom.

Aguado, R., Olazar, M., San Jose, M.J., Gaisan, B., Bilbao, J., 2002. Wax formation in the pyrolysis of poly ole fins in a conical spouted bed reactor. Energy & Fuels 16 (6), 1429–1437.

Al-Salem, S.M., Lettieri, P., Baeyens, J., 2009. Recycling and recovery routes of plastic solid waste (PSW): a review. Waste Management 29 (10), 2625–2643.

Balat, M., 2011. Fuels from biomass – an overview, In: Speight, J.G. (Ed.), The Biofuels Handbook Royal Society of Chemistry, London, United Kingdom. Part 1, (Chapter 3).

Basu, P., 2013. Biomass gasification, pyrolysis and torrefaction. In: Practical Design and Theory, second ed. Academic Press, Inc., New York.

Bhattacharya, S., Mizanur Rahman Siddique, A.H.M., Pham, H.-L., 1999. A study in wood gasification on low tar production. Energy 24, 285–296.

Boateng, A.A., Walawender, W.P., Fan, L.T., Chee, C.S., 1992. Fluidized-bed steam gasification of rice hull. Bioresource Technology 40 (3), 235–239.

Boateng, A.A., Daugaard, D.E., Goldberg, N.M., Hicks, K.B., 2007. Industrial & Engineering Chemistry Research 46, 1891–1897.

Brage, C., Yu, Q., Chen, G., Sjöström, K., 2000. Tar evolution profiles obtained from gasification of biomass and coal. Biomass and Bioenergy 18 (1), 87–91.

Brar, J.S., Singh, K., Wang, J., Kumar, S., 2012. Cogasification of coal and biomass: a review. International Journal of Financial Research 2012, 1–10, 2012.

Brigwater, A.V. (Ed.), 2003. Pyrolysis and Gasification of Biomass and Waste. CPL Press, Newbury, Berkshire, United Kingdom.

Calvin, M., 1980. Hydrocarbons from plants. Naturwissenschaften 67 (11), 525–533.

Cetin, E., Moghtaderi, B., Gupta, R., Wall, T.F., 2005. Biomass gasification kinetics: influences of pressure and char structure. Combustion Science and Technology 177 (4), 765–791.

Chadeesingh, R., 2011. The fischer-tropsch process. In: Speight, J.G. (Ed.), The Biofuels Handbook, The Royal Society of Chemistry, pp. 476–517. London, United Kingdom. Part 3, (Chapter 5).

Chen, G., Sjöström, K., Bjornbom, E., 1992. Pyrolysis/gasification of wood in a pressurized fluidized bed reactor. Industrial & Engineering Chemistry Research 31 (12), 2764–2768.

Collot, A.G., Zhuo, Y., Dugwell, D.R., Kandiyoti, R., 1999. Co-pyrolysis and cogasification of coal and biomass in bench-scale fixed-bed and fluidized bed reactors. Fuel 78, 667–679.

Crocker, M., Crofcheck, C., 2006. Reducing national dependence on imported oil. Energeia 17 (6) (Center for Applied Energy Research, University of Kentucky, Lexington, Kentucky).

Dasappa, S., Paul, P.J., Mukunda, H.S., Rajan, N.K.S., Sridhar, G., Sridhar, H.V., 2004. Biomass gasification technology – a route to meet energy needs. Current Science 87 (7), 908–916.

Deguchi, S., Mukai, S.A., Tsudome, M., Horikoshi, K., 2006. Facile generation of fullerene nanoparticles by hand-grinding. Advanced Materials 18, 729–732.

Demirbaş, A., 2004. Pyrolysis of municipal plastic wastes for recovery of gasoline-range hydrocarbons. Journal of Analytical and Applied Pyrolysis 72 (1), 97–102.

Demirbaş, A., 2006. Current technologies for biomass conversion into chemicals and fuels. Energy Sources Part A 28, 1181–1188.

Demirbaş, A., 2011. Production of fuels from crops. In: Speight, J.G. (Ed.), The Biofuels Handbook. Royal Society of Chemistry, London, United Kingdom. Part 2, (Chapter 1).

Ergudenler, A., Ghaly, A.E., 1993. Agglomeration of alumina sand in a fluidized bed straw gasifier at elevated temperatures. Bioresource Technology 43 (3), 259–268.

Erikkson, O., Lindgren, B.O., 1977. About the linkage between lignin and hemicelluloses in wood. Svensk Papperstidning 80 (2), 59–63.

Freudenberg, K., Neish, A.C., 1968. In: Springer, G.F., Kleinzeller, A. (Eds.), Constitution and Biosynthesis of Lignin. Springer-Verlag, New York.

Gabra, M., Pettersson, E., Backman, R., Kjellström, B., 2001. Evaluation of cyclone gasifier performance for gasification of sugar cane residue – Part 1: gasification of bagasse. Biomass and Bioenergy 21 (5), 351–369.

Garcia, L., French, R., Czernik, S., Chornet, E., 2000. Catalytic steam reforming of bio-oils for the production of hydrogen: effects of catalyst composition. Applied Catalysis A: General 201 (2), 225–239.

Hanaoka, T., Inoue, S., Uno, S., Ogi, T., Minowa, T., 2005. Effect of woody biomass components on air-steam gasification. Biomass and Bioenergy 28 (1), 69–76.

Hernandez, M.R., Garcia, A.N., Marcilla, A., 2007. Catalytic flash pyrolysis of HDPE in a fluidized bed reactor for recovery of fuel-like hydrocarbons. Journal of Analytical and Applied Pyrolysis 78 (2), 272–281.

Jayamurthy, M., Vasudevan, S., 1996. Methanol-to-Gasoline (MTG) conversion over ZSM-5: a temperature programmed surface reaction study. Catalysis Letters 36 (1–2), 111–114.

Jenkins, B.M., Ebeling, J.M., 1985. Thermochemical Properties of Biomass Fuels. California Agriculture, pp. 14–18 (May-June).

John, E.J., Singh, K., 2011. Production and properties of fuels from domestic and industrial waste. In: Speight, J.G. (Ed.), The Biofuels Handbook, The Royal Society of Chemistry, pp. 333–376. London, United Kingdom. Part 3, (Chapter 1).

Kaminsky, W., Zoriquetta, I.J.N., 2007. Catalytical and thermal pyrolysis of polyolefins. Journal of Analytical and Applied Pyrolysis 79 (1–2), 368–374.

Karhunen, P., Rummakko, P., Sipilä, J., Brunow, G., Pipeline, I., 1995. Dibenzodioxins: a novel type of linkage in softwood lignins. Tetrahedron Letters 36 (1), 167–170.

Kavalov, B., Peteves, S.D., 2005. Status and Perspectives of Biomass-To-Liquid Fuels in the European Union. European Commission. Directorate General Joint Research Centre (DG JRC). Institute for Energy, Petten, The Netherlands.

Kim, J., Park, C., Kim, T.H.L.M., Kim, S., Lee, S.W., 2003. Effects of various pretreatments for enhanced anaerobic digestion with waste activated sludge. Journal of Bioscience and Bioengineering 95, 271−275.

Ko, M.K., Lee, W.Y., Kim, S.B., Lee, K.W., Chun, H.S., 2001. Gasification of food waste with steam in fluidized bed. Korean Journal of Chemical Engineering 18 (6), 961−964.

Kumabe, K., Hanaoka, T., Fujimoto, S., Minowa, T., Sakanishi, K., 2007. Cogasification of woody biomass and coal with air and steam. Fuel 86, 684−689.

Lee, S., Shah, Y.T., 2013. Biofuels and Bioenergy. CRC Press, Taylor & Francis Group, Boca Raton, Florida.

Luque, R., Speight, J.G. (Eds.), 2015. Gasification for Synthetic Fuel Production: Fundamentals, Processes, and Applications. Woodhead Publishing, Elsevier, Cambridge, United Kingdom.

Lv, P.M., Xiong, Z.H., Chang, J., Wu, C.Z., Chen, Y., Zhu, J.X., 2004. An experimental study on biomass air-steam gasification in a fluidized bed. Bioresource Technology 95 (1), 95−101.

Machado, N.R.C.F., Calsavara, V., Guilherme, N., Astrath, C., Neto, A.M., Mauro Baesso, M.L., 2006. Hydrocarbons from ethanol using [Fe,Al]ZSM-5 zeolites obtained by direct synthesis. Applied Catalysis A: General 311, 193−198.

Marcilla, A., Beltran, M.I., Navarro, R., 2008. Evolution with the temperature of the compounds obtained in the catalytic pyrolysis of polyethylene over HUSY. Industrial & Engineering Chemistry Research 47 (18), 6896−6903.

Metzger, J.O., 2006. Production of liquid hydrocarbons from biomass. Angewandte Chemie International Edition 45, 696−698.

Molino, A., Chianese, S., Musmarra, D., 2016. Biomass gasification technology: the state of the art overview. Jounral of Energy Chemistry 25 (1), 10−25.

Mshandete, A., Bjornsson, L., Kivaisi, A.K., Rubindamayugi, M.S.T., Matthiasson, B., 2006. Effect of particle size on biogas yield from sisal fiber waste. Renewable Energy 31, 2385−2392.

Pakdel, H., Roy, C., 1991. Hydrocarbon content of liquid products and tar from pyrolysis and gasification of wood. Energy & Fuels 5, 427−436.

Pan, Y.G., Velo, E., Roca, X., Manyà, J.J., Puigjaner, L., 2000. Fluidized-bed cogasification of residual biomass/poor coal blends for fuel gas production. Fuel 79, 1317−1326.

Pearl, I.W., 1967. The Chemistry of Lignin. Marcel Dekker, Inc., New York.

Predel, M., Kaminsky, W., 2000. Pyrolysis of mixed poly-olefins in a fluidized-bed reactor and on a pyro-GC/MS to yield aliphatic waxes. Polymer Degradation and Stability 70 (3), 373−385.

Rajvanshi, A.K., 1986. Biomass gasification. In: Goswami, D.Y. (Ed.), Alternative Energy in Agriculture, vol. II. CRC Press, Boca Raton, Florida, pp. 83−102.

Ramroop Singh, N., 2011. Biofuel. In: Speight, J.G. (Ed.), The Biofuels Handbook. Royal Society of Chemistry, London, United Kingdom. Part 1, (Chapter 5).

Rapagnà, N.J., Latif, A., 1997. Steam gasification of almond shells in a fluidized bed reactor: the influence of temperature and particle size on product yield and distribution. Biomass and Bioenergy 12 (4), 281−288.

Rapagnà, N.J., Kiennemann, A., Foscolo, P.U., 2000. Steam-gasification of biomass in a fluidized-bed of olivine particles. Biomass and Bioenergy 19 (3), 187−197.

Sarker, M., Rashid, M.M., Rahman, M.S., Molla, M., 2012. A new kind of renewable energy: production of aromatic hydrocarbons naphtha chemical by thermal degradation of polystyrene (PS) waste plastic. American Journal of Climate Change 1, 145−153.

Sarkanen, K.V., Ludwig, C.H., 1971. In: Sarkanen, K.V., Ludwig, C.H. (Eds.), Lignin: Occurrence, Formation, Structure and Reactions. Wiley-Interscience, New York.

Scheirs, J., Kaminsky, W., 2006. Feedstock Recycling and Pyrolysis of Waste Plastics. John Wiley & Sons Inc., Chichester, United Kingdom.

Sjöström, E., 1993. Wood Chemistry: Fundamentals and Application. Academic Press, Orlando.

Sjöström, K., Chen, G., Yu, Q., Brage, C., Rosén, C., 1999. Promoted reactivity of char in cogasification of biomass and coal: synergies in the thermochemical process. Fuel 78, 1189−1194.

Smith, I.M., 2006. Management of FGD Residues. Center for Applied Energy Research, University of Kentucky, Lexington, Kentucky.

Speight, J.G., 2011a. The Refinery of the Future. Gulf Professional Publishing, Elsevier, Oxford, United Kingdom.

Speight, J.G., 2011b. An Introduction to Petroleum Technology, Economics, and Politics. Scrivener Publishing, Salem, Massachusetts.

Speight, J.G. (Ed.), 2011c. The Biofuels Handbook. The Royal Society of Chemistry, London, United Kingdom.

Speight, J.G., 2013a. The Chemistry and Technology of Coal, third ed. CRC Press, Taylor & Francis Group, Boca Raton, Florida.

Speight, J.G., 2013b. Coal-Fired Power Generation Handbook. Scrivener Publishing, Salem, Massachusetts.

Speight, J.G., 2014. The Chemistry and Technology of Petroleum, fifth ed. CRC Press, Taylor & Francis Group, Boca Raton, Florida.

Speight, J.G., Islam, M.R., 2016. Peak Energy − Myth or Reality. Scrivener Publishing, Beverly, Massachusetts.

Speight, J.G., 2017. Handbook of Petroleum Refining. CRC Press, Taylor & Francis Group, Boca Raton, Florida.

Speight, J.G., 2019. Biogas: Production and Properties. Nova Science publishers, New York.

Steinberg, M., Fallon, P.T., Sundaram, M.S., 1992. The flash pyrolysis and methanolysis of biomass (wood) for the production of ethylene, benzene, and methanol. In: von Herausgeg, R.C., Albright, L.F., Crynes, B.L., Nowak, S. (Eds.), Novel Production Methods for Ethylene, Light Hydrocarbons, and Aromatics. Marcel Dekker Inc., New York.

Vasudevan, P., Sharma, S., Kumar, A., 2005. Liquid fuels from biomass: an overview. Journal of Scientific & Industrial Research 64, 822−831.

Wright, L., Boundy, R., Perlack, R., Davis, S., Saulsbury, B., 2006. Biomass Energy Data Book: Edition 1. Office of Planning, Budget and Analysis. Energy Efficiency and Renewable Energy, United States Department of Energy, Oak Ridge, Tennessee. Contract No. DE-AC05-00OR22725. Oak Ridge National Laboratory.

Further reading

Adler, E., 1977. Lignin − past, present and future. Wood Science and Technology 11 (3), 169−218.

Perlin, J., 1989. A Forest Journey. The Role of Wood in the Development of Civilization. W. Norton & Co. Inc., New York.

Shah, S., Gupta, M.N., 2007. Lipase catalyzed preparation of biodiesel from Jatropha oil in a solvent free system. Process Biochemistry 42, 409−414.

Speight, J.G., 2012. Shale Oil Production Processes. Gulf Professional Publishing, Elsevier, Oxford, United Kingdom.

Chapter 8

Hydrocarbons from synthesis gas

1. Introduction

Synthesis gas (also referred to as *syngas*) is the name given to a gas mixture that contains varying amounts of carbon monoxide (CO) and hydrogen (H_2) generated by the gasification of a carbonaceous material. Examples include steam reforming of natural gas, crude oil residua, coal, and biomass. Synthesis gas is used as an intermediate in producing hydrocarbon derivatives via the Fischer-Tropsch process for use as fuels (Fig. 8.1).

Current commercial processes for synthesis gas and hydrogen production largely depend on fossil fuels both as the source of hydrogen and as the source of energy for the production processing. Fossil fuels are nonrenewable energy resources, but they provide a more economical path to hydrogen production in the near term (next three decades) and perhaps they will continue to play an important role in the mid-term (up to 50 years from now). Alternative processes need to be developed that do not depend on fossil hydrocarbon resources for either the hydrogen source or the energy source, and these

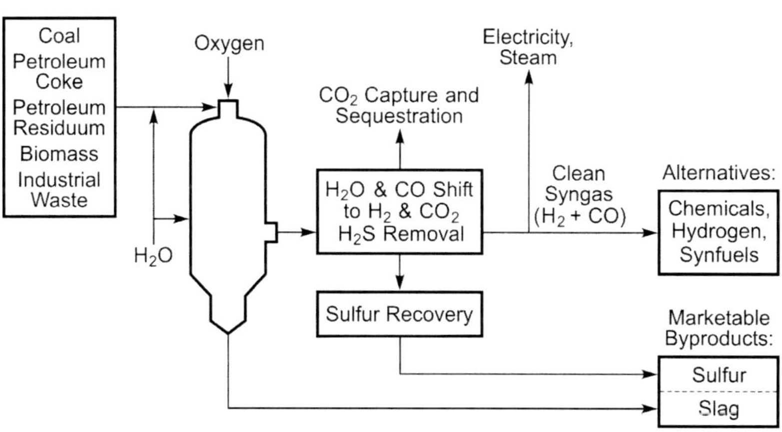

FIGURE 8.1 The gasification process can accommodate a variety of carbonaceous feedstocks.

Handbook of Industrial Hydrocarbon Processes. https://doi.org/10.1016/B978-0-12-809923-0.00008-4

processes need to be economical, environmentally friendly, and competitive. Efficient separation of the hydrogen from the gaseous products is also a major issue that must be addressed. In this respect, pressure swing adsorption (PSA) is used in current industrial practice. Furthermore, several types of membranes are being developed that, when incorporated into the separation process, will enable more efficient gas separation.

The process for producing synthesis gas involves three individual components: (i) synthesis gas generation, (ii) waste heat recovery, and (iii) gas processing. Within each of the above three listed systems are several options—synthesis gas can be generated to yield a range of compositions ranging from high-purity hydrogen to high-purity carbon monoxide. Two major routes can be utilized for high-purity gas production: (i) pressure swing adsorption and (ii) utilization of a cold box, where separation is achieved by distillation at low temperatures. In fact, both processes can also be used in combination as well. Unfortunately, both processes require high capital expenditure. However, to address these concerns, research and development is ongoing and successes can be measured by the demonstration and commercialization of technologies such as permeable membrane for the generation of high-purity hydrogen, which in itself can be used to adjust the H_2/CO ratio of the synthesis gas produced.

Both nonrenewable and renewable energy sources are important for production of synthesis gas and hydrogen. As energy carriers, hydrogen and synthesis gas can be produced from catalytic processing of various hydrocarbon fuels, alcohol fuels, and a variety of biofuels and biomass feedstocks. Gasification to produce synthesis gas can proceed from almost all carbonaceous organic materials such as coal, crude oil residua, as well as biomass and plastic waste (Speight, 2013a, 2013b, 2014b; Adhikari et al., 2015). The resulting synthesis gas burns cleanly into water vapor and carbon dioxide:

$$CO/H_2 + O_2 \rightarrow CO_2 + H_2O$$

Alternatively, synthesis gas may be converted efficiently to methane via the Sabatier reaction, or to a series of synthetic hydrocarbon derivatives via the Fischer-Tropsch reaction:

Sabatier reaction:

$$CO_2 + 4H_2 \rightarrow CH_4 + 2H_2O$$

Fischer-Tropsch reaction:

$$nCO + (2n+1)H_2 \rightarrow C_nH_{(2n+2)} + nH_2O$$

Inorganic components of the feedstock, such as metals and minerals, are trapped in an inert and environmentally safe form as char, which may have use as a fertilizer.

In principle, synthesis gas can be produced from any hydrocarbon feedstock. These include natural gas, naphtha, residual oil, crude oil coke, coal, and biomass. In fact, as the move to a bio-based economy (also known as a bioeconomy) continues understanding mechanisms and processing of synthesis gas from various sources will assist in the improvement of the relevant industrial processes. Briefly, the bio-based economy uses first-generation biomass (such as crops), second-generation biomass (such as crop refuge), and third-generation biomass (such as seaweed, algae). Several methods of processing are then used (in a biorefinery) to gather the most out of the biomass. This includes techniques such as (i) pyrolysis, (ii) fermentation, (iii) anaerobic digestion, and (iv) torrefaction.

Pyrolysis is the thermal decomposition of a feedstock, such as biomass, at elevated temperatures in an inert atmosphere. The process results in a change in the chemical composition of the feedstock and the process is irreversible. Pyrolysis is most commonly used in the treatment of organic materials and the products evolved during the process can also include hydrocarbon derivatives. In general, pyrolysis of organic substances produces volatile products and leaves a solid residue enriched in carbon. If the process is taken to completion (then it is referred to as carbonization), the products are gases, liquids, and a solid char. Applications of pyrolysis would convert biomass into synthesis gas, a liquid product (pyrolysis oil which may contain hydrocarbon derivatives), and biochar. Waste plastics can be converted to usable oil (that contains hydrocarbon constituents).

Fermentation is a metabolic process that produces chemical changes in organic feedstocks through the action of enzymes. The common, and perhaps most well-known example of the process, is the fermentation of sugars to produce ethanol, from which hydrocarbon derivatives can be produced. On the other hand, the anaerobic digestion process is generally used to produce the gaseous mixture known as biogas that also contains hydrocarbon products, particularly methane (Speight, 2019a).

The *anaerobic digestion process* is a collection of processes by which microorganisms break down feedstocks in the absence of oxygen. The process begins with bacterial hydrolysis of the feedstock after which other bacteria cause a series of reactions that result in the production of methane and carbon dioxide (Speight, 2019a). This biogas can be used directly as fuel, in combined heat and power gas engines or upgraded to natural gas-quality methane (sometimes referred to as *biomethane* as an indicator of the source of the gas).

In the torrefaction process—which is a mild form of pyrolysis (more correctly, thermal treatment at temperatures typically between 200°C and 320°C (390°F −610°F)—moisture and volatile materials are removed from biomass and the product is a solid form of the biomass that can be used as a solid fuel. During the torrefaction process, the biomass typically loses 20% w/w of its mass (bone dry basis) and 10% of its heating value, with no appreciable change in volume. This energy (the volatiles) can be used as a heating fuel for the torrefaction process. After the biomass is torrefied, it can be densified, usually into briquettes or pellets using conventional densification equipment, to increase its mass and energy density and to improve its hydrophobic properties. The final product may repel water and thus can be stored in moist air or rain without appreciable change in moisture content or heating value, unlike the original biomass. The product of the torrefaction process has been erroneously called biocoal and use of this term should be discontinued.

The lowest cost routes for synthesis gas production, however, are based on natural gas. The cheapest option is remote or stranded reserves. Current economic considerations dictate that the production of liquid fuels from synthesis gas translates into using natural gas as the hydrocarbon source. Nevertheless, the synthesis gas production operation in a gas-to-liquids plant amounts to greater than half of the capital cost of the plant. The choice of technology for synthesis gas production also depends on the scale of the synthesis operation. Synthesis gas production from solid fuels can require an even greater capital investment with the addition of feedstock handling and more complex synthesis gas purification operations. The greatest impact on improving gas-to-liquids plant economics is to decrease capital costs associated with synthesis gas production and improve thermal efficiency through better heat integration and utilization. Improved thermal efficiency can be obtained by combining the gas-to-liquids plant with a power generation plant to take advantage of the availability of low-pressure steam.

Synthesis gas is combustible and often used as a fuel source or as an intermediate for the production of other chemicals. Synthesis gas for use as a fuel is most often produced by gasification of coal or municipal waste mainly by the following paths:

$$C + O_2 \rightarrow CO_2$$

$$CO_2 + C \rightarrow 2CO$$

$$C + H_2O \rightarrow CO + H_2$$

When used as an intermediate in the large-scale, industrial synthesis of hydrogen and ammonia, it is also produced from natural gas (via the steam reforming reaction):

$$CH_4 + H_2O \rightarrow CO + 3H_2$$

The synthesis gas produced in large waste-to-energy gasification facilities is used as fuel to generate electricity.

The manufacture of gas mixtures of carbon monoxide and hydrogen has been an important part of chemical technology for approximately a century. Originally, such mixtures were obtained by the reaction of steam with incandescent coke and were known as *water gas*. Used first as a fuel, water gas soon attracted attention as a source of hydrogen and carbon monoxide for the production of chemicals, at which time it gradually became known as synthesis gas. Eventually, steam reforming processes, in which steam is reacted with natural gas (methane) or crude oil naphtha over a nickel catalyst, found wide application for the production of synthesis gas.

A modified version of steam reforming known as autothermal reforming is a combination of partial oxidation near the reactor inlet with conventional steam reforming further along the reactor which improves the overall reactor efficiency and increases the flexibility of the process. Partial oxidation processes using oxygen instead of steam also found wide application for synthesis gas manufacture, with the special feature that they could utilize low-value feedstocks such as high-boiling crude oil residua. In recent years, catalytic partial oxidation employing very short reaction times (milliseconds) at high temperatures (850–1000°C; 1560 to 1830°F) is providing still another approach to synthesis gas manufacture (Luque and Speight, 2015). Nearly complete conversion of methane, with close to 100% selectivity to H_2 and CO, can be obtained with a rhenium monolith under well-controlled conditions. Experiments on the catalytic partial oxidation of *n*-hexane conducted with added steam give much higher yields of H_2 than can be obtained in experiments without steam, a result of much interest in obtaining hydrogen-rich streams for fuel cell applications. The route of coal to synthetic automotive fuels, as practiced by SASOL (Couvaras, 1997; Jager, 1997) is technically proven and products with favorable environmental characteristics are produced. As is the case in essentially all coal conversion processes where air or oxygen is used for the utilization or partial conversion of the energy in the coal, the carbon dioxide burden is a drawback as compared to crude oil.

The uses of synthesis gas include use as a chemical feedstock and in gas-to-liquid processes (Mangone, 2002; 2019c), which use Fischer–Tropsch chemistry to make liquid fuels as feedstock for chemical synthesis; it is also being used in the production of fuel additives, including diethyl ether and methyl t-butyl ether (MTBE), acetic acid and its anhydride; synthesis gas could also make an important contribution to chemical synthesis through conversion to methanol. There is also the option in which stranded natural gas is converted to synthesis gas production followed by conversion to liquid fuels.

2. Coal gasification

Gasification is a process that converts carbonaceous materials, such as coal, crude oil, or biomass, into carbon monoxide and hydrogen by reacting the raw material at high temperatures with a controlled amount of oxygen (Speight, 2013a, 2013b; Luque and Speight, 2015). The resulting gas mixture is called synthesis gas or synthesis gas is itself a fuel. Gasification is a very efficient method for extracting energy from many different types of organic materials, and also has applications as a clean waste disposal technique. In the process, coal or coal char is converted to gaseous products by reaction with steam, oxygen, air, hydrogen, carbon dioxide, or a mixture of these.

The gasification of coal or a derivative (i.e., char produced from coal) is, essentially, the conversion of coal (by any one of a variety of processes) to produce combustible gases (Radovic et al., 1983; Radovic and Walker, 1984; Garcia and Radovic, 1986; Calemma and Radovic, 1991; Kristiansen, 1996; Luque and Speight, 2015). With the rapid increase in the use of coal from the 15th century onwards it is not surprising the concept of using coal to produce a flammable gas, especially the use of the water and hot coal, became common place. In fact, the production of gas from coal has been a vastly expanding area of coal technology and, as a result, the characteristics of rank, mineral matter, particle size, and reaction conditions are all recognized as having a bearing on the outcome of the process; not only in terms of gas yields but also on gas properties (Speight, 2013a, 2013b; Luque and Speight, 2015).

Gasification of coal has been, and continues to be, widely used on industrial scales to generate electricity. However, almost any type of organic carbonaceous material can be used as the raw material for gasification, such as wood, biomass, or even plastic waste. Thus, gasification may be an important technology for renewable energy. In particular biomass gasification is carbon neutral. Gasification has been considered for many years as an alternative to combustion of solid or liquid fuels. It is easier to clean gaseous mixtures than it is to clean solid or high-viscosity liquid fuels. Clean gas can be used in internal combustion–based power plant that would suffer from severe fouling or corrosion if solid or low-quality liquid fuels were burnt inside them. However, in the current context, the gasification of coal and, for that matter, any carbonaceous feedstock, offers an indirect route to hydrocarbon products.

The advantage of gasification is that using the synthesis gas is more efficient than direct combustion of the original fuel; more of the energy contained in the fuel is extracted. Synthesis gas may be burned directly in internal combustion engines, used to produce methanol and hydrogen, or converted via the Fischer-Tropsch process into synthetic fuel. Gasification can also begin with materials that are not otherwise useful fuels, such as biomass or organic waste. In addition, the high-temperature combustion refines out corrosive ash elements such as chloride and potassium, allowing clean gas production from otherwise problematic fuels.

Another advantage of gasification-based energy systems is that when oxygen is used in the gasifier in place of air, the carbon dioxide produced by the process is in a concentrated gas stream, making it easier and less expensive to separate and capture. Once the carbon dioxide is captured, it can be sequestered and prevented from escaping to the atmosphere, where it could otherwise potentially contribute to the *greenhouse effect*.

2.1 Chemistry

Gasification relies on chemical processes at elevated temperatures >700°C (>1290°F), which distinguishes it from biological processes such as anaerobic digestion that produce gas. In a gasifier, the carbonaceous material undergoes several different processes: (i) pyrolysis of the carbonaceous fuel, (ii) combustion, and (iv) gasification of the remaining char.

Pyrolysis (*devolatilization*) is the thermal degradation of an organic substance in the absence of air to produce char, pyrolysis oil, and synthesis gas, e.g., the conversion of wood to charcoal. The pyrolysis process occurs as the carbonaceous feedstock heats up. Volatiles are released and char is produced, resulting in up to 70% weight loss for coal. The process is very dependent on the properties of the carbonaceous material and determines the structure and composition of the char, which will then undergo gasification reactions.

The *combustion* process occurs as the volatile products and some of the char reacts with oxygen to form carbon dioxide and carbon monoxide, which provides heat for the subsequent gasification reactions:

$$2C + O_2 \rightarrow CO_2$$

Gasification is the decomposition of any one of several carbonaceous feedstocks into a synthesis gas by carefully controlling the amount of oxygen present, e.g., the conversion of coal into town gas. The gasification process occurs as the char reacts with carbon dioxide and steam to produce carbon monoxide and hydrogen:

$$C + H_2O \rightarrow H_2 + CO_2$$

In addition, the gas phase water gas shift reaction reaches equilibrium very fast at the temperatures in a gasifier. This removes carbon dioxide from the reactor and provides water for the gasification reaction;

$$CO + H_2O \leftrightarrow CO_2 + H_2$$

In practice, a limited amount of oxygen or air is introduced into the reactor to allow some of the organic material to be burned to produce carbon monoxide and energy, which drives a second reaction that converts further organic material to hydrogen and additional carbon monoxide.

Coal gasification chemistry, as represented by the equations, is reasonably simple and straightforward and, hence, coal gasification processes are reasonably efficient. For many years such processes were used to manufacture illuminating gas (coal gas) for gas lighting, before electric lighting became widely available. The simplest method, and the first used, was to heat coal in a retort in the absence of air, partially converting coal to gas with a residue of coke; the Scottish engineer William Murdock used this technique in pioneering the commercial gasification of coal in 1792. Murdock licensed his process to the Gas Light and Coke Company in 1813, and in 1816 the Baltimore Gas Company, the first coal gasification company in the United States, was established. The process of heating coal to produce coke and gas is still used in the metallurgical industry.

Currently, hydrogen is produced from coal by gasification and the subsequent processing of the resulting synthesis gas. In its simplest form, coal gasification works by first reacting coal with oxygen and steam under high pressures and temperatures to form a synthesis gas consisting primarily of carbon monoxide and hydrogen. This synthesis gas is cleaned of virtually all of its impurities and shifted to produce additional hydrogen. The clean gas is sent to a separation system to recover hydrogen.

The most complete conversion of coal or coke to gas that is feasible was achieved by reacting coal continuously in a vertical retort with air and steam. The gas obtained in this manner, called producer gas, has a relatively low heat content per unit volume of gas (100−150 Btu/ft^3). The development of a cyclic steam-air process in 1873 made possible the production of a gas of higher thermal content (300−350 Btu/ft^3), composed chiefly of carbon monoxide and hydrogen, and known as water gas. By adding oil to the reactor, the thermal content of gas was increased to 500 to 550 Btu/ft^3; this became the standard for gas distributed to residences and industry. Since 1940, processes have been developed to produce continuously a gas equivalent to water gas; this involves the use of steam and essentially pure oxygen as a reactant. A more recently developed process reacts coal with pure oxygen and steam at an elevated pressure (450 psi) to produce a gas that may be converted to synthetic natural gas.

2.2 Processes

Briefly, gasification is a process in which combustible materials are partially oxidized or partially combusted. The product of gasification is a combustible synthesis gas. Because gasification involves the partial, rather than complete, oxidization of the feed, gasification processes operate in an oxygen-lean environment. The process has grown from a predominately coal conversion process used for making *town gas* for industrial lighting to an advanced process for the production of multiproduct, carbon-based fuels from a variety of feedstocks such as crude oil viscous feedstocks, biomass, or other carbonaceous feedstocks (Fig. 8.1) (Kumar et al., 2009; Speight, 2013a, 2014a; 2014b; Luque and Speight, 2015).

The most common modern coal gasification process uses lump coal in a vertical retort. In the process, coal is fed at the top with air, and steam is introduced at the bottom and the gas (air and steam) rising up the retort heat the coal in its downward flow and react with the coal to convert it to gas. Ash is removed at the bottom of the retort. Using air and steam as reacting gases results in a producer gas; using oxygen and steam results in a water gas. Increasing operating pressure increases the productivity.

Two other processes currently in commercial use react finely powdered coal with steam and oxygen. One of these, the Winkler process, uses a fluidized bed in which the powdered coal is agitated with the reactant gases. The

Koppers-Totzek process operates at a much higher temperature, and the powdered coal is reacted while it is entrained in the gases passing through the reactor. The ash is removed as a molten slag at the bottom of the reactor.

As crude oil supplies decrease, the desirability of producing gas from coal may increase, especially in those areas where natural gas is in short supply. It is also anticipated that costs of natural gas will increase, allowing coal gasification to compete as an economically viable process. Research in progress on a laboratory and pilot-plant scale should lead to the invention of new process technology by the end of the century, thus accelerating the industrial use of coal gasification.

Thus, the products of coal gasification consist of carbon monoxide, carbon dioxide, hydrogen, methane, and some other gases in proportions dependent upon the specific reactants and conditions (temperatures and pressures) employed within the reactors, and the treatment steps which the gases undergo subsequent to leaving the gasifier. Similar chemistry can also be applied to the gasification of coke derived from crude oil and other sources. The reaction of coal or coal char with air or oxygen to produce heat and carbon dioxide could be called gasification, but it is more properly classified as combustion. The principal purposes of such conversion are the production of synthetic natural gas as a substitute gaseous fuel and synthesis gases for production of chemicals and plastics.

In all cases of commercial interest, gasification with steam, which is endothermic, is an important chemical reaction. The necessary heat input is typically supplied to the gasifier by combusting a portion of the coal with oxygen added along with the steam. From the industrial viewpoint, the final product is synthesis gas, medium-Btu gas, or substitute natural gas. Each of the gas types has potential industrial applications.

In the chemical industry, synthesis gas from coal is a potential alternative source of hydrogen and carbon monoxide. This mixture is obtained primarily from the steam reforming of natural gas, natural gas liquids, or other crude oil liquids. Fuel users in the industrial sector have studied the feasibility of using medium-Btu gas instead of natural gas or oil for fuel applications. Finally, the natural gas industry is interested in substitute natural gas, which can be distributed in existing pipeline networks.

The conversion of the gaseous products of coal gasification processes to synthesis gas, a mixture of hydrogen (H_2) and carbon monoxide (CO), in a ratio appropriate to the application, needs additional steps, after purification. The product gases—carbon monoxide, carbon dioxide, hydrogen, methane, and nitrogen—can be used as fuels or as raw materials for chemical or fertilizer manufacture.

2.3 Gasifiers

The focal point of any gasification-based system is the gasifier. A gasifier converts hydrocarbon feedstock into gaseous components by applying heat under pressure in the presence of steam.

A gasifier differs from a combustor in that the amount of air or oxygen available inside the gasifier is carefully controlled so that only a relatively small portion of the fuel burns completely. The *partial oxidation* process provides the heat and rather than combustion, most of the carbon-containing feedstock is chemically broken apart by the heat and pressure applied in the gasifier resulting in the chemical reactions that produce synthesis gas. However, the composition of the synthesis gas will vary because of dependence upon the conditions in the gasifier and the type of feedstock.

Minerals in the fuel (i.e., the rocks, dirt and other impurities which do not gasify) separate and leave the bottom of the gasifier either as an inert glasslike slag or other marketable solid products. Sulfur impurities in the feedstock are converted to hydrogen sulfide (H_2S) and carbonyl sulfide (COS), from which sulfur can be extracted, typically as elemental sulfur. Nitrogen oxides (NO_X), other potential pollutants, are not formed in the oxygen-deficient (reducing) environment of the gasifier. Instead, ammonia (NH_3) is created by nitrogen-hydrogen reactions and can be washed out of the gas stream.

Four types of gasifier are currently available for commercial use: (i) the countercurrent fixed bed gasifier, (ii) the cocurrent fixed bed gasifier, (iii) the fluidized bed gasifier, and (iv) the entrained flow gasifier (Speight, 2013a, 2013b; Luque and Speight, 2015).

The *countercurrent fixed bed gasifier* (*(updraft gasifier*) consists of a fixed bed of carbonaceous fuel (e.g., coal or biomass) through which the gasification agent (steam, oxygen, and/or air) flows in countercurrent configuration. The ash is either removed dry or as a slag. The slagging gasifiers require a higher ratio of steam and oxygen to carbon in order to reach temperatures higher than the ash fusion temperature. The nature of the gasifier means that the fuel must have high mechanical strength and must be noncaking so that it will form a permeable bed, although recent developments have reduced these restrictions to some extent. The throughput for this type of gasifier is relatively low. Thermal efficiency is high as the gas exit temperatures are relatively low. However, this means that tar and methane production is significant at typical operation temperatures, so product gas must be extensively cleaned before use or recycled to the reactor.

The *cocurrent fixed bed gasifier (downdraft gasifier)* is similar to the countercurrent type, but the gasification agent gas flows in cocurrent configuration with the fuel (downwards, hence the name downdraft gasifier). Heat needs to be added to the upper part of the bed, either by combusting small amounts of the fuel or from external heat sources. The produced gas leaves the gasifier at a high temperature, and most of this heat is often transferred to the gasification agent added in the top of the bed, resulting in energy efficiency on level with the countercurrent type. Since all tars must pass through a hot bed of char in this configuration, tar levels are much lower than the countercurrent type.

In the fluidized bed gasifier, the fuel is fluidized in oxygen (or air) and steam. The ash is removed dry or as high-density agglomerates. The temperatures are relatively low in dry ash gasifiers, so the fuel must be highly reactive; low-grade coals are particularly suitable. The agglomerating gasifiers have slightly higher temperatures, and are suitable for higher rank coals. Fuel throughput is higher than for the fixed bed, but not as high as for the entrained flow gasifier. The conversion efficiency is rather low, so recycle or subsequent combustion of solids is necessary to increase conversion. Fluidized bed gasifiers are most useful for fuels that form highly corrosive ash that would damage the walls of slagging gasifiers. A disadvantage of biomass feedstocks is that they generally contain high levels of corrosive ash.

In the *entrained flow gasifier* a dry pulverized solid, an atomized liquid fuel, or a fuel slurry is gasified with oxygen (much less frequent: air) in cocurrent flow. The gasification reactions take place in a dense cloud of very fine particles. Most coal is suitable for this type of gasifier because of the high operating temperatures and because the coal particles are well separated from one another. The high temperatures and pressures also mean that a higher throughput can be achieved but thermal efficiency is somewhat lower as the gas must be cooled before it can be cleaned with existing technology. The high temperatures also mean that tar and methane are not present in the product gas; however, the oxygen requirement is higher than for the other types of gasifiers. All entrained flow gasifiers remove the major part of the ash as a slag as the operating temperature is well above the ash fusion temperature. A smaller fraction of the ash is produced either as a very fine dry fly ash or as black fly ash slurry. Some fuels, in particular certain types of biomasses, can form slag that is corrosive for ceramic inner walls that serve to protect the gasifier outer wall. However, some entrained bed type of gasifiers do not possess a ceramic inner wall but have an inner water or steam cooled wall covered with partially solidified slag. These types of gasifiers do not suffer from corrosive slag. Some fuels have ashes with very high ash fusion temperatures. In this case mostly limestone is mixed with the fuel prior to gasification. Addition of limestone will usually suffice for lowering the fusion temperatures. The fuel particles must be much smaller than for other types of gasifiers. This means the fuel must be pulverized, which requires somewhat more energy than for the other types of gasifiers. By far the most energy consumption related to entrained bed gasification is not the milling of the fuel but the production of oxygen used for the gasification.

In Integrated Gasification Combined Cycle (IGCC) systems, the synthesis gas is cleaned of its hydrogen sulfide, ammonia, and particulate matter and is burned as fuel in a combustion turbine (much like natural gas is burned in a turbine). The combustion turbine drives an electric generator. And hot air from the combustion turbine can be channeled back to the gasifier or the air separation unit, while exhaust heat from the combustion turbine is recovered and used to boil water, creating steam for a steam turbine-generator.

The use of these two types of turbines—a combustion turbine and a steam turbine—in combination, known as a combined cycle, is one reason why gasification-based power systems can achieve unprecedented power generation efficiencies. Currently, commercially available gasification-based systems can operate at around 42% efficiencies; in the future, these systems may be able to achieve efficiencies approaching 60%. A conventional coal-based boiler plant, by contrast, employs only a steam turbine-generator and is typically limited to 33%–40% efficiency.

Higher efficiency means that less fuel is required to generate the rated power, resulting in better economics (which can mean lower costs to the consumer) and the formation of fewer greenhouse gases—a 60%-efficient gasification power plant can cut the formation of carbon dioxide by 40% compared to a typical coal combustion plant.

3. Gasification of crude oil fractions

The influx of viscous feedstocks such as heavy oil, extra heavy oil, and tar sand bitumen into refineries creates, and will continue to create, challenges but, at the same time, creates opportunities by improving the ability of refineries to handle viscous feedstocks thereby enhancing refinery flexibility to meet the increasingly stringent product specifications for refined fuels (Speight, 2013a, 2014a; 2014b). Upgrading viscous feedstocks is an increasingly prevalent means of extracting the maximum amount of liquid fuels from each barrel of crude oil that enters the refinery. Although solvent deasphalting processes and coking processes are used in refineries to upgrade viscous feedstocks to intermediate products (which are then processed further) to produce transportation fuels Speight (2014a), the integration of viscous feedstock processing units and gasification presents some unique synergies that will enhance the performance of the future refinery by preparing otherwise difficult to refine feedstocks to feedstocks that are suitable for the production of hydrocarbon products (Fig. 8.1) (Wallace et al., 1998; Furimsky, 1999; Penrose et al., 1999; Gray and Tomlinson, 2000; Abadie and Chamorro, 2009; Wolff and Vliegenthart, 2011; Speight, 2011a, 2011b, 2014a).

Gasification offers more scope for recovering products from waste than incineration. When waste is burnt in a modern incinerator the only practical product is energy, whereas the gases, oils and solid char from pyrolysis and gasification can not only be used as a fuel but also purified and used as a feedstock for petrochemicals and other applications. Many processes also produce a stable granulate instead of ash which can be more easily and safely utilized. In addition, some processes are targeted at producing specific recyclables such as metal alloys and carbon black. From waste gasification, in particular, it is feasible to produce hydrogen, which many see as an increasingly valuable resource.

A wide range of materials can be handled by gasification technologies and specific processes have been optimized to handle particular feedstock (for example, tire pyrolysis and sewage sludge gasification), while others have been designed to process mixed wastes. For example, recovering energy from agricultural and forestry residues, household and commercial waste, materials recycling (auto-shredder residue, electrical and electronic scrap, tires, mixed plastic waste, and packaging residues) are feasible processes.

Gasification is an appealing process for the utilization of relatively inexpensive feedstocks that might otherwise be declared as waste and sent to a landfill (where the production of methane—a so-called greenhouse gas—will be produced) or combusted which may not (depending upon the feedstock) be energy efficient. Overall, use of a gasification technology (Speight, 2013a, 2014b) with the necessary gas cleanup options can have a smaller environmental footprint and lesser effect on the environment than landfill operations or combustion of the waste. In fact, there are strong indications that gasification is a technically viable option for the waste conversion, including residual waste from separate collection of municipal solid waste. The process can meet existing emission limits and can have a significant effect on the reduction of landfill disposal using known gasification technologies (Arena, 2012; Speight, 2014b; Luque and Speight, 2015) or thermal plasma (Fabry et al., 2013).

In the gasification process, organic (carbonaceous) feedstocks are converted into carbon monoxide, carbon dioxide, and hydrogen by reacting the feedstock at high temperatures (>700°C, 1290°F), without combustion, with a controlled amount of oxygen and/or steam (Marano, 2003; Lee, 2007; Higman and Van der Burgt, 2008; Speight, 2013a, 2014b). Unconventional carbonaceous feedstocks include solids, liquids, and gases such as heavy oil, extra heavy oil, tar sand bitumen, residua, and biomass (Speight, 2014b). The gasification is not a single step process, but involves multiple subprocesses and reactions. The generated synthesis gas has wide range of applications ranging from power generation to chemicals production. The power derived from the gasification of carbonaceous feedstocks followed by the combustion of the product gas(es) is considered to be a source of renewable energy of derived gaseous products that are generated from a carbonaceous source (e.g., biomass) other than a fossil fuel. Speight, 2014b).

Indeed, the increasing interest in gasification technology reflects a convergence of changes in providing energy to the marketplace: (i) the maturity of gasification technology, and (ii) the extremely low emissions from integrated gasification combined cycle (IGCC) plants, especially air emissions, and (iii) the potential for control of greenhouse gases (Speight, 2014b). Another advantage of gasification is the use of synthesis gas is potentially more efficient as compared to direct combustion of the original fuel because it can be (i) combusted at higher temperatures, (ii) used in fuel cells, (iii) used to produce methanol and hydrogen, (iv) converted via the Fischer-Tropsch process into a range of synthesis liquid fuels suitable for use in gasoline engines or diesel engines (Keil, 1999; Chadeesingh, 2011; Serrano-Ruiz and Dumesic, 2012; Luque and Speight, 2015).

3.1 Feedstocks

The most common, and perhaps the best, feedstocks for steam reforming are low boiling saturated hydrocarbon derivatives that have a low sulfur content; including natural gas, refinery gas, liquefied petroleum gas (LPG), and low-boiling naphtha.

Natural gas is the most common feedstock for hydrogen production, since it meets all the requirements for reformer feedstock. Natural gas typically contains more than 90% methane and ethane with only a few percent of propane and higher boiling hydrocarbon derivatives (Speight, 2019b). Natural gas may (or most likely will) contain traces of carbon dioxide with some nitrogen and other impurities. Purification of natural gas, before reforming, is usually relatively straightforward (Speight, 2019b). Traces of sulfur must be removed to avoid poisoning the reformer catalyst; zinc oxide treatment in combination with hydrogenation is usually adequate.

Refinery gas, containing a substantial amount of hydrogen, can be an attractive steam reformer feedstock, since it is produced as a byproduct (Speight, 2014a, 2017). Processing of refinery gas will depend on its composition, particularly the levels of olefins and of propane and higher boiling hydrocarbon derivatives. Olefins, that can cause problems by forming coke in the reformer, are converted to saturated compounds in the hydrogenation unit. Higher-boiling hydrocarbon derivatives in refinery gas can also form coke, either on the primary reformer catalyst or in the preheater. If there is more than a few percent of C_3 and higher compounds, a promoted reformer catalyst should be considered, in order to avoid carbon deposits.

Refinery gases from different sources vary in suitability as hydrogen plant feed. Catalytic reformer off-gas (Speight, 2014a, 2017) for example, is saturated, very low in sulfur, and often has high hydrogen content. The process gases from a coking unit or from a fluid catalytic cracking unit are much less desirable because of the content of unsaturated constituents. In addition to olefins, these gases contain substantial amounts of sulfur that must be removed before the gas is used as feedstock. These gases are also generally unsuitable for direct hydrogen recovery, since the hydrogen content is usually too low. Hydrotreated off-gas lies in the middle of the range. It is saturated, so it is readily used as hydrogen plant feed. Content of hydrogen and higher-boiling hydrocarbon derivatives depends to a large extent on the upstream pressure. Sulfur removal will generally be required.

As hydrogen use has become more widespread in refineries, hydrogen production has moved from the status of a high-tech specialty operation to an integral feature of most refineries. This has been made necessary by the increase in hydrotreating and hydrocracking, including the treatment of progressively higher-boiling feedstocks (Speight, 2014a, 2017). The continued increase in hydrogen demand over the last several decades is a result of the conversion of crude oil to match changes in product slate and the supply of high-boiling, high-sulfur oil, and in order to make lower-boiling, cleaner, and more salable products. There are also many reasons other than product quality for using hydrogen in processes adding to the need to add hydrogen at relevant stages of the refining process and, most importantly according to the availability of hydrogen.

Hydrogen has historically been produced during catalytic reforming processes as a byproduct of the production of the aromatic compounds used in gasoline and in solvents. As reforming processes changed from fixed-bed to cyclic to continuous regeneration, process pressures have dropped and hydrogen production per barrel of reformate has tended to increase. However, hydrogen production as a byproduct is not always adequate to the needs of the refinery and other processes are necessary. Thus, hydrogen production by steam reforming or by partial oxidation of residua has also been used, particularly where heavy oil is available. Steam reforming is the dominant method for hydrogen production and is usually combined with pressure swing adsorption (PSA) to purify the hydrogen to greater than 99% by volume.

The gasification of residua and coke to produce hydrogen and/or power may become an attractive option for refiners. The premise that the gasification section of a refinery will be the *garbage can* for deasphalter residues, high-sulfur coke, as well as other refinery wastes is worthy of consideration.

Of the processes that are available for the production of the hydrogen, many can be considered dual processes insofar as they also produce carbon monoxide and, therefore, are considered as producers of synthesis gas. For example, most of the external hydrogen is manufactured by steam-methane reforming or by oxidation processes. Other processes, such as ammonia dissociation, steam-methanol interaction, or electrolysis, are also available for hydrogen production, but economic factors and feedstock availability assist in the choice between processing alternatives.

The processes described in this section are those gasification processes by which hydrogen is produced for use in other parts of the refinery.

3.2 Chemistry

In *steam reforming*, low-boiling hydrocarbon derivatives such as methane are reacted with steam to form hydrogen:

$$CH_4 + H_2O \rightarrow 3H_2 + CO \quad \Delta H_{298K} = +97{,}400 \text{ Btu/lb}$$

H is the heat of reaction. A more general form of the equation that shows the chemical balance for higher-boiling hydrocarbon derivatives is:

$$C_nH_m + nH_2O \rightarrow (n + m/2)H_2 + nCO$$

The reaction is typically carried out at approximately 815°C (1500°F) over a nickel catalyst packed into the tubes of a reforming furnace. The high temperature also causes the hydrocarbon feedstock to undergo a series of cracking reactions, plus the reaction of carbon with steam:

$$CH_4 \rightarrow 2H_2 + C$$

$$C + H_2O \rightarrow CO + H_2$$

Carbon is produced on the catalyst at the same time that hydrocarbon is reformed to hydrogen and carbon monoxide. With natural gas or similar feedstock, reforming predominates and the carbon can be removed by reaction with steam as fast as it is formed. When higher-boiling feedstocks are used, the carbon is not removed fast enough and builds up thereby requiring catalyst regeneration or replacement. Carbon buildup on the catalyst (when high-boiling feedstocks are employed) can be avoided by addition of alkali compounds, such as potash, to the catalyst thereby encouraging or promoting the carbon-steam reaction.

However, even with an alkali-promoted catalyst, feedstock cracking limits the process to hydrocarbon derivatives with a boiling point less than of 180°C (350°F). Natural gas, propane, butane, and low-boiling naphtha are most suitable. Prereforming, a process that uses an adiabatic catalyst bed operating at a lower temperature, can be used as a pretreatment to allow higher-boiling feedstocks to be used with lower potential for carbon deposition (coke formation) on the catalyst.

After reforming, the carbon monoxide in the gas is reacted with steam to form additional hydrogen (the *water-gas shift* reaction):

$$CO + H_2O \rightarrow CO_2 + H_2 \; \Delta H_{298K} = -16{,}500 \text{ Btu/lb}$$

This leaves a mixture consisting primarily of hydrogen and carbon monoxide that is removed by conversion to methane:

$$CO + 3H_2O \rightarrow CH_4 + H_2O$$

$$CO_2 + 4H_2 \rightarrow CH_4 + 2H_2O$$

The critical variables for steam reforming processes are (i) temperature, (ii) pressure, and (iii) the steam/hydrocarbon ratio. Steam reforming is an equilibrium reaction, and conversion of the hydrocarbon feedstock is favored by high temperature, which in turn requires higher fuel use. Because of the volume increase in the reaction, conversion is also favored by low pressure, which conflicts with the need to supply the hydrogen at high pressure. In practice, materials of construction limit temperature and pressure. On the other hand, and in contrast to reforming, shift conversion is favored by low temperature. The gas from the reformer is reacted over iron oxide catalyst at 315–370°C (600–700°F) with the lower limit being dictated activity of the catalyst at low temperature.

Hydrogen can also be produced by *partial oxidation* (POX) of hydrocarbon derivatives in which the hydrocarbon is oxidized in a limited or controlled supply of oxygen:

$$2CH_4 + O_2 \rightarrow CO + 4H_2 \; \Delta H_{298K} = -10{,}195 \text{ Btu/lb}$$

The shift reaction also occurs and a mixture of carbon monoxide and carbon dioxide is produced in addition to hydrogen. The catalyst tube materials do not limit the reaction temperatures in partial oxidation processes, and higher temperatures may be used that enhance the conversion of methane to hydrogen. Indeed, much of the design and operation of hydrogen plants involves protecting the reforming catalyst and the catalyst tubes because of the extreme temperatures and the sensitivity of the catalyst. In fact, minor variations in feedstock composition or operating conditions can have significant effects on the life of the catalyst or the reformer itself. This is particularly true of changes in molecular weight of the feed gas, or poor distribution of heat to the catalyst tubes.

Since the high temperature takes the place of a catalyst, partial oxidation is not limited to the lower-boiling feedstocks that are required for steam reforming. Partial oxidation processes were first considered for hydrogen production because of expected shortages of lower-boiling feedstocks and the need to have available a disposal method for higher-boiling, high-sulfur streams such as asphalt or crude oil coke.

Catalytic partial oxidation, also known as autothermal reforming, reacts oxygen with a low-boiling, low-density feedstock and by passing the resulting hot mixture over a reforming catalyst. The use of a catalyst allows the use of lower temperatures than in noncatalytic partial oxidation and which causes a reduction in oxygen demand.

The feedstock requirements for catalytic partial oxidation processes are similar to the feedstock requirements for steam reforming and low-boiling hydrocarbon derivatives from refinery gas to naphtha are preferred. The oxygen substitutes for much of the steam in preventing coking and a lower steam/carbon ratio is required. In addition, because a large excess of steam is not required, catalytic partial oxidation produces more carbon monoxide and less hydrogen than steam reforming. Thus, the process is more suited to situations where carbon monoxide is the more desirable product such as, for example, synthesis gas for chemical feedstocks.

3.3 Commercial processes

In spite of the use of low-quality hydrogen (that contain up to 40% by volume hydrocarbon gases), a high-purity hydrogen stream (95%–99% v/v hydrogen) is required for hydrodesulfurization, hydrogenation, hydrocracking, and petrochemical processes. Hydrogen, produced as a byproduct of refinery processes (principally hydrogen recovery from catalytic reformer product gases), often is not enough to meet the total refinery requirements, necessitating the manufacturing of additional hydrogen or obtaining supply from external sources.

3.3.1 Heavy residue gasification

High-boiling residua are gasified and the produced gas is purified to clean fuel (Speight, 2014a, 2017). As an example, solvent deasphalter residuum is gasified by partial oxidation method under pressure of approximately 570 psi and at a temperature between 1300 and 1500°C (2370 and 2730°F). The high temperature generated gas flows into the specially designed waste heat boiler, in which the hot gas is cooled and high pressure saturated steam is generated. The gas from the waste heat boiler is then heat exchanged with the fuel gas and flows to the carbon scrubber, where unreacted carbon particles are removed from the generated gas by water scrubbing.

The gas from the carbon scrubber is further cooled by the fuel gas and boiler feed water and led into the sulfur compound removal section, where hydrogen sulfide (H_2S) and carbonyl sulfide (COS) are removed from the gas to obtain clean fuel gas. This clean fuel gas is heated with the hot gas generated in the gasifier and finally supplied to the gas turbine at a temperature of 250–300°C (480–570°F).

The exhaust gas from the gas turbine, having a temperature of approximately 550–600°C (1020–1110°F), flows into the heat recovery steam generator consisting of five heat exchange elements. The first element is a superheater in which the combined stream of the high pressure saturated steam generated in the waste heat boiler and in the second element (high pressure steam evaporator) is super heated. The third element is an economizer, the fourth element is a low pressure steam evaporator, and the final or the fifth element is a deaerator heater. The off-gas from heat recovery steam generator having a temperature of approximately 130°C is emitted into the air via stack.

In order to decrease the nitrogen oxide (NO_X) content in the flue gas, two methods can be applied. The first method is the injection of water into the gas turbine combustor. The second method is to selectively reduce the nitrogen oxide content by injecting ammonia gas in the presence of de-NO_X catalyst that is packed in a proper position of the heat recovery steam generator. The latter is more effective that the former to lower the nitrogen oxide emissions to the air.

3.3.2 Hybrid gasification process

In the hybrid gasification process, a slurry of coal and residual oil is injected into the gasifier where it is pyrolyzed in the upper part of the reactor to produce gas and chars. The chars produced are then partially oxidized to ash. The ash is removed continuously from the bottom of the reactor.

In this process, coal and vacuum residue are mixed together into slurry to produce clean fuel gas. The slurry fed into the pressurized gasifier is thermally cracked at a temperature of 850–950°C (1560–1740°F) and is converted into gas, tar, and char. The mixture oxygen and steam in the lower zone of the gasifier gasify the char. The gas leaving the gasifier is quenched to a temperature of 450°C (840°F)in the fluidized bed heat exchanger, and is then scrubbed to remove tar, dust and steam at around 200°C (390°F).

The coal and residual oil slurry is gasified in the fluidized-bed gasifier. The charged slurry is converted to gas and char by thermal cracking reactions in the upper zone of the fluidized bed. The produced char is further gasified with steam and oxygen that enter the gasifier just below the fluidizing gas distributor. Ash is discharged from the gasifier and indirectly cooled with steam and then discharged into the ash hopper. It is burned with an incinerator to produce process steam. Coke deposited on the silica sand is removed in the incinerator.

3.3.3 Hydrocarbon gasification

The gasification of hydrocarbon derivatives to produce hydrogen is a continuous, noncatalytic process that involves partial oxidation of the hydrocarbon. Air or oxygen (with steam or carbon dioxide) is used as the oxidant at 1095–1480°C (2000–2700°F). Any carbon produced (2%–3% by weight of the feedstock) during the process is removed as a slurry in a carbon separator and pelletized for use either as a fuel or as raw material for carbon-based products.

3.3.4 Hypro process

The Hypro process is a continuous catalytic method for hydrogen manufacture from natural gas or from refinery effluent gases. The process is designed to convert natural gas:

$$CH_4 \rightarrow C + 2H_2$$

Hydrogen is recovered by phase separation to yield hydrogen of approximately 93% purity; the principal contaminant is methane.

3.3.5 Pyrolysis processes

There has been recent interest in the use of pyrolysis processes to produce hydrogen as well as hydrocarbon products. Specifically the interest has focused on the pyrolysis of methane (natural gas) and hydrogen sulfide.

Natural gas is readily available and offers a relatively rich stream of methane with lower amounts of ethane, propane, and butane also present. The thermocatalytic decomposition of natural gas hydrocarbon derivatives offers an alternate method for the production of hydrogen:

$$C_nH_m \rightarrow nC + (m/2)H_2$$

If a hydrocarbon fuel such as natural gas (methane) is to be used for hydrogen production by direct decomposition, then the process that is optimized to yield hydrogen production may not be suitable for production of high-quality carbon black byproduct intended for the industrial rubber market. Moreover, it appears that the carbon produced from high-temperature (850−950°C; 1560−1740°F) direct thermal decomposition of methane is a sootlike material with high tendency for the catalyst deactivation. Thus, if the object of methane decomposition is hydrogen production, the carbon byproduct may not be marketable as high-quality carbon black for rubber and tire applications.

Hydrogen sulfide decomposition is a highly endothermic process and equilibrium yields are poor. At temperatures less than 1500°C (2730°F), the thermodynamic equilibrium is unfavorable toward hydrogen formation. However, in the presence of catalysts such as platinum-cobalt (at 1000°C; 1830°F), disulfides of molybdenum or tungsten (Mo or W) at 800°C (1470°F), or other transition metal sulfides supported on alumina (at 500−800°C; 930 to 1470°F), decomposition of hydrogen sulfide proceeds rapidly. In the temperature range of approximately 800−1500°C (1470−2730°F), thermolysis of hydrogen sulfide can be treated simply:

$$H_2S \rightarrow H_2 + 1/xS_x \quad \Delta H_{298K} = +34{,}300 \text{ Btu/lb}$$

where $x = 2$. Outside this temperature range, multiple equilibria may be present depending on temperature, pressure, and relative abundance of hydrogen and sulfur.

Above approximately 1000°C (1830°F), there is a limited advantage to using catalysts since the thermal reaction proceeds to equilibrium very rapidly. The hydrogen yield can be doubled by preferential removal of either hydrogen or sulfur from the reaction environment, thereby shifting the equilibrium. The reaction products must be quenched quickly after leaving the reactor to prevent reversible reactions.

3.3.6 Shell gasification process

The Shell gasification process (*partial oxidation process*) is a flexible process for generating synthesis gas, principally hydrogen and carbon monoxide, for the ultimate production of high-purity high pressure hydrogen, ammonia, methanol, fuel gas, town gas, or reducing gas by reaction of gaseous or liquid hydrocarbon derivatives with oxygen, air, or oxygen-enriched air.

The most important step in converting high-boiling residue to industrial gas is the partial oxidation of the oil using oxygen with the addition of steam. The gasification process takes place in an empty, refractory-lined reactor at temperatures of approximately 1400°C (2550°F) and pressures between 29 and 1140 psi (196−7845 kPa). The chemical reactions in the gasification reactor proceed without catalyst to produce gas containing carbon amounting to some 0.5%−2% by weight, based on the feedstock. The carbon is removed from the gas with water, extracted in most cases with feed oil from the water and returned to the feed oil. The high reformed gas temperature is utilized in a waste heat boiler for generating steam. The steam is generated at 850 to 1565 psi (5884−10,787 kPa). Some of this steam is used as process steam and for oxygen and oil preheating. The surplus steam is used for energy production and heating purposes.

3.3.7 Steam-methane reforming

Steam-methane reforming is a catalytic process that involves a reaction between natural gas or other low-boiling hydrocarbon derivatives and steam. Steam-methane reforming is the benchmark process that has been employed over a period of several decades for hydrogen production. The process involves reforming natural gas in a continuous catalytic process in which the major reaction is the formation of carbon monoxide and hydrogen from methane and steam.

Steam reforming is an exothermic reaction that is carried out by passing a preheated mixture comprising methane (sometimes substituted by natural gas having high methane content) and steam through catalyst filled tubes. The products of the process are a mixture of hydrogen, carbon monoxide, and carbon dioxide. To maximize the conversion of the methane feed, primary and secondary reformers are often used—the *primary reformer* effects a 90%−92% v/v conversion of methane. Here, the hydrocarbon feed is partially reacted with steam at 900°C (1650°F) at 220 to 500 psi over a nickel-alumina catalyst to produce a synthesis gas in which the hydrogen/carbon monoxide (H_2/CO) ratio is on the order of 3:1. Any unconverted methane is reacted with oxygen at the top of a *secondary autothermal reformer* containing nickel catalyst in the lower region of the vessel.

Thus, the first reforming step catalytically reacts with methane (the chief chemical constituent of natural gas) to form hydrogen and carbon monoxide in an endothermic (heat-absorbing) reaction:

$$CH_4 + H_2O = CO + 3H_2$$

Higher molecular weight feedstocks can also be reformed to hydrogen:

$$C_3H_8 + 3H_2O \rightarrow 3CO + 7H_2$$

That is,

$$C_nH_m + nH_2O \rightarrow nCO + (0.5m + n)H_2$$

In the actual process, the feedstock is first desulfurized by passage through activated carbon, which may be preceded by caustic and water washes. The desulfurized material is then mixed with steam and passed over a nickel-based catalyst (730−845°C, 1350 to 1550°F and 400 psi). Effluent gases are cooled by the addition of steam or condensate to approximately 370°C (700°F), at which point carbon monoxide reacts with steam in the presence of iron oxide in a shift converter to produce carbon dioxide and hydrogen in which the carbon monoxide is then "shifted" with steam to form additional hydrogen and carbon dioxide in an exothermic (heat-releasing) reaction:

$$CO + H_2O = CO_2 + H_2$$

The carbon dioxide (usually by amine washing) leaving hydrogen is separated for its commercial use; the hydrogen is usually a high-purity (>99%) material.

Since the presence of any carbon monoxide or carbon dioxide in the hydrogen stream can interfere with the chemistry of the catalytic application, a third stage is used to convert these gases to methane:

$$CO + 3H_2 \rightarrow CH_4 + H_2O$$

$$CO_2 + 4H_2 \rightarrow CH_4 + 2H_2O$$

For many refiners, sulfur-free natural gas (CH_4) is not always available to produce hydrogen by this process. In that case, higher-boiling hydrocarbon derivatives (such as propane, butane, or naphtha) may be used as the feedstock to generate hydrogen.

The net chemical process for steam methane reforming is then given by:

$$CH_4 + 2H_2O \rightarrow CO_2 + 4H_2$$

Indirect heating provides the required overall endothermic heat of reaction for the steam-methane reforming.

One way of overcoming the thermodynamic limitation of steam reforming is to remove either hydrogen or carbon dioxide as it is produced, hence shifting the thermodynamic equilibrium toward the product side. The concept for sorption-enhanced methane steam reforming is based on in situ removal of carbon dioxide by a sorbent such as calcium oxide (CaO).

$$CaO + CO_2 \rightarrow CaCO_3$$

Sorption enhancement enables lower reaction temperatures, which may reduce catalyst coking and sintering, while enabling use of less expensive reactor wall materials. In addition, heat release by the exothermic carbonation reaction supplies most of the heat required by the endothermic reforming reactions. However, energy is required to regenerate the sorbent to its oxide form by the energy-intensive calcination reaction, i.e.,

$$CaCO_3 \rightarrow CaO + CO_2$$

Use of a sorbent requires either that there be parallel reactors operated alternatively and out of phase in reforming and sorbent regeneration modes, or that sorbent be continuously transferred between the reformer/carbonator and regenerator/calciner.

In autothermal (or secondary) reformers, the oxidation of methane supplies the necessary energy and is carried out either simultaneously or in advance of the reforming reaction. The equilibrium of the methane steam reaction and the water-gas shift reaction determines the conditions for optimum hydrogen yields. The optimum conditions for hydrogen production require: high temperature at the exit of the reforming reactor (800–900°C; 1470 to 1650°F), excess of steam (molar steam-to-carbon ratio of 2.5–3), and relatively low pressures (below 450 psi). Most commercial plants employ supported nickel catalysts for the process.

The steam-methane reforming process described briefly above would be an ideal hydrogen production process if it was not for the fact that large quantities of natural gas, a valuable resource, are required as both feed gas and combustion fuel. For each mole of methane reformed, more than 1 mole of carbon dioxide is coproduced and must be disposed. This can be a major issue as it results in the same amount of greenhouse gas emission as would be expected from direct combustion of natural gas or methane. In fact, the production of hydrogen as a clean burning fuel by way of steam reforming of methane and other fossil-based hydrocarbon fuels is not in environmental balance if in the process, carbon dioxide and carbon monoxide are generated and released into the atmosphere, although alternate scenarios are available.

Moreover, as the reforming process is not totally efficient, some of the energy value of the hydrocarbon fuel is lost by conversion to hydrogen but with no tangible environmental benefit, such as a reduction in emission of greenhouse gases. Despite these apparent shortcomings, the process has the following advantages: (i) produces 4 mol of hydrogen for each mole of methane consumed, (ii) feedstocks for the process (methane and water are readily available), (iii) the process is adaptable to a wide range of hydrocarbon feedstocks, (4) operates at low pressures, less than 450 psi, (v) requires a low steam/carbon ratio (2.5–3), (vi) good utilization of input energy (reaching 93%), (vii) can use catalysts that are stable and resist poisoning, and (viii) acceptable process kinetics.

Liquid feedstocks, either liquefied petroleum gas or naphtha, can also provide backup feed, if there is a risk of natural gas curtailments. The feed handling system needs to include a surge drum, feed pump, vaporizer (usually steam-heated) followed by further heating before desulfurization. The sulfur in liquid feedstocks occurs as mercaptans, thiophene derivatives, or higher boiling compounds. These compounds are stable and will not be removed by zinc oxide; therefore, a hydrogenation unit will be required. In addition, as with refinery gas, olefins must also be hydrogenated if they are present.

The reformer will generally use a potash-promoted catalyst to avoid coke buildup from cracking of the higher-boiling feedstock. If liquefied petroleum gas is to be used only occasionally, it is often possible to use a methane-type catalyst at a higher steam/carbon ratio to avoid coking. Naphtha will require a promoted catalyst unless a preformer is used.

3.3.8 Steam-naphtha reforming

Steam-naphtha reforming is a continuous process for the production of hydrogen from liquid hydrocarbon derivatives and is, in fact, similar to steam-methane reforming that is one of several possible processes for the production of hydrogen from low-boiling hydrocarbon derivatives other than ethane Murata et al., 1997; Muradov, 1998, 2000; Brandmair et al., 2003; Find et al., 2003). A variety of naphtha-types in the gasoline boiling range may be employed, including feeds containing up to 35% aromatics. Thus, following pretreatment to remove sulfur compounds, the feedstock is mixed with steam and taken to the reforming furnace (675–815°C, 1250 to 1500°F, 300 psi), where hydrogen is produced.

Thus, liquid feedstocks, either liquefied petroleum gas or naphtha can also provide backup feed for the steam-methane reformer, if there is a risk of natural gas curtailments (Breault, 2010). The feed handling system needs to include a surge drum, feed pump, vaporizer (usually steam-heated) followed by further heating before desulfurization. The sulfur in liquid feedstocks occurs as mercaptans, thiophene derivatives, or higher-boiling compounds. These compounds are stable and will not be removed by zinc oxide; therefore a hydrogenation unit will be required. In addition, as with refinery gas, olefins must also be hydrogenated if they are present.

Thus, steam-naphtha reforming is a continuous process for the production of hydrogen from liquid hydrocarbons and is, in fact, similar to steam-methane reforming that is one of several possible processes for the production of hydrogen from low-boiling hydrocarbons other than ethane (Murata et al., 1997; Muradov, 1998; 2000; Brandmair et al., 2003; Find et al., 2003). A variety of naphtha-types in the gasoline boiling range may be employed, including feeds containing up to 35% aromatics. Thus, following pretreatment to remove sulfur compounds, the feedstock is mixed with steam and taken to the reforming furnace (675–815°C, 1250 to 1500°F, 300 psi), where hydrogen is produced.

3.3.9 Texaco gasification process

The Texaco gasification process is a partial oxidation gasification process for generating synthetic gas, principally hydrogen and carbon monoxide. The characteristic of Texaco gasification process is to inject feedstock together with carbon dioxide, steam or water into the gasifier. Therefore, solvent deasphalted residua or crude oil coke rejected from any coking method can be used as feedstock for this gasification process. The produced gas from this gasification process can be used for the production of high-purity high-pressurized hydrogen, ammonia, and methanol. The heat recovered from the high temperature gas is used for the generation of steam in the waste heat boiler. Alternatively the less expensive quench type configuration is preferred when high-pressure steam is not needed or when a high degree of shift is needed in the downstream CO converter.

In the process, the feedstock, together with the feedstock carbon slurry recovered in the carbon recovery section, is pressurized to a given pressure, mixed with high-pressure steam, and then blown into the gas generator through the burner together with oxygen. The gasification reaction is a partial oxidation of hydrocarbon derivatives to carbon monoxide and hydrogen:

$$C_xH_{2y} + x/2O_2 \rightarrow xCO + yH_2$$

$$C_xH_{2y} + xH_2O \rightarrow xCO + (x + y)H_2$$

The gasification reaction is instantly completed, thus producing gas mainly consisting of hydrogen and carbon monoxide ($H_2 + CO = >90\%$). The high-temperature gas leaving the reaction chamber of the gas generator enters the quenching chamber linked to the bottom of the gas generator and is quenched to 200°C–260°C (390°F–500°F) with water.

3.4 Synthesis gas generation

The synthesis gas generation process is a noncatalytic process (principally hydrogen and carbon monoxide) for the ultimate production of high-purity hydrogen from gaseous or liquid hydrocarbon derivatives.

In this process, a controlled mixture of preheated feedstock and oxygen is fed to the top of the generator where carbon monoxide and hydrogen emerge as the products. Soot, produced in this part of the operation, is removed in a water scrubber from the product gas stream and is then extracted from the resulting carbon-water slurry with naphtha and transferred to a fuel oil fraction. The oil-soot mixture is burned in a boiler or recycled to the generator to extinction to eliminate carbon production as part of the process.

The soot-free synthesis gas is then charged to a shift converter where the carbon monoxide reacts with steam to form additional hydrogen and carbon dioxide at the stoichiometric rate of 1 mol of hydrogen for every mole of carbon monoxide charged to the converter. The reactor temperatures vary from 1095 to 1490°C (2000 –2700°F), while pressures can vary from approximately atmospheric pressure to approximately 2000 psi. The process has the capability of producing high-purity hydrogen although the extent of the purification procedure depends upon the use to which the hydrogen is to be put. For example, carbon dioxide can be removed by scrubbing with various alkaline reagents, while carbon monoxide can be removed by washing with liquid nitrogen or, if nitrogen is undesirable in the product, the carbon monoxide should be removed by washing with copper-amine solutions.

This particular partial oxidation technique has also been applied to a whole range of liquid feedstocks for hydrogen production. There is now serious consideration being given to hydrogen production by the partial oxidation of solid feedstocks such as crude oil coke (from both delayed and fluid-bed reactors), lignite, and coal, as well as crude oil residua.

The chemistry of the process, using naphthalene as an example, may be simply represented as the selective removal of carbon from the hydrocarbon feedstock and further conversion of a portion of this carbon to hydrogen:

$$C_{10}H_8 + 5O_2 \rightarrow 10CO + 4H_2$$

$$10CO + 10H_2O \rightarrow 10CO_2 + 10H_2$$

Although these reactions may be represented very simply using equations of this type, the reactions can be complex and result in carbon deposition on parts of the equipment thereby requiring careful inspection of the reactor.

4. Gasification of other feedstocks

Gasification offers more scope for recovering products from waste than incineration. When waste is burnt in a modern incinerator, the only practical product is energy, whereas the gases, oils, and solid char from pyrolysis and gasification can not only be used as a fuel but also purified and used as a feedstock for petrochemicals and other applications. Many processes also produce a stable granulate instead of ash which can be more easily and safely utilized. In addition, some processes are targeted at producing specific recyclables such as metal alloys and carbon black. From waste gasification, in particular, it is feasible to produce hydrogen, which many see as an increasingly valuable resource.

Gasification can be used in conjunction with gas engines (and potentially gas turbines) to obtain higher conversion efficiency than conventional fossil fuel energy generation. By displacing fossil fuels, waste pyrolysis and gasification can help meet renewable energy targets, address concerns related to global warming and contribute to achieving Kyoto Protocol commitments. Conventional incineration, used in conjunction with steam-cycle boilers and turbine generators, achieves lower efficiency.

Many of the processes fit well into a modern integrated approach to waste management. They can be designed to handle the waste residues and are fully compatible with an active program of composting for the waste fraction that is subject to decay and putrefaction.

This, by analogy with coal, the high temperature conversion of waste is a downdraft gasification process which gasifies the feed material within a controlled and limited oxygen supply. Combustion of the feed material is prevented by the limited oxygen supply. The temperature within the reactor reaches 2700°C (4890°F), at which point molecular dissociation takes place. The pollutants that were contained within the feed waste material such as dioxin derivatives, furan derivatives, as well as pathogens are completely cracked into harmless compounds.

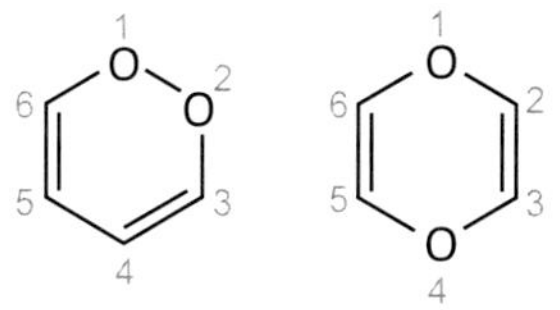

1,2-Dioxin 1,4-Dioxin

Another isomer, the 1,3-dioxin with the oxygen atoms in the 1-position and the 3-position also exists but is less common.

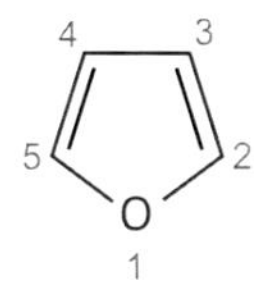

Furan

All metal components in the waste stream are converted into a castable iron alloy/pig iron for use in steel foundries. The mineral fraction is reduced to a nonleaching vitrified glass, which can used for road construction and/or further processed into a mineral wool for insulation. All of the organic material is fully converted to a fuel quality synthesis gas which can be used to produce electrical energy, heat, methanol, or used in the production of various other chemical compounds. The resultant synthesis gas, with a hydrogen/carbon monoxide ratio approximately equal to one, is also capable of being used for the production of Fischer-Tropsch hydrocarbon derivatives. Under certain conditions, heat from the reactor could be used for district heating, industrial steam production, or water desalination plants.

A wide range of materials can be handled by gasification technologies and specific processes have been optimized to handle particular feedstock (for example, tire pyrolysis and sewage sludge gasification), while others have been designed to process mixed wastes. For example, recovering energy from agricultural and forestry residues, household and commercial waste, materials recycling (auto-shredder residue, electrical and electronic scrap, tires, mixed plastic waste, and packaging residues) are feasible processes.

5. Fischer-Tropsch process

The Fischer-Tropsch process (often referred to as the Fischer-Tropsch reaction or Fischer-Tropsch chemistry) is used to produce synthetic hydrocarbon derivatives from gasified biomass. The process is actually a collection of chemical reactions that result in the conversion of synthesis gas (a mixture of carbon monoxide and hydrogen into liquid hydrocarbon derivatives. These reactions occur in the presence of metal catalysts and typically at temperatures on the order of 150–300°C (300–570°F) and pressures of 15 psi to pressure

TABLE 8.1 Summary of the catalysts for the conversion of synthesis gas to products.

Catalyst	Reaction type	Products
Ni	Methanation	Methane, CH_4
Cu/Cr/Zn oxide	Methanol synthesis	Methanol, CH_3OH
Fe, Co	Fischer-Tropsch synthesis	Hydrocarbons, C_nH_{2n+2}
Rh cluster	Glycol production	Glycol, CH_2OHCH_2OH

up to 500 psi using a variety of catalysts (Table 8.1) (Chadeesingh, 2011; Choudhury and Moholkar, 2013).

As a premier example of (what is often known as) C1 (C-one) chemistry, the Fischer-Tropsch process is an important reaction for the conversion of carbonaceous feedstocks (such as coal, crude oil resids, and biomass) by way of synthesis gas to liquid hydrocarbon derivatives (GTL, gas-to-liquids technology). In the usual implementation, carbon monoxide and hydrogen, the feedstocks for the Fischer-Tropsch process are produced by gasification of the feedstock after which the conversion of the gaseous product (synthesis gas) to hydrocarbon derivative is achieved. The Fischer-Tropsch process has received intermittent attention as a source of low-sulfur diesel fuel and to address the diminishing supply of crude oil—derived hydrocarbon derivatives.

In the process, carbonaceous material is gasified and the gas is processed to make purified synthesis gas which is then converted into gasoline-range and diesel-range hydrocarbon derivatives (Ryan, 1997). While biodiesel and ethanol production from biomass uses only parts of a plant, the production of liquids by the Fischer-Tropsch route uses the whole plant and the result is less land area is required per unit of energy produced.

Although the focus of this section is the production of hydrocarbon derivatives from synthesis gas, it is worthy of note that all or part of the clean synthesis gas can also be used (i) as chemical *building blocks* to produce a broad range of chemicals using processes well established in the chemical and petrochemical industry, (ii) as a fuel producer for highly efficient fuel cells (which run off the hydrogen made in a gasifier) or perhaps in the future, hydrogen turbines and fuel cell-turbine hybrid systems, and (iii) as a source of hydrogen that can be separated from the gas stream and used as a fuel or as a feedstock for refineries (which use the hydrogen to upgrade crude oil products). However, the decreasing availability and increased price of crude oil have renewed the worldwide interest in the production of liquid hydrocarbon derivatives from carbon monoxide and hydrogen using metal catalysts, also known as Fischer-Tropsch synthesis or Fischer-Tropsch process.

The synthesis of hydrocarbon derivatives from the hydrogenation of carbon monoxide over transition metal catalysts was discovered in 1902 when Sabatier and Sanderens produced methane from hydrogen and carbon monoxide mixtures passed over nickel, iron, and cobalt catalysts. In 1923, Fischer and Tropsch reported the use of alkalized iron catalysts to produce liquid hydrocarbon derivatives rich in oxygenated compounds.

The Fischer-Tropsch process (*Fischer-Tropsch synthesis*) is a series of catalyzed chemical reactions that convert a mixture of carbon monoxide and hydrogen into hydrocarbon derivatives. The process is a key component of gas-to-liquids (GTL) technology that produces liquid and solid hydrocarbon derivatives from coal, natural gas, biomass, or other carbonaceous feedstocks. Typical catalysts used are based on iron and cobalt and the hydrocarbon derivatives synthesized in the process are primarily liquid alkanes along with byproducts such as olefins, alcohols, and solid paraffins (waxes).

5.1 Chemistry

The synthesis of hydrocarbon derivatives from the hydrogenation of carbon monoxide was discovered in 1902 by Sabatier and Sanderens who produced methane by passing carbon monoxide and hydrogen over nickel (Ni), iron (Fe), and cobalt (Co) catalysts. At about the same time, the first commercial hydrogen from synthesis gas produced from steam methane reforming was commercialized. Haber and Bosch discovered the synthesis of ammonia from H_2 and N_2 in 1910 and the first industrial ammonia synthesis plant was commissioned in 1913. The production of liquid hydrocarbon derivatives and oxygenates from synthesis gas conversion over iron catalysts was discovered in 1923 by Fischer and Tropsch. Variations on this synthesis pathway were soon to follow for the selective production of methanol, mixed alcohols, and iso synthesis products. Another outgrowth of Fischer-Tropsch synthesis was the hydroformylation of olefins discovered in 1938.

The development of pressurized Fischer-Tropsch synthesis started in 1925 in Germany when Professor Franz Fischer, founding director of the Kaiser-Wilhelm Institute of Coal Research in Mälheim an der Ruhr, and his head of department, Dr. Hans Tropsch, applied for a patent describing a process to produce liquid hydrocarbon derivatives from carbon monoxide gas and hydrogen using metal catalysts.

The experiments took place in the laboratory of Franz Fischer at the Kaiser Wilhelm Institute for Coal Research, and developed to an industry with 600,000 tons per year in 1945. At those times strategic reasons for liquid fuel production from coal exceeded economic aspects. In the last decades the interest in Fischer-Tropsch synthesis has changed as a result of environmental demands, technological developments, and change in fossil energy reserves. A good example is the "oil-age" from 1955–1970 with plenty of cheap oil supply and as a result of only a marginal interest in Fischer-Tropsch synthesis.

High oil prices increase the focus at alternative fuels; likewise as carbon dioxide concentration concern arises, being related to global warming, the focus at new technologies rises. Today the driving forces are environmental concern, but also higher oil price, limited oil reserves, and increased focus at stranded gas.

Two main chemical characteristics of Fischer-Tropsch synthesis are the unavoidable production of a wide range of hydrocarbon products (olefins, paraffins, and oxygenated products) and the liberation of a large amount of heat from the highly exothermic synthesis reactions. Product distributions are influenced by temperature, feed gas composition (H_2/CO), pressure, catalyst type, and catalyst composition. Fischer-Tropsch products are produced in four main steps: (i) synthesis gas generation, (ii) gas purification, (iii) Fischer-Tropsch synthesis, and (iv) product upgrading. Depending on the types and quantities of Fischer-Tropsch products desired, either low (200−240°C; 390 to 465°F) or high temperature (300−350°C; 570 to 660°F) synthesis is used with either an iron (Fe) or cobalt catalyst (Co) (Van Berge, 1995).

The required gas mixture of carbon monoxide and hydrogen (synthesis gas) is created through a reaction of coke or coal with water steam and oxygen, at temperatures over 900°C. In the past, town gas and gas for lamps were a carbon monoxide-hydrogen mixture, made by gasifying coke in gas works. In the 1970s it was replaced with imported natural gas (methane). Coal gasification and Fischer-Tropsch hydrocarbon synthesis together bring about a two-stage sequence of reactions which allows the production of liquid fuels like diesel and petrol out of the solid combustible coal.

The Fischer-Tropsch synthesis is, in principle, a carbon chain building process, where methylene groups are attached to the carbon chain. The actual reactions that occur have been, and remain, a matter of controversy, as it has been in the last century since the 1930s.

$$(2n+1)H_2 + nCO \rightarrow C_nH_{(2n+2)} + nH_2O$$

Even though the overall Fischer-Tropsch process is described by the following chemical equation:

$$(2n+1)H_2 + nCO \rightarrow C_nH_{(2n+2)} + nH_2O$$

The initial reactants in the above reaction (i.e., CO and H_2) can be produced by other reactions such as the partial combustion of a hydrocarbon:

$$C_nH_{(2n+2)} + {}^1\!/_2\ nO_2 \rightarrow (n+1)H_2 + nCO$$

For example (when $n = 1$), methane (in the case of gas to liquids applications):

$$2CH_4 + O_2 \rightarrow 4H_2 + 2CO$$

Or by the gasification of any carbonaceous source, such as biomass:

$$C + H_2O \rightarrow H_2 + CO$$

The energy needed for this endothermic reaction is usually provided by the exothermic combustion reaction with air or oxygen:

$$2C + O_2 \rightarrow 2CO$$

The detailed behavior of these other reactions is not known and is a theme of controversy. The reactions are believed to be:

$$CO + 2H_2 \rightarrow [-CH_2-] + H_2O$$

$$2CO + H_2 \rightarrow [-CH_2-] + CO_2$$

$$2CO + H_2 \rightarrow [-CH_2-] + 2CO_2$$

$$CO_2 + 2H_2 \rightarrow [-CH_2-] + 2H_2O$$

These reactions are highly exothermic, and to avoid an increase in temperature, which results in lower-boiling hydrocarbon derivatives, it is important to have sufficient cooling, to secure stable reaction conditions. The total heat of reaction amounts to approximately 25% of the heat of combustion of the synthesis gas, and lays thereby a theoretical limit on the maximal efficiency of the Fischer-Tropsch process.

The reaction is dependent of a catalyst, mostly an iron or cobalt catalyst, where the reaction takes place. There is either a low- or high-temperature process (low-temperature Fischer-Tropsch, high-temperature Fischer-Tropsch), with temperatures ranging between 200°C and 240°C for low-temperature Fischer-Tropsch and 300–350°C for high-temperature Fischer-Tropsch. The high-temperature Fischer-Tropsch uses an iron catalyst and the low-temperature Fischer-Tropsch either an iron or a cobalt catalyst. The different catalysts include also nickel-based and ruthenium-based catalysts, which also have enough activity for commercial use in the process. But the availability of ruthenium is limited and the nickel-based catalyst has high activity but produces too much methane, and additionally the performance at high pressure is poor, due to production of volatile carbonyls. This leaves only cobalt and iron as practical catalysts, and this study will only consider these two. Iron is cheap, but cobalt has the advantage of higher activity and longer life, though it is on a metal basis 1000 times more expensive than iron catalyst.

In general, the product distribution of hydrocarbon derivatives formed during the Fischer-Tropsch process follows an Anderson-Schulz-Flory distribution:

$$Wn/n = (1\text{-}\alpha)2\alpha n\text{-}1$$

Wn is the weight fraction of hydrocarbon molecules containing n carbon atoms. α is the chain growth probability or the probability that a molecule will continue reacting to form a longer chain. In general, α is largely determined by the catalyst and the specific process conditions. The above equation reveals that methane will always be the largest single product; however, by increasing α close to one, the total amount of methane formed can be minimized compared to the sum of all of the various long-chain products. Increasing α increases the formation of long-chain, higher molecular weight hydrocarbon derivatives.

The very long-chain solid hydrocarbon derivatives are waxes, which are solid at room temperature. Therefore, for production of liquid transportation fuels it may be necessary to crack some of the Fischer-Tropsch products. In order to avoid this, some researchers have proposed using zeolites or other catalyst substrates with fixed sized pores that can restrict the formation of hydrocarbon derivatives longer than some characteristic size (usually $n < 10$). This way they can drive the reaction so as to minimize methane formation without producing lots of long-chain hydrocarbon derivatives. Such efforts have met with only limited success.

5.2 Catalysts

Catalysts play a pivotal role in synthesis gas conversion reactions. In fact, fuels and chemicals synthesis from synthesis gas does not occur in the absence of appropriate catalysts. The basic concept of a catalytic reaction is that reactants adsorb onto the catalyst surface and rearrange and combine into products that desorb from the surface. One of the fundamental functional differences between synthesis gas synthesis catalysts is whether or not the adsorbed carbon monoxide molecule dissociates on the catalyst surface. For FTS and higher alcohol synthesis, dissociation of the carbon monoxide is a necessary reaction condition. For methanol synthesis, the carbon-oxygen bond (of the carbon monoxide) bond remains intact. Hydrogen has two roles in catalytic synthesis gas synthesis reactions: (i) hydrogen is the reactant needed for carbon monoxide hydrogenation and (ii) the hydrogen is commonly used to reduce the metalized synthesis catalysts and activate the metal surface.

Various metals, including but not limited to iron, cobalt, nickel, and ruthenium, alone and in conjunction with other metals, can serve as Fischer-Tropsch catalysts. Cobalt is particularly useful as a catalyst for converting natural gas to high-boiling hydrocarbon derivatives suitable for the production of diesel fuel. Iron has the advantage of being readily available and relatively inexpensive but also has the disadvantage of greater water-gas shift activity. Ruthenium is highly active but quite expensive. Consequently, although ruthenium is not the economically preferred catalyst for commercial Fischer-Tropsch production, it is often used in low concentrations as a promoter with one of the other catalytic metals.

A variety of catalysts can be used for the Fischer-Tropsch process, but the most common are the transition metals cobalt, iron, and ruthenium. Nickel can also be used, but tends to favor methane formation (methanation). Cobalt seems to be the most active catalyst, although iron may be more suitable for low-hydrogen-content synthesis gases such as those derived from coal due to its promotion of the water-gas-shift reaction. In addition to the active metal the catalysts typically contain a number of promoters, including potassium and copper. Catalysts are supported on high-surface-area binders/supports such as silica (SiO_2), alumina (Al_2O_3), or the more complex zeolites. Cobalt catalysts are more active for Fischer-Tropsch synthesis when the feedstock is natural gas. Natural gas has a high hydrogen-to-carbon ratio, so the water-gas-shift is not needed for cobalt catalysts. Iron catalysts are preferred for lower-quality feedstocks such as crude oil residua, coal, or biomass.

Unlike the other metals used for this process (cobalt, Co; nickel, Ni; ruthenium, Ru) which remain in the metallic state during synthesis, iron catalysts tend to form a number of chemical phases, including various oxides and carbides during the reaction. Control of these phase transformations can be important in maintaining catalytic activity and preventing breakdown of the catalyst particles.

Fischer-Tropsch catalysts can lose activity as a result of (i) conversion of the active metal site to an inactive oxide site, (ii) sintering, (iii) loss of active area by carbon deposition, and (4) chemical poisoning. For example, Fischer-Tropsch catalysts are notoriously sensitive to poisoning by sulfur-containing compounds. The sensitivity of the catalyst to sulfur is greater for cobalt-based catalysts than for their iron counterparts. Some of these mechanisms are unavoidable and others can be prevented or minimized by controlling the impurity levels in the synthesis gas. By far the most abundant, important, and most studied FTS catalyst poison is sulfur. Other catalyst poisons include halides and nitrogen compounds (e.g., NH_3, NO_x, and HCN).

5.3 Reactors

Since its discovery the Fischer-Tropsch synthesis has undergone periods of rapid development and periods of inaction. Within 10 years of the discovery, German companies were building commercial plants. The construction of these plants stopped in 1940 but existing plants continued to operate during World War II.

Two types of reactors were used in the German commercial plants: (i) the parallel plate reactors and (ii) a variety of fixed-bed tubular reactors. For the parallel plate version, the catalyst bed was located in tubes fixed between the plates, which were cooled by steam/water that passed around the tubes within the catalyst bed. In another version, the reactor may be regarded as finned-tube in which large fins are penetrated by a large number of parallel or connected catalyst-filled tubes. Various designs were utilized for the tubular fixed bed reactor with the concentrically placed tubes being the preferred one. This type of reactor contained catalyst in the area between the two tubes with cooling water steam flowing through the inner tube and on the exterior of the outer tube.

Various types of reactors have been used to carry out Fischer-Tropsch reactions, including packed bed (also termed fixed-bed) reactors and gas-agitated multiphase reactors. Originally, the Fischer-Tropsch synthesis was carried out in packed bed reactors. Gas-agitated multiphase reactors, sometimes called "slurry reactors" or "slurry bubble columns," gained favor however, because the circulation of the slurry makes it much easier to control the reaction temperature in a slurry bed reactor than in a fixed bed reactor. Gas-agitated multiphase reactors operate by suspending catalytic particles in a liquid and feeding gas reactants into the bottom of the reactor through a gas distributor, which produces small gas bubbles. As the gas bubbles rise through the reactor, the reactants are absorbed into the liquid and diffuse to the catalyst particles where, depending on the catalyst system, they are typically converted to gaseous and liquid products. The gaseous products formed enter the gas bubbles and are collected at the top of the reactor. Liquid products are recovered from the suspending liquid using different techniques, including filtration, settling, and hydrocyclones.

Because the Fischer-Tropsch reaction is exothermic, temperature control is an important aspect of Fischer-Tropsch reactor operation. Gas-agitated multiphase reactors or slurry bubble column reactors (SBCRs) have very high heat transfer rates and therefore allow good thermal control of the reaction. On the other hand, because the desired liquid products are mixed with the suspending liquid, recovery of the liquid products can be relatively difficult. This difficulty is compounded by the tendency of the catalyst particles to erode in the slurry, forming fine catalyst particles that are also relatively difficult to separate from the liquid products. Fixed-bed reactors avoid the problems of liquid separation and catalyst separation, but do not provide the mixing of phases that allows good thermal control in slurry bubble column reactors.

Furthermore, Fischer-Tropsch reactors are typically sized to achieve a desired volume of production. When a fixed-bed reactor is planned, economies of scale tend to result in the use of long (tall) reactors. Because the Fischer-Tropsch reaction is exothermic, however, a thermal gradient tends to form along the length of the reactor, with the temperature increasing with distance from the reactor inlet. In addition, for most Fischer-Tropsch catalyst systems each 10 degree rise in temperature increases the reaction rate approximately 60%, which in turn results in the generation of still more heat. To absorb the heat generated by the reaction and offset the rise in temperature, a cooling liquid is typically circulated through the reactor.

Thus, for a given reactor system having a known amount of catalyst with a certain specific activity and known coolant temperature, the maximum flow rate of reactants through the reactor is limited by the need to maintain the catalyst below a predetermined maximum catalyst temperature at all points along the length of the catalyst bed and the need to avoid thermal runaway which can result in catalyst deactivation and possible damage to the physical

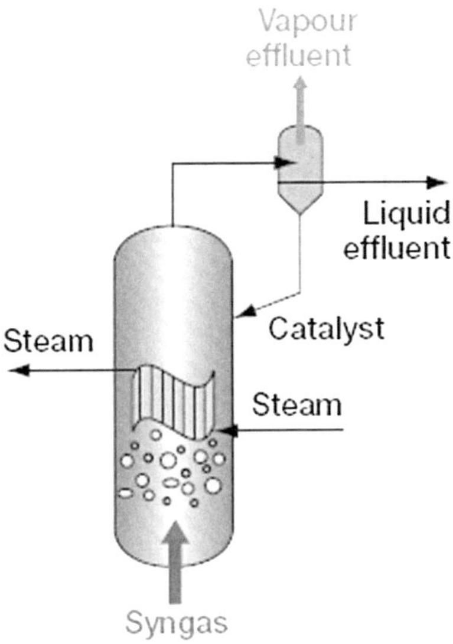

FIGURE 8.2 Schematic of a fluidized-bed reactor.

integrity of the reactor system. The net result is that it is unavoidable to operate most of the reactor at temperatures below the maximum temperature, with the corresponding low volumetric productivities over most of the reactor volume.

In the operation of a fluidized bed reactor, where solid catalyst is kept in suspension by synthesis gas (and gaseous products) (Fig. 8.2), liquid production must be limited or avoided. That is why this type of reactor can only be accommodated with HT-FT (above 350°C) with iron catalysts (cobalt catalyst selectivity toward methane is too detrimental at these temperatures). The corresponding catalyst selectivity (characterized by the Schultz-Flory [SF] coefficient for paraffins distribution:

$$a = Cn+1/Cn$$

Cn (is the number of moles of paraffin formed with n atoms of carbon) is in the order of 0.7, leading to more than 50% w/w production of C1 to C4 gases and less than 20% w/w of diesel and higher molecular weight hydrocarbon derivatives. This technology requires a diligent workup of the raw Fischer-Tropsch products to obtain valuable products like a relatively limited yield of gasoline, as well as olefins and specialty chemicals.

The low temperature Fischer-Tropsch process uses a cobalt catalyst in fixed-bed reactor technology (Fig. 8.3) for middle distillates (kerosene, diesel), naphtha, and waxes production. The catalyst is placed inside tubes. The synthesis gas passes through the tubes and the vapor-liquid products are recovered at the bottom of the reactor. This leads to a good final yield (after

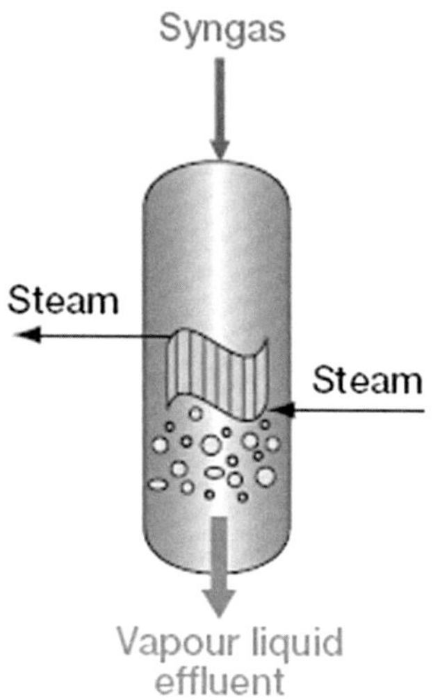

FIGURE 8.3 Schematic of a fixed-bed reactor.

product upgrading) of ultra-clean-diesel or middle distillates, with the possibility of producing lube base and waxes.

This fixed-bed technology has two important advantages: (i) the scale-up to industrial reactors is theoretically simple (multiplying the number of tubes) and (ii) there is normally no problem with liquid-solid separation, as the catalyst is fixed inside the tubes.

There are, however, mechanical constraints in the reactor design that should be noted with this technology (i.e., temperature gradient along the tubes, limitations in the thickness of the main tube sheets), and transfer limitations between the gas-liquid phase and the solid catalyst can lead to a limited capacity per single train: from 3000 to 6000 bpd per reactor of 7−8 m diameter. Also, catalyst continuous make-up is not feasible. For example, in the case of catalyst deactivation, the reactor must be shut down, tubes emptied and refilled (the estimated time is on the order of one-to-two weeks). This issue is particularly critical when the risk of having impurities/poisons in the synthesis gas is higher (coal, biomass). This is also why this type of technology is not combined with iron catalysts, which are known to deactivate much faster than cobalt-based catalysts.

The low temperature Fischer-Tropsch process using a cobalt catalyst in a slurry bubble column reactor (SBCR) technology (Fig. 8.4) for middle distillates (kerosene, diesel), naphtha, and waxes has been developed over the past 20 years. It is the most promising in terms of catalyst productivity, capacity per train, and operational flexibility. The reaction takes place in a three-phase slurry bubble column reactor, where the synthesis gas is contacted with solid catalyst mixed in the produced waxes. The obtained products are the same as in the fixed-bed case (mainly ultraclean diesel).

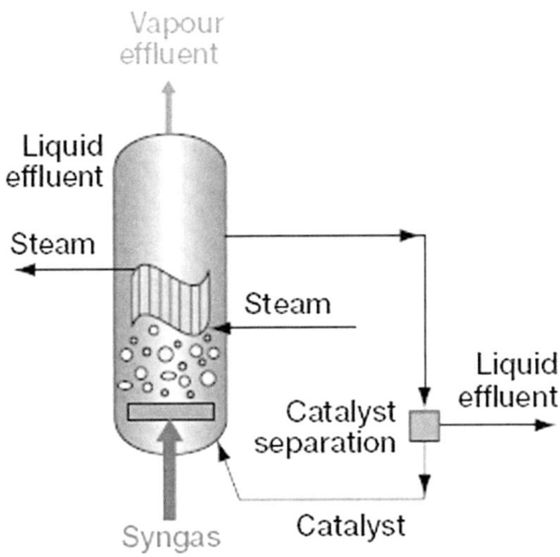

FIGURE 8.4 Schematic of a slurry bubble column reactor.

Compared to the fixed-bed technology, the slurry bubble column reactor has the following advantages: (i) higher capacity per train: at least 15,000 bpd per Fischer-Tropsch train (with reactors of up to 10 m diameter), (ii) easy isothermal operation in the reactor and (iii) the possibility of continuous catalyst make-up/withdrawal, allowing for a constant production to be maintained in case of catalyst deactivation or partial poisoning. However, some challenges must be solved, namely: (i) catalyst mechanical stress in a large reactor, and (ii) liquid-solid separation.

High-temperature circulating fluidized-bed reactors have been developed for gasoline and low-boiling olefin production. These reactors are known as Synthol reactors and operate at 350°C (660°F) and 375 psi (Fig. 8.5). The combined gas feed (fresh and recycled) enters at the bottom of the reactor and entrains catalyst that is flowing down the standpipe and through the slide valve. The high gas velocity carries the entrained catalyst into the reaction zone where heat is removed through heat exchangers. Product gases and catalyst are then transported into a large diameter catalyst hopper where the catalyst settles out and the product gases exit through a cyclone. These Synthol reactors have been successfully used for many years but there are a number of limitations. They are physically very complex reactors that involve circulation of large amounts of catalyst that leads to considerable erosion in particular regions of the reactor.

Carbon deposition is the most important mode of catalyst deactivation that can be impacted by addition of promoters to catalysts and reaction temperature and pressure. This mode of catalyst deactivation is largely unavoidable and FTS processes must be operated in a manner that the decreasing output from

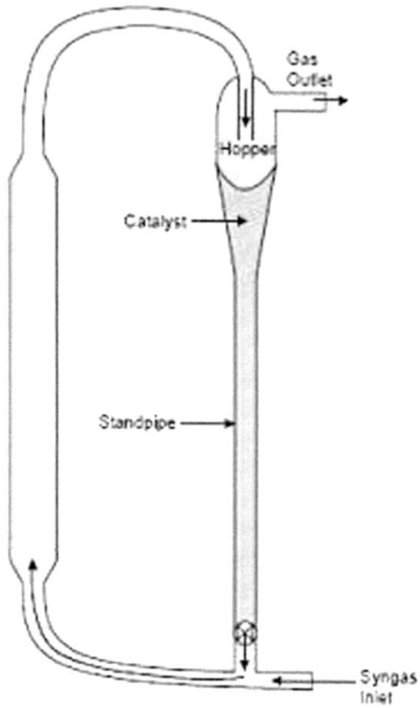

FIGURE 8.5 Schematic of a circulating fluidized-bed reactor (Synthol reactor).

coke deposition is balanced with the economic considerations of catalyst regeneration or replacement. In general, because of its high activity, the coke deposition rate is higher for an iron catalyst than a cobalt catalyst. Consequently, cobalt catalysts have longer lifetimes.

One of the more controllable modes of catalyst deactivation is that induced by poisoning of the active sites by impurities in the synthesis gas. By far the most abundant, important, and most studied catalyst poison is sulfur. Sulfur is present in both natural gas and coal and during steam reforming or gasification gets converted primarily to H_2S plus other organic sulfur compounds. Sulfur compounds rapidly deactivate both iron and cobalt catalysts, presumably by forming surface metal sulfides that do not have activity in the Fischer-Tropsch synthesis.

Ideally, there should be no sulfur in the synthesis gas. There is, however, always a small amount that gets through to the catalyst. There is really no safe sulfur level in FTS. Again, the level of gas cleaning required is based on economic considerations; namely how long the catalyst remains active versus the investment in gas cleaning. Other synthesis gas impurities are also known to poison the catalysts. Halide levels in synthesis gas should be less than 10 ppb and referenced nitrogen levels are 10 ppmv. Additionally, water oxidizes the iron and cobalt but the rate of oxidation is higher for the iron catalyst; water has an inhibiting effect on the iron catalyst because of its WGS activity.

Commercial processes are available to clean synthesis gas to meet these stringent contaminant requirements. The Rectisol process uses chilled methanol to scrub the raw synthesis gas. Ammonia, hydrogen sulfide, tar, and carbon dioxide are removed from synthesis gas to required levels. Other chemical absorption processes include potassium carbonate or alkanolamine (MEA: monoethanolamine or DEA: diethanolamine) for wet scrubbing. Fixed bed reactors containing ZnO are also used for sulfur polishing. Whether or not these gas-cleaning processes are economical will depend on the scale of the Fischer-Tropsch process.

5.4 Process parameters

For large-scale commercial Fischer-Tropsch reactors heat removal and temperature control are the most important design features to obtain optimum product selectivity and long catalyst lifetimes. Over the years, basically four Fischer-Tropsch reactor designs have been used commercially. These are the multitubular fixed bed, the slurry reactor, or the fluidized bed reactor (with either a fixed bed or a circulating bed).

The fixed-bed reactor consists of thousands of small tubes with the catalyst as surface-active agent in the tubes. Water surrounds the tubes and regulates the temperature by settling the pressure of evaporation. The slurry reactor is widely used and consists of fluid and solid elements, where the catalyst has no particularly position, but flows around as small pieces of catalyst together with the reaction components. The slurry and fixed-bed reactor are used in the low temperature Fischer-Tropsch process. The fluidized bed reactors are diverse, but characterized by the fluid behavior of the catalyst.

Sasol in South Africa uses coal and natural gas as a feedstock, and produces a variety of synthetic crude oil products. This process has received renewed attention in the quest to produce low sulfur diesel fuel in order to minimize the environmental impact from the use of diesel engines. The Fischer-Tropsch technology as applied at Sasol can be divided into two operating regimes: (i) the high-temperature Fischer-Tropsch process and (ii) the low-temperature Fischer-Tropsch process.

The high-temperature Fischer-Tropsch technology uses a fluidized catalyst at 300–330°C (570–635°F). Originally circulating fluidized bed units were used (Synthol reactors). Since 1989 a commercial scale classical fluidized bed unit has been implemented and improved upon.

The low-temperature Fischer-Tropsch technology has originally been used in tubular fixed bed reactors at 200–230°C (390–260°F). This produces a more paraffinic and waxy product spectrum than the high temperature technology. A new type of reactor (the Sasol slurry phase distillate reactor has been developed and is in commercial operation. This reactor uses a slurry phase system rather than a tubular fixed bed configuration and is currently the favored technology for the commercial production of synfuels.

The commercial Sasol Fischer-Tropsch reactors all use iron-based catalysts on the basis of the desired product spectrum and operating costs. Cobalt based catalysts have also been known since the early days of this technology and have the advantage of higher conversion for low temperature cases. Cobalt is not suitable for high temperature use due to excessive methane formation at such temperatures. For once-through maximum diesel production, cobalt has, despite its high cost, advantages and Sasol has also developed cobalt catalysts which perform very well in the slurry phase process.

The diesel produced by the slurry phase reactor has a highly paraffinic nature, giving a cetane number in excess of 70. The aromatic content of the diesel is typically below 3% and it is also sulfur-free and nitrogen-free. This makes it an exceptional diesel as such or it can be used to sweeten or to upgrade conventional diesels.

The Fischer-Tropsch process is an established technology and already applied on a large scale, although its popularity is hampered by high capital costs, high operation and maintenance costs, and the uncertain and volatile price of crude oil. In particular, the use of natural gas as a feedstock only becomes practical when using stranded gas, which is a source of natural gas that is far from major cities which are impractical to exploit with conventional gas pipelines and liquefied natural gas technology; otherwise, the direct sale of natural gas to consumers would become much more profitable.

Under most circumstances the production of synthesis gas by reforming natural gas will be more economical than from coal gasification, but site-specific factors need to be considered. In fact, any technological advance in this field (such as better energy integration or the oxygen transfer ceramic membrane reformer concept) will speed up the rate at which the synfuel technology will become common practice.

There are large coal reserves which may increasingly be used as a fuel source during oil depletion. Since there are large coal reserves in the world, this technology could be used as an interim transportation fuel if conventional oil were to become more expensive. Furthermore, combination of biomass gasification and Fischer-Tropsch synthesis is a very promising route to produce hydrocarbon products from renewable or green resources.

Often a higher concentration of some sorts of hydrocarbon derivatives is wanted, which might be achieved by changed reaction conditions. Nevertheless; the product range is wide and infected with uncertainties, due to lack of knowledge of the details of the process and of the kinetics of the reaction. Since the different products have quite different characteristics such as boiling point, physical state at ambient temperature, and thereby different use and ways of distribution, often only a few of the carbon chains is wanted. As an example the low-temperature Fischer-Tropsch is used when longer carbon chains are wanted, because lower temperature increases the portion of longer chains. But too low temperature is not wanted, because of reduced activity. When the desired products are shorter carbon chains, such as products from crude oil, the longer ones might be cracked into shorter chains.

The yield of diesel is therefore highly dependent on the chain growth probability, which again is dependent on pressure, temperature, feed gas composition, catalyst type, catalyst composition, and reactor design. The desire to increase the selectivity of some favorable products leads to a need of understanding the relation between reaction conditions and chain growth probability, which in turn request a mathematical expression for the growth probability in order to make a suitable model of the process. The different attempts to model the growth probability, a, have resulted in some models that are regarded in literature as appropriate to describe the product distribution. Two will be presented here to show the influence of temperature and partial pressure.

5.5 Refining Fischer-Tropsch products

The Fischer-Tropsch product stream typically contains hydrocarbon derivatives having a range of numbers of carbon atoms, including gases, liquids, and waxes. Depending on the molecular weight product distribution, different Fischer-Tropsch product mixtures are ideally suited to different uses. For example, Fischer-Tropsch product mixtures containing liquids may be processed to yield gasoline, as well as higher-boiling middle distillates. Hydrocarbon waxes may be subjected to additional processing steps for conversion to liquid and/or gaseous hydrocarbon derivatives. Thus, in the production of a Fischer-Tropsch product stream for processing to a fuel it is desirable obtain primarily hydrocarbon derivatives that are liquids and waxes (e.g., C_{5+} hydrocarbon derivatives).

Initially, the low-boiling gases in raw product are separated and sent to a gas cleaning operation. The higher boiling product is distilled to produce separate streams of naphtha, distillate, and wax.

The naphtha stream is first hydrotreated (Fig. 8.6), resulting in the production of hydrogen-saturated liquids (primarily paraffins), a portion of which is converted by isomerization from normal paraffins to iso-paraffins to boost

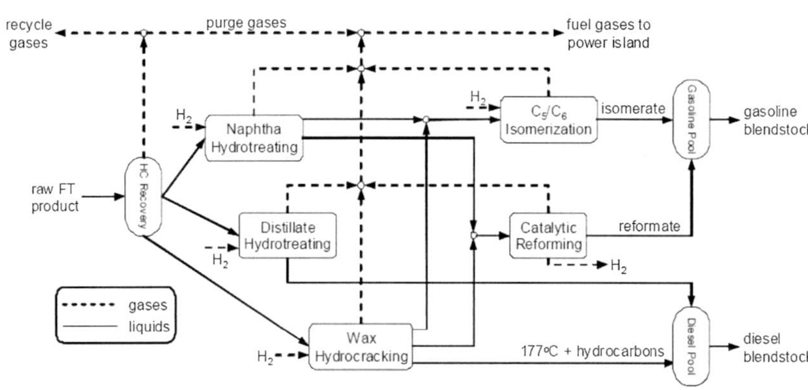

FIGURE 8.6 Illustration of the upgrading Scheme for the raw product from the Fischer-Tropsch process.

their octane value. Another fraction of the hydrotreated naphtha is catalytically reformed to provide some aromatic content to (and further boost the octane value of) the final gasoline blendstock. The distillate stream is also hydrotreated, resulting directly in a finished diesel blendstock. The wax fraction is hydrocracked into a finished distillate stream and naphtha streams that augment the hydrotreated naphtha streams sent for isomerization and for catalytic cracking.

Thus, conventional refinery processes (Speight, 2014a, 2017) can be used for upgrading of Fischer-Tropsch liquid and wax products. A number of possible processes for FT products are (i) wax hydrocracking, (ii) distillate hydrotreating, (iii) catalytic reforming, (iv) naphtha hydrotreating, (v) alkylation, and (vi) isomerization (Speight, 2014a, 2017). Fuels produced with by the Fischer-Tropsch synthesis are of a high quality due to a very low aromaticity and zero sulfur content.

The diesel fraction has a high cetane number resulting in superior combustion properties and reduced emissions. New and stringent regulations may promote replacement or blending of conventional fuels by sulfur and aromatic-free products. Also, other products besides fuels can be manufactured with Fischer-Tropsch in combination with upgrading processes, for example, ethylene, propylene, α-olefins, alcohols, ketones, solvents, and waxes. These valuable byproducts of the process have higher added values, resulting in an economically more attractive process economy.

References

Abadie, L.M., Chamorro, J.M., 2009. The economics of gasification: a market-based approach. Energies 2, 662–694.

Adhikari, U., Eikeland, M.S., Halvorsen, B.M., 2015. Gasification of biomass for production of syngas for biofuel. In: Proceedings. 56th SIMS, Linköping, Sweden. October 7–9, pp. 255–260. http://www.ep.liu.se/ecp/119/025/ecp15119025.pdf.

Arena, U., 2012. Process and technological aspects of municipal solid waste gasification a review. Waste Management 32, 625–639.

Brandmair, M., Find, J., Lercher, J.A., 2003. Combined autothermal reforming and hydrogenolysis of alkanes. In: Proceedings. DGMK Conference on Innovation in the Manufacture and Use of Hydrogen. Dresden, Germany. October 15–17, pp. 273–280.

Breault, R.W., 2010. Gasification processes old and new: a basic review of the major technologies. Energies 3 (2), 216–240.

Calemma, V., Radovic, L.R., 1991. On the gasification reactivity of Italian sulcis coal. Fuel 70, 1027.

Chadeesingh, R., 2011. The Fischer-Tropsch process. In: Speight, J.G. (Ed.), The Biofuels Handbook. The Royal Society of Chemistry, London, United Kingdom, pp. 476–517. Part 3, (Chapter 5).

Choudhury, H.A., Moholkar, V.S., 2013. Synthesis of liquid hydrocarbons by Fischer–Tropsch process using industrial iron catalyst. International Journal of Innovative Research in Science, Engineering and Technology 2 (8), 3493–3499.

Couvaras, G., 1997. Sasol's slurry phase distillate process and future applications. In: Monetizing Stranded Gas Reserves Conference, Houston, Texas. December.

Fabry, F., Rehmet, C., Rohani, V.-J., Fulcheri, L., 2013. Waste gasification by thermal plasma: a review. Waste and Biomass Valorization 4 (3), 421–439.

Find, J., Nagaoka, K., Lercher, J.A., 2003. Steam reforming of light alkanes in micro-structured reactors. In: Proceedings. DGMK Conference on Innovation in the Manufacture and Use of Hydrogen. Dresden, Germany. October 15–17, p. 257.

Furimsky, E., 1999. Gasification in a petroleum refinery of the 21st century. Oil & gas science and technology. Revue Institut Français du Pétrole 54 (5), 597–618.

Garcia, X., Radovic, L.R., 1986. Gasification reactivity of Chilean coals. Fuel 65, 292.

Gray, D., Tomlinson, G., 2000. Opportunities for petroleum coke gasification under Tighter sulfur limits for transportation fuels. In: Proceedings. 2000 Gasification Technologies Conference, San Francisco, California. October 8–11.

Higman, C., Van der Burgt, M., 2008. Gasification, second ed. Gulf Professional Publishing, Elsevier, Amsterdam, Netherlands.

Jager, B., 1997. Sasol's advanced Fischer-Tropsch processes. In: AIChE Spring Meeting, Houston, Texas. 9–13 March.

Keil, F.J., 1999. Methanol-to-Hydrocarbons: process technology. Microporous and Mesoporous Materials 29, 49–66.

Kristiansen, A., 1996. IEA Coal Research Report IEACR/86. Understanding Coal Gasification. International Energy Agency, London, United Kingdom.

Kumar, A., Jones, D.D., Hanna, M.A., 2009. Thermochemical biomass gasification: a review of the current status of the technology. Energies 2, 556–581.

Lee, S., 2007. Gasification of coal. In: Lee, S., Speight, J.G., Loyalka, S. (Eds.), Handbook of Alternative Fuel Technologies. CRC Press, Taylor & Francis Group, Boca Raton, Florida.

Luque, R., Speight, J.G. (Eds.), 2015. Gasification for Synthetic Fuel Production: Fundamentals, Processes, and Applications. Woodhead Publishing, Elsevier, Cambridge, United Kingdom.

Mangone, C., 2002. Gas to liquids conversions produce extremely pure base oils. Machinery Lubrication Magazine. Independent Lubricant Manufacturers Association (ILMA). November.

Marano, J.J., June 2003. Refinery Technology Profiles: Gasification and Supporting Technologies. Report prepared for the United States Department of Energy, National Energy Technology Laboratory. United States Energy Information Administration, Washington, DC.

Muradov, N.Z., 1998. CO_2-free production of hydrogen by catalytic pyrolysis of hydrocarbon fuel. Energy & Fuels 12 (1), 41–48.

Murata, K., Ushijima, H., Fujita, K., 1997. Process for Producing Hydrogen from Hydrocarbon. United States Patent 5,650,132.

Penrose, C.F., Wallace, P.S., Kasbaum, J.L., Anderson, M.K., Preston, W.E., 1999. Enhancing refinery profitability by gasification, hydroprocessing and power generation. In: Proceedings. Gasification Technologies Conference San Francisco, California. October. https://www.globalsyngas.org/uploads/eventLibrary/GTC99270.pdf.

Radovic, L.R., Walker Jr., P.L., Jenkins, R.G., 1983. Importance of carbon active sites in the gasification of coal chars. Fuel 62, 849.

Radovic, L.R., Walker Jr., P.L., 1984. Reactivities of chars obtained as residues in selected coal conversion processes. Fuel Processing Technology 8, 149.

Ryan III, T.W., 1997. Near ULEV emission level in a heavy-duty diesel engine using Fischer-Tropsch diesel fuel. In: Proceedings. Monetizing Stranded Gas Reserves Conference, Houston. December.

Serrano-Ruiz, J.C., Dumesic, J.A., 2012. Catalytic production of liquid hydrocarbon transportation fuels. In: Guczi, L., Erdöhelyi, A. (Eds.), Catalysis for Alternative Energy Generation. Springer Science + Business Media, New York, pp. 29–56 (Chapter 2).

Speight, J.G. (Ed.), 2011a. Biofuels Handbook. Royal Society of Chemistry, London, United Kingdom.

Speight, J.G., 2011b. The Refinery of the Future. Gulf Professional Publishing, Elsevier, Oxford, United Kingdom.

Speight, J.G., 2013a. The Chemistry and Technology of Coal, third ed. CRC Press, Taylor & Francis Group, Boca Raton, Florida.

Speight, J.G., 2013b. Coal-Fired Power Generation Handbook. Scrivener Publishing, Beverly, Massachusetts.

Speight, J.G., 2014a. The Chemistry and Technology of Petroleum, fifth ed. CRC Press, Taylor & Francis Group, Boca Raton, Florida.

Speight, J.G., 2014b. Gasification of Unconventional Feedstocks. Gulf Professional Publishing, Elsevier, Oxford, United Kingdom.

Speight, J.G., 2017. Handbook of Petroleum Refining. CRC Press, Taylor and Francis Group, Boca Raton, Florida.

Speight, J.G., 2019a. Biogas: Production and Properties. Nova Science publishers, New York.

Speight, J.G., 2019b. Natural Gas: A Basic Handbook, second ed. Gulf Publishing Company, Elsevier, Cambridge, Massachusetts.

Speight, J.G., 2019c. Handbook of Petrochemical Processes. CRC Press, Taylor and Francis Group, Boca Raton, Florida.

Van Berge, P.J., 1995. Cobalt as an alternative Fischer-Tropsch catalyst to iron for the production of middle distillates. In: Proceedings. 4th International Natural Gas Conversion Symposium, Kruger National Park, South Africa, November.

Wallace, P.S., Anderson, M.K., Rodarte, A.I., Preston, W.E., 1998. Heavy oil upgrading by the separation and gasification of asphaltenes. In: Proceedings. Presented at the Gasification Technologies Conference. San Francisco, California. October. https://www.globalsyngas.org/uploads/eventLibrary/gtc9817p.pdf.

Wolff, J., Vliegenthart, E., 2011. Gasification of heavy ends. Petroleum Technology Quarterly Q2, 1–5.

Further reading

Rostrup-Nielsen, J.R., Christiansen, L.J., Bak Hansen, J.-H., 1988. Activity of steam reforming catalysts: role and assessment. Applied Catalysis 43, 287–303.

Chapter 9

Chemical and physical properties of hydrocarbons

1. Introduction

Hydrocarbon derivatives (hydrocarbons) are the principal compounds of crude oil and natural gas and to some extent the major constituents of the gaseous and liquid products produced by the thermal decomposition of coal (Chapter 5), oil shale (Chapter 6), and biomass (Chapter 7). In many cases, the hydrocarbon derivatives produced from these sources have to be chemically altered to make useful products and materials. This conversion to saleable products is carried out by changes in the chemical and physical structure. Such differences in molecular structure, even though the empirical formula can remain the same, cause significant difference in the properties and behavior of hydrocarbon derivatives and fuels that are predominantly composed of hydrocarbon derivatives.

A hydrocarbon is any chemical compound that consists only of the elements carbon (C) and hydrogen (H) (Chapter 1). With the exception of methane (CH_4), all hydrocarbon derivatives contain a carbon-chain skeleton and have hydrogen atoms attached to the carbon skeleton. Most hydrocarbon derivatives are readily combustible (Chapter 10). Almost all usable supplies of hydrocarbon derivatives are currently obtained from crude oil and natural as. Hydrocarbon derivatives are the simplest organic compounds and contain only carbon and hydrogen but they can be straight-chain or branched-chain (Stoker, 2008; Klein, 2013) with the same empirical formula but showing difference in properties. The hydrocarbon derivatives can be divided into various homologous series (Chapter 1) and each member of such a series shows a definite relationship in its structural formula to the members preceding and following it, and there is generally some regularity in changes in physical properties of successive members of a series.

The alkane derivatives are a homologous series of saturated aliphatic hydrocarbon derivatives. The first and simplest member of this series is methane, CH_4; the series is sometimes called the methane series. Each successive member of a homologous series of hydrocarbon derivatives has one more carbon and two more hydrogen atoms in its molecule than the preceding member. The second alkane is ethane, C_2H_6, and the third is

Handbook of Industrial Hydrocarbon Processes. https://doi.org/10.1016/B978-0-12-809923-0.00009-6

propane, C_3H_8. Alkane derivatives have the general formula C_nH_{2n+2} (where n is an integer greater than or equal to 1). Other homologous series of hydrocarbon derivatives include the alkene derivatives and the alkyne derivatives (Chapter 1).

Hydrocarbon mixtures are composed of hydrocarbon derivatives, naphtha, benzine, and petroleum ether. Naphtha is a flammable liquid hydrocarbon mixture that typically boils over the range 0–200°C (32–390°F). Benzine (which should not be confused with benzene— an aromatic hydrocarbon) also known as petroleum ether is a hydrocarbon mixture of alkane derivatives, such as pentane, hexane, and heptane. Petroleum ether is obtained from crude oil refineries as the portion of the distillate which is intermediate between the low boiling naphtha and the higher boiling kerosene. It has a specific gravity of between 0.6 and 0.8 depending on its composition. Petroleum ether should not be confused with the class of organic compounds called ethers.

The physical properties of the unsaturated hydrocarbons (alkane derivatives and cycloalkanes derivatives) are similar to (but not the same as) the properties of the corresponding saturated hydrocarbons (alkene derivatives and alkyne derivatives). The molecules are nonpolar and thus relatively insoluble in water and any intermolecular bonds are weak van der Waal's bonds. Melting points and boiling points for the small molecules are fairly low. The higher the molecular weight of the hydrocarbon derivative, the higher the melting point and the boiling point.

Finally, the chemical and physical properties of hydrocarbon derivatives are also dictated by stereochemistry (Olah and Molnár, 2003). Both types of properties are related and the proportions of the stereoisomers serve to influence the chemical and/or physical properties.

2. Stereochemistry

Stereochemistry, a subdiscipline of chemistry, involves the study of the relative spatial arrangement of atoms within molecules. Stereochemistry is an important facet of chemistry and the study of stereochemical effects spans the entire range of chemical and physical properties (Eliel and Wilen, 1994; Eliel et al., 2001). Stereochemistry (molecular geometry) refers to the three-dimensional arrangement of a molecule. Organic molecules of the same chemical formula can have their atoms arranged differently in space, often leading to significantly different chemical properties.

Isomers are those types of compounds which have the same chemical formula but different atomic arrangements in space. Isomers can be divided into stereoisomers and structural isomers. Stereoisomers change their atomic arrangement as a result of changes in pressure or temperature. All bonds and types of bonds (single, double, triple) are conserved in the same original fashion, however. Structural isomers have atoms which change their position in a molecule. One example is a linear compound (where all of the carbon atoms are lined up in linear fashion), compared to the same chemical formula compound with a shorter linear structure and branching (chain isomerism). Functional groups can change their position (functional isomerism), or can differ from another isomer in the position of a double or triple bond (bond isomerism). The number of carbon atoms in a hydrocarbon determines how many forms that compound can take—or the number of possible isomers. The number of possible isomers in a compound rises as the number of carbon atoms it contains rises.

The molecular structure of the alkane derivatives directly affects their physical and chemical characteristics. It is derived from the electron configuration of carbon (Chapter 1), which has four valence electrons. The carbon atoms in alkane derivatives are always sp^3 hybridized, that is to say that the valence electrons are said to be in four equivalent orbitals derived from the combination of the 2s orbital and the three 2p orbitals. These orbitals, which have identical energies, are arranged spatially in the form of a tetrahedron, the angle of $\cos^{-1}(-1/3) \approx 109.47$ degrees between them.

An alkane molecule has only C—H and C—C single bonds. The former result from the overlap of a sp^3-orbital of carbon with the 1s-orbital of a hydrogen; the latter by the overlap of two sp^3-orbitals on different carbon atoms. The bond lengths amount to 1.09×10^{-10} m for a carbon-hydrogen (C—H) bond and 1.54×10^{-10} m for a carbon-carbon (C—C) bond.

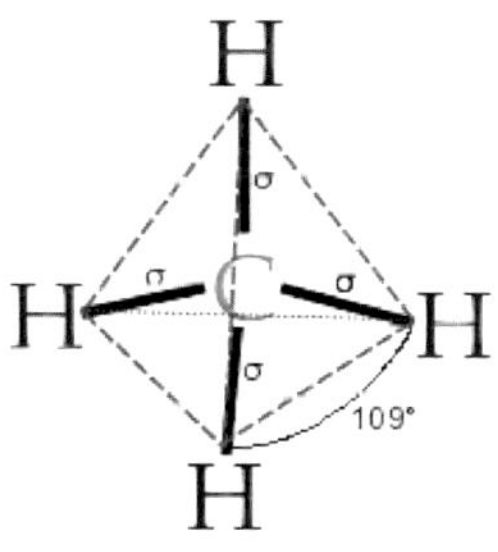

The tetrahedral structure of methane.

The spatial arrangement of the bonds is similar to that of four sp^3-orbitals—they are tetrahedrally arranged, with an angle of 109.47 degrees between them. Structural formulae that represent the bonds as being at right angles to one another, while both common and useful, do not correspond with the reality.

The structural formula and the bond angles are not usually sufficient to completely describe the geometry of a molecule. There is a further degree of freedom for each carbon-carbon bond: the torsion angle between the atoms or groups bound to the atoms at each end of the bond. The spatial arrangement described by the torsion angles of the molecule is known as its conformation.

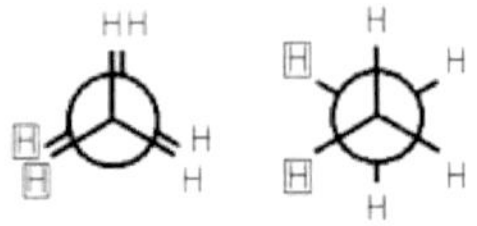

Newman projections of the two conformations of ethane: eclipsed on the left, staggered on the right.

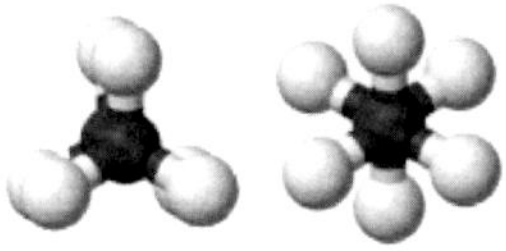

Ball-and-stick models of the two rotamers of ethane.

Ethane forms the simplest case for studying the conformation of alkane derivatives, as there is only one carbon-carbon bond. If one looks down the axis of the carbon-carbon bond, one will see the so-called Newman projection. The hydrogen atoms on both the front and rear carbon atoms have an angle of 120 degrees between them, resulting from the projection of the base of the tetrahedron onto a flat plane. However, the torsion angle between a given hydrogen atom attached to the front carbon and a given hydrogen atom attached to the rear carbon can vary freely between 0° and 360 degrees. This is a consequence of the free rotation about a carbon-carbon single bond. Despite this apparent freedom, only two limiting conformations are important: eclipsed conformation and staggered conformation. The two conformations, also known as rotamers, differ in energy: the staggered conformation is 12.6 kJ/mol lower in energy (more stable) than the eclipsed conformation (the least stable).

This difference in energy between the two conformations, known as the torsion energy, is low compared to the thermal energy of an ethane molecule at ambient temperature. There is constant rotation about the carbon-carbon single bond. The time taken for an ethane molecule to pass from one staggered conformation to the next, equivalent to the rotation of one methyl (CH_3) group by 120 degrees relative to the other, is of the order of 10−11 s.

The case of higher molecular weight alkane derivatives is more complex but based on similar principles, with the antiperiplanar conformation always being the most favored around each carbon-carbon bond. For this reason, alkane derivatives are usually shown in a zigzag arrangement in diagrams or in models. The actual structure will always differ somewhat from these idealized forms, as the differences in energy between the conformations are small compared to the thermal energy of the molecules: Alkane molecules have no fixed structural form, whatever the models may suggest. The geometry of acetylene is linear but the structure of ethylene and propylene are different because the two double bonded carbons are sp^2 hybridized and therefore are trigonal planar.

There is no free rotation of a double bond or triple. Therefore, many alkene derivatives and alkyne derivatives exhibit geometric isomerism. For example, cis-2-butene and trans-2-butene are geometric isomers— cis means on the same side, while trans means on opposite sides— and refer to (in the case of the butylenes) the relative position of the methyl groups.

3. Molecular weight

The *molecular mass* of a substance is the mass of one molecule of that substance, in unified atomic mass units (Drews, 1998; Speight, 2015). This is distinct from the *relative molecular mass* of a molecule, frequently referred to by the term *molecular weight*, which is the ratio of the mass of that molecule to 1/12 of the mass of carbon 12 and is a dimensionless number. Thus, it is incorrect to express relative molecular mass (molecular weight) in Daltons (Da) or kilo-Daltons (kDa) (unfortunately, the terms molecular weight and molecular mass have been confused on numerous websites, which often state that molecular weight was used in the past as another term for molecular mass).

For a well-defined molecular structure, the molecular weight is calculated from the atomic masses of the constituents. In the case of mixtures, the average molecular weight can be measured by the following methods: vapor pressure osmometry, freezing point depression, boiling point elevation, gel permeation chromatography, and nonfragmenting mass spectrometry (Speight, 2014, 2015). The different methods have advantages and drawbacks, which make them suitable for different molecular weight ranges. Vapor pressure osmometry, freezing point depression, and boiling point elevation are all based on the assumption that the change in the corresponding properties (vapor pressure, freezing point, and boiling point) in a pure solvent caused by introduction of a solute at low concentration is directly proportional to the concentration of the solute. Gel permeation chromatography, also known as size exclusion chromatography, takes advantage of the difference in elution time between molecules with different sizes. Nonfragmenting mass spectrometry principally provides detailed information of the hydrocarbon types, the formulae, and the concentration of all the components in a fraction.

Generally, hydrocarbon derivatives of low molecular weight, e.g., methane, ethane, and propane, are gases; those of intermediate molecular weight, e.g., hexane, heptane, and octane, are liquids; and those of high molecular weight, e.g., eicosane ($C_{20}H_{42}$) and polyethylene, are solids. Paraffin is a mixture of high molecular weight alkane derivatives.

4. Chemical properties

By way of recall (Chapter 1), the hydrocarbon family falls into five subgroups: (i) paraffin derivatives, (ii) cycloparaffin derivatives, also called naphthene derivatives, (iii) olefin derivatives, also called alkene derivatives, (iv) acetylene derivatives, also called alkyne derivatives, and (v) aromatic derivatives, also called arene derivatives. The chemical properties of hydrocarbon derivatives describe the *potential* of hydrocarbon derivatives to undergo chemical change or reaction by virtue of the hydrocarbon structure (Howard and Meylan, 1997; Yaws, 1999).

Chemical change results in the hydrocarbon yielding a product that may be entirely different in composition to the starting hydrocarbon—the exception is the isomerization reaction where a straight-chain hydrocarbon is converted to a branched-chain hydrocarbon. In such a case, the composition of the product is not changed over the composition of the starting material but the structure has been changed:

$$CH_3CH_2CH_2CH_2CH_3 \rightarrow CH_3CH_2CH(CH_3)CH_3$$

n-pentane, C_5H_{12} isopentane, C_5H_{12}

Thus, since a chemical change alters the composition of the original matter, the expected outcome is usually the presence of different elements or compounds at the end of the chemical change. The atoms in compounds are rearranged to make new and different compounds.

In the absence of a spark or a high-intensity light source, *alkane derivatives* are generally inert to chemical reactions. However, it must be recognized that alkane derivatives burst into flame in the presence of a spark. It doesn't matter whether the starting material is (i) the methane found in natural gas, (ii) the mixture of butane and isobutane used in disposable cigarette lighters, (iii) the mixture of pentane (C_5) to hexane (C_6) hydrocarbon derivatives in charcoal lighter fluid, or (iv) the complex mixture of hexane (C_6) to octane (C_8) hydrocarbon derivatives in naphtha or gasoline; once the reaction is ignited by a spark, these hydrocarbon derivatives burn to form carbon dioxide and water and give energy:

$$CH_4(g) + 2O_2(g) \rightarrow CO_2(g) + 2H_2O(g)$$

$$2C_4H_{10}(g) + 13O_2(g) \rightarrow 8CO_2(g) + 10H_2O(g)$$

$$C_5H_{12}(g) + 8O_2(g) \rightarrow 5\ CO_2(g) + 6\ H_2O(g)$$

$$2C_8H_{18}(l) + 25\ O_2(g) \rightarrow 16CO_2(g) + 18H_2O(g)$$

In the presence of light, or at high temperatures, alkane derivatives react with halogens to form alkyl halides in which the alkyl group is a derivative of the related hydrocarbon (Table 9.1). Reaction with chlorine gives an alkyl chloride and the ease with which chlorine can be introduced into the molecules of all the hydrocarbon types present in petroleum has resulted in the commercial production of a number of widely used compounds (Speight, 2019). With saturated hydrocarbon derivatives the reactions are predominantly substitution of hydrogen by chloride and are strongly exothermic, difficult to control, and inclined to become explosively violent:

$$RH + Cl_2 \rightarrow RCl + HCl$$

TABLE 9.1 General comparison of selected properties of alkane, alkene, and alkyne derivatives.

	Alkanes	Alkenes	Alkynes
General formula	CH_nH_{2n+2}	CH_nH_{2n}	CH_nH_{2n-2}
Naming	All the members end with *ane*	All the members end with *ene*	All the members end with *yne*
Physical state	Members having 1–4 carbon atoms per molecule are gases Members having 5–17 carbon atoms are liquids Higher members are solids at room temperature	Members having 2–4 carbon atoms per molecule are gases Members having 5–15 carbon atoms per molecule are liquids Higher members are solids	Members having 2–4 carbon atoms per molecule are gases Members having 5–13 carbon atoms per molecule are liquids Higher members are solids
Boiling points and melting points	Melting and boiling points increase with increase in molecular mass	Melting point and boiling point increase with the increase in molecular mass	Melting point and boiling point increase with the increase in molecular mass
Combustion	Undergo complete combustion Produce CO_2, H_2O, and heat	Burn with a sooty flame because of the higher percentage of carbon Produce CO_2, H_2O, and heat.	Burn with a sooty flame because of the higher percentage of carbon Produce CO_2, H_2O, and heat.

Moderately high temperatures are used, about 250–300°C (480–570°F), for the thermal chlorination of methane, but as the molecular weight of the paraffin increases the temperature may generally be lowered. A mixture of chlorinated derivatives is always obtained, and many variables, such as choice of catalyst, dilution of inert gases, and presence of other chlorinating agents (antimony pentachloride, sulfuryl chloride, and phosgene), have been tried in an effort to direct the path of the reaction.

Methane yields four compounds upon chlorination in the presence of heat or light:

$$CH_4 + Cl_2 \rightarrow CH_3Cl, CH_2Cl_2, CHCl_3, CCl_4$$

These compounds, known as chloromethane or methyl chloride, dichloromethane or methylene chloride, trichloromethane or chloroform, and tetrachloromethane or carbon tetrachloride, are used as solvents or in the production of chlorinated materials.

The successive substitution of methane hydrogens with chlorine produces a mixture of four chloromethanes: (i) methyl chloride, CH_3Cl, also known as monochloromethane, (ii) methylene dichloride, CH_2Cl_2, also known as dichloromethane, (iii) chloroform, $CHCl_3$, also known as trichloromethane, and (iv) carbon tetrachloride, CCl_4, also known as tetrachloromethane. Each of these four compounds has many industrial applications.

Methane is the most difficult alkane to chlorinate. The reaction is initiated by chlorine free radicals obtained via the application of heat (thermal) or light (hv). Thermal chlorination (more widely used industrially) occurs at approximately 350–370°C (660–700°F) and atmospheric pressure. A typical product distribution for a feedstock (methane-chlorine) ratio of 1.7 v/v is: methyl chloride: 58.7% v/v, methylene dichloride: 29.3% v/v, chloroform: 9.7% v/v, and carbon tetrachloride: 2.3% v/v.

The first step in the process involves the breaking of the chlorine-chlorine bond which forms two chlorine (Cl•) free radicals after which the chlorine free radical reacts with methane to form a methyl free radical (CH_3•) and hydrogen chloride. The methyl radical then reacts in a subsequent step with a chlorine molecule, forming methyl chloride and a chlorine radical:

$$Cl\bullet + CH_4 \rightarrow CH_3 + HCl$$

$$CH_3\bullet + Cl_2 \rightarrow CH_3Cl + Cl\bullet$$

The freshly generated chlorine radical Cl either attacks another methane molecule and repeats the above reaction, or it reacts with a methyl chloride molecule to form a chloromethyl free radical (•CH_2Cl, in which the free electron resides on the carbon atom) and hydrogen chloride:

$$Cl\bullet + CH_3Cl \rightarrow \bullet CH_2Cl$$

The chloromethyl free radical then attacks another chlorine molecule and produces dichloromethane along with a Cl atom:

$$\bullet CH_2Cl + Cl_2 \rightarrow CH_2Cl_2 + Cl\bullet$$

This formation of chlorine free radicals continues until all chlorine is consumed. Chloroform and carbon tetrachloride are formed in a similar way by reaction of dichloromethyl (•$CHCl_2$) and trichloromethyl (•CCl_3) free radicals with chlorine.

Product distribution among the products depends primarily on the mole ratio of the reactants in the feedstock. For example, the yield of methyl chloride (CH_3Cl) chloromethane could be increased to 80% v/v by increasing the methane-chlorine mole ratio to 10:1 at 450°C (840°F). If dichloromethane is the desired product, the methane-chlorine mole ratio is lowered and the monochloromethane recycled. Decreasing the methane-chlorine mole ratio generally increases polysubstitution and, hence, the yield of chloroform and carbon tetrachloride.

Similarly, the reaction of methane with bromine gives methyl bromide.

$$CH_4(g) + Br_2(l) \rightarrow CH_3Br(g) + HBr(g)$$

However, the ease of halogenation is influenced by the halogens—fluorine (fluorination) and chlorine—which are more electrophilic and are more aggressive halogenating agents. Bromine (bromination) is a weaker halogenating agent than both fluorine and chlorine and iodine (iodination) is the least reactive halogen. The facility of dehydrohalogenation follows the reverse trend: iodine is most easily removed from organic compounds, and organofluorine compounds are highly stable.

Alkene derivatives (olefin derivatives, C_nH_{2n}) are unsaturated compounds of carbon with hydrogen which contain one or two double bonds between atoms of carbon. They burn to form carbon soot and carbon dioxide and water. They are more reactive than alkane derivatives because of the fact that they contain double bonds.

Multiple bonds (double, triple bonds) are energetically less advantageous for atoms than corresponding single bonds. For this reason, the atoms in a compound will attempt to break multiple bonds to form single bonds, which are more advantageous energetically. This explains why compounds which contain double and triple bonds are so much more reactive than those which contain single bonds. The alkene derivatives include ethylene (C_2H_4), propylene (C_3H_6), butylene (C_4H_8), and pentylene (pentene, C_5H_{10}). Up to butylene, the alkene derivatives occur as gases. Up to hexadecene ($C_{16}H_{32}$) they are liquids, with higher alkene derivatives found in the solid state of matter. Hydrocarbon derivatives with double bonds make up the *alkene* family while hydrocarbon derivatives with triple bonds make up the *alkyne* family and there are similarities in physical properties (Table 9.1).

Open chain alkene derivatives with one double bond have the general formula, C_nH_{2n}, where n equals the number of carbon atoms. Open chain alkyne derivatives with one triple bond have the general formula C_nH_{2n-2}.

Like the alkane derivatives and other hydrocarbon derivatives, they are insoluble in water and are flammable. The most familiar alkene derivatives are ethylene and propylene. Ethyne (acetylene) is an important alkyne.

Unsaturated hydrocarbon derivatives such as alkene derivatives and alkyne derivatives are much more reactive than the parent alkane derivatives. They react rapidly with bromine, for example, to add a Br_2 molecule across the C=C double bond.

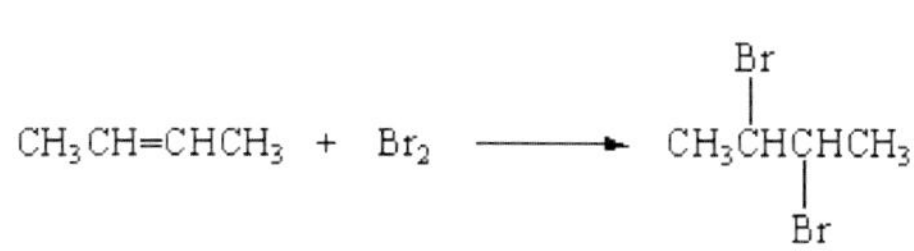

2,3-Dibromobutane

This reaction provides a way to test for alkene derivatives or alkyne derivatives. Solutions of bromine in carbon tetrachloride have an intense red-orange color. When bromine in carbon tetrachloride is mixed with a sample of an alkane, no change is initially observed. When it is mixed with an alkene or alkyne, the color of bromine rapidly disappears.

The reaction between 2-butene and bromine to form 2,3-dibromobutane is just one example of the addition reactions of alkene derivatives and alkyne derivatives. Hydrogen bromide (HBr) adds across a C=C double bond to form the corresponding alkyl bromide, in which the hydrogen ends up on the carbon atom that had more hydrogen atoms to begin with. Addition of HBr to 2-butene, for example, gives 2-bromobutane.

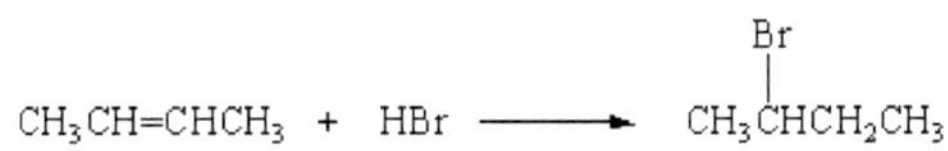

2-Bromobutane

H_2 adds across double (or triple bonds) in the presence of a suitable catalyst to convert an alkene (or alkyne) to the corresponding alkane.

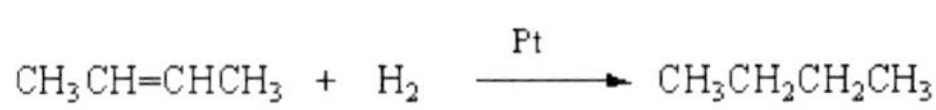

In the presence of an acid catalyst, it is even possible to add a molecule of water across a C=C double bond.

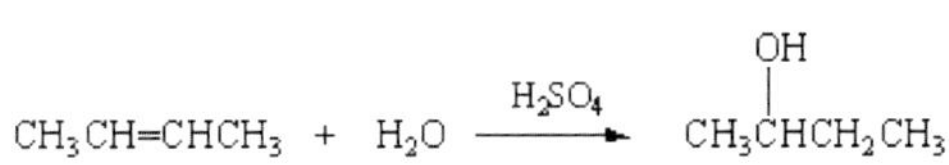

Addition reactions provide a way to add new substituents to a hydrocarbon chain and thereby produce new derivatives of the parent alkane derivatives.

In theory, two products can form when an unsymmetrical reagent such as HBr is added to an unsymmetrical C=C double bond. In practice, only one product is obtained. When HBr is added to 2-methylpropene, for example, the product is 2-bromo-2-methylpropane, not 1-bromo-2-methylpropane.

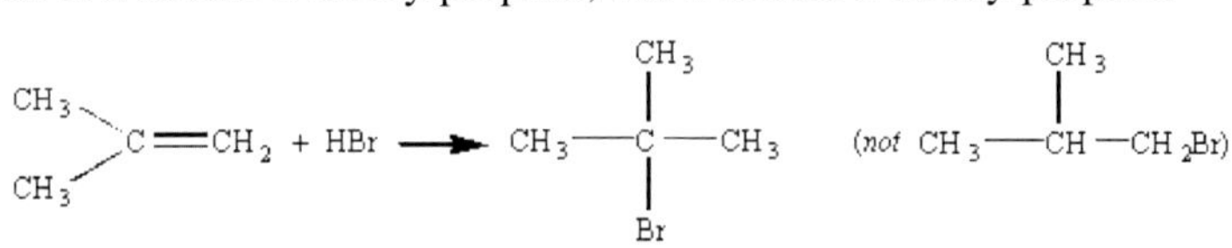

In 1870, after careful study of many examples of addition reactions, the Russian chemist Vladimir Markovnikov formulated a rule for predicting the product of these reactions. The Markovnikov rule states that the hydrogen atom adds to the carbon atom that already has the larger number of hydrogen atoms when HX adds to an alkene. Thus, water (H−OH) adds to propene to form the product in which the OH group is on the middle carbon atom.

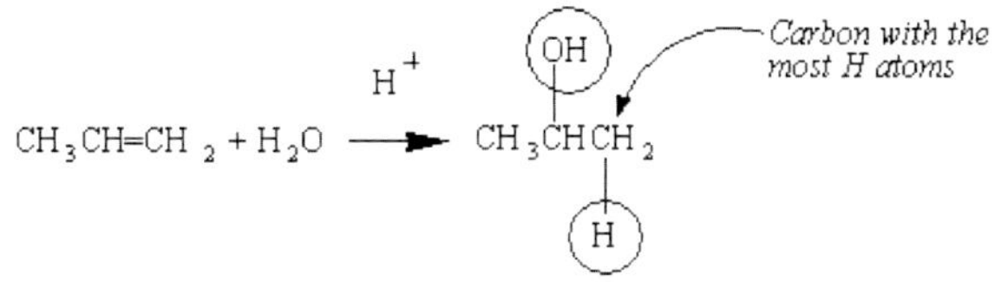

Alkyne derivatives (acetylene derivatives, C_nH_{2n-2}) are unsaturated hydrocarbon derivatives which contain one or more triple bonds between atoms of carbon. When they burn, they tend to form carbon soot. When oxygen is present during burning, high temperatures can be reached.

The simplest (lowest molecular weight) alkyne derivatives are acetylene (C_2H_2, HC≡CH), propyne (C_3H_4, $CH_3C{\equiv}CH$) and butyne (C_4H_6, $CH_3C{\equiv}CCH_3$ or $CH_3CH_2C{\equiv}CH$).

Cycloalkane derivatives (*cyclic alkane derivatives*) are differentiated from aliphatic hydrocarbon derivatives insofar as they contain a ring structure and form a homologous group of compounds. The first member of the series is cyclopentane followed by cyclohexane. Cycloalkane derivatives are saturated compounds and, like linear alkane derivatives, are not very reactive.

Aromatic hydrocarbon derivatives are derived from benzene. Group members have six free valence electrons which are distributed in a circle in the form of a charged cloud. Because of the presence of these valence electrons, we can predict that the reactivity of these aromatic compounds will be similar to other unsaturated hydrocarbon derivatives. However, benzene is much less reactive than other unsaturated hydrocarbon derivatives. Only at high temperatures and in the presence of a catalyst can benzene take on another hydrogen atom. When it does, cyclohexane is the resultant product.

5. Physical properties

Physical properties can be observed or measured without changing the composition of matter. Physical properties are used to observe and describe matter (Howard and Meylan, 1997; Yaws, 1999). The three states of matter are: solid, liquid, and gas. The melting point and boiling point are related to changes of the state of matter. All matter may exist in any of three physical states of matter. A physical change takes place without any changes in molecular composition. The same element or compound is present before and after the change. The same molecule is present throughout the changes. Physical changes are related to physical properties since some changes require a change in the three-dimensional structure of the molecule.

Physical properties that are of interest in the current context include: (i) boiling point, (ii) density and specific gravity, (iii) dew point, (iv) flash point and ignition temperature, (v) melting point, and (vi) vapor density. These properties are listed in alphabetical order rather than attempt to assign importance to any individual property. The properties present indications the behavior of hydrocarbon derives as determined by application of standard test method (Speight, 2015).

The physical properties of alkene derivatives are similar to those of the alkane derivatives. The boiling points of straight-chain alkenes increase with increasing molar mass, just as with alkanes. For molecules with the same number of carbon atoms and the same general shape, the boiling points usually differ only slightly, just as would be expected for chemicals in which whose molar mass differs by only one to two hydrogen atoms (i.e., $RCH_2CH{=}CH_2$ compared to $RCH_2CH_2CH_3$). Like other hydrocarbons, the alkene derivatives are insoluble in water but soluble in organic solvents.

5.1 Boiling point

The *boiling point* of an organic compound is the temperature at which the vapor pressure of the liquid equals the environmental pressure surrounding the liquid. Thus, the normal boiling point of a pure hydrocarbon is the temperature

TABLE 9.2 Selected properties of lower molecular weight alkane derivatives.

Name	Molecular formula	Structural formula	Boiling point (°C)	Melting point (°C)	Density (g/ml, 20°C)
Methane	CH_4	CH_4	−161.5	−182.5	
Ethane	C_2H_6	CH_3CH_3	−88.6	−183.3	
Propane	C_3H_8	$CH_3CH_2CH_3$	−42.1	−189.7	
Butane	C_4H_{10}	$CH_3(CH_2)_2CH_3$	−0.5	−138.4	
Pentane	C_5H_{12}	$CH_3(CH_2)_3CH_3$	36.1	−129.7	0.626
Hexane	C_6H_{14}	$CH_3(CH_2)_4CH_3$	68.7	−95.3	0.659
Heptane	C_7H_{16}	$CH_3(CH_2)_5CH_3$	98.4	−90.6	0.684
Octane	C_8H_{18}	$CH_3(CH_2)_6CH_3$	125.7	−56.8	0.703
Nonane	C_9H_{20}	$CH_3(CH_2)_7CH_3$	150.8	−53.5	0.718
Decane	$C_{10}H_{22}$	$CH_3(CH_2)_8CH_3$	174.1	−29.7	0.730

TABLE 9.3 Common alkyl groups[a].

Group	Carbon atoms	Formula
Methyl	1	CH_3^-
Ethyl	2	$CH_3CH_2^-$
Propyl	3	$CH_3CH_2CH_2^-$
Butyl	4	$CH_3CH_2CH_2CH_2^-$
Pentyl	5	$CH_3CH_2CH_2CH_2CH_2^-$
Hexyl	6	$CH_3CH_2CH_2CH_2CH_2CH_2^-$
Heptyl	7	$CH_3CH_2CH_2CH_2CH_2CH_2CH_2^-$
Octyl	8	$CH_3CH_2CH_2$ $CH_2CH_2CH_2CH_2CH_2^-$
Nonyl	9	$CH_3CH_2CH_2CHCH_2CH_2CH_2CH_2CH_2^-$
Decyl	10	$CH_3CH_2CH_2CH_2CH_2CH_2CH_2CH_2CH_2CH_2^-$

[a]*For convenience, only the n-isomers are shown.*

at which the substance changes state from liquid to vapor at a pressure of 1 atm (14.7 psi) (Table 9.2) from which the common alkyl groups are derived (Table 9.3). This property is known and tabulated for all the pure components normally considered in petroleum engineering except for the high boiling substances that do not have a boiling point at atmospheric pressure. For the distillation fractions of a hydrocarbon fuel, the normal boiling point is usually chosen to be a certain average value of the temperature interval of the distillation fraction.

The boiling points of organic compounds can give important clues to other physical properties and structural characteristics. A liquid boils when its vapor pressure is equal to the atmospheric pressure. Vapor pressure is determined by the kinetic energy of molecules. Kinetic energy is related to temperature and the mass and velocity of the molecules (K.E. $= 1/2\ mv^2$). When the temperature reaches the boiling point, the average kinetic energy of the liquid particles is sufficient to overcome the forces of attraction that hold molecules in the liquid state.

Vapor pressure is caused by equilibrium between molecules in the gaseous state and molecules in the liquid state. When molecules in the liquid state have sufficient kinetic energy, they may escape from the surface and turn into a gas. Molecules with the most independence in individual motions achieve sufficient kinetic energy (velocities) to escape as gases at lower temperatures. The vapor pressure will be higher (more gas molecules are present) and therefore the compound will boil at a lower temperature.

In each homologous series of hydrocarbon derivatives, the boiling points increase with molecular weight and structure also has a marked influence since it is a general rule that branched paraffin isomers have lower boiling points than the corresponding *n*-alkane. In any given series, steric effects notwithstanding, there is an increase in boiling point with an increase in carbon number of the alkyl side chain. This particularly applies to alkyl aromatic compounds where alkyl-substituted aromatic compounds can have higher boiling points than polycondensed aromatic systems.

The boiling points of hydrocarbon fuels are rarely, if ever, distinct temperatures; it is, in fact, more correct to refer to the boiling ranges of the various fuels. To determine these ranges, the crude oil is tested in various methods of distillation, either at atmospheric pressure or at reduced pressure. In general, the limiting molecular weight range for distillation at atmospheric pressure without thermal degradation is 200–250, whereas the limiting molecular weight range for conventional vacuum distillation is 500–600.

Alkane derivatives experience intermolecular van der Waals forces. Stronger intermolecular van der Waals forces give rise to greater boiling points of alkane derivatives. There are two determinants for the strength of the van der Waals forces: (i) the number of electrons surrounding the molecule, which increases with the molecular weight of the alkane, and (ii) the surface area of the molecule.

Under standard conditions (STP), alkane derivatives from methane (CH_4) to butane (C_4H_{10}) are gaseous; from pentane (C_5H_{12}) to $C_{17}H_{36}$ they are liquids; and after $C_{18}H_{38}$ they are solids. As the boiling point of alkane derivatives is primarily determined by weight, it should not be a surprise that the boiling point has almost a linear relationship with the size (molecular weight) of the molecule. As a general rule, the boiling point rises 20–30°C for each carbon added to the chain; this rule applies to other homologous series.

A straight-chain alkane will have a boiling point higher than a branched-chain alkane due to the greater surface area in contact, which results in greater van der Waals forces between adjacent molecules. For example, compare iso-butane (2-methylpropane) and n-butane (butane), which boil at −12 and 0°C, respectively, and 2,2-dimethylbutane and 2,3-dimethylbutane which boil at 50°C (122°F) and 58°C (136°F), respectively. For the latter case, two molecules of 2,3-dimethylbutane can associate with each other better than the cross-shaped 2,2-dimethylbutane, hence the greater van der Waals forces. On the other hand, cycloalkane derivatives tend to have higher boiling points than their linear counterparts due to the locked conformations of the molecules, which give a plane of intermolecular contact.

The boiling points of alkene derivatives, like the boiling points of the alkane derivatives, increase with molecular weight (Table 9.4). Branched-chain alkene derivatives have lower boiling points than the corresponding straight-chain alkene derivatives. However, the boiling point of each alkene is

TABLE 9.4 Boiling points of alkene derivatives.

Alkene	Boiling point (°C)
Ethylene	−104
Propylene	−47
Trans-2-butene	0.9
Cis-2-butene	3.7
1-Pentene	30
Trans-2-pentene	36
Cis-2-pentene	37
1-Heptene	115
3-Octene	122
3-Nonene	147
5-Decene	170

very similar to that of the alkane with the same number of carbon atoms (Table 9.2, Table 9.4). Ethylene, propylene, and the various butenes are gases at room temperature. The higher boiling alkene derivatives are liquids. In each case, the alkene has a boiling point which is slightly lower than the boiling point of the corresponding alkane. The only attractions involved are van der Waals dispersion forces, and these depend on the shape of the molecule and the number of electrons it contains. Each alkene has two fewer electrons than the alkane with the same number of carbons.

The melting points of alkene derivatives also increase with molecular weight (Table 9.5). Generally, alkene derivatives have similar melting points to that of corresponding alkane derivatives. However, melting points of alkene derivatives depend on the packaging of the molecules—*cis* isomers are packaged in a U-bending shape and, therefore, display lower melting points compared to their respective *trans* isomers. In keeping with the general trend of alkane derivatives and alkene derivatives (and to no surprise), the boiling points and melting points of *alkyne derivatives* increase as the number of carbon atoms (i.e., molecular weight) increases (Table 9.6).

However, alkyne derivatives have higher boiling points than alkane derivatives or alkene derivatives, with its increased number of weakly held Π electrons because of the triple bond. Because of these weakly held electrons, the electric field is more easily distorted, producing stronger attractive forces between molecules. This holds the molecules together at higher temps preventing vaporization.

TABLE 9.5 Melting points of alkene derivatives.

Alkene	Melting point (°C)
Ethene	−169
Propene	−185
Butene	−138
1-Pentene	−165
Trans-2-pentene	−135
Cis-2-pentene	−180
1-Heptene	−119
3-Octene	−101.9
3-Nonene	−81.4
5-Decene	−66.3

TABLE 9.6 Physical properties of alkyne derivatives.

Name	Formula	Melting point (°C)	Boiling point (°C)	Density (20°C)
Acetylene	HC≡CH	−82	−75	
Propyne	HC≡CCH3	−101.5	−23	
1-Butyne	$HC≡CCH_2CH_3$	−122	91	
2-Butyne	$CH_3C≡CCH_3$	−24	27	0.694
1-Pentyne	$HC≡C(CH_2)2CH_3$	−98	40	0.695
2-Pentyne	$CH3C≡CCH_2CH_3$	−101	55	0.714
1-Hexyne	$HC≡C(CH_2)3CH_3$	−124	72	0.719
1-Heptyne	$HC≡C(CH_2)4CH_3$	−80	100	0.733
1-Octyne	$HC≡C(CH_2)5CH_3$	−70	126	0.747
1-Nonyne	$HC≡C(CH_2)6CH_3$	−65	151	0.763
1-Decyne	$HC≡C(CH_2)7CH_3$	−35	182	0.770

Cycloalkane derivatives are similar to alkane derivatives in their general physical properties, but they have higher boiling points, melting points, and densities than alkane derivatives. This is due to stronger London forces because the ring shape allows for a larger area of contact. Containing only carbon-carbon (C–C) and carbon-hydrogen (C–H) bonds, unreactivity of cycloalkane derivatives with little or no ring strain (see below) are comparable to noncyclic alkane derivatives.

The London dispersion force is the weakest intermolecular force. The London dispersion force is a temporary attractive force that results when the electrons in two adjacent atoms occupy positions that make the atoms form temporary dipoles. This force is sometimes called an induced dipole-induced dipole attraction. London forces are the attractive forces that cause nonpolar substances to condense to liquids and freeze into solids when the temperature is lowered sufficiently.

Because of the constant motion of the electrons, an atom or molecule can develop a temporary (instantaneous) dipole when its electrons are distributed unsymmetrically about the nucleus. A second atom or molecule, in turn, can be distorted by the appearance of the dipole in the first atom or molecule (because electrons repel one another) which leads to an electrostatic attraction between the two atoms or the two molecules.

5.2 Cloud point and pour point

The relationship of *cloud point*, *pour point*, *melting point*, and *freezing point* to one another varies widely from one hydrocarbon fuel to another, depending upon the composition of the fuel in terms of the hydrocarbon derivative present in the fuel. Hence, the significance of the data also varies for the different types of product. In general, cloud, melting, and freezing points are of more limited value and each is of narrower range of application than the pour point.

The *cloud point* of hydrocarbon fuel is the temperature at which paraffin wax or other solidifiable compounds present in the oil appear as a haze when the sample is chilled under definitely prescribed conditions (ASTM D2500; ASTM D6751). To determine the *cloud point* and the *pour point* (ASTM D97, ASTM D512, ASTM D1835, ASTM D524, ASTM D5501, ASTM D975) the oil is contained in a glass test tube fitted with a thermometer and immersed in one of three baths containing coolants. The sample is dehydrated and filtered at a temperature higher than the above anticipated cloud point. It is then placed in a test tube and cooled progressively in coolants held at -1 to $+2$°C (30–35°F), -18 to -20°C (-4 to 0°F) and -32 to -35°C (-26 to -31°F), respectively. The sample is inspected for cloudiness at temperature intervals of 1°C (2°F). If conditions or oil properties are such that reduced temperatures are required to determine the pour point, alternate tests are available that accommodate the various types of samples. Related to the *cloud point*, the wax appearance temperature or wax appearance point is also determined (ASTM D6751).

The *pour point* of crude oil hydrocarbon derivatives or a crude oil product is determined using this same technique (ASTM D97), and it is the lowest temperature at which the oil pours or flows. It is actually 2°C (3°F) above the temperature at which the oil ceases to flow under these definitely prescribed conditions when it is chilled without disturbance at a standard rate. To determine the pour point, the sample is first heated to 46°C (115°F) and cooled in air to 32°C (90°F) before the tube is immersed in the same series of coolants as used for the determination of the *cloud point*. The sample is inspected at temperature intervals of 2°C (3°F) by withdrawal and holding horizontal for 5 s until no flow is observed during this time interval.

Cloud and pour points are useful for predicting the temperature at which the observed viscosity of oil deviates from the true (Newtonian) viscosity in the low-temperature range. They are also useful for identification of oils or when planning the storage of oil supplies, as low temperatures may cause handling difficulties with some oils.

The pour point of a crude oil was originally applied to crude oil that had high wax content. More recently, the pour point, like the viscosity, is determined principally for use in pumping arid pipeline design calculations. Difficulty occurs in these determinations with waxy crude oils that begin to exhibit irregular flow behavior when wax begins to separate. These crude oils possess viscosity relationships that are difficult to predict in pipeline operation. In addition, some waxy crude oils are sensitive to heat treatment that can also affect their viscosity characteristics. This complex behavior limits the value of viscosity and pour point tests on waxy crude oils. At the present time, multimile crude oil pipelines and the increasing production of waxy crude oils make it necessary to make an assessment of the pumpability of a wax-containing crude oil through a given system because it may be a matter of some difficulty that can often only be resolved after field trials.

Alkene derivatives contain a carbon-carbon double bond, which affects the physical properties of alkene derivatives relative to the physical properties of alkane derivatives. At room temperature, alkene derivatives exist in all three phases, solid, liquids, and gases. Melting and boiling points of alkene derivatives are similar to that of alkane derivatives but isomers of the *cis*-alkene derivatives have lower melting points than that of *trans*-isomers. Alkene derivatives display a weak dipole-dipole interaction due to the electron-attracting sp^2 carbon. The physical properties of alkene derivatives are comparable with those of alkane derivatives but the physical state does depend on the molecular weight. The lower molecular weight alkene derivatives (ethylene, propylene, and butylene) are gases while linear alkene derivatives of approximately 5 to 16 carbons are liquids, and higher alkene derivatives are waxy solids.

Cis-isomers and *trans*-isomers often have different physical properties. Differences between isomers, in general, arise from the differences in the shape of the molecule or the overall dipole moment of the molecule. This difference can be small as in the case of the boiling point of straight-chain alkene derivatives, such as 2-pentene which is 37°C (98°F), in the *cis*-isomer and 36°C (96°F) in the *trans*-isomer.

5.3 Density and specific gravity

Density is the mass of a unit volume of material at a specified temperature and has the dimensions of grams per cubic centimeter (a close approximation to grams per milliliter). *Specific gravity* (also called the *relative density*) is the ratio of the mass of a volume of the substance to the mass of the same volume of water and is dependent on two temperatures, those at which the masses of the sample and the water are measured. When the water temperature is 4°C (39°F), the specific gravity is equal to the density in the centimeter-gram-second (cgs) system, since the volume of 1 g of water at that temperature is, by definition, 1 mL. Thus the density of water, for example, varies with temperature, and its specific gravity at equal temperatures is always unity. The standard temperatures for a specific gravity in the crude oil industry in North America are 60/60°F (15.6/15.6°C).

Specific gravity usually meant to imply the relative density of the substance with respect to the density of water. If the specific gravity of a hydrocarbon is less than one then it is less dense than water (Table 9.2), or the reference chemical. Conversely, if the specific density if greater than one, it is denser than the reference. If the relative density is exactly one then the densities are equal; that is, equal volumes of the two substances have the same mass. If the reference material is water then a substance with a relative density (or specific gravity) less than one will float in water.

Temperature and pressure must be specified for both the sample and the reference pressure is typically 14.7 psi (1 atm) but, for correctness, the pressure should be stated. Where it is not it is more usual to specify the density directly. Temperatures for both sample and reference vary from industry to industry. In British brewing practice the specific gravity as specified above is multiplied by 1000. Specific gravity is commonly used in industry as a simple means of obtaining information about the concentration of solutions of various materials such as brine, sugar solutions, and acids. The *density* and *specific gravity* of crude oil and hydrocarbon fuels (ASTM D70; ASTM D71; ASTM D287; ASTM D1217; ASTM D1298; ASTM D1480; ASTM D1481; ASTM D1555; ASTM D1657; ASTM D4052) are two properties that have found wide use in the industry for preliminary assessment of the character and quality of crude oil.

In the early years of the crude oil industry, and hydrocarbon properties, density was the principal specification for crude oil and refinery products; it was used to give an estimation of the gasoline and, more particularly, the kerosene present in the crude oil. However, the derived relationships between the density of crude oil and its fractional composition were valid only if they were applied to a certain type of crude oil and lost some of their significance when applied to different types of crude oil. Nevertheless, density is still used to give a rough estimation of the nature of crude oil and crude oil products. Although density and specific gravity are used extensively, the API (American Petroleum Institute) gravity is the preferred property. This property was derived from the Baumé scale:

$$\text{Degrees Baumé} = (140/\text{sp gr @ } 60/60^{\circ}\text{F}) - 130$$

However, a considerable number of hydrometers calibrated according to the Baum scale were found at an early period to be in error by a consistent amount, and this led to the adoption of the equation:

$$\text{Degrees API} = (141.5/\text{sp gr @ } 60/60^{\circ}\text{F}) - 131.5$$

The specific gravity of crude oil usually ranges from approximately 0.8 (45.3 API) for the lighter crude oils to over 1.0 (less than 10 API) for heavy crude oil and bitumen.

Specific gravity is influenced by the chemical composition of crude oil, but quantitative correlation is difficult to establish. Nevertheless, it is generally recognized that increased amounts of aromatic compounds result in an increase in density, whereas an increase in saturated compounds results in a decrease in density. Indeed, it is also possible to recognize certain preferred trends between the density of crude oil and one or other of the physical properties. For example, an approximate correlation exists between the density (API gravity) and sulfur content, Conradson carbon residue, viscosity, and nitrogen content (Speight, 2017).

Density, specific gravity, or API gravity may be measured by means of a hydrometer (ASTM D287; ASTM D1298; ASTM D1657, IP 160), a pycnometer (ASTM D70, ASTM D1217, ASTM D1480, and ASTM D1481), or by means of a digital density meter (ASTM D4052) and a digital density analyzer (ASTM D5002). The pycnometer method (ASTM D70; ASTM D1217; ASTM D1480; ASTM D1481) for determining density is reliable, precise, and requires relatively small test samples. However, because of the time required, other methods such as using the hydrometer (ASTM D1298, ASTM D1298), the density meter (ASTM D4052), and the digital density analyzer (ASTM D5002) are often preferred. However, surface tension effects can affect the displacement method and the density meter method loses some of its advantage when measuring the density of heavy oil and bitumen.

The pycnometer method (ASTM D70, ASTM D1217, ASTM D1480, and ASTM D1481) is routinely used to measure the density of samples being charged to a distillation flask, where volume charge is needed, but the volume is not conveniently measured. The volume may be found weighing the sample, and determining the sample density. It is also used in routine measurements of material properties. It is worthy of note that even a small amount of solids in the sample will influence its measured density. For example, 1% by weight solids in the sample can raise the density by 0.007 g/cm^3.

The densimeter method (ASTM D4052) uses an instrument that measures the total mass of a tube by determining its natural frequency of vibration. This frequency is a function of the dimensions and the elastic properties of the tube, and the weight of the tube and contents. Calibration with water and air provides data for the determination of the instrument constraints which allow conversion of the natural frequency of vibration to sample density.

The variation of density with temperature, effectively the coefficient of expansion, is a property of great technical importance, since most crude oil products are sold by volume and specific gravity is usually determined at the prevailing temperature (21°C, 70°F) rather than at the standard temperature (15.6°C, 60°F). The tables of gravity corrections (ASTM D1555) are based on an assumption that the coefficient of expansion of all crude oil products is a function (at fixed temperatures) of density only. However, not all of these methods are suitable for measuring the density (or specific gravity) of heavy oil and bitumen although some methods lend themselves to adaptation.

The API gravity of a feedstock (ASTM D287) is calculated directly from the specific gravity. The specific gravity of bitumen shows a fairly wide range of variation. The largest degree of variation is usually due to local conditions that affect material close to the faces, or exposures, occurring in surface oil sand beds. There are also variations in the specific gravity of the bitumen found in beds that have not been exposed to weathering or other external factors. The range of specific gravity usually varies over the range of the order of 0.995−1.04.

A very important property of the Athabasca bitumen (which also accounts for the success of the hot water separation process) is the variation in density (specific gravity) of the bitumen with temperature. Over the temperature range 30−130°C (85−265°F) the bitumen is lighter than water. Flotation of the bitumen (with aeration) on the water is facilitated, hence the logic of the hot water separation process (Speight, 2014).

The density of the *alkane derivatives* usually increases with increasing number of carbon atoms, but remains less than that of water (Table 9.6). Hence, alkanes form the upper layer in an alkane-water mixture. The density of the *alkene derivatives* is higher than the density of the corresponding alkane derivatives. Again, density usually increases with increasing number of carbon atoms, but all alkene derivatives have the density smaller than 1. The density of the cycloalkane derivatives is higher than the density of the corresponding alkane derivatives (Table 9.7).

5.4 Dew point

The dew point is the temperature to which air must be cooled to become saturated with water vapor. When further cooled, the airborne water vapor will condense to form liquid water (dew). The measurement of the dew point is related to humidity—for example, a higher dew point means there will be

TABLE 9.7 Physical properties of cycloalkane derivatives compared to the properties of the corresponding (matching number of carbon atoms) alkane derivatives.

Hydrocarbon	Melting point, °C	Boiling point, °C	Density, @20°C
Cyclopropane	−127	−33	
Propane	−167	−42	
Cyclobutane	−80	13	
Butane	−138	0	
Cyclopentane	−94	49	0.746
Pentane	−130	36	0.626
Cyclohexane	6.5	81	0.778
Hexane	−95	69	0.659
Cycloheptane	−12	118	0.810
Heptane	−90	−98	0.684
Cyclooctane	14	149	0.830
Octane	−57	126	0.703

more moisture in the air. In addtion, when air cools to its dew point through contact with a surface that is colder than the air, water will condense on the surface. When the temperature is below the freezing point of water, the dew point is called the frost point because frost is formed rather than dew.

Briefly, the frost point is similar to the dew point in that it is the temperature to which a given amount of humid air must be cooled, at constant atmospheric pressure (14.7 psi) for water vapor to be deposited on a surface as ice crystals without passing through an intermediate liquid phase. The frost point for a given amount of air is always higher than the dew point, as the stronger bonding between water molecules on the surface of ice requires higher temperature to break. On the other hand, as a reversal of the frost point, sublimation is the transition of a substance directly from the solid phase to the gas phase without passing through the intermediate liquid phase. Sublimation is an endothermic process that occurs at temperatures and pressures below the triple point of a substance in the phase diagram which corresponds to the lowest pressure at which the substance can exist as a liquid. Sublimation has also been used as a generic term to describe a solid-to-gas transition (sublimation) followed by a gas-to-solid transition (deposition). While a transition from liquid to gas is described as evaporation if it occurs below the boiling point of the liquid, and as boiling if it occurs at the boiling point, there is no such distinction within the solid-to-gas transition, which is always described as sublimation.

The hydrocarbon dew point (HDP) is a function of the composition of the gas mixture and is strongly influenced by the concentration of the higher molecular weight hydrocarbon derivatives, especially C_{6+}. The presence of higher molecular weight hydrocarbon derivatives will increase the HDP and failure to include them in a hydrocarbon dew point calculation will under-predict the hydrocarbon dew point.

For most pipeline conditions, the HDP temperature at a given pressure increases as the concentration of heavier hydrocarbon derivatives increases. Thus, the potential to form liquids at certain pipeline conditions exists for gases rich in C_{6+}. Processing of the gas stream primarily removes or extracts higher molecular weight hydrocarbon derivatives and thus reduces the HDP of a given mixture. The level of hydrocarbon removal directly impacts the HDP.

The HDP curve is plotted as a function of gas pressure (P) and temperature (T) (Fig. 9.2). The left-hand side of the curve is the bubble point line and divides the single-phase liquid region from the two-phase gas-liquid region. The right-hand side of the curve is the dew point line and divides the two-phase gas-liquid region and the single-phase gas region. The bubble point and dew point lines intersect at the critical point, where the distinction between gas and liquid properties disappears. Note that two dew point temperatures are possible at a given pressure (P3) and two dew point pressures are possible at a given temperature (T3). This phase envelope phenomenon provides for behavior known as retrograde condensation. The retrograde phenomenon occurs when liquids form at a given temperature when the pressure is lowered (see red arrow).

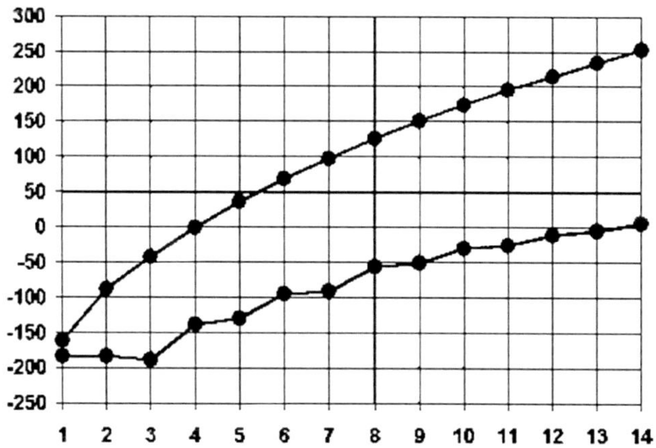

FIGURE 9.1 Melting points (*lower line*) and boiling points (*upper line*) of the C_1–C_{14} alkane derivatives.

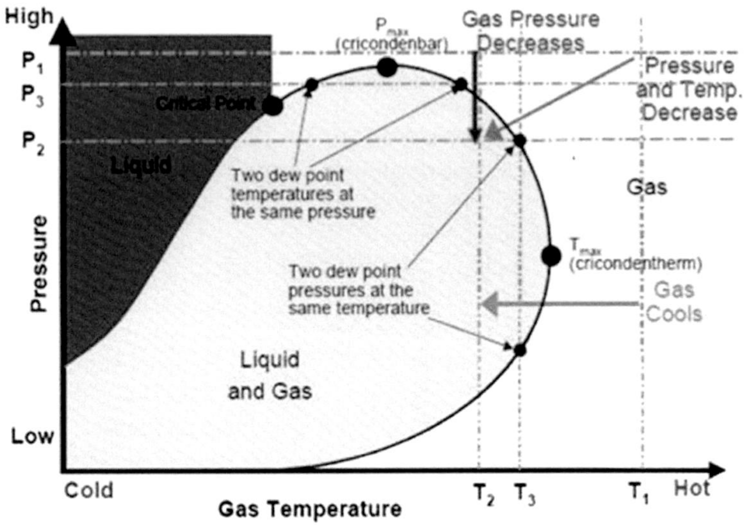

FIGURE 9.2 Hydrocarbon dew point curve for natural gas.

The word *retrograde* indicates a phenomenon that is contradictory to the phase behavior of pure components, which condense with increasing pressure and or decreasing temperature. The maximum pressure at which phase change occurs (P_{max}) is called the *cricondenbar*, and the maximum temperature (T_{max}) at which phase change l occurs from is called the *cricondentherm*.

For unprocessed and processed gas mixtures there are variations in the HDP (Fig. 9.3). The unprocessed HDP curve has a higher cricondentherm temperature than the processed HDP curve and illustrates the impact of

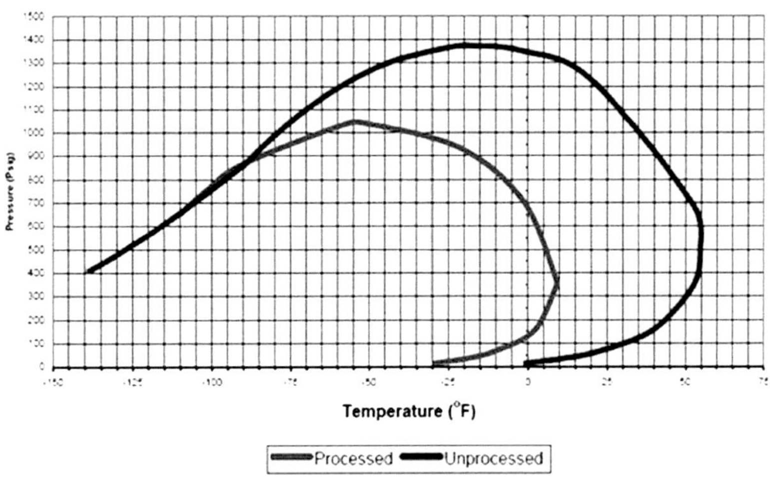

FIGURE 9.3 Contrast between unprocessed (black line) and processed (*gray line*) natural gas.

processing on the HDP. The significance of the HDP curve for gas transmission and distribution operations lies in the potential transition from the single-phase gas region to the two-phase gas-liquid region.

5.5 Flash point and ignition temperature

The flash point of a volatile liquid is the lowest temperature at which it can vaporize to form an ignitable mixture in air (Table 9.8). At the flash point, the vapor may cease to burn when the source of ignition is removed. The flash point is often used as a descriptive characteristic of liquid fuel, and it is also used to describe liquids that are not normally used as fuels but are flammable liquids and/or combustible liquids. There are various international standards for defining each, but most agree that liquids with a flash point less than 43°C (109°F) are flammable, while those having a flash point above this temperature are combustible.

The *fire point* is a slightly higher temperature and is the temperature at which the vapor continues to burn after being ignited. Neither the flash point

TABLE 9.8 Flash points, autoignition temperature, and flammability limits for various hydrocarbons.

			Flammable limits	
	Flash point	Autoignition temperature	Upper	Lower
Hydrocarbon	°C	°C	(vol % at 25°C)	
Methane	−188	630	5.0	15.0
Ethane	−135	515	3.0	12.4
Propane	−104	450	2.1	9.5
n-Butane	−74	370	1.8	8.4
n-Pentane	−49	260	1.4	7.8
n-Hexane	−23	225	1.2	7.4
n-Heptane	−3	225	1.1	6.7
n-Octane	14	220	0.95	6.5
n-Nonane	31	205	0.85	-
n-Decane	46	210	0.75	5.6
n-Dodecane	74	204	0.60	-
n-Tetradecane	99	200	0.50	-

nor the fire point is related to the temperature of the ignition source or of the burning liquid, which are much higher. The flash point is not to be confused with the autoignition temperature, which does not require an ignition source.

The *ignition temperature* is the minimum temperature to which a substance must be heated before it will spontaneously burn independently of the source of heat. The autoignition temperature (*kindling point*) of a substance (Table 9.8) is the lowest temperature at which it will spontaneously ignite in a normal atmosphere without an external source of ignition, such as a flame or spark. This temperature is required to supply the activation energy needed for combustion. The temperature at which a chemical will ignite decreases as the pressure increases or as oxygen concentration increases. It is usually applied to a combustible fuel mixture. Autoignition temperatures of liquid chemicals are typically measured using a 500 mL flask placed in a temperature controlled oven in accordance with a standard test procedure (ASTM E659).

The flash point of a volatile liquid is the lowest temperature at which the liquid can vaporize to form an ignitable mixture in air. Measuring the flash point of a liquid requires an ignition source. At the flash point, the vapor may cease to burn when the source of ignition is removed. The flash point is not to be confused with the autoignition temperature, which does not require an ignition source. The fire point, a slightly higher temperature than the flash point, is the temperature at which the vapor continues to burn after being ignited. Neither the flash point nor the fire point is related to the temperature of the ignition source or of the burning liquid, which are much higher.

The flash point is often used as a descriptive characteristic of hydrocarbon derivatives (Table 9.8) and hydrocarbon fuels (Table 9.9) and it is also used to describe other liquids, including those that are not normally used as fuels. Flash point refers to both flammable liquids and combustible liquids. There are various international standards for defining each, but most agree that liquids with a flash point less than 43°C (109°F) are flammable, while those having a flash point above this temperature are combustible.

There are two basic types of flash point measurement: open cup and closed cup. The best known example of the open cup method is the Cleveland Open

TABLE 9.9 Examples of the flash points of hydrocarbon fuels.

Fuel	Flash point	Autoignition temperature
Gasoline	<−40°C (−40°F)	246°C (475°F)
Diesel fuel	>62°C (143°F)	210°C (410°F)
Jet fuel	>60°C (140°F)	210°C (410°F)
Kerosene	>38°C −72°C (100°F −162°F)	220°C (428°F)

Cup (COC). In the open-cup method, the sample is contained in an open cup which is heated, and at intervals a flame is brought over the surface. The measured flash point will actually vary with the height of the flame above the liquid surface, and at sufficient height the measured flash point temperature will coincide with the fire point.

There are two types of closed cup testers: nonequilibrium, such as Pensky-Martens where the vapors above the liquid are not in temperature equilibrium with the liquid, and equilibrium where the vapors are deemed to be in temperature equilibrium with the liquid. In both these types the cups are sealed with a lid through which the ignition source can be introduced. Closed cup testers normally give lower values for the flash point than open cup (typically 5–10°C) and are a better approximation to the temperature at which the vapor pressure reaches the lower flammability limit (LFL).

The flash point is an empirical measurement rather than a fundamental physical parameter. The measured value will vary with equipment and test protocol variations, including temperature ramp rate (in automated testers), time allowed for the sample to equilibrate, sample volume, and whether the sample is stirred. Methods for determining the flash point of a liquid are specified in many standards. For example, testing by the Pensky-Martens closed cup method is detailed in ASTM D93. Determination of flash point by an alternate closed cup method is detailed in ASTM D3828, ASTM D3278.

Gasoline is designed for use in an engine which is driven by a spark and the fuel should be premixed with air within its flammable limits and heated above its flash point, then ignited by the spark plug. The fuel should not preignite in the hot engine. Therefore, gasoline is required to have a low flash point and a high autoignition temperature (Table 9.8).

Diesel fuel flash points vary between 52 and 96°C (126–204°F). Diesel is designed for use in a high compression engine in which air is compressed until it has been heated above the autoignition temperature of the fuel. The diesel fuel is then injected as a high-pressure spray, keeping the fuel-air mix within the flammable limits of diesel. There is no ignition source and, therefore, diesel is required to have a high flash point and a low autoignition temperature (Table 9.9).

Gasoline is designed for use in a spark ignition engine and the hydrocarbon fuel is premixed with air within its flammable limits and heated above its flash point, then ignited by the spark plug. The hydrocarbon fuel should not preignite in the hot engine and, therefore, gasoline is required to have a low flash point and a high autoignition temperature.

The flash point of diesel fuel varies between 52 and 96°C (126–204°F). Diesel is designed for use in a high compression engine in which air is compressed until it has been heated above the autoignition temperature of the diesel fuel. Then the fuel is injected as a high-pressure spray, keeping the fuel-air mix within the flammable limits of diesel. There is no ignition source and, therefore, diesel is required to have a high flash point and a low autoignition temperature.

The flash point of jet fuel also varies considerably. Both Jet A and Jet A-1 have flash points between 38 and 66°C (100–150°F).

5.6 Melting point

The *melting point* of a solid is the temperature at which the vapor pressure of the solid and the liquid are equal. At the melting point, the solid and liquid phases exist in equilibrium. The melting point (sometimes referred to as the liquefaction point) is the point where a chemical changes from the solid state to the liquid state and at the melting point the solid and liquid phases exist in equilibrium. The melting point of a substance depends on the pressure and is usually specified at a standard pressure such as 1 atm (14.7 psi). When the change of phase is reversed (that is from the liquid phase to the solid phase), the temperature at which this phase change occurs is referred to as the freezing point or the crystallization point. Because of the ability of some substances to supercool, the freezing point is not always considered to be a characteristic property of a substance.

The melting points of the alkane derivatives follow a similar trend to boiling points of alkane derivatives (Table 9.2, Fig. 9.1) for the same reason as outlined above. That is (all other things being equal), the larger the molecule the higher the melting point. There is one significant difference between boiling points and melting points. Solids have more rigid and fixed structure than liquids. This rigid structure requires energy to break down. Thus the stronger, better put together solid structures will require more energy to break apart. For alkane derivatives, the odd-numbered alkane derivatives have a lower trend in melting points than even-numbered alkane derivatives (Fig. 9.1). Even-numbered alkane derivatives pack well in the solid phase, forming a well-organized structure, which requires more energy to break apart. The odd-number alkane derivatives pack less well and so the looser organized solid packing structure requires less energy to break apart.

The melting points of branched-chain alkane derivatives can be either higher or lower than those of the corresponding straight-chain alkane derivatives, again depending on the ability of the alkane in question to packing well in the solid phase. This is particularly true for iso-alkane derivatives (2-methyl isomers), which often have melting points higher than those of the linear analogs.

Hydrocarbon derivatives that are liquids at ambient temperature, and problems that may arise from solidification during normal use are not common. Nevertheless, the *melting point* is a test (ASTM D87 and ASTM D127) that is widely used by suppliers of wax and wax consumers; it is particularly applied to the highly paraffinic or crystalline waxes. Quantitative prediction of the melting point of pure hydrocarbon derivatives is difficult, but the melting point tends to increase qualitatively with the molecular weight and symmetry of the molecule.

Unsubstituted and symmetrically substituted compounds (e.g., benzene, cyclohexane, *p*-xylene, and naphthalene) melt at higher temperatures relative to the paraffin compounds of similar molecular weight: the unsymmetrical isomers generally melt at lower temperatures than the aliphatic hydrocarbon derivatives of the same molecular weight.

Unsaturation affects the melting point principally by its alteration of symmetry; thus ethane (−172°C, −278°F) and ethylene (−169.5°C, −273°F) differ only slightly, but the melting points of cyclohexane (6.2°C, 21°F) and cyclohexene (−104°C, −155°F) contrast strongly. All types of highly unsymmetrical hydrocarbon derivatives are difficult to crystallize; asymmetrically branched aliphatic hydrocarbon derivatives as low as octane and most substituted cyclic hydrocarbon derivatives comprise the greater part of the lubricating fractions of crude oil, crystallize slowly, if at all, and on cooling merely take the form of glasslike solids.

Although the melting points of crude oil and crude oil products are of limited usefulness, except to estimate the purity or perhaps the composition of waxes, the reverse process, *solidification*, has received attention in crude oil chemistry. In fact, solidification of crude oil and crude oil products has been differentiated into four categories, namely, *freezing point*, *congealing point*, *cloud point*, and *pour point*.

Crude oil becomes more or less a plastic solid when cooled to sufficiently low temperatures. This is due to the congealing of the various hydrocarbon derivatives that constitute the oil. The cloud point of a crude oil or a crude oil product is the temperature at which paraffin wax or other solidifiable compounds present in the oil appear as a haze when the oil is chilled under definitely prescribed conditions (ASTM D2500; ASTM D6751). As cooling is continued, all crude oils and crude oil products become more and more viscous and flow becomes slower and slower. The pour point of crude oil or crude oil product is the lowest temperature at which the oil pours or flows under definitely prescribed conditions when it is chilled without disturbance at a standard rate (ASTM D97).

The solidification characteristics of a hydrocarbon or hydrocarbon fuel depend on its grade or kind. For grease, the temperature of interest is that at which fluidity occurs, commonly known as the *dropping point*. The dropping point of grease is the temperature at which the grease passes from a plastic solid to a liquid state and begins to flow under the conditions of the test (ASTM D56 and ASTM D2650). For another type of plastic solid, including petrolatum and microcrystalline wax, both *melting point* and *congealing point* are of interest.

The *melting point* of wax is the temperature at which the wax becomes sufficiently fluid to drop from the thermometer; the *congealing point* is the temperature at which melted petrolatum ceases to flow when allowed to cool under definitely prescribed conditions (ASTM D93). For paraffin wax, the *solidification temperature* is of interest. For such purposes, the *melting point* is the temperature at which the melted paraffin wax begins to solidify, as shown by the minimum rate of temperature change, when cooled under prescribed conditions. For pure or essentially pure hydrocarbon derivatives, the solidification temperature is the freezing point, the temperature at which a hydrocarbon passes from a liquid to a solid state.

5.7 Vapor density

Vapor density is the density of a vapor in relation to that of air—hydrogen may also be used as the standard of comparison. In the case of air (which is commonly used in relation to hydrocarbon derivatives and hydrocarbon fuels), the vapor density is the mass of a specified volume of the substance divided by mass of same volume of air and air is given an arbitrary vapor density of one. With this definition, the vapor density would indicate whether a gas is denser (greater than one) or less dense (less than one) than air.

The vapor density has implications for container storage and personnel safety—if a container can release a dense gas, its vapor could sink and, if flammable, collect until it is at a concentration sufficient for ignition. Even if not flammable, it could collect in the lower floor or level of a confined space and displace air, possibly presenting a smothering hazard to individuals entering the lower part of that space.

5.8 Use of the data

Thus, this chapter has been devoted to presentation of the various methods of hydrocarbon analysis and there is no doubt that the analysis of petroleum products is used to ensure that the products match the specifications that are required for their sale and use. In the modern era, as a result of the expansion of computer simulators and advanced analytical tools accompanied by the availability of more accurate analytical data, the fields of petroleum analysis and petroleum product analysis have expanded even further.

The data derived from any one, or more, of the evaluation techniques described in this chapter give an indication of the characteristics of the hydrocarbon as well as options for use or other form of processing as well as for the prediction of product properties. Other properties may also be required for more detailed evaluation of the hydrocarbon and for comparison between hydrocarbons.

However, proceeding from the raw evaluation data to use may not always be the preferred step. Further evaluation of the hydrocarbon is usually necessary in order to develop accurate and realistic relationships using the data obtained from the test methods. After that, there may be other test methods that can also play an important role to assist in the various *tweaks* that are needed to produce a saleable product with the necessary properties. It is, of course, a matter of choosing the relevant and meaningful properties to meet the nature of the task.

However, as analytical procedures evolved and data become available, it became obvious that each piece of data not only reflected the test method applied to the sample and the exact steps followed in the analysis of a sample depend, among other factors, on the interpretation by the analyst of the written test method and the unique characteristics of the test sample—not all samples of petroleum or of a specific petroleum product are the same or even equivalent. Thus, the procedure and the resulting data will (i) reflect the background of the investigator, (ii) reflect the ability of the investigator to interpret the results in relation to the application, (iii) be dependent upon the analytical facilities available, and (iv) be shown the required levels of accuracy and precision. As a result, it has been necessary to support the objectives of standard organizations (such as ASTM International), which provide a forum for the exchange of information related to test methods, discussion of any issues, and cooperative development of modified and/or standard test methods.

Indeed, whatever the source, hydrocarbon fuels are complex mixtures of hydrocarbon compounds, ranging from low molecular weight, low-boiling volatile organic compounds to high molecular weight, high-boiling constituents (even nonvolatile under achievable temperatures). Furthermore, the composition of petroleum products varies depending upon (i) the source of the crude oil—crude oil is derived from a variety of source materials which vary greatly in chemical composition, and (ii) the refining processes used to produce the hydrocarbon product.

When the data from the various analytical tests and investigations have been collected, the obvious question that comes to mind relates to use of the data. For the most part, the data are used to determine whether or not a crude oil may produce a certain hydrocarbon product and whether or not the product can meet the specifications for use. However since the early days of analysis, there has been a growing tendency to use the analytical data as a means of more detailed and accurate projections of (i) the refinery process by which the hydrocarbon is produced, (ii) the yield of the hydrocarbon product, (iii) properties of the hydrocarbon product, and (iv) predictability of the behavior of the product. For example, predictive methods to determine the composition of the hydrocarbon product is vital to determine if the product meets specifications set by the federal government or by local authorities.

Thus, the data derived from any one or more of the analytical techniques applied to a hydrocarbon product not only give an indication of the product characteristics but also whether or not the product is suitable (i.e., meets specification) for the proposed use. Indeed, the use of physical properties for hydrocarbon evaluation and yields and, in some cases, product properties has continued in process research laboratories to the present time and will continue for some time.

Thus, using the data derived from the application of a test method, it is possible to assess hydrocarbon quality and acquire a degree of predictability of performance during use. However, knowledge of the basic concepts of hydrocarbon production will help the analyst understand the production and, to a large extent, the anticipated properties of the product, which in turn is related to storage, sampling, and handling the products. In addition, there are many instances in which interrelationships of the specification data enable properties to be predicted from the measured properties with as good precision as can be obtained by a single test. It would be possible to examine in this way the relationships between all the specified properties of a hydrocarbon product and to establish certain key properties from which the remainder could be predicted, but this would be a tedious task.

An alternative approach to that of picking out the essential tests in a specification (especially a hydrocarbon mixture such as a fuel) using regression analysis is to examine the specifications as a whole, and to use the necessary component. This is termed *principal components analysis*. And in this method a set of data as points in *multidimensional* space (*n-dimensional*, corresponding to *n* original tests) is examined to determine the direction that accounts for the biggest variability in the data (*first principal component*). The process is repeated until *n* principal components are evaluated, but it must be determined which are of practical importance since some principal components may be due to experimental error. The number of significant principal components shows the number of independent properties being measured by the tests considered.

Having established the number of independent properties for such a product, it is also necessary to examine the basis for making the specification more efficient. In the long-term, it might be possible to obtain new tests of a fundamental nature to replace existing tests. In the short-term, selecting the best of the existing tests to define product quality will be beneficial.

In summary, hydrocarbon product analysis is a complex discipline involving a variety of standard test methods, some of which have been mentioned above, and which needs a multidimensional approach. No single technique should supersede the other without adequate testing, along with an explanation of the data that are obtained with adequate interpretation.

References

ASTM D56, 2019. Standard Test Method for Flash Point by Tag Closed Cup Tester. Annual Book of Standards. ASTM International, West Conshohocken, Pennsylvania.

ASTM D70, 2019. Standard Test Method for Density of Semi-solid Bituminous Materials (Pycnometer Method). Annual Book of Standards. ASTM International, West Conshohocken, Pennsylvania.

ASTM D71, 2019. Standard Test Method for Relative Density of Solid Pitch and Asphalt (Displacement Method). Annual Book of Standards. ASTM International, West Conshohocken, Pennsylvania.

ASTM D87, 2019. Standard Test Method for Melting Point of Petroleum Wax (Cooling Curve). Annual Book of Standards. ASTM International, West Conshohocken, Pennsylvania.

ASTM D93, 2019. Standard Test Methods for Flash Point by Pensky-Martens Closed Cup Tester. Annual Book of Standards. ASTM International, West Conshohocken, Pennsylvania.

ASTM D97, 2019. Standard Test Method for Pour Point of Petroleum Products. Annual Book of Standards. ASTM International, West Conshohocken, Pennsylvania.

ASTM D127, 2019. Standard Test Method for Drop Melting Point of Petroleum Wax, Including Petrolatum. Annual Book of Standards. ASTM International, West Conshohocken, Pennsylvania.

ASTM D287, 2019. Standard Test Method for API Gravity of Crude Petroleum and Petroleum Products (Hydrometer Method). Annual Book of Standards. ASTM International, West Conshohocken, Pennsylvania.

ASTM D512, 2019. Standard Test Methods for Chloride Ion in Water. Annual Book of Standards. ASTM International, West Conshohocken, Pennsylvania.

ASTM D524, 2019. Standard Test Method for Ramsbottom Carbon Residue of Petroleum Products. Annual Book of Standards. ASTM International, West Conshohocken, Pennsylvania.

ASTM D975, 2019. Standard Specification for Diesel Fuel Oils. Annual Book of Standards. ASTM International, West Conshohocken, Pennsylvania.

ASTM D1217, 2019. Standard Test Method for Density and Relative Density (Specific Gravity) of Liquids by Bingham Pycnometer. Annual Book of Standards. ASTM International, West Conshohocken, Pennsylvania.

ASTM D1298, 2019. Standard Test Method for Density, Relative Density, or API Gravity of Crude Petroleum and Liquid Petroleum Products by Hydrometer Method. Annual Book of Standards. ASTM International, West Conshohocken, Pennsylvania.

ASTM D1480, 2019. Standard Test Method for Density and Relative Density (Specific Gravity) of Viscous Materials by Bingham Pycnometer. Annual Book of Standards. ASTM International, West Conshohocken, Pennsylvania.

ASTM D1481, 2019. Standard Test Method for Density and Relative Density (Specific Gravity) of Viscous Materials by Lipkin Bicapillary Pycnometer. Annual Book of Standards. ASTM International, West Conshohocken, Pennsylvania.

ASTM D1555, 2019. Standard Test Method for Calculation of Volume and Weight of Industrial Aromatic Hydrocarbons and Cyclohexane [Metric]. Annual Book of Standards. ASTM International, West Conshohocken, Pennsylvania.

ASTM D1657, 2019. Standard Test Method for Density or Relative Density of Light Hydrocarbons by Pressure Hydrometer. Annual Book of Standards. ASTM International, West Conshohocken, Pennsylvania.

ASTM D1835, 2019. Standard Specification for Liquefied Petroleum (LP) Gases. Annual Book of Standards. ASTM International, West Conshohocken, Pennsylvania.

ASTM D2500, 2019. Standard Test Method for Cloud Point of Petroleum Products. Annual Book of Standards. ASTM International, West Conshohocken, Pennsylvania.

ASTM D2650, 2019. Standard Test Method for Chemical Composition of Gases by Mass Spectrometry. Annual Book of Standards. ASTM International, West Conshohocken, Pennsylvania.

ASTM D3828, 2019. Standard Test Method for pH of Activated Carbon. Annual Book of Standards. ASTM International, West Conshohocken, Pennsylvania.

ASTM D4052, 2019. Standard Test Method for Density, Relative Density, and API Gravity of Liquids by Digital Density Meter. Annual Book of Standards. ASTM International, West Conshohocken, Pennsylvania.

ASTM D5002, 2019. Standard Test Method for Density, Relative Density, and API Gravity of Crude Oils by Digital Density Analyzer. Annual Book of Standards. ASTM International, West Conshohocken, Pennsylvania.

ASTM D6751, 2019. Standard Specification for Biodiesel Fuel Blend Stock (B100) for Middle Distillate Fuels. Annual Book of Standards. ASTM International, West Conshohocken, Pennsylvania.

ASTM E659, 2019. Standard Test Method for Autoignition Temperature of Liquid Chemicals. Annual Book of Standards. ASTM International, West Conshohocken, Pennsylvania.

Drews, A.W. (Ed.), 1998. Manual on Hydrocarbon Analysis. American Society for Testing and Materials, West Conshohocken, Pennsylvania.

Eliel, E.L., Wilen, S.H., 1994. Stereochemistry of Organic Compounds. John Wiley & Sons Inc., New York.

Eliel, E.L., Wilen, S.H., Doyle, M.P., 2001. Basic Organic Stereochemistry. John Wiley & Sons Inc., New York.

Howard, P.H., Meylan, 1997. W.M. Handbook of Physical Properties of Organic Chemicals. CRC Press, Taylor & Francis Group, Boca Raton, Florida.

Klein, D., 2013. Organic Chemistry, second ed. John Wiley & Sons Inc., Hoboken, New Jersey.

Olah, G.A., Molnár, A., 2003. Hydrocarbon Chemistry, second ed. John Wiley & Sons Inc., Hoboken, New Jersey.

Speight, J.G., 2014. The Chemistry and Technology of Petroleum, fifth ed. CRC Press, Taylor & Francis Group, Boca Raton, Florida.

Speight, 2015. Handbook of Petroleum Product Analysis, second ed. John Wiley & Sons Inc., Hoboken, New Jersey.

Speight, J.G., 2017. Lange's Handbook of Chemistry, seventeenth ed. McGraw Hill, New York.

Speight, J.G., 2019. Handbook of Petrochemical Processes. CRC Press, Taylor & Francis Group, Boca Raton, Florida.

Stoker, H.S., 2008. General, Organic, and Biological Chemistry. Florence Kentucky.

Yaws, C.L., 1999. Chemical Properties Handbook. McGraw-Hill, New York.

Further reading

ASTM D5502, 2019. Standard Test Method for Density and Relative Density of Crude Oils by Digital Density Analyzer. Annual Book of Standards. ASTM International, West Conshohocken, Pennsylvania.

Chapter 10

Combustion of hydrocarbons

1. Introduction

Combustion (*burning*) is the sequence of exothermic chemical reactions between a hydrocarbon and an oxidant accompanied by the production of heat and conversion of chemical species (Glassman, 1996). The release of heat can result in the production of light, usually in the form of a flame. Hydrocarbon derivatives of interest often include organic compounds (especially hydrocarbon derivatives) in the gas, liquid, or solid phase. For the most part, combustion involves a mixture of hot gases and is the result of a chemical reaction, primarily between oxygen in air and a hydrocarbon (or a hydrocarbon fuel). In addition to other products, the combustion reaction produces carbon dioxide (CO_2), steam (H_2O), light, and heat.

The burning of any substance, in gaseous, liquid, or solid form: in this broad definition, combustion includes fast exothermic chemical reactions, generally in the gas phase but not excluding the reaction of solid carbon with a gaseous oxidant. Flames represent combustion reactions that can propagate through space at subsonic velocity and are accompanied by the emission of light. The flame is the result of complex interactions of chemical and physical processes whose quantitative description must draw on a wide range of disciplines, such as chemistry, thermodynamics, fluid dynamics, and molecular physics. In the course of chemical reaction, energy is released in the form of heat, atoms, and free radicals, all highly reactive intermediates of the combustion reactions.

The physical processes involved in combustion are primarily transport processes: transport of mass and energy and, in systems with flow of the reactants, transport of momentum. The reactants in the chemical reaction are normally a hydrocarbon and an oxidant. In practical combustion systems the chemical reactions of the major chemical species, carbon and hydrogen in the hydrocarbon and oxygen in the air, are fast at the prevailing high temperatures (greater than 930°C, 1,700°F) because the reaction rates increase exponentially with temperature. In contrast, the rates of the transport processes that exhibit much smaller dependence on temperature are, therefore, lower than those of the chemical reactions.

Thus in most practical flames the rate of evolution of the main combustion products, carbon dioxide and water, and the accompanying heat release

Handbook of Industrial Hydrocarbon Processes. https://doi.org/10.1016/B978-0-12-809923-0.00010-2

depends on the rates at which the reactants are mixed and heat is being transferred from the flame to the fresh hydrocarbon-oxidant mixture injected into the flame. However, this generalization cannot be extended to the production and destruction of minor species in the flame, including those of trace concentrations of air pollutants such as nitrogen oxides, polycyclic aromatic hydrocarbon derivatives, soot, carbon monoxide, and sub-micrometer-size inorganic particulate matter.

Combustion applications are wide ranging with respect to the fields in which they are used and to their thermal input, extending from a few watts for a candle to hundreds of megawatts for a utility boiler. Combustion is the major mode of hydrocarbon utilization in domestic and industrial heating, in production of steam for industrial processes and for electric power generation, in waste incineration, and in propulsion in internal combustion engines, gas turbines, or rocket engines.

Thus, during combustion, new chemical substances (exhaust gases) are created from the hydrocarbon and the oxidizer. When a hydrocarbon-based fuel (such as gasoline) burns, the exhaust includes water and carbon dioxide. However, the exhaust gases can also include chemical combinations from the oxidizer alone. For example, if the gasoline is burned in air (21% v/v oxygen and 78% v/v nitrogen), the exhaust gases can also include nitrogen oxides (NO_X). The temperature of the exhaust gases is high because of the heat that is transferred to the exhaust during combustion. Because of the high temperatures, exhaust usually occurs as a gas, but there can be liquid (tar and other high boiling products) or solid (soot, carbon).

Finally, the specific energy content of a hydrocarbon is the heat energy obtained when a certain quantity of the hydrocarbon is burned. It is sometimes called the heat of combustion (Table 10.1). The *heat of combustion* (*energy content*) of a hydrocarbon or, for that matter any substance, natural gas is the amount of energy that is obtained from the burning of a volume of the hydrocarbon, typically measured in British thermal units (Btu). One Btu is the quantity of heat required to raise the temperature of one pound of water by 1 degree Fahrenheit at atmospheric pressure.

The *lower heating value* (LHV) or *higher heating value* (HHV) of a hydrocarbon (or a fuel) is an important consideration for use of the hydrocarbon (or the fuel). Whenever a hydrocarbon is burned one product of combustion is water. The quantity of water produced is dependent upon the amount of hydrogen in the hydrocarbon. Due to high combustion temperatures, this water takes the form of steam which stores a small fraction of the energy released during combustion as the latent heat of vaporization; in simple terms, as heat energy stored in the vaporized state of water. The total amount of heat liberated during the combustion of a unit of fuel, the high heating value (or the high calorific value, HCV) includes the latent heat stored in the vaporized water. In some applications it is possible to condense this vapor back to its liquid state and recover a proportion of this energy.

TABLE 10.1 Lower heating value for selected alkane derivatives, isoparaffin derivatives, and naphthene derivatives.

	MJ/kg	MJ/L	BTU/lb	kJ/mol
Alkane derivatives				
Methane	50.009	6.9	21,504	802.34
Ethane	47.794	—	20,551	1437.2
Propane	46.357	25.3	19,934	2044.2
Butane	45.752	—	19,673	2659.3
Pentane	45.357	28.39	21,706	3272.6
Hexane	44.752	29.30	19,504	3856.7
Heptane	44.566	30.48	19,163	4465.8
Octane	44.427	—	19,104	5074.9
Nonane	44.311	31.82	19,054	5683.3
Decane	44.240	33.29	19,023	6294.5
Undecane	44.194	32.70	19,003	6908.0
Dodecane	44.147	33.11	18,983	7519.6
Isoparaffin derivatives				
Isobutane	45.613	—	19,614	2651.0
Isopentane	45.241	27.87	19,454	3264.1
2-Methylpentane	44.682	29.18	19,213	6850.7
2,3-Dimethylbutane	44.659	29.56	19,203	3848.7
2,3-Dimethylpentane	44.496	30.92	19,133	4458.5
2,2,4-Trimethylpentane	44.310	30.49	19,053	5061.5
Naphthene derivatives				
Cyclopentane	44.636	33.52	19,193	3129.0
Methylcyclopentane	44.636?	33.43?	19,193?	3756.6?
Cyclohexane	43.450	33.85	18,684	3656.8
Methylcyclohexane	43.380	33.40	18,653	4259.5

The amount of heat available from a fuel after the latent heat of vaporization, the lower heating value (or the lower calorific value, LCV), is deducted from the higher heating value and it is this, that is available when the hydrocarbon is burned. The energy input into a gas engine should be defined using the LHV of the fuel. The LHVs of the hydrocarbon (or the fuel) determines the flow rate required when going into a combustor (such as a gasoline engine) because the total quantity of energy input necessary for the engine to produce a specific output power is defined and fixed. Hence the gas flow rate has to be such in order to provide the required energy input.

However, the energy content of a hydrocarbon mixture or any fuel is, in the latter case of a fuel that is a mixture of various constituents, variable because fuel mixture will have variations in the amount and types of energy constituents (for example, natural gas contains varying amounts of gases (methane, ethane, propane, and butane and the higher the content of noncombustible constituents in the mixture, the lower the energy content (Btu)). In addition, the volume mass of energy gases which are present in a natural gas accumulation also influences the Btu value of natural gas. The more carbon atoms in a hydrocarbon, the higher the Btu value. The gross heats of combustion (Q) of hydrocarbon mixtures can be calculated with a reasonable degree of accuracy by the equation:

$$Q = 12,400 - 2,100d^2$$

In this equation, d is the 60/60°F specific gravity and deviation of the calculated value from the measured value is typically less than 1%.

Two different values of specific heat energy exist for the same batch of hydrocarbon: (i) the high heat of combustion (gross heat of combustion) and (ii) the low heat of combustion (net heat of combustion). The high value is obtained when, after the combustion, the water in the exhaust is in liquid form. For the low value, the exhaust has all the water in vapor form (steam). Since water vapor gives up heat energy when it changes from vapor to liquid, the high value is larger since it includes the latent heat of vaporization of water. The difference between the high and low values is significant and is typically on the order of 8%–10% (Table 10.2). This accounts for most of the apparent discrepancy in the heat value of gasoline.

Finally, the *methane number* of a hydrocarbon or a hydrocarbon mixture provides an indication of the *knock tendency* of the hydrocarbon or the hydrocarbon mixture (ASTM D8221). The methane numbers is a product of the different constituent gases within the hydrocarbon mixture, particularly the proportions of methane, ethane, propane, and butane. Methane, which has a high knock resistance, is given an index value of 100. Hydrogen, which burns quickly relative to methane, has a low knock resistance and is given the index value of 0. If a gas mixture has a methane number of 80, the knock resistance is equivalent to that of a gas comprised of 80% v/v methane and 20% v/v hydrogen. There are gas constituents which have a higher methane number than 100; therefore, it is also possible for a gas composite to have a higher methane number than 100—for example, biogas often has a methane number in excess of 100.

TABLE 10.2 Higher (HHV) and lower (LHV) heating values of selected hydrocarbons and common hydrocarbon fuels.

Hydrocarbon fuel	HHV, MJ/kg	HHV Btu/lb	HHV kJ/mol	LHV MJ/kg
Methane	55.50	23,900	889	50.00
Ethane	51.90	22,400	1560	47.80
Propane	50.35	21,700	2220	46.35
Butane	49.50	20,900	2877	45.75
Pentane				45.35
Gasoline	47.30	20,400		44.40
Kerosene	46.20	20,400		43.00
Benzene	41.80	18,000	3270	

2. Combustion chemistry

Combustion chemistry often involves a sequence of free radical reactions. For example, solid fuels such as coal and wood first undergo endothermic pyrolysis to produce gaseous hydrocarbon derivatives (or other fuels) where the combustion of these materials supplies the heat required to produce more of them. Combustion of a hydrocarbon (or any such fuel) in air is always exothermic because the bond in the oxygen molecule (O_2) is much weaker than other double bonds or pairs of single bonds, and therefore the formation of the stronger bonds in the combustion products carbon dioxide and water results in the release of energy.

Uncatalyzed combustion in air requires fairly high temperatures. Complete combustion is stoichiometric with respect to the hydrocarbon. Generally, the chemical equation for the stoichiometric combustion of a hydrocarbon in oxygen is:

$$C_xH_y + zO_2 \rightarrow xCO_2 + y/2H_2O$$

For example:

$$C_3H_8 + 5O_2 \rightarrow 3CO_2 + 4H_2O$$

On the other hand, if the stoichiometric combustion takes place using air as the oxygen source, the nitrogen present in the air can be added to the equation (although it does not react) to show the stoichiometric composition of the fuel in air and the composition of the resultant flue gas.

Thermodynamically, the chemical equilibrium of combustion in air is overwhelmingly on the side of the products. However, complete combustion is almost impossible to achieve, since the chemical equilibrium is not necessarily reached, or may contain unburnt products such as carbon monoxide and hydrogen and even carbon (char or soot) or mineral ash. The oxidants for combustion have high oxidation potential and include, for example, atmospheric or pure oxygen.

In order to initiate, and maintain, the combustion process, the properties of the hydrocarbon (such as flash point, fire point, and ignition temperature) are extremely important. Since heat is both required to start combustion and is itself a product of combustion, it is easy to visualize why combustion takes place very rapidly. Furthermore, once combustion commences, it is not necessary to provide the heat source because the heat of combustion maintains the combustion process (Warnatz et al., 1996). Thus, for combustion to occur three items are necessary: (i) the hydrocarbon to be burned, (ii) a source of oxygen, and (iii) a source of heat.

2.1 General principles

Hydrocarbon derivatives are currently the main source of electrical energy and heat sources (such as home heating) because of the energy produced when burnt. Often this energy is used directly as heat such as in home heaters, which use either crude oil or natural gas. The hydrocarbon is burnt and the heat is used to heat water, which is then circulated. A similar principle is used to create electric energy in power plants.

Common properties of hydrocarbon derivatives are the facts that they produce steam, carbon dioxide, and heat during combustion and that oxygen is required for combustion to take place. The combustion of hydrocarbon derivatives follows the general equation:

$$\text{Hydrocarbon} + \text{Oxygen} \rightarrow \text{Carbon dioxide} + \text{Water}$$

For example, when the combustion of methane occurs the products are carbon dioxide (CO_2), water (H_2O), and energy:

$$CH_4[g]+2O_2[g] \rightarrow CO_2[g]+2H_2O[g] + \text{energy}$$

One molecule of methane (in the gaseous state) reacts with two oxygen molecules (also in the gaseous state) to form a carbon dioxide molecule, two water molecules (usually given off as steam or water vapor during the reaction), and energy.

However, Since different hydrocarbons have different ratios of hydrogen to carbon, they produce different ratios of water to carbon dioxide. In general, the longer and more complex the molecule, the greater the ratio of carbon to hydrogen. For this reason, combustion of equal amounts of different hydrocarbons will yield different quantities of carbon dioxide, depending on the ratio of carbon to hydrogen in molecules of each. Since coal contains the longest and most complex hydrocarbon molecules, burning coal releases more

TABLE 10.3 Energy density and specific energy of the C1 to C12 alkane derivatives.

Hydrocarbon	Density, kg/m^3	Energy density, MJ/m^3	Specific energy, MJ/kg
Methane (CH_4)	423	−23,529	−55.6
Ethane (C_2H_6)	545	−28,246	−51.8
Propane (C_3H_8)	585	−29,449	−50.3
Butane (C_4H_{10})	601	−29,729	−49.5
Pentane (C_5H_{12})	621	−30,223	−48.7
Hexane (C_6H_{14})	655	−31,633	−48.3
Heptane (C_7H_{16})	680	−32,690	−48.1
Octane (C_8H_{18})	698	−33,433	−47.9
Nonane (C_9H_{20})	714	−34,086	−47.7
Decane ($C_{10}H_{22}$)	726	−34,581	−47.6
Undecane ($C_{11}H_{24}$)	737	−35,039	−47.5
Dodecane ($C_{12}H_{26}$)	745	−35,378	−47.5

carbon dioxide than burning the same mass of oil or natural gas which also changes the energy density of each of these fuels. The energy density is the amount of energy stored in a given system or region of space per unit volume—the term may also be used for energy per unit mass but in this case it is more correct to use *specific energy* (Table 10.3). Thus:

$$\text{Energy density} = \text{the energy per unit volume of a fuel } (E_d) = E/V$$

$$\text{Specific energy} = \text{energy per unit mass of a fuel } (E_s) = E/m$$

Natural gas is the cleanest burning fossil fuel. Coal (Speight, 2013) and crude oil (Speight, 2014), the other fossil fuels, are more chemically complicated than natural gas, and when combusted, release a variety of potentially harmful air pollutants. Burning methane releases only carbon dioxide and water. Since purified, processed, refined natural gas is methane (Chapters 2 and 4) (Mokhatab et al., 2006; Speight, 2014, 2019), the combustion of natural gas releases fewer byproducts than other fossil fuels.

Another example is propane:

$$C_3H_8 + 5O_2 \rightarrow 4H_2O + 3CO_2 + \text{Energy}$$

Burning of hydrocarbon derivatives is an example of exothermic chemical reaction.

Overall, the products of stoichiometric combustion of a hydrocarbon are carbon dioxide and water. For reaction thermochemistry calculations, it is usually assumed that the diatomic nitrogen (N_2) does not react:

$$C_xH_y + a_{stoich}(O_2 + 3.76N_2) \rightarrow x\ CO_2 + \frac{y}{2}H_2O + 3.76\ a_{stoich}N_2$$

For stoichiometric combustion of hydrocarbon derivatives:

$$a_{stoich} = x + \frac{y}{4}$$

And the equivalence ratio is:

$$\Phi = \frac{(F/A)}{(F/A)_{stoich}} = \frac{AF_{stoich}}{AF} = \frac{a_{stoich}}{a} \quad \Phi > 1\ rich, \quad \Phi < 1\ lean$$

Examples of stoichiometric combustion ($\Phi = 1$) are:

$$\begin{aligned}
&methane: \ CH_4 + 2(O_2 + 3.76\ N_2) \rightarrow CO_2 + 2H_2O + 7.52\ N_2\\
&ethylene: \ C_2H_4 + 3(O_2 + 3.76\ N_2) \rightarrow 2CO_2 + 2H_2O + 11.28\ N_2\\
&propane: \ C_3H_8 + 5(O_2 + 3.76\ N_2) \rightarrow 3CO_2 + 4H_2O + 18.8\ N_2\\
&octane: \ C_8H_{18} + 12.5(O_2 + 3.76\ N_2) \rightarrow 8CO_2 + 9H_2O + 47\ N_2
\end{aligned}$$

For lean combustion, major reaction products are carbon dioxide and water. Excess oxygen is present in the air and it does not all react, so the products of lean combustion will contain oxygen:

$$C_xH_y + a(O_2 + 3.76N_2) \rightarrow b\ CO_2 + c\ H_2O + d\ O_2 + 3.76a\ N_2$$

Atom Balances :

$$C: x = b$$

$$H: y = 2c$$

$$O: 2a = 2b + c + 2d \Rightarrow d = a - b - (c/2)$$

$$N: 2 \times 3.76a = 2 \times 3.76a$$

Example:

$$C_3H_8 + 8(O_2 + 3.76N_2) \rightarrow 3CO_2 + 4H_2O + 3O_2 + 30.08N_2$$

For rich combustion, major reaction products are carbon monoxide, hydrogen, carbon dioxide, and water. The atom balances alone are not sufficient to determine the composition of the products.

$$C_xH_y + a(O_2 + 3.76N_2) \rightarrow b\ CO_2 + c\ H_2O + d\ H_2 + e\ CO + 3.76a\ N_2$$

Atom Balances :

$$C: x = b + e$$

$$H: y = 2c + 2d$$

$$O: 2a = 2b + c + e$$

$$N: 2 \times 3.76a = 2 \times 3.76a$$

The chemical structure of the hydrocarbon constituents of crude oil involves the presence of hydrocarbon chains of different lengths and the different hydrocarbon derivatives are separated by fractional distillation to produce gasoline, jet fuel, kerosene, and other hydrocarbon derivatives. The general formula for these alkane derivatives is C_nH_{2n+2}. For example 2,2,4-trimethylpentane (iso-octane, C_8H_{18}), which is widely used as a measure of the octane rating for gasoline, and it reacts with oxygen exothermally:

$$2C_8H_{18(l)} + 25O_{2(g)} \rightarrow 16CO_{2(g)} + 18H_2O_{(l)} + 10.86\ \text{MJ}$$

Incomplete combustion of crude oil or gasoline results in production of potentially toxic byproducts. Too little oxygen results in carbon monoxide. Combustion in air (which contains mostly nitrogen) results in nitric oxides:

$$C_8H_{18(l)} + 12.5O_{2(g)} + N_{2(g)} \rightarrow 6CO_{2(g)} + 2CO_{(g)} + 2NO_{(g)} + 9H_2O_{(l)} + \text{heat}$$

Generally, the chemical equation for the stoichiometric burning of hydrocarbon in oxygen is:

$$C_xH_y + \left(x + \frac{y}{4}\right)O_2 \rightarrow xCO_2 + \left(\frac{y}{2}\right)H_2O$$

For example, the burning of propane is represented by:

$$C_3H_8 + 5O_2 \rightarrow 3CO_2 + 4H_2O$$

Generally, the chemical equation for the stoichiometric incomplete combustion of a hydrocarbon in oxygen is:

$$(z)C_xH_y + \left(z\left(\frac{x}{2} + \frac{y}{4}\right)\right)O_2 \rightarrow z \cdot xCO + \left(\frac{z \cdot y}{2}\right)H_2O$$

For example, the incomplete combustion of propane is represented by:

$$2C_3H_8 + 7O_2 \rightarrow 2C + 2CO + 8H_2O + 2CO_2$$

If the combustion takes place using air as the oxygen source, the nitrogen can be added to the equation, as and although it does not react, to show the composition of the flue gas:

$$C_xH_y + \left(x + \frac{y}{4}\right)O_2 + 3.76\left(x + \frac{y}{4}\right)N_2 \rightarrow xCO_2 + \left(\frac{y}{2}\right)H_2O + 3.76\left(x + \frac{y}{4}\right)N_2$$

For example, the burning of propane is then represented by:

$$C_3H_8 + 5O_2 + 18.8N_2 \rightarrow 3CO_2 + 4H_2O + 18.8N_2$$

Nitrogen may also oxidize when there is an excess of oxygen. The reaction is thermodynamically favored only at high temperatures.

Diesel engines are run with an excess of oxygen to combust small particles that tend to form with only a stoichiometric amount of oxygen, necessarily producing nitrogen oxide emissions. Both the United States and European

Union are imposing limits to nitrogen oxide emissions, which necessitate the use of a special catalytic converter or treatment of the exhaust with a chemical, such as urea.

Combustion of a liquid hydrocarbon in an oxidizing atmosphere actually happens in the gas phase—it is the vapor that burns, not the liquid. Therefore, a liquid will normally catch fire only above a certain temperature: its flash point. The flash point of a liquid hydrocarbon is the lowest temperature at which it can form an ignitable mix with air. It is also the minimum temperature at which there is enough evaporated hydrocarbon in the air to start combustion.

2.2 Slow combustion

Slow combustion (*smoldering*) is the slow, low-temperature, flameless form of combustion, sustained by the heat evolved when oxygen directly attacks the surface of a condensed-phase hydrocarbon. It is a slow, low-temperature, flameless form of combustion of a condensed fuel that poses safety and environmental hazards and allows novel technological application but the fundamentals remain mostly unknown. The terms filtering combustion, smoking problem, deep seated fires, hidden fires, peat or peatland fires, lagging fires, low oxygen combustion, in-situ combustion, fireflood, and underground gasification all refer to smoldering combustion phenomena. Also, smoldering combustion is a typically incomplete combustion reaction (Rein, 2009). Solid materials that can sustain a smoldering reaction include coal, cellulose, wood, cotton, tobacco, peat, coal duff (coal fines), humus, synthetic foams, charring polymers including polyurethane. Common examples of smoldering phenomena are the initiation of residential fires on upholstered furniture by weak heat sources (e.g., a cigarette, a short-circuited wire), and the persistent combustion of biomass behind the flaming front of wildfires.

The term *smoldering* is sometimes inappropriately used to describe a nonflaming response of condensed phase organic materials to an external heat flux. Any organic material, when subjected to a sufficient heat flux, will degrade, gasify, and give off smoke. There usually is little or no oxidation involved in this gasification process, and thus it is endothermic. This process is more appropriately referred to as *forced pyrolysis*, not smoldering.

Smoldering is a branch of solid hydrocarbon combustion quite distinct in many aspects from flaming, but equally diverse and complex. Unfortunately it has not been studied nearly to the same extent as flaming. This is quite apparent in the lack of quantitative guidelines that can be provided here for estimating the behavior of realistic smolder propagation processes, smolder detection, toxic gas production, and the transition to flaming. The experimental data provided can be readily used for closely analogous situations. They must be used cautiously for dissimilar conditions. The reader should always bear in mind the strong role that the oxygen supply rate has on the smolder process. The other very important factor is the relative direction of movement of oxygen supply and smolder propagation. This can be somewhat obscure in many realistic configurations. The actual chemical nature of the hydrocarbon is relatively secondary, at least with regard to smolder rate. It may be important for toxic gas production rates, but the data here are quite limited.

Smoldering poses safety and environmental hazards and allows novel technological application but its fundamentals remain mostly unknown to the scientific community. Smoldering is the leading cause of deaths in residential fires and a source of safety concern in space and commercial flights. Smoldering wildfires destroy large amounts of biomass and cause great damage to the soil, contributing significantly to atmospheric pollutant and greenhouse gas emissions. Subsurface fires in coal mines and seams burn for very long periods of time, making them the oldest continuously burning fires on Earth.

The effects of smoldering fires on the landscape can range from the small scale (pockets of burning in superficial layers or the root of a single tree), to the large scale (burning of a hill-top or the destruction of the root network of a complete forest stand). In general, smoldering fires have a severe impact on the local soil system, because the burning hydrocarbon is the organic portion of the soil itself. The prolonged heating from the slowly propagating fire can kill roots, seeds, and plant stems and the affected layers of soil sustain large losses of biomass. This coupled with expositing of underlying layers increase the likelihood of long-term damage and erosion.

The fundamental difference between smoldering and flaming combustion is that smoldering occurs on the surface of the solid rather than in the gas phase. Smoldering is a surface phenomenon but can propagate to the interior of a porous hydrocarbon (such as certain types of hydrocarbon wax) if it is permeable to flow. The characteristic temperature and heat released during smoldering are low compared to those in the flaming combustion [i.e., approximately 600°C (1110°F) compared to up to 1500°C (2730°F)]. Smoldering propagates in a creeping fashion, around 0.1 mm/s, which is approximately 10 times slower than flames spread over a solid. In spite of its weak combustion characteristics, smoldering is a significant fire hazard. Smoldering emits toxic gases (such as carbon monoxide) at a higher yield than flaming fires and leaves behind a significant amount of solid residue. The emitted gases are flammable and could later be ignited in the gas phase, triggering the transition to flaming combustion.

Hydrocarbon derivatives (even the solid hydrocarbon derivatives with the exception of certain types of hydrocarbon wax) are less likely to smolder. However many materials can sustain a smoldering reaction, including coal and biomass fuels. Smoldering fuels are generally porous, permeable to flow and formed by aggregates (particulates, grains, fibers or of cellular structure). These aggregates facilitate the surface reaction with oxygen by allowing gas flow through the hydrocarbon and providing a large surface area per unit volume. They also act as thermal insulation, reducing heat losses.

The transition process from smolder to flaming in the above bedding and upholstery fires is essentially spontaneous. At ambient conditions both smoldering and flaming are possible in many such systems. In the domain of overlapping smolder and flaming potential there is a hysteresis in the spontaneous transition between these two combustion modes. The enhanced air supply presumably accelerates local char oxidation, heating the char to the point where it can ignite pyrolysis gases. Such a mechanism is plausible but it has not been demonstrated to be operable where the chimney effect may not develop so readily.

Transition to flaming (fast exothermic gas-phase reactions) requires both a mixture of gases and air that are within their flammability limits and a sufficient heat source to ignite this mixture. Furthermore, these two requirements must be realized at the same locus in space and at the same time. Any factor that either enhances the net rate of heat generation or decreases the net rate of heat loss will move the smoldering material toward flaming ignition by increasing both local temperature and rate of pyrolysis gas generation. Such factors include an enhanced oxygen supply, an increase in scale (which usually implies lesser surface heat losses per unit volume of smoldering material), or an increasingly concave smolder front geometry, which reduces radiative losses to the surroundings and enhances gaseous hydrocarbon concentration buildup.

2.3 Rapid combustion

Rapid combustion is a form of combustion in which large amounts of heat and light energy are released, which often results in a fire. This is used in a form of machinery such as internal combustion engines. This form of combustion is the result of a rapid release of heat. Sometimes, a large volume of gas is liberated in combustion besides the production of heat and light. The sudden evolution of large quantities of gas from the hydrocarbon causes excessive pressure and if the gas cannot dissipate quickly enough, then *extremely rapid combustion* and explosions occur. As an example, fire is the rapid combustion (rapid oxidation) of the hydrocarbon derivatives in a hydrocarbon fuel in the chemical process of combustion, releasing heat, light, and various reaction products. The *flame* is the visible portion of the fire and consists of glowing hot gases. Depending on the substances alight, and any impurities outside, the color of the flame and the intensity of the fire might vary.

Fire in its most common form can result in conflagration, which has the potential to cause physical damage through burning. The positive effects of fire include stimulating growth and maintaining various ecological systems. The negative effects of fire include decreased water purity, increased soil erosion, increase in atmospheric pollutants, and an increased hazard to human life.

Rapid combustion (fire) commences when a flammable and/or a combustible material (such as a hydrocarbon), in combination with a sufficient quantity of an oxidizer (such as oxygen gas or another oxygen-rich compound), is exposed to a source of heat or ambient temperature above the flash point for the hydrocarbon/oxidizer mix, and is able to sustain a rate of rapid oxidation that produces a chain reaction. Fire cannot exist without all of these elements in place and in the right proportions (though as previously stated, another strong oxidizer can replace oxygen). For example, a flammable hydrocarbon liquid will start burning only if the hydrocarbon and oxygen are in the right proportions. Some hydrocarbon-oxygen mixes may require a catalyst, a substance that is not directly involved in any chemical reaction during combustion, but which enables the reactants to combust more readily. Once ignited, a chain reaction must take place whereby combustion can sustain its own heat by the further release of heat energy in the process of combustion and may propagate, provided there is a continuous supply of an oxidizer and hydrocarbon.

Fire can be extinguished by removing any one of the necessary elements that contribute to the combustion reaction: (i) turning off the hydrocarbon (gas) supply, which removes the hydrocarbon source, (ii) covering the flame completely, which smothers the flame as the combustion both uses the available oxidizer (the oxygen in the air) and displaces it from the area around the flame with carbon dioxide, (iii) application of fire retardant—not always water—which removes heat and retards the combustion reaction until the rate of combustion is too slow to maintain the chain reaction.

In contrast, fire is intensified by increasing the overall rate of combustion. Methods to do this include balancing the input of hydrocarbon and oxidizer to stoichiometric proportions, increasing the input of the hydrocarbon and the oxidizer in this balanced mix, increasing the ambient temperature so the heat of the fire is better able to sustain combustion, or providing a catalyst—a nonreactant medium in which the hydrocarbon and oxidizer can more readily react.

Turbulent combustion is a form of *rapid combustion* that results in a turbulent flame and is the most used for industrial application (such as gas turbines and gasoline engines) because the turbulence helps the mixing process between the hydrocarbon and oxidizer.

2.4 Complete and incomplete combustion

During complete combustion, the reactant burns in oxygen, producing a limited number of products. When a hydrocarbon burns in oxygen, the reaction will only yield carbon dioxide and water. When elements are burned, the products are primarily the most common oxides. Carbon will yield carbon dioxide, nitrogen will yield nitrogen dioxide, and sulfur will yield sulfur dioxide.

In most industrial applications and in fires, air is the source of oxygen (O_2). Nitrogen does not take part in combustion, but at high temperatures, some nitrogen will be converted to nitrogen oxides (NO_x).

$$CH_4+O_2+N_2 \rightarrow CO_2+2H_2O + N_2 + CO + NOx + heat$$

Incomplete combustion occurs when there isn't enough oxygen to allow the hydrocarbon to react completely with the oxygen to produce carbon dioxide and water, and also when the combustion is quenched by a heat sink such as a solid surface or flame trap.

Complete or incomplete combustion (an indicator of the combustion efficiency) is a calculation of how well the equipment is burning a specific hydrocarbon, shown in percent. Complete combustion efficiency would extract all the energy available in the hydrocarbon. However 100% combustion efficiency is not realistically achievable. Common combustion processes produce efficiencies from 10% to 95%. Combustion efficiency calculations assume complete hydrocarbon combustion and are based on three factors: (i) the chemistry of the hydrocarbon, (ii) the net temperature of the stack gases, and (iii) the percentage of oxygen or CO_2 by volume after combustion.

Combustion efficiency relates to the part of the reactants that combine chemically. Combustion efficiency increases with increasing temperature of the reactants; increasing time that the reactants are in contact; increasing vapor pressures, surface areas, and stored chemical energy. One way of increasing the temperature of the reactants and their vapor pressures is to preheat them by circulating them around the combustion chamber and throat before being injected into the combustion chamber. The specific heat of combustion is a chemical property that refers to the amount of energy that can theoretically be extracted from a hydrocarbon at 100% combustion efficiency. The heating value is a more realistic term and does not include the condensation of the water vapor produced. It is thus more easily applied to real combustion processes.

Air preheating is one method used in steel works, for instance, to increase combustion efficiency. This uses the heat in the flue gases to heat one of a pair of chambers and the inlet air passes through the other one. The use of the chambers is switched as soon as one chamber has reached temperature, so the air passes through the heated chamber. This is one of the simplest and best methods of increasing combustion efficiency in this kind of process; such preheaters are standard equipment these days for larger systems.

In this same context, hydrocarbon efficiency is a form of thermal efficiency of a process that converts chemical potential energy contained in a carrier hydrocarbon into kinetic energy. Overall hydrocarbon efficiency may vary per device, which in turn may vary per application, and this spectrum of variance is often illustrated as a continuous energy profile.

2.5 Spontaneous combustion

Spontaneous combustion is the phenomenon in which a hydrocarbon (or a chemical substance) substance unexpectedly bursts into flame without apparent cause. In ordinary combustion, the hydrocarbon is deliberately heated to its ignition point to make it burn. During this process, many hydrocarbon derivatives undergo slow oxidation that, like the rapid oxidation of burning, releases heat. If the heat so released cannot escape, the temperature of the hydrocarbon rises until ignition takes place. Spontaneous combustion often occurs in piles of hydrocarbon-soaked (oily) rags and can constitute a serious fire hazard.

Fires started by spontaneous combustion are caused by the following mechanisms: (i) spontaneous heating, (ii) pyrophoricity, and (iii) hypergolic reactions.

Spontaneous heating is the slow oxidation of an element or compound which causes the bulk temperature of the element or compound to rise without the addition of an external heat source. Spontaneous heating may be the result of direct oxidation of hydrocarbon derivatives (for example, oils and solvents) or it may occur because of the action of microorganisms in organic materials. Saturated hydrocarbon derivatives (such as alkane derivatives) do not have a tendency for spontaneous combustion whereas unsaturated hydrocarbon derivatives do have a tendency for spontaneous combustion.

Pyrophoric substances ignite instantly upon exposure to air (atmospheric oxygen). A pyrophoric substance may be a solid, liquid, or gas. Most materials are not pyrophoric unless they are in a very finely divided state. Although there are some pyrophoric liquids and gases, most pyrophoric materials are metals. *Pyrophoricity* is a special case of a *hypergolic* reaction because the oxidizing agent is restricted to atmospheric oxygen. Whereas pyrophoricity is concerned only with the spontaneous combustion of a material when exposed to air (atmospheric oxygen), a hypergolic reaction describes the ability of a material to spontaneously ignite or explode upon contact with *any* oxidizing agent.

Some hydrocarbon derivatives are capable of spontaneous heating and ignition under proper conditions. Spontaneous heating of hydrocarbon derivatives usually involves a combustible liquid hydrocarbon in contact with combustible materials. An example of this would be combustible rags impregnated with oils or solvents. Whether or not spontaneous heating leads to ignition depends on several items: (i) the rate at which heat is generated and removed from the material being oxidized, (ii) the ignition temperature of the fibrous combustible material, hydrocarbon, or any gases liberated by oxidation, (iii) the specific area (cm^2/g, defined below) of the hydrocarbon exposed to an oxidizer, and (iv) the amount of moisture present in the atmosphere and the fibrous material.

For spontaneous ignition to occur, the rate of heat being generated through oxidation must exceed the rate of heat removal by conduction, convection, and radiation (thermal). As the temperature of the material begins to rise, the rate of heat generation will often increase. The result is a *runaway reaction* which ultimately causes ignition. If the rate of heat removal exceeds the rate of generation, the material will cool and will not ignite. The rate of heat removal may be increased through physical contact with a thermally conductive surface, by rotating piles of combustibles to cool hot spots, and by circulating inert gases through the piles to cool hot spots and displace oxygen.

The ignition temperature of the materials is obviously of concern and varies widely among materials. Much more stringent controls must be placed on materials which have lower ignition temperatures and those which liberate explosive gases. Although most materials with high ignition temperatures are of lesser concern, some are more explosive than those with lower ignition temperatures.

The specific area of a combustible substance is a measure of the surface area of the material exposed to an oxidizing atmosphere per gram of material and is expressed in units of cm^2/g. Materials which have a high specific area are more prone to heat and ignite spontaneously. For example, combustible liquids on fibrous material pose a spontaneous fire hazard because the fibers of the material allow the liquid to spread out over a larger surface area, allowing more contact with oxygen and, therefore, porous combustible materials are more likely to ignite than tightly packed solid materials.

It is important to keep potentially spontaneously heating compounds as dry as possible. High ambient temperatures compound moisture problems. As the ambient temperature rises, the rate of spontaneous heat generation will also rise. High ambient temperatures also reduce the rate of heat removal, bringing the hydrocarbon closer to its ignition temperature.

Spontaneous combustion may occur in piles of moist organic material where heat is generated in the early stages by the respiration of bacteria, molds, and microorganisms. High moisture content is required for vigorous activity, and heating is generally controlled by maintaining the moisture content below a predetermined level. This type of heating can only raise the material to the temperature range of 50°C–75°C (122°F–167°F), where the living organisms die.

Beyond this point, oxidation reactions must take over if ignition is to occur. The existence of biological heating requires careful control of moisture, air supply, and nearby combustible or flammable materials. If a *hot spot* in a pile of organic material comes in contact with a highly flammable liquid or gas, a fire or explosion may occur. Heat generated by biological action may also act as a catalyst for other reactions which occur only at elevated temperatures.

The likelihood of biological heating may be reduced by the following measures: (i) provide adequate ventilation of the organic material to remove moisture, heat, and dust particles, (ii) limit the storage time of the organic material using a "first in, first out" rule of thumb, and (iii) circulate large quantities of organic materials to disperse areas of localized heating.

3. Process parameters

A combustion reaction is a type of redox reaction in which a combustible material combines with an oxidizer to form oxidized products and generate heat (exothermic reaction). Combustion in oxygen is a radical chain reaction where many distinct radical intermediates participate. The high energy required for initiation is due to the structure of the di-oxygen molecule. The lowest-energy configuration of the di-oxygen molecule is a stable, relatively unreactive diradical in a triplet spin state. Bonding can be described with three bonding electron pairs and two antibonding electrons, where the spins are aligned, such that the molecule has nonzero total angular momentum. Most hydrocarbon derivatives, on the other hand, are in a single state, with paired spins and zero total angular momentum. Interaction between the two is quantum mechanically a forbidden transition, i.e., possible with a very low probability.

To initiate combustion, energy is required to force di-oxygen into a spin-paired state, or singlet oxygen, which is extremely reactive. The energy is supplied as heat and the reaction produces heat, which means as long as the hydrocarbon is provided the reaction is self-perpetuating. The combustion of hydrocarbon derivatives is thought to be initiated by hydrogen atom abstraction (not proton abstraction) from the hydrocarbon to oxygen, to give a hydroperoxide radical (HOO). This reacts further to give a hydroperoxide, which breaks up to give hydroxyl radicals. There are a great variety of these processes that produce hydrocarbon radicals and oxidizing radicals.

Oxidizing species include singlet oxygen, hydroxyl, monatomic oxygen, and the hydroperoxyl radical. Such intermediates are short-lived and cannot be isolated. However, nonradical intermediates are stable and are produced in incomplete combustion. An example is acetaldehyde (CH_3CHO) produced in the combustion of ethanol. An intermediate in the combustion of carbon and hydrocarbon derivatives, carbon monoxide, is of special importance because it is a poisonous gas, but also economically useful for the production of synthesis gas (syngas). Solid hydrocarbon derivatives also undergo a number of pyrolysis reactions that produce gaseous hydrocarbon derivatives. These reactions are endothermic and require constant energy input from the combustion reactions. Lack of oxygen results in the pyrolysis products being emitted as thick black smoke.

Generally, the chemical equation for stoichiometric combustion of a hydrocarbon in oxygen is:

$$C_xH_y + \left(x + \frac{y}{4}\right)O_2 \rightarrow xCO_2 + \left(\frac{y}{2}\right)H_2O$$

For example:

$$C_3H_8 + 5O_2 \rightarrow 3CO_2 + 4H_2O$$

On the other hand, the chemical equation for the stoichiometric incomplete combustion of a hydrocarbon is:

$$(z)C_xH_y + \left(z\left(\frac{x}{2}+\frac{y}{4}\right)\right)O_2 \rightarrow z \cdot xCO + \left(\frac{z \cdot y}{2}\right)H_2O$$

For example:

$$2C_3H_8 + 7O_2 \rightarrow 2C + 2CO + 8H_2O + 2CO_2$$

If combustion takes place using air as the oxygen source, the nitrogen can be added to the equation, as and although it does not react, to show the composition of the flue gas:

$$C_xH_y + \left(x+\frac{y}{4}\right)O_2 + 3.76\left(x+\frac{y}{4}\right)N_2 \rightarrow xCO_2 + \left(\frac{y}{2}\right)H_2O + 3.76\left(x+\frac{y}{4}\right)N_2$$

For example:

$$C_3H_8 + 5O_2 + 18.8N_2 \rightarrow 3CO_2 + 4H_2O + 18.8N_2$$

Nitrogen may also oxidize when there is an excess of oxygen—in fact, the reaction is thermodynamically favored only at high temperatures.

Assuming perfect combustion conditions, such as complete combustion under adiabatic conditions (i.e., no heat loss or gain), the combustion temperature can be determined. The formula that yields this temperature is based on the first law of thermodynamics and requires that the heat of combustion is used entirely for heating the hydrocarbon, the combustion air or oxygen, and the combustion product gases (flue gases). In the case of hydrocarbon derivatives burned in air, the combustion temperature depends on all of the following: (i) the heating value, (ii) the stoichiometric air/hydrocarbon ratio, λ, (iii) the specific heat capacity of the hydrocarbon and air, and (iv) the air and hydrocarbon inlet temperatures.

The heating value (calorific value) of a hydrocarbon derivative is the amount of heat released during the combustion of a specified amount of the hydrocarbon. The calorific value is a characteristic for each substance. The heat of combustion for hydrocarbon derivatives and hydrocarbon fuels is expressed as the HHV, LHV, or GHV. For example, the higher heating value (HHV, gross calorific value, gross energy, upper heating value) is determined by bringing all the products of combustion back to the original precombustion temperature, and in particular condensing any vapor produced. This is the same as the thermodynamic heat of combustion since the enthalpy change for the reaction assumes a common temperature of the compounds before and after combustion, in which case the water produced by combustion is liquid. On the other hand, the quantity known as the lower heating value (LHV, net calorific value) is determined by subtracting the heat of vaporization of the water vapor from the higher heating value. The energy required to vaporize the water therefore is not realized as heat.

Finally, the gross heating value accounts for water in the exhaust leaving as vapor, and includes liquid water in the hydrocarbon prior to combustion. This value is important for complex fuels such as wood or coal, which will usually contain some amount of water prior to burning.

A common method of relating the higher heating value (HHV) to the lower heating value (LHV) is:

$$HHV = LHV + hv\ x\ (nH_2O,out/nfuel,in)$$

where hv is the heat of vaporization of water, nH_2O,out is the moles of water vaporized, and nfuel,in is the number of moles of fuel combusted. Most applications which burn hydrocarbon derivatives produce water vapor which is not used, thus, wasting the heat content. In such applications, the lower heating value is the applicable measure. This is particularly relevant for natural gas where the high content of hydrocarbon derivatives produces considerable amounts of water.

Both the higher heating value and the lower heating value can be expressed in terms of the hydrocarbon on an as-received (AR) basis in which all of the moisture is included, or on a moisture-free (MF), or on a moisture- and ash-free basis (MAF), all of which are commonly used for indicating the heating values of coal: (i) as-received (AR) indicates that the hydrocarbon heating value has been measured with all moisture and ash forming minerals present, (ii) moisture-free (MF, dry) indicates that the hydrocarbon heating value has been measured after the hydrocarbon has been dried of all inherent moisture but still retains its ash forming minerals, and (iii) moisture- and ash-free (MAF) or dry and ash-free (DAF), which indicates that the hydrocarbon heating value has been measured in the absence of inherent moisture and ash forming minerals.

The adiabatic combustion temperature (adiabatic flame temperature) increases for higher heating values and inlet air and hydrocarbon temperatures and for stoichiometric air ratios approaching one. Most commonly, the adiabatic combustion temperatures for oil is approximately around 2150°C (3990°F) (for inlet air and hydrocarbon at ambient temperatures and for $\lambda = 1.0$) for coals and approximately 2000°C (3630°F) for natural gas.

In power plants steam generators, the more common way of expressing the usage of more than the stoichiometric combustion air is percent excess combustion air. For example, excess combustion air of 15% means that 15% more than the required stoichiometric air is being used.

3.1 Air-hydrocarbon ratio

The air-hydrocarbon ratio (air-hydrocarbon ratio, λ) is the mass ratio of air to the mass of the hydrocarbon present during combustion.

$$AFR = mair/mfuel$$

When all the hydrocarbon is combined with all the free oxygen, typically within the combustion chamber of a vehicle, the reaction is chemically balanced and this air-hydrocarbon ratio is a stoichiometric relationship. The air-hydrocarbon ratio is an important measure for antipollution and performance tuning reasons.

Most practical air-hydrocarbon ratio devices actually measure the amount of residual oxygen (for lean mixes) or unburned hydrocarbon derivatives (for rich mixtures) in the exhaust gas. Lambda (λ) is the ratio of actual air-hydrocarbon ratio to stoichiometry for a given mixture. Lambda of 1.0 is at stoichiometry, rich mixtures are less than 1.0, and lean mixtures are greater than 1.0. There is a direct relationship between lambda and the air-hydrocarbon ratio and to calculate the air-hydrocarbon ratio from a given value of lambda, multiply the measured lambda by the stoichiometric air-hydrocarbon ratio for the hydrocarbon. Alternatively, to recover lambda from an air-hydrocarbon ratio, divide the air-hydrocarbon ratio by the stoichiometric air-hydrocarbon ratio for that hydrocarbon:

$$\lambda = \frac{AFR}{AFR_{stoich}}$$

For low-boiling hydrocarbon liquids, such as naphtha and gasoline, the stoichiometric air/hydrocarbon mixture is approximately 15 times the mass of air to hydrocarbon. Any mixture less than 14.7 to 1 is considered to be a rich mixture, whereas more than 14.7 to 1 is a lean mixture—assuming a perfect (ideal) gasoline consisting of solely n-heptane and iso-octane). In reality, gasoline is more complex than a simple two-component mixture and the stoichiometric ratio is altered.

Lean mixtures produce hotter combustion gases than does a stoichiometric mixture, so much so that pistons can melt as a result. Rich mixtures produce cooler combustion gases than does a stoichiometric mixture, primarily due to the excessive amount of carbon which oxidizes to form carbon monoxide, rather than carbon dioxide. The chemical reaction oxidizing carbon to form carbon monoxide releases significantly less heat than the similar reaction to form carbon dioxide.

Carbon monoxide retains significant potential chemical energy because it is a fuel whereas carbon dioxide is not, being the result of complete combustion of a hydrocarbon or carbonaceous fuel. Lean mixtures, when consumed in an internal combustion engine, produce less power than does the stoichiometric mixture. Similarly, rich mixtures return poorer fuel efficiency than the stoichiometric mixture. The mixture for the best fuel efficiency is slightly different from the stoichiometric mixture.

3.2 Equivalence ratio

The equivalence ratio (ø) of a system is defined as the ratio of the hydrocarbon-to-oxidizer ratio to the stoichiometric hydrocarbon-to-oxidizer ratio:

$$\phi = \frac{\text{fuel} - \text{to} - \text{oxidizer ratio}}{(\text{fuel} - \text{to} - \text{oxidizer ratio})_{st}} = \frac{m_{fuel}/m_{ox}}{(m_{fuel}/m_{ox})_{st}} = \frac{n_{fuel}/n_{ox}}{(n_{fuel}/n_{ox})_{st}}$$

In this equation, m represents the mass, n represents number of moles, suffix st stands for stoichiometric conditions.

The advantage of using equivalence ratio over hydrocarbon-to-oxidizer ratio is that it does not have the same dependence as hydrocarbon-to-oxidizer ratio on the units being used. For example, hydrocarbon-to-oxidizer ratio based on mass of the hydrocarbon and the oxidizer is not same as defined by the number of moles. This is not the case for equivalence ratio. For example, in the case of a mixture of 1 mole of ethane (C_2H_6) and 1 mole of oxygen (O_2), the hydrocarbon-to-oxidizer ratio of this mixture based on the mass of hydrocarbon and air:

$$\frac{m_{C_2H_6}}{m_{O_2}} = \frac{1 \cdot (2 \cdot 12 + 6 \cdot 1)}{1 \cdot (2 \cdot 16)} = \frac{30}{12} = 0.938$$

The hydrocarbon-to-oxidizer ratio of this mixture based on the number of moles of hydrocarbon and air is:

$$\frac{n_{C_2H_6}}{n_{O_2}} = \frac{1}{1} = 1.$$

The two values are not equal and to compare it to the equivalence ratio, the hydrocarbon-to-oxidizer ratio of ethane and oxygen mixture needs to be determined from the stoichiometric reaction of ethane and oxygen:

$$C_2H_6 + \frac{7}{2}O_2 \rightarrow 2CO_2 + 3H_2O$$

This gives the hydrocarbon-to-oxidizer based on mass:

$$(\text{fuel} - \text{to} - \text{oxidizer ratio based on mass})_{st} = \left(\frac{m_{C_2H_6}}{m_{O_2}}\right)_{st}$$

$$= \frac{1 \cdot (2 \cdot 12 + 6 \cdot 1)}{3.5 \cdot (2 \cdot 16)} = \frac{30}{112} = 0.268$$

$$(\text{fuel} - \text{to} - \text{oxidizer ratio based on number of moles})_{st} = \left(\frac{n_{C_2H_6}}{n_{O_2}}\right)_{st}$$

$$= \frac{1}{3.5} = 0.286$$

From which the equivalence ratio of the mixture can be determined:

$$\phi = \frac{m_{C_2H_6}/m_{O_2}}{(m_{C_2H_6}/m_{O_2})_{st}} = \frac{0.938}{0.268} = 3.5$$

The equivalently is then:

$$\phi = \frac{n_{C_2H_6}/n_{O_2}}{(n_{C_2H_6}/n_{O_2})_{st}} = \frac{1}{0.286} = 3.5$$

Another advantage of using the equivalence ratio is that ratios greater than one always represent excess hydrocarbon in the hydrocarbon-oxidizer mixture than would be required for complete combustion (stoichiometric reaction) irrespective of the hydrocarbon and oxidizer being used, while ratios less than one represent a deficiency of hydrocarbon or equivalently excess oxidizer in the mixture. This is not the case if one uses hydrocarbon-to-oxidizer ratio, which will take different values for different mixtures.

It should be noted that equivalence ratio is related to λ (defined previously) as follows:

$$\phi = \frac{1}{\lambda}$$

4. Combustion of hydrocarbons

Combustion of a *liquid hydrocarbon* in an oxidizing atmosphere actually happens in the gas phase—it is the vapor that burns, not the liquid. Therefore, a liquid will normally catch fire only above the flash point of the liquid. The flash point of a liquid hydrocarbon is the lowest temperature at which it can form an ignitable mix with air. It is also the minimum temperature at which there is enough evaporated hydrocarbon in the air to start combustion.

Combustion of a *solid hydrocarbon* consists of three relatively distinct but overlapping phases: (i) the preheating phase, when the unburned hydrocarbon is heated up to its flash point and then to the fire point; flammable gases start being evolved in a process similar to distillation, (ii) the distillation phase or gaseous phase, when the mix of evolved flammable gases with oxygen is ignited; energy is produced in the form of heat and flames are often visible; during this phase, heat transfer from the combustion to the solid maintains the evolution of flammable vapors, and (iii) the charcoal phase or solid phase, when the output of flammable gases from the material is too low for persistent presence of flame and the charred hydrocarbon does not burn rapidly anymore but glows and smolders.

4.1 Gaseous hydrocarbons

Natural gas (methane) is the cleanest burning fossil fuel. Crude oil–based fuels (such as fuel oil and resid) and coal are more chemically complicated than natural gas, and when combusted release a variety of potentially harmful chemicals into the air, whereas combustion of methane releases only carbon dioxide and water vapor into the air:

$$CH_4[g] + 2O_2[g] \rightarrow CO_2[g] + 2H_2O[g]$$

Natural gas is one of the major combustion fuels used throughout the country. It is mainly used to generate industrial and utility electric power, produce industrial process steam and heat, and heat residential and commercial space. Natural gas, as supplied to the consumer, is mostly methane and the gross heating value of natural gas is approximately 1020 Btu/ft^3 but usually varying from 950 to 1050 Btu/ft^3.

There are three major types of boilers used for natural gas combustion in commercial, industrial, and utility applications: (i) water tube boilers, (ii) firetube boilers, and (iii) cast iron boilers.

Water tube boilers are designed to pass water through the inside of heat transfer tubes while the outside of the tubes is heated by direct contact with the hot combustion gases and through radiant heat transfer. The water tube design is the most common in utility and large industrial boilers. Water tube boilers are used for a variety of applications, ranging from providing large amounts of process steam, to providing hot water or steam for space heating, to generating high-temperature, high-pressure steam for producing electricity. Furthermore, water tube boilers can be distinguished either as field erected units or packaged units.

Firetube boilers are designed such that the hot combustion gases flow through tubes, which heat the water circulating outside of the tubes. These boilers are used primarily for space heating systems, industrial process steam, and portable power boilers. Firetube boilers are almost exclusively packaged units. The two major types of firetube units are Scotch Marine boilers and the older firebox boilers.

In *cast iron boilers*, as in firetube boilers, the hot gases are contained inside the tubes and the water being heated circulates outside the tubes. However, the units are constructed of cast iron rather than steel. These boilers are used to produce either low-pressure steam or hot water, and are most commonly used in small commercial applications.

Natural gas is also combusted in residential boilers and furnaces. Residential boilers and furnaces generally resemble firetube boilers with flue gas. The emissions from natural gas–fired boilers and furnaces include nitrogen oxides (NO_x), carbon monoxide (CO), and carbon dioxide (CO_2), methane (CH_4), nitrous oxide (N_2O), volatile organic compounds (VOCs), trace amounts of sulfur dioxide (SO_2), and particulate matter (PM).

Nitrogen oxide is formed by three fundamentally different mechanisms. The principal mechanism of NO_x formation in natural gas combustion is thermal NO_x. The thermal NO_x mechanism occurs through the thermal dissociation and subsequent reaction of nitrogen (N_2) and oxygen (O_2) molecules in the combustion air. Most NO_x formed through the thermal NO_x mechanism occurs in the high temperature flame zone near the burners. The formation of thermal NO_x is affected by three furnace-zone factors: (i) oxygen concentration, (ii) peak temperature, and (iii) time of exposure at peak temperature. As these three factors increase, NO_x emission levels increase. The emission trends due to changes in these factors are fairly consistent for all types of natural gas–fired boilers and furnaces. Emission levels vary considerably with the type and size of combustor and with operating conditions (e.g., combustion air temperature, volumetric heat release rate, load, and excess oxygen level).

The second mechanism of NO_X formation, called prompt NO_X, occurs through early reactions of nitrogen molecules in the combustion air and hydrocarbon radicals from the hydrocarbon. Prompt NO_X reactions occur within the flame and are usually negligible when compared to the amount of NO_X formed through the thermal NO_X mechanism. However, prompt NO_X levels may become significant with ultra-low-NO_X burners. The third mechanism of NO_X formation, called hydrocarbon NO_X, stems from the evolution and reaction of hydrocarbon-bound nitrogen compounds with oxygen. Due to the characteristically low hydrocarbon nitrogen content of natural gas, NO_X formation through the hydrocarbon NO_X mechanism is insignificant.

The rate of CO emissions from boilers depends on the efficiency of natural gas combustion. Improperly tuned boilers and boilers operating at off-design levels decrease combustion efficiency resulting in increased CO emissions. In some cases, the addition of NO_X control systems such as low NO_X burners and flue gas recirculation (FGR) may also reduce combustion efficiency, resulting in higher carbon monoxide emissions relative to uncontrolled boilers.

The rate of emissions of volatile organic compounds from boilers and furnaces also depends on combustion efficiency. Emissions of volatile organic compounds are minimized by combustion practices that promote high combustion temperatures, long residence times at those temperatures, and turbulent mixing of hydrocarbon and combustion air. Trace amounts of volatile organic compounds in the natural gas hydrocarbon (e.g., formaldehyde and benzene) may also contribute to emissions of volatile organic compounds if they are not completely combusted in the boiler.

Emissions of sulfur dioxide from natural gas-fired boilers are low because pipeline quality natural gas typically has sulfur levels of 2000 grains per million cubic feet. However, sulfur-containing odorants are added to natural gas for detecting leaks, leading to small amounts of SO_2 emissions. Boilers combusting unprocessed natural gas may have higher sulfur oxide emissions due to higher levels of sulfur in the raw natural gas.

Because natural gas is a gaseous hydrocarbon, filterable emissions particulate matter are typically low. Particulate matter from natural gas combustion has been estimated to be less than 1 μm in size and has filterable and condensable fractions. Particulate matter in natural gas combustion is usually larger molecular weight hydrocarbon derivatives that are not fully combusted. Increased emissions of particulate matter may result from poor air/fuel mixing or maintenance problems.

Carbon monoxide, methane, and nitrous oxide emissions are all produced during natural gas combustion. In properly tuned boilers, nearly all of the fuel carbon (99.9%) in natural gas is converted to carbon dioxide during the combustion process—this conversion is relatively independent of boiler or combustor type. Hydrocarbon carbon not converted to carbon dioxide results in emissions of methane, carbon monoxide, and volatile organic compounds and is due to incomplete combustion. Even in boilers operating with poor combustion efficiency, the amounts of methane, carbon monoxide, and volatile organic compounds produced is insignificant compared to carbon dioxide levels.

Formation of nitrous oxide (N_2O) during the combustion process is affected by two furnace-zone factors. Nitrous oxide emissions are minimized when combustion temperatures are kept high (above 800°C, 1475°F) and excess oxygen is kept to a minimum (less than 1%).

Methane emissions are highest during low-temperature combustion or incomplete combustion, such as the start-up or shut-down cycle for boilers. Typically, conditions that favor formation of nitrous oxide also favor emissions of methane.

4.2 Liquid hydrocarbons

The ignition and combustion of liquid hydrocarbon derivatives play a major role in the operation of the gasoline engine (internal combustion engine), the diesel engine (compression engine), gas turbines, and industrial burners. Liquid hydrocarbon derivatives are usually burned as sprays of small liquid droplets, the droplets first evaporating to produce a cloud of vapor which then burns in the gas phase.

The purpose of a gasoline car engine (*the internal combustion engine*) is to convert gasoline into motion. In the engine, the gasoline ignites and is converted to motion energy. The diesel engine is also an internal combustion engine but the hydrocarbon fuel (diesel fuel) is not ignited by a spark but by an increase in pressure which creates sufficient heat to ignite the fuel thereby commencing the generation of energy. The principle behind any reciprocating internal combustion engine is the amount of energy is released in the form of expanding gas from the hydrocarbon fuel. The engine creates a cycle that allows energy release in the form of controlled explosions that occur many times per second.

Almost all cars currently use a *four-stroke combustion cycle* to convert gasoline into motion (the Otto cycle, invented by Nikolaus Otto in 1867): (i) intake stroke, (ii) compression stroke, (iii) combustion stroke, and (iv) exhaust stroke. To commence the cycle, the piston starts at the top of the cylinder, the intake valve opens, and the piston moves down the cylinder to allow the engine to take in a cylinder-full of air and gasoline (*intake stroke*). The piston moves back up to compress this hydrocarbon/air mixture (*compression stroke*) and when the piston reaches the top of its stroke, the spark plug emits a spark to ignite the gasoline (*explosion stroke*). Once the piston hits the bottom of its stroke, the exhaust valve opens and the exhaust gases leave the cylinder to go out the tailpipe (*exhaust stroke*) after which the engine is ready for the next cycle, which commences with another intake of air and hydrocarbon. The motion that comes out of an internal combustion engine is rotational because the linear motion of the pistons is converted into rotational motion by the crankshaft, which transmits the energy to the wheels.

The steam engine in old-fashioned trains and steam boats is the best example of an *external combustion engine*. The fuel (coal, wood, oil) in a steam engine burns outside the engine to create steam, and the steam creates motion inside the engine. Internal combustion is a lot more efficient insofar as it requires less fuel per mile than the external combustion engine. Furthermore, an internal combustion engine is much smaller in size than an equivalent external combustion engine, thereby reducing the overall weight of the vehicle to be transported. An aircraft using an external combustion engine would not make it of the ground— the weight/fuel ratio (i.e., weight/energy ratio) would be too high.

4.3 Solid hydrocarbons

The combustion of solid fuels is typically restricted to the combustion of resid, coal, coke, and the like. In the loose terminology that is often used, such fuels are often (incorrectly) referred to as hydrocarbons, but they are not true hydrocarbon and are more correctly referred to as hydrocarbonaceous fuel. A hydrocarbonaceous fuel is a fuel that contains not only carbon and hydrogen but also other elements such as sulfur, nitrogen, and metals that are chemically bound within, or chemically associated with, the fuel.

The combustion of such fuels consists of three phases: (i) the preheating phase in which the unburned fuel is heated up to the flash point and then the fire point at which time flammable gases start being evolved in a process similar to dry distillation, (ii) the distillation phase or gaseous phase in which the mixture of evolved flammable gases with oxygen is ignited and heat transfer from the combustion to the solid maintains the evolution of flammable vapors, and (iii) the charcoal phase or solid phase in which the production of flammable gases from the fuel is too low for the continued presence of a flame and the charred fuel does not burn rapidly but just glows and later only smolders.

4.4 Nonhydrocarbons

For the purpose of this text nonhydrocarbons are those materials that contain elements other than carbon and hydrogen. The nonhydrocarbons that occur in crude oils and crude oil products may be small in quantity but some of them have considerable influence on product quality. In many cases they have noxious or harmful effects and must be removed or converted to less harmful compounds during the refining process. The most common occurring nonhydrocarbons are sulfur, nitrogen, and oxygen. There may also be small amounts of vanadium, nickel, sodium, and potassium. Any of these nonhydrocarbon compounds can pass into the higher-boiling fuels, such as the various fuel oil grades and special grades that are known as the residual fuel oils.

4.4.1 Fuel oil

Fuel oil is classified or graded in several ways (Tables 10.4 and 10.5) but in the United States was formally divided into two main types: *distillate fuel oil* and *residual fuel oil*, each of which was a blend of two or more refinery streams. Distillate fuel oils are vaporized and condensed during a distillation process and thus have a definite boiling range and do not contain high-boiling constituents. A fuel oil that contains any amount of the residue from crude distillation of thermal cracking is a residual fuel oil. The terms *distillate fuel oil* and *residual fuel oil* are losing their significance, since fuel oil is now made for specific uses and may be either distillates or residuals or mixtures of the two. The terms *domestic fuel oil*, *diesel fuel oil*, and *heavy fuel oil* are more indicative of the uses of fuel oils.

All of the fuel oil classes described here are refined from crude oil and may be categorized as either a distillate fuel or a residual fuel depending on the method of production (Speight, 2014). Fuel oil No. 1 and fuel oil No. 2 are distillate fuels which consist of distilled process streams. Residual fuel oil, such as fuel oil No. 4, is composed of the residuum remaining after distillation or cracking, or blends of such residues with distillates. Diesel fuel is approximately similar to fuel oil used for heating (fuel oil No. 1, fuel oil No. 2, and fuel oil No. 4).

All fuel oils consist of complex mixtures of aliphatic and aromatic hydrocarbons, the relative amounts depending on the source and grade of the fuel oil. The aliphatic alkanes (paraffins) and cycloalkane constituents (naphthene constituents) are hydrogen saturated and compose as much as 90% w/w of the fuel oil. Aromatic constituents (e.g., benzene) and olefin constituents compose up to 20% v/v and 1% v/v, respectively, of the fuel oils. Fuel oil No. 1 (straight-run kerosene) is a distillate which consists primarily of hydrocarbons in the C_9 to C_{16} range while fuel oil No. 2 is a higher-boiling, usually blended, distillate with hydrocarbons in the C_{11} to C_{20} range.

Diesel fuels predominantly contain a mixture of C_{10} to C_{19} hydrocarbons, which include aliphatic hydrocarbons (approximately 65% v/v), olefin hydrocarbons (up to 2% v/v), and aromatic hydrocarbons (up to 35% v/v). Jet fuels are based primarily on straight-run kerosene, as well as additives. All of the above fuel oils contain less than 5% v/v polycyclic aromatic hydrocarbons. Fuel No. 4 (also known as *marine diesel fuel*) is less volatile than diesel fuel No. 2 and may contain up to 15% v/v residual (high-boiling) streams, in addition to 5%–10% v/v polycyclic aromatic hydrocarbon constituents.

In some countries, residual fuel oil is also known as *heavy fuel oil* which consists primarily of the residue from distillation or cracking units in the

TABLE 10.4 General grading of fuel oil.

Fuel oil	Description
Number 1	A volatile distillate oil intended for vaporizing pot-type burners. Other names include coal oil, stove oil, and range oil.
Number 2	A distillate home heating oil. Sometimes known as Bunker A.
Number 3	A distillate oil for burners requiring low-viscosity fuel. ASTM merged this grade into the number 2 specification, and the term has been rarely used since the mid-20th century.
Number 4	A commercial heating oil for burner installations not equipped with preheaters. May be obtained from the heavy (vacuum) gas oil fraction.
Number 5	A residual-type industrial heating oil requiring preheating to 77°C–104°C (171°F–219°F) for proper atomization at the burners. Sometimes known as Bunker B. May be obtained from the heavy (vacuum) gas oil fraction or may be a blend of residual oil with enough number 2 oil to adjust viscosity until it can be pumped without preheating.
Number 6	A high-viscosity residual oil requiring preheating to 104°C–127°C (219°F–261°F). May contain various undesirable impurities, including 2% water and 0.5% mineral soil. May be known as residual fuel oil (RFO) or Bunker C.

refinery. Historically, fuel oils were based on residua from the atmospheric distillation column and were known as straight-run fuels. However, the increasing demand for transportation fuels such as gasoline, kerosene, and diesel fuel has led to an increased value for the atmospheric residue as a feedstock for vacuum distillation and for cracking processes. As a consequence, most heavy fuel oils are currently based on vacuum residua from thermal and catalytic cracking operations. These fuels differ in character from straight-run fuels in that the density and mean molecular weight are higher—the atomic hydrogen-carbon is lower indicating a higher degree of aromaticity. The density of some heavy fuel oils can be above that of water (>1.000) which has environmental implications in the event of a spillage into water systems.

TABLE 10.5 Other nomenclature of the various fuel oils.

Name	Other name	Other name	Type	Chain length
No. 1 fuel oil	No. 1 distillate	No. 1 diesel fuel	Distillate	9–16
No. 2 fuel oil	No. 2 distillate	No. 2 diesel fuel	Distillate	10–20
No. 3 fuel oil	No. 3 distillate	No. 3 diesel fuel	Distillate	
No. 4 fuel oil	No. 4 distillate	No. 4 residual fuel oil	Distillate/ residual	12–70
No. 5 fuel oil	No. 5 residual fuel oil	Heavy fuel oil	Residual	12–70
No. 6 fuel oil	No. 6 residual fuel oil	Heavy fuel oil	Residual	20–70

Residual fuel oil (and/or heavy fuel oil) is typically more complex in composition and impurities than distillate fuel oil. Therefore, a specific composition cannot always be determined—the sulfur content in residual fuel oil has been reported to vary up to 5% w/w. Residual fuel oils are complex mixtures of high molecular weight compounds having a typical boiling range from 350°C to 650°C (660°F −1200°F). They consist of aromatic, aliphatic, and naphthenic hydrocarbons, typically having carbon numbers from C_{20} to C_{50}, together with asphaltene constituents and smaller amounts of heterocyclic compounds containing sulfur, nitrogen, and oxygen. They have chemical characteristics similar to liquid asphalt and hence, are considered to be stabilized suspensions of asphaltene constituents in an oily medium.

Residual fuel oil also contains organometallic compounds from their presence in the original crude oil—the most important of which are nickel and vanadium. The metals (especially vanadium) are of particularly major significance for fuels burned in both diesel engines and boilers because when combined with sodium (perhaps from brine contamination from the reservoir or remaining after the refinery dewatering/desalting process) and other metallic compounds in critical proportions can lead to the formation of high melting point ash which is corrosive to engine parts. Other elements that occur in heavy fuel oils include iron, potassium, aluminum, and silicon—the latter two metals are mainly derived from refinery catalyst fines.

Distillate fuel oil is a product of the distillation process and has a definite boiling range; it does not contain high-boiling oils or asphaltic components.

Heavy fuel oil includes a variety of oils ranging from distillates to residual oils that must be heated to 260°C (500°F) or higher before they can be used. The quality and performance requirements for fuel oils differ widely although general quality limitations for various fuel grades are used to serve as guides in the manufacture, sale, and purchase of the oils. These quality definitions are often referred to as *specifications* or *classifications* but more precise specifications of quality requirements such as the vapor pressure (ASTM D323) and metals content (ASTM D5184; ASTM D5708; ASTM D5863), may be required for any given application (ASTM D396).

Fuel oil, therefore, in its various categories has an extensive range of applications and the choice of a standard procedure to be used for assessing or controlling fuel quality must, of necessity, depend both upon the type of fuel and its ultimate use. Specifications for both middle distillate heating fuels and transportation fuels are similar; as a consequence, it is often possible for refiners to satisfy the performance requirements of both applications with the same process stream or blend of process streams.

While fuel oil is typically in the liquid state, it is not usually a pure hydrocarbon mixture and does contain nonhydrocarbon constituents. Two major categories of fuel oil are burned by combustion sources: distillate oils and residual oils. These oils are further distinguished by grade numbers, with Nos. 1 and 2 being distillate oils; Nos. 5 and 6 being residual oils; and No. 4 being either distillate oil or a mixture of distillate and residual oils. No. 6 fuel oil is sometimes referred to as Bunker C.

Distillate oils are more volatile and less viscous than residual oils. They have negligible nitrogen and ash contents and typically contain less than 0.3% sulfur (by weight). Distillate oils are used mainly in domestic and small commercial applications, and include kerosene and diesel fuels. Being more viscous and less volatile than distillate oils, the heavier residual oils (Nos. 5 and 6) may need to be heated for ease of handling and to facilitate proper atomization. Because residual oils are produced from the residue remaining after the lower boiling fractions (gasoline, kerosene, and distillate oils) have been removed from the crude oil, they contain significant quantities of ash, nitrogen, and sulfur. Residual oils are used mainly in utility, industrial, and large commercial applications.

The major boiler configurations for fuel oil combustion are (i) water tube boiler, (ii) firetube boiler, and (iii) cast iron boiler (see previous section). Boilers are classified according to design and orientation of heat transfer surfaces, burner configuration, and size. These factors can all strongly influence emissions as well as the potential for controlling emissions. In addition to the three categories of boilers, another type of heat transfer configuration used on smaller boilers is the *tubeless design*. This design incorporates nested pressure vessels with water in between the shells. Combustion gases are fired into the inner pressure vessel and are then sometimes recirculated outside the second vessel.

Emissions from fuel oil combustion depend on the grade and composition of the fuel, the type and size of the boiler, the firing and loading practices used, and the level of equipment maintenance. Because the combustion characteristics of distillate and residual oils are different, their combustion can produce significantly different emissions. In general, the baseline emissions of criteria and noncriteria pollutants are those from uncontrolled combustion sources, which are sources without add-on air pollution control equipment or other combustion modifications designed for emission control.

Particulate emissions from fuels oil combustion may be categorized as either (i) filterable or (ii) condensable. Filterable emissions are generally considered to be the particulate matter that is trapped by a glass fiber filter, which traps particulate matter larger 0.3 microns from passing through the filter. Condensable particulate matter is material that is emitted in the vapor state which later condenses to form homogeneous and/or heterogeneous aerosol particles. The condensable particulates emitted from boilers used for fuel oil combustion is primarily inorganic in nature. Filterable particulate matter emissions depend predominantly on the grade of fuel fired. Combustion of lower-boiling distillate oil results in significantly lower particulate matter formation than does the combustion of higher boiling residual oil. Among residual oils, firing of No. 4 or No. 5 oil usually produces less particulate matter than does the firing of heavier No. 6 oil.

In general, filterable particulate matter emissions depend on the completeness of combustion as well as on the oil ash content. The particulate matter emitted by distillate oil–fired boilers primarily comprises carbonaceous particles resulting from incomplete combustion of oil and is not correlated to the ash or sulfur content of the oil. However, particulate matter emissions from residual oil burning are related to the oil sulfur content because low-sulfur No. 6 oil, either from naturally low-sulfur crude oil or desulfurized by one of several processes, exhibits substantially lower viscosity and reduced asphaltene content, reduced mineral matter content, and reduced sulfur content, which results in better atomization and more complete combustion.

Boiler load can also affect filterable particulate emissions in units firing No. 6 oil. At low load (50% of maximum rating) conditions, particulate emissions from utility boilers may be lowered by 30%–40% and by as much as 60% from small industrial and commercial units. At very low load conditions (approximately 30% of maximum rating), proper combustion conditions may be difficult to maintain and particulate emissions may increase significantly.

Sulfur oxides (SO_X) are generated during oil combustion from the oxidation of sulfur contained in the fuel—the emissions of sulfur oxides from conventional combustion systems are predominantly in the form of sulfur dioxide. Uncontrolled emission of sulfur oxides are almost entirely dependent on the sulfur content of the fuel and are not affected by boiler size, burner design, or grade of fuel being fired. On average, more than 95% of the fuel sulfur is converted to sulfur dioxide, approximately 1%−5% is further oxidized to sulfur trioxide (SO_3), and 1%−3% is emitted as particulate matter containing sulfur (usually as sulfates). Sulfur trioxide reacts readily with water vapor (both in the atmosphere and in flue gases) to form a sulfuric acid mist.

Oxides of nitrogen (NO_X—a mixture of nitric oxide (NO) and nitrogen dioxide (NO_2)) formed in combustion processes are due either to thermal fixation of atmospheric nitrogen in the combustion air (*thermal* NO_x), or to the conversion of chemically bound nitrogen in the fuel (*fuel* NO_x). Fuel nitrogen conversion is the more important NO_X-forming mechanism in residual oil boilers. It can account for 50% of the total nitrogen oxide emissions from residual oil firing. The percent conversion of fuel nitrogen to nitrogen oxides typically varies from 20% to 90% w/w of nitrogen in the fuel oil. On the other hand, thermal fixation is the dominant NO_X-forming mechanism in units firing distillate fuel oils, primarily because of the negligible nitrogen content in these lower-boiling oils. Distillate fuel oil−fired boilers are usually smaller and have lower heat release rates and the quantity of thermal NO_X formed in them is less than that of larger units which typically burn residual fuel oil.

A number of variables influence how much NO_X is formed by these two mechanisms. One important variable is firing configuration. NOx emissions from tangentially (corner) fired boilers are, on an average, less than those of horizontally opposed units. Also important are the firing practices employed during boiler operation. Low excess air (LEA) firing, flue gas recirculation (FGR), staged combustion (SC), reduced air preheat (RAP), low NO_X burners (LNBs), burning oil/water emulsions (OWE), or some combination thereof may result in NO_X reductions of 5%−60%. Load reduction (LR) can likewise decrease production of nitrogen oxides, which may also be reduced from 0.5% to 1% for each percentage reduction in load from full-load operation. Most of these variables, with the exception of excess air, only influence the emissions of nitrogen oxides of large fuel oil−fired boilers. Low excess air-firing is possible in many small boilers, but the resulting reductions of nitrogen oxide emissions are less significant.

The rate of *carbon monoxide* (CO) emissions from combustion sources depends on the oxidation efficiency of the fuel. By controlling the combustion process carefully, carbon monoxide emissions can be minimized. Smaller boilers, heaters, and furnaces tend to emit more of these pollutants than larger combustors because smaller units usually have a higher ratio of heat transfer surface area to flame volume than the larger combustors. This leads to reduced flame temperature and combustion intensity and, therefore, to lower combustion efficiency.

The presence of *carbon monoxide* in the exhaust gases of combustion systems results principally from incomplete fuel combustion. Several conditions can lead to incomplete combustion, including (i) insufficient oxygen availability, (ii) poor fuel/air mixing, (iii) cold-wall flame quenching, (iv) reduced combustion temperature, (v) decreased combustion gas residence time, and (vi) load reduction (i.e., reduced combustion intensity). Since various combustion modifications for NO_X reduction can produce one or more of the above conditions, the possibility of increased carbon monoxide emissions is an environmental concern as well as energy efficiency, and operational aspects of the boiler.

Small amounts of *organic compounds* are emitted from fuel oil combustion. As with carbon monoxide emissions, the rate at which organic compounds are emitted depends, to some extent, on the combustion efficiency of the boiler. Therefore, any combustion modification which reduces the combustion efficiency will most likely increase the concentrations of organic compounds in the flue gases.

Total organic compounds (TOCs) include volatile organic compounds, semivolatile organic compounds, and condensable organic compounds. Emissions of volatile organic compounds are primarily characterized by the criteria pollutant class of unburned vapor phase hydrocarbon derivatives. Unburned hydrocarbon emissions can include essentially all vapor phase organic compounds emitted from a combustion source. These are primarily emissions of aliphatic, oxygenated, and low molecular weight aromatic compounds which exist in the vapor phase at flue gas temperatures. These emissions include all alkane, alkene, aldehyde, carboxylic acid, and substituted benzene derivatives (e.g., benzene, toluene, xylene, ethyl benzene—BTEX). Monitoring and mitigating the emissions of the BTEX group (comprising benzene, toluene, ethylbenzene, and xylene and often also expressed as total xylenes, *ortho* + *meta* + *para*) is necessary for both ambient applications, and industrial health and safety applications:

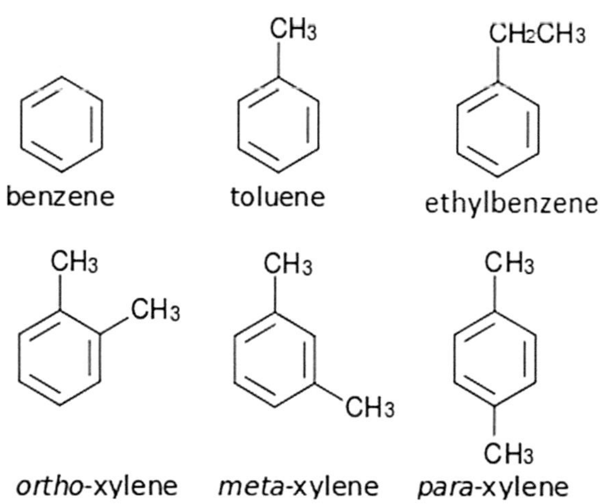

Benzene (C_6H_6) is a clear, colorless, flammable liquid that is found in ambient air as a result of burning fuels, such as crude oil–based fuels, coal, and wood. Benzene is common in unleaded fuel, where it is added as a substitute for lead, allowing smoother running. Toluene ($C_6H_5CH_3$), also known as methylbenzene, is a colorless liquid which is widely used in industrial processes as a solvent. In nonindustrial uses, toluene can be found in gasoline as an octane booster and in glues, solvents, and resins. Ethylbenzene ($C_6H_5CH_2CH_3$) is a colorless liquid which is widely used in industrial processes for the manufacture of styrene, which is then used for polystyrene manufacture and also as a solvent in inks, dyes, and in gasoline. Xylene (C_8H_{10}) is the term used to describe the three isomers of dimethyl benzene; *m*-xylene, *p*-xylene, and *o*-xylene. Usually concentrations of each are added together as total xylenes. Xylene is refined from crude oil, and is a clear, greasy liquid widely used in the production of plastic bottles and polyester clothing and as a solvent with a range of applications from circuit board cleaning to thinning paints and varnishes.

The amount of BTEX in emissions often gives an indication of volatile organic compound (VOC) emissions from a range of sources. For example, benzene is commonly present in gasoline, vehicle exhaust, and burning of solid and liquid fuels. In an urban environment, benzene is usually present from these sources. The remainder of the BTEX suite is found in gasoline but the largest sources are industrial emissions. In areas with significant manufacturing industries, especially those using glues, solvents, and dyes, higher concentrations of toluene, ethylbenzene, and xylenes may be present. Monitoring BTEX will allow quantification of these, and can be used to identify health and safety concerns, demonstrate compliance with local regulations, and even assess if there is a leak in storage or processing infrastructure.

The remaining organic emissions are composed largely of compounds emitted from combustion sources in a condensed phase. These compounds can almost exclusively be classed into a group known as polycyclic organic matter (POM) or polynuclear aromatic hydrocarbon derivatives (PNA or PAH). There are also polynuclear aromatic hydrocarbon-nitrogen analogs.

Formaldehyde is also formed and emitted during combustion of hydrocarbon-based fuel oils. Formaldehyde is present in the vapor phase of the flue gas and is subject to oxidation and decomposition at high temperatures encountered during combustion. Thus, larger units with efficient combustion (resulting from closely regulated air-fuel ratios, uniformly high combustion chamber temperatures, and relatively long gas retention times) have lower formaldehyde emission rates than do smaller, less efficient combustion units.

Trace elements are also emitted from the combustion of fuel oil and, as expected, the quantity of trace elements emitted from the boiler depends on (i) the composition of the fuel oil, (ii) the combustion temperature, and (iii) the fuel feed mechanism. The temperature determines the degree of volatilization of specific compounds contained in the fuel. The fuel feed mechanism affects the separation of emissions into bottom ash and fly ash. In general, the quantity of any given metal emitted depends on the physical and chemical properties of the element itself; concentration of the metal in the fuel; the combustion conditions; and the type of particulate control device used, and its collection efficiency as a function of particle size.

Some trace metals concentrate in certain waste particle streams from a combustor (bottom ash, collector ash, flue gas particulate). By understanding trace metal partitioning and concentration in fine particulate, it is possible to postulate the effects of combustion controls on incremental trace metal emissions. For example, several NO_X controls for boilers reduce peak flame temperatures. If combustion temperatures are reduced, fewer metals will initially volatilize, and fewer will be available for subsequent condensation and enrichment on fine particulate matter. Therefore, for combustors with particulate controls, lower volatile metal emissions should result due to improved particulate removal.

The *greenhouse gases* carbon dioxide, methane, and nitrous oxide are all produced during fuel oil combustion. Nearly all of the fuel carbon (99%) in fuel oil is converted to carbon dioxide during the combustion process. Although the formation of carbon monoxide acts to reduce carbon dioxide emissions, the amount of carbon monoxide produced is insignificant compared to the amount of carbon dioxide produced. The majority of the fuel carbon not converted to carbon dioxide is due to incomplete combustion in the fuel stream.

Formation of nitrous oxide (N_2O) during the combustion process is governed by a complex series of reactions and its formation is dependent upon many factors. Formation of nitrous oxide is minimized when combustion temperatures are kept high (above 800°C, 1475°F) and excess air is kept to a minimum (less than 1% v/v). Emissions can vary widely from unit to unit—even from the same unit at different operating conditions.

Methane emissions vary with the type of fuel and firing configuration, but are highest during periods of incomplete combustion or low-temperature combustion, such as the start-up or shut-down cycle for oil-fired boilers. Typically, conditions that favor formation of nitrous oxide also favor emissions of methane.

4.4.2 Coal

Coal combustion is used in a range of applications which vary from domestic fires to large industrial furnaces and utility boilers. While, for reasons of economy, the oxidant is usually air, the coal may be in any degree of dispersion. In fact, coal combustion provides the majority of consumable energy to the world and despite the continuing search for alternate sources of energy (whether or not they are other fossil fuels or nonfossil fuels), coal appears to be so firmly entrenched that there is little doubt that coal combustion will remain important into the 21st century, particularly where a convenient method of storing energy is required, as for example in transport applications.

A major concern in the present day combustion of coal is the performance of the process in an environmentally acceptable manner through the use of a variety of environmentally acceptable technologies such as the use of a low-sulfur coal or through the use of postcombustion cleanup of the off-gases (Speight, 2019). Thus, there is a marked trend in the modern research to more efficient methods of coal combustion. In fact, the ideal would be a combustion system that is able to accept any coal without a precombustion treatment or without the need for postcombustion treatment or without emitting objectionable amounts of sulfur and nitrogen oxides and particulates.

Coal combustion is a complex science because of the variety of physical and chemical properties of coal (Field et al., 1967; Morrison, 1986; Heitmann, 1993). In addition, it is not only the amount of energy available from coal combustion but also other aspects such as fuel handling, ash removal, emissions, and environmental control techniques that are of extreme importance.

Combustion occurs, chemically, by initiation and propagation of a self-supporting exothermic reaction. The physical processes involved in combustion are principally those which involve the transport of matter and the transport of energy. The conduction of heat, the diffusion of chemical species, and the bulk flow of the gas all follow from the release of chemical energy in an exothermic reaction. Thus, combustion phenomena arise from the interaction of chemical and physical processes. The first requirement, somewhat difficult with coal because of its molecular complexity, is that the overall stoichiometry of the reaction must always be established. For these purposes, coal is usually represented by carbon which can react with oxygen in two ways, producing either carbon monoxide or carbon dioxide.

In direct combustion, coal is burned (i.e., the carbon and hydrogen in the coal are oxidized into carbon dioxide and water) to convert the chemical energy of the coal into thermal energy after which the sensible heat in the products of combustion then can be converted into steam that can be external work or directly into shaft horsepower (e.g., in a gas turbine). In fact, the combustion process actually represents a means of achieving the complete oxidation of coal.

Coal combustion may be simply represented as the staged oxidation of coal carbon to carbon dioxide with any reactions of the hydrogen in the coal being considered to be of secondary importance. The stoichiometric reaction equations are quite simple, but there is a confusing variation of hypotheses related to the sequential reaction mechanism which is caused to be a great extent by the heterogeneous nature (solid and gaseous phases) of the reaction. But, for the purposes of this text, the chemistry will remain simple as shown in the above equations. Other types of combustion systems may be rate-controlled due to the onset of the Boudouard reaction.

In more general terms, the combustion of carbonaceous materials (which contain hydrogen and oxygen as well as carbon) involves a wide variety of reactions between the many reactants, intermediates, and products. The reactions occur simultaneously and consecutively (in both forward and reverse directions) and may at times approach a condition of equilibrium. Furthermore, there is a change in the physical and chemical structure of the fuel particle as it burns.

Coal quality and/or rank have an impact, often significant, on combustion, especially on many areas of power plant operation. The parameters of rank, mineral matter content (ash content), sulfur content, and moisture content are regarded as determining factors in combustibility as it relates to both heating value and ease of reaction. Thus, lower rank coals (though having lower heat content) may be more reactive than higher rank coals, thus implying that rank does not influence coal combustibility. At the same time, anthracites (with a low volatile matter content) are generally more difficult to burn than bituminous coals.

The lower the rank of a coal the greater the wettability with water, but the higher the rank the greater the wettability with tar or pitch. High moisture content is associated with a high unit surface area of the coal (especially for retained moisture after drying) and coals also become harder to grind as the percentage of volatiles decreases). Lignite usually serves as the more extreme example of low-grade fuel of high moisture content and the problems encountered in lignite combustion are often applicable to other systems (Nowacki, 1980). Lignite gives up its moisture more slowly than harder coals but the higher volatile content tends to offset the effect of high moisture. For the combustion of pulverized material, it appears essential to dry lignite and brown coals to 15%–20% moisture; the lowest possible ash and moisture contents are desired as well as high grindability, high heat content, and high fusion temperature.

Finally, since coal quality can be affected by oxidation or weathering, the question is raised related to the effects of oxidation and weathering on combustion and whether oxidized or weathered coal could maintain a self-sustaining flame in an industrial boiler. The inhibition of volatile matter release due to changes in the char morphology, because of reduced thermoplasticity of coal—as a result of the oxidation/weathering—suggests that this may not be the case. One option for managing coal quality for power generation is to blend one particular coal with others until a satisfactory feedstock is achieved. This is similar to current crude oil refinery practice where one refinery actually accepts a blend of various crude oils and operates on the basis of average feedstock composition. The exact nature of the coal combustion process is difficult to resolve but can be generally formulated as two processes: (i) the degradation of hydrogen, and (ii) the degradation of carbon.

Combustion actually occurs on the surface with the oxidant being adsorbed there prior to reaction. However, the initial reaction at (on) the surface is not necessarily the rate-determining step; the process involves a sequence of reactions, any one of which may control the rate. The initial step is the transfer of reactant (i.e., oxygen) through the layer of gas adjacent to the surface of the particle. The reactant is then adsorbed and reacts with the solid after which the gaseous products diffuse away from the surface. If the solid is porous, much of the available surface can only be reached by passage of the oxidant along the relatively narrow pores and this may be a rate controlling step. Rate control may also be exercised by: (a) adsorption and chemical reaction, which are considered as chemical reaction control; and (b) pore diffusion, by which the products diffuse away from the surface. This latter phenomenon is seldom a rate-controlling step. In general, rate control will occur if the surface reaction is slow compared with the diffusion processes; while diffusion shows less-marked temperature dependence, reaction control predominates at low temperatures but diffusion control is usually more important at higher temperatures.

On a chemical basis, hydrogen degradation outweighs the slower-starting carbon degradation in the early, or initial, stage of combustion. But, at the same time, the carbon monoxide/carbon dioxide ratio is decreased. After the initial stages of combustion, during which volatile material is evolved (which is also combustible), a nonvolatile carbonaceous residue (coke, char), which can comprise up to 90% of the original mass of the coal, remains.

During the combustion of the coke, three different zones (regimes) of combustion can be distinguished. In the first zone (I), the rate of diffusion to and away from the surface is very fast compared with the rate of the surface reaction; such phenomena are observed at low temperatures. At much higher temperatures, the rate at which oxygen molecules are transported from the bulk gas to the external surface is slow enough to be rate controlling (Zone III); the observed rate can be equated to the molar flux of oxygen to unit area of external surface. Finally (Zone II; intermediate between I and III), the oxygen transport to the external surface is rapid but diffusion into the pores before reaction is relatively slow.

The complex nature of coal as a molecular entity (Speight, 2013) has resulted in treatments of coal combustion being confined to the carbon in the system and, to a lesser extent, the hydrogen but it must be recognized that the system is extremely complex. Even with this simplification, there are several principal reactions that are considered to be an integral part of the overall combustion of coal.

In summary, it is more appropriate to consider the combustion of coal (which contains carbon, hydrogen, nitrogen, oxygen, and sulfur) as involving a variety of reactions between (i) the reactants; (ii) the intermediate, or transient, species; and (iii) the products. The reactions can occur both simultaneously and consecutively (in both forward and reverse directions) and may even approach steady state (equilibrium) conditions and, there is a change in the physical and chemical structure of the fuel particle during the process.

The ignition of coal has been described as occurring in just a few hundredths of a second with the onset of burning in less than half a second. It burns to carbon dioxide at distances close to the surface. Water is evaporated in the initial stages and the ignition is propagated through a dry bed.

For coal, the ignition temperatures are usually of the order of 700°C (1290°F), but may be as low as 600°C (1110°F) or as high as 800°C (1470 degrees), depending on volatiles evolved. In fact, ignition temperatures depend on rank and generally range from 150°C to 300°C (300°F–570°F) for lignite to 300°C–600°C (570°F–1110°F) for anthracite with some dependence on particle size being noted.

4.5 Formation of particulate matter

When pseudohydrocarbon fuels are used in place of pure hydrocarbon fuels (such as hydrocarbon themselves, gasoline, diesel, and the like), combustion processes can (depending upon the properties of the fuel) emit large quantities of particles to the atmosphere. Particles formed in combustion systems fall roughly into two categories. The first category, referred to as ash, comprises particles derived from noncombustible constituents (primarily mineral inclusions) in the fuel and from atoms other than carbon and hydrogen (heteroatoms) in the organic structure of the fuel. The second category consists of carbonaceous particles that are formed by pyrolysis of the fuel molecules.

Particles produced by combustion sources are generally complex chemical mixtures that often are not easily characterized in terms of composition. The particle sizes vary widely, and the composition may be a strong function of particle size.

Ash is derived from noncombustible material introduced in the combustor along with the fuel and from inorganic constituents in the fuel itself. Such fuels include high-boiling fuel oil (sometimes referred to as residual fuel oil), crude oil residua, and coal—all of which are often referred to *erroneously* as hydrocarbon fuels but not pure hydrocarbon fuels and contain heteroatoms such as nitrogen, oxygen, sulfur, and mineral matter.

The ash produced in coal combustion, for example, arises from mineral inclusions in the coal as well as from heteroatoms, which are present in the coal molecules. High-boiling fuel oils produce much less ash than coals since noncombustible material such as mineral inclusions are not always present in such fuel oils and heteroatoms are the only source of ash. Ash particles produced in residua combustion and coal combustion have long been controlled by cleaning the flue gases with electrostatic precipitators. Most of the mass of particulate matter is removed by such devices, so ash received relatively little attention as an air pollutant.

Residua and coal are complex, heterogeneous, and variable substances containing, in addition to the hydrocarbonaceous molecular species, dispersed mineral matter. The chemical and physical properties of the mineral matter vary considerably, but the mineral matter does eventually form ash particles as the carbon bums out. The ash particles that are entrained in the combustion gases are called *fly ash*.

Volatile fractions originally present in the feedstock or formed by pyrolysis are vaporized, and the particle may burst open from the internal evolution of such gases. As the carbon is consumed, mineral constituents come into contact with one another, forming larger ash agglomerates. Since the temperature of the combustion process is generally high enough that the ash melts, these agglomerates coalesce to form large droplets of molten ash on the surface of the burning char. The fragmentation of the char limits the degree of agglomeration of the ash within a single fuel particle, so a number of ash residue particles are produced from each parent feedstock particle.

To understand the vaporization of ash during coal combustion, it is necessary to examine the thermodynamics and chemistry of the ash and the transport of the volatilized ash from the surface of the particle. Some components of the ash are highly volatile; examples include sodium, potassium, and arsenic. Volatile ash constituents may vaporize completely during combustion unless inhibited by diffusional resistances, either in transport through the porous structure of the char or to the surfaces of the mineral inclusions. The vaporization process can be a direct transformation from the condensed phase to the vapor phase or it may involve the production of volatile suboxides or elemental forms from the original oxides. The former mechanism may dominate for the more volatile ash constituents, but there is evidence that the reduction reactions play an important role in the vaporization of species with relatively low vapor pressures.

4.6 Char and coke

The carbonaceous char residue that remains after a high-boiling fuel is devolatilized burns slowly by surface reactions. If the char particle is too large, mixing in the combustion is poor, or heat is transferred too quickly, char particles may not be fully consumed. High-boiling fuel oils may produce similar carbonaceous particles (*coke*) which are relatively large and account for the majority of particulate mass emitted from boilers fired with high-boiling fuel oil. They are hard *cenospheres*, porous carbonaceous shells containing many blowholes. Fuel impurities tend to concentrate in these cenospheres and coke particles are formed by liquid-phase pyrolysis of high-boiling fuel oil droplets.

The combustion of individual millimeter-sized droplets of heavy fuel oil involves two combustion times: (i) the droplet burning time, which corresponds to the time required for coke formation, and (ii) the time required for the coke particle to burn. During the first 60% of the droplet burning phase, combustion was relatively quiescent. In the final stages of droplet burning, the droplet deforms and finally appears to froth just prior to forming a small coke cenosphere. Small droplets may be ejected during the latter violent phase of coke formation. Immature coke particles were found to be tarry and soluble in organic solvents.

Typically, the coke particles account for approximately 3% w/w of the residual component of the fuel oil—even when the residual oil was diluted with 60% v/v of low-boiling distillate. Coke particle formation appears to be almost unavoidable in the combustion of heavy fuel oil. Emission rates would be reduced substantially by reducing the time required in coke combustion. Improved atomization or dilution of the heavy fuel oil with a lower boiling (lower molecular weight) component would decrease the initial size of the coke particles, thereby reducing the combustion time.

4.7 Soot

Carbonaceous particles can also be produced in the combustion of gaseous fuels and from the volatilized components of liquid or solid fuels. The particles (*soot*) formed by this route differ markedly from the char and coke discussed previously. Most commonly, soot particles are agglomerates of small, roughly spherical particles but while the size and morphology of the clusters can vary widely, the small spheres differ little from one source to another. The structural similarity between soot particles and the inorganic particles produced from volatilized ash suggests a common origin. However, the genesis of soot is much less well understood that that of the inorganic particles due to the extreme complexity of hydrocarbon chemistry in the flame, as well as to the fact that soot particles can burn if exposed to oxygen at high temperatures.

Soot forms in a flame as the result of a chain of events that begins with pyrolysis and oxidative pyrolysis of the fuel into small molecules, followed by chemical reactions that build up larger molecules that eventually get big enough to become very small particles. Soot formation is favored when the molar ratio of carbon to oxygen approaches 1.0, as suggested by the stoichiometry. In premixed flames the critical carbon/oxygen ratio for soot formation is found to be smaller than 1.0 and is closer to 0.5. The lower carbon/oxygen ratio suggests that an appreciable amount of the carbon is tied up in stable molecules such as carbon dioxide.

The propensity to form soot (as measured by the critical carbon/oxygen ratio at which soot formation begins) is a complex function of flame type, temperature, and the properties of the fuel oil. There is general agreement that the rank ordering of the soot-forming tendency of fuel components is:

Naphthalene hydrocarbons > Benzene hydrocarbons > Aliphatic hydrocarbons

However, the order of soot-forming tendencies of the aliphatic hydrocarbon derivatives (alkane, alkene, and alkyne derivatives) varies dramatically with flame type and flame temperature. In premixed flames, soot formation appears to be determined by a competition between the rate of pyrolysis and growth of soot precursors and the rate of oxidative attack on these precursors. As the temperature increases, the oxidation rate increases faster than the pyrolysis rate, and soot formation decreases.

The difference between the soot-forming tendencies of aromatic hydrocarbon derivatives and aliphatic hydrocarbon derivatives is thought to result from different routes of formation. Aliphatic hydrocarbon derivatives appear to form soot primarily through formation of acetylene and polyacetylenes, but at a relatively slow rate. Aromatic hydrocarbon derivatives can form soot by a similar process, but there is a more direct route involving ring condensation or polymerization reactions that build on the existing aromatic structure. The fragmentation of aromatics should occur primarily at high temperature, but such reactions may not be important. In flames, fuel pyrolysis generally begins at relatively low temperature as the fuel approaches the flame front, so the soot inception process may be completed well before temperatures are high enough to initiate the competitive reactions.

References

ASTM D323, 2019. Standard Test Method for Vapor Pressure of Petroleum Products (Reid Method). Annual Book of Standards. ASTM International West Conshohocken, Pennsylvania.

ASTM D396, 2019. Standard Specification for Fuel Oils. Annual Book of Standards, ASTM International West Conshohocken, Pennsylvania. Annual Book of Standards, ASTM International West Conshohocken, Pennsylvania.

ASTM D5184, 2019. Standard Test Methods for Determination of Aluminum and Silicon in Fuel Oils by Ashing, Fusion. In: Inductively Coupled Plasma Atomic Emission Spectrometry, and Atomic Absorption Spectrometry. Annual Book of Standards, ASTM International West Conshohocken, Pennsylvania.

ASTM D5708, 2019. Standard Test Methods for Determination of Nickel, Vanadium, and Iron in Crude Oils and Residual Fuels by Inductively Coupled Plasma (ICP) Atomic Emission Spectrometry. Annual Book of Standards, ASTM International West Conshohocken, Pennsylvania.

ASTM D5863, 2019. Standard Test Methods for Determination of Nickel, Vanadium, Iron, and Sodium in Crude Oils and Residual Fuels by Flame Atomic Absorption Spectrometry. Annual Book of Standards, ASTM International West Conshohocken, Pennsylvania.

ASTM D8221, 2019. Standard Practice for Determining the Calculated Methane Number (MNC) of Gaseous Fuels Used in Internal Combustion Engines. Annual Book of Standards, ASTM International West Conshohocken, Pennsylvania.

Field, M.A., Gill, D.W., Morgan, B.B., Hawksley, P.G.W., 1967. Combustion of Pulverized Coal. British Coal Utilization Research Association, Leatherhead, Surrey.

Glassman, I., 1996. Combustion, third ed. Academic Press Inc., New York.

Heitmann, H.-G., 1993. Handbook of Power Plant Chemistry. CRC Press Inc., Boca Raton, Florida.

Mokhatab, S., Poe, W.W., Speight, J.G., 2006. Handbook of Natural Gas Transmission and Processing. Elsevier, Amsterdam, Netherlands.

Morrison, G.F., 1986. Understanding Pulverized Coal Combustion. Report No. ICTIS/TR34. IEA Coal Research. International Energy Agency, London.

Nowacki, P., 1980. Lignite Technology. Noyes Data Corporation, Park Ridge, New Jersey.

Rein, G., 2009. Smoldering combustion phenomena in science and technology. International Review of Chemical Engineering 1, 3–18.

Speight, J.G., 2013. The Chemistry and Technology of Coal, third ed. CRC Press, Taylor and Francis Group, Boca Raton, Florida.

Speight, J.G., 2014. The Chemistry and Technology of Petroleum, fifth ed. CRC Press, Taylor and Francis Group, Boca Raton, Florida.

Speight, J.G., 2019. Natural Gas: A Basic Handbook, second ed. Gulf Publishing Company, Elsevier, Cambridge, Massachusetts.

Warnatz, J., Maas, U., Dibble, R.W., 1996. Combustion: Physical and Chemical Fundamentals, fourth ed. Springer-Verlag, Berlin, Germany.

Chapter 11

Reactions of hydrocarbons

1. Introduction

Hydrocarbon derivatives and hydrocarbon fuels (gas, liquid, and solid) are one of the Earth's most important energy resources. The predominant use of hydrocarbon derivatives (individually or as fuels) is as a combustible fuel source. However, in addition, hydrocarbon fuels can be harnessed to create mechanical energy through combustion (Chapter 10).

Hydrocarbon mixtures are produced in refineries by distillation from natural gas (Chapter 4) and crude oil (Chapter 3) as well as by thermal decomposition of higher boiling predominantly hydrocarbon fractions (such as gas oil) (Parkash, 2003; Gary et al., 2007; Speight, 2014, 2017; Hsu and Robinson, 2017). For example, naphtha is obtained from crude oil refineries as the lowest boiling portion of the distillate from which crude oil is manufactured. Naphtha is also produced by fluid catalytic cracking of higher boiling feedstocks and has a density (specific gravity) on the order of 0.6 and 0.8 depending on the composition of the naphtha (Parkash, 2003; Gary et al., 2007; Speight, 2014, 2017; Hsu and Robinson, 2017).

The individual hydrocarbon derivatives differ both in the total number of carbon and hydrogen atoms in their molecules and in the proportion of hydrogen to carbon and can be divided into various homologous series (Chapter 1). Each member of such a series shows a definite relationship in its structural formula to the members preceding and following it, and there is generally some regularity in changes in physical properties of successive members of a series.

The alkane derivatives are a homologous series of saturated aliphatic hydrocarbon derivatives. The first and simplest member of this series is methane (CH_4); the series is sometimes called the methane series. Each successive member of a homologous series of hydrocarbon derivatives has one more carbon and two more hydrogen atoms in its molecule than the preceding member. The second alkane is ethane (C_2H_6) and the third is propane (C_3H_8). Alkane derivatives have the general formula C_nH_{2n+2} (where n is an integer greater than or equal to 1). Generally, alkane derivatives of low molecular weight (such as methane, ethane, and propane) are gases while the alkane derivatives of intermediate molecular weight (e.g., hexane, heptane, and octane) are liquids, and

Handbook of Industrial Hydrocarbon Processes. https://doi.org/10.1016/B978-0-12-809923-0.00011-4

the higher molecular weight alkane derivatives (those above heptadecane ($C_{17}H_{36}$)) are solids. Other homologous series of hydrocarbon derivatives include the alkene derivatives (RC=CR) and the alkyne derivatives (RC≡CR). The various alkyl derivatives of benzene are sometimes referred to as the benzene series. The hydrocarbon derivatives differ in thermal activity.

Methane and ethane are gaseous at ambient temperatures and pressures (STP) and cannot be readily liquefied by pressure alone. Propane, which is also gaseous at STP, is however easily liquefied, and exists in propane bottles mostly as a liquid. Butane, also a gas at standard temperature and pressure (STP) is so easily liquefied that it provides a safe, volatile fuel for small pocket lighters. Pentane is a clear liquid at room temperature, commonly used in chemistry and industry as a powerful nearly odorless solvent of waxes and high molecular weight organic compounds, including greases. Hexane is also a widely used nonpolar, nonaromatic solvent, as well as a significant fraction of gasoline.

The five-carbon through ten-carbon alkane derivatives, alkene derivatives, and isomeric cycloalkane derivatives are the top components of naphtha and kerosene and specialized industrial solvent mixtures. With the progressive addition of carbon units, the simple nonring structured hydrocarbon derivatives have higher viscosities, lubricating indices, boiling points, and solidification temperatures; color may become more prominent but that is usually because of impurities.

However, even though the thermal and catalytic decomposition of hydrocarbon derivatives can be represented by simple equations (as presented below), it must always be remembered that the complexity of the individual reactions occurring in an extremely complex mixture and the *interference* of the products with those from other components of the mixture is unpredictable. Or the *interference* of secondary and tertiary products with the course of a reaction and, hence, with the formation of primary products may also be a cause for concern. Hence, caution is advised when applying the data from model compound studies to the behavior of crude oil, especially the molecularly complex heavy oils. These have few, if any, parallels in organic chemistry.

An understanding of the chemical properties of the alkane derivatives is an important aspect of organic chemistry because of the many products derived from the hydrocarbon derivatives. Thus, it is the purpose of this chapter to present the simper reactions—other more focused and complete texts are available for the readers who wish to gather more details (Brückner, 2002; Carey, 2006; Klein, 2013).

2. Thermal reactions

Thermal decomposition (*thermolysis, noncatalytic thermal decomposition*) is a chemical reaction in which a compounds decomposes under the influence of heat into at least two other (lower molecular weight) products. Typically the reaction is endothermic as heat is required to break chemical bonds in the compound undergoing decomposition. The *decomposition temperature* of a substance is the temperature at which the substance decomposes into at least two other (lower molecular weight) products.

There are various theories relating to the thermal decomposition of organic molecules and this area of crude oil technology has been the subject of study for several decades. The relative reactivity of crude oil constituents can be assessed on the basis of bond energies, but the thermal stability of an organic molecule is dependent upon the bond strength of the weakest bond. And even though the use of bond energy data is a method for predicting the reactivity or the stability of specific bonds under designed conditions, the reactivity of a particular bond is also subject to its environment. Thus, it is not only the reactivity of the constituents of crude oil that are important in processing behavior, it is also the stereochemistry of the constituents as they relate to one another that is also of some importance (Speight, 2014). It must be appreciated that the stereochemistry of organic compounds is often a major factor in determining reactivity and properties (Eliel and Wilen, 1994).

In the present context, it is necessary to recognize that most hydrocarbon derivatives decompose thermally at temperatures above approximately 650°F (340°C), so the high boiling points of many crude oil constituents cannot be measured directly and must be estimated from other measurements. And in the present context, it is as well that hydrocarbon derivatives decompose at elevated temperatures. Thereby lies the route to many modern products. For example, in a crude oil refinery, the highest value products are transportation fuels: (i) gasoline—boiling range: 35–220°C, 95–425°F, (ii) jet fuel—boiling range 175–290°C (350 to 550°F), and (iii) diesel fuel, boiling range 175–370°C, 350–700°F. These boiling ranges are not always precise to the degree and are subject to variation and depend upon the process used for their production. In winter, gasoline will typically (in cold regions) have butane added to the mix (to facilitate cold starting) thereby changing the boiling range to 0–220°C (32–425°F). The fuels are produced by thermal decomposition of a variety of hydrocarbon derivatives, high molecular weight paraffin derivatives included. Less than one-third of a typical crude oil distills in these ranges and thus the goal of refining chemistry might be stated simply as the methods by which crude oil is converted to these fuels. It must be recognized that refining involves a wide variety of chemical reactions, but the production of liquid fuels is the focus of a refinery.

In the current context, thermal decomposition (thermal decomposition, thermolysis) is a chemical reaction in which a hydrocarbon (or any chemical compound) breaks up into at least two other substances when heated. The reaction is usually endothermic as heat is required to break chemical bonds in the hydrocarbon. The *decomposition temperature* is the temperature at which the hydrocarbon decomposes into smaller substances or into its constituent atoms. In some case, the decomposition temperature is noted as the temperature at which the rate of thermal decomposition becomes noticeable and measurable.

On the other hand, thermal depolymerization is a process using pyrolysis for the reduction of the molecular weight of high molecular weight hydrocarbon derivatives. It may not be depolymerization insofar as it is not always the reverse of polymerization. For example, under the influence of heat, high molecular weight organic constituents of crude oil decompose into lower molecular weight usable hydrocarbon derivatives with a maximum molecular weight (chain length) that is dependent on (i) temperature, (ii) pressure, (iii) residence time, and (iv) the presence of an added material such as hydrogen.

The term *cracking* applies to the decomposition of crude oil constituents that is induced by elevated temperatures (>350°C, >660°F) whereby the higher molecular weight constituents of crude oil are converted to lower molecular weight products. Thermal decomposition reactions involve carbon-carbon bond rupture and are thermodynamically favored at high temperature.

2.1 Thermal decomposition

Thermal decomposition (cracking) of hydrocarbon derivatives is the major process in petrochemical industry for the production of low-boiling olefin derivatives. The process converts hydrocarbon feedstock into more valuable products, by means of highly endothermic reactions. The performance of thermal decomposition processes is influenced to a great extent by the feedstock composition and degree of saturation, as the product yield depends on the conversion level and extent of reaction.

Thus, thermal decomposition is a phenomenon by which higher boiling (higher molecular weight) constituents in crude oil are converted into lower boiling (lower molecular weight) products. However, certain products may interact with one another to yield products having higher molecular weights than the constituents of the original feedstock. Some of the products are expelled from the system as, say, gases, gasoline-range materials, kerosene-range materials, and the various intermediates that produce other products such as coke. Materials that have boiling ranges higher than gasoline and kerosene may (depending upon the refining options) be referred to as *recycle* stock, which is recycled in the thermal decomposition equipment until conversion is complete.

Thermal decomposition of the lower molecular weight hydrocarbon derivatives is used in the petrochemical industry. For example, the chief use of ethane is in the chemical industry in the production of ethylene by steam cracking (parkas). When diluted with steam and briefly heated to very high temperatures (900°C, 1650°F, or higher), heavy hydrocarbon derivatives break down into lower molecular weight products. Ethane is favored for ethylene production because the steam cracking of ethane is fairly selective for ethylene, while the steam cracking of heavier hydrocarbon derivatives yields a product mixture poorer in ethylene, and richer in higher molecular weight such as propylene and butadiene as well as aromatic hydrocarbon derivatives.

Generally, the thermal decomposition of a higher molecular weight alkane produces a lower molecular weight alkane (relative to the molecular weight of the starting alkane) plus a low molecular weight alkene (relative to the molecular weight of the starting alkane):

$$RCH_2CH_2CH_2CH_2R^1 \rightarrow RCH_2CH_3 + R^1CH{=}CH_2$$

R and R^1 may or may not be equal alkyl moieties. The reaction is, of course, much more complex than illustrated above due to molecular factors such as chain length, branching, and stereochemistry. Secondary reactions of the two primary products (RCH_2CH_3 and $R^1CH{=}CH_2$) complicate the ultimate product slate even further.

Using *n*-decane as the starting alkane, the primary products, for example, are often considered to be n-octane and ethylene:

$$CH_3(CH_2)_8CH_3 \rightarrow CH_3(CH_2)_6CH_3 + CH_2{=}CH_2$$

Unless they were allowed to escape from the reaction vessel, further reactions of the octane and ethylene would produce lower molecular weight product that may result in high yield of methane, carbon, and hydrogen.

Furthermore, there are several other potential reactions that can occur that lead to a variety of products, for example:

$$CH_3(CH_2)_8CH_3 \rightarrow CH_3(CH_2)_5CH_3 + CH_2{=}CHCH_3$$

$$CH_3(CH_2)_8CH_3 \rightarrow CH_3(CH_2)_4CH_3 + CH_2{=}CHCH_2CH_3$$

$$CH_3(CH_2)_8CH_3 \rightarrow CH_3(CH_2)_3CH_3 + CH_2{=}CH(CH_2)_2CH_3$$

$$CH_3(CH_2)_8CH_3 \rightarrow CH_3(CH_2)_2CH_3 + CH_2{=}CH(CH_2)_3CH_3$$

$$CH_3(CH_2)_8CH_3 \rightarrow CH_3CH_2CH_3 + CH_2{=}CH(CH_2)_4CH_3$$

$$CH_3(CH_2)_8CH_3 \rightarrow CH_3CH_3 + CH_2{=}CH(CH_2)_5CH_3$$

$$CH_3(CH_2)_8CH_3 \rightarrow CH_4 + CH_2{=}CH(CH_2)_8CH_3$$

The products are dependent on temperature and residence time, and the simple reactions shown above do not consider the potential for isomerization of the products such as, for example, the conversion of butene ($CH_3CH_2CH{=}CH_2$ or $CH_3CH{=}CHCH_3$) to iso-butylene [$(CH_3)_2C{=}CH_2$]. Other products include naphtha as well as higher boiling products often referred to as *thermal tar*.

In the crude oil industry, thermal decomposition is the process by which high molecular weight hydrocarbon molecules are thermally decomposed into usable products. This is achieved by using high pressures and temperatures without a catalyst, or lower temperatures and pressures in the presence of a catalyst. The source of the large hydrocarbon molecules is often the naphtha fraction or the gas oil fraction from the fractional distillation of crude oil. These fractions are obtained from the distillation process as liquids, but are re-vaporized before thermal decomposition. In the process, the high molecular weight hydrocarbon derivatives are decomposed in a random manner to produce mixtures of lower molecular weight hydrocarbon derivatives, some of which have carbon-carbon double bonds.

In thermal decomposition, high temperatures (typically in the range of 450–750°C) and pressures (up to about 70 atm) are used to break the high molecular weight hydrocarbon derivatives into lower molecular weight hydrocarbon derivatives. Thermal decomposition gives mixtures of products containing high proportions of hydrocarbon derivatives with double bonds—alkene derivatives. Thermal decomposition does not involve ionic intermediates but the carbon-carbon bonds are broken so that each carbon atom ends up with a single electron free radical.

Two general types of reaction occur during thermal decomposition: (i) the decomposition of large molecules into small molecules (primary reactions), and (ii) reactions by which some of the primary products interact to form higher molecular weight materials (secondary reactions). In the first instance, i.e., the primary reactions, examples are the thermal decomposition of butane to methane and propylene or to ethane and ethylene:

$$CH_3CH_2CH_2CH_3 \rightarrow CH_4 + CH_3CH{=}CH_2$$

Butane Methane Propene

$$CH_3CH_2CH_2CH_3 \rightarrow CH_3CH_3 + CH_2{=}CH_2$$

Butane Ethane Ethylene

In the second instance, i.e., the secondary reaction, examples are the thermal decomposition of butane to methane and propylene followed by reactions by which some of the primary products interact to form higher molecular weight materials (secondary reactions) such as the recombination of ethylene with another ethylene molecule to produce butylene-1:

$$CH_2{=}CH_2 + CH_2{=}CH_2 \rightarrow CH_3CH_2CH{=}CH_2$$

Ethylene + Ethylene Butylene-1

This type of reaction can occur with a variety of olefin derivatives and can be represented generally as:

$$RCH = CH_2 + R^1CH{=}CH_2 \rightarrow \text{Cracked residuum} + \text{Coke} + \text{Other products}$$

Thermal decomposition is a *free radical* chain reaction; a free radical is an atom or group of atoms possessing an unpaired electron. Free radicals are very reactive, and it is their mode of reaction that actually determines the product distribution during thermal decomposition. Free radical reacts with a hydrocarbon by abstracting a hydrogen atom to produce a stable end product and a new free radical. Free radical reactions are extremely complex, and it is hoped that these few reaction schemes illustrate potential reaction pathways. Any of the preceding reaction types are possible, but it is generally recognized that the prevailing conditions and those reaction sequences that are thermodynamically favored determine the product distribution. One of the significant features of hydrocarbon free radicals is their resistance to isomerization, for example, migration of an alkyl group and, as a result, thermal decomposition does not produce any degree of branching in the products other than that already present in the feedstock.

Data obtained from the thermal decomposition of pure compounds indicate certain decomposition characteristics that permit predictions to be made of the product types that arise from the thermal decomposition of various feedstocks. For example, normal (straight-chain) paraffin derivatives (n-paraffin derivatives) are believed to form, initially, higher molecular weight material, which subsequently decomposes as the reaction progresses. Other paraffinic materials and terminal olefin derivatives are produced. An increase in pressure inhibits the formation of low molecular weight gaseous products and therefore promotes the formation of higher molecular weight materials.

Branched paraffin derivatives react somewhat differently to the normal paraffin derivatives during thermal decomposition processes and produce substantially higher yields of olefin derivatives having one fewer carbon atom that the parent hydrocarbon. Cycloparaffin derivatives (naphthene derivatives) react differently to their noncyclic counterparts and are somewhat more stable. For example, cyclohexane produces hydrogen, ethylene, butadiene, and benzene: Alkyl-substituted cycloparaffin derivatives decompose by means of scission of the alkyl chain to produce an olefin and a methyl or ethyl cyclohexane.

The main feature of the thermal decomposition of aromatic hydrocarbon derivatives is the aromatic ring, which is generally stable at moderate thermal decomposition temperatures (350–500°C, 660 to 930°F). Alkylated aromatic hydrocarbon derivatives, like the alkylated naphthene derivatives, are more prone to dealkylation than to ring destruction. For example, the hydrodealkylation reaction can be used to convert toluene ($C_6H_5CH_3$) to benzene (C_6H_6). In this hydrogen-intensive process, toluene is mixed with hydrogen, then passed over a chromium oxide, molybdenum oxide, or platinum oxide catalyst at 500–600°C (930–1110°F) and up to 1000 psi pressure (higher temperatures can be used instead of a catalyst) whereupon toluene undergoes dealkylation to benzene and methane:

$$C_6H_5CH_3 + H_2 \rightarrow C_6H_6 + CH_4$$

This irreversible reaction is accompanied by an equilibrium side reaction that produces biphenyl (diphenyl) at higher temperature:

$$2\ C_6H_6 \leftrightarrow H_2 + C_6H_5C_6H_5$$

If the raw material stream contains much nonaromatic components (paraffin derivatives or naphthene derivatives), those are likely decomposed to lower hydrocarbon derivatives such as methane, which increases the consumption of hydrogen.

Where a petrochemical complex has similar demands for benzene and xylene, then toluene disproportionation is a suitable alternate to the toluene hydrodealkylation. In the process, two toluene molecules are reacted and the methyl groups rearranged from one toluene molecule to the other, yielding one benzene molecule and one xylene molecule.

$$2C_6H_5CH_3 \rightarrow C_6H_6 + CH_3C_6H_4CH_3$$

Xylene isomers, ethyl benzene and propyl benzene, are decomposed to benzene and other products to varying degrees at 500°C (930°F) over a silica-alumina catalyst. As the size of the alkyl group increases, the ease of thermal decomposition becomes greater and the selectivity of the bond cleavage, as evidenced by the yield of benzene, remains high.

The olefin derivatives formed in the thermal decomposition of alkyl aromatic hydrocarbon derivatives can undergo further reactions so that, depending upon the reactions parameters, the product will contain a variety of hydrocarbon derivatives quite different from the structure of the original substituent alkyl group. However, ring destruction of the benzene derivatives occurs above 500°C (930°F), but condensed aromatic hydrocarbon derivatives may undergo ring destruction at somewhat lower temperatures (450°C, 840°F).

Thermal decomposition of hydrocarbon distillates in a refinery is complex because of the number of constituents that make up these fractions. Naphtha, the lowest boiling distillate, is the most convenient example. Other examples are available from various process descriptions. Naphtha refers to a number of different flammable liquid mixtures of hydrocarbon derivatives boiling below 200°C (390°F). It is a broad term covering the lightest and most volatile fraction of the liquid hydrocarbon derivatives in crude oil.

Full range naphtha is defined as the fraction of hydrocarbon derivatives in crude oil boiling between 30°C (86°F) and 200°C (390°F) and consists of a complex mixture of hydrocarbon molecules generally having between 5 and 12 carbon atoms. *Light naphtha* is the fraction boiling between 30°C (86°F) and 90°C (195°F) and consists of molecules with five to six carbon atoms. *Heavy naphtha* boils between 90°C (195°F) and 200°C (390°F) and consists of molecules with 6–12 carbons. Naphtha is flammable and has a density on the order of 0.7–0.75.

Naphtha is used primarily as feedstock for producing high octane gasoline (via the catalytic reforming process). It is also used in the petrochemical industry for producing olefin derivatives in steam cracking units and in the chemical industry for solvent (cleaning) applications.

2.2 Steam cracking

Steam cracking is a refinery (petrochemical) process in which saturated hydrocarbon derivatives (alkane derivatives) are thermally decomposed into lower molecular weight, often unsaturated, hydrocarbon derivatives (olefin derivatives). It is the principal industrial method for producing the lower molecular weight olefin derivatives, including ethylene and propylene.

In the steam cracking process, a gaseous or liquid hydrocarbon feed is diluted with steam and then briefly heated in a furnace. Typically, the reaction temperature is in excess of 900°C (1650°F) and the residence time of the feedstock in the reaction zone may only be a few tenths of a second before the feedstock/product steam is being quenched by contact with a colder fluid stream. The products produced in the reaction depend on the composition of the feed, the hydrocarbon to steam ratio, and on the thermal decomposition temperature and residence time.

Lower molecular weight feedstocks (such as ethane, propane, butane, or low boiling naphtha) give product streams rich in the lower molecular weight olefin derivatives including ethylene, propylene, and butadiene ($CH_2{=}CHCH{=}CH_2$). Higher molecular weight hydrocarbon feedstocks (full range naphtha and high boiling naphtha) also yield products rich in aromatic hydrocarbon derivatives and hydrocarbon derivatives suitable for inclusion in gasoline.

A higher thermal decomposition temperature (*higher severity*) favors the production of ethylene and benzene, whereas a lower thermal decomposition temperature (lower severity) produces relatively higher amounts of propylene, butanes, and butylenes, as well as low boiling liquid products. The process also results in the slow deposition of coke on the reactor walls. This degrades the effectiveness of the reactor, so reaction conditions are designed to minimize this. Nonetheless, a steam cracking furnace can usually only run for a few months at a time between decoking operations.

A variety of chemical reactions take place during steam cracking, most of them based on free radical chemistry. The major types of reactions that take place, with examples, include: (i) initiation reactions, (ii) hydrogen abstraction reactions, (iii) radical decomposition reactions, (iv) radical addition reactions, and (v) termination reactions.

Initiation reactions are those reactions in which a single molecule breaks apart into two free radicals. Only a small fraction of the feed molecules actually undergo initiation, but these reactions are necessary to produce the free radicals that drive the rest of the reactions. In steam cracking, initiation usually involves breaking a chemical bond between two carbon atoms, rather than the bond between a carbon atom and a hydrogen atom:

$$CH_3CH_3 \rightarrow 2\ CH_3\bullet$$

Hydrogen abstraction reactions are those reactions in which a free radical removes a hydrogen atom from another molecule, turning the second molecule into a free radical.

$$CH_3\bullet + CH_3CH_3 \rightarrow CH_4 + CH_3CH_2\bullet$$

Radical decomposition reactions are those reactions in which a free radical breaks apart into two molecules, one an alkene, the other a free radical. This is the process that results in the alkene products of steam cracking:

$$CH_3CH_2\bullet \rightarrow CH_2{=}CH_2 + H\bullet$$

Radical addition reactions are those reactions in which a radical reacts with an alkene to form a single, larger free radical. These processes are involved in forming the aromatic products that result when heavier feedstocks are used:

$$CH_3CH_2\bullet + CH_2{=}CH_2 \rightarrow CH_3CH_2CH_2CH_2\bullet$$

Termination reactions are those reactions in which two free radicals react with each other to produce products that are not free radicals. Two common forms of termination are *recombination*, where the two radicals combine to form one larger molecule, and *disproportionation*, where one radical transfers a hydrogen atom to the other, giving an alkene and an alkane:

$$CH_3\bullet + CH_3CH_2\bullet \rightarrow CH_3CH_2CH_3$$

$$CH_3CH_2\bullet + CH_3CH_2\bullet \rightarrow CH_2{=}CH_2 + CH_3CH_3$$

When used as feedstock in petrochemical steam crackers, naphtha is heated in the presence of water vapor and the absence of oxygen or air until the hydrocarbon molecules fall apart. The primary products of the thermal decomposition process are olefin derivatives (ethylene, propylene, butenes, and butadiene). When naphtha is used as a feedstock in catalytic reforming, the primary products are aromatic hydrocarbon derivatives including benzene, toluene, and xylene isomers. The olefin derivatives are used as feedstocks for derivative units that produce plastics (such as polyethylene and polypropylene for example) and industrial chemicals. The aromatic hydrocarbon derivatives are used for octane boosting in fuel blending as well as polyethylene terephthalate feedstock as well as paint solvents and coating solvents.

2.3 Thermal reforming

Thermal reforming is a crude oil refining process using heat (but no catalyst) to effect molecular rearrangement of a low-octane naphtha to form high-octane motor gasoline. The process is carried out at higher temperature when noncyclic hydrocarbon derivatives are converted to high octane number olefin derivatives and aromatic hydrocarbon derivatives.

In the process, a feedstock, such as 200°C (390°F) end-point naphtha is heated to 510−595°C (950−1100°F) in a furnace much the same as a thermal decomposition furnace, with pressures from 400 to 1000 psi. As the heated naphtha leaves the furnace, it is cooled or quenched by the addition of cold naphtha. The quenched, reformed material then enters a fractional distillation tower where any heavy products are separated. The remainder of the reformed material leaves the top of the tower to be separated into gases and reformate. The higher octane number of the product (*reformate*) is due primarily to the thermal decomposition of higher molecular weight paraffin derivatives into higher-octane olefin derivatives.

Thermal reforming is in general less effective than catalytic processes and has been largely supplanted. As it was practiced, a single-pass operation was employed at temperatures in the range of 540–760°C (1000–1140°F) and pressures in the range 500–1000 psi. Octane number improvement depended on the extent of conversion but was not directly proportional to the extent of thermal decomposition-per-pass.

The amount and quality of *reformate* are dependent on the temperature. A general rule is the higher the reforming temperature, the higher the octane number of the product but the yield of reformate is relatively low. For example, naphtha with an octane number of 35 when reformed at 515°C (960°F) yields 92.4% of 56 octane reformate; when reformed at 555°C (1,030°F) the yield is 68.7% of 83 octane reformate. However, high conversion is not always effective since coke production and gas production usually increase.

Modifications of the thermal reforming process due to the inclusion of hydrocarbon gases with the feedstock are known as *gas reversion* and *polyforming*. Thus, olefinic gases produced by thermal decomposition and reforming can be converted into liquids boiling in the gasoline range by heating them under high pressure. Since the resulting liquids (polymers) have high octane numbers, they increase the overall quantity and quality of gasoline produced in a refinery.

The gases most susceptible to conversion to liquid products are olefin derivatives with three and four carbon atoms. These are propylene ($CH_3CH{=}CH_2$), which is associated with propane in the C_3 fraction, and butylene ($CH_3CH_2CH{=}CH_2$ and/or $CH_3CH{=}CHCH_3$) and *iso*-butylene [$(CH_3)_2C{=}CH_2$], which are associated with butane ($CH_3CH_2CH_2CH_3$) and *iso*-butane [$(CH_3)_2CHCH_3$] in the C_4 fraction. When the C_3 and C_4 fractions are subjected to the temperature and pressure conditions used in thermal reforming, they undergo chemical reactions that result in a small yield of gasoline. When the C_3 and C_4 fractions are passed through a thermal reformer in admixture with naphtha, the process is called *naphtha-gas reversion* or *naphtha polyforming*.

These processes are essentially the same but differ in the manner in which the gases and naphtha are passed through the heating furnace. In gas reversion, the naphtha and gases flow through separate lines in the furnace and are heated independently of one another. Before leaving the furnace, both lines join to form a common soaking section where the reforming, polymerization, and other reactions take place. In naphtha reforming, the C_3 and C_4 gases are premixed with the naphtha and pass together through the furnace. Except for the gaseous components in the feedstock, both processes operate in much the same manner as thermal reforming and produce similar products.

3. Catalytic decomposition

Refining processes involve the use of various thermal and catalytic processes to higher molecular weight constituents to lower-boiling products (Parkash, 2003; Gary et al., 2007; Speight, 2014, 2017; Hsu and Robinson, 2017). This efficiency translates into a strong economic advantage, leading to widespread use of conversion processes in refineries today. However, in order to understand the principles of catalytic cracking, understanding the principles of adsorption and reaction on solid surfaces is valuable (Samorjai, 1994; Masel, 1995).

In the catalytic decomposition process (catalytic cracking process, heterolysis), the alkane is brought into contact with the catalyst at a temperature of about 500°C and moderately low pressures. The process involves the presence of acid catalysts (usually solid acids such as silica-alumina and zeolites), which promote a heterolytic (asymmetric) breakage of bonds yielding pairs of ions of opposite charges, usually a carbocation and the very unstable hydride anion.

Catalytic cracking is the thermal decomposition of crude oil constituents, hydrocarbon derivatives, in the presence of a catalyst. Thermal decomposition has been superseded by catalytic thermal decomposition as the process for gasoline manufacture. Indeed, gasoline produced by catalytic thermal decomposition is richer in branched paraffin derivatives, cycloparaffin derivatives, and aromatic hydrocarbon derivatives, which all serve to increase the quality of the gasoline. Catalytic thermal decomposition also results in production of the maximum amount of butenes and butanes (C_4H_8 and C_4H_{10}) rather that ethylene and ethane (C_2H_4 and C_2H_6).

Zeolites as the catalysts, which are complex aluminosilicates, are large lattices of aluminum, silicon, and oxygen atoms carrying a negative charge. They are, of course, associated with positive ions such as sodium ions (Na^+). The zeolites used in catalytic thermal decomposition are chosen to give high percentages of hydrocarbon derivatives with between 5 and 10 carbon atoms—particularly useful for gasoline. The reaction also produces high proportions of branched alkane derivatives and aromatic hydrocarbon derivatives like benzene.

The zeolite catalyst has sites which can remove hydrogen from an alkane together with the two electrons which bound it to the carbon. That leaves the carbon atom with a positive charge (carbonium ion, carbocation). Rearrangement of these ions leads to the various products of the reaction.

Carbon-localized free radicals and cations are both highly unstable and undergo processes of chain rearrangement, carbon-carbon bond scission in a position beta (i.e., thermal decomposition), and intra and intermolecular hydrogen transfer or hydride transfer. In both types of processes, the corresponding reactive intermediates (radicals, ions) are permanently regenerated, and thus they proceed by a self-propagating chain mechanism. The chain of reactions is eventually terminated by radical or ion recombination.

Catalytic cracking processes evolved in the 1930s from research on crude oil and coal liquids. The crude oil work came to fruition with the invention of acid thermal decomposition. The work to produce liquid fuels from coal, most notably in Germany, resulted in metal sulfide hydrogenation catalysts. In the 1930, a catalytic cracking catalyst for crude oil that used solid acids as catalysts was developed using acid-treated clay minerals.

Clay minerals are a family of crystalline aluminosilicate solids, and the acid treatment develops acidic sites by removing aluminum from the structure. The acid sites also catalyze the formation of coke, and Houdry developed a moving bed process that continuously removed the cooked beads from the reactor for regeneration by oxidation with air.

Although thermal decomposition is a free radical (neutral) process, catalytic cracking is an ionic process involving carbonium ions, which are hydrocarbon ions having a positive charge on a carbon atom. The formation of carbonium ions during catalytic cracking can occur by: (i) addition of a proton from an acid catalyst to an olefin and (ii) abstraction of a hydride ion (H^-) from a hydrocarbon by the acid catalyst or by another carbonium ion. However, carbonium ions are not formed by cleavage of a carbon-carbon bond.

In essence, the use of a catalyst permits alternate routes for cracking reactions, usually by lowering the free energy of activation for the reaction. The acid catalysts first used in catalytic thermal decomposition were amorphous solids composed of approximately 87% silica (SiO_2) and 13% alumina (Al_2O_3) and were designated low-alumina catalysts. However, this type of catalyst is now being replaced by crystalline aluminosilicates (zeolites) or molecular sieves.

The first catalysts used for catalytic thermal decomposition were acid-treated clay minerals, formed into beads. In fact, clay minerals are still employed as catalyst in some thermal decomposition processes (Parkash, 2003; Gary et al., 2007; Speight, 2014, 2017; Hsu and Robinson, 2017). Clay minerals are a family of crystalline aluminosilicate solids, and the acid treatment develops acidic sites by removing aluminum from the structure. The acid sites also catalyze the formation of coke, and the development of a moving bed process that continuously removed the cooked beads from the reactor reduced the yield of coke; clay regeneration was achieved by oxidation with air.

Clay minerals are natural compounds of silica and alumina, containing major amounts of the oxides of sodium, potassium, magnesium, calcium, and other alkali and alkaline earth metals. Iron and other transition metals are often found in natural clay minerals, substituted for the aluminum cations. Oxides of virtually every metal are found as impurity deposits in clay minerals. Clay minerals are layered crystalline materials. They contain large amounts of water within and between the layers. Heating the clay minerals above 100°C (212°F) can drive out some or all of this water; at higher temperatures, the clay structures themselves can undergo complex solid-state reactions. Such behavior makes the chemistry of clay minerals a fascinating field of study in its own right. Typical clay minerals include kaolinite, montmorillonite, and illite (Keller, 1985). They are found in most natural soils and in large, relatively pure deposits, from which they are mined for applications ranging from adsorbents to paper making.

Once the carbonium ions are formed, the modes of interaction constitute an important means by which product formation occurs during catalytic thermal decomposition, for example, isomerization either by hydride ion shift or by methyl group shift, both of which occur readily. The trend is for stabilization of the carbonium ion by *movement* of the charged carbon atom toward the center of the molecule, which accounts for the isomerization of α-olefin derivatives to internal olefin derivatives when carbonium ions are produced. Cyclization can occur by internal addition of a carbonium ion to a double bond which, by continuation of the sequence, can result in aromatization of the cyclic carbonium ion.

Like the paraffin derivatives, naphthene derivatives do not appear to isomerize before thermal decomposition. However, the naphthenic hydrocarbon derivatives (from C_9 upward) produce considerable amounts of aromatic hydrocarbon derivatives during catalytic thermal decomposition. Reaction schemes similar to that outlined here provide possible routes for the conversion of naphthene derivatives to aromatic hydrocarbon derivatives. Alkylated benzene derivatives undergo nearly quantitative dealkylation to benzene without apparent ring degradation below 500°C (930°F). However, polymethlybenzene derivatives undergo disproportionation and isomerization with very little benzene formation.

Like the thermal decomposition (noncatalytic) reactions, catalytic thermal decomposition reactions can be represented by simple reaction schemes. However, questions have arisen as to how the thermal decomposition of paraffin derivatives is initiated. Several hypotheses for the initiation step in catalytic thermal decomposition of paraffin derivatives have been proposed. The Lewis site mechanism is the most obvious, as it proposes that a carbenium ion is formed by the abstraction of a hydride ion from a saturated hydrocarbon by a strong Lewis acid site: a tricoordinated aluminum species. On Brønsted sites a carbenium ion may be readily formed from an olefin by the addition of a proton to the double bond or, more rarely, via the abstraction of a hydride ion from a paraffin by a strong Brønsted proton. This latter process requires the formation of hydrogen as an initial product. This concept was, for various reasons that are of uncertain foundation, often neglected.

It is therefore not surprising that the earliest thermal decomposition mechanisms postulated that the initial carbenium ions are formed only by the protonation of olefin derivatives generated either by thermal decomposition or present in the feed as an impurity. For a number of reasons this proposal was not convincing, and in the continuing search for initiating reactions it was even proposed that electrical fields associated with the cations in the zeolite are responsible for the polarization of reactant paraffin derivatives, thereby activating them for thermal decomposition. More recently, however, it has been convincingly shown that a penta-coordinated carbonium ion can be formed on the alkane itself by protonation, if a sufficiently strong Brønsted proton is available.

Coke formation is considered, with just cause to a malignant side reaction of normal carbenium ions. However, while chain reactions dominate events occurring on the surface, and produce the majority of products, certain less desirable biomolecular events have a finite chance of involving the same carbenium ions in a bimolecular interaction with one another. Of these reactions, most will produce a paraffin and leave carbene/carboid-type species on the surface. This carbene/carboid-type species can produce other products but the most damaging product will be one which remains on the catalyst surface and cannot be desorbed and results in the formation of coke, or remains in a noncoke form but effectively blocks the active sites of the catalyst.

A general reaction sequence for coke formation from paraffin derivatives involves oligomerization, cyclization, and dehydrogenation of small molecules at active sites within zeolite pores:

Alkanes → alkenes.

Alkenes → oligomers.

Oligomers → naphthenes.

Naphthenes → aromatics.

Aromatics → coke.

Whether or not these are the true steps to coke formation can only be surmised. The problem with this reaction sequence is that it ignores sequential reactions in favor of consecutive reactions. And it must be accepted that the chemistry leading up to coke formation is a complex process, consisting of many sequential and parallel reactions.

There is a complex and little understood relationship between coke content, catalyst activity, and the chemical nature of the coke. For instance, the atomic hydrogen/carbon ratio of coke depends on how the coke was formed; its exact value will vary from system to system. And it seems that catalyst decay is not related in any simple way to the hydrogen-to-carbon atomic ratio of the coke or to the total coke content of the catalyst or any simple measure of coke properties. Moreover, despite many and varied attempts, there is currently no consensus as to the detailed chemistry of coke formation. There is, however, much evidence and good reason to believe that catalytic coke is formed from carbenium ions which undergo addition, dehydrogenation and cyclization, and elimination side reactions in addition to the main-line chain propagation processes.

3.1 Fluid catalytic cracking

Fluid catalytic cracking is a commonly used process Sadeghbeigi (2000), and a modern oil refinery will typically include a fluid catalytic cracking unit (*cat cracker*, FCC unit) particularly at refineries where demand for gasoline is high.

The process was first used in the 1940s and employs powdered catalysts. During the Second World War, fluid catalytic cracking provided Allied Forces with plentiful supplies of gasoline. Initial process implementations were based on an alumina catalyst and a reactor where the catalyst particles were suspended in a rising flow of feed hydrocarbon derivatives in the fluidized bed reactor.

In newer process designs, thermal decomposition takes place using a very active zeolite-based catalyst in a short-contact time vertical or upward sloped pipe called the *riser* (hence *riser pipe cracking*).

In the process, preheated feed is sprayed into the base of the riser via feed nozzles where it contacts hot fluidized catalyst at 665−760°C (1230 to 1400°F). The hot catalyst vaporizes the feed and catalyzes the thermal decomposition reactions that decompose the high molecular weight oil into lower boiling components. The catalyst-hydrocarbon mixture flows upward through the riser for just a few seconds and then the mixture is separated via cyclones. The catalyst-free hydrocarbon derivatives are routed to a main fractionator for separation into hydrocarbon gases, naphtha (the precursor to gasoline), kerosene (a precursor to diesel), and, light cycle oils (also used for in diesel production as well as jet fuel), and heavy fuel oil.

During the process, the thermal decomposition catalyst is deactivated (*spent*) by reactions which deposit coke on the catalyst and greatly reduce activity and selectivity. The spent catalyst is disengaged from the cracked hydrocarbon vapors and sent to a stripper where it is contacted with steam to remove hydrocarbon derivatives remaining in the catalyst pores. The spent catalyst then flows into a fluidized-bed regenerator where air (or in some cases air plus oxygen) is used to burn off the coke to restore catalyst activity and also provide the necessary heat for the next reaction cycle, thermal decomposition being an endothermic reaction. The *regenerated catalyst* is sent to the base of the riser and the process is repeated.

The naphtha produced in the fluid catalytic cracking unit has a relatively high octane number but is less chemically stable compared to other gasoline components due to the presence of olefin derivatives, which are also responsible for the formation of deposits in storage tanks, fuel lines, and injectors. The hydrocarbon gases from the fluid catalytic cracking unit are an important source of propylene and butylenes as well as iso-butane, which are essential feedstocks for the alkylation process (which produces high octane gasoline components).

3.2 Hydrocracking

Hydrocracking is a refining technology in which the outcome is the conversion of a variety of feedstocks to a range of products and units to accomplish this goal can be found at various points in a refinery (Ancheyta and Speight, 2007).

The history of the process goes back to the late 1920s when it was realized that there was a need for gasoline of a higher quality than that obtained by catalytic cracking, and this led to the development of the hydrocracking process. One of the first plants to use hydrocracking was commissioned for the commercial hydrogenation of brown coal at Leuna in Germany. Tungsten sulfide was used as a catalyst in this one-stage unit, in which high reaction pressures, 2900 to 4350 psi, were applied. The catalyst displayed a very high hydrogenation activity: the aromatic feedstock, coal and heavy fractions of oil, containing sulfur, nitrogen and oxygen, were virtually completely converted into paraffin derivatives and iso-paraffin derivatives. In 1939, Imperial Chemical Industries in Britain developed the second-stage catalyst for a plant that contributed largely to Britain's supply of aviation gasoline in the subsequent years.

During World War II, two-stage processes were applied on a limited scale in Germany, Britain, and USA. In Britain, feedstocks were creosote from coal tar and gas oil from crude oil. In the USA, Standard Oil of New Jersey operated a plant at Baton Rouge, producing gasoline from a Venezuelan kerosene/light gas oil fraction. Operating conditions in those units were comparable: approximate reaction temperature 400°C (750°F) and reaction pressures of 2900−4350 psi. After the war, commercial hydrocracking was very expensive but by the end of the 1950s, the process had become economic. The development of improved catalyst made it possible to operate the process at considerably lower pressure, that is, 1000 to 2200 psi. This in turn resulted in a reduction in equipment wall thickness, whereas simultaneously, advances were made in mechanical engineering, especially in the field of reactor design and heat transfer. These factors, together with the availability of relatively low cost hydrogen from steam reforming process, brought hydrocracking back on the refinery scene. The first units of the second generation were built in USA to meet the demand for conversion of surplus fuel oil in the gasoline-oriented refineries.

The older hydrogenolysis type of hydrocracking practiced in Europe during and after World War II used tungsten sulfide (WS_2) or molybdenum sulfide (MoS) as catalysts. These processes required high reaction temperatures and operating pressures, sometimes in excess of about 3000 psi (20,684 kPa) for continuous operation. The modern hydrocracking processes were initially developed for converting refractory feedstocks to gasoline and jet fuel; process and catalyst improvements and modifications have made it possible to yield products from gases and naphtha to furnace oils and catalytic cracking feedstocks. The zeolites most frequently used in commercial hydrocracking catalysts are partially dealuminated and low-sodium, or high-silica, Type Y zeolites in hydrogen or rare-earth forms. Other zeolites and mixtures of zeolites also are used. The zeolites often are imbedded in a high-surface-area amorphous matrix, which serves as a binder. The metals can reside inside the zeolite and on the amorphous matrix.

The concept of hydrocracking allows the refiner to produce products having a lower molecular weight with higher hydrogen content and a lower yield of coke. In summary, hydrocracking facilities add flexibility to refinery processing and to the product slate. Hydrocracking is more severe than hydrotreating there being the intent, in hydrocracking processes, to convert the feedstock to lower boiling products rather than to treat the feedstock for heteroatom and metals removal only.

Hydrocracking is an extremely versatile process that can be utilized in many different ways, and one of the advantages of hydrocracking is its ability to break down high boiling aromatic stocks produced by catalytic cracking or coking. To take full advantage of hydrocracking, the process must be integrated in the refinery with other process units. In gasoline production, for example, the hydrocracker product must be further processed in a catalytic reformer as it has a high naphthene content and relatively low octane number. The high naphthene content makes the hydrocracker gasoline an excellent feed for catalytic reforming, and good yields of high octane number gasoline can be obtained.

If high molecular weight hydrocarbon fractions are *pyrolyzed*, that is, if no hydrogenation occurs, progressive thermal decomposition and condensation reactions generally lead to the final products. These products are usually:

1. Gaseous and low boiling liquid compounds of high hydrogen content.
2. Liquid material of intermediate molecular weight with a hydrogen-carbon atomic ratio differing more or less from that of the original feedstock, depending on the method of operation.
3. Material of high molecular weight, such as coke, possessing a lower hydrogen-carbon atomic ratio than the starting material.

Highly aromatic or refractory recycle stocks or gas oils that contain varying proportions of highly condensed aromatic structures (for example, naphthalene and phenanthrene) usually crack, in the absence of hydrogen, to yield intractable residues and coke.

The mechanism of hydrocracking is basically similar to that of catalytic cracking, but with concurrent hydrogenation. The catalyst assists in the production of carbonium ions via olefin intermediates and these intermediates are quickly hydrogenated under the high-hydrogen partial pressures employed in hydrocracking. The rapid hydrogenation prevents adsorption of olefin derivatives on the catalyst and, hence, prevents their subsequent dehydrogenation, which ultimately leads to coke formation so that long on-stream times can be obtained without the necessity of catalyst regeneration.

One of the most important reactions in hydrocracking is the partial hydrogenation of polycyclic aromatic hydrocarbon derivatives followed by rupture of the saturated rings to form substituted monocyclic aromatic hydrocarbon derivatives. The alkyl side chains may then be split off to give *iso*-paraffin derivatives. It is desirable to avoid excessive hydrogenation activity of the catalyst so that the monocyclic aromatic hydrocarbon derivatives become hydrogenated to naphthene derivatives; furthermore, repeated hydrogenation leads to loss in octane number, which increases the catalytic reforming required to process the hydrocracked naphtha.

Side chains composed of three or four carbon atoms are relatively easily removed from an aromatic ring during catalytic cracking, but the reaction of aromatic rings with shorter side chains appears to be quite different. For example, hydrocracking single-ring aromatic hydrocarbon derivatives containing four or more methyl groups produces largely *iso*-butane and benzene. It may be that successive isomerization of the feed molecule adsorbed on the catalyst occurs until a four-carbon side chain is formed, which then breaks off to yield *iso*-butane and benzene. Overall, coke formation is very low in hydrocracking since the secondary reactions and the formation of the precursors to coke are suppressed as the hydrogen pressure is increased.

The products from hydrocracking are composed of either saturated or aromatic compounds; no olefin derivatives are found. In making gasoline, the lower paraffin derivatives formed have high octane numbers; for example, the five- and six-carbon number fractions have leaded research octane numbers of 99–100. The remaining gasoline has excellent properties as a feed to catalytic reforming, producing a highly aromatic gasoline that is capable of a high octane number. Both types of gasoline are suitable for premium-grade motor gasoline. Another attractive feature of hydrocracking is the low yield of gaseous components, such as methane, ethane, and propane, which are less desirable than gasoline. When making jet fuel, more hydrogenation activity of the catalysts is used, since jet fuel contains more saturates than gasoline.

While whole families of catalysts are required depending on feed available and the desired product slate or product character, the number of process stages is also important to catalysts choice. Generally, the refinery utilizes one of three options. Thus, depending on the feedstock being processed and the type of plant design employed (*single-stage* or *two-stage*), flexibility can be provided to vary product distribution among the following principal end products:

Hydrocracking adds that flexibility and offers the refiner a process that can handle varying feeds and operate under diverse process conditions. Utilizing different types of catalysts can modify the product slate produced. Reactor design and number of processing stages play a role in this flexibility.

3.3 Catalytic reforming

Like thermal reforming, *catalytic reforming* converts low-octane gasoline into high-octane gasoline (reformate). Although thermal reforming can produce reformate with a research octane number in the range 65–80 depending on the yield, catalytic reforming produces reformate with octane numbers of the order of 90–95. Catalytic reforming is conducted in the presence of hydrogen over hydrogenation-dehydrogenation catalysts, which may be supported on alumina or silica-alumina. Depending on the catalyst, a definite sequence of reactions takes place, involving structural changes in the charge stock. The catalytic reforming process was commercially nonexistent in the United States before 1940. The process is really a process of the 1950s and showed phenomenal growth in 1953–1959 time period. As a result, thermal reforming is now somewhat obsolete.

Catalytic reformer feeds are saturated (i.e., not olefinic) materials; in the majority of cases the feed may be a straight-run naphtha, but other byproduct low-octane naphtha (e.g., coker naphtha) can be processed after treatment to remove olefin derivatives and other contaminants. Hydrocarbon naphtha that contains substantial quantities of naphthene derivatives is also a suitable feed. The process uses a precious metal catalyst (platinum supported by an alumina base) in conjunction with very high temperatures to reform the paraffin and naphthene constituents into high-octane components.

Sulfur is a poison to the reforming catalyst, which requires that virtually all the sulfur must be removed from the heavy naphtha by a hydrotreating process prior to reforming.

Several different types of chemical reactions occur in the reforming reactors: paraffin derivatives are *isomerized* to branched chains and to a lesser extent to naphthene derivatives, and naphthene derivatives are *converted* to aromatic hydrocarbon derivatives. Overall, the reforming reactions are endothermic. The resulting product stream (*reformate*) from catalytic reforming has an RON from 96 to 102 depending on the reactor severity and feedstock quality. The dehydrogenation reactions which convert the saturated naphthene derivatives into unsaturated aromatic hydrocarbon derivatives produce hydrogen, which is available for distribution to other refinery hydroprocesses.

The catalytic reforming process consists of a series of several reactors (Fig. 11.1) which operate at temperatures of approximately 480°C (900°F).

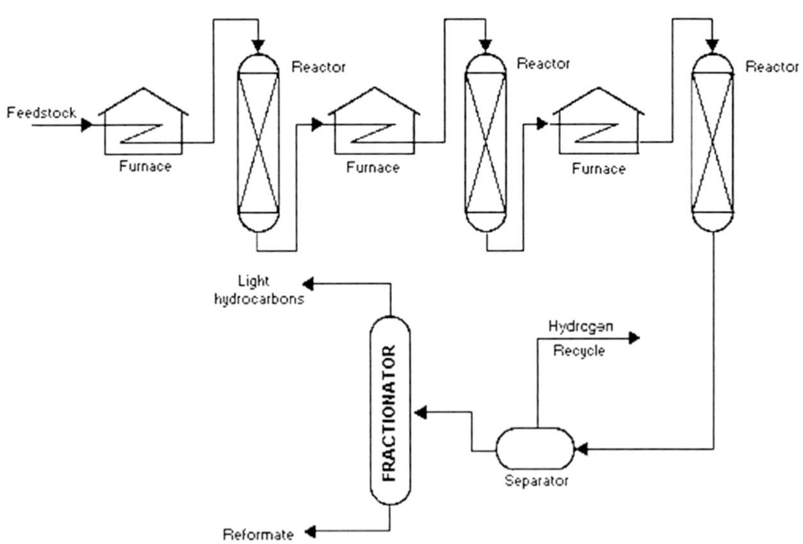

FIGURE 11.1 Catalytic reforming process. (Osha Technical Manual. Section IV, Chapter 2. Petroleum refining processes).

The hydrocarbon derivatives are reheated by direct-fired furnaces in between the subsequent reforming reactors. As a result of the very high temperatures, the catalyst becomes deactivated by the formation of coke (i.e., essentially pure carbon) on the catalyst which reduces the surface area available to contact with the hydrocarbon derivatives.

Catalytic reforming is usually carried out by feeding a naphtha (after pretreating with hydrogen if necessary) and hydrogen mixture to a furnace where the mixture is heated to the desired temperatures 450−520°C (840−965°F), and then passed through fixed-bed catalytic reactors at hydrogen pressures of 100−1000 psi. Normally two (or more than one) reactors are used in series, and reheaters are located between adjoining reactors to compensate for the endothermic reactions taking place. Sometimes as many as four or five are kept on-stream in series while one or more are being regenerated. The on-stream cycle of any one reactor may vary from several hours to many days, depending on the feedstock and reaction conditions.

The product issuing from the last catalytic reactor is cooled and sent to a high-pressure separator where the hydrogen-rich gas is split into two streams: one stream goes to recycle, and the remaining portion represents excess hydrogen available for other uses. The excess hydrogen is vented from the unit and used in hydrotreating, as a fuel, or for manufacture of chemicals (e.g., ammonia). The liquid product (reformate) is stabilized (by removal of light ends) and used directly in gasoline or extracted for aromatic blending stocks for aviation gasoline.

The commercial processes available for use can be broadly classified as the *moving-bed*, *fluid-bed*, and *fixed-bed* types. The fluid-bed and moving-bed processes use mixed nonprecious metal oxide catalysts in units equipped with separate regeneration facilities. Fixed-bed processes use predominantly platinum-containing catalysts in units equipped for cycle, occasional, or no regeneration.

There are several types of catalytic reforming process configurations that differ in the manner that they accommodate the regeneration of the reforming catalyst. Catalyst regeneration involves burning off the coke with oxygen. The semiregenerative process is the simplest configuration but does require that the unit be shut down for catalyst regeneration in which all reactors (typically four) are regenerated. The cyclic configuration utilizes an additional swing reactor that enables one reactor at a time to be taken offline for regeneration while the other four remain in service. The continuous catalyst regeneration (CCR) configuration is the most complex configuration and enables the catalyst to be continuously removed for regeneration and replaced after regeneration. The benefits to the more complex configurations are that operating severity may be increased as a result of higher catalyst activity but this does come at an increased capital cost for the process.

Although subsequent olefin reactions occur in thermal reforming, the product contains appreciable amounts of unstable unsaturated compounds. In the presence of catalysts and of hydrogen (available from dehydrogenation reactions), hydrocracking of paraffin derivatives to yield two lower paraffin derivatives occurs. Olefin derivatives that do not undergo dehydrocyclization are also produced. The olefin derivatives are hydrogenated with or without isomerization, so that the end product contains only traces of olefin derivatives.

The addition of a hydrogenation-dehydrogenation catalyst to the system yields a dual-function catalyst complex. Hydrogen reactions—hydrogenation, dehydrogenation, dehydrocyclization, and hydrocracking—take place on the one catalyst, and thermal decomposition, isomerization, and olefin polymerization take place on the acid catalyst sites.

Under the high-hydrogen partial pressure conditions used in catalytic reforming, sulfur compounds are readily converted into hydrogen sulfide, which, unless removed, builds up to a high concentration in the recycle gas. Hydrogen sulfide is a reversible poison for platinum and causes a decrease in the catalyst dehydrogenation and dehydrocyclization activities. In the first catalytic reformers the hydrogen sulfide was removed from the gas cycle stream by absorption in, for example, diethanolamine. Sulfur is generally removed from the feedstock by use of a conventional desulfurization over cobalt-molybdenum catalyst. An additional benefit of desulfurization of the feed to a level of <5 ppm sulfur is the elimination of hydrogen sulfide (H_2S) corrosion problems in the heaters and reactors.

Organic nitrogen compounds are converted into ammonia under reforming conditions, and this neutralizes acid sites on the catalyst and thus represses the activity for isomerization, hydrocracking, and dehydrocyclization reactions. Straight-run materials do not usually present serious problems with regard to nitrogen, but feeds such as coker naphtha may contain around 50 ppm nitrogen and removal of this quantity may require high-pressure hydrogenation (800–1000 psi) over nickel-cobalt-molybdenum on an alumina catalyst.

The yield of gasoline of a given octane number and at given operating conditions depends on the hydrocarbon types in the feed. For example, high-naphthene stocks, which readily give aromatic gasoline, are the easiest to reform and give the highest gasoline yields. Paraffinic stocks, however, which depend on the more difficult isomerization, dehydrocyclization, and hydrocracking reactions, require more severe conditions and give lower gasoline yields than the naphthenic stocks. The end point of the feed is usually limited to about 190°C (375°F), partially because of increased coke deposition on the catalyst as the end point during processing at about 15°C (27°F). Limiting the feed end point avoids redistillation of the product to meet the gasoline end-point specification of 205°C (400°F), maximum.

Dehydrogenation is a main chemical reaction in catalytic reforming, and hydrogen gas is consequently produced in large quantities. The hydrogen is recycled through the reactors where the reforming takes place to provide the atmosphere necessary for the chemical reactions and also prevents the carbon from being deposited on the catalyst, thus extending its operating life. An excess of hydrogen above whatever is consumed in the process is produced, and as a result, catalytic reforming processes are unique in that they are the only crude oil refinery processes to produce hydrogen as a byproduct.

4. Hydrogenation

The purpose of hydrogenating hydrocarbon derivatives, especially in the crude oil industry is to convert constituents to (i) improve existing crude oil products or develop new products or even new uses, (ii) convert inferior or low-grade materials into valuable products, and (iii) transform higher molecular weight constituents into liquid fuels. The distinguishing feature of the hydrogenating processes is that, although the composition of the feedstock is relatively unknown and a variety of reactions may occur simultaneously, the final product may actually meet all the required specifications for its particular use (Parkash, 2003; Gary et al., 2007; Speight, 2014; Hsu and Robinson, 2017; Speight, 2017).

Hydrogenation processes for the conversion of crude oil and crude oil products may be classified as *nondestructive* and *destructive*. Nondestructive, or simple, hydrogenation is generally used for the purpose of improving product (or even feedstock) quality without appreciable alteration of the boiling range. Treatment under such mild conditions is often referred to as *hydrotreating* or *hydrofining* and is essentially a means of eliminating nitrogen, oxygen, and sulfur as ammonia, water, and hydrogen sulfide, respectively. On the other hand, the latter process (*hydrogenolysis* or *hydrocracking*) is characterized by the rupture of carbon-carbon bonds and is accompanied by hydrogen saturation of the fragments to produce lower boiling products. Such treatment requires rather high temperatures and high hydrogen pressures, the latter to minimize coke formation. Many other reactions, such as isomerization, dehydrogenation, and cyclization, can occur under these conditions (Parkash, 2003; Gary et al., 2007; Speight, 2014, 2017; Hsu and Robinson, 2017).

The chemistry of the hydroprocesses can be defined by the type of catalysis employed, either (i) metal catalysis or acid (ii) catalysis:

Metal catalysis:

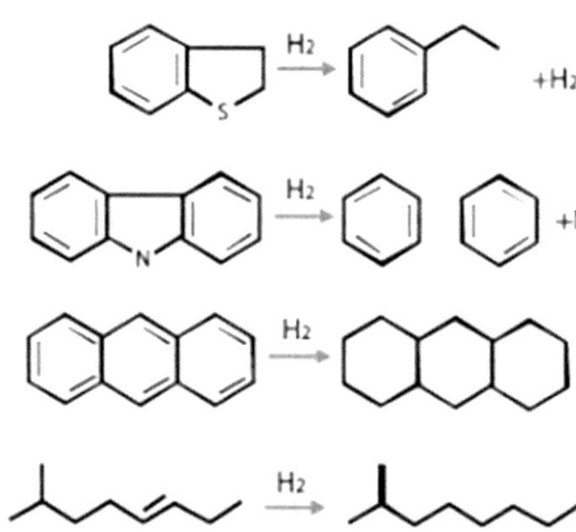

Acid catalysis:

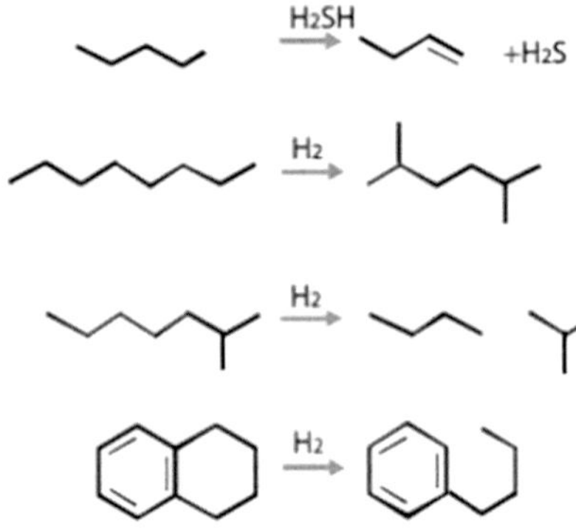

One advantage with catalytic processing is the flexibility it offers in feedstock processing. Heavier feedstocks require more extensive processing which can be accomplished by higher conversion severity in the hydrocracking unit. This may result in lower yields of some products but the converted products from the hydrocracking unit are primarily higher-value transportation fuel such as high-quality diesel and jet fuel.

Hydrotreating, as practiced in the crude oil industry, is a common form of hydrogenation. It is generally recognized that the higher the hydrogen content of a crude oil product, especially the fuel products, the better is the quality of the product. This knowledge has stimulated the use of hydrogen-adding processes in the crude oil refinery. Hydrotreating (i.e., hydrogenation without simultaneous occurrence of thermal decomposition) (Chapter 4) is used for saturating olefins or for converting aromatics to naphthenes as well as for heteroatom removal. Under atmospheric pressure, olefins can be hydrogenated up to approximately 500°C (930°F), but beyond this temperature dehydrogenation commences. The presence of hydrogen changes the nature of the products (especially the decreasing coke yield) by preventing the buildup of precursors that are incompatible in the liquid medium and form coke (Parkash, 2003; Gary et al., 2007; Speight, 2014, 2017; Hsu and Robinson, 2017).

In contrast to the visbreaking process, in which the general principle is the production of products for use as fuel oil, hydroprocessing is employed to produce a slate of products for use as hydrocarbon fuels. Nevertheless, the decomposition of asphaltene constituents is, again, an issue, and just as models consisting of large polynuclear aromatic systems are inadequate to explain the chemistry of visbreaking, they are also of little value for explaining the chemistry of hydrocracking.

A wide variety of metals are active hydrogenation catalysts; those of most interest are nickel, palladium, platinum, cobalt, iron, nickel-promoted copper, and copper chromite. Special preparations of the first three are active at room temperature and atmospheric pressure. The metallic catalysts are easily poisoned by sulfur-containing and arsenic-containing compounds, and even by other metals. To avoid such poisoning, less effective but more resistant metal oxides or sulfides are frequently employed, generally those of tungsten, cobalt, chromium, or molybdenum.

Alternatively, catalysts poisoning can be minimized by mild hydrogenation to remove nitrogen, oxygen, and sulfur from feedstocks in the presence of more resistant catalysts, such as cobalt-molybdenum-alumina (Co-Mo-Al_2O_3). The reactions involved in nitrogen removal are somewhat analogous to those of the sulfur compounds and follow a stepwise mechanism to produce ammonia and the relevant substituted aromatic compound.

Hydrotreating catalysts consist of metals impregnated on a porous alumina support. Almost all of the surface area is found in the pores of the alumina (200–300 m^2/g) and the metals are dispersed in a thin layer over the entire alumina surface within the pores. This type of catalyst does display a huge catalytic surface for a small weight of catalyst. Cobalt (Co), molybdenum (Mo), and nickel (Ni) are the most commonly used metals for desulfurization catalysts. The catalysts are manufactured with the metals in an oxide state. In the active form they are the sulfide state, which is obtained by sulfiding the catalyst either prior to use or with the feed during actual use. Any catalyst that exhibits hydrogenation activity will catalyze hydrodesulfurization to some extent. However, the Group VIB metals (chromium, molybdenum, and tungsten) are particularly active for desulfurization, especially when promoted with metals from the iron group (iron, cobalt, nickel).

Hydrodesulfurization and demetallization occur simultaneously on the active sites within the catalyst pore structure (Ancheyta and Speight, 2007; El-Gendy and Speight, 2016). Sulfur and nitrogen occurring in residua are converted to hydrogen sulfide and ammonia in the catalytic reactor and these gases are scrubbed out of the reactor effluent gas stream. The metals in the feedstock are deposited on the catalyst in the form of metal sulfides and cracking of the feedstock to distillate produces a laydown of carbonaceous material on the catalyst; both events poison the catalyst and activity or selectivity suffers. The deposition of carbonaceous material is a fast reaction that soon equilibrates to a particular carbon level and is controlled by hydrogen partial pressure within the reactors. On the other hand, metal deposition is a slow reaction that is directly proportional to the amount of feedstock passed over the catalyst.

Removal of sulfur from a feedstock results in a gradual increase in catalyst activity, returning almost to the original activity level. As with ammonia, the concentration of the hydrogen sulfide can be used to control precisely the activity of the catalyst. Nonnoble metal-loaded zeolite catalysts have an inherently different response to sulfur impurities since a minimum level of hydrogen sulfide is required to maintain the nickel-molybdenum and nickel-tungsten in the sulfide state. Alternatively, catalysts poisoning can be minimized by mild hydrogenation to remove nitrogen, oxygen, and sulfur from feedstocks in the presence of more resistant catalysts, such as cobalt-molybdenum-alumina (Co-Mo-Al_2O_3). The reactions involved in nitrogen removal are somewhat analogous to those of the sulfur compounds and follow a stepwise mechanism to produce ammonia and the relevant substituted aromatic compound.

In the presence of a catalyst— typically platinum, palladium, nickel, or rhodium— hydrogen can be added across a triple or a double bond to take an alkyne to an alkene or an alkene to an alkane. In practice, it is difficult to isolate the alkene product of this reaction, though a poisoned catalyst (i.e., a catalyst with fewer available reactive sites) can be used to do so. As the hydrogen is immobilized on the surface of the catalyst, the triple or double bonds are hydrogenated in a *syn* fashion in which the hydrogen atoms add to the same side of the molecule.

5. Dehydrogenation

Dehydrogenation is a chemical reaction that involves the elimination of hydrogen (H_2) and is the reverse of hydrogenation. Dehydrogenation reactions are of major industrial importance, especially in the production of constituents for the manufacture of high octane gasoline.

The dehydrogenation of low boiling alkane derivatives has become of great industrial importance because it presents an alternative method for obtaining alkene derivatives (e.g., butenes, butadiene) from low cost saturated hydrocarbon feedstocks.

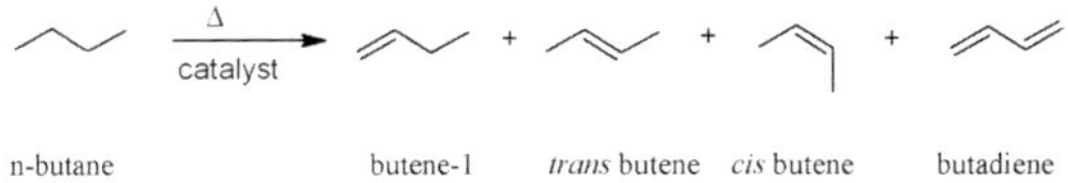

The reaction is strongly endothermic, and therefore heat must be supplied to maintain the reaction temperature. When the detached hydrogen is immediately oxidized (oxidative dehydrogenation), the conversion of reactants to products is increased because the equilibrium concentration is shifted toward the products and the added exothermic oxidation reaction supplies the needed heat of reaction. On the other hand, excess hydrogen is sometimes added to a dehydrogenation reaction in order to diminish the complete breakup of the molecule into many fragments.

Within the context of the current text, dehydrogenation reactions fall into two major classes: (i) aromatization in which six-membered cycloalkane derivatives rings can be aromatized in the presence of hydrogenation catalysts (such as sulfur, selenium, or quinones), and (ii) dehydrogenation of alkane derivatives and alkene derivatives in which n-pentane and iso-pentane can be converted to pentene and isoprene using chromium (III) oxide as a catalyst at 500°C (930°F). Oxidation, such as the conversion of alcohols to ketones or aldehydes and the dehydrogenation of amines to nitriles are other examples that are not considered here.

The most important reactions pertain to hydrocarbon derivatives, principally paraffin derivatives to olefin derivatives and diolefin derivatives, or alkyl aromatic hydrocarbon derivatives to alkenyl aromatic hydrocarbon derivatives (e.g., ethylbenzene to styrene).

Dehydrogenation, in general, concerns the abstraction of hydrogen from a compound to produce a less saturated analog. Although dehydrogenation can be used on compounds that contain heteroatoms (e.g., alcohols to aldehydes), by and large, the most important reactions pertain to hydrocarbon derivatives, principally paraffin derivatives to olefin derivatives and diolefin derivatives, or alkyl aromatic hydrocarbon derivatives to alkenyl aromatic hydrocarbon derivatives (e.g., ethylbenzene to styrene). Dehydrogenation can be effected thermally or catalytically.

Dehydrogenation can be effected thermally or catalytically. Thermal dehydrogenation is best exemplified by the pyrolysis of hydrocarbon derivatives to produce olefin derivatives, usually in the presence of steam, in pyrolysis furnaces or steam crackers at elevated temperatures. By controlling temperatures and residence times, thermal pyrolysis can be surprisingly selective, as is the case, for example, in the conversion of ethane to ethylene (approximately 80%–90% w/w yield) or in the production of alpha-olefin derivatives (α-olefin derivatives) by the thermal decomposition of wax. In general, however, thermal pyrolysis has limited selectivity patterns. For example, the thermal pyrolysis of naphtha will typically yield only about 29%–33% w/w ethylene and 14%–17% w/w propylene. But these yields are a strong function of the feedstock composition, and significantly higher yields of ethylene can be obtained if the feed naphtha is rich in *n*-paraffin derivatives.

Dehydrogenation can be effected thermally or catalytically. Thermal dehydrogenation is best exemplified by the pyrolysis of hydrocarbon derivatives to produce olefin derivatives, usually in the presence of steam, in pyrolysis furnaces or steam crackers at elevated temperatures. By controlling temperatures and residence times, thermal pyrolysis can be surprisingly selective but, in general thermal decomposition has limited selectivity patterns. Thus, for example, the thermal pyrolysis of a naphtha will typically yield only about 29%–33% w/w ethylene and 14%–17% w/w propylene. The yields are a strong function of the feedstock composition, and significantly higher yields of ethylene can be obtained if the feed naphtha is rich in *n*-paraffin derivatives.

The common primary reactions of alkane pyrolysis are dehydrogenation and carbon bond scission. The extent of one or the other varies with the starting material and operating conditions, but because of its practical importance, methods have been found to increase the extent of dehydrogenation and, in some cases, to render it almost the only reaction.

For example, at 550°C (1025°F) n-butane loses hydrogen to produce butene-1 and butene-2. The development of selective catalysts, such as chromic oxide (chromia, Cr_2O_3) on alumina (Al_2O_3) has rendered the dehydrogenation of paraffin derivatives to olefin derivatives particularly effective, and the formation of higher molecular weight material is minimized. Higher selectivity can be obtained when the dehydrogenation reactions proceed catalytically, but always subject to per-pass conversion limitations imposed by thermodynamic equilibria, so that recycle of unconverted hydrocarbon derivatives is usually required.

Catalytic paraffin dehydrogenation for the production of olefin derivatives has been in commercial use since the late 1930s, while catalytic paraffin oxydehydrogenation for olefin production has not yet been commercialized. However, there are some interesting recent developments worthy of further research and development. During World War II, catalytic dehydrogenation of butanes over a chromia-alumina catalyst was performed for the production of butenes, which were then dimerized to octenes and hydrogenated to octanes to yield high-octane aviation fuel. More recently, platinum or modified platinum catalysts have been used instead of chromia-alumina catalysts. Important aspects in dehydrogenation entail the approach to equilibrium or near-equilibrium conversions while minimizing sidereactions and coke formation.

Thermal dehydrogenation is best exemplified by the pyrolysis of hydrocarbon derivatives to produce olefin derivatives, usually in the presence of steam, in pyrolysis furnaces or steam crackers at elevated temperatures. The dehydrogenation of light alkane derivatives (C_3-C_5) to corresponding alkene derivatives (olefin derivatives) is studied in the project. Alkene derivatives can be used as raw materials in etherification, alkylation, or polymerization processes. Dehydrogenation catalysts used in the industrial processes are supported chromium oxide and platinum metal. Supported vanadium oxide and molybdenum oxide have been studied as potential new catalysts.

Alkane dehydrogenation is an endothermic reaction that requires relatively high temperatures and low pressures. The equilibrium reaction is shown in general form in the equation:

$$C_nH_{2n+2} \leftrightarrow C_nH_{2n} + H_2$$

Commercial processes for the catalytic dehydrogenation of propane and butanes attain per-pass conversions in the range of 30%–60%, while the catalytic dehydrogenation of C_{10}-C_{14} paraffin derivatives typically operates at conversion levels of 10%–20%.

Oxydehydrogenation employs catalysts containing vanadium and, more recently, platinum. Oxydehydrogenation at approximately 1000°C (1830°F) and very short residence time over Pt and Pt-Sn catalysts can produce ethylene in higher yields than in steam cracking. However, there are a number of issues related to safety and process upsets that need to be addressed. Important objectives in oxydehydrogenation are attaining high selectivity to olefin derivatives with high conversion of paraffin derivatives and minimizing potentially dangerous mixtures of paraffin and oxidant. More recently, the use of carbon dioxide as an oxidant for ethane conversion to ethylene has been investigated as a potential way to reduce the negative impact of dangerous oxidant-paraffin mixtures and to achieve higher selectivity.

Industrial processes use temperatures above 500°C (930°F) but the high temperatures also favor sidereactions such as thermal decomposition and deactivation of the catalysts through coke formation. Hence, in all processes the catalyst undergoes dehydrogenation and regeneration. For example, the thermal pyrolysis of naphtha will typically yield only about 29%–33% w/w ethylene and 14%–17% w/w propylene. But these yields are a strong function of the feedstock composition, and significantly higher yields of ethylene can be obtained if the feed naphtha is rich in *n*-paraffin derivatives (Vora and Pujadó, 2005).

Higher selectivity can be obtained when the dehydrogenation reactions proceed catalytically, but always subject to per-pass conversion limitations imposed by thermodynamic equilibria, so that recycle of unconverted hydrocarbon derivatives is usually required. This entry concerns mostly the application of catalytic dehydrogenation. Catalytic paraffin dehydrogenation for the production of olefin derivatives has been in commercial use since the late 1930s, while catalytic paraffin oxydehydrogenation for olefin production has not yet been commercialized. However, there are some interesting recent developments worthy of further research and development. During World War II, catalytic dehydrogenation of butanes over a chromia-alumina catalyst was performed for the production of butenes, which were then dimerized to octenes and hydrogenated to octanes to yield high-octane aviation fuel. More recently, platinum or modified platinum catalysts have been used instead of chromia-alumina catalysts. Important aspects in dehydrogenation entail the approach to equilibrium or near-equilibrium conversions while minimizing sidereactions and coke formation.

Naphthene derivatives are somewhat more difficult to dehydrogenate, and cyclopentane derivatives form only aromatic hydrocarbon derivatives if a preliminary step to form the cyclohexane structure can occur. Alkyl derivatives of cyclohexane usually dehydrogenate at 480–500°C (895–930°F), and polycyclic naphthene derivatives are also quite easy to dehydrogenate thermally. In the presence of catalysts, cyclohexane and its derivatives are readily converted into aromatic hydrocarbon derivatives; reactions of this type are prevalent in catalytic cracking and reforming. Benzene and toluene are prepared by the catalytic dehydrogenation of cyclohexane to benzene:

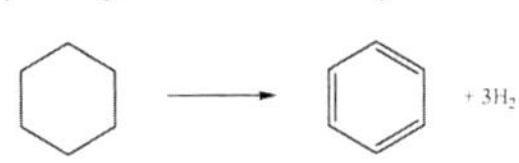

Similarly, methylcyclohexane can be dehydrogenated to toluene over a variety of catalysts (Boufaden et al., 2016).

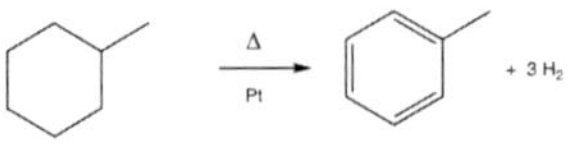

Polycyclic naphthene derivatives can also be converted to the corresponding aromatic hydrocarbon derivatives by heating at 450°C (840°F) in the presence of a chromia-alumina (Cr_2O_3-Al_2O_3) catalyst.

Alkyl aromatic compounds also dehydrogenate to various products. For example, the majority of industrial production of styrene follows from the dehydrogenation of ethylbenzene:

$$C_6H_5CH_2CH_3 \rightarrow C_6H_5CH{=}CH_2 + H_2$$

Sidereactions are:

$$C_6H_5CH_2CH_3 \rightarrow C_6H_6 + C_2H_4$$

$$C_6H_5CH_2CH_3 \rightarrow 8C + 5H_2$$

$$C_6H_5CH{=}CH_2 + 2H_2 \rightarrow C_6H_5CH_3 + CH_4$$

This dehydrogenation process involves the catalytic reaction of ethylbenzene. Fresh ethylbenzene is mixed with a recycle stream and vaporized. Steam is then added before feeding the effluent into a train of two–four reactors. This process involves a highly endothermic reaction carried out in the vapor phase over a solid catalyst. Steam is used to provide heat of this reaction, to prevent excessive coking or carbon formation, to shift equilibrium of the reversible reaction toward the products, and to clean the catalyst of any carbon that does form. The reactors are run adiabatically with steam added before each stage with typical yields of 88%–94%.

Crude styrene from the reactors is then fed into a distillation train. Because of the possibility of polymerization of the styrene during distillation, small residence time, avoidance of high temperature, and addition of inhibitor are necessary. A large proportion of styrene is produced with an unsupported iron (Fe^{3+}) oxide catalyst promoted with a potassium compound. Potassium carbonate or hydroxide and chromium oxides are added to the catalyst to improve reaction selectivity. Other alkylbenzene derivatives can be dehydrogenated similarly; *iso*-propyl benzene yields α-methyl styrene.

6. Dehydrocyclization

Dehydrocyclization is any process involving both dehydrogenation and cyclization, as is used in crude oil refining. An example is a reaction in which an alkane is converted into an aromatic hydrocarbon and hydrogen, such as the conversion of heptane to toluene and hydrogen as can occur in the reforming unit:

$$CH_3(CH_2)_5CH_3 \rightarrow C_6H_5CH_3 + 4H_2$$

Catalytic aromatization involving the loss of 1 mol of hydrogen followed by ring formation and further loss of hydrogen has been demonstrated for a variety of paraffin derivatives. Thus, n-hexane can be converted to benzene, and octane is converted to ethyl benzene and to *o*-xylene (*ortho*-xylene):

$$CH_3(CH_2)_4CH_3 \rightarrow C_6H_6 + 4H_2$$

$$CH_3(CH_2)_6CH_3 \rightarrow C_6H_5CH_2CH_3 + + 4H_2$$

$$CH_3(CH_2)_6CH_3 \rightarrow 1{,}2\text{-}C_6H_4(CH_3)_2 + 4H_2$$

Conversion takes place at low pressures, even atmospheric, and at temperatures above 300°C (570°F), although 450–550°C (840–1020°F) is the preferred temperature range.

The catalysts are metals (or their oxides) of the titanium, vanadium, and tungsten groups and are generally supported on alumina; the mechanism is believed to be dehydrogenation of the paraffin to an olefin, which in turn is cyclized and dehydrogenated to the aromatic hydrocarbon. In support of this, olefin derivatives can be converted to aromatic hydrocarbon derivatives much more easily that the corresponding paraffin derivatives.

The term *dehydrocyclization of paraffin derivatives* is usually considered to be synonymous with the conversion of paraffin derivatives to yield aromatic hydrocarbon derivatives. Such a reaction can be effected by the catalytic action of platinum or oxides of metals from Group VI, deposited on alumina or other carriers. This reaction may be named Ca-dehydrocyclization.

The direct catalytic conversion of alkane derivatives into aromatic hydrocarbon derivatives has found potentially important industrial applications. Supported platinum catalysts were found active for the aromatization of alkane derivatives; the drawbacks of these catalysts were their deactivation with time on stream and the existence of simultaneous parallel reactions. A bifunctional mechanism which involves both the metal and the acid sites of the support and a monofunctional mechanism involving only the metallic sites operate over, respectively, platinum supported on acidic support and platinum supported on nonacidic support (Mériaudeau and Naccache, 1997).

Over monofunctional Pt catalysts two possible mechanisms prevail: 1,6 ring closure on the Pt surface involving primary and secondary C-H bond rupture, followed by dehydrogenation of the cycloalkane derivatives into aromatic hydrocarbon derivatives (1,5 ring closure to a lesser extent also contributes to aromatic production); or dehydrogenation of the alkane derivatives into olefin derivatives, dienes, and trienes followed by thermal ring closure.

Zeolites were found most suitable as support for preparing catalysts more active and more selective in the alkane aromatization. In addition, catalysts based on noble metals supported on zeolite appeared more resistant against deactivation by coke. In this review the aromatization of hexane, heptane, and octane over Pt-zeolite catalysts is discussed in detail.

Zeolite structure also affects the aromatic product distribution, in particular when the alkane contains more than seven carbon atoms. It is shown how Pt on medium-pore zeolites such as in ZSM-5, the silicalite centers will favor the aromatization of C_8 alkane isomers into ethylbenzene-styrene with respect to other C_8 aromatic hydrocarbon derivatives. Aromatization of light alkane derivatives, C_2-C_5, requires the increase of the hydrocarbon chain length up to six carbon atoms and higher, followed by cyclization reaction.

In the presence of platinized charcoal, paraffin derivatives can also be converted into hydrocarbon derivatives with a five-membered ring (Kasanski and Liberman, 1959). n-Pentane is converted to cyclopentane with greater difficulty than paraffin derivatives of higher molecular weight, and this conversion requires higher temperature. At 310°C (590°F), n-hexane yields methyl cyclopentane and benzene; n-octane gives a mixture of n-propyl cyclopentane and 1-methyl-2-ethyl cyclopentane and 1% toluene. Branched-chain paraffin derivatives are converted more readily; 3-ethylpentane yields ethyl cyclopentane, and isooctane gives 1,1,3-trimethyl cyclopentane. This reaction may be called 6-dehydrocyclization. n-propyl-, sec-butyl, and iso-butyl benzene derivatives are converted to indane and 2- and 3-methyl indane derivatives.

The direct catalytic conversion of alkane derivatives into aromatic hydrocarbon derivatives has found potentially important industrial applications (Mériaudeau and Naccache, 1997). Initially only alkane derivatives with six and more carbon atoms in the chain were concerned. Supported platinum catalysts were found active for the aromatization of alkane derivatives; the drawbacks of these catalysts were their deactivation with time on stream and the existence of simultaneous parallel reactions. Much discussion has been published on the aromatization of C_{6+} alkane derivatives.

A bifunctional mechanism which involves both the metal and the acid sites of the support and a monofunctional mechanism involving only the metallic sites operate over, respectively, Pt supported on acidic support and Pt supported on nonacidic support. In the present review the mechanisms proposed for the aromatization of alkane derivatives are described. Over monofunctional Pt catalysts two possible mechanisms prevail: 1,6 ring closure on the Pt surface involving primary and secondary C-H bond rupture, followed by dehydrogenation of the cycloalkane derivatives into aromatic hydrocarbon derivatives (1,5 ring closure to a lesser extent also contributes to aromatic production); or dehydrogenation of the alkane derivatives into olefin derivatives, dienes, and trienes followed by thermal ring closure. Zeolites were found most suitable as support for preparing catalysts more active and more selective in the alkane aromatization.

In addition, catalysts based on noble metals supported on zeolite appeared more resistant against deactivation by coke. In this review the aromatization of hexane, heptane, and octane over Pt-zeolite catalysts is discussed in detail. Comparisons between different zeolite structures and different dehydrogenation sites are given. In particular a critical analysis of the results and interpretation concerning Pt-KL catalysts strongly suggests that the exceptional high selectivity toward aromatization of *n*-hexane exhibited by Pt-KL could not be explained by only the nest or constraint effect exerted by the channel dimension and morphology, not only by the terminal cracking properties or the partial electron transfer from the zeolite support to the Pt particlesor by the Pt particle size. Zeolite structure also affects the aromatic product distribution, in particular when the alkane contains more than seven carbon atoms.

Aromatization of light alkane derivatives, C_2-C_5, requires the increase of the hydrocarbon chain length up to six carbon atoms and higher, followed by cyclization reaction. Recently new processes to convert C_2-C_5 alkane derivatives into aromatic hydrocarbon derivatives have become available and these processes use bifunctional catalysts possessing a dehydrogenating and an acid function. The catalysts consist of a metal ion or metal oxide supported on a microporous acid solid. In this review we analyze the results concerning mainly platinum supported on pentasil-type zeolite. It is shown that although Pt has better dehydrogenating properties as compared with gallium and zinc, the efficiency of catalysts based on Pt-ZSM-5 for light alkane aromatization is less because undesirable reactions such as hydrogenolysis and ethene (olefin derivatives) hydrogenation occur on the platinum surface, resulting in the production of unreactive alkane derivatives. These drawbacks could be partially suppressed by alloying Pt and by increasing the reaction temperature.

Catalytic reforming is a chemical process (Fig. 11.1) used to convert naphtha, which typically has a low octane rating, into a high-octane liquid product (reformates), which is a component of high-octane gasoline. Basically, the process rearranges or restructures the hydrocarbon constituents of the naphtha to aromatic products. In so doing, the process produces significant amounts of byproduct hydrogen gas for use in a number of the other processes involved in a refinery. Other byproducts are small amounts of methane, ethane, propane, and butane(s).

There are a good many chemical reactions that occur in the catalytic reforming process, all of which occur in the presence of a catalyst and a high partial pressure of hydrogen. Depending upon the type or version of catalytic reforming used as well as the desired reaction severity, the reaction conditions range from temperatures of about 495–525°C and from pressures of about 75–650 psi. The commonly used catalytic reforming catalysts contain noble metals such as platinum and/or rhenium, which are very susceptible to poisoning by sulfur and nitrogen compounds in the feedstock. Therefore, the naphtha feedstock to a catalytic reformer is always preprocessed in a hydrodesulfurization unit which removes both the sulfur and the nitrogen compounds.

The four major catalytic reforming reactions are (i) dehydrogenation of naphthene derivatives to aromatic derivatives, (ii) isomerization of *n*-paraffin derivatives to iso-paraffin derivatives, (iii) dehydrogenation and aromatization of paraffin derivatives to aromatic hydrocarbon derivatives, and (iv) hydrocracking of paraffin derivatives into lower molecular weight products.

The *dehydrogenation* of naphthene derivatives to produce aromatic hydrocarbon derivatives using the conversion of methylcyclohexane to toluene as an example:

Methylcyclohexane → Toluene + 3 H_2

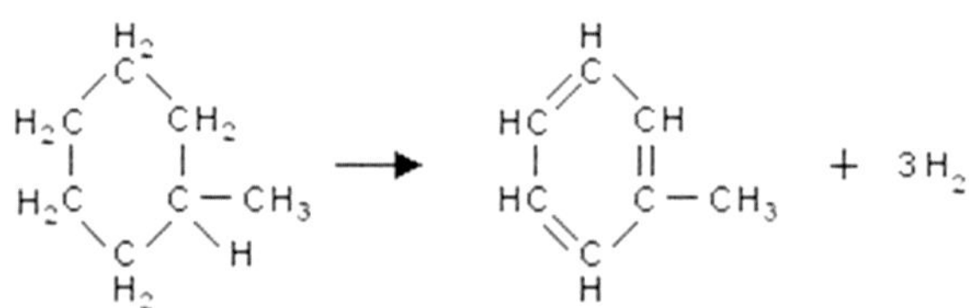

The isomerization of *n*-paraffin derivatives to iso-paraffin derivatives as, for example, in the conversion of *n*-octane to 2,5-dimethyl hexane:

n-Octane → 2,5-Dimethylhexane

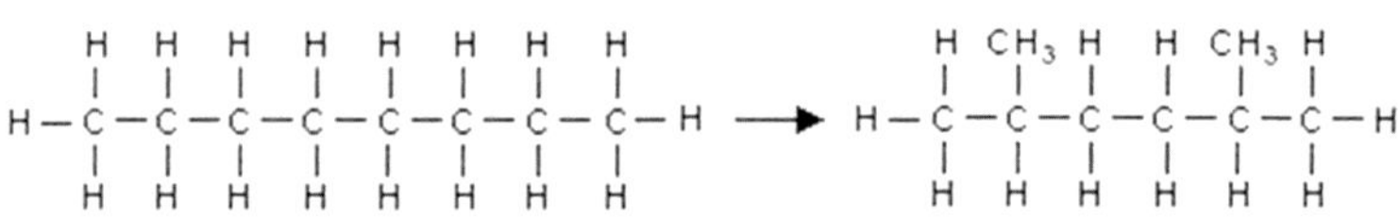

The dehydrogenation and aromatization of paraffin derivatives to aromatic hydrocarbon derivatives (dehydrocyclization) as in the conversion of *n*-heptane to toluene:

n-Heptane → Toluene + 4 H_2

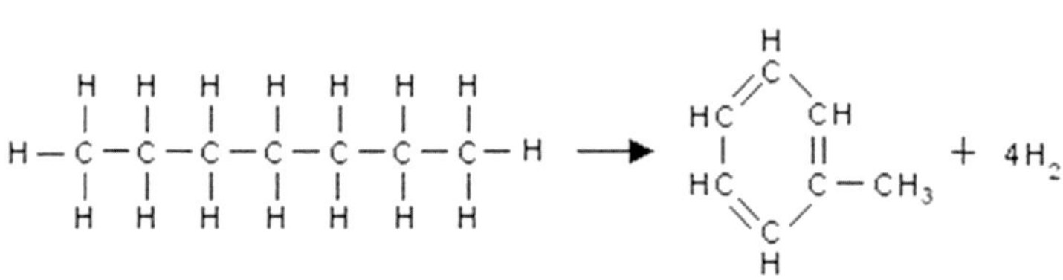

The hydrocracking of paraffin derivatives into lower molecular weight products, such as the thermal decomposition *n*-heptane to iso-pentane and ethane:

n-Heptane + H_2 → Isopentane + Ethane

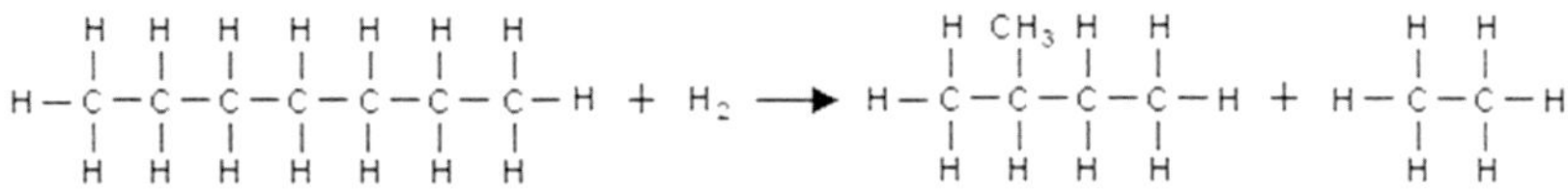

Hydrocracking paraffin derivatives is the only one of the above four major reforming reactions that consume hydrogen. The isomerization of normal paraffin derivatives does not consume or produce hydrogen. However, both the dehydrogenation of naphthene derivatives and the dehydrocyclization of paraffin derivatives produce hydrogen.

The overall net production of hydrogen in the catalytic reforming of crude oil naphtha ranges from approximately 300 to 1200 ft^3 of hydrogen gas per barrel (4.2 ft^3) of liquid naphtha feedstock. In many crude oil refineries, the net hydrogen produced in catalytic reforming supplies a significant part of the hydrogen used elsewhere in the refinery (for example, in hydrodesulfurization processes).

7. Chemical reactions

The reactions of hydrocarbon derivatives are many and varied. However, there are differences between the reactions of saturated hydrocarbon and unsaturated hydrocarbon derivatives. Saturated hydrocarbons form four single covalent bonds with other atoms on its each carbon atom in the molecule and are single bonded with a greater amount of hydrogen. Unsaturated hydrocarbon derivatives are usually double bonded (some contain triple bonds), and thus it do not contain the maximum amount of hydrogen.

Alkane derivatives contain strong carbon-carbon single bonds and strong carbon-hydrogen (nonpolar) bonds As a result, there is no portion of the molecule that carries any significant amount of positive or negative charge, which is required for other ionic and polar molecules to be attracted to it. Therefore alkane derivatives generally do not react with ionic compounds such as most laboratory acids, bases, oxidizing agents, or reducing agents:

$$\underset{\text{butane}}{CH_3CH_2CH_2CH_3} + OH^- \longrightarrow \text{no reaction}$$

$$+ H^+ \longrightarrow \text{no reaction}$$

$$+ MnO_4^- \longrightarrow \text{no reaction}$$

Neither positive ions nor negative ions are attracted to a nonpolar molecule. However, the unsaturated hydrocarbon derivatives do offer varying degrees of reactivity.

It is not the purpose of this section to present a comprehensive review of the chemical reactions of hydrocarbon derivatives but to present an overview of the types of reactions of hydrocarbon derivatives that can be of use for the commercial production of saleable products.

7.1 Alkylation

Alkylation is the transfer of an alkyl group from one molecule to another. The alkyl group may be transferred as an alkyl carbocation, a free radical, a carbanion, or a carbene. An alkyl group is a piece of a molecule with the general formula C_nH_{2n+1}, where n is the integer depicting the number of carbons linked together. For example, a methyl group ($n = 1$, CH_3) is a fragment of a methane molecule (CH_4). Alkylating agents use selective alkylation by adding the desired aliphatic carbon chain to the previously chosen starting molecule. This is one of many known chemical syntheses. Alkyl groups can also be removed in a process known as dealkylation.

Carbon alkylation is a process for the formation of carbon-carbon bonds and for this types of reaction, the electrophilic nature of alkyl halides is enhanced by the presence of a Lewis acid, such as aluminum trichloride ($AlCl_3$). The carbon alkylation reaction can also be effected by alkene derivatives in the presence of acids.

Nitrogen alkylation is an important process for the formation of carbon-nitrogen bonds. Amines are readily alkylated and the rate of alkylation follows the order tertiary amine < secondary amine < primary amine. As an example, a tertiary amine can be converted into a quaternary ammonium salt by reaction with an alkyl (hailed the Menshutkin reaction):

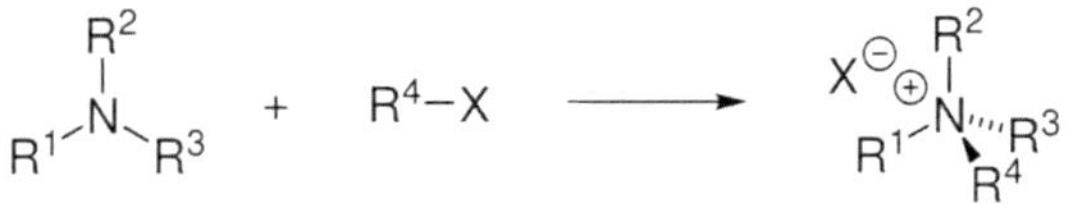

Alcohol derivatives can also be alkylated to yield ethers derivatives: for example, using an alkyl halide (the Williamson ether synthesis)

$$ROH + R^1X \rightarrow ROR^1$$

Alcohols are also good alkylating agents in the presence of suitable acid catalysts. For example, most methyl amines are prepared by alkylation of ammonia with methanol. The alkylation of phenols is particularly straightforward, since it is subject to fewer competing reactions.

$$C_6H_5OH + NaOH \rightarrow C_6H_5O^- \ Na^+ + H_2O$$

$$C_6H_5O^- \ Na^+ + (CH_3)_2SO_4 \rightarrow C_6H_5OCH_3 + CH_3SO_4^- + Na^+$$

Thiol derivatives (RSH) are also readily alkylated to yield thioether derivatives (RSR^1). The reaction is typically conducted in the presence of a base or using the conjugate base of the thiol.

On an industrial scale, the typical route for the alkylation of benzene uses ethylene ($CH_2{=}CH_2$) with a zeolite (ZSM 5) catalyst:

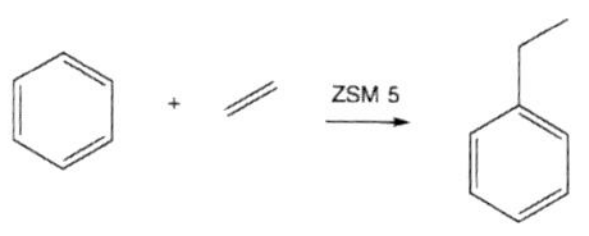

In addition.

7.2 Halogenation

Halogenation is a chemical reaction that involves the addition of one or more halogens to a compound or material. The pathway and stoichiometry of the reaction depends on the structural features and functional groups of the hydrocarbon, as well as on the specific halogen. Unlike the complex transformations of combustion, the halogenation of an alkane appears to be a simple substitution reaction in which a carbon-hydrogen bond is broken and a new carbon-halogen bond is formed. For example:

$$CH_4 + Cl_2 \rightarrow CH_3Cl + HCl$$

One complication is that all the hydrogen atoms of an alkane may undergo substitution, resulting in a mixture of products:

$$4CH_4 + 10Cl_2 \rightarrow CH_3Cl + CH_2Cl_2 + CHCl_3 + CCl_4 + 10HCl$$

The relative amounts of the various products depend on the proportion of the two reactants used. In the case of methane, a large excess of the hydrocarbon favors formation of methyl chloride as the chief product; whereas, an excess of chlorine favors formation of chloroform and carbon tetrachloride.

In halogenation reactions, the reactivity of the halogens decreases in the following order: $F_2 > Cl_2 > Br_2 > I_2$; the reactions may be conducted in either the gaseous or liquid phase. When alkane derivatives having a higher molecular weight than ethane are halogenated, isomeric products are formed. Thus chlorination of propane gives both 1-chloropropane (*n*-propyl chloride) and 2-chloropropane (iso-propyl chloride) as the monochlorinated products:

$$2CH_3CH_2CH_3 + Cl_2 \rightarrow CH_3CH_2CH_2Cl + CH_3CHClCH_3 + 2HCl$$

1-Chloropropane 2-Chloropropane

Thus, the halogenation of alkane derivatives typically occurs by a substitution reaction, in which one or more of the hydrogen atoms of the alkane derivative is replaced with a different atom or group of atoms. The carbon-

carbon bonds remain intact in these reactions, and the hybridization of the carbon atoms does not change. For example, the reaction between ethane and molecular chlorine is a substitution reaction:

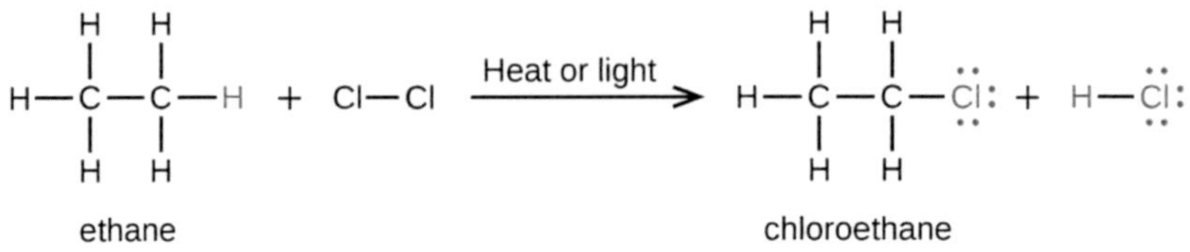

Thus, saturated hydrocarbons typically do not add halogens but undergo free radical halogenation, involving substitution of hydrogen atoms by halogen. The chemistry of the halogenation of alkane derivatives is usually determined by the relative weakness of the available carbon-hydrogen bonds. The preference for reaction at tertiary and secondary positions results from greater stability of the corresponding free radicals and the transition state leading to them. Free radical halogenation is used for the industrial production of chlorinated methane derivatives, which can lead to a variety of chloromethane derivatives:

$$CH_4 + Cl_2 \rightarrow CH_3Cl + HCl$$

$$CH_3Cl + Cl_2 \rightarrow CH_2Cl_2 + HCl$$

$$CH_2Cl_2 + Cl_2 \rightarrow CHCl_3 + HCl$$

$$CHCl_3 + Cl_2 \rightarrow CCl_4 + HCl$$

The original mixture of a colorless gas (CH_4) and a green gas (Cl_2) would produce steamy fumes of hydrogen chloride (HCl) and a mist of organic liquids (mixture of the chlorinated methane). All of the organic products are liquid at room temperature with the exception of the chloromethane (CH_3Cl) which is a gas. When a higher molecular weight alkane derivative is used in place of methane, the reaction in much more complex and, in addition, rearrangement often accompanies such free radical reactions. In fact, most reactions that occur with alkene derivatives are addition reactions in which, as the name implies, during an addition reaction a compound is added to the molecule across the double bond. The result is loss of the double bond (or alkene structure), and the formation of the alkane structure.

For example, unsaturated hydrocarbons (alkene derivatives and alkyne derivatives) readily react with halogens by addition reactions with halogens: halogen atoms (X) are added across the double (C=C) or triple (C≡C) bond. As an example:

$$CH_3CH{=}CHCH_3 + X_2 \rightarrow CH_3CHXCHXCH_3$$

In this equation, X_2 is the halogen.

The most commonly encountered halogens in the halogenated hydrocarbon products are fluorine and chlorine, but sometimes bromine or iodine occur, or combinations of any of these.

In the *hydrohalogenation* reaction, an alkene derivative reacts with a molecule that contains one hydrogen and one halogen—hydrogen chloride (HCl) and hydrogen bromide (HBr) are common hydrohalogen derivatives that are used in this reaction type. In hydrohalogenation, the hydrohalogen is a polar molecule, unlike the nonpolar molecules observed in the halogenation and hydrogenation reactions. In the case of the hydrohalogen, the end of the molecule containing hydrogen is partially positive, while the end of the molecule containing the halogen is partially negative. Thus, when the negatively charged electron from the alkene double bond attacks the hydrohalogen, it will preferentially attack the hydrogen side of the molecule, since the electron will be attracted to the partial positive charge. The halogen will then form the negatively charged anion observed in the intermediate structure and attach second during the addition reaction. The final product is a haloalkane. In the simplest case, the reaction is:

$$CH_2{=}CH_2 + HCl \rightarrow CH_3CH_2Cl$$

Aromatic compounds are subject to electrophilic halogenation in which the halogen atom (X) can replace a hydrogen atom from the aromatic ring:

$$RC_6H_5 + X_2 \rightarrow RC_6H_4X + HX$$

This reaction works only for chlorine and bromine and is carried in the presence of a Lewis acid such as ferric chloride (iron trichloride, $FeCl_3$) The role of the Lewis acid is to polarize the halogen-halogen bond, making the halogen molecule more electrophilic. Industrially, this is achieved by treating the aromatic compound with the halogen in the presence of iron metal. When the halogen is introduced into the reaction vessel, it reacts with iron thereby generating FeX_3 in catalytic amounts.

In the case of iodine, oxidizing conditions must be used in order to perform iodination. The iodination reaction is a reversible process and, thus, the products have to be removed from the reaction medium in order to drive the reaction forward, This can be achieved by conducting the reaction in the presence of an oxidizer that oxidizes the hydrogen iodide to iodine thereby removing the hydrogen iodide from the reaction and generating more iodine that can further react. However, because fluorine is an extremely reactive halogen, the protocol described above would not be efficient as the aromatic molecule would react destructively with fluorine. Therefore, other methods, such as the Balz-Schiemann reaction, are often employed to prepare fluorinated aromatic compounds.

The Balz-Schiemann reaction is a chemical reaction in which a primary aromatic amine is transformed to an aromatic fluoride via a diazonium tetrafluoroborate intermediate. Both thermal and photolytic decomposition of the diazonium intermediate proceed through an arene cation, as evidenced by the equal product ratios in both cases. Thus;

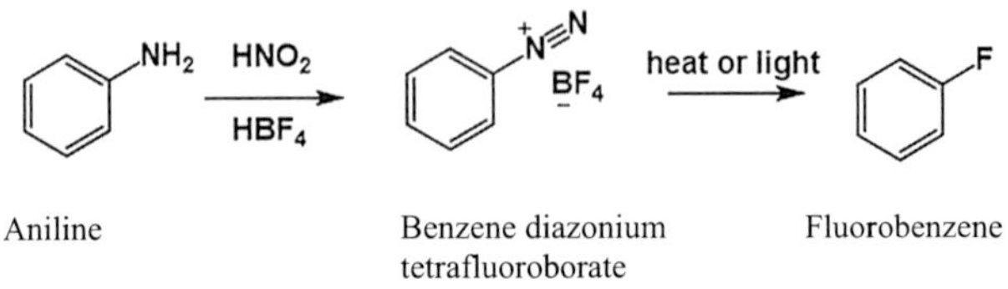

However, it should be noted (and caution observed) that the large-scale thermal decomposition of the diazonium salts is potentially explosive.

7.3 Hydration

The hydration reaction involves the addition of water to a molecule while dehydration means the removal or elimination of water from a molecule. In the simplest cases, the reactions are:

Hydration:

$$CH_2{=}CH_2 + H_2O \rightarrow CH_3CH_2OH$$

Dehydration:

$$CH_3CH_2OH \rightarrow CH_2{=}CH_2 + H_2O$$

Similar to the hydrohalogenation reaction (above), water is also a polar molecule and is split into two groups (H^+ and OH^-) to be added across the double bond of the alkene derivative. Similar to the hydrohalogenation reaction, the hydrogen adds first, as it carries the partial positive charge. The hydroxyl group forms the negative anion intermediate and is then added to the carbocation to form the final product, which is an alcohol.

In more complex molecules, hydrohalogenation and hydration reactions can lead to the formation of more than one possible product. For example, if 2-methylpropene [$(CH_3)_2C{=}CH_2$] reacts with water to form the alcohol, two possible products can form but the addition reaction is not random. One of the products is the major product (being produced in higher abundance) while the other product is the minor product.

Major product.

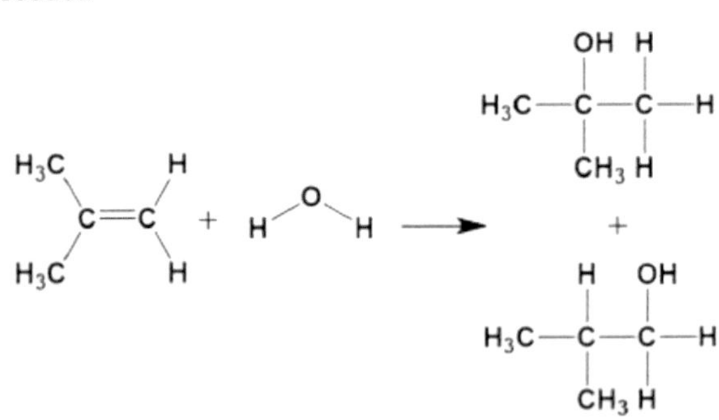

In these types of reactions, the Markovnikov rule can be used to predict which product will be the major product insofar as in addition reactions with HX (where X is a halogen or the OH group from water), the hydrogen always attaches to the carbon that already has the most hydrogens, and the halogen or hydroxyl group attaches the carbon with the fewer hydrogen atoms.

In summary, alkene derivatives and alkyne derivatives can also be halogenated with the halogen adding across the double or triple bond, in a similar fashion to hydrogenation. The halogenation of an alkene results in a dihalogenated alkane derivative, while the halogenation of an alkyne can produce a tetrahalogenated alkane derivative.

7.4 Oxidation

The most common oxidation reactions of hydrocarbon derivatives are the oxidation reactions that occur during combustion in which the final products are carbon dioxide and water (Chapter 10). Alkane derivatives burn in the presence of oxygen, a highly exothermic oxidation-reduction reaction that produces carbon dioxide and water. As example, gasoline is a liquid mixture of continuous- and branched-chain alkane derivatives, each containing from five to nine carbon atoms, plus various additives to improve its performance as a fuel. Kerosene, diesel oil, and fuel oil are primarily mixtures of alkane derivatives with higher molecular masses. Examples of the various combustion reactions of hydrocarbons are the combustion of cyclohexane, cyclohexene, and toluene:

cyclohexane

$$C_6H_{12} + 9\,O_2 \longrightarrow 6\,CO_2 + 6\,H_2O + \text{energy}$$

cyclohexene

$$C_6H_{10} + 8.5\,O_2 \longrightarrow 6\,CO_2 + 5\,H_2O + \text{energy}$$

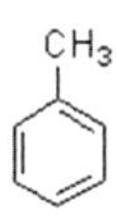

toluene

$$C_7H_8 + 9\,O_2 \longrightarrow 7\,CO_2 + 4\,H_2O + \text{energy}$$

The combustion of carbon compounds, especially hydrocarbons, has been the most important source of heat energy for human civilizations throughout recorded history. The practical importance of this reaction cannot be denied, but the massive and uncontrolled chemical changes that take place in combustion make it difficult to deduce mechanistic paths.

In terms of the oxidation using chemical agents, potassium permanganate ($KMnO_4$) does not react with alkane derivatives but does react with (and oxidize) unsaturated hydrocarbon derivatives such as alkene derivatives to form vicinal diols—osmium tetroxide, OsO_4, exhibits a similar reactivity with unsaturated hydrocarbon derivatives (Lee and Chen, 1989). Potassium permanganate will also react with alkyl benzene derivatives (such as toluene) to form benzoic acid derivatives.

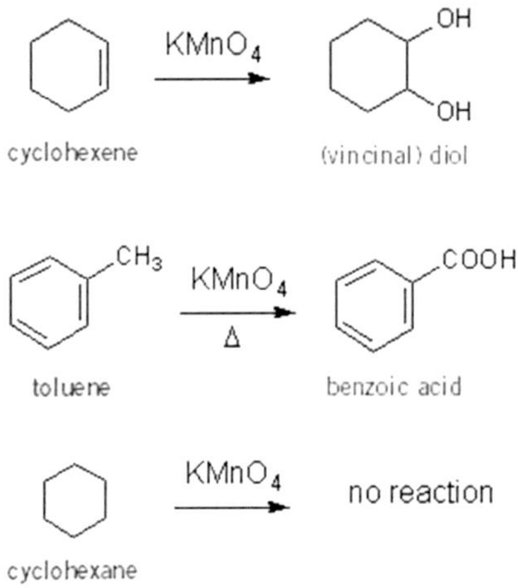

Alkene derivatives and aromatic derivatives also react with sulfuric acid in different ways. Sulfuric acid adds to cyclohexene via an acid catalyzed addition but adds to toluene via electrophilic aromatic substitution:

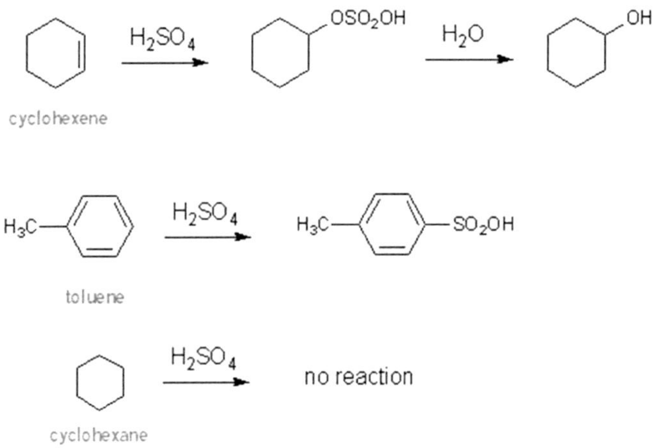

Oxidation of alkyne derivatives by strong oxidizing agents such as potassium permanganate ($KMnO_4$) or ozone (O_3) will yield a pair of carboxylic acids.

7.5 Polymerization

The polymerization of hydrocarbon derivatives results in the formation of a polymer which is a large molecule (macromolecule) that is composed of many repeated subunits. In some cases, different hydrocarbon derivatives may be used as the feedstocks for the formation of the polymer. The hydrocarbon subunits may be organized along the backbone in a variety of ways (Fig. 11.2). In fact, polymerization reactions are considered to be the most important commercial reactions of alkenes. In these reactions, monomer molecules are assembled into a high molecular weight product (the polymer). For example, ethylene (a gas) is used to manufacture plastics (solids):

$$nCH_2 = CH_2\ H\text{-}(CH_2CH_2)_n\text{-}H$$

Polyethylene (often referred to as polythene) is the most common polymer. Ethylene is a stable molecule that polymerizes only upon contact with catalysts. The conversion is highly exothermic. The most common catalysts consist of titanium(III) chloride ($TiCl_3$) while the catalyst prepared by depositing chromium (VI) oxide (CrO_3) on silica is another common catalyst.

The transition metal catalyzed polymerization of olefin derivatives is, essentially, a reaction in which the carbon-carbon π-bond in α-olefins are reformed as a carbon-carbon σ-bonds, linking the olefins together in long hydrocarbon chains. This reaction is exothermic and for typical catalysts, the activation energy is low, making this a very easy-to-accomplish reaction:

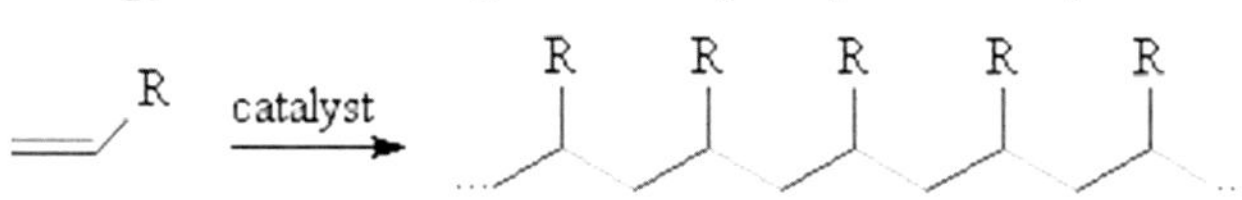

FIGURE 11.2 Variation in polymer structure.

In the process, R is typically H in which case the polymer is polyethylene. If R is methyl, the polymer is polypropylene.

7.6 Rearrangement

A rearrangement reaction is a specific organic reaction that causes the alteration of the structure to form an isomer. With alkene structures, rearrangement reactions often result in the conversion of a *cis*-isomer into the *trans*-conformation. For example, in the *cis*-conformation the methyl groups are on the same side of the molecule while in the *trans*-conformation the methyl groups are on the opposite sides of the molecule.

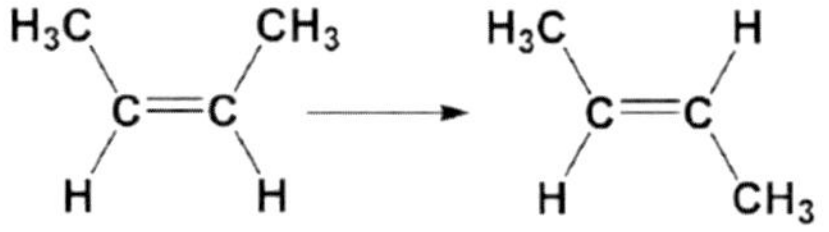

7.7 Substitution

Alkene derivatives are highly reactive and typically undergo addition reactions rather than substitutions reactions. The exception is the benzene ring in which the double-bonded structure of the benzene ring gives this molecule a resonance structure such that all of the carbon atoms in the ring share a continually rotating partial bond structure. Thus, the overall structure is very stable compared to other alkene derivatives and benzene rings do not readily undergo addition reactions. They behave more similarly to alkane structure and lack chemical reactivity.

However, one of the few types of reactions that a benzene ring will undergo is a substitution reaction in which an atom or group of atoms is replaced by another atom or group of atoms. Halogenation is a common substitution reaction that occurs with benzene ring structures. In the diagram below, notice that the hydrogen atom is substituted by one of the bromine atoms.

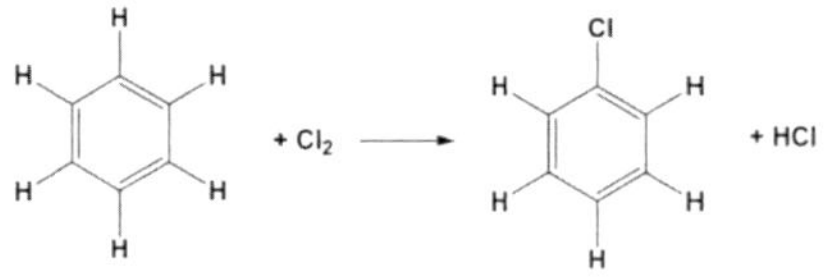

References

Ancheyta, J., Speight, J.G., 2007. Hydroprocessing of Heavy Oils and Residua. CRC Press, Taylor & Francis Group, Boca Raton, Florida.

Boufaden, N., Akkari, R., Pawelec, B., Fierro, J.L.G., Said Zina, M., Ghorbel, A., 2016. Dehydrogenation of methylcyclohexane to toluene over partially reduced silica-supported Pt-Mo catalysts. Journal of Molecular Catalysis 420, 96–106.

Brückner, R., 2002. Advanced Organic Chemistry: Reaction Mechanisms. Academic Press Inc., New York.

Carey, F.A., 2006. Organic Chemistry, sixth ed. McGraw-Hill, New York.

El-Gendy, N.S., Speight, J.G., 2016. Handbook of Refinery Desulfurization. CRC Press, Taylor & Francis Group, Boca Raton, Florida.

Eliel, E., Wilen, S., 1994. Stereochemistry of Organic Compounds. John Wiley & Sons Inc., New York.

Gary, J.G., Handwerk, G.E., Kaiser, M.J., 2007. Petroleum Refining: Technology and Economics, fifth ed. CRC Press, Taylor & Francis Group, Boca Raton, Florida.

Hsu, C.S., Robinson, P.R. (Eds.), 2017. Handbook of Petroleum Technology. Springer International Publishing AG, Cham, Switzerland.

Kasanski, B.A., Liberman, A.L., 1959. Catalytic dehydrocyclization of paraffinic hydrocarbons. In: Proceedings. 5th World Petroleum Congress, New York. May 30−June 5.

Klein, D.R., 2013. Organic Chemistry, second ed. John Wiley & Sons Inc., Hoboken, New Jersey.

Lee, D.G., Chen, T., 1989. Oxidation of hydrocarbons. 18. Mechanism of the reaction between permanganate and carbon-carbon double bonds. Journal of the American Chemical Society 111 (19), 7534−7538.

Masel, R.I., 1995. Principles of Adsorption and Reaction on Solid Surfaces. John Wiley & Sons Inc., New York.

Mériaudeau, P., Naccache, C., 1997. Dehydrocyclization of alkanes over zeolite-supported metal catalysts: monofunctional or bifunctional route. Catalysis Reviews 39 (1&2), 5−48.

Parkash, S., 2003. Refining Processes Handbook. Gulf Professional Publishing, Elsevier, Amsterdam, Netherlands.

Sadeghbeigi, R., 2000. Fluid Catalytic Cracking Handbook, second ed. Gulf Publishing, Houston, Texas.

Samorjai, G.A., 1994. Introduction to Surface Chemistry and Catalysis. John Wiley & Sons Inc., New York.

Speight, J.G., 2014. The Chemistry and Technology of Petroleum, fifth ed. CRC Press, Taylor and Francis Group, Boca Raton, Florida.

Speight, J.G., 2017. Handbook of Petroleum Refining. CRC Press, Taylor and Francis Group, Boca Raton, Florida.

Vora, B.V., Pujadó, P.R., 2005. In: Lee, S. (Ed.), Encylopedia of Chemisl Processing. CRC Press, Taylor & Francis Group, Boca Raton, Florida.

Chapter 12

Petrochemicals

1. Introduction

Petrochemicals are chemical products derived from crude oil, although many of the same chemical compounds are also obtained from other fossil fuels such as coal and natural gas or from renewable sources such as corn, sugar cane, and other types of biomass (Matar and Hatch, 2001; Meyers, 2005; Speight, 2011). In the current context of *petrochemicals*, this chapter focuses on organic hydrocarbon compounds that are not burned as a fuel and are generally known as crude oil products (also known as *petroleum products*) (Chapter 3) (Parkash, 2003; Gary et al., 2007; Speight, 2014, 2017, 2019a; Hsu and Robinson, 2017).

Petrochemical production relies on multiphase processing of crude oil and the associated gas. Key raw materials in the petrochemical industry include products of crude oil refining (primarily gases and naphtha). Petrochemical goods include ethylene, propylene, and benzene; source monomers for synthetic rubbers, and inputs for technical carbon. Petrochemicals are used for production of several feedstocks and monomers and monomer precursors. The monomers after polymerization process create several polymers which are ultimately used to produce gels, lubricants, elastomers, plastics, and fibers.

Petrochemical and petroleum products are the second level products being derived from crude oil after several refining processes. Crude oil is the basic component to produce all petrochemical and crude oil components after a long process of refinement in oil refineries. The major hydrocarbon products initially produced from crude oil by refining are (i) liquefied petroleum gas, (ii) naphtha, (iii) kerosene (iv) fuel oil, (v) lubricating oil, and (vi) paraffin wax. The lower boiling products (Table 12.1) are (after purification) ready for use as petrochemical feedstocks (Parkash, 2003; Gary et al., 2007; Speight, 2014, 2017, 2019a; Hsu and Robinson, 2017).

On the basis of chemical structure, petrochemicals are categorized into three categories of petrochemical products (i) olefin derivatives, (ii) aromatic derivatives, and (iii) synthesis gas (Table 12.2). Ethylene and propylene, the major part of olefin derivatives, are the basic source in preparation of several industrial chemicals and plastic products whereas butadiene is used to prepare synthetic rubber. Benzene, toluene, and the xylene isomers are major components of aromatic chemicals. These aromatic petrochemicals are used in

Handbook of Industrial Hydrocarbon Processes. https://doi.org/10.1016/B978-0-12-809923-0.00012-6

TABLE 12.1 Petrochemical feedstocks from natural gas and crude oil.

Starting material	Process	Product
Natural gas	Refining	Methane
		Ethane
		Propane
		Butane
Crude oil	Distillation	Light ends
		Methane
		Ethane
		Propane
		Butane
	Catalytic cracking	Ethylene
		Propylene
		Butylene derivatives
		Higher boiling olefins
	Coking	Ethylene
		Propylene
		Butylene derivatives
		Higher boiling olefins

manufacturing of secondary products like synthetic detergents, polyurethanes, plastic, and synthetic fibers. Synthesis gas comprises of carbon monoxide and hydrogen which are basically used to produce ammonia and methanol which are further used to produce other chemical and synthetic substances.

Petrochemical intermediates are generally produced by chemical conversion of primary petrochemicals to form more complicated derivative products (Parkash, 2003; Gary et al., 2007; Speight, 2014, 2017, 2019a; Hsu and Robinson, 2017). Petrochemical derivative products can be made in a variety of ways: directly from primary petrochemicals; through intermediate products which still contain only carbon and hydrogen; and, through intermediates which incorporate chlorine, nitrogen, or oxygen in the finished derivative. In some cases, they are finished products; in others, more steps are needed to arrive at the desired composition.

Some typical petrochemical intermediates are: (1) vinyl acetate for paint, paper, and textile coatings, (2) vinyl chloride for polyvinyl chloride (PVC),

TABLE 12.2 The three major categories of petrochemical products.

Category	Comments
Olefins	Example are ethylene ($CH_2{=}CH_2$) and propylene ($CH_3CH{=}CH_2$). Important sources of industrial chemicals and plastics; butadiene ($CH_2{=}CHCH{=}CH_2$) is used in making synthetic rubber.
Aromatics	Examples are benzene, toluene, and the xylene isomers. Benzene is a raw material for dyes and synthetic detergents. Benzene and toluene are raw materials for isocyanates. Xylenes are used in the manufacture of plastics and synthetic fibers.
Synthesis gas	A mixture of carbon monoxide and hydrogen. Used in the Fischer-Tropsch process to produce gasoline-range and diesel-range hydrocarbons. Also used in the manufacture of methanol and dimethyl ether.

(3) ethylene glycol for polyester textile fibers, and (4) styrene which is important in rubber and plastic manufacturing. Of all the processes used, one of the most important is polymerization. It is used in the production of plastics, fibers, and synthetic rubber, the main finished petrochemical derivatives.

The production of petrochemicals commences with refining of crude oil and natural gas. Crude oil refining begins with the distillation, or fractionation, of crude oils into separate fractions of hydrocarbon groups (Parkash, 2003; Gary et al., 2007; Speight, 2014, 2017, 2019a; Hsu and Robinson, 2017). The resultant products (26) are directly related to the characteristics of the crude oil being processed. Most of these products of distillation are further converted into more usable products by changing their physical and molecular structures through cracking, reforming, and other conversion processes. These products are subsequently subjected to various treatment and separation processes, such as extraction, hydrotreating, and sweetening, in order to produce finished products. Whereas the simplest refineries are usually limited to atmospheric and vacuum distillation, integrated refineries incorporate fractionation, conversion, treatment, and blending with lubricant, high-boiling fuel oils, and asphalt manufacturing; they may also include petrochemical processing.

It is during the refining process that other products are also produced. These products include the gases dissolved in the crude oil that are released during distillation as well as the gases produced during the various refining processes that provide fodder for the petrochemical industry. The gas (often referred to as *refinery gas* or *process gas*) varies in composition and volume, depending on the origin of the crude oil and on any additions (i.e., other crude oils blended into the refinery feedstock) to the crude oil made at the loading point. It is not uncommon to reinject low-boiling hydrocarbon derivatives such as propane and butane into the crude oil before dispatch by tanker or pipeline. This results in a higher vapor pressure of the crude, but it allows one to increase the quantity of low-boiling products obtained at the refinery. Since low-boiling gases in most crude oil markets command a premium, while in the oil field itself propane and butane may have to be reinjected or flared, the practice of *spiking* crude oil with liquefied petroleum gas is becoming fairly common. These gases are recovered by distillation (Fig. 12.1). In addition to distillation, gases produced in the various thermal processes, such as *catalytic cracking* (Fig. 12.2), are also available (Parkash, 2003; Gary et al., 2007; Speight, 2014, 2017, 2019a; Hsu and Robinson, 2017).

In processes such as coking or visbreaking, a variety of gases are produced. Another group of refining operations that contributes to gas production is that of the *catalytic cracking processes*. Both catalytic and thermal cracking processes result in the formation of unsaturated hydrocarbon derivatives, particularly

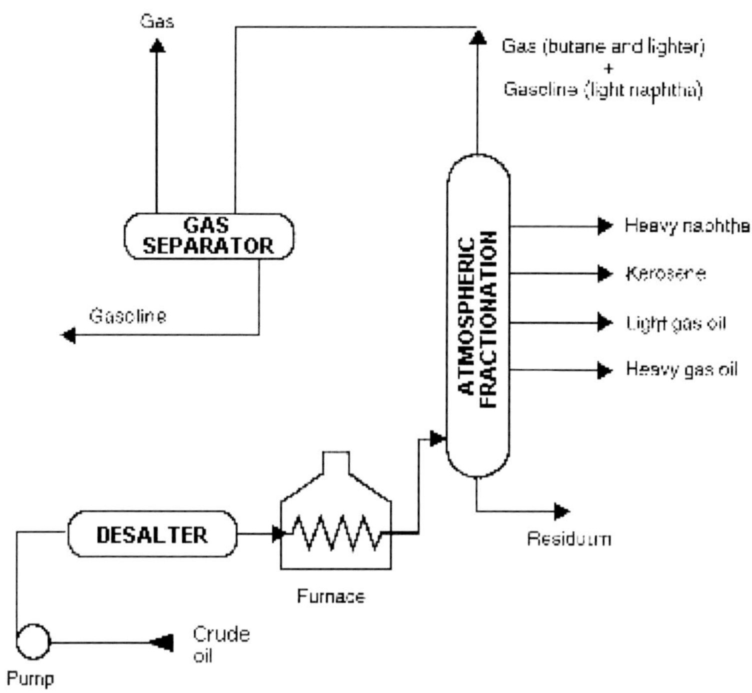

FIGURE 12.1 Gas recovery by distillation (OSHA Technical Manual. Section IV, Chapter 2. petroleum refining processes).

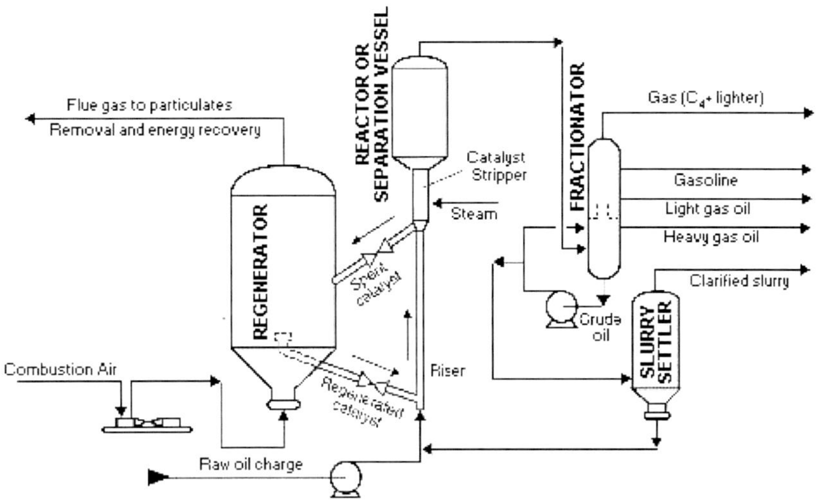

FIGURE 12.2 Gas production during catalytic cracking (OSHA Technical Manual. Section IV, Chapter 2. petroleum refining processes).

ethylene ($CH_2{=}CH_2$), but also propylene (propene, $CH_3CH{=}CH_2$), *iso*-butylene [*iso*-butene, $(CH_3)_2C{=}CH_2$], and the *n*-butenes ($CH_3CH_2CH{=}CH_2$ and $CH_3CH{=}CHCH_3$) in addition to hydrogen (H_2), methane (CH_4), and smaller quantities of ethane (CH_3CH_3), propane ($CH_3CH_2CH_3$), and butanes [$CH_3CH_2CH_2CH_3$, $(CH_3)_3CH$]. Diolefin derivatives such as butadiene ($CH_2{=}CHCH{=}CH_2$) are also present. A further source of refinery gas is *hydrocracking*, a catalytic high-pressure pyrolysis process in the presence of fresh and recycled hydrogen. The feedstock is again high-boiling gas oil or residual fuel oil, and the process is mainly directed at the production of additional middle distillates and gasoline. Since hydrogen is to be recycled, the gases produced in this process again have to be separated into lower-boiling streams and higher-boiling streams, any surplus recycle gas and the liquefied petroleum gas from the hydrocracking process are both saturated.

In a series of *reforming processes* (Parkash, 2003; Gary et al., 2007; Speight, 2014, 2017, 2019a; Hsu and Robinson, 2017), commercialized under names such as *Platforming*, paraffin and naphthene (cyclic, nonaromatic) hydrocarbon derivatives are converted in the presence of hydrogen and a catalyst into aromatics, or isomerized to more highly branched hydrocarbon derivatives. Catalytic reforming processes thus not only result in the formation of a liquid product of higher octane number, but also produce substantial quantities of gases. The latter are rich in hydrogen, but also contain hydrocarbon derivatives from methane to butanes, with a preponderance of propane ($CH_3CH_2CH_3$), *n*-butane ($CH_3CH_2CH_2CH_3$), and *iso*-butane [$(CH_3)_3CH$].

The composition of the process gas varies in accordance with reforming severity and reformer feedstock. All catalytic reforming processes require substantial recycling of a hydrogen stream. Therefore, it is normal to separate reformer gas into a propane ($CH_3CH_2CH_3$) and/or a butane stream [$CH_3CH_2CH_2CH_3$ plus $(CH_3)_3CH$], which becomes part of the refinery liquefied petroleum gas production, and a lower-boiling gas fraction, part of which is recycled. In view of the excess of hydrogen in the gas, all products of catalytic reforming are saturated, and there are usually no olefin gases present in either gas stream.

In many refineries, naphtha in addition to other finery gases is also used as the source of petrochemical feedstocks. In the process, naphtha crackers convert naphtha feedstock (produced by various process) (Table 12.3) into ethylene, propylene, benzene, toluene, and the xylene isomers as well as other byproducts in a two-step process of cracking and separating. In some cases, a combination of naphtha, gas oil, and liquefied petroleum gas may be used. The feedstock, typically naphtha, is introduced into the pyrolysis section of the naphtha where it is cracked in the presence of steam. The naphtha is converted into lower boiling fractions, primarily ethylene and propylene. The hot gas effluent from the furnace is then quenched to inhibit further cracking and to condense higher molecular weight products. The higher molecular weight products are subsequently processed into fuel oil, low boiling cycle oil, and pyrolysis gas byproducts. The pyrolysis gas stream can then be fed to the aromatics plants for benzene and toluene production.

The cooled gases are then compressed, treated to remove acid gases, dried over a desiccant, and fractionated into separate components at low temperature through a series of refrigeration processes (Mokhatab et al., 2006; Speight, 2014, 2019b). Hydrogen and methane are removed by way of a compression/expansion process after which the methane is distributed to other process as deemed

TABLE 12.3 Naphtha production.

Process	Primary product	Secondary process	Secondary product
Atmospheric distillation	Naphtha		High boiling naphtha
			High boiling naphtha
	Gas oil	Catalytic cracking	Naphtha
	Gas oil	Hydrocracking	Naphtha
Vacuum distillation	Gas oil	Catalytic cracking	Naphtha
		Hydrocracking	Naphtha
	Residuum	Coking	Naphtha
		Hydrocracking	Naphtha

appropriate or fuel gas. Hydrogen is collected and further purified in a pressure swing unit for use in the hydrogenation process. Polymer grade ethylene and propylene are separated in the cold section after which the ethane and propane streams are recycled back to the furnace for further cracking while the mixed butane (C_4) stream is hydrogenated prior to recycling back to the furnace for further cracking. The refinery gas (or the process gas) stream and the products of naphtha cracking are the source of a variety of *petrochemicals*. Thus, petrochemicals are chemicals derived from crude oil and natural gas and, for convenience of identification can be divided into two groups: (1) primary petrochemicals (Fig. 12.3) (2) intermediates and derivatives.

Primary petrochemicals include olefin derivatives (ethylene, propylene, and butadiene); aromatics (benzene, toluene, and xylenes); and methanol. *Petrochemical intermediates* are generally produced by chemical conversion of primary petrochemicals to form more complicated derivative products. Petrochemical derivative products can be made in a variety of ways: directly from primary petrochemicals; through intermediate products which still contain only carbon and hydrogen; and, through intermediates which incorporate chlorine, nitrogen, or oxygen in the finished derivative. In some cases, they are finished products; in others, more steps are needed to arrive at the desired composition.

In the strictest sense petrochemicals is any of a large group of chemicals manufactured from crude oil and natural gas as distinct from fuels and other products (Speight, 2014, 2019b), derived from crude oil and natural gas and used

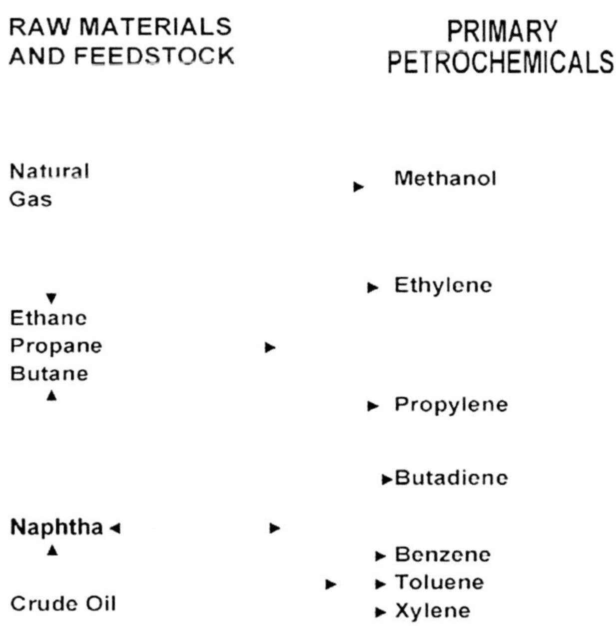

FIGURE 12.3 Raw materials and primary petrochemicals.

for a variety of commercial purposes. The definition has been broadened to include the whole range of organic chemicals. In many instances, a specific chemical included among the petrochemicals may also be obtained from other sources, such as coal, coke, or vegetable products. For example, materials such as benzene and naphthalene can be made from either crude oil or coal, while ethyl alcohol may be of petrochemical or vegetable origin. This makes it difficult to categorize a specific substance as, strictly speaking, petrochemical or nonpetrochemical.

The *chemical industry* is the *chemical process industry* by which a variety of chemicals are manufactured. The chemical process industry is, in fact, subdivided into other categories: (i) chemicals and allied products in which chemicals are manufactured from a variety of feedstocks and may then be put to further use, (ii) rubber and miscellaneous products which focus on the manufacture of rubber and plastic materials, and (iii) crude oil refining and related industries. Thus, the petrochemical industry falls under the subcategory of *petroleum and related industries*.

The crude oil era was ushered in by the discovery of crude oil at Titusville, Pennsylvania, in 1859. But the production of chemicals from natural gas and crude oil has been a recognized industry only since the early twentieth century. Nevertheless, the petrochemical industry has made quantum leaps in the production of a wide variety of chemicals (Chenier, 1992), with those being based on starting feedstocks from crude oil being termed *petrochemicals*.

Thermal cracking processes (Parkash, 2003; Gary et al., 2007; Speight, 2014, 2017, 2019a; Hsu and Robinson, 2017) developed for crude oil refining, starting in 1913 and continuing the next two decades, were focused primarily on increasing the quantity and quality of gasoline components. As a byproduct of this process, gases were produced that included a significant proportion of lower molecular weight olefin derivatives, particularly ethylene ($CH_2{=}CH_2$), propylene ($CH_3CH{=}CH_2$), and butylenes (butenes, $CH_3CH{=}CHCH_3$ and $CH_3CH_2CH{=}CH_2$). Catalytic cracking (Parkash, 2003; Gary et al., 2007; Speight, 2014, 2017, 2019a; Hsu and Robinson, 2017), introduced in 1937, is also a valuable source of propylene and butylene, but it does not account for a very significant yield of ethylene, the most important of the petrochemical building blocks. Ethylene is polymerized to produce polyethylene or, in combination with propylene, to produce copolymers that are used extensively in food-packaging wraps, plastic household goods, or building materials. Prior to the use of crude oil and natural gas as sources of chemicals, coal was the main source of chemicals (Speight, 2013). Petrochemical products include such items as plastics, soaps and detergents, solvents, drugs, fertilizers, pesticides, explosives, synthetic fibers and rubbers, paints, epoxy resins, and flooring and insulating materials. Petrochemicals are found in products as diverse as aspirin, luggage, boats, automobiles, aircraft, polyester clothes, and recording discs and tapes.

The petrochemical industry has grown with the crude oil industry (Goldstein, 1949; Steiner, 1961; Hahn, 1970) and is considered by some to be a mature industry. However, as is the case with the latest trends in changing crude oil types, it must also evolve to meet changing technological needs. The manufacture of chemicals or chemical intermediates from a variety of raw materials is well established (Wittcoff and Reuben, 1996). And the use of crude oil and natural gas is an excellent example of the conversion of such raw materials to more valuable products. The individual chemicals made from crude oil and natural gas are numerous and include industrial chemicals, household chemicals, fertilizers, and paints, as well as intermediates for the manufacture of products, such as synthetic rubber and plastics.

Petrochemicals are generally considered chemical compounds derived from crude oil either by direct manufacture or indirect manufacture as byproducts from the variety of processes that are used during the refining of crude oil. Gasoline, kerosene, fuel oil, lubricating oil, wax, asphalt, and the like are excluded from the definition of petrochemicals, since they are not, in the true sense, chemical compounds but are in fact intimate mixtures of hydrocarbon derivatives.

The classification of materials as petrochemicals is used to indicate the source of the chemical compounds, but it should be remembered that many common petrochemicals can be made from other sources, and the terminology is therefore a matter of source identification.

The starting materials for the petrochemical industry are obtained from crude oil in one of two general ways. They may be present in the raw crude oil and, as such, are isolated by physical methods, such as distillation or solvent extraction. On the other hand, they may be present, if at all, in trace amounts and are synthesized during the refining operations. In fact, unsaturated (olefin) hydrocarbon derivatives, which are not usually present in crude oil, are nearly always manufactured as intermediates during the various refining sequences.

The manufacture of chemicals from crude oil is based on the ready response of the various compound types to basic chemical reactions, such as oxidation, halogenation, nitration, dehydrogenation addition, polymerization, and alkylation. The low molecular weight paraffin derivatives and olefin derivatives, as found in natural gas and refinery gases, and the simple aromatic hydrocarbon derivatives have so far been of the most interest because they can be readily isolated and dealt with. A wide range of compounds is possible, many are being manufactured and we are now progressing to the stage in which a sizable group of products is being prepared from the higher molecular weight fractions of crude oil.

The various reactions of asphaltene constituents indicate that these materials may be regarded as containing chemical functions and are therefore different and are able to participate in numerous chemical or physical conversions to, perhaps, more useful materials. The overall effect of these modifications is the production of materials that either affords good-grade aromatic cokes comparatively easily or the formation of products bearing functional groups that may be employed as a nonfuel material.

For example, the sulfonated and sulfomethylated materials and their derivatives have satisfactorily undergone tests as drilling mud thinners, and the results are comparable to those obtained with commercial mud thinners. In addition, these compounds may also find use as emulsifiers for the in situ recovery of heavy oils. These are also indications that these materials and other similar derivatives of the asphaltene constituents, especially those containing such functions as carboxylic or hydroxyl, readily exchange cations and could well compete with synthetic zeolites. Other uses of the hydroxyl derivatives and/or the chloroasphaltenes include high-temperature packing or heat transfer media.

Reactions incorporating nitrogen and phosphorus into the asphaltene constituents are particularly significant at a time when the effects on the environment of many materials containing these elements are receiving considerable attention. Here we have potential slow-release soil conditioners that only release the nitrogen or phosphorus after considerable weathering or bacteriological action. One may proceed a step further and suggest that the carbonaceous residue remaining after release of the heteroelements may be a benefit to humus-depleted soils, such as the gray-wooded and solonetzic soils. It is also feasible that coating a conventional quick-release inorganic fertilizer with a water-soluble or water-dispersible derivative will provide a slower release fertilizer and an organic humuslike residue. In fact, variations on this theme are multiple.

Nevertheless, the main objective in producing chemicals from crude oil is the formation of a variety of well-defined chemical compounds that are the basis of the petrochemical industry. It must be remembered, however, that ease of separation of a particular compound from crude oil does not guarantee its use as a petrochemical building block. Other parameters, particularly the economics of the reaction sequences, including the costs of the reactant equipment, must be taken into consideration.

For the purposes of this text, there are four general types of petrochemicals: (1) aliphatic compounds, (2) aromatic compounds, (3) inorganic compounds, and (4) synthesis gas (carbon monoxide and hydrogen). Synthesis gas is used to make ammonia (NH_3) and methanol (methyl alcohol, CH_3OH). Ammonia is used primarily to form ammonium nitrate (NH_4NO_3), a source of fertilizer. Much of the methanol produced is used in making formaldehyde ($HCH{=}O$). The rest is used to make polyester fibers, plastics, and silicone rubber.

An aliphatic petrochemical compound is an organic compound that has an open chain of carbon atoms, be it normal (straight), e.g., *n*-pentane ($CH_3CH_2CH_2CH_2CH_3$) or branched, e.g., *iso*-pentane [2-methylbutane, $CH_3CH_2CH(CH_3)CH_3$], or unsaturated. The unsaturated compounds, olefin derivatives, include important starting materials such as ethylene ($CH_2{=}CH_2$), propylene ($CH_3CH{=}CH_2$), butene-1 ($CH_3CH_2CH_2{=}CH_2$), *iso*-butene (2-methylpropene [$CH_3(CH_3)C{=}CH_2$], and butadiene ($CH_2{=}CHCH{=}CH_2$).

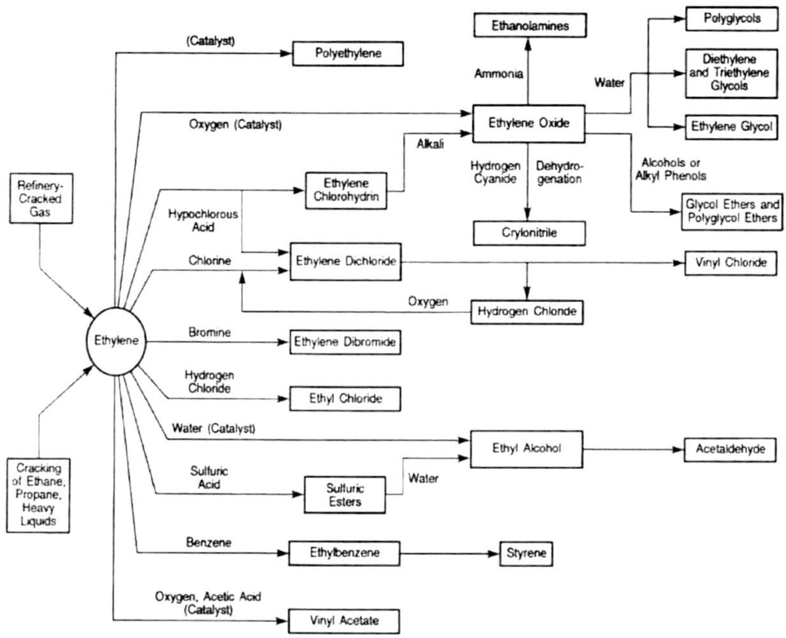

FIGURE 12.4 Chemicals from ethylene.

Ethylene is the hydrocarbon feedstock used in greatest volume in the petrochemical industry (Fig. 12.4). From ethylene, for example, are manufactured ethylene glycol, used in polyester fibers and resins and in antifreezes; ethyl alcohol, a solvent and chemical reagent; polyethylene, used in film and plastics; styrene, used in resins, synthetic rubber, plastics, and polyesters; and ethylene dichloride, for vinyl chloride, used in plastics and fibers. Propylene is also an important source of petrochemicals (Fig. 12.5) and is used in making such products as acrylics, rubbing alcohol, epoxy glue, and carpets. Butadiene is used in making synthetic rubber, carpet fibers, paper coatings, and plastic pipes. An aromatic petrochemical is also an organic chemical compound but one that contains, or is derived from, the basic benzene ring system.

Petrochemicals are made, or recovered from, the entire range of crude oil fractions, but the bulk of petrochemical products are formed from the lower boiling (C_1 to C_4) hydrocarbon gases as raw materials. These materials generally occur in natural gas, but they are also recovered from the gas streams produced during refinery, especially cracking, operations. Refinery gases are also particularly valuable because they contain substantial amounts of olefin derivatives that, because of the double bonds, are much more reactive then the saturated (paraffin) hydrocarbon derivatives. Also important as raw materials are the aromatic hydrocarbon derivatives (benzene, toluene, and xylene), that are obtained in rare cases from crude oil and, more likely, from the various product

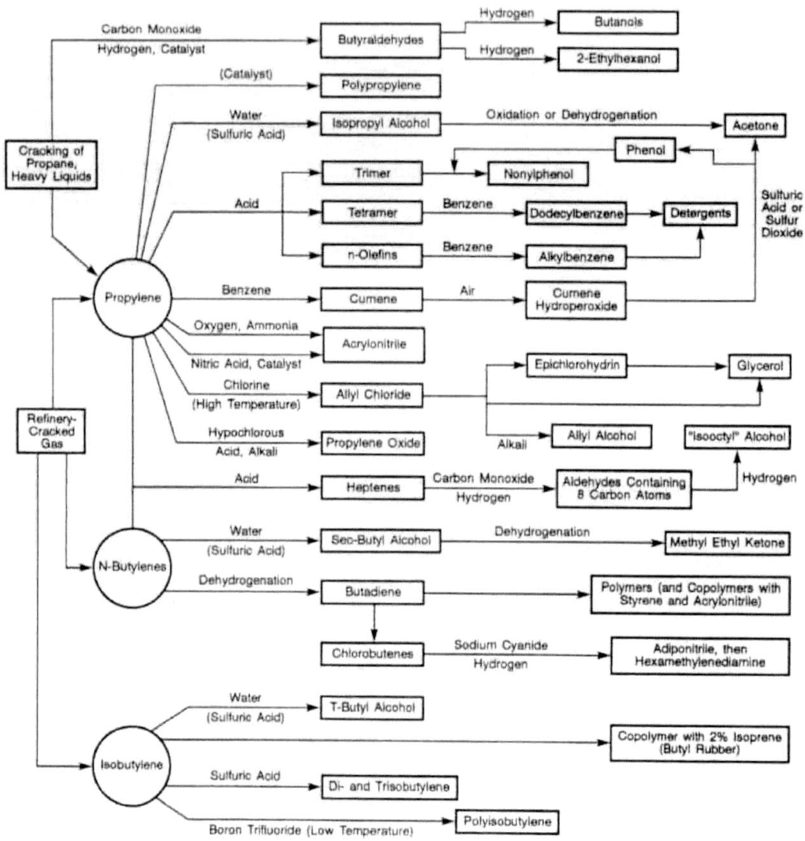

FIGURE 12.5 Chemicals from propylene.

streams. By means of the catalytic reforming process, nonaromatic hydrocarbon derivatives can be converted to aromatics by dehydrogenation and cyclization.

A highly significant proportion of these basic petrochemicals are converted into plastics, synthetic rubbers, and synthetic fibers. Together these materials are known as polymers, because their molecules are high molecular weight compounds made up of repeated structural units that have combined chemically. The major products are polyethylene, polyvinyl chloride, and polystyrene, all derived from ethylene, and polypropylene, derived from monomer propylene. Major raw materials for synthetic rubbers include butadiene, ethylene, benzene, and propylene. Among synthetic fibers the polyesters, which are a combination of ethylene glycol and terephthalic acid (made from xylene), are the most widely used. They account for approximately one-half of all synthetic fibers. The second major synthetic fiber is nylon, its most important raw material being benzene. Acrylic fibers, in which the major raw material is the propylene derivative acrylonitrile, make up most of the remainder of the synthetic fibers.

2. Chemicals from paraffin hydrocarbons

By way of recall, paraffin derivatives are straight-chain or branched-chain hydrocarbon derivatives having the chemical formula C_nH_{2n+2}.

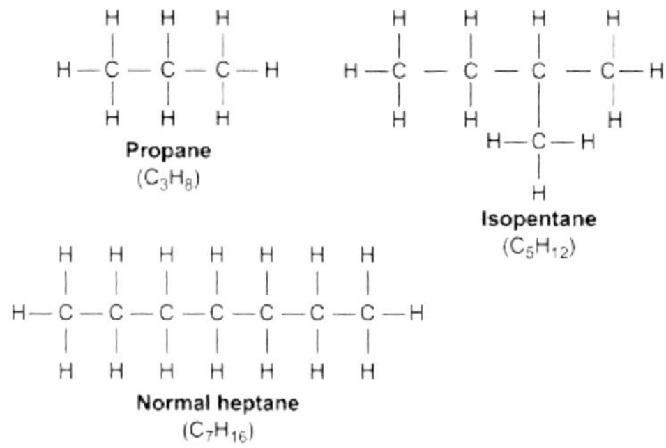

For values of $n < 5$, the paraffin derivatives (methane, ethane, propane, and the butane isomers) are gaseous at normal temperatures and pressures. For values of $n = 5$ (pentane, C_5H_{12}) to $n = 15$ (pentadecane, $C_{15}H_{32}$) the paraffin derivatives are liquid at normal temperatures and pressures (hexadecane is borderline with a melting point of 18°C (64°F) and for values of $n > 16$ they grade from low melting solid to high melting waxes (Table 12.4). Two types of paraffin isomers increase in molecular weight along the series by the addition of methylene (CH_2) groups. One series consists of straight-chain molecules (n-paraffin derivatives or normal paraffin derivatives), the other of branched-chain molecules. Using heptane (C_7H_{16}) as the example:

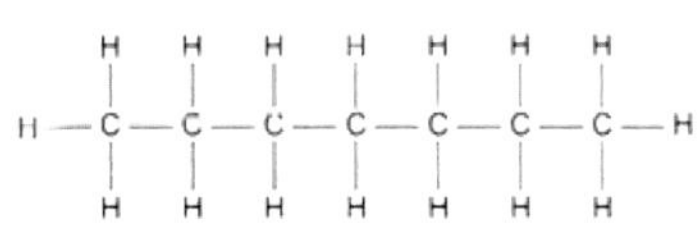

n-Heptane

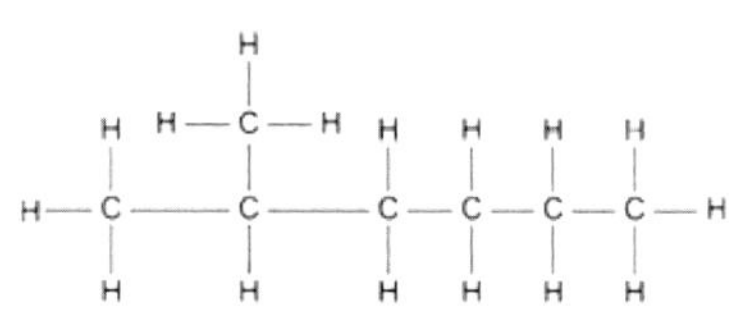

iso-heptane (2-methyl hexane)

Individual members of the paraffin series (C_nH_{2n+2}) have been recorded up to $C_{78}H_{158}$. For a given molecular weight, the normal paraffin derivatives have higher boiling points than do equivalent weight *iso*-paraffin derivatives. It is generally true that only paraffin hydrocarbon derivatives from methane (CH_4) through propane (C_3H_8) are used as starting materials for specific chemicals syntheses. This is because the higher members of the series are less easy to fractionate from crude oil in pure form, and also because the number of compounds formed in each particular chemical treatment makes the separation of individual products quite difficult.

TABLE 12.4 Examples of the paraffin series.

Alkane	Formula	Boiling point, °C	Melting point, °C	Density [kg/m^3] (at 20°C)
Methane	CH_4	−162	−182	0.656 (gas)
Ethane	C_2H_6	−89	−183	1.26 (gas)
Propane	C_3H_8	−42	−188	2.01 (gas)
Butane	C_4H_{10}	0	−138	2.48 (gas)
Pentane	C_5H_{12}	36	−130	626 (liquid)
Hexane	C_6H_{14}	69	−95	659 (liquid)
Heptane	C_7H_{16}	98	−91	684 (liquid)
Octane	C_8H_{18}	126	−57	703 (liquid)
Nonane	C_9H_{20}	151	−54	718 (liquid)
Decane	$C_{10}H_{22}$	174	−30	730 (liquid)
Undecane	$C_{11}H_{24}$	196	−26	740 (liquid)
Dodecane	$C_{12}H_{26}$	216	−10	749 (liquid)
Pentadecane	$C_{15}H_{32}$	270	10	769 (liquid)
Hexadecane	$C_{16}H_{34}$	287	18	773 (liquid)
Heptadecane	$C_{17}H_{36}$	303	22	777 (solid)
Icosane	$C_{20}H_{42}$	343	37	789 (solid)
Triacontane	$C_{30}H_{62}$	450	66	810 (solid)
Tetracontane	$C_{40}H_{82}$	525	82	817 (solid)
Pentacontane	$C_{50}H_{102}$	575	91	824 (solid)
Hexacontane	$C_{60}H_{122}$	625	100	829 (solid)

The chemical reactions of hydrocarbon derivatives have been presented elsewhere (Chapter 11), and it is the purpose of this chapter to show the methods by which these reactions are incorporated into processes for the production of petrochemicals.

2.1 Alkylation, transalkylation, and dealkylation

Alkylation chemistry contributes to the efficient utilization of C_4 olefin derivatives generated in the cracking operations (Parkash, 2003; Gary et al., 2007; Speight, 2014, 2017, 2019a; Hsu and Robinson, 2017). The process is used in crude oil refineries to convert *iso*-butane and low molecular weight alkene derivatives (primarily a mixture of propylene and butylene) into alkylate which is used as a high octane blend stock for gasoline manufacture. The octane number of the alkylate depends mainly upon the kind of alkenes used and upon operating conditions. For example, *iso*-octane is produced by combining butylene with *iso*-butane and (by definition) has an octane rating of 100. There are other products in the alkylate effluent, however, so the octane rating of the product will vary accordingly.

Thus, iso-butane can be added to butylene derivatives (and other low-boiling olefin derivatives) to give a mixture of highly branched octanes (e.g., heptane derivatives) by a process called alkylation. The reaction is thermodynamically favored at low temperatures (<20°C), and thus very powerful acid catalysts are employed. Typically, sulfuric acid (85%–100%), anhydrous hydrogen fluoride, or a solid sulfonic acid is employed as the catalyst in these processes. The first step in the process is the formation of a carbocation by combination of an olefin with an acid proton:

$$(CH_3)_2C{=}CH_2 + H^+ \rightarrow (CH_3)_3C^+$$

Step 2 is the addition of the carbocation to a second molecule of olefin to form a dimer carbocation. The extensive branching of the saturated hydrocarbon results in high octane. In practice, mixed butenes are employed (*iso*-butylene, 1-butene, and 2-butene), and the product is a mixture of isomeric octanes that has an octane number of 92–94.

For aromatic alkylation, an alkyl halide (such as the chloroalkane derivative, RCl) in the presence of an aluminum chloride catalyst accomplishes the reaction (March, 1985; Stefanidakis and Gwyn, 1993):

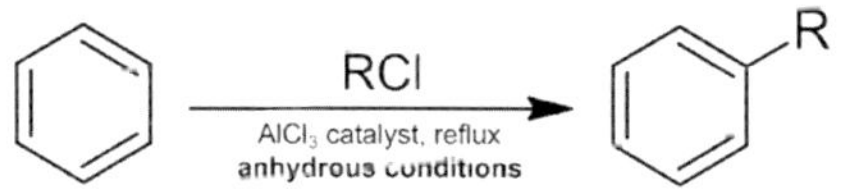

On an industrial scale, the typical route for the alkylation of benzene uses ethylene ($CH_2{=}CH_2$) with a zeolite (ZSM 5) catalyst:

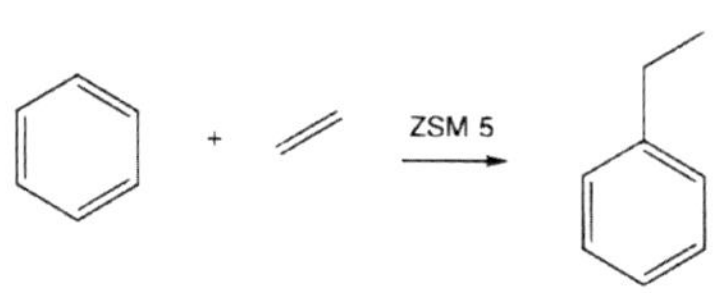

Diethylbenzene derivatives arise as sideproducts of the alkylation of benzene with ethylene. Since there is only a limited market for diethylbenzene, much of the diethylbenzene product is recycled by transalkylation to yield ethylbenzene:

$$C_6H_4(C_2H_5)_2 + C_6H_6 \rightarrow 2C_6H_5C_2H_5$$

Also, in the refining industry, the typical route to 2,4-dimethyl pentane involves the acid-catalyzed reaction of propylene with iso-butane using a Brønsted acid catalyst:

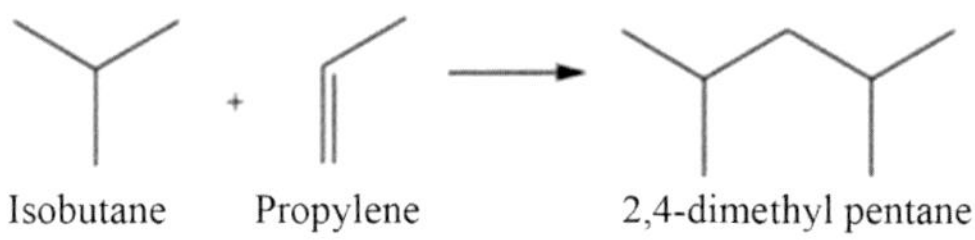

The Brønsted acid catalyst, which can include solid acid such as zeolite derivative, protonates the alkene derivative (in this case, propene) to produce a carbocation, which alkylates the isobutane. The product (typically referred to as alkylate) is composed of a mixture of high-octane, branched-chain paraffinic hydrocarbon derivatives (predominantly *iso*-heptane, in the case of *iso*-butane as the feedstock). The alkylate product is a premium blending stock in the manufacture of gasoline because it has exceptional antiknock properties and is clean burning (Parkash, 2003; Gary et al., 2007; Speight, 2014, 2017, 2019a; Hsu and Robinson, 2017).

Once of considerable importance but now of lesser importance in the modern refinery is the Friedel-Crafts alkylation of benzene (and many other aromatic rings) using an alkyl halide in the presence of a strong Lewis acid catalyst. Similarly, the Friedel-Crafts acylation reaction is a related example of electrophilic aromatic substitution. The reaction involves the acylation of benzene (or many other aromatic rings) with an acyl chloride (acid chloride) using a strong Lewis acid catalyst such as aluminum chloride ($AlCl_3$) or iron (ferric) chloride ($FeCl_3$):

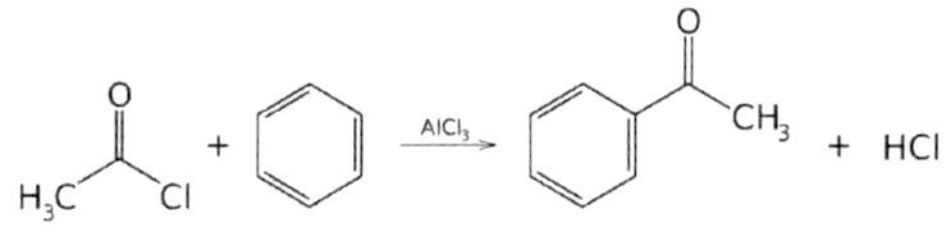

On the other hand, transalkylation is a chemical reaction which involves the transfer of an alkyl group from one organic compound to another. The reaction is used predominantly for the transfer of methyl and ethyl groups between benzene rings and the reaction is of particular use in the petrochemical industry for the manufacture of *p*-xylene, styrene, and other aromatic compounds (Speight, 2019a). Transalkylation (using a zeolite catalyst) can convert toluene into benzene and the relatively valuable xylene isomers (Table 12.5).

TABLE 12.5 Properties of the xylene isomers.

Common name	Xylene	*o*-Xylene	*m*-Xylene	*p*-Xylene
Systematic name	Dimethylbenzene	1,2-Dimethylbenzene	1,3-Dimethylbenzene	1,4-Dimethylbenzene
Other names	Xylol	*o*-Xylol; *o*-Xylene	*m*-Xylol; *m*-Xylene	*p*-Xylol; *p*-Xylene
Molecular formula	C_8H_{10}			
Density and phase	0.864 g/mL, liquid	0.88 g/mL, liquid	0.86 g/mL, liquid	0.86 g/mL, liquid
Solubility in water	Practically insoluble			
Soluble in nonpolar solvents such as aromatic hydrocarbons				
Melting point	−47.4°C (−53.3°F)	−25°C (−13°F)	−48°C (−54°F)	13°C (55°F)
Boiling point	138.5°C (281.3°F)	144°C (291°F	139°C (282°F)	138°C (280°F)
Flash point	30°C (86°F)	17°C (63°F)	25°C (77°F)	25°C (77°F)

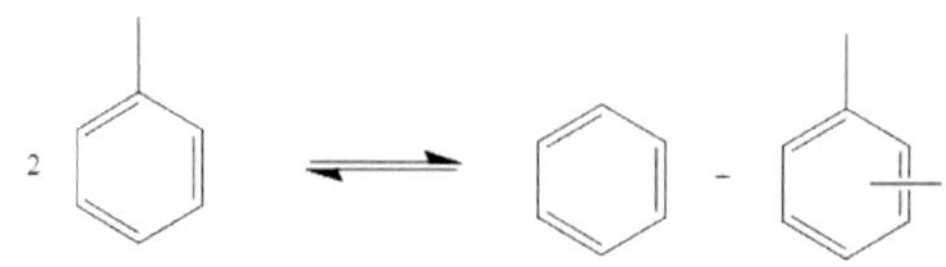

Toluene Benzene Xylene isomers

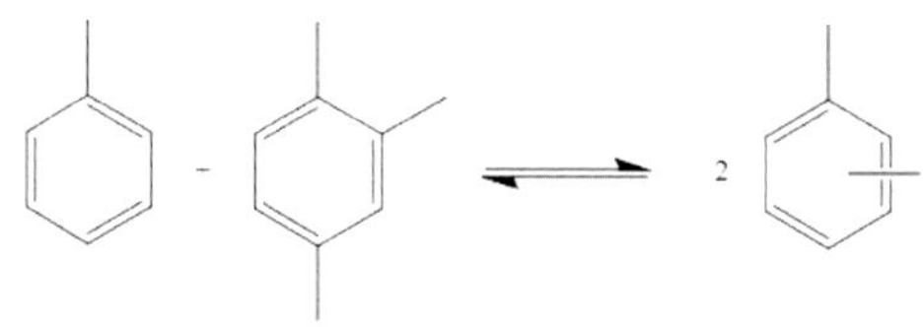

Toluene Trimethylbenzene Xylene isomers

Several other important chemicals are produced by alkylation using readily available feedstocks, such as (i) ethylbenzene, $C_6H_5CH_2CH_3$, which is a precursor to styrene, $C_6H_5CH{=}CH_2$, (ii) cumene, $C_6H_5CH(CH_3)_2$, which is a precursor to phenol, C_6H_5OH, and acetone, CH_3COCH_3, and (iii) linear alkyl benzene sulfonate derivatives such as sodium dodecylbenzene sulfonate, which is obtained by the alkylation of benzene with dodecene, is a precursor to linear alkylbenzene sulfonate detergents:

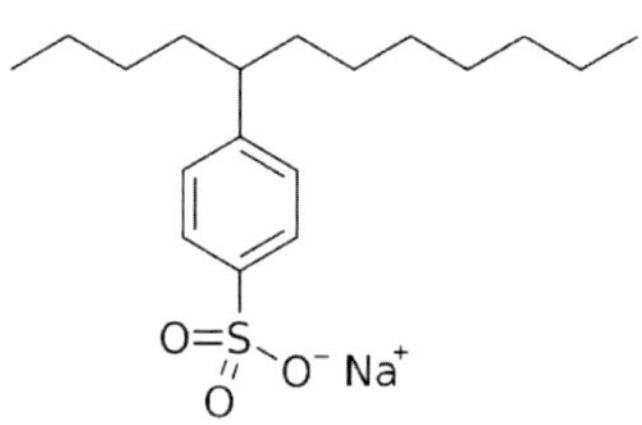

Sodium dodecylbenzene sulfonate

Hydrodealkylation—the converse reaction to alkylation—is a chemical reaction that often involves reacting an aromatic hydrocarbon derivative—such as toluene— in the presence of hydrogen to form a simpler aromatic hydrocarbon that is a lesser number of alkyl groups such as, for example, the conversion of 1,2,4-trimethylbenzene to the xylene isomers (Griesbaum et al., 2002):

Trimethylbenzene Xylene isomers

This process typically requires high temperature and high pressure, as well as the presence of a catalyst. The catalyst is predominantly composed of transition metals, such as chromium or molybdenum as used in the refining industry and in the petrochemical industry (Parkash, 2003; Gary et al., 2007; Speight, 2014, 2017, 2019a; Hsu and Robinson, 2017).

The hydrodealkylation of toluene produces benzene; methane is the other product:

In the process, toluene is mixed with hydrogen, and then passed over a chromium, molybdenum, or platinum oxide catalyst at 500°C −600°C (930°F to 1112°F) and 600 to 900 psi pressure. Sometimes, higher temperatures are used instead of a catalyst (at the similar reaction condition).

2.2 Halogenation

The ease with which chlorine can be introduced into the molecules of all the hydrocarbon types present in crude oil has resulted in the commercial production of a number of widely used compounds. With saturated hydrocarbon derivatives the reactions are predominantly substitution of hydrogen by chloride and are strongly exothermic, difficult to control, and inclined to become explosively violent:

$$RH + Cl_2 \rightarrow RCl + HCl$$

Moderately high temperatures are used, approximately 250°C−300°C (480°F−570°F) for the thermal chlorination of methane, but as the molecular weight of the paraffin increases the temperature may generally be lowered. A mixture of chlorinated derivatives is always obtained, and many variables, such as choice of catalyst, dilution of inert gases, and presence of other chlorinating agents (antimony pentachloride, sulfuryl chloride, and phosgene), have been tried in an effort to direct the path of the reaction. Methane yields four compounds upon chlorination in the presence of heat or light:

$$CH_4 + Cl_2 \rightarrow CH_3Cl, CH_2Cl_2, CHCl_3, CCl_4$$

These compounds, known as chloromethane or methyl chloride, dichloromethane or methylene chloride, trichloromethane (chloroform), and tetrachloromethane (carbon tetrachloride), are used as solvents or in the production of chlorinated materials.

Other examples of the chlorination reaction include the formation of ethyl chloride by the chlorination of ethane:

$$CH_3CH_3 + Cl_2 \rightarrow CH_3CH_2Cl + HCl$$

Ethyl chloride (CH_3CH_2Cl) is also prepared by the direct addition of hydrogen chloride (HCl) to ethylene ($CH_2{=}CH_2$) or by reacting ethyl ether ($CH_3CH_2OCH_2CH_3$) or ethyl alcohol (CH_3CH_2OH) with hydrogen chloride. The chlorination of *n*-pentane and *iso*-pentane does not take place in the liquid or vapor phase below 100°C (212°F) in the absence of light or a catalyst, but above 200°C (390°F) it proceeds smoothly by thermal action alone. The hydrolysis of the mixed chlorides obtained yields all the isomeric amyl (C_5) alcohols except *iso*-amyl alcohol. Reaction with acetic acid produces the corresponding amyl acetates, which find wide use as solvents.

The alkyl chloride obtained on substituting an equivalent of one hydrogen atom by a chloride atom in kerosene is used to alkylate benzene or naphthalene for the preparation of a sulfonation stock for use in the manufacture of detergents and antirust agents. Similarly, paraffin wax can be converted to a hydrocarbon monochloride mixture, which can be employed to alkylate benzene, naphthalene, or anthracene. The product finds use as a pour-point depressor effective for retarding wax crystal growth and deposition in cold lubricating oils.

2.3 Nitration

Hydrocarbon derivatives that are usually gaseous (including normal and *iso*-pentane) react smoothly in the vapor phase with nitric acid to give a mixture of nitro-compounds, but there are sidereactions, mainly of oxidation. Only mononitro-derivatives are obtained with the lower paraffin derivatives as high temperatures, and they correspond to those expected if scission of a C-C and C-H bond occurs.

Ethane, for example, yields nitromethane and nitroethane,

$$CH_3CH_3 + HNO_3 \rightarrow CH_3CH_2NO_2 + CH_3NO_2$$

Propane yields nitromethane, nitroethane, 1-nitropropane, and 2-nitropropane:

The nitro-derivatives of the lower paraffin derivatives are colorless and noncorrosive and are used as solvents or as starting materials in a variety of syntheses. For example, treatment with inorganic acids and water yields fatty acids (RCO_2H) and hydroxylamine (NH_2OH) salts and condensation with an aldehyde ($RCH = O$) yields nitro-alcohols [$RCH(NO_2)OH$].

2.4 Oxidation

The oxidation of hydrocarbon derivatives and hydrocarbon mixtures has received considerable attention, but the uncontrollable nature of the reaction and the mixed character of the products have made resolution of the reaction sequences extremely difficult. Therefore it is not surprising that, except for the preparation of mixed products having specific properties, such as fatty acids, hydrocarbon derivatives higher than pentanes are not employed for oxidation because of the difficulty of isolating individual compounds.

Methane undergoes two useful reactions at 90°C (195°F) in the presence of iron oxide (Fe_3O_4) as a catalyst:

$$CH_4 + H_2O \rightarrow CO + 3H_2$$

$$CO + H_2O \rightarrow CO_2 + H_2$$

Alternatively, partial combustion of methane can be used to provide the required heat and steam. The carbon dioxide produced then reacts with methane at 900°C (1650°F) in the presence of a nickel catalyst:

$$CH_4 + 2O_2 \rightarrow O_2 + 2H_2O$$

$$CO_2 + CH_4 \rightarrow 2CO + 2H_2$$

$$CH_4 + H_2O \rightarrow CO + 3H_2$$

Methanol (methyl alcohol, CH_3OH) is the second major product produced from methane. Synthetic methanol has virtually completely replaced methanol obtained from the distillation of wood, its original source material. One of the older trivial names used for methanol was wood alcohol. The synthesis reaction takes place at 350°C (660°F) and 4400 psi in the presence of ZnO as a catalyst:

$$2CH_4 + O_2 \rightarrow 2CH_3OH$$

Most of the methanol is then oxidized by oxygen from air to formaldehyde (sometimes referred to as methanal):

$$2CH_3OH + O_2 \rightarrow 2CH_2O + 2H_2O$$

Formaldehyde is used to produce synthetic resins either alone or with phenol, urea, or melamine; other uses are minor.

By analogy to the reaction with oxygen, methane reacts with sulfur in the presence of a catalyst to give the carbon disulfide used in the rayon industry:

$$CH_4 + 4S(g) \rightarrow CS_2 + 2H_2S$$

The major nonpetrochemical use of methane is in the production of hydrogen for use in the Haber synthesis of ammonia. Ammonia synthesis requires nitrogen, obtained from air, and hydrogen. The most common modern source of the hydrogen consumed in ammonia production, approximately 95% of it, is methane.

When propane and butane are oxidized in the vapor phase, without a catalyst, at 270°C–350°C (520°F–660°F) and at 50 to 3000 psi, a wide variety of products is obtained, including C_1 to C_4 acids, C_2 to C_7 ketones, ethylene oxide, esters, formals, acetals, and others.

Cyclohexane is oxidized commercially and is somewhat selective in its reaction with air at 150°C–250°C (300°F–480°F) in the liquid phase in the presence of a catalyst, such as cobalt acetate. Cyclohexanol derivatives are the initial products, but prolonged oxidation produces adipic acid. On the other hand, oxidation of cyclohexane and methylcyclohexane over vanadium pentoxide at 450°C–500°C (840°F–930°F) affords maleic acid and glutaric acid.

The preparation of carboxylic acids from crude oil, particularly from paraffin wax, for esterification to fats or neutralization to form soaps has been the subject of a large number of investigations. Wax oxidation with air is comparatively slow at low temperature and normal pressure, very little reaction taking place at 110°C (230°F), with a wax melting at 55°C (130°F) after 280 h. At higher temperatures the oxidation proceeds more readily; maximum yields of mixed alcohol and high molecular weight acids are formed at 110°C–140°C (230°F–285°F) at 60 to 150 psi; higher temperatures (140°C–160°C, 285°F –320°F) result in more acid formation:

$$\text{Paraffin wax} \rightarrow \underset{\text{Alcohol}}{ROH} + \underset{\text{acid}}{RCO_2H}$$

Acids from formic (HCO_2H) to that with a 10-carbon atom chain [$CH_3(CH_2)_9CO_2H$] have been identified as products of the oxidation of paraffin wax. Substantial quantities of water-insoluble acids are also produced by the oxidation of paraffin wax, but apart from determination of the average molecular weight (ca. 250), very little has been done to identify individual numbers of the product mixture.

Aromatic derivatives are more resistant to oxidation than the paraffin hydrocarbon derivatives, and higher temperatures are necessary; the oxidations are carried out in the vapor phase over a catalyst, generally supported vanadium oxide. *Ortho*-xylene is oxidized by nitric acid to phthalic anhydride, *m*-xylene to *iso*-phthalic acid, and *p*-xylene with nitric acid to terephthalic acid. These acid products are used in the manufacture of fibers, plastics, plasticizers, and the like.

Phthalic anhydride is also produced in good yield by the air oxidation of naphthalene at 400°C–450°C (750°F–840°F) in the vapor phase at approximately 25 psi over a fixed-bed vanadium pentoxide catalyst. Terephthalic acid is produced in a similar manner from *p*-xylene, and an intermediate in the process, *p*-toluic acid, can be isolated because it is slower to oxidize than the *p*-xylene starting material.

Of the xylenes, *o*-xylene is used to produce phthalic anhydride and other compounds. Another xylene, *p*-xylene is used in the production of polyesters in the form of terephthalic acid or its methyl ester (Figure 8.1). Terephthalic acid is produced from *p*-xylene by two reactions in four steps. The first of these is oxidation with oxygen at 190°C (375°F):

$$CH_3C_6H_4CH_3 + O_2 \rightarrow HOOCC_6H_4CH_3$$

This is followed by formation of the methyl ester at 150°C (302°F):

$$HOOCC_6H_4CH_3 + CH_3OH \rightarrow HOOCC_6H_4CH_3$$

Repetition of these steps gives methyl diester of terephthalic acid:

$$CH_3OOCC_6H_4CH_3 + O_2 \rightarrow CH_3OOCC_6H_4CCOOH$$

$$CH_3OOCC_6H_4CCOOH + CH_3OH \rightarrow CH_3OOCC_6H_4CCOOCH_3$$

This diester, $CH_3OOCC_6H_4CCOOCH_3$, when polymerized with ethylene glycol at 200°C (390°F), yields the polymer after loss of methanol to give a monomer. The polymerization step requires a catalyst.

In the process to produce terephthalic acid, the crude acid is produced by the catalytic oxidation of *p*-xylene with air in the liquid phase using acetic acid as a solvent. The feedstock mix (*p*-xylene, solvent, and catalyst) is continuously fed, with compressed air, to the e-column oxidizer which operates at moderate temperature. The oxidizer product is purified in a separation step in which the impurities are removed from the product by exchanging the reaction liquor with lean solvent from the solvent recovery system. The reactor overhead—mainly reaction water, acetic acid, and nitrogen—is sent to the solvent recovery system where water is separated from the solvent by distillation. The offgas is sent to a regenerative thermal oxidation unit for further cleaning. To produce polymer-grade terephthalic acid, the crude acid is purified in a postoxidation step, at elevated temperature. The post oxidizers serve as reactors to increase the conversion of the partially oxidized compounds to terephthalic acid.

2.5 Thermolysis

Although there are relatively unreactive organic molecules, paraffin hydrocarbon derivatives are known to undergo thermolysis when treated under high-temperature, low-pressure vapor-phase conditions. The cracking chemistry of crude oil constituents has been extensively studied (Oblad et al., 1979; Albright and Crynes, 1976). Cracking is the major process for generating ethylene and the other olefin derivatives that are the reactive building blocks of the petrochemical industry (Chenier, 1992). In addition to thermal cracking, other very important processes that generate sources of hydrocarbon raw materials for the petrochemical industry include catalytic reforming, alkylation, dealkylation, isomerization, and polymerization.

Cracking reactions involve the cleavage of carbon-carbon bonds with the resulting redistribution of hydrogen to produce smaller molecules. Thus cracking of crude oil or crude oil fractions is a process by which larger molecules are converted into smaller, lower boiling molecules. In addition, cracking generates two molecules from one, with one of the product molecules saturated (paraffin) and the other unsaturated (olefin).

At the high temperatures of refinery crackers (usually >500°C, 950°F), there is a thermodynamic driving force for the generation of more molecules from fewer molecules; that is, cracking is favored. Unfortunately, in the cracking process certain products interact with one another to produce products of increased molecular weight over that in the original feedstock. Thus some products are taken off from the cracker as useful low boiling products (olefin derivatives, gasoline, and others), but other products include higher boiling oil and coke.

3. Chemicals from olefin hydrocarbons

Olefin derivatives (C_2H_{2n}) are the basic building blocks for a host of chemical syntheses (Chenier, 1992). These unsaturated materials enter into polymers, and rubbers and with other reagents react to form a wide variety of useful compounds, including alcohols, epoxides, amines, and halides.

Olefin derivatives present in gaseous products of catalytic cracking processes offer promising source materials. Cracking paraffin hydrocarbon derivatives and heavy oils also produces olefin derivatives. For example, cracking ethane, propane, butane, and other feedstock such as gas oil, naphtha, and residua produces ethylene. Propylene is produced from thermal and catalytic cracking of naphtha and gas oils, as well as propane and butane.

As far as can be determined, the first large-scale petrochemical process was the sulfuric acid absorption of propylene ($CH_3CH{=}CH_2$) from refinery cracked gases to produce isopropyl alcohol [$(CH_3)_2CHOH$]. The interest, then, in thermal reactions of hydrocarbon derivatives has been high since the 1920s when alcohols were produced from the ethylene and propylene formed during crude oil cracking. The range of products formed from crude oil pyrolysis has widened over the past six decades to include the main chemical building blocks. These include ethane, ethylene, propane, propylene, the butanes, butadiene, and aromatics. Additionally, other commercial products from thermal reactions of crude oil include coke, carbon, and asphalt.

Ethylene manufacture via the steam cracking process is in widespread practice throughout the world. The operating facilities are similar to gas oil cracking units, operating at temperatures of 840°C (1550°F) and at low pressures (24 psi, atmospheric pressure = 14.7 psi). Steam is added to the vaporized feed to achieve a 50-50 mixture, and furnace residence times are only 0.2–0.5 s. Ethane extracted from natural gas is the predominant feedstock for ethylene cracking units. Propylene and butylene are largely derived from catalytic cracking units and from cracking a naphtha or low boiling gas oil fraction to produce a full range of olefin products.

Virtually all propene or propylene is made from propane, which is obtained from natural gas stripper plants or from refinery gases:

$$CH_3CH_2CH_3 \rightarrow CH_3\text{-}CH{=}CH_2 + H_2$$

The uses of propene include gasoline (80%), polypropylene, *iso*-propanol, trimers, and tetramers for detergents, propylene oxide, cumene, and glycerin.

Two butenes or butylenes (1-butene, $CH_3CH_2CH{=}CH_2$, and 2-butene, $CH_3CH{=}CHCH_3$) are industrially significant. The latter has end uses in the production of butyl rubber and polybutylene plastics. On the other hand, 1-butene is used in the production of 1,3-butadiene ($CH_2{=}CHCH\text{-}CH_2$) for the synthetic rubber industry. Butenes arise primarily from refinery gases or from the cracking of other fractions of crude oil.

Butadiene can be recovered from refinery streams as butadiene, butenes, or butanes; the latter two on appropriate heated catalysts dehydrogenate to give 1,3-butadiene:

$$CH_2{=}CHCH_2CH_3 \rightarrow CH_2{=}CHCH{=}CH_2 + H_2$$

$$CH_3CH_2CH_2CH_3 \rightarrow CH_3{=}CHCH{=}CH_2$$

An alternative source of butadiene is ethanol, which on appropriate catalytic treatment also gives the compound di-olefin:

$$2C_2H_5OH \rightarrow CH_2{=}CHCH{=}CH_2 + 2H_2O$$

Olefin derivatives containing more than four carbon atoms are in little demand as petrochemicals and thus are generally used as fuel. The single exception to this is 2-methyl-1,3-butadiene or isoprene, which has a significant use in the synthetic rubber industry. It is more difficult to make than 1,3-butadiene. Some is available in refinery streams, but more is manufactured from refinery stream 2-butene by reaction with formaldehyde:

$$CH_3CH{=}CHCH_3 + HCHO \rightarrow CH_2{=}CH(CH_3)CH{=}CH_2 + H_2O$$

3.1 Ester formation

Esters (RCO_2R') are formed directly by the addition of acids to olefin derivatives, mercaptans by the addition of hydrogen sulfide to olefin derivatives, sulfides by the addition of mercaptans to olefin derivatives, and amines by the addition of ammonia and other amines to olefin derivatives.

More specifically, ester is a chemical compounds that is derived from an organic acid or an inorganic acid in which at least one hydroxyl (-OH) group is replaced by an alkyl group. Usually, esters are derived from a carboxylic acid and an alcohol and are considered high-quality solvents for a broad array of plastics, plasticizers, resins, and lacquers, as well as a class of synthetic lubricants.

The classic synthesis of ester is by the Fischer esterification reaction, which involves treating a carboxylic acid with an alcohol in the presence of a dehydrating agent (sulfuric acid is a typical catalyst for this reaction):

$$RCO_2H + R'OH \rightleftharpoons RCO_2R' + H_2O$$

Esters are also formed by the reaction of alcohols with acyl chlorides and acid anhydrides:

$$RCOCl + R'OH \rightarrow RCO_2R' + HCl$$

$$(RCO)_2O + R'OH \rightarrow RCO_2R' + RCO_2H$$

Since acyl chlorides and acid anhydrides also react with water, anhydrous conditions are preferred. The analogous acylation of amines produces amides.

Transesterification involves changing one ester into another one, is widely practiced, especially in the production of biodiesel (Speight, 2011):

$$RCO_2R' + CH_3OH \rightarrow RCO_2CH_3 + R'OH$$

Biodiesel production is the process of producing the biodiesel by the chemical reactions of transesterification and involves the reaction of vegetable fats or animal fats with short-chain alcohols (typically methanol or ethanol). In the process, the added alcohol is deprotonated with a base to make it a stronger nucleophile. The acid or base are not consumed by the transesterification reaction and are, thus are not reactants, but catalysts.

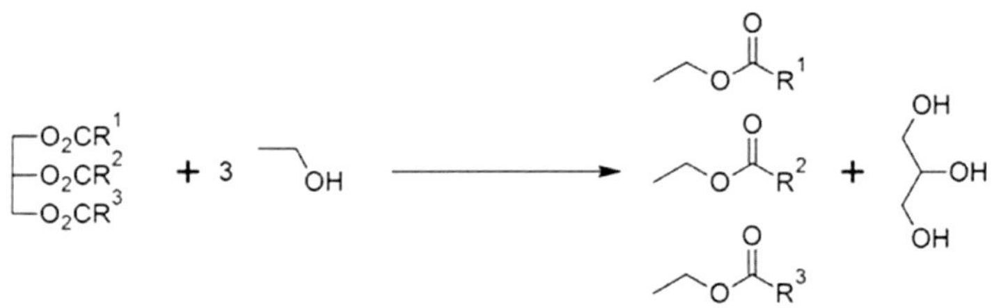

Common catalysts for transesterification include sodium hydroxide (NaOH), potassium hydroxide (KOH), and sodium methoxide ($CH3O^- Na^+$).

3.2 Halogenation

Generally, at ordinary temperatures, chlorine reacts with olefin derivatives by addition. Thus, ethylene is chlorinated to 1,2-dichloroethane (dichloroethane) or to ethylene dichloride:

$$H_2C{=}CH_2 + Cl_2 \rightarrow H_2ClCCH_2Cl$$

There are some minor uses for ethylene dichloride, but approximately 90% of it is cracked to vinyl chloride, the monomer of polyvinyl chloride (PVC):

$$H_2ClCCH_2Cl \rightarrow HCl + H_2C{=}CHCl$$

At slightly higher temperatures, olefin derivatives and chlorine react by substitution of a hydrogen atom by a chlorine atom. Thus, in the chlorination of propylene, a rise of 50°C (90°F) changes the product from propylene dichloride [$CH_3CH(Cl)CH_2Cl$] to allyl chloride ($CH_2{=}CHCH_2Cl$).

3.3 Hydroxylation

The earliest method for conversion of olefin derivatives into alcohols involved their absorption in sulfuric acid to form esters, followed by dilution and hydrolysis, generally with the aid of steam. In the case of ethyl alcohol, the direct catalytic hydration of ethylene can be employed. Ethylene is readily absorbed in 98%–100% sulfuric acid at 75°C–80°C (165°F–175°F), and both ethyl and diethyl sulfate are formed; hydrolysis takes place readily on dilution with water and heating.

The direct hydration of ethylene to ethyl alcohol is practiced over phosphoric acid on diatomaceous earth or promoted tungsten oxide under 100 psi pressure and at 300°C (570°F):

$$CH_2{=}CH_2 + H_2O \rightarrow C_2H_5OH$$

Ethylene with a high degree of purity is required in direct hydration than in the acid absorption process and the conversion per pass is low, but high yields are possible by recycling. Propylene and the normal butenes can also be hydrated directly.

Ethylene, produced from ethane by cracking, is oxidized in the presence of a silver catalyst to ethylene oxide:

$$2H_2C{=}CH_2 + O_2 \rightarrow C_2H_4O$$

The vast majority of the ethylene oxide produced is hydrolyzed at 100°C to ethylene glycol:

$$C_2H_4O + H_2O \rightarrow HOCH_2CH_2OH$$

Approximately 70% of the ethylene glycol produced is used as automotive antifreeze and much of the rest is used in the synthesis of polyesters.

Of the higher alkenes, one of the first alcohol syntheses practiced commercially was that of isopropyl alcohol from propylene. Sulfuric acid absorbs propylene more readily than it does ethylene, but care must be taken to avoid polymer formation by keeping the mixture relatively cool and using acid of approximately 85% strength at 300 to 400 psi pressure; dilution with inert oil may also be necessary. Acetone is readily made from isopropyl alcohol, either by catalytic oxidation or by dehydrogenation over metal (usually copper) catalysts. Secondary butyl alcohol is formed on absorption of 1-butene or 2-butene by 78%–80% sulfuric acid, followed by dilution and hydrolysis. Secondary butyl alcohol is converted into methyl ethyl ketone by catalytic oxidation or dehydrogenation.

There are several methods for preparing higher alcohols. One method in particular, the so-called Oxo reaction, involves the direct addition of carbon monoxide (CO) and a hydrogen (H) atom across the double bond of an olefin to form an aldehyde ($RCH = O$), which in turn is reduced to the alcohol (RCH_2OH). Hydroformylation (the Oxo reaction) is brought about by contacting the olefin with synthesis gas (1:1 carbon monoxide-hydrogen) at 75°C −200°C (165°F−390°F) and 1500 to 4500 psi over a metal catalyst, usually cobalt. The active catalyst is held to be cobalt hydrocarbonyl $HCO(CO)_4$, formed by the action of the hydrogen on dicobalt octacarbonyl $[Co_2(CO)_8]$.

A wide variety of olefin derivatives enter the reaction, those containing terminal unsaturated being the most active. The hydroformylation is not specific; the hydrogen and carbon monoxide add across each side of the double bond. Thus propylene gives a mixture of 60% *n*-butyraldehyde and 40% *iso*-butyraldehyde. Terminal ($RCH = CH_2$) and nonterminal ($RCH = CHR'$) olefin derivatives, such as 1-pentene and 2-pentene, give essentially the same distribution of straight-chain and branched-chain C_6 aldehydes, indicating that rapid isomerization takes place. Simple branched structures add mainly at the terminal carbon; *iso*-butylene forms 95% *iso*-valeraldehyde and 5% trimethyl acetaldehyde. Commercial application of the synthesis has been most successful in the manufacture of *iso*-octyl alcohol from a refinery C_3-C_4 copolymer, decyl alcohol from propylene trimer, and tridecyl alcohol from propylene tetramer. Important outlets for the higher alcohols lie in their sulfonation to make detergents and the formation of esters with dibasic acids for use as plasticizers and synthetic lubricants.

The hydrolysis of ethylene chlorohydrin ($HOCH_2CH_2Cl$) or the cyclic ethylene oxide produces ethylene glycol ($HOCH_2CH_2OH$). The main use for this chemical is for antifreeze mixtures in automobile radiators and for cooling aviation engines; considerable amounts are used as ethylene glycol dinitrate in low-freezing dynamite. Propylene glycol is also made by the hydrolysis of its chlorohydrin or oxide.

Glycerin can be derived from propylene by high-temperature chlorination to produce alkyl chloride, followed by hydrolysis to allyl alcohol and then conversion with aqueous chloride to glycerol chlorohydrin, a product that can be easily hydrolyzed to glycerol (glycerin). Glycerin has found many uses over the years; important among these are as solvent, emollient, sweetener, in cosmetics, and as a precursor to nitroglycerin and other explosives.

3.4 Oxidation

The most striking industrial olefin oxidation process involves ethylene, which is air oxidized over a silver catalyst at 225°C−325°C (435°F−615°F) to give pure ethylene oxide in yields ranging from 55% to 70%.

Analogous higher olefin oxides can be prepared from propylene, butadiene, octene, dodecene, and styrene via the chlorohydrin route or by reaction with peracetic acid. Acrolein is formed by air oxidation or propylene over a supported cuprous oxide catalyst or by condensing acetaldehyde and formaldehyde. When acrolein and air are passed over a catalyst, such as cobalt molybdate, acrylic acid is produced or if acrolein is reacted with ammonia and oxygen over molybdenum oxide, the product is acrylonitrile. Similarly, propylene may be converted to acrylonitrile.

Acrolein and acrylonitrile are important starting materials for the synthetic materials known as acrylates; acrylonitrile is also used in plastics, which are made by copolymerization of acrylonitrile with styrene or with a styrene-butadiene mixture.

Oxidation of the higher olefin derivatives by air is difficult to control, but at temperatures between 350°C and 500°C (660°F and 930°F) maleic acid is obtained from amylene and a vanadium pentoxide catalyst; higher yields of the acid are obtained from hexene, heptene, and octene.

3.5 Polymerization

The polymerization of ethylene under pressure (1500 to 3000 psi) at 110°C–120°C (230°F–250°F) in the presence of a catalyst or initiator, such as a 1% solution of benzoyl peroxide in methanol, produces a polymer in the 2000 to 3000 molecular weight range. Polymerization at 15,000 to 30,000 psi and 180°C–200°C (355°F–390°F) produces a wax melting at 100°C (212°F) and 15,000 to 20,000 molecular weight but the reaction is not as straightforward as the equation indicates, since there are branches in the chain. However, considerably lower pressures can be used over catalysts composed of aluminum alkyls (R_3Al) in presence of titanium tetrachloride ($TiCl_4$), supported chromic oxide (CrO_3), nickel (NiO), or cobalt (CoO) on charcoal, and promoted molybdena-alumina (MoO_2-Al_2O_3), which at the same time give products more linear in structure. Polypropylenes can be made in similar ways, and mixed monomers, such as ethylene-propylene and ethylene-butene mixtures, can be treated to give high molecular weight copolymers of good elasticity. Polyethylene has excellent electrical insulating properties; its chemical resistance, toughness, machinability, light weight, and high strength make it suitable for many other uses.

Lower molecular weight polymers, such as the dimers, trimers, and tetramers, are used as such in motor gasoline. The materials are normally prepared over an acid catalyst. Propylene trimer (dimethyl heptene derivatives) and tetramer (trimethyl nonene derivatives) are applied in the alkylation of aromatic hydrocarbon derivatives for the production of alkyl-aryl sulfonate detergents and also as olefin-containing feedstocks in the manufacture of C_{10} and C_{13} oxo-alcohols. Phenol is alkylated by the trimer to make nonylphenol, a chemical intermediate for the manufacture of lubricating oil detergents and other products.

Iso-butylene also forms several series of valuable products; the di- and tri-*iso*-butylenes make excellent motor and aviation gasoline components; they can also be used as alkylating agents for aromatic hydrocarbon derivatives and phenols and as reactants in the oxo-alcohol synthesis. Poly(*iso*butylene) derivatives in the viscosity range of 55,000 SUS (38°C, 100°F) have been employed as viscosity index improvers in lubricating oils. Butene-1 ($CH_3CH_2CH{=}CH_2$) and butene-2 ($CH_3CH{=}CHCH_3$) participate in polymerization reactions by the way of butadiene ($CH_2{=}CHCH{=}CH_2$), the dehydrogenation product, which is copolymerized with styrene (23.5%) to form GR-S rubber, and with acrylonitrile (25%) to form GR-N rubber:

Derivatives of acrylic acid (butyl acrylate, ethyl acrylate, 2-ethylhexyl acrylate, and methyl acrylate) can be homopolymerized using peroxide initiators or copolymerized with other monomers to generate acrylic or acryloid resins.

4. Chemicals from aromatic hydrocarbons

Aromatic compounds (sometime referred to as arenes) are those compounds that contain one or more benzene rings or similar ring structures (Table 12.6) (March, 1985), many of which occur in crude oil and crude oil products

TABLE 12.6 Representative single-ring aromatic compounds.

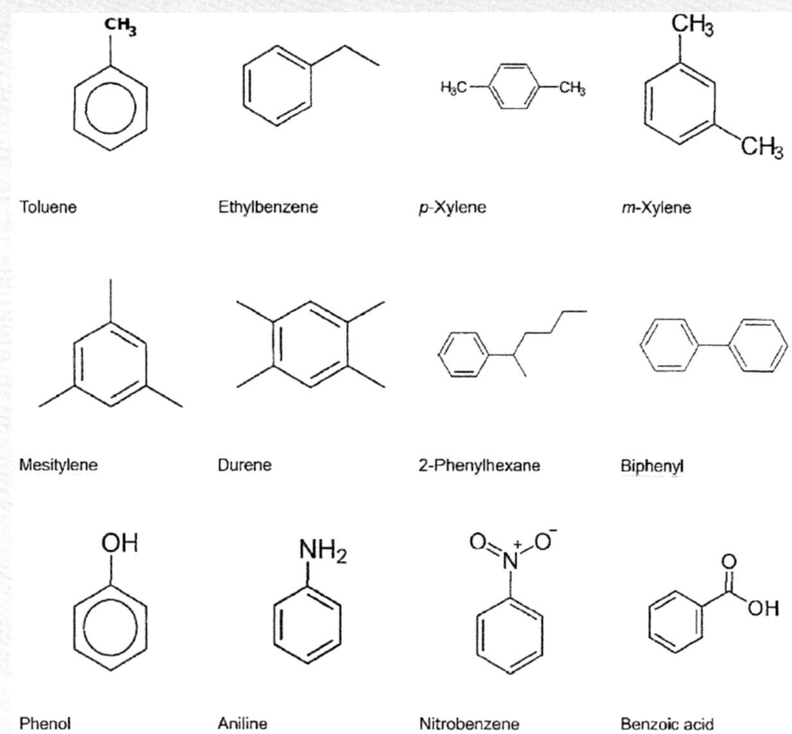

(Parkash, 2003; Gary et al., 2007; Speight, 2014, 2017; Hsu and Robinson, 2017). The majority of the aromatic compounds for petrochemical use are produced in various refinery streams and are then separated into fractions, of which the most significant constituents are benzene (C_6H_6), methylbenzene or toluene ($C_6H_5CH_3$), and the dimethylbenzene or xylene derivatives ($CH_3C_6H_4CH_3$) with the two-ring condensed aromatic compound naphthalene ($C_{10}H_8$) also being a source of petrochemicals.

Briefly, aromatic compounds are those containing one or more benzene rings or similar ring structures. The majority are taken from refinery streams which contain them and separated into fractions, of which the most significant fractions are benzene (C_6H_6), methylbenzene or toluene ($C_6H_5CH_3$), and the dimethylbenzenes or xylenes ($CH_3C_6H_4CH_3$) with the two-ring condensed aromatic compound naphthalene ($C_{10}H_8$) also being a source of petrochemical. In the traditional chemical industry, aromatics such as benzene, toluene, and xylene were made from coal during the course of carbonization in the production of coke and town gas. A much larger volume of these chemicals are now made as refinery byproducts. A further source of supply is the aromatic-rich liquid fraction produced in the cracking of naphtha or low-boiling gas oils during the manufacture of ethylene and other olefin derivatives.

Aromatic compounds are valuable starting materials for a variety of chemical products (Chenier, 1992). Reforming processes have made benzene, toluene, xylene, and ethylbenzene economically available from crude oil sources. They are generally recovered by extractive or azeotropic distillation, by solvent extraction (with water-glycol mixtures or liquid sulfur dioxide), or by adsorption. Naphthalene and methylnaphthalenes are present in catalytically cracked distillates. A substantial part of the benzene consumed is now derived from crude oil, and it has many chemical uses.

Aromatic compounds, such as benzene, toluene, and the xylenes, are major sources of chemicals (Fig. 12.6). For example, benzene is used to make styrene ($C_6H_5CH{=}CH_2$), the basic ingredient of polystyrene plastics, as well as paints, epoxy resins, glues, and other adhesives. The process for the manufacture of styrene proceeds through ethylbenzene, which is produced by reaction of benzene and ethylene at 95°C (203°F) in the presence of a catalyst:

$$C_6H_6 + CH_2{=}CH_2 \rightarrow C_6H_5CH_2CH_3$$

In the presence of a catalyst and superheated steam ethylbenzene dehydrogenates to styrene:

$$C_6H_5CH_2CH_3 \rightarrow C_6H_5CH{=}CH_2 + H_2$$

Toluene is usually added to the gasoline pool or used as a solvent, but it can be dealkylated to benzene by catalytic treatment with hydrogen:

$$C_6H_5CH_3 + H_2 \rightarrow C_6H_6 + CH_4$$

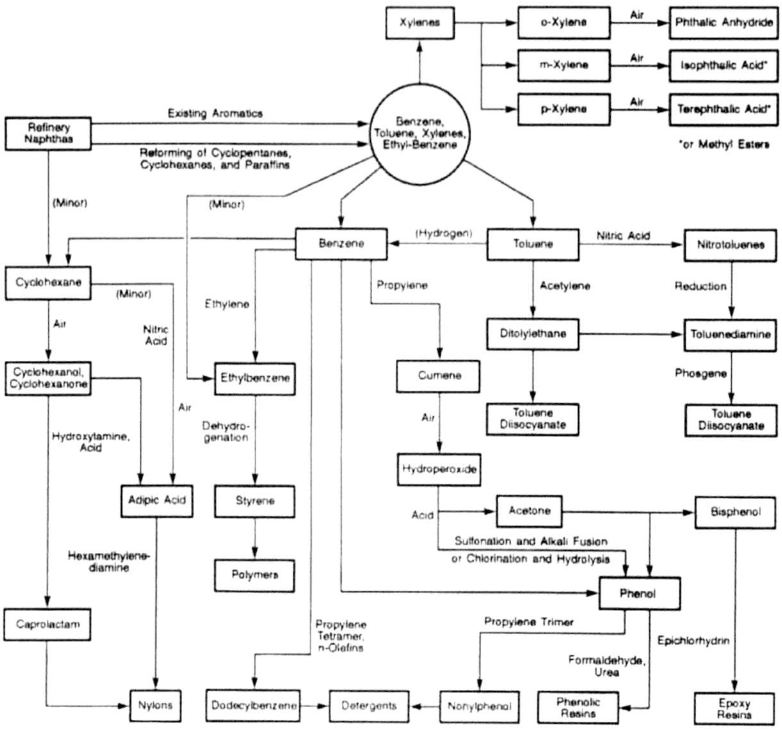

FIGURE 12.6 Chemicals from benzene, toluene, and the xylenes.

Similar processes are used for dealkylation of methyl-substituted naphthalene. Toluene is also used to make solvents, gasoline additives, and explosives.

Toluene is usually in demand as a source of trinitrotoluene (TNT) but has fewer chemical uses than benzene. Alkylation of toluene with ethylene, followed by dehydrogenation, yields α-methylstyrene [$C_6H_5C(CH_3) = CH_2$], which can be used for polymerization. Alkylation of toluene with propylene tetramer yields a product suitable for sulfonation to a detergent-grade surface-active compound.

Of the xylene isomers, *o*-xylene is used to produce phthalic anhydride and other compounds. Another xylene, *p*-xylene is used in the production of polyesters in the form of terephthalic acid or its methyl ester. Terephthalic acid is produced from *p*-xylene by two reactions in four steps. The first of these is oxidation with oxygen at 190°C (375°F):

$$CH_3C_6H_4CH_3 + O_2 \rightarrow HOOCC_6H_4CH_3$$

This is followed by formation of the methyl ester at 150°C (302°F):

$$HOOCC_6H_4CH_3 + CH_3OH \rightarrow HOOCC6H_4CH_3$$

Repetition of these steps gives the methyl diester of terephthalic acid: This diester, $CH_3OOCC_6H_4CCOOCH_3$, when polymerized with ethylene glycol at 200°C (390°F), yields the polymer after loss of methanol to give a monomer. The polymerization step requires a catalyst.

Aromatics are more resistant to oxidation than the paraffin hydrocarbon derivatives, and higher temperatures are necessary; the oxidations are carried out in the vapor phase over a catalyst, generally supported vanadium oxide. *Ortho*-xylene is oxidized by nitric acid to phthalic anhydride, *m*-xylene to *iso*-phthalic acid, and *p*-xylene with nitric acid to terephthalic acid. These acid products are used in the manufacture of fibers, plastics, plasticizers, and the like.

Phthalic anhydride is also produced in good yield by the air oxidation of naphthalene at 400°C–450°C (750°F–840°F) in the vapor phase at approximately 25 psi over a fixed-bed vanadium pentoxide catalyst. Terephthalic acid is produced in a similar manner from *p*-xylene, and an intermediate in the process, *p*-toluic acid, can be isolated because it is slower to oxidize than the *p*-xylene starting material.

5. Chemicals from acetylene

Acetylene is the simplest member of alkyne hydrocarbon derivatives. In the first half of the 20th century acetylene was the most important of all starting materials for organic synthesis. Acetylene is a colorless, combustible gas with a distinctive odor. When acetylene is liquefied, compressed, heated, or mixed with air, it becomes highly explosive. As a result special precautions are required during its production and handling. The most common use of acetylene is as a raw material for the production of various organic chemicals including 1,4-butanediol, which is widely used in the preparation of polyurethane and polyester plastics. The second most common use is as the fuel component in oxy-acetylene welding and metal cutting. Some commercially useful acetylene compounds include acetylene black, which is used in certain dry-cell batteries, and acetylenic alcohols, which are used in the synthesis of vitamins.

Acetylene was discovered in 1836, when Edmund Davy was experimenting with potassium carbide. One of his chemical reactions produced a flammable gas, which is now known as acetylene. In 1859, Marcel Morren successfully generated acetylene when he used carbon electrodes to strike an electric arc in an atmosphere of hydrogen. The electric arc tore carbon atoms away from the electrodes and bonded them with hydrogen atoms to form acetylene molecules. He called this gas *carbonized hydrogen*.

By the late 1800s, a method had been developed for making acetylene by reacting calcium carbide with water. This generated a controlled flow of acetylene that could be combusted in air to produce a brilliant white light. Carbide lanterns were used by miners and carbide lamps were used for street illumination before the general availability of electric lights. In 1897, Georges Claude and A. Hess noted that acetylene gas could be safely stored by dissolving it in acetone. Nils Dalen used this new method in 1905 to develop long-burning, automated marine and railroad signal lights. In 1906, Dalen went on to develop an acetylene torch for welding and metal cutting.

Between 1960 and 1970, when worldwide acetylene production peaked, it served as the primary feedstock for a wide variety of commodity and specialty chemicals. Advances in olefin derivatives technology are related to the safety aspects of acetylene use, but mostly loss of cost-competitiveness, reduced and effectively limited the importance of acetylene. Now, with the current rise in crude oil prices, acetylene is finding a new place in the chemical industry.

Acetylene is the only petrochemical produced in significant quantity which contains a triple bond, and is a major intermediate species. The usefulness of acetylene is partly due to the variety of additional reactions which its triple bond undergoes, and partly due to the fact that its weakly acidic hydrogen atoms are replaceable by reaction with strong bases to form acetylide salts. However, acetylene is not easily shipped, and as a consequence its consumption is close to the point of origin.

However, acetylene was largely replaced by olefin feedstocks, such as ethylene and propylene, because of its high cost of production and the safety issues of handling acetylene at high pressures. Its use has largely been eliminated, except for the continued, and in some instances, growing production of vinyl chloride monomer, 1,4-butanediol, and carbon black. Up until the 1970s, acetylene was a basic chemical raw material used for the production of a wide range of chemicals (Fig. 12.7).

Currently, there are several routes to acetylene. Hydrocarbon derivatives are the major feedstocks in the United States and Western Europe, either in the form of natural gas in partial oxidation processes or as byproducts in ethylene production. However, coal is becoming an ever increasing source of acetylene in countries with plentiful and cheap coal supplies, such as China, for the

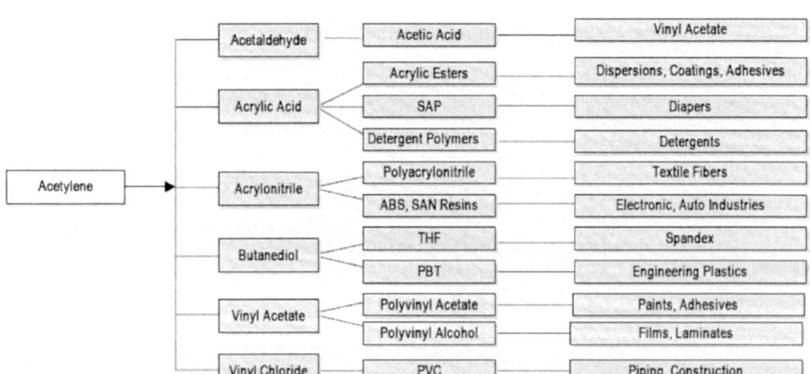

FIGURE 12.7 Chemicals from acetylene and end uses.

production of vinyl chloride and this source of lower cost acetylene may prove to be the impetus for returning acetylene to its place as a major chemical feedstock, especially in respect of the current and projected high oil prices and improvements in the safety, cost, and environmental protection of the calcium carbide process for the production of acetylene.

The resurgence of the use of acetylene for chemicals production will depend upon the relative cost of acetylene versus the more commonly used feedstocks. The technologies for the chemicals production are well known and have been improved since the heyday of acetylene. More importantly, the process technology to produce acetylene has been greatly improved and optimized, and now can offer attractive competitiveness in the right situations.

The classic commercial route to acetylene, first developed in the late 1800s, is the calcium carbide route in which lime is reduced by carbon (in the form of coke) in an electric furnace to yield calcium carbide. During this process a considerable amount of heat is produced, which is removed to prevent the acetylene from exploding. This reaction can occur via wet or dry processes depending on how much water is added to the reaction process. The calcium carbonate is first converted into calcium oxide and the coal into coke. The two are then reacted with each other to form calcium carbide and carbon monoxide:

$$CaO + 3C \rightarrow CaC_2 + CO$$

The calcium carbide is then hydrolyzed to produce acetylene:

$$CaC_2 + 2H_2O \rightarrow C_2H_2 + Ca(OH)_2$$

Acetylene can also be manufactured by the partial oxidation (partial combustion) combustion of methane with oxygen. The process employs a homogeneous gas phase hydrogen halide catalyst other than hydrogen fluoride to promote the pyrolytic oxidation of methane. The homogeneous gas phase catalyst employed can also consist of a mixture of gaseous hydrogen halide and gaseous halogen, or a halogen gas.

The electric arc or plasma pyrolysis of coal can also be used to produce acetylene. The electric arc process involves a 1 megawatt arc plasma reactor which utilizes a DC electric arc to generate and maintain a hydrogen plasma. The coal is then fed into the reactor and is heated to a high temperature as it passes through the plasma. It is then partially gasified to yield acetylene, hydrogen, carbon monoxide, hydrogen cyanide, and several hydrocarbon derivatives.

Acetylene can also be produced as a byproduct of ethylene steam cracking. The use of acetylene as a commodity feedstock decreased due to the competition of cheaper, more readily accessible and workable olefin derivatives when these olefin derivatives were produced from low cost crude oil products. With the rising cost of crude oil, natural gas, and the associated olefin derivative feedstocks (such as naphtha, ethane, propane, etc.) the olefin derivatives prices are no longer low enough to preclude the possibility of using acetylene. Additionally, regional shortages of these olefin derivatives and their feedstocks have forced the search for alternate routes to the commodity chemicals.

Acetylene is used as a special fuel gas (oxyacetylene torches) and as a chemical raw material. Historically, acetylene has been used to produce many important chemicals, such as (listed alphabetically): acetaldehyde, acrylate esters, acrylonitrile, 1,4-butynediol, 1,2-dichloroethane, polyacetylene, and polydiacetylene vinyl acetate, vinyl chloride monomer, and vinyl ether (Table 12.7). Based on its availability, its many uses and prospective uses, acetylene is definitely an interesting possibility going forward, if available at competitive cost.

6. Chemicals from natural gas

Natural gas can be used as a source of hydrocarbon derivatives (e.g., ethane and propane) that are of higher molecular weight than methane and that are important chemical intermediates. The preparation of chemicals and chemical intermediates from methane (natural gas) should not be restricted to those described here but should be regarded as some of the building blocks of the petrochemical industry (Fig. 12.8) (Lowenheim and Moran, 1975; Sasma and Hedman, 1984).

The availability of hydrogen from catalytic reforming operations has made its application economically feasible in a number of crude oil–refining operations. Previously, the chief sources of largescale hydrogen (used mainly for ammonia manufacture) were the cracking of methane (or natural gas) and the reaction between methane and steam. In the latter, at 900°C to 1000°C (1650°F to 1830°F) conversion into carbon monoxide and hydrogen results:

$$CH_4 + H_2O \rightarrow CO + 3H_2$$

If this mixture is treated further with steam at 500°C over catalyst, the carbon monoxide present is converted into carbon dioxide and more hydrogen is produced:

$$CO + H_2O \rightarrow H_2 + CO_2$$

The reduction of carbon monoxide by hydrogen is the basis of several syntheses, including the manufacture of methanol and higher alcohols—the Fischer-Tropsch reaction (Chapter 8). Indeed, the synthesis of hydrocarbon derivatives by the Fischer-Tropsch reaction has received considerable attention:

$$nCO + 2nH_2 \rightarrow (CH_2)_n + nH_2O$$

TABLE 12.7 Chemicals from acetylene[a].

Chemical	Comments
Acetaldehyde	Water added to acetylene produces acetaldehyde, used as a solvent and flavoring in food, cosmetics, and perfumes.
Acrylate esters	Acetylene reacts with carbon monoxide and alcohol forming acrylate esters, used in the manufacture of plexiglass and safety glasses.
Acrylonitrile	Hydrogen cyanide added to acetylene produces acrylonitrile, used as an intermediate in the production of nitrile rubbers, acrylic fibers, and insecticides.
1,4-Butynediol	Formaldehyde added to acetylene produces 1, 4-butynediol, which is then hydrogenated to 1, 4-butanediol and used as a chain extender for polyurethane. These resins include urethane foams for cushioning material, carpet underlay and bedding, insulation in refrigerated appliances and vehicles, sealants, caulking and adhesives.
1,2-Dichloroethane	Chlorine added to acetylene forms 1,2-dichloroethylene, used primarily as a feedstock for vinyl chloride monomer, which, in turn, is the monomer for the widely used plastic, polyvinyl chloride.
Polyacetylene	Acetylene can polymerize forming polyacetylene. The delocalized electrons of the alternating single and double bonds between carbon atoms give polyacetylene its conductive properties. Doping of polyacetylene makes this polymer a better conductor. Polyacetylene is used in rechargeable batteries that could be used in electric cars and could also replace copper wires in aircraft because of its light weight.
Polydiacetylene	Polydiacetylene is also a polymer of the future. It behaves as a photoconductor and could be used for time-temperature indicators or monitoring of irradiation.
Vinyl acetate	Acetic acid added to acetylene forms vinyl acetate, used as an intermediate in polymerized form for films and lacquers.
Vinyl chloride monomer	First produced by reacting acetylene with hydrogen chloride. Acetylene-based technology predominated until the early 1950s.

Continued

TABLE 12.7 Chemicals from acetylene[a].—cont'd

Chemical	Comments
	Due to the high energy input needed in the acetylene-based process and the hazards of handling acetylene, the ethylene-based route has become the predominant one. The acetylene-based route does have its advantages, such as in countries where there is a shortage of ethylene cracker feedstock.
Vinyl ether	Alcohol added to acetylene yields vinyl ether that can be used as an anesthetic.

[a]*Listed alphabetically.*

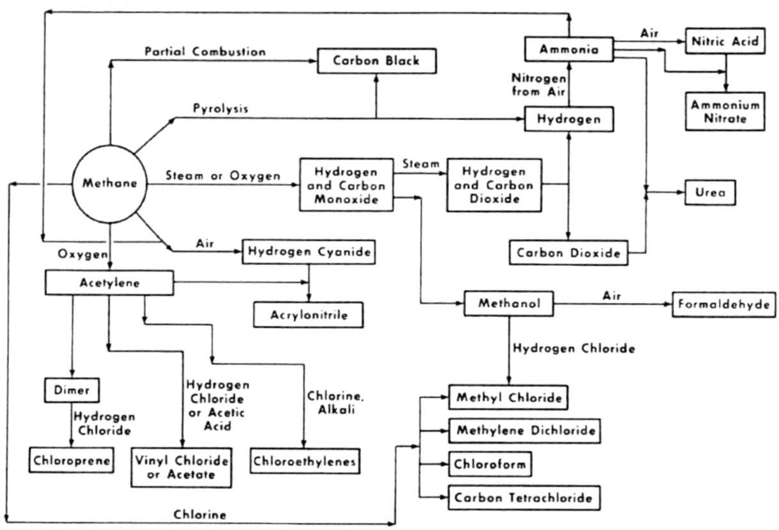

FIGURE 12.8 Chemicals from methane.

This occurs in the temperature range 200°C–350°C (390°F–660°F), which is sufficiently high for the water-gas shift to take place in presence of the catalyst:

$$CO + H_2O \rightarrow CO_2 + H_2$$

The major products are olefin derivatives and paraffin derivatives, which together with some oxygen-containing organic compounds in the product mix may be varied by changing the catalyst or the temperature, pressure, and carbon monoxide-hydrogen ratio.

The hydrocarbon derivatives formed are mainly aliphatic, and on a molar basis methane is the most abundant; the amount of higher molecular weight hydrocarbon derivatives usually decreases gradually with increased molecular weight. *Iso*-paraffin formation is more extensive over zinc oxide (ZnO) or thoria (ThO_2) at 400°C–500°C (750°F–930°F) and at higher pressure. Paraffin waxes are formed over ruthenium catalysts at relatively low temperatures (170°C–200°C, 340°F–390°F), high pressures (1500 psi), and with a carbon monoxide-hydrogen ratio. The more highly branched product made over the iron catalyst is an important factor in a choice for the manufacture of automotive fuels. On the other hand, a high-quality diesel fuel (paraffin character) can be prepared over cobalt.

Secondary reactions play an important part in determining the final structure of the product. The olefin derivatives produced are subjected to both hydrogenation and double-bond shifting toward the center of the molecule; *cis* and *trans* isomers are formed in approximately equal amounts. The proportion of straight-chain molecules decreases with rise in molecular weight, but even so they are still more abundant than branched-chain compounds up to approximately C_{10}. The small amount of aromatic hydrocarbon derivatives found in the product covers a wide range of isomer possibilities. In the C_6 to C_9 range, benzene, toluene, ethylbenzene, xylene, *n*-propyl- and *iso*-propylbenzene, methyl ethylbenzenes, and trimethylbenzenes have been identified; naphthalene derivatives and anthracene derivatives are also present.

In addition, acetylene is prepared by passing methane through an electric arc. When methane is made to react with chlorine (gas), various chloromethane derivatives are produced: chloromethane (CH_3Cl), dichloromethane (CH_2Cl_2), chloroform ($CHCl_3$), and carbon tetrachloride (CCl_4). However, the use of these chemicals is declining—acetylene may be replaced by less costly substitutes, and the chloromethane derivatives are used less often because of health and environmental concerns. It must be recognized that there are many other options for the formation of chemical intermediates and chemicals from methane by indirect routes, i.e., where other compounds are prepared from methane which are then used as further sources of petrochemical products.

7. Chemicals from synthesis gas

Synthesis gas is a mixture of carbon monoxide (CO) and hydrogen (H_2) that is the beginning of a wide range of chemicals (Fig. 12.9). The production of synthesis gas, i.e., mixtures of carbon monoxide and hydrogen—the Fischer-Tropsch reaction—has been known since the 19th century (Chapter 8). In fact, synthesis gas may be produced from a variety of feedstocks, such as from natural gas (CH_4) by the steam reforming process. The first step in the process is to ensure that the methane feedstock is free from hydrogen sulfide. The purified gas is then mixed with steam and introduced to the first reactor (primary reformer). The reactor is constructed from vertical stainless steel tubes lined in a refractory furnace. The steam to natural gas ratio varies from

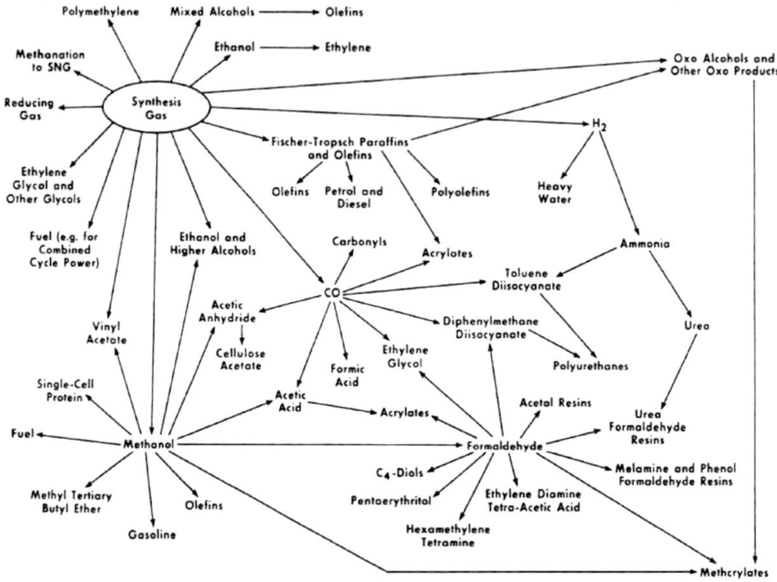

FIGURE 12.9 Production of chemicals from synthesis gas.

four to five depending on natural gas composition (natural gas may contain ethane and heavier hydrocarbon derivatives) and the pressure used.

It is only with the commercialization of the Fischer-Tropsch reaction that the importance of synthesis gas has been realized. The thermal cracking (pyrolysis) of crude oil or fractions thereof was an important method for producing gas in the years following its use for increasing the heat content of water gas. Many water-gas sets operations converted into oil-gasification units; some have been used for base-load city gas supply but most find use for peak-load situations in the winter.

In addition to the gases obtained by distillation of crude oil, further gaseous products are produced during the processing of naphtha and middle distillate to produce gasoline. Hydrodesulfurization processes involving treatment of naphtha, distillates, and residual fuels and from the coking or similar thermal treatment of vacuum gas oils and residual fuel oils also produce gaseous products. The chemistry of the oil-to-gas conversion has been established for several decades and can be described in general terms although the primary and secondary reactions can be truly complex. The composition of the gases produced from a wide variety of feedstocks depends not only on the severity of cracking but often to an equal or lesser extent on the feedstock type. In general terms, gas heating values are on the order of 950−1350 Btu/ft^3.

A second group of refining operations which contribute to gas production are the catalytic cracking processes, such as fluid-bed catalytic cracking, and other variants, in which heavy gas oils are converted into gas, naphtha, fuel oil, and coke. The catalysts will promote steam-reforming reactions that lead to a product gas containing more hydrogen and carbon monoxide and fewer unsaturated hydrocarbon products than the gas product from a noncatalytic process. The resulting gas is more suitable for use as a medium heat-value gas than the rich gas produced by straight thermal cracking. The catalyst also influences the reactions rates in the thermal cracking reactions, which can lead to higher gas yields and lower tar and carbon yields.

Almost all crude oil fractions can be converted into gaseous fuels, although conversion processes for the higher-boiling fractions require more elaborate technology to achieve the necessary purity and uniformity of the manufactured gas stream. In addition, the thermal yield from the gasification of higher-boiling feedstocks is invariably lower than that of gasifying low-boiling naphtha or liquefied petroleum gas since, in addition to the production of synthesis gas components (hydrogen and carbon monoxide) and various gaseous hydrocarbon derivatives, heavy feedstocks also yield some tar and coke.

Synthesis gas can be produced from heavy oil by partially oxidizing the oil:

$$[2CH]_{crude\ oil} + O_2 \rightarrow 2CO + H_2$$

The initial partial oxidation step consists of the reaction of the feedstock with a quantity of oxygen insufficient to burn it completely, making a mixture consisting of carbon monoxide, carbon dioxide, hydrogen, and steam.

Success in partially oxidizing high boiling feedstocks depends mainly on details of the burner design. The ratio of hydrogen to carbon monoxide in the product gas is a function of reaction temperature and stoichiometry and can be adjusted, if desired, by varying the ratio of carrier steam to oil fed to the unit.

Many chemicals are produced from synthesis gas as a consequence of the high reactivity associated with hydrogen and carbon monoxide gases, the two constituents of synthesis gas. The reactivity of this mixture was demonstrated during World War II, when it was used to produce alternative hydrocarbon fuels using Fischer-Tropsch technology (Speight, 2013, 2019a).

Synthesis gas is also an important building block for aldehydes from olefin derivatives. The catalytic hydroformylation reaction (Oxo reaction) is used with many olefin derivatives to produce aldehydes and alcohols of commercial importance. The two major chemicals based on synthesis gas are ammonia and methanol. Each compound is a precursor for many other chemicals. From ammonia, urea, nitric acid, hydrazine, acrylonitrile, methylamines, and many other minor chemicals are produced. Each of these chemicals is also a precursor to many other chemicals.

References

Albright, L.F., Crynes, B.L., 1976. Industrial and Laboratory Pyrolysis. In: Symposium Series No. 32. Am. Chem. Soc., Washington, DC.

Chenier, P.J., 1992. Survey of Chemical Industry, second revised ed. VCH Publishers Inc., New York.

Gary, J.G., Handwerk, G.E., Kaiser, M.J., 2007. Petroleum Refining: Technology and Economics, fifth ed. CRC Press, Taylor & Francis Group, Boca Raton, Florida.

Goldstein, R.F., 1949. The Petrochemical Industry. E. & F. N. Spon, London.

Griesbaum, K., Behr, A., Biedenkapp, D., Voges, H.-W., Garbe, D., Paetz, C., Collin, G., Mayer, D., Höke, H., 2002. Hydrocarbons. In: Ullmann's Encyclopedia of Industrial Chemistry 2002. Wiley-VCH, Weinheim, Germany.

Hahn, A.V., 1970. The Petrochemical Industry: Market and Economics. McGraw-Hill, New York.

Hsu, C.S., Robinson, P.R. (Eds.), 2017. Handbook of Petroleum Technology. Springer International Publishing AG, Cham, Switzerland.

Lowenheim, F.A., Moran, M.K., 1975. Industrial Chemicals. John Wiley & Sons, New York.

Matar, S., Hatch, L.F., 2001. Chemistry of Petrochemical Process, second ed. Butterworth-Heinemann, Woburn, Massachusetts.

March, J., 1985. Advanced Organic Chemistry: Reactions, Mechanisms and Structure, third ed. John Wiley & Sons Inc., Hoboken, New Jersey.

Meyers, R.A., 2005. Handbook of Petrochemicals Production Processes. McGraw-Hill, New York.

Mokhatab, S., Poe, W.A., Speight, J.G., 2006. Handbook of Natural Gas Transmission and Processing. Elsevier, Amsterdam, Netherlands.

Oblad, A.G., Davis, H.B., Eddinger, R.T., 1979. Thermal Hydrocarbon Chemistry. In: Advances in Chemistry Series No. 183. Am. Chem. Soc., Washington, DC.

Parkash, S., 2003. Refining Processes Handbook. Gulf Professional Publishing, Elsevier, Amsterdam, Netherlands.

Sasma, M.E., Hedman, B.A., 1984. Proceedings. International Gas Research Conference. Gas Research Institute, Chicago Illinois.

Speight, J.G., 2013. The Chemistry and Technology of Coal, third ed. CRC Press, Taylor and Francis Group, Boca Raton, Florida.

Speight, J.G. (Ed.), 2011. The Biofuels Handbook. The Royal Society of Chemistry, London, United Kingdom.

Speight, J.G., 2014. The Chemistry and Technology of Petroleum, fifth ed. CRC Press, Taylor and Francis Group, Boca Raton, Florida.

Speight, J.G., 2017. Handbook of Petroleum Refining. CRC Press, Taylor and Francis Group, Boca Raton, Florida.

Speight, J.G., 2019a. Handbook of Petrochemical Processes. CRC Press, Taylor and Francis Group, Boca Raton, Florida.

Speight, J.G., 2019b. Natural Gas: A Basic Handbook, second ed. Gulf Publishing Company, Elsevier, Cambridge, Massachusetts.

Stefanidakis, G., Gwyn, J.E., 1993. Alkylation. In: McKetta, J.J. (Ed.), Chemical Processing Handbook. CRC Press, Taylor & Francis Group, Boca Raton, Florida, pp. 80–138.

Steiner, H., 1961. Introduction to Petroleum Chemicals. Pergamon Press, New York.

Wittcoff, H.A., Reuben, B.G., 1996. Industrial Organic Chemicals. John Wiley & Sons Inc., New York.

Chapter 13

Pharmaceuticals

1. Introduction

The modern pharmaceutical industry can trace its beginnings to local apothecaries—now called chemists in the United Kingdom and pharmacists in the United States—who expanded from their traditional role distributing botanical drugs such as morphine and quinine to wholesale manufacture in the mid-1800s. By the late 1880s, German dye manufacturers had perfected the purification of individual organic compounds from coal tar and other mineral sources and had also established fundamental methods in organic chemical synthesis. The development of synthetic chemical methods allowed scientists to systematically vary the structure of chemical substances, and growth in the emerging science of pharmacology expanded their ability to evaluate the biological effects of these structural changes. It is from these early beginning and the recognition of the wealth of chemical that could be produced from crude oil that led to the rapid expansion of the medicines from crude oil industry as an extension of the petrochemical industry Purdy (1967).

From the previous chapters, it is obvious that petrochemicals play many roles in modern life because they are used to create resins, films, and plastics. In addition, petrochemicals also play a major role in the production of medicines because they are used to produce chemicals such as (i) phenol and cumene that are used to create a substance that is essential for manufacturing penicillin—an extremely important antibiotic—and aspirin, (ii) petrochemical resins that are used to purify medicines, speeding up the manufacturing process, (iii) resins made from petrochemicals, which are used in the manufacture of medicines including treatments for aids, arthritis, and cancer, (iv) plastics and resins that are used to make devices such as artificial limbs and skin, and (v) plastics that are used to make a wide range of medical equipment including bottles, disposable syringes, and much more (Hess et al., 2011).

Thus, it would be remiss not to mention the role of petrochemical intermediates in the manufacture of pharmaceutical products. Petrochemical solutions and petrochemicals are the second-phase products and solutions that originate from crude oil, following a number of refining methods (Parkash, 2003; Gary et al., 2007; Speight, 2014, 2017, 2019; Hsu and Robinson, 2017). Crude oil is the fundamental ingredient that offers petrochemical products and

Handbook of Industrial Hydrocarbon Processes. https://doi.org/10.1016/B978-0-12-809923-0.00013-8

by-products after an extensive procedure of refining that takes place in various oil refineries.

Petrochemicals play an important role in the production of medicines. For example, most medicines contain two types of ingredients: (i) the active ingredient that is composed of one or more compounds manufactured synthetically or extracted and purified from plant or animal sources and is the chemical that reacts with your body to produce a therapeutic effect and (ii) the inactive ingredients that are typically the additives present in the medication, which are normally inactive/inert and which may have been added as preservatives, flavoring agents, coloring, sweeteners, and sorbents.

Also, for the purposes of this chapter, there are two general definitions that are used: (i) a medicine or medication, which is a chemical that is available as an over-the-counter purchase at a pharmacy and (ii) a drug, which is available only by prescription from an authorized person. Over-the-counter medicine (also known as OTC or medicine) is a nonprescription medicine. All of these terms refer to medicine that you can buy without a prescription. They are safe and effective when you follow the directions on the label and as directed by your health care professional. Examples of the former (over-the-counter medicines) are based on benzene and naphthalene (Table 13.1) through

TABLE 13.1 Examples of readily available over-the-counter medications based on benzene.

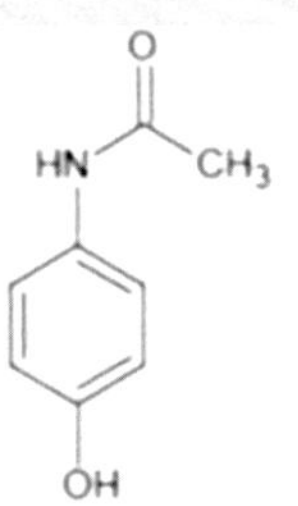

Acetaminophen

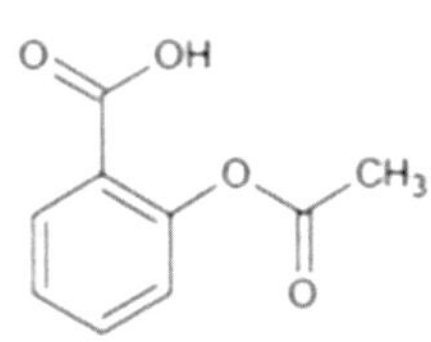

Aspirin

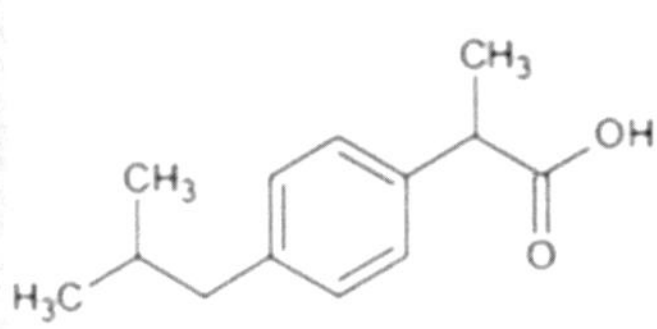

Naproxen

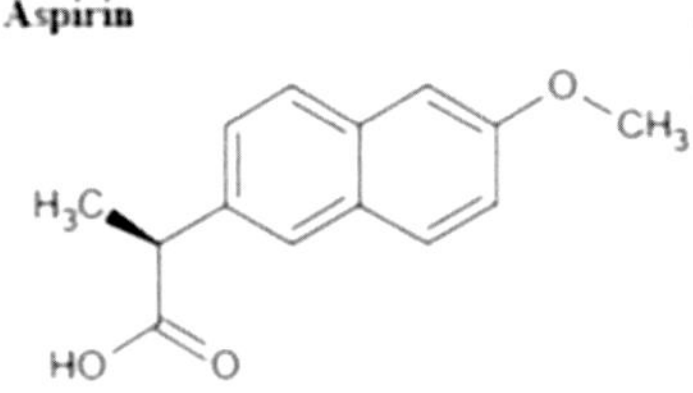

Ibuprofen

published synthesis, while the latter (i.e., medicines that are available only by prescription) are not included in the subject of this chapter.

In addition, many synthetic routes to medicines are not published because of proprietary issues as well as dangerous-to-health issues. There are also the questions of nomenclature that can be troublesome as well as confusing. Because of proprietary issues, even over-the-counter medications have names that often bear no relationship to the actual chemical for industrial usage. In all cases, where possible, the trade name and the chemical name of the medication are presented.

A word of caution should be added here. Although relatively easy to obtain, over-the-counter medications can still carry a risk, even though they do not require a prescription. There is the possibility of side effects, interactions with other medications, or harm due to excessive doses. All patients should consult with their doctor, pharmacist, or other health care provider if they have additional questions concerning use of over-the-counter medications.

Thus, medications (usually referred to as *drugs*) that change behavior patterns are not included in this chapter. It is not the purpose of this chapter to produce methods by which drugs (especially harmful medications, often referred to as *drugs*) can be synthesized but to present to the reader a section of the published synthetic methods that result in the production of commonly used medications. For this, it will also point out the starting materials or other constituents that originated from petrochemical processes.

A medicine is a chemical substance that has known biological effects on humans or other animals, used in the treatment, cure, mitigation, prevention, or diagnosis of disease or used to enhance physical or mental well-being. Medicines may be used for a limited duration or on a regular basis for chronic disorders and are generally taken to cure and/or relieve any symptoms of an illness or medical condition or may be used as prophylactic medicines. One or more of the constituents of the medicine usually interacts with either normal or abnormal physiological process in a biological system and produces a desired and positive biological action. However, if the effect causes harm to the body, the medicine is classified as a poison and is no longer a medication. The medications can treat different types of diseases such as infectious diseases, noninfectious diseases, and nondiseases (alleviation of pain, prevention of pregnancy, and anesthesia). Many of the modern medications are prepared from petrochemical starting materials (Table 13.2).

Petrochemicals have contributed to the development of many medications for diverse indications. While most US pharmaceutical companies have reduced or eliminated in-house natural product groups, new paradigms and new enterprises have evolved to carry on a role for natural products in the pharmaceutical industry. Many of the reasons for the decline in popularity of natural products are being addressed by the development of new techniques for screening and production. This chapter aims to inform pharmacologists of

TABLE 13.2 Selection of common hydrocarbon products from crude oil used in the pharmaceutical industry.

Chemical	Processes		
Benzene	C		
Cyclohexane	C		
n-Heptane	C	F	B
n-Hexane	C	F	B
Toluene	C	F	B
Xylene isomer	C		

C, Chemical synthesis, *F*, Fermentation, *B*, Biological or natural extraction.

current strategies and techniques that make petrochemicals a continuing and viable strategic choice for use in medication synthesis programs.

As early as 1500 B.C., the use of asphalt for medicinal purposes and (when mixed with beer) as a sedative for the stomach has been recorded. It is also recorded in the code of Hammurabi that hot asphalt was to be poured over the ear of a miscreant as a form of punishment. In more modern times, medicinal oil (sometimes referred to paraffin oil) distilled from crude oil was prescribed to lubricate the alimentary tract where coal dust was likely to collect. From these humble beginnings, crude oil has, through the production of petrochemicals, become a major contributor to the pharmaceutical industry. For example, the first analgesics and antipyretics, exemplified by phenacetin and acetanilide, were simple chemical derivatives of aniline and p-nitrophenol, both of which were by-products from coal tar and not from crude oil. An extract from the bark of the white willow tree had been used for centuries to treat various fevers and inflammation. The active principle in white willow, salicin or salicylic acid, had a bitter taste and irritated the gastric mucosa, but a simple chemical modification was much more palatable. This was acetylsalicylic acid, better known as aspirin, the first drug that could be generally administered for a variety of ailments. At the start of the 20th century, the first of the barbiturate family of drugs entered the pharmacopoeia leading to the start of the evolution of the modern pharmaceutical industry (Mahdi et al., 2006; Fuster and Sweeney, 2011; Jones, 2011; Wick, 2012; Aronson, 2013).

Hydrocarbon derivatives are a heterogeneous group of naturally occurring and manifested organic chemicals that are primarily composed of carbon and hydrogen molecules (Forbes, 1958a, 1958b; 1959; Guthrie, 1960; Warren, 2006; Speight, 2014). Hydrocarbon derivatives are quite abundant in modern society; their use includes fuels, paints, paint and spot removers, dry cleaning solutions, lamp oil, lubricants, rubber cement, and solvents. In addition, many volatile substances that contain hydrocarbon derivatives (such as glue and propellants) are commonly abused for their euphoric effects.

The hydrocarbon derivatives can be derived from either crude oil or from wood. Crude oil distillates include kerosene, gasoline, and naphtha, while wood-derived hydrocarbon derivatives include turpentine and pine oil. The length of the chains as well as the degree of branching determines the phase of the hydrocarbon at room temperature; most are liquid, but some short-chain hydrocarbon derivatives (e.g., butane) are gas at room temperature, while other long-chain hydrocarbon derivatives (e.g., waxes) are solid at room temperature. Hydrocarbon derivatives can be classified as being aliphatic, in which the carbon moieties are arranged in a linear or branched chain, or aromatic, in which the carbon moieties are arranged in a ring (Chapter 1) (Clayden et al., 2001). Halogenated hydrocarbon derivatives are a subgroup of aromatic hydrocarbon derivatives, in which one of the hydrogen molecules is substituted by a halogen group. The most important halogenated hydrocarbon derivatives include carbon tetrachloride, trichloroethylene, tetrachloroethylene, trichloroethane, chloroform, and methylene chloride.

A *pharmaceutical drug* (*medicine*, *medication*) is any chemical substance intended for use in the medical diagnosis, treatment, cure, or prevention of disease. On the other hand, a drug (a chemical which is a subcategory of *pharmaceuticals*) is (i) a chemical substance that affects the processes of the mind or body or (ii) a substance used recreationally for its effects on the central nervous system, such as a narcotic. In this respect, a *designer drug* is a new drug of abuse similar in action to an older abused drug and usually created by making a small chemical modification in the older one, while a *mind-altering drug* is a drug that produces an altered state of consciousness. These are *not* the subject of this text. Medications can be classified in various ways, such as, for example, by (i) chemical properties, (ii) mode of administration, (iii) biological system affected, or (iv) therapeutic effects.

Because hydrocarbon derivatives are the simplest organic compounds containing only carbon and hydrogen, they can be straight chain, branched chain, or cyclic molecules (Chapter 1) but generally offer little in the way of pharmaceutical properties. Nevertheless, there are those hydrocarbon derivatives that do have pharmaceutical properties. Thus, for the purposes of this chapter and in the context of this book, medications are classified as (i) hydrocarbon derivatives and (ii) nonhydrocarbon, with the focus of this chapter being on the hydrocarbon medications. The definition and interpretation of hydrocarbon derivatives and nonhydrocarbon derivatives as used here in this chapter is the same as the definition and interpretation given elsewhere in this text (Chapter 1).

The pharmaceutical industry includes the manufacture, extraction, processing, purification, and packaging of chemical materials to be used as medications for humans or animals (Gad, 2008). Pharmaceutical manufacturing is divided into two major stages: the production of the active ingredient or medicine (primary processing or manufacture) and secondary processing, the conversion of the active medicines into products suitable for administration. The products are available as tablets, capsules, liquids (in the form of solutions, suspensions, emulsions, gels, or injectables), creams (usually oil-in-water emulsions), ointments (usually water-in-oil emulsions), and aerosols, which contain inhalable products or products suitable for external use. Propellants used in aerosols include chlorofluorocarbons, which are being phased out. Recently, butane has been used as a propellant in externally applied products.

The major manufactured groups include (i) antibiotics such as penicillin, streptomycin, tetracyclines, chloramphenicol, and antifungals, (ii) other synthetic drugs, including sulfa drugs, antituberculosis drugs, antileprotic drugs, analgesics, anesthetics, and antimalarials, (iii) vitamins, (iv) synthetic hormones, (v) glandular products, (vi) drugs of vegetable origin such as quinine, strychnine and brucine, emetine, and digitalis, (vii) glycosides, and (viii) vaccines.

Other pharmaceutical chemicals such as calcium gluconate, ferrous salts, nikethamide, glycerophosphates, chloral hydrate, saccharin, antihistamines (including meclozine and buclozine), tranquilizers (including meprobamate and chloropromoazine), antifilarials, diethyl carbamazine citrate, and oral antidiabetics, including tolbutamide and chloropropamide and surgical sutures and dressings. The main pharmaceutical groups manufactured include (i) proprietary ethical products or prescription-only medicines, which are usually patented products, (ii) general ethical products, which are basically standard prescription-only medicines made to a recognized formula that may be specified in standard industry reference books, and (iii) over-the-counter, or nonprescription, products. For those readers interested in the synthesis of medications available by prescription, there are descriptions available (Karaman, 2015; Flick et al., 2017).

The principal manufacturing steps are (i) preparation of process intermediates, (ii) introduction of functional groups, (iii) coupling and esterification, (iv) separation processes such as washing and stripping; and (v) purification of the final product. Additional product preparation steps include granulation; drying; tablet pressing, printing, and coating; filling; and packaging.

Finally, it is not the purpose of this chapter to show preference for any type of medication, but it is the purpose to show the methods by which selected over-the-counter medicines can be produced from hydrocarbon derivatives.

2. History

The earliest written documents indicate that the use of drugs such as herbs, powders, and poultices had a place in religion and mysticism as well as medicine (Table 13.3) (Forbes, 1958a, 1958b; 1959; Guthrie, 1960; Purdy,

TABLE 13.3 Brief time line of the use of drugs.

4000 BC	The Sumerians use opium, suggested by the fact that they have an ideogram for it with the meaning joy or rejoicing.
3500 BC	Earliest historical record of the production of alcohol—the description of a brewery in an Egyptian papyrus.
3000 BC	Approximate date of the supposed origin of the use of tea in China.
2500 BC	Earliest historical evidence of the eating of poppy seeds among the Lake Dwellers in Switzerland.
2000 BC	Earliest record of prohibitionist teaching, by an Egyptian priest, who writes forbids a pupil to enter a tavern where beer is sold.
300 BC	Theophrastus (371–287 BC), Greek naturalist and philosopher, records what has remained as the earliest reference to the use of poppy juice.
350 AD	Earliest mention of tea in a Chinese dictionary.
1000 AD	Opium is widely used in China and the Far East.

1967; Bough and Trammel, 2006). In the period 3000 to 4000 BC, the Chinese documented the use of herbal medicine to cure illness in humans and valuable animals and made early discoveries related to the medicinal values of herbs—many of which are still recognized in modern pharmacy. Sumerian clay tablets from 2100 BC (recovered for the level Ur III, approximately 2050 BC) contain pharmacologic descriptions involving ingredients such as salt, saltpeter, thyme, seeds, roots, and bark. Early Hindus used snake root to treat mental disorders, and Egyptians used opium to treat gastrointestinal disorders.

Hippocrates (460–375 BC) a Greek physician (after whom the Hippocratic Oath is named) believed that there was limited use for drugs. He noted that sick people generally got well even if drugs were not used. However, the scientific basis for medicine was formed shortly after his time by the Greek philosopher Aristotle (384–322 BC) based his ideas on biology-related observations and systematic classifications and recorded much of what was known about natural science at the time, including similarities and differences between the biology of humans and animals. His student Theophrastus, known as the father of botany, systematically classified medicinal plants. Dioscorides (40–90 AD), from Asia Minor, worked with medicinal plants as well as drugs from mineral and animal sources and recorded drug names, sources, identification, preparation, dosage, and usage. His work established a structure used and developed for future pharmacopeias. Also from Asia Minor, Galen (130–200 AD) practiced and taught pharmacy and medicine. His contributions focused on the correct compounding methods and are still useful in the modern world.

During the Middle Ages, much emphasis was placed on combining multiple ingredients in medicines so that they could be used for any ailment. However, the Middle Ages produced little advancement in the area of pharmacy in Europe. However, during this time, the Arabian scientists and medical doctors contributed to drug knowledge by recording new information related to the preparation of drug and various medications. In 1498, the first official *pharmacopeia* was published in Florence, Italy. The goal was to provide a source for uniform pharmaceutical standards. In 1606, the Society of Apothecaries of London was formed. At that time, an apothecary was similar to a modern pharmacist, preparing and selling medicinal substances. When King James I granted a charter to the society in 1617, he created the first official organization of pharmacists in the Anglo-Saxon world. During the 18th century, pharmaceutical and medical services were provided in the America (which would become the United States) by governors, religious leaders, and educators. These men used imported drugs as well as drugs derived from local plants. In 1821, the Philadelphia College of Pharmacy was founded; it was the first association of pharmacists in America. As the development of drugs continued, pharmaceutical education developed with a stronger focus on chemistry and standardization.

Scientists began developing biological compounds in the late 1700s and throughout the 1800s. The first diseases these drugs affected were smallpox, diphtheria, and tetanus. Louis Pasteur (1822–95), who is responsible for numerous scientific achievements, discovered that weakened forms of microbes could be used as immunizations for more virulent forms of microbes. His work led to the development of vaccines for chicken cholera, anthrax, and swine erysipelas as well as modern rabies vaccines for humans and dogs. In 1903, the first US government inspection and licensure policies were implemented for those manufacturing viruses, serums, toxins, and analogous products. The Pure Food and Drug Act, passed in 1906, gave the US government the ability to enforce United States Pharmacopeia (USP) standards and to bring action against those who adulterated or misbranded drugs. This act was prompted by the exposure of popular patent medicines for humans and animals as largely ineffective—and sometimes harmful—concoctions. Until the 1920s, in some medical schools, materia medica (diluted pharmacy courses) was taught, and the term materia medica has since been replaced by the term *pharmacology*, which was the early study of compounding and preparing drugs, usually from natural sources.

The introduction of chemotherapy in 1936 and overall drug industry growth after World War II kept the momentum going. As these changes occurred, a greater emphasis was placed on pharmacology in the medical curriculum. Unfortunately, the veterinary field lagged behind in drug development because of economic factors as well as the fact that the profession was much smaller. After 1950, scientific exploration in the veterinary drug industry began to increase, and although economic and societal factors still contribute to slower progress in this area, significant growth has occurred. During the 20th century and into the 21st century, remarkable changes have occurred in the production and use of drugs.

3. Hydrocarbon pharmaceuticals

This section deals with the synthesis of the bulk fractions that have been used and, in some countries, continue to be used as medications as well the individual molecular active ingredients of medications and their usage in drug formulations to deliver the prescribed dosage. Formulation is also referred to as galenical production. A galenical is a simple cure in the form of a vegetable or herbal remedy as prescribed as by Galen (Aelius Galenus or Claudius Galenus or better known to the Western World as Galen of Pergamon, 129 to 217 AD), a Greek physician, surgeon, and philosopher in the Roman Empire.

The crude oil industry is first encountered in the archaeological record near Hit (Tuttul), which is now Iraq. Hit is on the banks of the Euphrates river and is the site of an oil seep known locally as The Fountains Of Pitch. There, the bitumen was quarried for use as mortar between building stones as early as 6000 years ago and was also used as a waterproofing agent for baths, pottery, and boats. The Babylonians caulked their ships with bitumen, and in Mesopotamia around 4000 BC, bitumen was used as caulking for ships, a setting for jewels and mosaics, and an adhesive to secure weapon handles. On the human side of bitumen use, the Egyptians used it for embalming while the ancient Persians, 10th Century Sumatrans, and pre-Columbian natives of the Americas believed that crude oil had medicinal benefits. Although it is not a true hydrocarbon, it is a hydrocarbonaceous material, which means it contains other atoms in addition to carbon and hydrogen—the bitumen (in the Bible it is referenced as *slime*) is not the same as the refinery product known as asphalt (Parkash, 2003; Gary et al., 2007; Speight, 2014; Hsu and Robinson, 2017; Speight, 2014, 2014, 2017, 2019). Bitumen is a natural-occurring material that occurs in tar sand formations and that has seeped from crude oil formation.

Typically, the bitumen, which has been referenced in ancient texts, is unless recovered from a tar sand formation, equivalent to an atmospheric residuum insofar as it is found as a seepage on the surface and is crude oil from which the more volatile constituents have escaped by evaporation. The bitumen obtained from the area of Hit (Tuttul) in Iraq (Mesopotamia) or as blocks floating on the Dead Sea are examples of such occurrences (Abraham, 1945; Forbes, 1958a, 1958b, 1959). Typically, asphalt is produced from crude oil as the treated (usually air-blown) vacuum residuum (Parkash, 2003; Gary et al., 2007; Speight, 2011, 2014; Hsu and Robinson, 2017; Speight, 2017).

Surface manifestations of bitumen are found in Middle Eastern countries as seepages from rocks. This bitumen has been extensively employed for a variety of uses, including in medicine. The historical evidence on the medicinal uses of bitumen spans at least 3000 years and, while many of the attributes of bitumen as a drug in antiquity are not based on medical evidence, certain treatments using bitumen may have been confirmed by modern medicine. For example, the application of bitumen of asphalt as a therapy for skin diseases, in humans and in animals, has been borne out in modern times by extensive experimentation. The nature of the active ingredient or ingredients in the bitumen has not been investigated as yet no constituents have been identified with any degree of certainty. Also, it has long been thought, for instance, that bitumen from what is now Iraq and Syria was exported to Egypt for embalming purposes from at least the early Ptolemaic period—the accession of Soter after the death of Alexander the Great in 323 BC and which ended with the death of Cleopatra and the Roman conquest of Egypt in 30 BC.

Furthermore, when going further back into history, it has become evident that bitumen was used widely in the Middle East, especially in the Zagros mountains of Iran (Connan, 1999). Ancient people from northern Iraq, southwest Iran, and the Dead Sea area extensively used this ubiquitous natural resource until the Neolithic period (7000 to 6000 BC). Evidence of earlier use has been recently documented in the Syrian Desert near El Kowm, where bitumen-coated flint implements, dated to 40,000 BC (Mousterian period), have been unearthed. This discovery at least proves that bitumen was used by Neanderthal populations as hafting material to fix handles to their flint tools. Numerous testimonies, proving the importance of this crude oil–based material in Ancient civilizations, were brought to notice by the excavations conducted in the Near East as of the beginning of the century.

The early records show that bitumen was largely used in Mesopotamia and Elam as mortar in the construction of palaces (e.g., the Darius Palace in Susa), temples, ziggurats (e.g., the so-called Tower of Babel in Babylon), terraces (e.g., the famous Hanging Gardens of Babylon), and exceptionally for roadway coating (e.g., the processional way of Babylon). Since Neolithic times, bitumen served to waterproof containers (baskets, earthenware jars, storage pits), wooden posts, palace grounds (e.g., in Mari and Haradum), reserves of lustral waters, bathrooms, and palm roofs. Mats, sarcophagi, coffins, and jars, used for funeral practices, were often covered and sealed with bitumen. Reed and wood boats were also caulked with bitumen. Bitumen was also a widespread adhesive in antiquity and served to repair broken ceramics and fix eyes and horns on statues (e.g., at Tell al-'Ubaid around 2500 BC). Decorations with stones, shells, mother of pearl, on palm trees, cups, ostrich eggs, musical instruments (such as the lyre that, reputedly, belonged to the Queen) and other items, such as rings, jewelry, and games, have been excavated from the Royal tombs in Ur (Connan, 1999).

Bitumen was also considered as a powerful remedy in medical practice, especially as a disinfectant and insecticide, and was used by the ancient Egyptians to prepare mixtures to embalm the corpses of their dead. Recent geochemical studies on more than 20 balms from Egyptian mummies from the Intermediate, Ptolemaic, and Roman periods have revealed that these balms are composed of various mixtures of bitumen, conifer resins, grease, and beeswax. The physician Ibn al-Baitar described as a preservative for embalming the dead, in order that the dead bodies might remain in the state in which they were buried and neither decay nor change. In addition, the historical records show that bitumen was used since ancient times for cosmetic, art, and the caulk of boats and was reputed to be useful to cure varying pulmonary, digestive, ear–nose-throat troubles, and even to set fractured bones (Bourée et al., 2011).

In medicine, Muslim physicians used crude oil and bitumen for pleurisy and dropsy—the patient was given *bituminous water* to drink—and for various skin ailments and wounds. There is also fragmentary evidence that hot bitumen was used to cauterize the wound resulting from a severed limb—as a side note, medieval physicians used fire as the cauterizing agent. Whether or not the bitumen-treated patient survived is not clear. Another law of the time suggests the use of hot bitumen as a *curative* agent not in the sense of a medicinal cure but as a punishment. The hot bitumen was to be poured over the head of the miscreant. The record do not show if the miscreant survived as a bald person after the bitumen was removed or if the miscreant actually survived the treatment.

For example, an early mention of the use of bitumen as a punishment appears in orders that Richard I of England (also known as Richard the Lionheart) issued to his navy when he set out of the Holy Land in 1189: (quote) *Concerning the lawes and ordinances appointed by King Richard for his navie the forme thereof was this … item, a thiefe or felon that hath stolen, being lawfully convicted, shal have his head shorne, and boyling pitch poured upon his head, and feathers or downe strawed upon the same whereby he may be knowen, and so at the first landing-place they shall come to, there to be cast up* (end quote) (Hakluyt, 1582).

In other literature, the name *shilajit* occurs frequently and is the Sanskrit name for Asphaltum (bitumen, also called *mineral pitch, vegetable asphalt, shilajita, guj, kalmadam, perangyum, rel-yahudi*, and *silaras*) that refers to a curative agent as an analgesic, antiinflammatory, antibacterial, cholagogic, diuretic, wound cleaner, expectorant, respiratory stimulant, and general health medicine, amongst a host of other effects (Jonas, 2005).

From that time, the ancient literature acts as a record of the use of crude oil. In fact, it was the Persian scientist Ibn Sina (who lived approximately from 980 to 1037 AD), known in the West as Avicenna, who discussed medicinal crude oil in his enormously influential encyclopedia of medicine. The translation of this work into Latin spread that knowledge into Europe, where it reached Constantinus Africanus (who lived approximately from 1020 to 1087 AD), who may have been the first Latin writer to use the word *petroleum*—the word was also used by Georgius Agricola (Georg Bauer) in his work entitled *De Natura Fossilium* (published 1546). From that time, there was a tradition of employing crude oil in medicine, which included concoctions recommended for eye diseases, reptile bites, respiratory problems, hysteria, and epilepsy. Mixing crude oil and the ashes of cabbage stalks was recommended for the treatment of scabies, and a preparation of crude oil was prescribed to warm the brain by applying it to the forehead. Marco Polo (who lived approximately from 1254 to 1324) reported that bitumen was used in the Caspian Sea region to treat camels for mange, and the first oil exported from Venezuela (in 1539) was intended as a gout treatment for the Holy Roman Emperor Charles V (reigned 1519–56).

The native North Americans collected oil for medicines, and European settlers found its presence in the water supplies a contamination, but they learned to collect it to use as fuel in their lamps. Native Americans also traded crude oil that they obtained from oil seeps in upstate New York among other places. The Seneca tribe traded oil for so long that all crude oil was referred to as Seneca Oil, which was reputed to have great medicinal value. In fact, in 1901, a crude oil technology text was published, in which it was noted that crude oil was an excellent remedy for diphtheria (Purdy, 1957). The members of the Seneca tribe also used crude oil for body paint and for ceremonial fires.

Several historical factors evolved to change the use of crude oil. The kerosene lamp, invented in 1854, ultimately created the first large-scale demand for crude oil. Kerosene first was made from coal but by the late 1880s most was derived from crude oil. However, in 1859, at Titusville, Pennsylvania, Colonel Edwin Drake drilled the first successful well through rock and produced crude oil. However, bulk oil products from crude oil still find a variety of uses in health and human service roles (i.e., cosmetics), and, because of the imperative of these products, a brief discussion of the various types of products and their roles within the various human communities is also included here—the oil products being considered to be bulk petrochemical products.

In fact, mineral oil and petrolatum are crude oil by-products used in many creams and topical pharmaceuticals. Tar (also called resid, asphalt, pitch), for psoriasis and dandruff, is also produced from crude oil. Most pharmaceuticals are complex organic compounds that have their basis in smaller, simpler precursor organic molecules that are crude oil by-products.

3.1 Mineral oil

The term *mineral oil* (sometime referred to as white oil because of the overall absence of constituents that darken the color of the oil) is used in two different senses: (i) for crude oil (petroleum) as naturally occurring in geological formations and (ii) for a refined by-product of the distillation of crude oil (Parkash, 2003; Gary et al., 2007; Speight, 2014; Hsu and Robinson, 2017; Speight, 2017, 2019). It is the second meaning that is implied here by the use of the term; mineral oils should not be confused with essential oils, which are concentrated, hydrophobic liquids containing volatile aroma compounds and are isolated from (biological) plants.

The first use of the term *mineral oil* was 1771, and before the late 19th century, the chemical science to determine such makeup was unavailable. White oils are highly refined, odorless, tasteless, and have excellent color stability. They are chemically and biological stable and do not support bacterial growth. The inert nature of mineral makes it easy to work with as they lubricate sooth, soften, and hold in moisture in formulations. These oils are used in a variety of product lines such as antibiotics, baby oils, lotions, creams, shampoos, sunscreens, and tissues.

White oils are manufactured from highly refined base oils and consist of saturated paraffin derivatives and cycloparaffin derivatives. The refinement process ensures complete removal of aromatic derivatives, sulfur compounds, and nitrogen compounds. The technologies employed result in products that are highly stable over time besides being hydrophobic, colorless, odorless, and tasteless. White mineral oils are extensively used as bases for pharmaceuticals and personal care products. The inertness of the product offers properties such as good lubricity, smoothness, and softness and resistance to moisture in the formulations. The products are also used in the polymer processing and plastic industry such as polystyrene, polyolefin, and thermoplastic elastomers. The oil controls the melt flow behavior of the finished polymer besides providing release properties. Very often the oils also impart improvement in physical characteristics of the finished product.

Mineral oil is used to designate for liquid by-products in the distillation of crude oil to produce naphtha and other products (Parkash, 2003; Gary et al., 2007; Speight, 2014; Hsu and Robinson, 2017; Speight, 2017, 2019). Mineral oil in this sense is a transparent, colorless, and composed mainly of alkane derivatives (typically 15−40 carbons) and cyclic paraffins. It has a density of around 0.8 g/cm and is currently considered to be of relatively low value. It is, however, still available in some drug stores and can be purchased as low-density crude and high-density grades of crude oil.

In the refining process, the feedstock is hydrotreated and the hydrotreated feedstock exits hydrotreater and is conducted to fractionating column (Parkash, 2003; Gary et al., 2007; Speight, 2014; Hsu and Robinson, 2017; Speight, 2017, 2019). Low-boiling constituents, especially hydrogen sulfide and ammonia, are removed, and the hydrotreated product is then conducted to a second hydrotreater where it is hydrotreated using process parameters that may be the same or different from the hydrotreating conditions in the first hydrotreater. The product from the second hydrotreater is sent to a catalytic dewaxing unit after which the dewaxed product exits dewaxing unit and is sent to a hydrofinishing unit. The product is analyzed for the $Cn{:}C_p$ (naphthene carbon/paraffin carbon) ratio. When the desired $Cn{:}C_p$ ratio is attained typically in the range 0.45–0.65, the medicinal white product is finished.

Mineral oil is any of various colorless, odorless, mixtures of higher molecular weight alkane derivatives from a mineral source, particularly as a distillate from crude oil. The name *mineral oil* by itself is imprecise, having been used for many specific types pf oils over the past several centuries. Other names, similarly imprecise and more physically description rather than chemical descriptive, include *white oil*, *paraffin oil*, *liquid paraffin* (a highly refined medical grade), *paraffinum liquidum* (Latin), and *liquid petroleum*. The product popularly called *baby oil* is a mineral oil to which scented ingredients (perfumes) have been added. More specifically, there are three basic classes of refined mineral oils: (i) paraffin oils, based on n-alkane derivatives, (ii) naphthenic oils, based on cycloalkane derivatives, and (iii) aromatic oils, based on aromatic hydrocarbon derivatives.

Mineral oil with added fragrance is marketed as *baby oil* in the United States, Canada, and Great Britain. While baby oil is primarily marketed as a generic skin ointment, other applications exist in common use. It is often used to alleviate mild eczema (and diaper rash), particularly when the use of corticosteroid cream is not desirable. Mineral or baby oil can also be employed in small quantities (2–three drops daily) to clean inside ears. Over a period of a few weeks, the mineral oil softens dried or hardened earwax so that a gentle flush of water can remove the debris. In the case of a damaged or perforated eardrum, however, mineral oil should not be used, as oil in the middle ear can promote ear infections.

During the middecades of the 20th century, mineral oil was taken orally as a lubricative for the alimentary tract and was particularly in common use by coal miners who ingested a large amount of coal dust during their work. In most countries, the use of mineral oil as laxative is considered obsolete mainly due to its potentially harmful effects on the lungs if accidently aspirated. Furthermore, the oil may be absorbed to a small percentage into internal tissue and cause adverse reactions to the body. In addition, mineral oil temporarily coats the intestines and prevents the uptake of certain essential vitamins and nutrients.

3.2 Paraffin oil

Paraffin oil or liquid paraffin oil is obtained in the process of crude oil distillation (Parkash, 2003; Gary et al., 2007; Speight, 2014, 2017, 2019; Hsu and Robinson, 2017). It is a colorless and odorless oil that is used for varied purposes. In some cases, *paraffin oil* and *mineral oil* are synonymous terms. In other cases, there are subtle, often undetectable differences in composition and properties that can only be determined by careful and detailed analysis of the two.

Liquid paraffin oil is a mineral oil and is a by-product of crude oil distillation. It is transparent, colorless, odorless, and tasteless oil, which is mainly composed of high-boiling alkane derivatives. Liquid paraffin (high-boiling mineral oil) is a mixture of higher molecular weight alkane derivatives and has a number of names, including nujol, adepsine oil, alboline, glymol, medicinal paraffin, or saxol. It has a density of approximately 0.8 g/cm^3. It is not soluble in water and is known to have low reactivity. Paraffin oil and paraffin wax have found a wide range of industrial, medical, and cosmetic uses in the modern times. Liquid paraffin oil usually comes in two forms, high-boiling liquid paraffin oil and low-boiling liquid paraffin oil.

Remembering that there is high-boiling paraffin oil and power boiling paraffin oil (kerosene range), liquid paraffin oil has found numerous applications—from manufacturing candles to the production of cosmetics or beauty products. Several of the most noteworthy uses of liquid paraffin oil are

- As a fuel in burning lamps and used as a fuel in many parts of the world; in this case, the oil is usually a high-boiling kerosene fraction and should not be used for medicinal purposes.
- As a laxative—this oil is not absorbed by the intestinal tract.
- In the manufacture of penicillin and is an important ingredient in many medicated creams, ointments, and balms.
- In the production of paints, dyes, pigments, wax, polythene, and insecticides.
- As a solvent and lubricant in the industrial sector.
- In the textile industry—mainly for spinning, weaving, and lubricating the sewing machines.
- In the cosmetic industry as well for the preparation of a number of solid and liquid brilliantine, moisturizers, cold cream, and lotions, as well as in makeup products such as lipstick, lip balm, and foundation cream.
- In skin treatment, especially in treating diaper rash and eczema and to preserve unstable or reactive substances.

Liquid paraffin (medicinal) used to aid problems of the gastrointestinal tract and it passes through the tract without itself being taken into the body. In the food industry, where paraffin oil may be called *wax*, it can be used as a lubricant in mechanical mixing, applied to baking tins to ensure that loaves are easily released when cooked and as a coating for fruit or other items requiring a “shiny” appearance for sale.

Paraffin oil (boiling in the kerosene boiling range) can pose certain health hazards, especially if it is inhaled or ingested and also due to repeated or prolonged skin exposure. Inhalation of paraffin oil can irritate the respiratory tract, and cause cough, shortness of breath, and occasionally, lead to hydrocarbon pneumonitis. On the other hand, prolonged skin exposure to this oil can cause skin irritation, which can lead to contact dermatitis, especially in individuals who already have skin disorders or diseases. Ingestion of paraffin oil can cause upset of the intestinal tract.

Paraffin oil, which has not been highly refined, is often considered as a carcinogen or cancer causing agent. Therefore, adequate precaution is required, while using paraffin oil. Ideally, liquid paraffin oil should be stored in a cool and well-ventilated place, in a tightly closed container. As some paraffin oil is highly inflammable, be sure to keep it away from any source of heat or ignition and also out of direct sunlight. Lastly, while using this oil for various purposes, be sure to follow the instructions mentioned in the label of the product, regarding the handling and storage of liquid paraffin oil.

3.3 Petroleum jelly

Crude oil products generally defined collectively as petrolatum have a long history in medical applications and that heritage continues as pharmaceutical grade petrolatum constituents are common components in a variety of balms, ointments, creams, moisturizers, haircare products, and other products where a virtually odorless additive that helps retain (and even lock-in) moisture is desired.

Petroleum jelly is a mixture of hydrocarbon derivatives, having a melting point usually close to human body temperature, approximately 37°C (99°F). Petroleum jelly is typically composed of paraffin wax, microcrystalline, wax, and mineral oil in varying amounts. The composition of highly refined constituents and their physical properties vary considerably according to the origin of the raw material and the refining methods. The solid or liquid elements of the hydrocarbon derivatives may contain 16 to 60 carbon atoms with significantly different molecular weights; therefore, the possible structures are extremely varied and their number practically infinite.

Vaseline is a brand name for petroleum jelly-based products that include plain (unaltered petroleum jelly) and a selection of skin creams, soaps, lotions, cleansers, and deodorants to provide various types of skincare and protection by minimizing friction or reducing moisture loss or by functioning as a grooming aid. It is believed that the use of petroleum jelly comes from a product known as *rod wax* that was used by early oil workers in Titusville Pennsylvania to heal cuts and burns. In many countries, the word *vaseline* (*vasenol* in some countries) is used as generic for petroleum jelly.

Petrolatum, a related product to petroleum jelly although the names are often used interchangeably, is a by-product of crude oil refining with a melting point close to body temperature—body temperature ranges from 36.1°C (97°F) to 37.2°C (99°F); in older adults, the typical *body temperature* is lower than 36.2°C (98.6°F). Petrolatum softens on application and forms a water-repellant film around the applied area, creating an effective barrier against the evaporation of the natural moisture from the skin and foreign particles or microorganisms that may cause infection. Petrolatum is odorless and colorless, and it has an inherently long shelf life. These qualities make petrolatum a popular ingredient in skincare products and cosmetics.

Petroleum jelly has been, and continues to be, manufactured from the highest-boiling crude oil refinery fraction (resid). However, because of the occurrence of cancer-forming polynuclear aromatic derivatives (as well as other constituents that are risky to health) in resids, number of cleanup (purification) steps are required to meet the stringent requirements of a product used for direct skin and mouth contact. Although not a comprehensive list, these cleanup steps can include propane deasphalting, hydrogenation, solvent dewaxing, and fixed-bed adsorption using adsorbents such as bauxite and carbon. In the simplest process, paraffin wax is introduced into a reaction vessel after which microcrystalline wax (i.e., wax with a very fine crystalline structure) is added. The mixture is melted with continuous mixing, and the temperature is maintained between 120 and 130°C (248–266°F). Liquid paraffin is added with continuous stirring (150–200 rpm) at constant temperature, so that ingredients are mixed together to form emulsion or gel after which the mass is cooled.

Briefly, bauxite is a complex mineral that is often claimed to be alumina (Al_2O_3) but which, in reality, consists mostly of the aluminum minerals gibbsite [$Al(OH)_3$], boehmite (γ-AlO(OH), and diaspore (α-AlO(OH), mixed with the two oxides of iron, namely goethite and hematite, as well as the aluminum clay mineral kaolinite, as well as small amounts of anatase (TiO_2) and ilmenite ($FeTiO_3$ or $FeO.TiO_2$). Petroleum jelly can also be produced by way of synthesis gas in which the process for conversion of synthesis gas to hydrocarbon products is adapted to produce higher molecular weight paraffin derivatives (Abhari, 2010). Thus, petroleum jelly is a subtle balance of liquid and solid hydrocarbon derivatives. The crystalline structure of the substances in its composition is one of the basic qualitative elements. The role of the amorphous solid hydrocarbon derivatives is, in fact, to retain in a sufficiently dense fibrous mesh, oily hydrocarbon derivatives of a generally high molecular weight.

Petroleum jelly is flammable only when heated to the liquid state at which point the fumes will combust but the liquid does not combust, not the liquid itself, so a wick material such as leaves, bark, or small twigs is needed to ignite petroleum jelly. Petroleum jelly is colorless or has a pale yellow color (when not highly distilled), translucent, and devoid of taste and smell when pure. It does not oxidize on exposure to the air and is not readily acted on by chemical reagents and is insoluble in water. It is soluble in dichloromethane (CH_2Cl_2), chloroform ($CHCl_3$), benzene (C_6H_6), diethyl ether ($CH_3CH_2OCH_2CH_3$), and carbon disulfide (CS_2).

According to the requirements of the International Nomenclature of Cosmetic Ingredients (INCI), which lists and assigns the INCI names of cosmetic ingredients, there are two possible designations depending on the manufacturing method of the petroleum jelly: (i) if the product is manufactured by blending paraffin oil, wax, and mineral paraffin, the INCI name of the mixture is composed of all the INCI names of the ingredients (*paraffinum liquidum [and] cera microcristallina [and] paraffin*) or (ii) if the product is manufactured by directly refining the crude oil or its derivatives of crude oil, the INCI name is *petrolatum*.

3.4 Paraffin wax

Paraffin wax is a white or colorless soft, solid wax that is composed of a complex mixture of hydrocarbon derivatives with the following general properties: (i) nonreactive, (ii) nontoxic, (iii) water barrier, and (iv) colorless. Paraffin wax is characterized by a clearly defined crystal structure and has the tendency to be hard and brittle with a melting point typically in the range 50–70°C (122–158°F). On a more specific basis, petroleum wax is of two general types: (i) paraffin wax in crude oil distillates and (ii) microcrystalline wax in crude oil residua. The melting point of wax is not directly related to its boiling point because waxes contain hydrocarbon derivatives of different chemical nature. Nevertheless, waxes are graded according to their melting point and oil content.

In the process for wax manufacture known as wax sweating (Parkash, 2003; Gary et al., 2007; Speight, 2011, 2014; Hsu and Robinson, 2017; Speight, 2017), a cake of slack wax (paraffin wax from a solvent dewaxing operation) is slowly warmed to a temperature at which the oil in the wax and the lower melting waxes become fluid and drip (or sweat) from the bottom of the cake, leaving a residue of higher melting wax. However, wax sweating can be carried out only when the residual wax consists of large crystals that have spaces between them, through which the oil and lower melting waxes can percolate; it is therefore limited to wax obtained from low-boiling paraffin distillate.

Wax recrystallization, like wax sweating, separates slack wax into fractions, but instead of using the differences in melting points, it makes use of the different solubility of the wax fractions in a solvent, such as the ketone used in the dewaxing process. When a mixture of ketone and slack wax is heated, the slack wax usually dissolves completely, and if the solution is cooled slowly, a temperature is reached at which a crop of wax crystals is formed. These crystals will all be of the same melting point, and if they are removed by filtration, a wax fraction with a specific melting point is obtained. If the clear filtrate is further cooled, a second crop of wax crystals with a lower melting point is obtained. Thus, by alternate cooling and filtration, the slack wax can be subdivided into a large number of wax fractions, each with different melting points.

Microcrystalline wax (sometimes also called *micro wax* or *microwax*) is produced from a combination of high-boiling lube distillates and residual oils and differs from paraffin wax in that the microcrystalline was has a less well-defined crystalline structure and is darker color. The physical properties of microcrystalline wax is affected significantly by the oil content (Kumar et al.,

2007), and by achieving the desired level of oil content, wax of the desired physical properties and specifications can be produced. Deep deoiling of microcrystalline wax is comparatively difficult compared with paraffin wax (macrocrystalline wax) as the oil remains occluded in these and is difficult to separate by sweating. Also as wax and residual oil have similar boiling ranges, separation by distillation is difficult. However, these waxes can be deoiled by treatment with solvents at lower temperature that have high oil miscibility and poor wax solubility, and these have been used extensively to separate.

Paraffin wax is mostly used for relief of discomfort and pain in following conditions such as bursitis, eczema, psoriasis, dry flaky skin, stiff joints, fibromyalgia, tired sore muscles, inflammation, and arthritis. Paraffin wax is often used in skin-softening salon and spa treatments on the hands, cuticles, and feet because this type of wax is colorless, tasteless, and odorless. It can also be used to provide pain relief to sore joints and muscles. Paraffin wax is often used as lubrication, electrical insulation, and to make candles and crayons. Cosmetically, paraffin wax is often applied to the hands and feet. The wax is a natural emollient, helping make skin supple and soft. When applied to the skin, it adds moisture and continues to boost the moisture levels of the skin after the treatment is complete. It can also help open pores and remove dead skin cells. This may help make the skin look fresher and feel smoother and give comfort to the user.

3.5 Steroids

The term *steroid* is applied to a group of naturally occurring or synthetic fat-soluble organic compounds (lipids), whose structure is chemically based on the hydrocarbon sterane nucleus. Sterane, the parent compound of steroids, is a hydrocarbon based on the 17 carbon atom four-ring perhydrocyclopentanophenanthrene ring system (fully hydrogenated cyclopentanoperhydrophenanthrene ring) (Fig. 13.1). The sterane structure constitutes the core of all nonhydrocarbon sterols and steroids. The characteristic base structure of a sterane (the degraded and saturated version of a steroid) (Fig. 13.1) has three cyclohexane rings and one cyclopentane ring and a side chain emerging from C_{17}. Sterane is the hypothetical parent molecule for any steroid hormone. The name was originally conceived to achieve forms

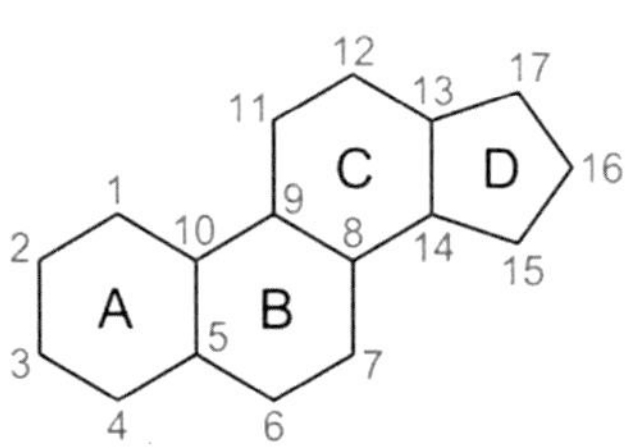

FIGURE 13.1 Numbering of the Sterane ring system and carbon system when the typical Alkyl side-chain is included.

of systemic nomenclature, but it is now supplanted by the fundamental structural variants such as gonane, estrane, cholestane, and pregnane.

Gonane is the fundamental tetracyclic unit (Fig. 13.1) with no methyl groups at C-10 and C-13 and with no side chain at C-17 steroid nucleus. *Gonane* exists as either of two isomers, known as 5α-gonane and 5β-gonane.

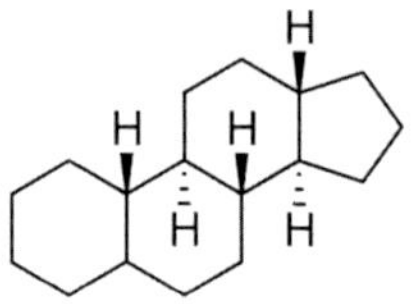

Gonane

Estrane is a sterane derivative; estrenes are estrane derivatives containing a double bond.

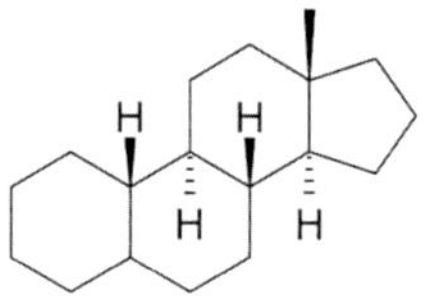

Estrane

Cholestane is a saturated hydrocarbon 27-carbon sterane that serves as the basis for many organic molecules. Derivatives are classified in two families: (i) sterols (with an alcohol group) and (ii) cholestenes (with a double bond). Some steroids, such as cholesterol, are both a sterol and a cholestene.

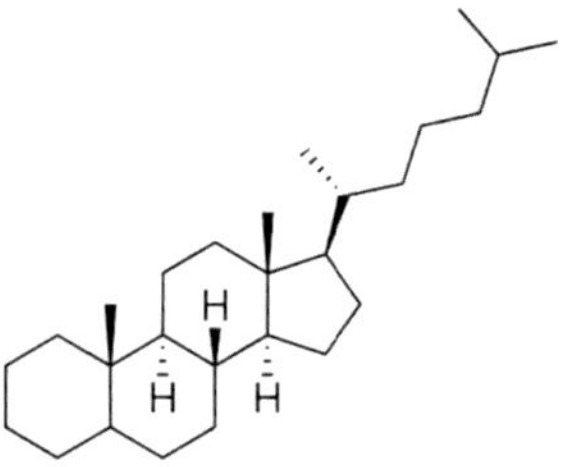

Cholestane

Pregnane is the parent hydrocarbon for two series of steroids stemming from 5α-pregnane (originally allopregnane) and 5β-pregnane (17β-ethyletiocholane):

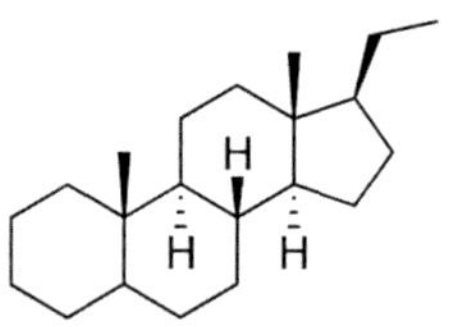

Pregnane

5β-Pregnane is the parent of the progesterones, pregnane alcohols, ketones, and several adrenocortical hormones and is found largely in urine as a metabolic product of 5β-pregnane compounds.

During diagenesis and catagenesis, the biological stereospecificity of sterols, particularly at C-5, C-14, C-17, and C-20, is usually lost, and a large range of isomers is generated (Fig. 13.2) The term *alphabetabeta sterane* (sometimes the word *alphabeta*) is commonly used as shorthand to denote steranes with the 5-alpha(H), 14-beta(H), and 17-beta(H) configuration, while *alphaalphaalpha* sterane would denote 5-alpha(H), 14-alpha(H), and 17-alpha(H) stereochemistry. The notation 14-alpha(H) indicates that the hydrogen is located below the plane of the paper, whereas in 14-beta(H) it is above the plane.

In steranes, if no other carbon number is cited, *S* and *R* always refer to the stereochemistry at C-20. The prefix *nor*, as for example in 24-norcholestane, indicates that the molecule is formally derived for the parent structure by loss of the indicated carbon atom, i.e., C-24 is removed from cholestane

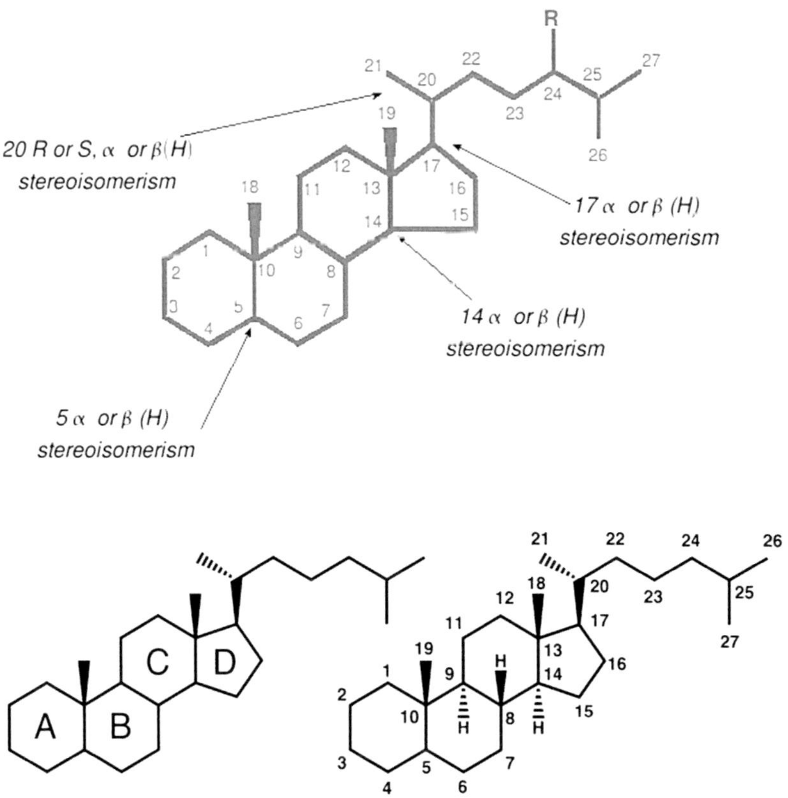

FIGURE 13.2 Sterane nomenclature and stereoisomerism.

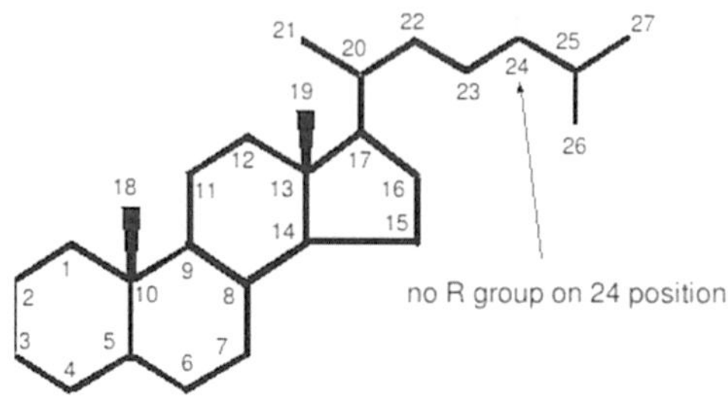

FIGURE 13.3 Norcholestane, a C_{27} to C_{30} sterane without the R group on its chain.

(Fig. 13.3). The term *desmethylsteranes* is sometimes used to distinguish steranes that do not possess an additional alkyl group at ring A, i.e., at carbon atoms C-1 to C-4.

Diasteranes (Fig. 13.4) are rearranged steranes that have no biological precursors and are most likely formed during diagenesis and catagenesis. Steranes may be rearranged into diasteranes during diagenesis. Thus, the diasterane/sterane ratio may be a signal of the maturity of the source rock. Norcholestane, shown above, a cholestane with one carbon missing, has some interesting uses as a biomarker. Only three series of these C_{26} steranes are known: 21-, 24- and 27-norcholestane. 24-norcholestane has a particular source or depositional environment meaning, whereas 21- and 27- are markers for maturity.

Sterane finds some use as a drug (the generic equivalent is the nonhydrocarbon prednisolone) but offers more information when considered as a biomarker in determining the origin of crude oil. Biomarkers (*molecular fossils*) from ancient sediments, crude oil source rocks, and crude oil are of uppermost importance for organic geochemists to characterize and identify oils, establish correlations, and develop paleoenvironmental interpretations (Fleck et al., 2000).

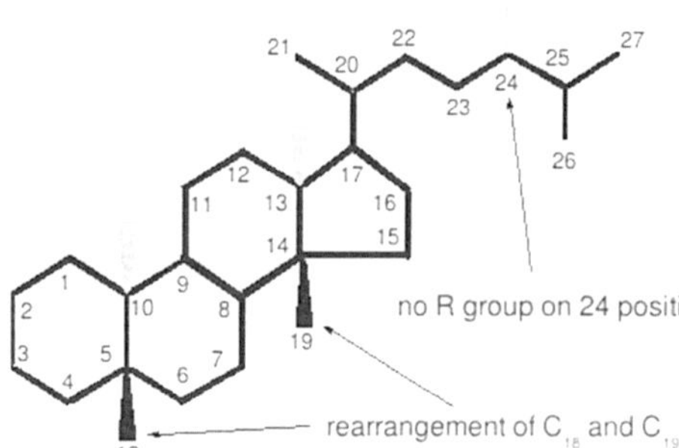

FIGURE 13.4 Diasterane.

For the most common sterane markers used in studies dealing with ancient sediments, four major classes of sterols are considered as precursors and derive from eukaryotic organisms. They contain 27 carbon atoms (e.g., cholesterol found in animals, algae or plankton), 28 carbon atoms (e.g., ergosterol found in fungi), 29 carbon atoms (sitosterol, stigmasterol found in vascular plants and some algae), and 30 carbon atoms (sterols from marine derived biomass). In addition to the variability in the organic sources, transformation of the biomass in the water column and the sediments as well as early diagenetic processes modify the initial structure of the precursor molecules, leading to the formation of steranes. Among them, the regular steranes are the most widely used in organic geochemistry. Especially, the relative proportions on the C27, C28, and C29 steranes are used for the assessment of organic input to the sediments and of paleoenvironmental conditions of deposition.

One of the environments in which crude oil is believed to be formed is a lacustrine environment (in addition to the marine environment). The lacustrine environment is usually characterized by a higher relative concentration in C28 steranes (Huang and Meinschein, 1979). A low concentration of this steranes in an environmental sample set suggests the absence of typically fresh aquatic organisms (the absence of a true lacustrine environment is also supported by geological evidences); this is probably because of shallow fresh water conditions (in opposition to deep lacustrine), related to swamp type environments). Foliage fall and turnover of plants are the dominant source of plant debris. These are utilized in the food web by heterotrophs. C27 sterols can thus originate from organisms living onto the plant debris, i.e., variable invertebrates (Huang and Meinschein (1979), and/or from the microbial degradation of C28 and C29 sterols side chains (Murohisa and Iida, 1993). To unravel all these possibilities and improve the paleoenvironmental assessment, correlation of organic information with the geological and biological context is necessary Volkman (2008), and it is necessary to adjust the paleoenvironmental interpretation of steranes by considering geological and biological information.

In fact, biomarkers add complementary information to the fossil palynomorph record (Schwark and Empt, 2006). Steranes are important constituents of eukaryotic cell membranes and are preserved in sediments as steranes. C_{28}- and C_{29}-steranes are indicators for the presence of green and C_{27}-steranes for the presence of red algae, respectively. The relative abundance of steranes allows the investigation of the fossil record for Paleozoic algal diversification and evolution.

For example, a sharp increase of the C_{28}/C_{29}-sterane ratio from <0.55 to >0.70 at the Devonian/Carboniferous boundary implies a fundamental change in the green algae assemblage from more primitive, mainly C_{29}-sterane–producing algae to modern C_{28}-sterane–producing algae. A pronounced but short-lived rise in the C_{28}-sterane content occurs, which is attributed to an episodic increase in prasinophytes. The gradual radiation of algae may have been triggered by frequent mass extinctions in the Upper Devonian culminating with the massive decline of acritarchs at the D/C boundary. The coeval rise in the C_{28}/C_{29}-sterane ratio indicates the presence of a nonencysting algal group and coincides with the global augmentation of numerous filamentous Codiacea (Siphonales) and the rise of euspondyle and metaspondyle Dasycladales.

A steroid is characterized by its sterane core to which nonhydrocarbon functional groups may be incorporated or attached. The core is a carbon structure of four fused rings: three cyclohexane rings and one cyclopentane ring (Fig. 13.1). The steroids vary by the functional groups attached to these rings and the oxidation state of the rings. The sterane core of steroids is composed of 17 carbon atoms bonded together to form four fused rings: three cyclohexane rings (designated as rings A, B, and C) and one cyclopentane ring (the D ring) (Fig. 13.1). The steroids vary by the functional groups attached to these rings and by the oxidation state of the rings. Sterols are forms of steroids, with a hydroxyl group at position 3 and a skeleton derived from cholestane (Fig. 13.2). Many hormones, body constituents, and drugs are steroids—not necessarily hydrocarbon derivatives but based on the sterane core.

All the corticosteroid hormones of the adrenal cortex (glucocorticoids or mineralocorticoids), all the sex hormones (sex hormones are found in higher quantities in one sex than in the other; male sex hormones are androgens, which include testosterone; female sex hormones are estrogens and progesterone), all vitamins of the Vitamin D group (calciferol), the bile acids (ursodeoxycholic acid and analogues), cardiac aglycones, sterols such as cholesterol, toad poison, saponins, some carcinogenic hydrocarbon derivatives, and some corticosteroid drugs such as prednisone are all steroids. Synthetic chemical analogues of many of the naturally occurring steroids are vital in medicine. Both natural and synthetic steroids are used to treat many disorders and play a vital role in the normal functioning of the body.

Steroidal drugs may be of three types—anabolic, androgenic, and *corticosteroids*. *Anabolic steroids* are chemically derived from testosterone. Many attempts were made to separate the anabolic effects of the hormones from their androgenic effects, but with little success. Thus, anabolic compounds may cause androgenic side effects, especially when used for extended periods. Anabolic effects are seen as the growth or thickening of the tissues of the nonreproductive tract of the body, including the skeletal muscles, bones, the larynx, and vocal chords, and a decrease in body fat. *Androgenic steroid* effects are seen in the growth of the male reproductive tract and the development of male secondary sexual characteristics. Medically, anabolic steroids were given for osteoporosis in women, but it not recommended nowadays.

Cortico steroid is a generic name for the group of hormones that have a cortisone-like action. They are man-made steroids that mimic the activity of cortisone. Cortisone is produced naturally in the body and is involved in regulating inflammation, thus dealing with injury. Thus, corticosteroids are not the same as anabolic steroids. Corticosteroids are used in the treatment of many diseases such as asthma, eczema, allergies, arthritis, colitis, and kidney disease.

Anabolic steroids control or contribute to the large muscle mass of males because of the nitrogen-retaining effects of androgen. They may have a property of protein building and when taken lead to an increase in muscle bulk and strength. *Anabolic steroids* were developed in the late 1930s to treat hypogonadism—a condition in which the testes do not produce sufficient testosterone for normal growth, development, and sexual functioning. The primary medical uses of anabolic steroids are to treat delayed puberty, some types of impotence, and wasting of the body caused by HIV infection or other diseases. Around the same time, scientists discovered that these compounds could facilitate the growth of skeletal muscles in laboratory animals, which led to their use first by bodybuilders and weightlifters and then by athletes in a variety of other sports.

Anabolic steroids are illegal without a prescription but steroidal supplements can be bought over-the-counter legally. Such supplements are more commonly called dietary supplements, though they are not food products. Steroidal supplements contain dehydroepiandrosterone and/or androstenedione. If large quantities of steroidal supplements substantially increase testosterone levels in the body, they might most likely produce the same side effects as anabolic steroids.

Medically, anabolic steroids may be used for many purpose, including (i) stimulation of protein anabolism in debilitating illness and in acute renal failure, (ii) promotion of growth in children with pituitary dwarfism and other growth disorders, (iii) retention of nitrogen and calcium may benefit patients with osteoporosis and patients receiving corticosteroid therapy, and (iv) stimulation of bone marrow function in hypoplastic anemia. However, when abused, anabolic steroids can have serious side effects. Athletes and bodybuilders aiming to improve their strength, stamina, speed, or body size have always abused them. Steroids appear to work by decreasing the amount of time necessary for recovering between bouts of exercise. Because of this, trainees can exercise more often, or more intensely, without overwhelming the ability of the body to adapt or overtraining.

It is important to understand that using steroids does not increase skill, agility, and performance. These are determined by many factors, including genetics, body size, age, sex, diet, and how hard the athlete trains. Anabolic steroids are not legal in organized sports. Most professional and amateur sports organizations and medical associations ban *anabolic steroids*. Athletes who test positive for steroids will be suspended or disqualified and may lose their chance to compete in their sport.

Cholesterol contains a hydroxyl group that also provides slightly hydrophilic features to a substance that otherwise is structured like a hydrocarbon and hence a lipo-soluble substance.

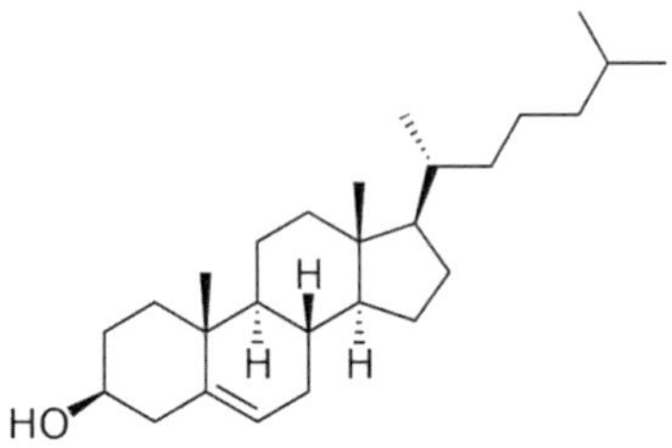

Cholesterol

This specific feature increases the water-retaining capacity of wool fat (*adeps lanae*) in which it is contained. Lanolin is the term for a mixture of wool fat (65 g), paraffin oil (15 g), and water (20 g) and a frequent component of W/O emulsions in pharmaceutical skin ointments. In the cosmetic field, *wool fat* and *lanolin* are synonyms. Cholesterol has excellent skin-protecting effects and is a component of the natural skin barrier.

Cholesterol is a main component for the human metabolism. It is transported in the blood stream with the help of lipoproteins whose main components are proteins and phosphatidylcholine. Chylomicrons that can be imagined as minuscule emulsion-like droplets help to transport the cholesterol assimilated with the daily nutrition from the small intestines via the lymphatic system into the blood vessels. A significant product of the cholesterol metabolism is *pregnenolone*, a gestagen that is the base substance for bile acids and steroid hormones.

Plant sterols (*phytosterols*) are structurally related to cholesterol and can therefore replace the animal cholesterol in skincare creams. This explains the excellent skincare characteristics of avocado oil, which is rich in phytosterols. The biosynthesis of cholesterol in the human body starts with activated acetic acid (acetyl-CoA) via the terpenes geraniol (monoterpene), farnesol (sesquiterpene), and squalene (triterpene). Squalene is a significant refattening ingredient of the human sebum and metabolized into lanosterine, which is a precursor of cholesterol, is also contained in wool fat and has similar emulsifying properties in creams.

Progesterone, which forms from pregnenolone, is the base substance for *androgens* (such as testosterone) and also for the *estrogens* (such as estrone and estradiol). In contrast to androgens, estrogens have an aromatic ring. This leads to the fact that the hydroxy group located right at the ring has phenolic characteristics. This specific feature is the reason for its structural resemblance to plant isoflavones (polyphenols), which are also called *phytohormones.*

Soybean and red clover−based phytohormones are mainly used in anti-aging products and skincare products for the blemished skin. Contrary to phytohormones, steroid hormones and extracts containing steroid hormones are banned in many European countries.

The *glucocorticoids* include *cortisol* (*hydrocortisone*):

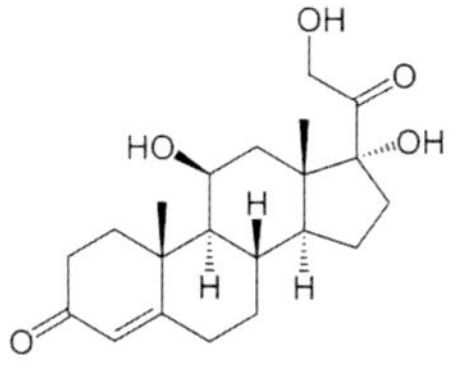

Cortisol

The biosynthesis of cortisol and cortisone from progesterone occurs in the adrenal cortex. Cortisone as such is inactive; cortisol, however, has manifold physiological effects. Orally taken inactive cortisone is transformed in the liver into active cortisol.

Cortisol is characterized by its antiinflammatory and immune-suppressive effects and is applied in ointments against all kinds of allergies and skin reactions. The skin condition frequently improves within a few days already. A disadvantage though is the atrophic skin condition developing after a long-term use. The skin becomes thinner and more permeable for externally affecting irritants and allergens. All in all, the skin becomes more sensible to relapses. To reduce these and other side effects, a whole series of artificial corticoids has been developed in addition to hydrocortisone. Another source for the technical manufacturing of cortisol besides the phytosterol *sitosterol* is the herbal *diosgenin*. Diosgenin belongs to the group of herbal *saponins* with a steroidal ring system. It is also a base substance for the industrially produced progesterone.

Like bile acids, saponins also are surface active and have formerly been used for cleansing purposes. In India and other Asian regions, the fruits of the wash nut tree (soap nut) with their specifically high saponin content are used still used in the modern world. Unlike the anionic emulsifying bile acids, the cleansing effect of saponins results from the glycosidic linkage of water-soluble sugar residues with the steroidal ring system. That is why saponins can be compared with nonionic emulsifiers like modern-day sugar tensides (which are medications that decrease the pore sizes in the skin), which are used for facial cleansing.

Cardiac glycosides have a similar glycosidic steroidal structure as saponins. The main active agent digitoxin is extracted from the leaves of the purple foxglove (*Digitalis purpurea*). Also related to saponins are the *steroidal alkaloids* of the solanum family. The most famous representative here is solanine, which occurs in potatoes and has a low toxic effect.

In connection with steroids, vitamin D_3 is worth mentioning as it is formed from 7-dehydrocholesterol, which is a prestage of cholesterol. 7-Dehydrocholesterol occurs in the stratum spinosum and stratum basale of the skin and is transformed into vitamin D_3 by influence of ultraviolet light. During this process, one of the four steroidal rings is opened. The vitamin is also assimilated with the daily nutrition. The more important this is, the less the skin exposed to sunlight and the more the sunscreens used. A major source for the vitamin is the consumption of fish especially of those with high fat content such as herring, salmon, and mackerel.

3.6 Carotenoids and vitamins

Carotenoids are organic pigments that are naturally occurring in the chloroplasts and chromoplasts of plants and some other photosynthetic organisms such as like algae, various types of fungi, and various types of bacteria. There are several hundred known carotenoids; they are split into two classes: (i) carotenes, pure hydrocarbon derivatives, and (ii) xanthophylls, which contain oxygen. However, in contrast to the steroids where the true hydrocarbon derivatives play a limited pharmaceutical role, the carotenoid hydrocarbon derivatives have a much greater role as pharmaceuticals.

3.6.1 Hydrocarbon carotenoids

Hydrocarbon carotenoids (carotenes) fall into a group of hydrocarbon compounds having the formula $C_{40}Hx$, (where x is variable) which are synthesized by plants but cannot be made by animals. Carotene is an orange photosynthetic pigment important for photosynthesis. Carotenes are all colored to the human eye. They are responsible for the orange color of the carrot, for which this class of chemicals is named, and for the colors of many other fruits and vegetables. Carotenes are also responsible for the orange (but not all of the yellow) colors in dry foliage. They also (in lower concentrations) impart the yellow coloration to milk fat and butter.

β-Carotene is composed of two retinyl groups, and it can be stored in the liver and body fat and converted to retinal as needed, thus making it a form of vitamin A for humans and some other mammals.

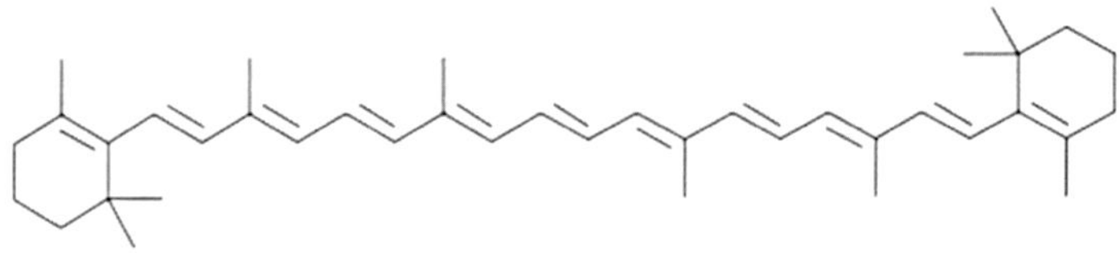

β-Carotene

α-Carotene and γ-carotene, due to their single retinyl group (β-ionone ring), also have some vitamin A activity (though less than β-carotene), as does the xanthophyll carotenoid β-cryptoxanthin. All other carotenoids, including lycopene, have no β-ring and thus no vitamin A activity (although they may have antioxidant activity and thus biological activity in other ways).

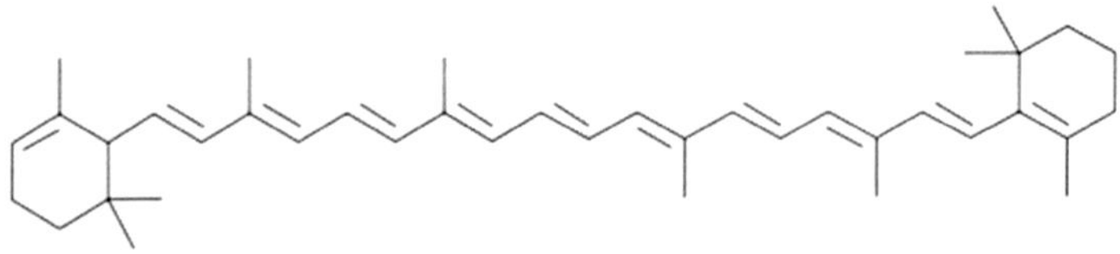

α-Carotene

The two ends of the β-carotene molecule are structurally identical (β-rings). Specifically, the group of nine carbon atoms at each end forms a β-ring. The α-carotene molecule has a β-ring at one end; the other end is called an ***ε-ring*** (there is no such designation as an *α-ring*). These and similar names for the ends of the carotenoid molecules form the basis of a systematic naming scheme, according to which

- α-carotene is β,ε-carotene;
- β-carotene is β,β-carotene;
- γ-carotene (with one β-ring and one uncyclized end that is labeled *psi*) is β,ψ-carotene;
- δ-carotene (with one ε ring and one uncyclized end) is ε,ψ-carotene;
- ε-carotene is ε,ε-carotene;
- lycopene is ψ,ψ-carotene.

Probably, the most well-known carotenoid is the compound that gives this group its name: carotene, which is found in carrots and also apricots. Crude palm oil, however, is the richest source of carotenoids in nature in terms of retinol (provitamin A) equivalent. The Vietnamese gac fruit contains the highest known concentration of the carotenoid lycopene:

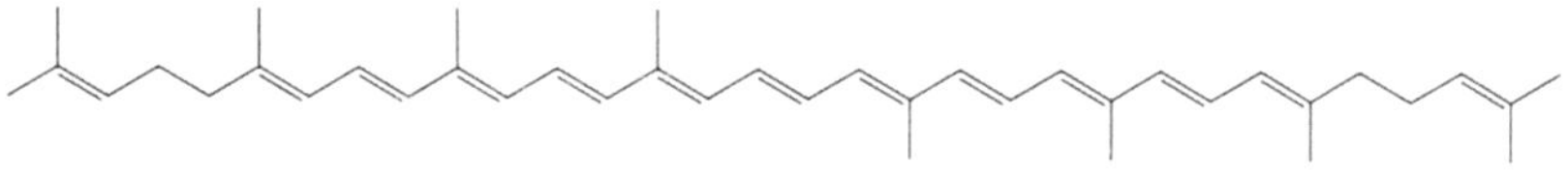

Lycopene

Lycopene is a bright red carotene and carotenoid pigment and phytochemical found in tomatoes and other red fruits and vegetables, such as carrots, watermelons, and papayas (but not strawberries or cherries). Although lycopene is chemically a carotene, it has no vitamin A activity. In plants, algae, and other photosynthetic organisms, lycopene is an important intermediate in the biosynthesis of many carotenoids, including β-carotene, responsible for yellow, orange, or red pigmentation, photosynthesis, and photoprotection. Like all carotenoids, lycopene is a polyunsaturated hydrocarbon (an unsubstituted alkene). Structurally, lycopene is a tetraterpene assembled from eight isoprene units, composed entirely of carbon and hydrogen, and is insoluble in water.

The 11 conjugated double bonds in lycopene give it its deep red color and are responsible for its antioxidant activity.

3.6.2 Nonhydrocarbon carotenoids

The nonhydrocarbon carotenoids are important components of the light harvesting in plants, expanding the absorption spectra of photosynthesis. The major carotenoids in this context are lutein, violaxanthin, and neoxanthin:

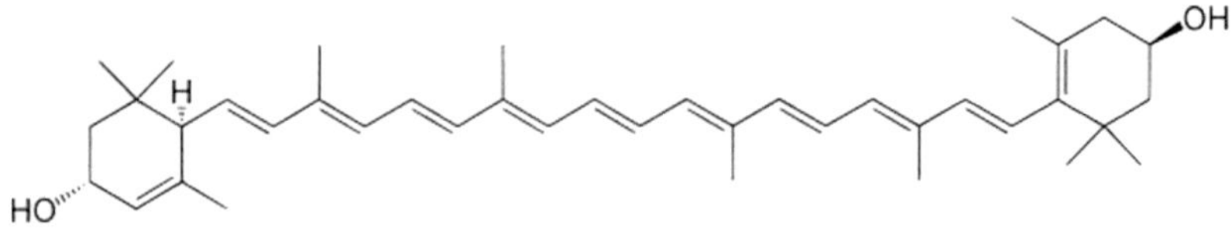

Lutein

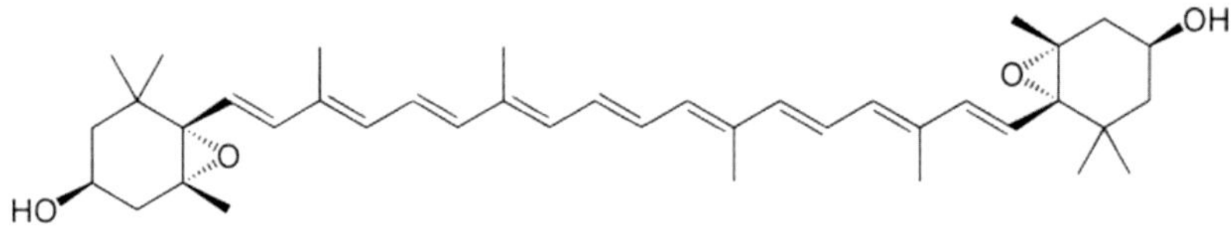

Violaxanthin

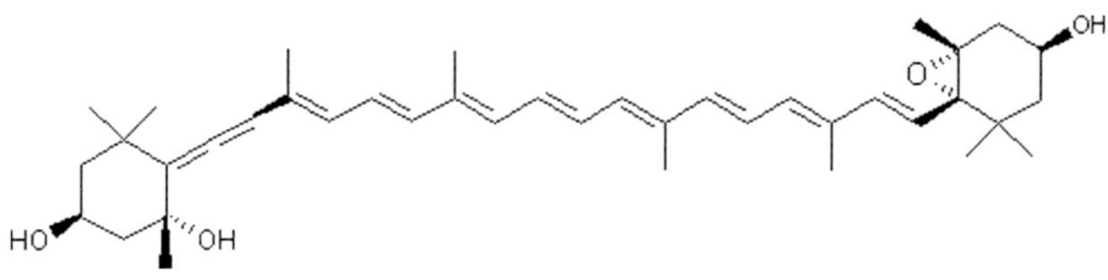

Neoxanthin

Additionally, there is considerable evidence that indicates a photo-protective role of xanthophylls preventing damage by dissipating excess light. In mammals, carotenoids exhibit immunomodulatory actions, likely related to their anticarcinogenic effects. Carotenoids generally absorb blue light, and they serve two key roles in plants and algae: they absorb light energy for use in photosynthesis, and they protect chlorophyll from photodamage. In humans, four carotenoids (β-carotene, α-carotene, γ-carotene, and β-cryptoxanthin) have vitamin activity and can also act as antioxidants.

Carotenoids belong to the category of tetraterpenoids (i.e., they contain 40 carbon atoms)—structurally they are in the form of a polyene chain that is sometimes terminated by rings.

Xanthophylls are not pure hydrocarbon derivatives and often yellow, hence their class name:

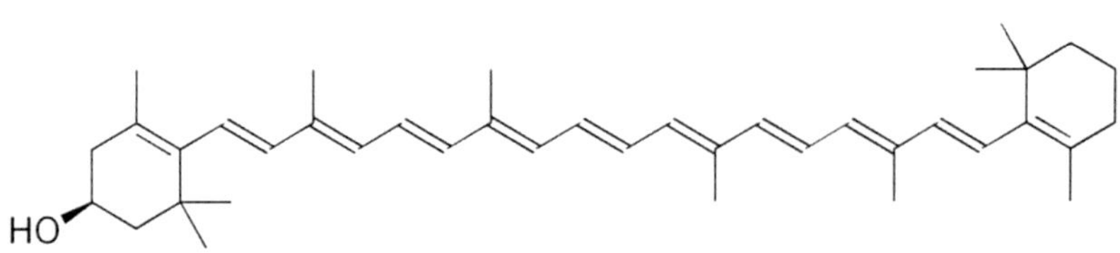

Cryptoxanthin

The carbon–carbon double bonds interact with each other through conjugation, which allows electrons in the molecule to move freely across these areas of the molecule. As the number of double bonds increases, electrons associated with conjugated systems have more room to move and require less energy to change states. This causes the range of energies of light absorbed by the molecule to decrease. As more frequencies of light are absorbed from the short end of the visible spectrum, the compounds acquire an increasingly red appearance.

In photosynthetic organisms, specifically flora, carotenoids play a vital role in the photosynthetic reaction center. They either participate in the energy transfer process or protect the reaction center from autooxidation. In humans, carotenoids have been linked to oxidation-preventing mechanisms. Carotenoids have many physiological functions. Given their structure, carotenoids are efficient free-radical scavengers, and they enhance the vertebrate immune system. There are several dozen carotenoids in foods people consume, and most carotenoids have antioxidant activity.

Humans and animals are incapable of synthesizing carotenoids and must obtain them through their diet, yet they are common and often in ornamental features. For example, the pink color of flamingos and salmon and the red coloring of lobsters are due to carotenoids.

The most common carotenoids include lycopene and the vitamin A precursor β-carotene. In plants, the xanthophyll lutein is the most abundant carotenoid and its role in preventing age-related eye disease is currently under investigation. Lutein and the other carotenoid pigments found in mature leaves are often not obvious because of the presence of chlorophyll. However, when chlorophyll is not present, as in young foliage and also dying deciduous foliage (such as autumn leaves), the yellows, reds, and oranges of the carotenoids are predominant. For the same reason, carotenoid colors often predominate in ripe fruits (e.g., oranges, tomatoes, bananas), after being unmasked by the disappearance of chlorophyll.

4. Pharmaceuticals based on hydrocarbons

The pharmaceutical industry includes the manufacture, extraction, processing, purification, and packaging of chemical materials to be used as medications for humans or animals. Pharmaceutical manufacturing is divided into two major stages: (i) the production of the active ingredient or drug (primary processing or manufacture) and (ii) secondary processing, the conversion of the active medicines into products suitable for administration (Gad, 2008). However, before a medication can be manufactured at any scale, much work goes into the actual formulation of the medicine. Formulation development scientists must evaluate a compound for uniformity, stability, and many other factors. After the evaluation phase, a solution must be developed to deliver the medication in its required form such as solid, semisolid, immediate, or controlled release, tablet, and capsule.

In the pharmaceutical industry, a wide range of excipients may be blended together to create the final blend used to manufacture the solid dosage form. The range of materials that may be blended (excipients, API) presents a number of variables that must be addressed to achieve products of acceptable blend uniformity. These variables may include the particle size distribution (including aggregates or lumps of material), particle shape (spheres, rods, cubes, plates, and irregular), presence of moisture (or other volatile compounds), and particle surface properties (roughness, cohesiveness).

The following sections present the published synthetic routes for several over-the-counter (nonprescription) medications that start from a hydrocarbon. There are many other medications that commence production from a simple hydrocarbon—too many for inclusion in this book—but the production methods and starting materials for the medications presented below are used examples. These are listed alphabetically rather than by preference or by stated use or effect.

4.1 Acetaminophen

Acetaminophen (paracetamol) is an analgesic and fever-reducing medicine similar in effect to aspirin.

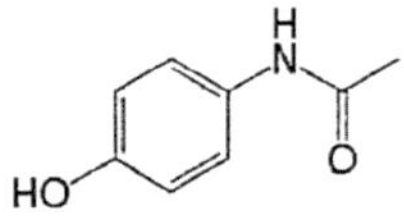

It is an active ingredient in many over-the-counter medicines, including Tylenol and Midol. Introduced in the early 1900s, acetaminophen is a coal tar derivative that acts by interfering with the synthesis of prostaglandins and other substances necessary for the transmission of pain impulses.

In keeping with the content of this book, the dominant chemical pathway for the production of acetaminophen commences with the production of phenol from benzene. The cumene process uses benzene and propene as feedstock and involves the partial oxidation of cumene (isopropyl benzene) via the Hock rearrangement:

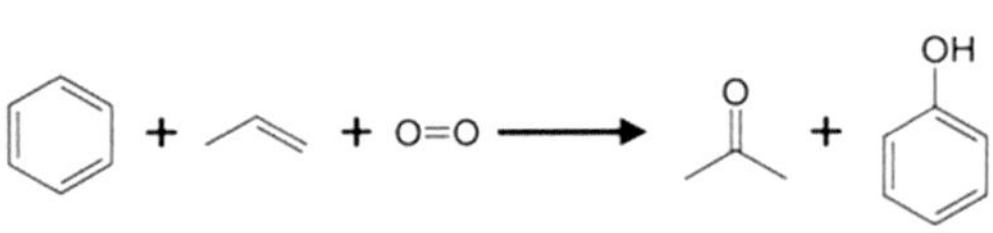

Compared with most other processes, the cumene process uses relatively mild synthesis conditions and relatively inexpensive raw materials; acetone

(CH_3COCH_3) is produced as a by-product. The phenol is then used as the starting material; p-aminophenol (4-aminophenol) is produced from phenol by nitration followed by reduction with iron. Alternatively, the partial hydrogenation of nitrobenzene affords phenylhydroxylamine, which rearranges primarily to 4-aminophenol:

$$C_6H_5NO_2 + 2H_2 \rightarrow C_6H_5NHOH + H_2O$$

$$C_6H_5NHOH \rightarrow HOC_6H_4NH_2$$

The p-aminophenol can also be produced from nitrobenzene by electrolytic conversion to phenylhydroxylamine, which, under the reaction conditions, spontaneously rearranges to 4-aminophenol. p-Aminophenol is a white powder that is moderately soluble in alcohols and can be recrystallized from hot water. Also, it is the final intermediate in the industrial synthesis of paracetamol by treatment with acetic anhydride:

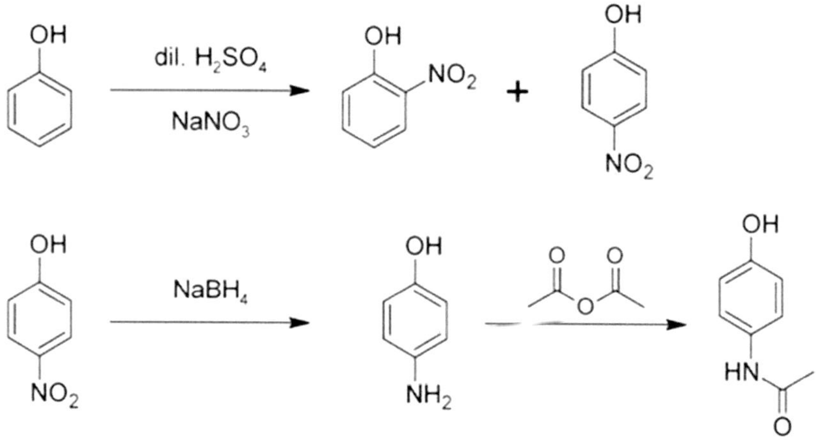

Alternatively, acetaminophen can be produced from p-aminophenol by the reaction with acetic anhydride:

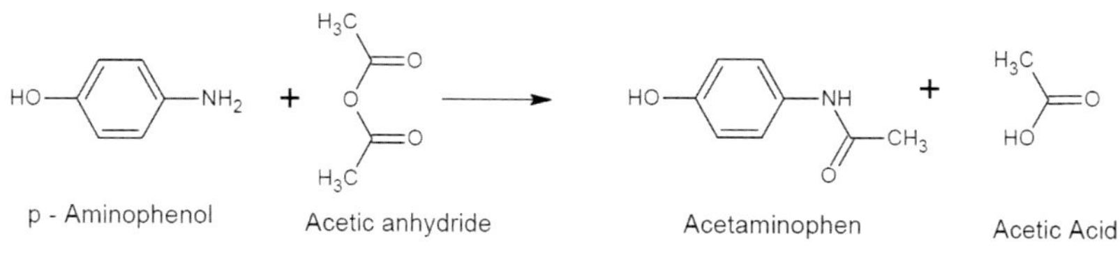

4.2 Aleve

The active constituent of Aleve is Naproxen sodium, which is an antiinflammatory compound. Naproxen is used to treat a variety of inflammatory conditions and symptoms that are due to excessive inflammation, such as pain and fever—naproxen has fever-reducing (antipyretic) properties in addition to its antiinflammatory activity.

The starting point for the synthesis is naphthalene, from which 2-naphthol is produced. Traditionally, 2-naphthol (2-hydroxynaphthalene, b-hydroxynaphthalene, also known as β-naphthol and sometime—but rarely—as 2-naphthalenol) is produced by a two-step process that begins with the sulfonation of naphthalene in sulfuric acid (Booth, 2005) after which the sulfonic acid group is then cleaved in molten sodium hydroxide and neutralization of the sodium salt with acid gives 2-naphthol. 2-Naphthol can also be produced by a method analogous to the cumene process.

$$C_{10}H_8 + H_2SO_4 \rightarrow C_{10}H_7SO_3H + H_2O$$

$$C_{10}H_7SO_3H + 3\ NaOH \rightarrow C_{10}H_7O^-\ Na^+ + Na_2SO_3 + 2H_2O$$

$$C_{10}H_7O^-\ Na^+ + H^+OH^-\ C_{10}H_7OH + NaOH$$

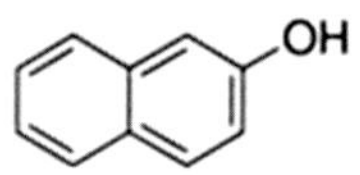

2-Naphthol

2-Naphthol

Neutralization of the product with acid gives 2-naphthol, which can also be produced from naphthalene by a method analogous to the cumene process. After which the naproxen has been produced starting from 2-naphthol (β-naphthol)—a constituent of coal tar or which can be prepared by the following series of reactions:

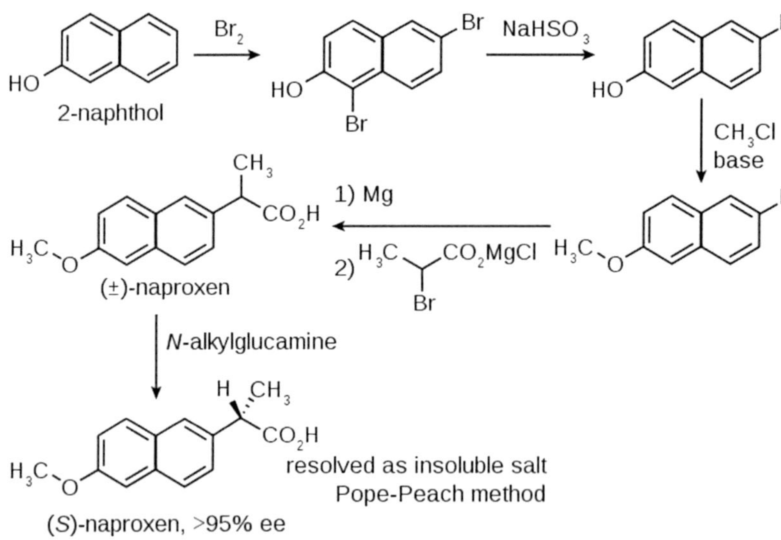

2-Naphthol is not a product that is isolated from crude oil. It is prepared from naphthalene (see above), but it can also be isolated from the products of the thermal decompositon of coal and some types of biomass. 2-Naphthol is also the base from which certain dyestuffs can be manufactured (Speight, 2019).

4.3 Aspirin

Acetylsalicylic acid commonly known as aspirin is a widely used drug. The analgesic, antipyretic, and antiinflammatory properties make it a powerful and effective drug to relive symptoms of pain, fever, and inflammation.

In the current content, salicylic acid can be synthesized from benzene via phenol by a three-step process.

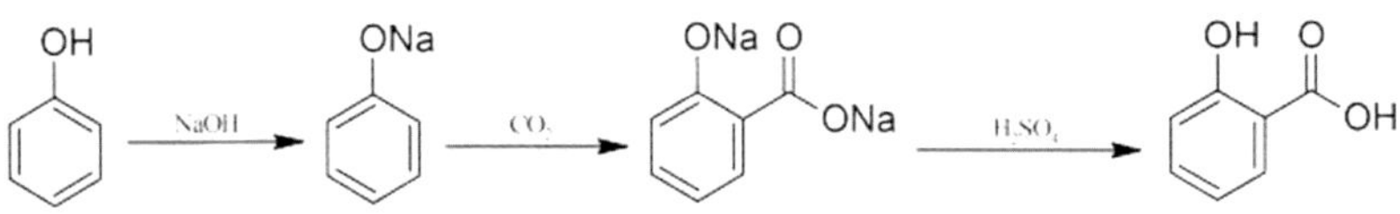

The synthesis of aspirin may be achieved in one simple step, O-acetylation of salicylic acid, which is incorporated into many undergraduate synthetic chemistry laboratory courses. The purity of the product as a pharmaceutical is crucial.

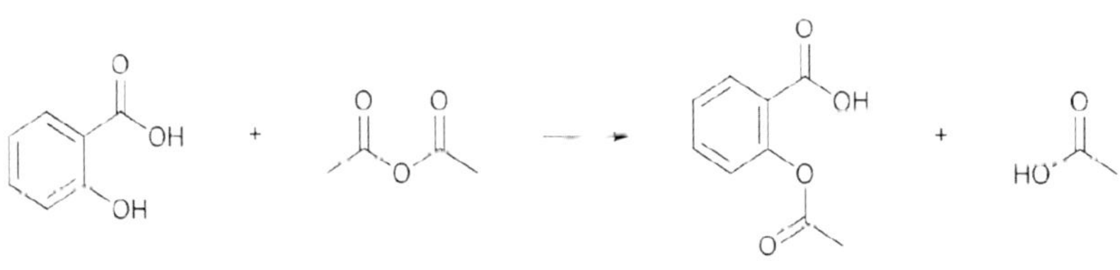

4.4 Cepacol

The main ingredient of Cepacol is benzocaine, which is commonly used as a topical pain reliever or in cough drops.

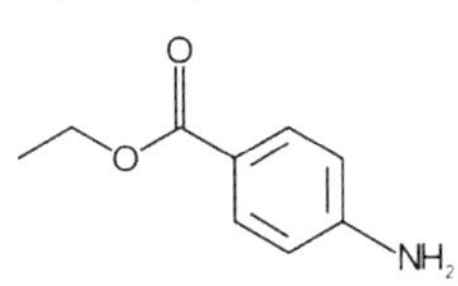

Benzocaine

It is the active ingredient in many over-the-counter anesthetic ointments such as products for oral ulcers.

Benzocaine is the ethyl ester of p-aminobenzoic acid and can be prepared by the reaction of p-aminobenzoic acid with ethanol or via the reduction of ethyl p-nitrobenzoate. Benzocaine is sparingly soluble in water; it is more soluble in dilute acids and very soluble in ethanol, chloroform, and ethyl ether, and it can be synthesized from toluene by a three-step process.

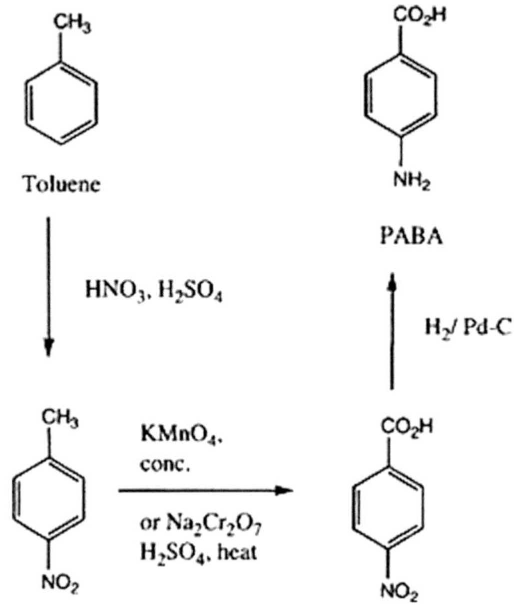

4.5 Ibuprofen

Ibuprofen is a medication in the nonsteroidal antiinflammatory drug (nonsteroidal antiinflammatory drug, NSAID) class (NSAID class) that is used for treating pain, fever, and inflammation. Since the introduction of the drug in 1969, ibuprofen has become one of the most common painkillers in the world. Ibuprofen in an NSAID, and like other drugs of its class it possesses analgesic, antipyretic, and antiinflammatory properties. While ibuprofen is a relatively simple molecule, there is still sufficient structural complexity to ensure that a large number of different synthetic approaches are possible.

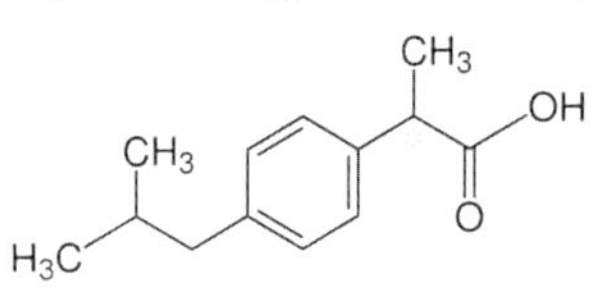

Ibuprofen

Ibuprofen is typically found in many over-the-counter drugs, such as Motrin, Advil, Potrin, and Nuprin. In other words, it often comes in capsules, tablets, or powder form. Comparing with that of aspirin, for example, Ibuprofen is somewhat short-lived and relatively mild. However, it is known to have an antiplatelet (non—blood clotting) effect.

The starting hydrocarbon for the production of ibuprofen is cumene, which is produced from benzene by the Friedel—Crafts alkylation of benzene with propylene. The original route for manufacturing of cumene was by alkylation of benzene in the liquid phase using sulfuric acid as a catalyst, and because of the complicated neutralization and recycling steps required, together with corrosion problems, this process has been largely replaced. As an alternative, solid phosphoric acid supported on alumina (Al_2O_3) was used as the catalyst:

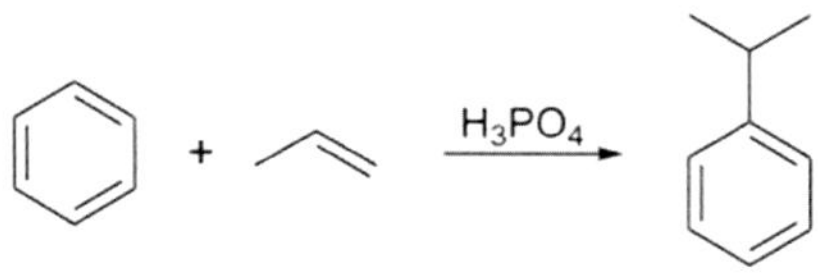

Since the mid-1990s, commercial production has switched to zeolite-based catalysts. The by-products are predominantly poly-isopropyl benzene derivatives. In 1976, an improved cumene process that uses aluminum chloride as a catalyst was developed. The addition of two equivalents of propylene to the reaction produces di-isopropyl benzene which, by transalkylation, with benzene yields the desired product (Vora et al., 2003).

Two of the most popular ways to obtain Ibuprofen are the Boot process and the Hoechst process. The Boot process is an older commercial process, and the Hoechst process is a newer process. The Boot process requires six steps, while the Hoechst process, with the assistance of catalysts, is completed in only three steps.

The Boot process:

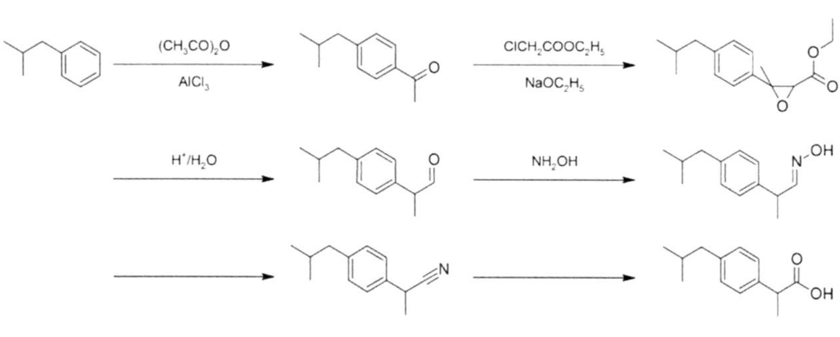

The Hoechst process:

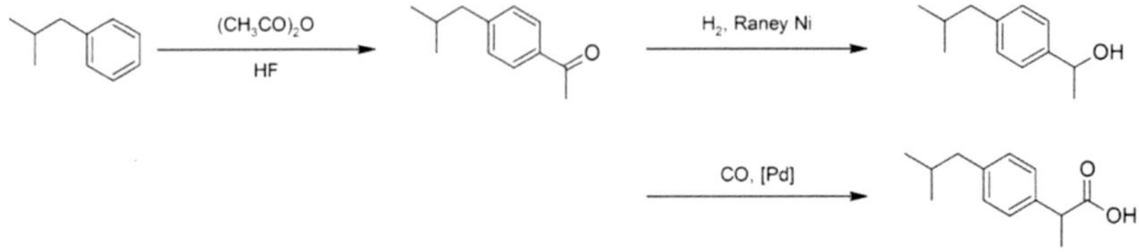

The starting material, cumene (isopropyl benzene, 2-phenylpropane, or 1-methylethyl benzene), for both of these processes is produced by the gas-phase reaction (Friedel–Crafts alkylation) of benzene by propylene. In the process, benzene and propylene are compressed together to a pressure on the order of 450 psi 250°C (482°F) in presence of a Lewis acid catalyst (such as an aluminum halide—a phosphoric acid (H_3PO_4) catalyst is often favored over an aluminum halide catalyst). Cumene is a colorless, volatile liquid with a gasoline-like odor. It is a natural component of coal tar and crude oil and also can be used as a blending component in gasoline.

4.6 Kaopectate

Kaopectate is an orally taken medication used for the treatment of mild indigestion, nausea, and stomach ulcers. The active ingredients have varied over time and are different between the United States and Canada. The original active ingredients were kaolinite (a layered clay mineral that has the approximate chemical composition $Al_2Si_2O_5(OH)_4$) and pectin (a structural heteropolysaccharide contained in the primary cell walls of terrestrial plants). In the United States, the active ingredient is now bismuth subsalicylate, which has the empirical chemical formula of $C_7H_5BiO_4$, and it is a colloidal substance obtained by hydrolysis of bismuth salicylate ($Bi(C_6H_4(OH)CO_2)_3$).

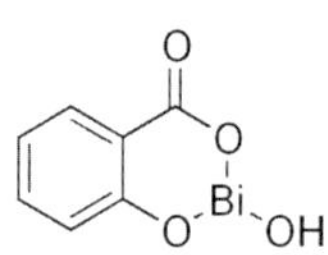

Bismuth subsalicylate

Bismuth subsalicylate is also the active ingredient in Pepto-Bismol and displays antiinflammatory action (due to salicylic acid) and is used to relieve the discomfort that arises from an upset stomach due to overindulgence in food and drink, including heartburn, indigestion, nausea, gas, and fullness (Tables 13.4, 13.5).

As stated previously, salicylic acid (or as a precursor to the acid, sodium salicylate) is produced commercially by treating sodium phenate (the sodium salt of phenol—phenol is a well-known petrochemical starting material) with carbon dioxide at high pressure (1500 psi) and high temperature (117°C,

TABLE 13.4 Examples of hydrocarbon solvents used in the pharmaceutical industry.

Solvent		*Use*	
Benzene	C		
Cyclohexane	C		
o-Dichlorobenzene (1,2-dichlorobenzene)	C		
n-Heptane	C	F	B
n-Hexane	C	F	B
Petroleum naphtha	C	F	B
Toluene	C	F	B
Xylene isomers	C		

C, Chemical synthesis, *F*, Fermentation, *B*, Biological or natural extraction.

242°F) (the Kolbe-Schmitt reaction) after which acidification of the product with sulfuric acid yield gives salicylic acid:

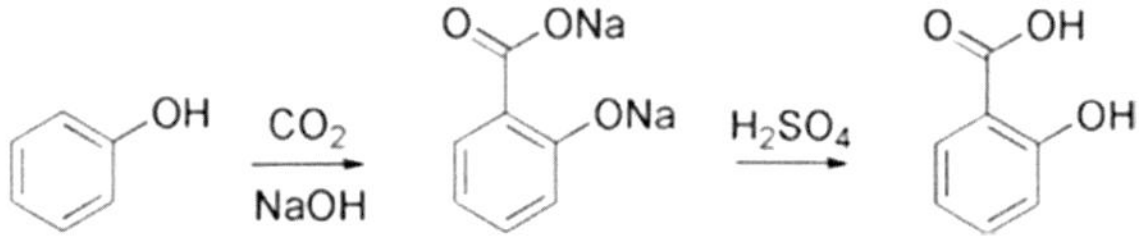

Salicylic acid can also be prepared by the hydrolysis of acetylsalicylic acid (aspirin) or by the hydrolysis of methyl salicylate (oil of wintergreen) with a strong acid or base. Another method for the production of salicylic acid involves biosynthesis from phenylalanine.

Salicylic acid is also used in the production of other pharmaceuticals, including 4-amino-salicylic acid and sandulpiride—the latter is an antipsychotic of the benzamide class that is used mainly in the treatment of psychosis associated with schizophrenia and depressive disorders. Other derivatives include methyl salicylate that is used as a liniment to soothe joint and muscle pain and choline salicylate that is used topically to relieve the pain of mouth ulcers.

4.7 Tylenol

The active constituent of Tylenol is acetaminophen that is an analgesic and fever-reducing medicine similar in effect to aspirin. It is an active ingredient in many over-the-counter medicines, including Tylenol and Midol. Introduced in the early 1900s, acetaminophen is a coal tar derivative that acts by interfering

TABLE 13.5 Physical properties of selected polynuclear aromatic compounds.

	Naphthalene	Anthracene	Phenanthrene	Benzo(a)pyrene
Molecular weight (g/mole)	128.16	178.23	178.23	252.3
Melting point (°C)	80.28	216.4	100.5	179
Boiling point (°C)	217.95	340	338	310–312
Solubility (aqueous, (mg/L)	30	0.065	1.28	3.8×10^{-3}
Vapor pressure (Torr)	0.082	5.63×10^{-6}	1.250×10^{-04}	5.25×10^{-9}
Henry's constant (atm-m^3/mol)	4.27×10^{-04}	1.8×10^{-6}	2.800×10^{-04}	5.53×10^{-07}
Molar volume (cm^3/mole)	148	197	199	263
Heat of vaporization (kJ/mol)	43.2	52.4	52.7	71.7
Molecular volume (Angstroms^3)	L 26.9	170.3	169.5	228.6
Molecular surface area (Angstroms^2)	L 55.8	202.2	198	225.6

with the synthesis of prostaglandins and other substances necessary for the transmission of pain impulses.

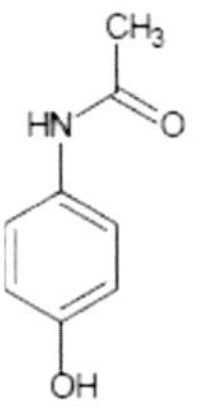

Acetaminophen

The preparation of acetaminophen involves treating an amine with an acid anhydride to form an amide. In this case, p-aminophenol, the amine, is treated with acetic anhydride to form acetaminophen (p-acetamidophenol), the amide:

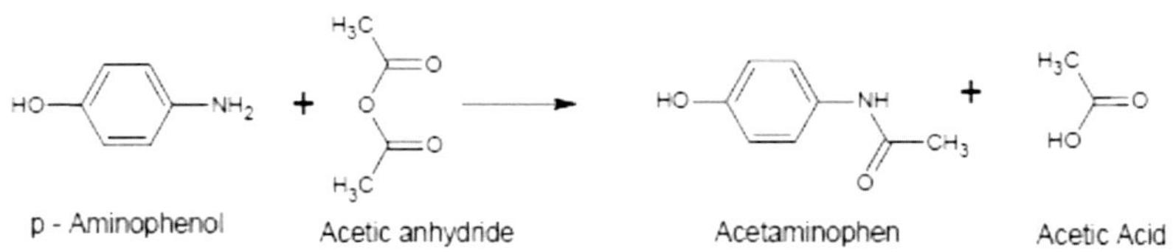

References

Abhari, R., 2010. Process for Producing Synthetic Petroleum Jelly. United States Patent 7,851,663. December 14.

Aronson, S.M., 2013. A tree-bark and its pilgrimage through history. Rhode Island Medical Journal 96 (2), 10–11.

Booth, G., 2005. Naphthalene derivatives. In: Ullmann's Encyclopedia of Industrial Chemistry. Wiley-VCH, Weinheim, Germany.

Bough, M., Trammel, H.L., 2006. Veterinary Technician. May, pp. 273–275.

Bourée, P., Blanc-Valleron, M.M., Ensaf, M., Ensaf, A., 2011. Use of bitumen in medicine throughout the ages. Histoire des Sciences Medicales 45 (2), 119–125.

Clayden, J., Greeves, N., Warren, S., Wothers, P., 2001. Organic Chemistry. Oxford University Press, Oxford, England.

Connan, J., 1999. Use and trade of bitumen in antiquity and prehistory: molecular archaeology reveals secrets of past civilizations. Philosophical Transactions of the Royal Society of London B: Biological Sciences 354 (1379), 33–50.

Fleck, S., Michels, R., Faure, P., Schlepp, L., Elie, M., Ashkan, S., Landais, P., 2000. Goldschmidt 2000, Oxford, UK. Journal of Conference Abstracts 5 (2), 403. September 3rd–8th, 2000.

Flick, A.C., Ding, H.X., Leverett, C.A., Kyne Jr., R.E., Liu, K.K.-C., Fink, S.J., O'Donnell, C.J., 2017. Synthetic approaches to the new drugs approved during 2015. Journal of Medicinal Chemistry 60, 6480–6515.

Forbes, R.J., 1958a. A History of Technology, vol. V. Oxford University Press, Oxford, England, p. 102.

Forbes, R.J., 1958b. Studies in Early Petroleum Chemistry. E. J. Brill, Leiden, The Netherlands.

Forbes, R.J., 1959. More Studies in Early Petroleum Chemistry. E.J. Brill, Leiden, The Netherlands.
Fuster, V., Sweeny, J.M., 2011. Aspirin: a historical and contemporary therapeutic overview. Circulation 123 (7), 768−778.
Gad, S.C. (Ed.), 2008. Pharmaceutical Manufacturing Handbook: Regulations and Quality. Wiley-Interscience, John Wiley & Sons Inc, Hoboken, New Jersey.
Gary, J.G., Handwerk, G.E., Kaiser, M.J., 2007. Petroleum Refining: Technology and Economics, fifth ed. CRC Press, Taylor & Francis Group, Boca Raton, Florida.
Guthrie, V., 1960. Petrochemical Products Handbook. McGraw-Hill, New York.
Hakluyt, R., 1582. Divers Voyages Touching the Discoverie of America and the Islands Adjacent unto the Same, Made First of All by Our Englishmen and Afterwards by the Frenchmen and Britons: With Two Mappes Annexed Hereunto. Thomas Dawson for T. Woodcocke, London, England (now: United Kingdom).
Hess, J., Bednarz, D., Bae, J., 2011. Petroleum and health care: evaluating and managing health care's vulnerability to petroleum supply shifts. American Journal of Public Health 101 (9), 1568−1579.
Hsu, C.S., Robinson, P.R. (Eds.), 2017. Handbook of Petroleum Technology. Springer International Publishing AG, Cham, Switzerland.
Huang, W.-Y., Meinschein, W.G., 1979. Geochimica et Cosmochimica Acta 43, 739−745. Speight, J.G. 2007. The Chemistry and Technology of Petroleum. 4th Edition. CRC Press, Taylor and Francis Group, Boca Raton, Florida.
Jones, A.W., 2011. Early drug discovery and the rise of pharmaceutical chemistry. Drug Testing and Analysis 3 (6), 337−344.
Karaman, R., 2015. Commonly Used Drugs − Uses, Side Effects, Bioavailability & Approaches to Improve it. Nova Biomedical, Nova Publishers, New York.
Kumar, S., Nautiyal, S.P., Agrawal, K.M., 2007. Physical properties of petroleum waxes 1: effect of oil content. Petroleum Science and Technology 25, 1531−1537.
Mahdi, J.G., Mahdi, A.J., Bowen, I.D., 2006. The historical analysis of aspirin discovery, its relation to the willow tree and antiproliferative and anticancer potential. Cell Proliferation 39 (2), 147−155.
Murohisa, T., Iida, M., 1993. Journal of Fermentation and Bioengineering 75, 13−17.
Parkash, S., 2003. Refining Processes Handbook. Gulf Professional Publishing, Elsevier, Amsterdam, Netherlands.
Purdy, G.A., 1967. Petroleum − Prehistoric to Petrochemicals. Copp Clark Publishing Co., Toronto, Ontario, Canada.
Schwark, L., Empt, P., 2006. Paleogeography, Paleoclimatology, Paleoecology 240 (1−2), 225−236.
Speight, J.G., 2014. The Chemistry and Technology of Petroleum, fifth ed. CRC-Taylor and Francis Group, Boca Raton, Florida.
Speight, J.G., 2017. Handbook of Petroleum Refining. CRC Press, Taylor & Francis Group, Boca Raton, Florida.
Speight, J.G., 2019. Handbook of Petrochemical Processes. CRC Press, Taylor & Francis Group, Boca Raton, Florida.
Volkman, J.K., 2008. In: Fleet, A.J., Kelts, K., Talbot, M.R. (Eds.), Lacustrine Petroleum Source Rocks, vol. 40. Special Publication, Geological Society, Oxford, pp. 103−122.
Vora, B.V., Kocal, J.A., Barger, P.T., Schmidt, R.J., Johnson, J.A., 2003. Alkylation. In: Kirk-Othmer Encyclopedia of Chemical Technology. John Wiley and Sons Inc., Hoboken, New Jersey.

Warren, J.K., 2006. Evaporites: Sediments, Resources and Hydrocarbons. Springer, Berlin, Germany.

Wick, J.Y., 2012. Aspirin: a history, a love story. The Consultant Pharmacist 27 (5), 322–329.

Further reading

Dias, J.R., 1987a. Handbook of Polycyclic Hydrocarbons: Part A, Benzenoid Hydrocarbons. Elsevier, Amsterdam, Netherlands.

Dias, J.R., 1987b. Handbook of Polycyclic Hydrocarbons: Part B: Polycyclic Isomers and Heteroatom Analogs of Benzenoid Hydrocarbons. Elsevier, Amsterdam, Netherlands.

FR, 2001. Federal Register: October 25, 2001 Rules and Regulations, 66 (207), pp. 53951–53957.

Harvey, R.G., 1991. Polycyclic Aromatic Hydrocarbons: Chemistry and Carcinogenicity. Cambridge University Press, Cambridge, England.

Mascal, M., Dutta, S., 2011. Synthesis of ranitidine (zantac) from cellulose-derived 5- (chloromethyl) furfural. Electronic Supplementary Material for Green Chemistry (The Royal Society of Chemistry).

Solomons, T.W.G., Fryhle, C.B., 2004. Organic Chemistry, eighth ed. John Wiley & Sons Inc., Hoboken, New Jersey.

Wise, S.A., 2003. Large (C>24) polycyclic hydrocarbon chemistry and analysis. Polycyclic Aromatic Compounds 23 (1), 109–111.

Chapter 14

Monomers, polymers, and plastics

1. Introduction

The list of hydrocarbon derivatives produced by the petrochemical industry includes, but is not limited to, (i) synthesis gas−based products including ammonia, methanol, and their derivatives, (ii) ethylene and derivatives, (iii) propylene, including on-purpose and methanol-based routes, and derivatives, (iv) C4 monomers, aromatic derivatives; oxides, glycol derivatives, and polyol derivatives, (v) chlor-alkali, ethylene dichloride, vinyl chloride monomer, and polyvinyl chloride, (vi) polyolefins—solution, slurry, and gas phase; alpha olefins and poly alpha olefins; polyethylene terephthalate (PET)—bottles and fiber; polystyrene—general purpose, high impact, and expandable, (vii) styrene derivatives—such as acrylonitrile butadiene styrene, acrylonitrile styrene, and acrylonitrile styrene acrylate, and (viii) specialty polymers including polyoxymethylene, superabsorbent polymers, and poly(methylmetacrylate); and nylon 6, 6-6, and intermediates (Matar and Hatch, 2001; Gary et al., 2007; Speight, 2014; Hsu and Robinson, 2017; Speight, 2017, 2019).

Thus, the ascent of polymer technology during the 20th century is due in no small part to the availability of starting materials that became available through the evolving and expanding petrochemical industry (Table 14.1). In fact, a high proportion of all petrochemicals are used for the production of polymers, the most important building blocks being ethylene, propylene, and butadiene (Table 14.2) (Matar and Hatch, 2001). These three petrochemicals can be polymerized directly, but an important part of their production is used to create more complex monomers through different ways of information into a polymer (Table 14.3). Ethylene is the progenitor of most vinyl monomers; hence, the need for an almost endless pressure on ethylene supply is particularly high. In fact, the C_2 and C_3 building blocks can be combined with benzene to form another set of monomers and intermediates, particularly valuable for constructing the complex repeat units noted in the last section. Other chemicals are also produced, such as plasticizers that are then added in a subsequent stage to polymers to modify their properties. But first, the relevant definitions.

Handbook of Industrial Hydrocarbon Processes. https://doi.org/10.1016/B978-0-12-809923-0.00014-X

TABLE 14.1 Polymers from hydrocarbon derivatives.

Polymer	Origin
Polyethylene	Derived from ethylene ($CH_2{=}CH_2$). Ethylene is derived from natural gas, from overhead gases in refinery, and from crackers.
Polypropylene	Derived from propylene ($CH_3CH{=}CH_2$). Propylene has almost same origin as ethylene. Used for making clothes various other plastics.
Rubber	Various synthetic rubbers such as polyacrylate rubber and ethylene-acrylate rubber. Derived from various petroleum-derived chemicals, such as 1,3-butadiene ($CH_2{=}CHCH{=}CH_2$) and acrylic acid ($CH_2{=}CHCO_2H$).
Polyesters	Produced from terephthalic acid that is derived from p-xylene (1,4-$HO_2CC_6H_4CO_2H$). p-Xylene ($H_3CC_6H_4CH_3$) has its origin from various aromatic compounds found in crude oil. Refining separates benzene derivatives.

A monomer is the original molecular form from which a polymer (and plastic product) is produced. A polymer (which may also be referred to as a macromolecule) consists of repeating molecular units that are usually held together by covalent bonds (Ali et al., 2005). Polymerization is the process of covalently bonding the low molecular weight monomers into a high molecular weight polymer.

Monomers are the basic molecular form which polymers and plastics are produced. Polymers consist of repeating molecular units that are usually joined by covalent bonds. Polymerization is the process of covalently bonding the low molecular weight monomers into a high molecular weight polymer. A polymer may also be referred to as a macromolecule (Ali et al., 2005). For a molecule to be a monomer, it must be at least bifunctional insofar as it has the capacity to interlink with other monomer molecules. While not truly bifunctional in the sense that they contain to functional groups, olefins have the ability to act as bifunctional molecules though the extra pair of electrons in the double bond.

A polymer may be a natural or synthetic macromolecule comprising repeating units of a smaller molecule (monomers). The terms polymer and plastic are often used interchangeably, but polymers are a much larger class of molecules that includes plastics, in addition to many other materials, such as cellulose, amber, and natural rubber. Examples of hydrocarbon polymers include polyethylene and synthetic rubber (Schroeder, 1983).

TABLE 14.2 Selected hydrocarbon addition polymers.

Name	*Formula*	*monomer*	*Physical properties*	*Uses*
Polyethylene: low density (LDPE)	$-(CH_2-CH_2)_n-$	Ethylene $CH_2{=}CH_2$	Soft, waxy solid	Film wrap, plastic bags
Polyethylene: high density (HDPE)	$-(CH_2-CH_2)_n-$	Ethylene $CH_2{=}CH_2$	Rigid, translucent solid	Electrical insulation, bottles, toys
Polypropylene: (PP) different grades	$-[CH_2-CH(CH_3)]_n-$	Propylene $CH_2{=}CHCH_3$	Atactic: soft, elastic solid Isotactic: hard, strong solid	Similar to LDPE, carpet, upholstery
Polystyrene (PS)	$-[CH_2-CH(C_6H_5)]_n-$	Styrene $CH_2{=}CHC_6H_5$	Hard, rigid, clear solid Soluble in organic solvents	Toys, cabinets, packaging (foamed)
Cis-polyisoprene: natural rubber	$-[CH_2-CH{=}C(CH_3)-CH_2]_n-$	Isoprene $CH_2{=}CH-C(CH_3)=CH_2$	Soft, sticky solid	Requires vulcanization for practical use

TABLE 14.3 Description of the various types of copolymers.

Statistical copolymers	Also called random copolymers. Here the monomeric units are distributed randomly, and sometimes unevenly, in the polymer chain: ~ABBAAABAABBBABAABA~.
Alternating copolymers	Here the monomeric units are distributed in a regular alternating fashion, with nearly equimolar amounts of each in the chain: ~ABABABABABABABAB~.
Block copolymers	Instead of a mixed distribution of monomeric units, a long sequence or block of one monomer is joined to a block of the second monomer: ~AAAAA-BBBBBBB~AAAAAAA~BBB~.
Graft copolymers	The side chains of a given monomer are attached to the main chain of the second monomer: ~AAAAAAA(BBBBBBB~)AAAAAAA(BBBB~)AAA~.

In the current context, a monomer is a low molecular weight hydrocarbon molecule—typically produced from crude oil (Parkash, 20013; Gary et al., 2007; Speight, 2014; Hsu and Robinson, 2017; Speight, 2017, 2019)—that has the potential of chemically bonding to other monomers of the same species to form a polymer. The lower molecular weight compounds built from monomers are also referred to as dimers (two monomer units), trimers (three monomer units), tetramers (four monomer units), pentamers (five monomer units), octamers (eight monomer units), continuing up to very high numbers of monomer units in the product.

The structure of monomer units in the polymer is retained by the chemical bonds between adjacent atoms, thereby conferring on the polymer configuration. However, there can be many different configurations for a given set of atoms of a particular type. Different isomers of the monomer unit, which have different properties, confer different properties on the polymer. This structural configuration of the monomer is an important structural feature and plays a major role as the complexity of the monomer increases and is a major determinant of the structure and properties of the polymer chains. In addition to structural isomerism in the monomer, which can be represented simply by the position of the double bond in butylene and is shown as butylene-1 and butylene-2, there is also a second type of isomerism.

CH3CH2CH=CH2	CH3CH=CHCH3
butylene-1	butylene-2
1-butene	2-butene

This type is isomerism (*geometrical isomerism*) occurs with various monomers and is present in both natural rubber and butadiene rubber (BR). In these cases, the single double bond in the final polymer can exist in two ways: a *cis* form and a *trans* form. In the *cis* form, the pendant methyl group appears on the same side as the lone hydrogen atom while in the trans form the pendant methyl group appears on the opposite side of the lone hydrogen atom.

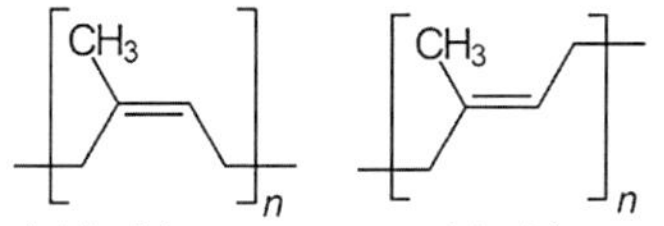

cis-1,4-polyisoprene trans-1,4-polyisoprene

Similarly, commencing with 2-butylene, the final polymer may have the methyl group on the same side of the final product or on alternate sides of the final product.

Because of the variations in monomer structure, the chemical structure of many polymers is rather complex because the polymerization reaction does not necessarily produce identical molecules. In fact, a polymeric material typically consists of a distribution of molecular sizes and sometimes also of shapes.

The properties of polymers are strongly influenced by details of the chain structure. The structural parameters that determine properties of a polymer include the overall chemical composition and the sequence of monomer units in the case of copolymers, the stereochemistry or tacticity of the chain, and geometric isomerization in the case of diene-type polymers. However, the properties of a specific polymer can often be varied by means of (i) controlling the molecular weight, (ii) the nature of the end groups, (iii) the manufacturing process, and (iv) any cross-linking that occurs within the molecular structure of the polymer. Therefore, it is possible to classify a single polymer in more than one category. For example, some polymers nylon can be produced as fibers in the crystalline forms or as plastics in the less crystalline forms. Also, certain polymers can be processed to act as plastics or elastomers.

2. Polymerization

Polymerization is the process by which polymers are manufactures, and during the polymerization process, some chemical groups may be lost from each monomer, and the polymer does not always retain the chemical properties or the reactivity of the monomer unit (Rudin, 1999; Braun et al., 2001; Carraher, 2003; Odian, 2004). Generally, polymerization is a relatively simple process, but the ways in which monomers are joined together vary and it is more convenient to have more than one system of describing polymerization. Polymerization occurs via a variety of reaction mechanisms that vary in complexity due to functional groups present in reacting compounds.

Polymerization reactions can occur in bulk (without solvent) in solution, emulsion, suspension, or a gas-phase process. Interfacial polymerization is also used with reactive monomers, such as acid chlorides. Polymers obtained by the *bulk technique* are usually pure due to the absence of a solvent. The purity of the final polymer depends on the purity of the monomers. Heat and viscosity are not easily controlled, as in other polymerization techniques, due to absence of a solvent, suspension, or emulsion medium. This can be overcome by carrying the reaction to low conversion and strong agitation. Outside cooling can also control the exothermic heat.

In *solution polymerization,* an organic solvent dissolves the monomer. Solvents should have low chain transfer activity to minimize chain transfer reactions that produce low molecular weight polymers. The presence of a solvent makes heat and viscosity control easier than in bulk polymerization. Removal of the solvent may not be necessary in certain applications such as coatings and adhesives.

Emulsion polymerization is widely used to produce polymers in the form of emulsions, such as paints and floor polishes. It also used to polymerize many water insoluble vinyl monomers, such as styrene and vinyl chloride. In emulsion polymerization, an agent emulsifies the monomers. Emulsifying agents should have a finite solubility. They are either ionic, as in the case of alkylbenzene sulfonates, or nonionic, as polyvinyl alcohol. Water is extensively used to produce emulsion polymers with a sodium stearate emulsifier. The emulsion concentration should allow micelles of large surface areas to form. The micelles absorb the monomer molecules activated by an initiator (such as a sulfate ion radical). X-ray and light-scattering techniques show that the micelles start to increase in size by absorbing the macromolecules. For example, in the free radical polymerization of styrene, the micelles increased to 250 times their original size.

In *suspension polymerization,* the monomer is first dispersed in a liquid, such as water, and mechanical agitation keeps the monomer dispersed. Initiators should be soluble in the monomer and stabilizers, such as talc or polyvinyl alcohol, prevent polymer chains from adhering to each other, and keep the monomer dispersed in the liquid medium. As a result, the final polymer appears in a granular form. Suspension polymerization produces polymers more pure than those from solution polymerization due to the absence of chain transfer reactions. As in a solution polymerization, the dispersing liquid helps control the heat of the reaction.

Interfacial polymerization is mainly used in polycondensation reactions with very reactive monomers. One of the reactants, usually an acid chloride, dissolves in an organic solvent (such as benzene or toluene), and the second reactant, a diamine or a diacid, dissolves in water. This technique produces polycarbonates, polyesters, and polyamides. The reaction occurs at the interface between the two immiscible liquids, and the polymer is continuously removed from the interface.

During the polymerization process, the structure of monomer units in the polymer is retained by the chemical bonds between adjacent atoms, thereby conferring on the polymer the configuration. However, there can be many different configurations for a given set of atoms of a particular type. Different isomers of the monomer unit, which have different properties, confer different properties on the polymer. This structural configuration of the monomer is an important structural feature and plays a major role as the complexity of the monomer increase and is a major determinant of the structure and properties of the polymer chains.

One system of separating polymerization processes is related to the amount of the original molecule (the monomer) that is left when the monomers bond. In *addition polymerization*, monomers are added together with their structure unchanged. Alkene derivatives that are relatively stable due to σ bonding between carbon atoms form polymers through relatively simple radical reactions.

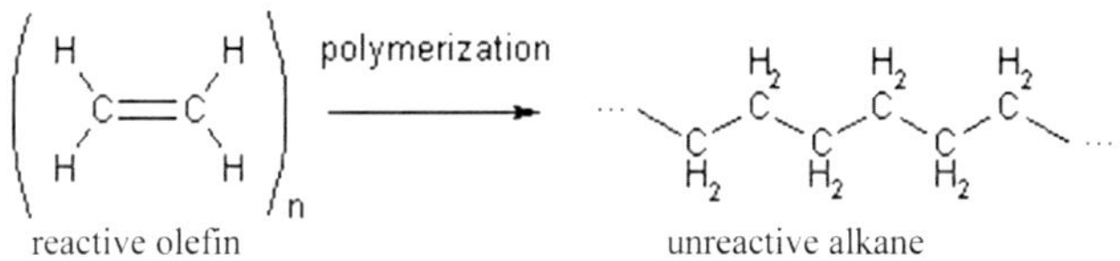

The chain terminating group can be a hydrogen atom (H) or any nonreactive (in this case) hydrocarbon moiety. On the other hand, condensation polymerization results in a polymer that is less massive than the two or more monomers that form the polymer because not all of the original monomer is incorporated into the polymer. Water is one of the common molecules chemically eliminated during condensation polymerization.

Polymers such as polyethylene are generally referred to as homopolymer as they consist of repeated long chains or structures of the same monomer unit, compared with polymers where polymerization occurs via a variety of reaction mechanisms that vary in complexity due to functional groups present in reacting compounds and their inherent steric effects. In more straightforward polymerization, alkene, which is relatively stable due to σ-bonding between carbon atoms, forms polymers through relatively simple radical reactions.

For hydrocarbon polymers, chain-growth polymerization (or addition polymerization) involves the linking together of molecules incorporating double or triple chemical bonds. These unsaturated monomers (the identical molecules that make up the polymers) have extra internal bonds that are able to break and link up with other monomers to form the repeating chain. Chain-growth polymerization is involved in the manufacture of polymers such as polyethylene and polypropylene.

All the monomers from which addition polymers are made are alkene derivatives or functionally substituted alkene derivatives. The most common and thermodynamically favored chemical transformations of alkene derivatives are addition reactions, and many of these addition reactions are known to proceed in a stepwise fashion by way of reactive intermediates:

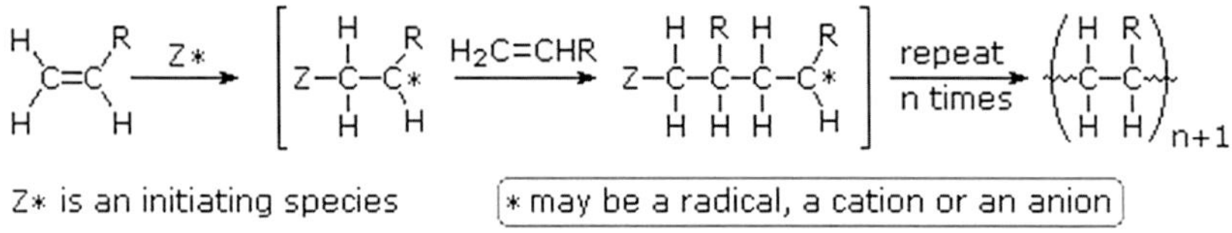

In principle, once initiated, a radical polymerization might be expected to continue unchecked, producing a few extremely long chain polymers. In practice, larger numbers of moderately sized chains are formed, indicating that chain-terminating reactions must be taking place. The most common termination processes are *radical combination* and *disproportionation* (Fig. 14.1). In both types of termination, two reactive radical sites are removed by simultaneous conversion to stable product(s). Because the concentration of radical species in a polymerization reaction is small relative to other reactants (e.g., monomers, solvents, and terminated chains), the rate at which these

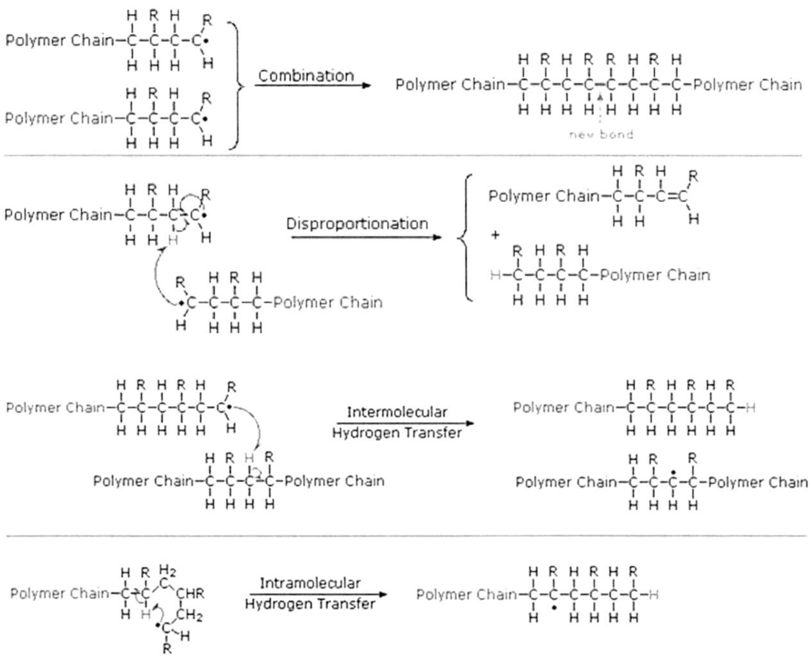

FIGURE 14.1 Examples of chain termination reactions.

radical–radical termination reactions occurs is very small, and most growing chains achieve moderate length before termination.

The relative importance of these terminations varies with the nature of the monomer undergoing polymerization. For acrylonitrile and styrene, combination is the major process. However, methyl methacrylate and vinyl acetate are terminated chiefly by disproportionation. Another reaction that diverts radical chain-growth polymerizations from producing linear macromolecules is *chain transfer* (Fig. 14.1), in which a carbon radical from one location is moved to another by an intermolecular or intramolecular hydrogen atom transfer.

Chain transfer reactions are especially prevalent in the high pressure radical polymerization of ethylene, which is the method used to make low-density polyethylene (LDPE). The primary radical at the end of a growing chain is converted to a more stable secondary radical by hydrogen atom transfer. Further polymerization at the new radical site generates a side chain radical, and this may in turn lead to creation of other side chains by chain transfer reactions. As a result, the morphology of LDPE is an amorphous network of highly branched macromolecules.

In the free radial polymerization of ethylene, the Π-bond (pi-bond) is broken, and the two electrons rearrange to create a new propagating center. The form this propagating center takes depends on the specific type of addition mechanism. There are several mechanisms through which this can be initiated. The free radical mechanism was one of the first methods to be used. Free radicals are very reactive atoms or molecules that have unpaired electrons. Taking the polymerization of ethylene as an example, the free radical mechanism can be divided in to three stages: chain initiation, chain propagation, and chain termination:

The free radical addition polymerization of ethylene must take place at high temperatures and pressures, approximately 300°C (570°F) and 29,000 psi. While most other free radical polymerizations do not require such extreme temperatures and pressures, they do tend to lack control. One effect of this lack of control is a high degree of branching. Also, as termination occurs randomly, when two chains collide, it is impossible to control the length of individual chains. A newer method of polymerization similar to free radical, but allowing more control, involves the Ziegler–Natta catalyst, especially with respect to polymer branching.

As alkene derivatives can be formed in somewhat straightforward reaction mechanisms, they form useful compounds such as polyethylene when undergoing radical reactions. Polymers such as polyethylene are generally referred to as homopolymers as they consist of repeated long chains or structures of the

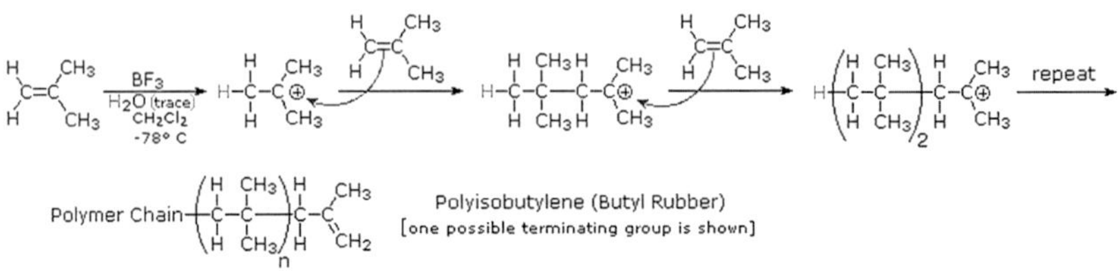

FIGURE 14.2 Cationic chain growth polymerization.

same monomer unit, whereas polymers that consist of more than one type of monomer are referred to as copolymers.

Polymerization of isobutylene (2-methylpropene) by traces of strong acids is an example of cationic polymerization (Fig. 14.2). The polyisobutylene product is a soft rubbery solid (Tg = −70°C), which is used for inner tubes. This process is similar to radical polymerization, and chain growth ceases when the terminal carbocation combines with a nucleophile or loses a proton, giving a terminal alkene.

Hydrocarbon monomers bearing cation stabilizing groups, such as alkyl, phenyl, or vinyl, can be polymerized by cationic processes. These are normally initiated at low temperature in methylene chloride solution. Strong acids, such as perchloric acid ($HClO_4$), or Lewis acids containing traces of water (Fig. 14.2) serve as initiating reagents. At low temperatures, chain transfer reactions are rare in such polymerizations, so the resulting polymers are cleanly linear (unbranched).

An example of anionic chain-growth polymerization is the treatment of a cold tetrahydrofuran solution of styrene with 0.001 equivalents of n-butyl lithium, which causes an immediate polymerization (Fig. 14.3). Species that have been used to initiate anionic polymerization include alkali metals, alkali amides, alkyl lithium compounds, and various electron sources. Chain growth may be terminated by water or carbon dioxide, and chain transfer seldom occurs. Only monomers having anion stabilizing substituents, such as phenyl, cyano, or carbonyl, are good substrates for this polymerization technique. Many of the resulting polymers are largely isotactic in configuration and have high degrees of crystallinity.

An example of ring opening polymerization is the process by which epoxy resins are produced when epichlorohydrin is reacted with a diphenol

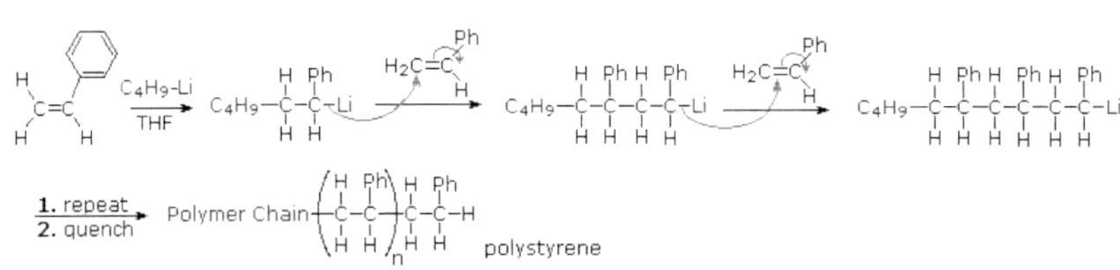

FIGURE 14.3 Anionic chain growth polymerization.

derivative. Bisphenol A is the diphenol generally used. The reaction, a ring opening polymerization of the epoxide ring, is catalyzed with strong bases such as sodium hydroxide. A nucleophilic attack of the phenoxy ion displaces a chloride ion and opens the ring. The linear polymer formed is cured by cross-linking either with an acid anhydride, which reacts with the −OH groups, or by an amine, which opens the terminal epoxide rings. Cresol derivatives and other bisphenol derivatives are also used for producing epoxy resins.

Finally, the use of Ziegler−Natta catalysts provides a stereospecific catalytic polymerization procedure (discovered by Karl Ziegler and Giulio Natta in the 1950s). Their catalysts permit the synthesis of unbranched, high molecular weight polyethylene (high-density polyethylene [HDPE]), laboratory synthesis of natural rubber from isoprene, and configurational control of polymers from terminal alkene derivatives, such as propylene (e.g., pure isotactic and syndiotactic polymers). In the case of ethylene, rapid polymerization occurs at atmospheric pressure and moderate to low temperature, giving a stronger (more crystalline) HDPE than that from radical polymerization (LDPE).

Ziegler−Natta catalysts are prepared by reacting specific transition metal halides with organometallic reagents such as alkyl aluminum, lithium, and zinc reagents. The catalyst formed by reaction of triethylaluminum with titanium tetrachloride has been widely studied, but other metals (e.g., vanadium and zirconium) have also proven effective (Fig. 14.4). Other catalysts have been suggested, with changes to accommodate the heterogeneity or homogeneity of the catalyst. Polymerization of propylene through action of the titanium catalyst gives an isotactic product; whereas, vanadium-based catalyst gives a syndiotactic product.

3. Polymers

A polymer is a large molecular (macromolecule) composed of repeating structural units (monomers) typically connected by covalent chemical bonds

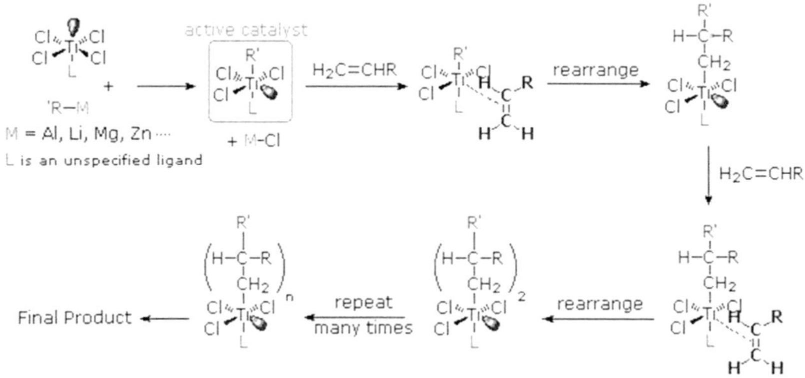

FIGURE 14.4 Mechanism of Ziegler−Natta catalysis.

(Rudin, 1999; Braun et al., 2001; Carraher, 2003; Odian, 2004). While polymer in the present context refers to hydrocarbon polymers, the term actually refers to a large class of natural and synthetic materials with a wide variety of properties, many of which are not true hydrocarbon derivatives. Polymers are formed by chemical reactions in which a large number of monomers are joined sequentially, forming a chain. In many polymers, only one monomer is used. In others, two or three different monomers may be combined. Hydrocarbon derivatives (alkene derivatives) are prevalent in the formation of addition polymers but do not usually participate in the formation of condensation polymers.

Before the early 1920s, chemists doubted the existence of molecules having molecular weights greater than a few thousand. This limiting view was challenged by Hermann Staudinger, a German chemist with experience in studying natural compounds such as rubber and cellulose. In contrast to the prevailing rationalization of these substances as aggregates of small molecules, Staudinger proposed they were made up of high molecular weight molecules composed of 10,000 or more atoms. He formulated a polymeric structure for rubber, based on a repeating monomer isoprene. Recognition that polymers make up many important natural materials was followed by the creation of synthetic analogs having a variety of properties. Indeed, applications of these materials as fibers, flexible films, adhesives, resistant paints, and tough but low-density solids have transformed modern society.

Polymers are classified by the characteristics of the reactions by which they are formed. If all atoms in the monomers are incorporated into the polymer, the polymer is called an addition polymer (Table 14.2). If some of the atoms of the monomers are released into small molecules, such as water, the polymer is called a condensation polymer. Most addition polymers are made from monomers containing a double bond between carbon atoms and are typical of polymers formed from olefins (alkene derivatives), and most commercial addition polymers are polyolefins. Condensation polymers are made from monomers that have two different groups of atoms that can join together to form, for example, ester or amide links. Polyesters are an important class of commercial polymers, as are polyamides (nylon).

The term polymer in popular usage suggests plastic but actually refers to a large class of natural and synthetic materials with a wide range of properties. A simple example is polyethylene (a polymer composed of a repeating ethylene unit) in which the range of properties varies depending on the number of ethylene units that make up the polymer. It is produced by the addition polymerization of ethylene (CH2=CH2). The properties of polyethylene depend on the manner in which ethylene is polymerized. When catalyzed by organometallic compounds at moderate pressure (220−450 psi), the product is HDPE. Under these conditions, the polymer chains grow to very great length, and molecular weight on the order of many hundreds of thousands are recorded. HDPE is hard, tough, and resilient. Most HDPE is used in the manufacture of containers, such as milk bottles and laundry detergent jugs.

When ethylene is polymerized at high pressure (15,000 to 30,000 psi), elevated temperatures (190–210°C, 380 to 410°F), and catalyzed by peroxide derivatives, the product is LDPE. This form of polyethylene has molecular weights on the order of 20,000–40,000. LDPE is relatively soft, and most of it is used in the production of plastic films, such as those used in sandwich bags.

Polypropylene is produced by the addition polymerization of propylene (CH3CH=CH2). The molecular structure is similar to that of polyethylene, but has a methyl group (–CH3) on alternate carbon atoms of the chain. The molecular weight falls in the range from 50,000 to 200,000. Polypropylene is slightly more brittle than polyethylene but softens at a temperature of approximately 40°C (104°F). Polypropylene is used extensively in the automotive industry for interior trim, such as instrument panels, and in food packaging, such as yogurt containers. It is formed into fibers of very low absorbance and high stain resistance used in clothing and home furnishings, especially carpeting.

Briefly, and by way of explanation, a fiber is often used as a polymer with a length-to-diameter ratio of at least 100 (Browne and Work, 1983). Fibers (synthetic or natural) are polymers with high molecular symmetry and strong cohesive energies between chains that result usually from the presence of polar groups. Fibers possess a high degree of crystallinity characterized by the presence of stiffening groups in the polymer backbone and of intermolecular hydrogen bonds. Also, they are characterized by the absence of branching or irregularly space-dependent groups that will otherwise disrupt the crystalline formation. Fibers are normally linear and drawn in one direction to make them long, thin, and threadlike, with great strength along the fiber. These characteristics permit formation of this type of polymers into long fibers suitable for textile applications. Typical examples of fibers include polyesters, nylons, and acrylic polymers, such as polyacrylonitrile, and naturally occurring polymers, such as cotton, wool, and silk.

Styrene ($C_6H_5CH{=}CH_2$) polymerizes readily to form polystyrene, a hard, highly transparent polymer. The molecular structure is similar to that of polypropylene, but with the methyl groups of polypropylene replaced by phenyl (C_6H_5) groups. A large portion of production goes into packaging. The thin, rigid, transparent containers in which fresh foods, such as salads, are packaged are made from polystyrene. Polystyrene is readily foamed or formed into beads. These foams and beads are excellent thermal insulators and are used to produce home insulation and containers for hot foods. Styrofoam is a trade name for foamed polystyrene.

When rubber is dissolved in styrene before it is polymerized, the polystyrene produced is much more impact resistant. This type of polystyrene is used extensively in home appliances, such as the interior of refrigerators and air conditioner housing. Natural polymeric materials such as natural rubber have been in use for centuries, and a variety of other natural polymers exist, such as cellulose, which is the main constituent of wood and paper. In spite of claim to the contrary, coal is not a polymer—coal is a macromolecular carbonaceous natural-occurring material that has no regularity of structure and hence does not have the chemical or physical character of a polymer (Speight, 2013, 2015).

Polypropylene is also a common polymer and is based on the propylene monomer but, in contrast to polyethylene, has a methyl group attached to the hydrocarbon backbone:

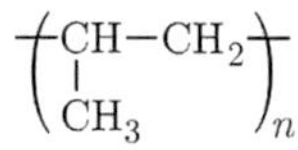

Polypropylene

Natural rubber is an elastomer (an elastic hydrocarbon polymer) that was originally derived from latex, a milky colloidal suspension found in the sap of some plants. It is useful directly in this form, but the invention of vulcanization in which natural rubber was heated with, sulfur forming cross-links between polymer chains (vulcanization), improves elasticity and durability.

Isoprene (2-methyl-1,3-butadiene) is a common organic compound with the formula $CH_2{=}C(CH_3)CH{=}CH_2$:

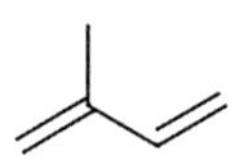
Isoprene

Isoprene is present under standard conditions as a colorless liquid and is the monomer of natural rubber as well as a precursor to an immense variety of other naturally occurring compounds. Natural rubber is a polymer of isoprene—most often cis-1,4-polyisoprene—with a molecular weight of 100,000 to 1,000,000. Typically, a few percent of other materials, such as proteins, fatty acids, resins, and inorganic materials, are found in high-quality natural rubber. Some natural rubber sources called gutta percha are composed of trans-1,4-polyisoprene, a structural isomer that has similar, but not identical, properties.

This type of isomerism (geometrical isomerism) occurs with various monomers and is present in polymers such as natural rubber and BR. In these cases, the single double bond in the final polymer can exist in two ways: a cis form and a trans form.

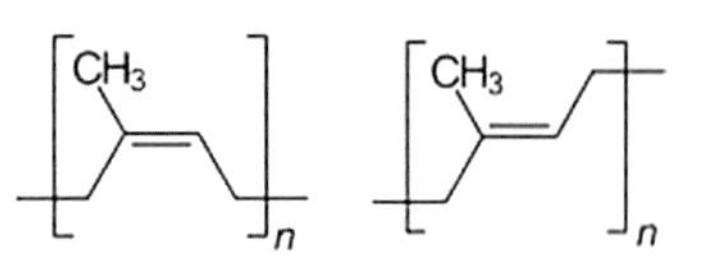

cis-1,4-polyisoprene trans-1,4-polyisoprene

Cis-1,4-polyisoprene Trans-1,4-polyisoprene

The pendant methyl group appears on the same side as the lone hydrogen atom or on the opposite side of the lone hydrogen atom. Similarly, commencing with 2-butylene, the final polymer may have the methyl group (i) on the same side of the final product or (ii) on alternate sides of the final product. Thus, because of the variations in monomer structure, the chemical structure of many polymers is rather complex because the polymerization reaction does not necessarily produce identical molecules. In fact, a polymeric material typically consists of a distribution of molecular sizes and sometimes also of shapes.

Synthetic rubber is any type of artificial elastomer, invariably a polymer. An elastomer is a material with the mechanical (or material) property that it can undergo much more elastic deformation under stress than most materials and still return to its previous size without permanent deformation. Synthetic rubber serves as a substitute for natural rubber in many cases, especially when improved material properties are required.

Synthetic rubber can be made from the polymerization of a variety of monomers including isoprene (2-methyl-1,3-butadiene), 1,3-butadiene, and iso-butylene (methylpropene) with a small percentage of isoprene for cross-linking. These and other monomers can be mixed in various desirable proportions to be copolymerized for a wide range of physical, mechanical, and chemical properties. The monomers can be produced pure, and the addition of impurities or additives can be controlled by design to give optimal properties. Polymerization of pure monomers can be better controlled to give a desired proportion of cis and trans double bonds.

3.1 Chain length

The physical properties of a polymer are strongly dependent on the size or length of the polymer chain. For example, as chain length is increased, melting and boiling temperatures increase quickly (http://en.wikipedia.org/wiki/Polymer-cite_note-PP 5-12). Impact resistance also tends to increase with chain length, as does the viscosity, or resistance to flow, of the polymer in its melt state. Chain length is related to melt viscosity in the approximate ratio of 1:10, which is a 10-fold increase in polymer chain length and results in a viscosity increase of over 1000 times.

Increasing chain length furthermore tends to decrease chain mobility, increase strength and toughness, and increase the glass transition temperature (Tg). This is a result of the increase in chain interactions such as van der Waals attractions and entanglements that come with increased chain length. These interactions tend to fix the individual chains more strongly in position and resist deformations and matrix breakup, both at higher stresses and higher temperatures.

3.2 Copolymers

Polymers containing a mixture of different repeat units are known as copolymers (e.g., styrene-ethylene copolymer). The synthesis of macromolecules composed of more than one monomeric repeating unit has been explored as a means of controlling the properties of the resulting material. In this respect, it is useful to distinguish several ways in which different monomeric units might be incorporated in a polymeric molecule. The examples presented (Table 14.3) refer to a two component system (A and B) in which the relationships of the monomers are varied.

Most direct copolymerization processes of equimolar mixtures of different monomers give statistical copolymers, if one monomer is much more reactive a nearly homopolymer of that monomer. Radical polymerization gives a statistical copolymer. In cases where the relative reactivity is different, the copolymer composition can sometimes be controlled by continuous introduction of a biased mixture of monomers into the reaction.

A growing number of commercial polymers are actually composed of different types of unit attached together by chemical covalent bonds (copolymers) and can comprise just two different units (binary copolymers) or three different units (ternary copolymers) and so on. This allows manipulation of the polymer properties to gain just the right combination of properties for a specific application. Monomers within a copolymer may be organized along the backbone in a variety of ways (Table 14.3): (i) alternating copolymers possess regularly alternating monomer residues, (ii) periodic copolymers have monomer residue types arranged in a repeating sequence, (iii) random copolymers that have monomer residues arranged in no particular order, (iv) block copolymers have two or more homopolymer subunits linked by covalent bonds, and (v) graft or grafted copolymers that contain side chains that have a different composition or configuration than the main chain.

Graft copolymers are a special type of branched copolymer in which the side chains are structurally distinct from the main chain (Table 14.3). However, the individual chains of a graft copolymer may be homopolymers or copolymers. Block copolymers are made up of blocks of different monomers and can be in the form of a diblock copolymer, which contains two different chemical blocks; there are also triblock copolymers, tetrablock copolymers, and multiblock copolymers. The most powerful strategy to prepare block copolymers is the chemoselective stepwise coupling between polymeric precursors and heterofunctional linking agents.

One of the best known examples of property modification using a copolymer involves polystyrene. In the homopolymer form (Fig. 14.5, line 1), polystyrene is a rigid (hence, brittle), transparent thermoplastic that finds little application for stressed applications in its original state. Polystyrene also shows a glass transition temperature (Tg) of approximately 97°C (approximately 206°F) and is not suitable for use in manufacturing containers to hold

FIGURE 14.5 Illustration of the variations in polymer structure. Key: (1) Regular polymer—also called a homogeneous polymer. (2) Alternating copolymer. (3) Random copolymer. (4) Block copolymer. (5) Graft copolymer.

boiling water. The properties can be changed by copolymerizing styrene with acrylonitrile to produce a styrene–acrylonitrile (SAN) polymer, where the styrene and acrylonitrile units alternate along the backbone chain of the material (Fig. 14.5, line 2). Other variations in polymer structure are also possible (Fig. 14.5, lines 3 to 5) and include polymers such as (i) the random copolymer, in which there is no order to the backbone, (ii) the block copolymer, in which the monomers occur in cells with no definitive number of monomers per cell, and (iii) the graft copolymer, in which polymer cells from different monomers are grafted on to the side of the main polymer chain.

The glass transition temperature of the SAN polymer is approximately 107°C (225°F), which is above the boiling point of water. Furthermore, if rubber (polybutadiene polymer) chains are grafted onto the main backbone polystyrene chain (to produce high-impact polystyrene), the graft copolymer so formed is much tougher, owing to molecular segregation of the rubber chains. This reduces the stiffness of the copolymer compared with the parent polystyrene, and the bulk material is ductile and tough. The benefits of both a high glass transition temperature and toughness are achieved by use of an acrylonitrile-butadiene-styrene terpolymer of the three component repeat units, which toughens the otherwise brittle product by adding rubber particles (rubber-toughening).

In copolymers, the different repeat units added to the original polymer are always covalently bonded and are locked into the structure, so the composition is highly variable. As an example, if the amount of rubber component in a copolymer is increased, the properties change from those of a plastic to that of a reinforced rubber. In fact, copolymerization of butadiene and styrene was employed at a very early stage in the development of synthetic rubber during the last World War after it was discovered that the stiffness of polybutadiene rubber could be improved by copolymerization styrene (approximately 24% w/w) to give a random copolymer of the two units, known as styrene-butadiene rubber (SBR).

In the 1960s, there came the discovery that another way of putting the units together was possible, and this involved an alternate route to use a mixture of monomers. The units were added sequentially and a block copolymer was the result, in which the properties are different to those of the random copolymer because each type of chain segregates together to form minute domains. Such materials retain thermoplastic behavior yet behave as cross-linked rubbers, and the styrene-butadiene-styrene (SBS) block copolymer was the first commercial thermoplastic elastomer (TPE).

3.3 Glass transition temperature

The two most important transitions exhibited by polymers are (i) the glass transition temperature, T_g, and (ii) the crystalline melting temperature, T_m. The glass transition temperature (vitrification temperature) is the temperature at which a liquid transforms into a glass, which usually occurs on rapid cooling. It is a dynamic phenomenon occurring between two distinct states of matter (liquid and glass), each with different physical properties. On cooling through the temperature range of glass transition (glass transformation range), without forming any long-range order or significant symmetry of atomic arrangement, the liquid contracts more continuously at approximately the same rate as above the melting point until there is a decrease in the thermal expansion coefficient. The crystalline melting temperature, which is related to the glass transition temperatures, is the temperature at which the crystalline domains lose their structure, or melt. As crystallinity increases, so does the crystalline melting temperature.

The glass transition temperature often depends on the history of the sample, particularly previous heat treatment, mechanical manipulation, and annealing. It is sometimes interpreted as the temperature above which significant portions of polymer chains are able to slide past each other in response to an applied force. The introduction of relatively large and stiff substituents (such as benzene rings) will interfere with this chain movement, thus increasing the glass transition temperature. The introduction of low molecular weight molecular compounds (plasticizers) into the polymer matrix increases the interchain spacing, allowing chain movement at lower temperatures, with a resulting decrease in the glass transition temperature.

Many glass transition temperatures (Table 14.4) are mean values because the glass transition temperature depends on the cooling rate and molecular weight distribution and could be influenced by additives. Note also that for a semicrystalline material, such as polyethylene that is 60%−80% crystalline at room temperature, the quoted glass transition refers to what happens to the amorphous part of the material on cooling.

The glass transition temperature, Tg, is lower than melting temperature, Tm, due to supercooling and depends on the time scale of observation which must be defined by convention. One approach is to agree on a standard cooling

TABLE 14.4 Glass transition temperatures of various polymers.

Material	T_g (°C)
Tire rubber	−70
Polypropylene (atactic)	−20
Poly(vinyl acetate) (PVAc)	30
Polyethylene terephthalate (PET)	70
Poly(vinyl alcohol) (PVA)	85
Poly(vinyl chloride) (PVC)	80
Polystyrene	95
Polypropylene (isotactic)	0
Poly-3-hydroxybutyrate (PHB)	15
Polymethylmethacrylate (atactic)	105

rate of 10 K/min. Another approach is by requiring a viscosity of 1013 P. Otherwise, it is more correct to present or discuss a glass transformation range as the base point has not been standardized.

3.4 Molecular weight

A common means of expressing the length of a chain is the degree of polymerization, which defines or quantifies the number of monomers incorporated into the chain and therefore gives an indication of the molecular weight of the polymer. Because synthetic polymerization techniques typically yield a polymer product including a range of molecular weights, the weight is often expressed statistically to describe the distribution of chain lengths present in the same. Common examples are the number average molecular weight (Mn) and the weight average molecular weight (Mw). The ratio of these two values is the polydispersity index, which is commonly used to express the extent of the molecular weight distribution. A final measurement is the contour length, which can be understood as the length of the chain backbone in its fully extended state. The molecular weights of polymers are of the most difficult measurements to make, and the reliability of the data may also be questioned.

Two experimentally determined values are common: (i) M_n, the number average molecular weight, which is calculated from the mole fraction distribution of different sized molecules in a sample, and (ii) M_w, the weight average molecular weight, which is calculated from the weight fraction distribution of different sized molecules.

Because larger molecules in a sample weigh more than smaller molecules, the weight average Mw is necessarily skewed to higher values and is always greater than Mn. As the weight dispersion of molecules in a sample narrows, Mw approaches Mn, and in the unlikely case that all the polymer molecules have identical weights (a pure monodisperse sample), the ratio Mw/Mn becomes unity.

3.5 Phase separation

Polymer mixtures (solutions) as a rule are viscous, and the timescales required to effect phase separation can be much longer than for analogous solutions in which small (nonpolymeric molecules) are dissolved (Mannella et al., 2016). The propensity for phase separation is related to the small entropy of mixing characteristic of polymer mixtures. Four manifestations of this small mixing entropy are (i) the overwhelming majority of polymer pairs (blends) are immiscible, (ii) polymer solutions and blends that are miscible tend to phase separate at elevated temperatures, (iii) a ternary blend may phase separate even when the corresponding pairs are completely miscible, and (iv) an AB copolymer may be miscible with the homopolymer even though the corresponding homopolymer pairs are immiscible. Although the specific reasons for these four effects differ, they have small mixing entropy as the common ingredient and primary influence.

Block copolymers can microphase separate to form periodic nanostructures, as in the SBS block copolymer. Because of incompatibility between the blocks, block copolymers undergo separation but, because the chemical blocks are covalently bonded to each other, they cannot separate macroscopically. In microphase separation, the blocks form nanometer-sized structures. Depending on the relative lengths of each block, several morphologies can be obtained. In diblock copolymers, sufficiently different block lengths lead to nanometer-sized spheres of one block in a matrix of the second. Using less different block lengths, a hexagonally packed cylinder geometry can be obtained. Blocks of similar length form layers (lamellae), and between the cylindrical and lamellar phase is the gyroid phase.

3.6 Polymer degradation

Polymer degradation is a change in the properties of the polymer, such tensile strength, color, shape, and molecular weight, or of a polymer-based product under the influence of one or more environmental factors, such as heat, light, chemicals, or any other applied force. Degradation is often due to a change in the chemical and/or physical structure of the polymer chain, which in turn leads to a decrease in the molecular weight of the polymer. These changes may be undesirable, such as changes during use, or desirable, as in biodegradation or deliberately lowering the molecular weight of a polymer. Such changes occur primarily because of the effect of these factors on the chemical composition of the polymer. The susceptibility of a polymer to degradation depends on its structure. Epoxies and chains containing aromatic functionality are especially susceptible to ultraviolet degradation, while hydrocarbon-based polymers are susceptible to thermal degradation and are often not ideal for high temperature applications.

The degradation of polymers to form smaller molecules may proceed by random scission or specific scission. The degradation of polyethylene occurs by random scission—a random breakage of the bonds within the polymer. When heated above 450 C (840°F), polyethylene degrades to form a mixture of hydrocarbon derivatives. Other hydrocarbon polymers, such as poly-α-methylstyrene, undergo specific chain scission with breakage occurring only at the ends, and such polymers depolymerize (unzip) to produce the constituent monomer.

While the degradation process represents failure of the polymer to perform in service, the process can be useful from the viewpoints of understanding the structure of a polymer or recycling/reusing the polymer waste to prevent or reduce environmental pollution by conversion to useful hydrocarbon derivatives (Sarker et al., 2011; Abbas and Mohamed, 2015; Pundhir, and Gagneja, 2016). The sorting of polymer waste for recycling purposes may be facilitated by the knowledge of the degradation process and assist in recycling.

3.7 Properties

The properties of polymers are strongly influenced by details of the chain structure. The structural parameters that determine properties of a polymer include the overall chemical composition and the sequence of monomer units in the case of copolymers, the stereochemistry or the relative stereochemistry of the stereocenters in the polymer chain, and geometric isomerization in the case of diene-type polymers. Thus, the properties of a specific polymer can often be varied by means of controlling molecular weight, end groups, processing, and cross-linking. Therefore, it is possible to classify a single polymer in more than one category. For example, some polymers nylon can be produced as fibers in the crystalline forms or as plastics in the less crystalline forms. Also, certain polymers can be processed to act as plastics or elastomers. Synthetic fibers are long-chain polymers characterized by highly crystalline regions resulting mainly from secondary forces (e.g., hydrogen bonding). They have a much lower elasticity than plastics and elastomers. They also have high tensile strength, a low density (i.e., light weight), and low moisture absorption.

The attractive forces between polymer chains play a large part in determining the properties of a polymer. Because polymer chains are so long, these interchain forces are amplified far beyond the attractions between conventional molecules. Different side groups on the polymer can lend the polymer to ionic bonding or hydrogen bonding between its own chains. These stronger forces typically result in higher tensile strength and higher crystalline melting points.

The intermolecular forces in polymers can be affected by dipoles in the monomer units. Polymers containing nonhydrocarbon groups can form hydrogen bonds between adjacent chains, which can result in the high tensile strength and melting point of the polymers. Other nonhydrocarbon groups can have dipole–dipole bonding between the nonhydrocarbon functions. However, dipole bonding is not as strong as hydrogen bonding, and the melting points of such polymers will be lower than hydrogen-bonded polymers, but the dipole-bonded polymers will have greater flexibility. In a true hydrocarbon polymer such as polyethylene, the salutation is different.

Ethylene has no permanent dipole. The attractive forces between polyethylene chains arise from weak van der Waals forces. Molecules are often pictures (rightly or wrongly) as being surrounded by a cloud of negative electrons. As two polymer chains approach, their electron clouds repel one another. This has the effect of lowering the electron density on one side of a polymer chain, creating a slight positive dipole on this side. This charge is enough to attract the second polymer chain. However, because van der Waals forces are weak, polyethylene can have a lower melting temperature compared with other polymers.

3.8 Repeat unit placement

The microstructure of a polymer (sometimes called configuration) relates to the physical arrangement of monomer residues along the backbone of the polymer. These are the elements of polymer structure that require the breaking of a covalent bond to change. Structure has a strong influence on the other properties of a polymer. For example, two samples of natural rubber may exhibit different durability, even though their molecules comprise the same monomers.

An important microstructural feature determining polymer properties is the polymer architecture. The simplest polymer architecture is a linear chain: a single backbone with no branches. A related unbranching architecture is a ring polymer. A branched polymer molecule is composed of a main chain with one or more substituent side chains or branches.

Branching of polymer chains affects the ability of chains to slide past one another by altering intermolecular forces, in turn affecting bulk physical polymer properties. Long chain branches may increase polymer strength, toughness, and the glass transition temperature due to an increase in the number of entanglements per chain. The effect of such long-chain branches on the size of the polymer in solution is characterized by the branching index. On the other hand, random length and atactic short chains may reduce polymer strength due to disruption of organization and may likewise reduce the crystallinity of the polymer.

Three factors that influence the degree of crystallinity are (i) chain length, (ii) chain branching, and (iii) interchain bonding. The importance of the first two factors is illustrated by the differences between LDPE and HDPE. Increased crystallinity is associated with an increase in rigidity, tensile strength, and opacity (due to light scattering). Amorphous polymers are usually less rigid, weaker, and more easily deformed. They are often transparent.

LDPE is composed of smaller and more highly branched chains that do not easily adopt crystalline structures. This material is therefore softer, weaker, less dense, and more easily deformed than HDPE. Generally, mechanical properties such as ductility, tensile strength, and hardness rise and eventually level off with increasing chain length. On the other hand, HDPE is composed of very long unbranched hydrocarbon chains. These pack together easily in crystalline domains that alternate with amorphous segments, and the resulting material, while relatively strong and stiff, retains a degree of flexibility. LDPE has significant numbers of both long and short branches, is quite flexible, and is used in applications such as plastic films, whereas HDPE has a very low degree of branching, is quite stiff, and is used in applications such as milk jugs.

In contrast, natural rubber is a completely amorphous polymer and the potentially useful properties of raw latex rubber are limited by temperature dependence; however, these properties can be modified by chemical change. The cis-double bonds in the hydrocarbon chain provide planar segments that stiffen, but do not straighten, the chain. If these rigid segments are completely removed by hydrogenation, the chains lose all constraints, and the product is a low melting paraffin-like semisolid of little value. If instead, the chains of rubber molecules are slightly cross-linked by sulfur atoms (vulcanization—discovered by Charles Goodyear in 1839), the desirable elastomeric properties of rubber are substantially improved. At 2%–3% cross-linking, a useful soft rubber is produced, which no longer suffers stickiness and brittleness problems on heating and cooling. At 25%–35% cross-linking, a rigid hard rubber product is formed.

Dendrimers are a special case of polymer where every monomer unit is branched. This tends to reduce intermolecular chain entanglement and crystallization. Alternatively, dendritic polymers are not perfectly branched but share similar properties to dendrimers due to their high degree of branching.

The architecture of the polymer is often physically determined by the functionality of the monomers from which it is formed. This property of a monomer is defined as the number of reaction sites at which may form chemical covalent bonds. The basic functionality required for forming even a linear chain is two bonding sites. Higher functionality yields branched or even cross-linked or networked polymer chains.

An effect related to branching is chemical cross-linking—the formation of covalent bonds between chains, which tends to increase strength and the glass transition temperature (Tg). Among other applications, this process is used to strengthen rubbers in a process known as vulcanization (cross-linking by sulfur). Car tires, for example, are highly cross-linked to reduce the leaking of air out of the tire and to toughen their durability. Eraser rubber, on the other hand, is not cross-linked to allow flaking of the rubber and prevent damage to the paper.

A cross-link suggests a branch point from which four or more distinct chains emanate. A polymer molecule with a high degree of cross-linking is referred to as a polymer network. Sufficiently, high cross-link concentrations may lead to the formation of an infinite network, also known as a gel, in which networks of chains are of unlimited extent—essentially all chains are linked into one molecule.

In the saturated hydrocarbon derivatives, it is not possible to form distinct isomers with just three or less carbon atoms linked together (Chapter 1). There is only one way in which one carbon and four hydrogen atoms can be linked together, the single compound being methane. A similar situation holds for ethane and propane, but with butane, two possible structures can be formed: n-butane, which has a linear structure, and isobutane, where the central carbon atom is linked to three adjacent carbons rather than one.

Their physical properties are slightly different, for example, their boiling points differ by 10°C (18°F), but otherwise, they are very similar compounds. The number of possible isomeric structures increases with carbon number: there are 3 isomers for pentane, five for hexane, nine for heptane, and as the number of possible structures increases, their properties diverge: the boiling points of the heptanes range over 20°C and the melting points by no less than 110°C (230°F).

The increase in isomeric structures is so rapid that at C30, there are no less than 4,111,846,763 theoretically possible compounds. So with even the lowest molecular mass polyethylene, there is an almost infinite number of isomers. Fortunately, the situation is simplified enormously by the way polymerization occurs and in fact there is relatively few chemically distinct polyethylene polymers. The concept of a distinct molecular formula is redundant with most commercial polymers as chain lengths are very variable even within a single sample, but the idea of branching is important for polyethylene in particular.

In addition, the formulas for 2- and 3-methyl pentane show that a single methyl group (CH_3) can occur in two different positions along an essentially linear carbon–carbon chain. The methyl group is a very simple kind of branch along the chain, and it is easy to extend the idea to much larger molecules. Thus, LDPE is a polymer based on a linear backbone chain with the repeat unit (CH_2CH_2) but is in addition branched with very long chains at infrequent points along the main chain (approximately 1 in 1000). Branching is caused during polymerization at high pressure by growth sometimes starting from an initiation point in a chain rather than at the end. An alternative way of making polyethylene is at low pressure using a special catalyst, and this usually results in a linear chain without branching (HDPE).

However, it is easy to polymerize a mixture of ethylene with a higher alkene such as hex-1-ene, so that the new units copolymerize together to form a chain where a certain proportion of the chains have tails or short branches along the linear sequence. This and similar copolymers are generically part of the polyethylene family and are known as linear low-density polyethylene.

This kind of structural variation is important because it affects the properties of the polymers, as their names indicate. Thus, branches along the chain hinder crystallization of the chains, resulting in a less dense and lower modulus material. LDPE typically has a density of 0.92, while HDPE has a higher density of 0.96.

A second type of isomerism occurs with diene monomers and is present in both natural rubber and BRs because the single double bond in the final polymer can exist in two ways: a cis form and a trans form. The two parts of the chain in which this single repeat unit sits lie on the opposite side of the double bond:

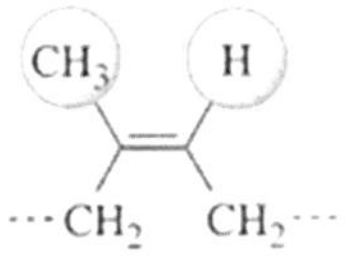

cis- polyisoprene

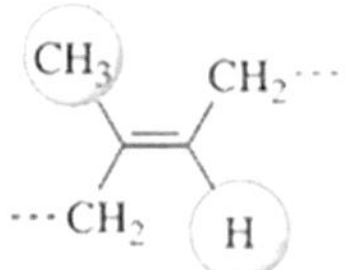

trans- polyisoprene

A final type of isomeric variation occurs as a result of the three-dimensional structure of some polymers. It is possible because a four-valent atom like carbon can exist in two different forms when the subsidiary groups or atoms attached to the carbon are all different (*asymmetric carbon atom*).

The carbon atom in a vinyl polymer to which is attached the pendant side group (i.e., every alternate carbon atom in the main chain) is another example of an asymmetric carbon atom, which gives rise to *tacticity*. The methyl groups in polypropylene, for example, can occur all on one side (*isotactic*), on alternate sides (*syndiotactic*), or be placed at random (*atactic*). The properties of each type of polymer are quite different to one another, primarily because *isotactic* and *syndiotactic* polypropylene have ordered chains and so can crystallize, but atactic chains are quite irregular and cannot crystallize. Isotactic polypropylene is the common form of the commercial material, although atactic polypropylene is used as a binder for paper for example.

Another type of polymer structure relates to the addition of adding monomer units to a growing chain in reverse rather than in their normal position. Monomer molecules, having a particular shape in space, will usually approach a growing chain end to minimize any spatial interaction, and a regular chain structure results from head-to-tail joints (structure a, below). A defective joint can sometimes occur, however, when heads combine to form a head-to-head joint (structure b, below).

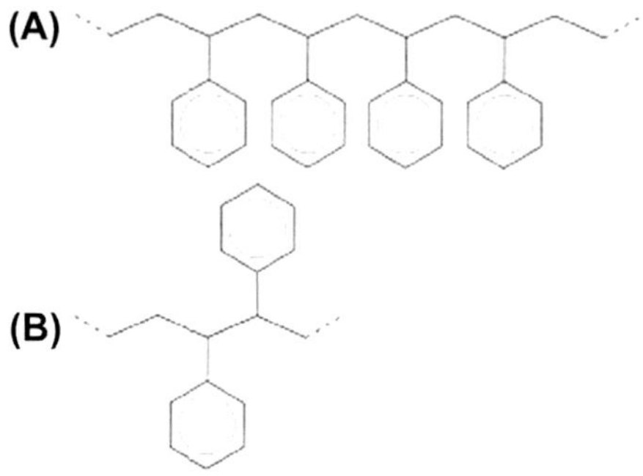

Structure b has a high potential to change the properties of the polymer as it may induce weakness in the bonding. For example, the change on bonding may cause degradation (thermal or oxidative) at this point.

3.9 Structure

Polymer properties are broadly divided into several classes based on the scale at which the property is defined as well as on its physical basis. The most basic property of a polymer is the identity of its constituent monomers. A second set of properties, known as microstructure, essentially describes the arrangement of these monomers within the polymer at the scale of a single chain. These basic structural properties play a major role in determining bulk physical properties of the polymer, which describe how the polymer behaves as a continuous macroscopic material. Chemical properties, at the nanoscale, describe how the chains interact through various physical forces. At the macroscale, they describe how the bulk polymer interacts with other chemicals and solvents.

The identity of the monomer units comprising a polymer is the first and most important attribute. Polymer nomenclature is generally based on the type of monomer residues comprising the polymer. Polymers that contain only a single type of repeat unit are known as homopolymer (e.g., polystyrene):

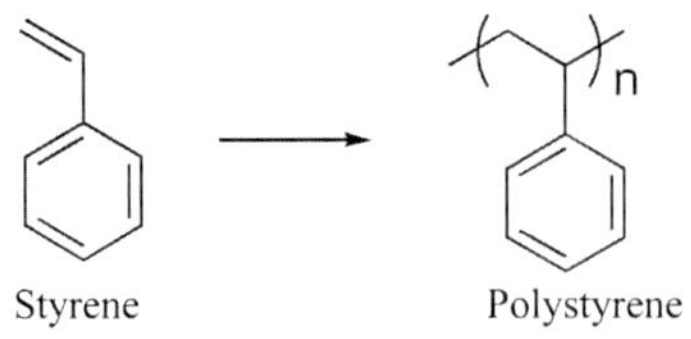

The repeating structural unit of most simple polymers not only reflects the monomer(s) from which the polymers are constructed but also provides a concise means for drawing structures to represent these macromolecules. For polyethylene, ethylene is the monomer, and the corresponding linear polymer is HDPE. HDPE is composed of macromolecules in which n ranges from 10,000 to 100,000 (molecular weight $2 \times 10^5 3 \times 10^6$):

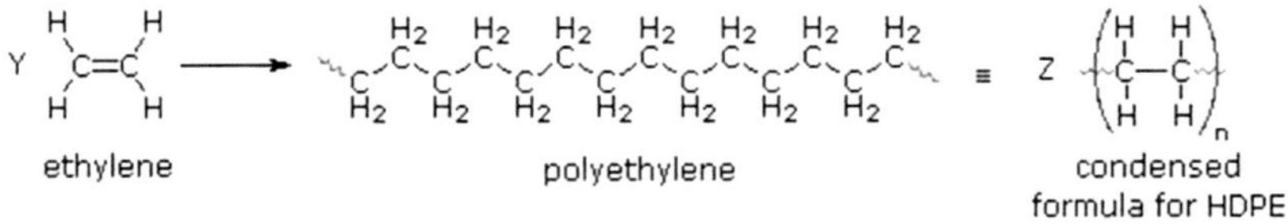

If Y and Z represent moles of monomer and polymer respectively, Z is approximately 105 Y. The two open bonds remaining at the ends of the long chain of carbons are normally not specified because the atoms or groups found there depend on the chemical process used for polymerization. Unlike simpler pure compounds, most polymers are not composed of identical molecules. The HDPE molecules, for example, are all long carbon chains, but the lengths may vary by thousands of monomer units. Because of this, polymer molecular weights are usually given as averages.

Symmetrical monomers such as ethylene can join together in only one way. Mono-substituted monomers, on the other hand, may join together in two organized ways, described in the following diagram, or in a third random manner. Most monomers of this kind, including propylene and styrene, join in a head-to-tail fashion, with some randomness occurring from time to time:

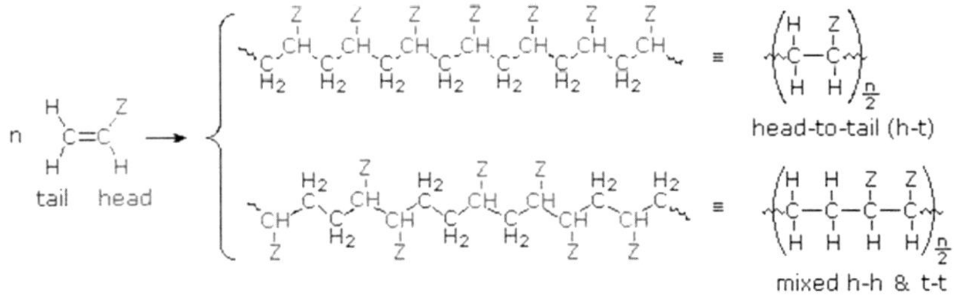

If the polymer chain (above) is drawn in the plane of the paper (above), each of the substituent groups (Z) will necessarily be located above or below the plane defined by the carbon chain. Consequently, configurational isomers of such polymers can be identified. If all of the substituents lie on one side of the chain, the configuration is isotactic, but if the substituents alternate from one side to another in a regular manner, the configuration is syndiotactic. A random arrangement of substituent groups is referred to as atactic:

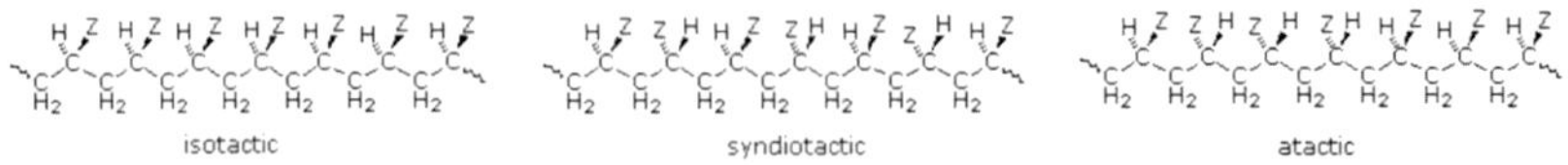

Many common hydrocarbon polymers, such as polystyrene, are atactic as normally prepared. Customized catalysts that effect stereoregular polymerization of polypropylene and some other monomers have been developed, and the improved properties associated with the increased crystallinity of these products have made this an important field of investigation.

The properties of a given polymer vary considerably with its stereo-configuration. For example, atactic polypropylene is employed mainly as a component of adhesives or as a soft matrix for composite materials. In contrast, isotactic polypropylene is a high melting solid (ca. 170°C) that can be molded or machined into structural components and used as a solid construction material.

4. Plastics

Plastic is the general common term for a wide range of synthetic organic (usually solid) materials produced and used in the manufacture of industrial products (Jones and Simon, 1983; Austin, 1984; Lokensgard, 2010). Plastics are the polymeric materials with properties in chemical structure; the demarcation between fibers and plastics may sometimes be blurred. A plastic is also any organic material with the ability to flow into a desired shape when heat and pressure are applied to it and to retain the shape when they are withdrawn. Plastics are typically polymers of high molecular weight and may contain other substances to improve performance and/or reduce costs.

Polymers such as polypropylene and polyamides can be used as fibers and plastics by a proper choice of processing conditions. Plastics can be extruded as sheets or pipes, painted on surfaces, or molded to form countless objects. A typical commercial plastic resin may contain two or more polymers in addition to various additives and fillers. Additives and fillers are used to improve some property such as the processability, thermal or environmental stability, and mechanical properties of the final product.

The first man-made plastic was created by Alexander Parkes who publicly demonstrated it at the 1862 Great International Exhibition in London. The material called Parkesine was an organic material derived from cellulose that once heated could be molded and retained its shape when cooled.

The common word plastic should not be confused with the technical adjective plastic, which is applied to any material which undergoes a permanent change of shape (plastic deformation) when strained beyond a certain point. A plastic is a type of polymer—all plastics are polymers but not all polymers are plastics. Polymers can be fibers, elastomers, or adhesives, and they are a wide group of solid composite materials that are largely organic, usually based on synthetic resins or modified polymers of natural origin, and possess appreciable mechanical strength.

A plastic exhibits plasticity and the ability to be deformed or undergo change of shape under pressure, temperature, or both. At a suitable stage in their manufacture, plastics can be cast, molded, or polymerized directly.

Resins are basic building materials that constitute the greater bulk of plastics. Resins undergo polymerization reactions during the development of plastics. Plastics are formed when polymers are blended with specific external materials in a process known as compounding. The important compounding ingredients include plasticizers, stabilizers, chelating agents, and antioxidants.

Hydrocarbon plastics are plastics based on resins made by the polymerization of monomers composed of carbon and hydrogen only.

A plastic is made up principally of a binder together with plasticizers, fillers, pigments, and other additives. The binder gives a plastic its main characteristics and usually its name. Binders may be natural materials, e.g., cellulose derivatives, casein, or milk protein but are more commonly synthetic resins. In either case, the binder materials consist of polymers. Cellulose derivatives are made from cellulose, a naturally occurring polymer; casein is also a naturally occurring polymer. Synthetic resins are polymerized, or built up, from small simple molecules called monomers. Plasticizers are added to a binder to increase flexibility and toughness. Fillers are added to improve particular properties, e.g., hardness or resistance to shock. Pigments are used to impart various colors. Virtually any desired color or shape and many combinations of the properties of hardness, durability, elasticity, and resistance to heat, cold, and acid can be obtained in a plastic.

Plastic deformation is observed in most materials including metals, soils, rocks, concrete, and plastics. However, the physical mechanisms that cause plastic deformation can vary widely. At the crystal scale, plasticity in metals is usually a consequence of dislocations, and although in most crystalline materials such defects are relatively rare, there are also materials where defects are numerous and are part of the very crystal structure, in such cases plastic crystallinity can result. In brittle materials, plasticity is caused predominantly by slippage at microcracks.

Plastics are so durable that they will not rot or decay as do natural products such as those made of wood. As a result, great amounts of discarded plastic products accumulate in the environment as waste. It has been suggested that plastics could be made to decompose slowly when exposed to sunlight by adding certain chemicals to them. Plastics present the additional problem of being difficult to burn. When placed in an incinerator, they tend to melt quickly and flow downward, clogging the grate of the incinerator; they also emit harmful fumes.

4.1 Classification

There are two types of plastics: thermoplastics and thermosetting polymers. Thermoplastics will soften and melt if enough heat is applied; examples among the truly hydrocarbon derivatives polymers are polyethylene and polystyrene. Thermosetting polymers can melt and take shape once; after they have solidified, they remain solid.

Thermoset plastics harden during the molding process and do not soften after solidifying. During molding, these resins acquire three-dimensional cross-linked structure with predominantly strong covalent bonds that retain their strength and structure even on heating. However, on prolonged heating, thermoset plastics get charred. In the softened state, these resins harden quickly with pressure assisting the curing process. Thermoset plastics are usually harder, stronger, and more brittle than thermoplastics and cannot be reclaimed from wastes. These resins are insoluble in almost all inorganic solvents.

Thermoplastics, when compounded with appropriate ingredients, can usually withstand several heating and cooling cycles without suffering any structural breakdown. Examples of commercial thermoplastics are polystyrene, polyolefins (e.g., polyethylene and polypropylene), nylon, poly(vinyl chloride), and poly(ethylene terephthalate). Thermoplastics are used for a wide range of applications, such as film for packaging, photographic, and magnetic tape, beverage and trash containers, and a variety of automotive parts and upholstery. Advantageously, waste thermoplastics can be recovered and refabricated by application of heat and pressure.

Thermosets are polymers whose individual chains have been chemically linked by covalent bonds during polymerization or by subsequent chemical or thermal treatment during fabrication. The thermosets usually exist initially as liquids called prepolymers; they can be shaped into desired forms by the application of heat and pressure. Once formed, these cross-linked networks resist heat softening, creep, and solvent attack and cannot be thermally processed or recycled. Such properties make thermosets suitable materials for composites, coatings, and adhesive applications. Principal examples of thermosets include epoxies, phenol formaldehyde resins, and unsaturated polyesters. Vulcanized rubber used in the tire industry is also an example of thermosetting polymers. Thermosetting polymers are usually insoluble because the cross-linking causes a tremendous increase in molecular weight. At most, thermosetting polymers only swell in the presence of solvents, as solvent molecules penetrate the network. The designation of a material as thermoplastic reflects the fact that above the glass transition temperature, the material may be shaped or pressed into molds, spun or cast from melts, or dissolved in suitable solvents for later fashioning. The polymers that are characterized by a high degree of cross-linking resist deformation and solution once their final morphology is achieved. Such polymers (thermosets) are usually prepared in molds that yield the desired object, and these polymers, once formed, cannot be reshaped by heating.

Plastics can be classified by chemical structure, namely the molecular units (the monomers) that make up the backbone and side chains of the polymer. Plastics can also be classified by the chemical process used in their synthesis, such as condensation, polyaddition, and cross-linking. Other classifications are based on qualities that are relevant for manufacturing or product design and include classes such as the thermoplastic and thermoset, elastomers, structural, conductive, and biodegradable. Plastics can also be classified by various physical properties, such as density, tensile strength, glass transition temperature, and resistance to various chemical products.

The use of plastics is constrained chiefly by their organic chemistry, which seriously limits their hardness, density, and their ability to resist heat, organic solvents, oxidation, and ionizing radiation. In particular, most plastics will melt or decompose when heated above 200°C (390°F).

4.2 Chemical structure

Common thermoplastics range in molecular weight from 20,000 to 500,000 while thermosets have higher, almost indefinable molecular weights. The molecular chains are made up of many repeating monomer units, and each plastic will have several thousand repeating units.

In the current context, the plastics are composed of polymers of hydrocarbon units with hydrocarbon moieties attached to the hydrocarbon backbone, which is that part of the chain in which a large number of repeat units together are linked together. To customize the properties of a plastic, different molecular groups are attached to the backbone. This fine tuning of the properties of the polymer by repeating the molecular structure of the unit has allowed plastics to become such an indispensable part of 21st century world.

Some plastics are partially crystalline and partially amorphous giving them both a melting point and one or more glass transitions (temperatures above which the extent of localized molecular flexibility is substantially increased). The so-called semicrystalline hydrocarbon plastics include polyethylene and polypropylene. Many plastics are completely amorphous, such as polystyrene and its copolymers, and all thermosets.

4.3 Properties

A *thermoplastic* (*thermosoftening plastic*) is a polymer that turns to a liquid when heated and freezes to a very glassy state when cooled sufficiently. Most hydrocarbon-based thermoplastics are high molecular weight polymers whose chains associate through weak van der Waals forces (polyethylene) or even stacking of aromatic rings (polystyrene). Thermoplastic polymers differ from thermosetting polymers as they can, unlike thermosetting polymers, be remelted and remolded. Many thermoplastic materials are additional polymers that result from vinyl chain-growth polymers such as polyethylene and polypropylene.

4.3.1 Chemical properties

Many applications require that plastics retain critical properties, such as strength, toughness, or appearance, during and after exposure to natural environmental conditions. Furthermore, the rapid growth of the use of plastics in major appliances has forced an examination of how best to manage this material once these products have reached the end of service. Integrated resource management requires that alternatives be developed to best utilize the material value of this postconsumer plastic.

Because the value of recovered materials will be determined by composition, the value over time changes as the composition of refrigerators changes. Any recycling process developed for plastics recovered should not only accommodate materials used 15–20 years ago but also be adaptable for the effective reclamation of the recovery of plastics.

Some of the environmental effects that may damage plastic materials are as follows:

Corrosion of metallic materials takes place via an electrochemical reaction at a specific corrosion rate. However, plastics do not have such specific rates. They are usually completely resistant to a specific corrodent or they deteriorate rapidly. Polymers are attacked either by chemical reaction or solvation. Solvation is the penetration of the polymer by a corrodent, which causes softening, swelling, and ultimate failure. Corrosion of plastics can be classified in the following ways as to attack mechanism: (i) disintegration or degradation of a physical nature due to absorption, permeation, solvent action, or other factors, (ii) oxidation, where chemical bonds are attacked, (iii) hydrolysis, where ester linkages are attacked, (iv) radiation, (v) thermal degradation involving depolymerization and possibly repolymerization, and (vi) dehydration, which is less common.

The absorption of *UV light*, mainly from sunlight, degrades polymers in two ways. First, the UV light adds thermal energy to the polymer as in heating, causing thermal degradation. Second, the UV light excites the electrons in the covalent bonds of the polymer and weakens the bonds. Hence, the plastic becomes more brittle.

Some plastics that are originated from natural products, or plastics that have natural products mixed with them, are potentially susceptible to *degradation* by microorganisms. This is not a desired property in the use stage of the plastic product. However, at the end of their life cycle, disposal of plastics become an important issue.

Oxidation is a degradation phenomenon when the electrons in a polymeric bond are so strongly attracted to another atom or molecule (here, oxygen) outside the bond that the polymer bond breaks. The results of oxidation are loss of mechanical and physical properties, embrittlement, and discoloration.

Environmental stress cracking occurs when the plastic is exposed to hostile environment conditions and mechanical stresses at the same time. It is different from polymer degradation because stress cracking does not break polymer bonds. Instead, it breaks the secondary linkages between polymers. These are broken when the mechanical stresses cause minute cracks in the polymer, and they propagate rapidly under harsh environmental conditions.

Nevertheless, the plastic material would not fail that fast if exposed to either the damaging environment or the mechanical stresses separately.

Crazing: In some cases, an environmental chemical embrittles the plastic material even when there is no mechanical stress applied. Cracks may also appear when the plastic part is stressed (usually in tensile) with no apparent environmental solvent present. These phenomena are called crazing and differ from environmental stress cracking in both the direction of the cracks and the extent of the cracking. The crack direction in environmental stress cracking is in the direction of molecular orientation in the part, while in crazing the cracks are much more numerous in a small area but are much shorter than environmental stress cracks.

Permeation is molecular migration through microvoids either in the polymer or between polymer molecules. Permeability is a measure of how easily gases or liquids can pass through a material. All materials are somewhat permeable to chemical molecules, but plastic materials tend to be an order of magnitude greater in their permeability than metals. However, not all polymers have the same rate of permeation. In fact, some polymers are not affected by permeation.

4.3.2 Electrical properties

Resistivity of a material is the resistance that a material presents to the flow of electrical charge.

Dielectric strength is the voltage that an insulating material can withstand before breakdown occurs. It usually depends on the thickness of the material and on the method and conditions of the test.

Arc resistance is the property that measures the ease of formation of a conductive path along the surface of a material, rather than through the thickness of the material as is done with dielectric strength.

Dielectric constant or permittivity is a measure of how well the insulating material will act as a dielectric capacitor. This constant is defined as the capacitance of the material in question compared (by ratio) with the capacitance of a vacuum. A high dielectric constant indicates that the material is highly insulating.

Dissipation factor of a material measures the tendency of the material to dissipate internally generated thermal energy (i.e., heat) resulting from an applied alternating electric field.

4.3.3 Mechanical properties

Plastics have the characteristics of a viscous liquid and a spring-like elastomer, traits known as *viscoelasticity*. These characteristics are responsible for many of the characteristic material properties displayed by plastics. Under mild loading conditions, such as short-term loading with low deflection and small loads at room temperature, plastics usually react like springs, returning to their original shape after the load is removed. Under long-term heavy loads or elevated temperatures, many plastics deform and flow similar to high viscous liquids, although still solid.

Creep is the deformation that occurs over time when a material is subjected to constant stress at constant temperature. This is the result of the viscoelastic behavior of plastics. *Stress relaxation* is another viscoelastic phenomenon. It is defined as a gradual decrease in stress at constant temperature. *Recovery* is the degree to which a plastic returns to its original shape after a load is removed. *Specific gravity* is the ratio of the weight of any volume to the weight of an equal volume of some other substance taken as the standard at a stated temperature. For plastics, the standard is water. *Water absorption* is the ratio of the weight of water absorbed by a material to the weight of the dry material. Many plastics are *hygroscopic*, meaning that over time they absorb water.

Tensile strength at break is a measure of the stress required to deform a material before breakage. It is calculated by dividing the maximum load applied to the material before its breaking point by the original cross-sectional area of the test piece.

Tensile modulus (modulus of elasticity) is the slope of the line that represents the elastic portion of the stress–strain graph. *Elongation at break* is the increase in the length of a tension specimen, usually expressed as a percentage of the original length of the specimen. *Compressive strength* is the maximum compressive stress a material is capable of sustaining. For materials that do not fail by a shattering fracture, the value depends on the maximum allowed distortion. *Flexural strength* is the strength of a material in bending expressed as the tensile stress of the outermost fibers of a bent test sample at the instant of failure. *Flexural modulus* is the ratio, within the elastic limit, of stress to the corresponding strain. *Izod impact* is one of the most common ASTM tests for testing the impact strength of plastic materials. It gives data to compare the relative ability of materials to resist brittle fracture as the service temperature decreases.

The *coefficient of thermal expansion* is the change in unit length or volume resulting from a unit change in temperature. Commonly used unit is 10^{-6} cm/cm/°C. *Thermal conductivity* is the ability of a material to conduct heat, a physical constant for the quantity of heat that passes through a unit cube of a material in a unit of time when the difference in temperature of two faces is 1°C (1.8°F). The *limiting oxygen index* is a measure of the minimum oxygen level required to support combustion of the polymer.

4.3.4 Optical properties

Light transmission: Plastics differ greatly in their ability to transmit light. The materials that allow light pass through them are called transparent. Many plastics do not allow any light to pass through. These are called opaque materials. Some plastic materials have light transmission properties between transparent and opaque. These are called translucent.

Surface reflectance: The reflection of light off the surface of a plastic part determines the amount of gloss on the surface. The reflectance is dependent on a property of materials called the index of refraction, which is a measure of the change in direction of an incident ray of light as it passes through a surface boundary. If the index of refraction of the plastic is near the index of air, light will pass through the boundary without significant change in direction. If the index of refraction between the air and the plastic is large, the ray of light will significantly change direction causing some of the light to be reflected back toward its source.

5. Synthetic rubber

Synthetic rubber (an elastomer) is a long-chain polymer with special chemical and physical as well as mechanical properties. These materials have chemical stability, high abrasion resistance, strength, and good dimensional stability. Many of these properties are imparted to the original polymer through cross-linking agents and additives. An important property of elastomeric materials is their ability to be stretched at least twice their original length and to return back to nearly their original length when released.

Natural rubber is a polymer of isoprene—most often cis-1,4-polyisoprene—with a molecular weight of 100,000 to 1,000,000. Typically, a few percent of other materials, such as proteins, fatty acids, resins, and inorganic materials, are found in high-quality natural rubber. Some natural rubber sources called gutta percha (i.e., trees of the genus *Palaquium* in the family Sapotaceae and the rigid natural latex produced from the sap of these trees, particularly from *Palaquium gutta*) are composed of trans-1,4-polyisoprene, a structural isomer that has similar, but not identical, properties.

Isoprene (2-methyl-1,3-butadiene) is a common organic compound with the formula $CH_2{=}C(CH_3)CH{=}CH_2$:

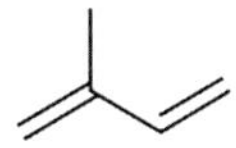

Isoprene

Under standard conditions, isoprene is a colorless liquid and is the monomer of natural rubber as well as a precursor to an immense variety of other naturally occurring compounds.

Synthetic rubber is any type of artificial elastomer, invariably a polymer. An elastomer is a material with the mechanical (or material) property that it can undergo much more elastic deformation under stress than most materials and still return to its previous size without permanent deformation. Synthetic rubber serves as a substitute for natural rubber in many cases, especially when improved material properties are required.

Synthetic rubber can be made from the polymerization of a variety of monomers including isoprene (2-methyl-1,3-butadiene), 1,3-butadiene, and isobutylene (methylpropene) with a small percentage of isoprene for cross-linking. These and other monomers can be mixed in various desirable proportions to be copolymerized for a wide range of physical, mechanical, and chemical properties. The monomers can be produced pure and the addition of impurities or additives can be controlled by design to give optimal properties. Polymerization of pure monomers can be better controlled to give a desired proportion of cis and trans double bonds.

Natural rubber is an elastomer constituted of isoprene units. These units are linked in a cis-1,4 configuration that gives natural rubber the outstanding properties of high resilience and strength. Natural rubber occurs as a latex (water emulsion) and is obtained from *Hevea brasiliensis,* a tree that grows in Malaysia, Indonesia, and Brazil. Charles Goodyear was the first to discover that the latex could be vulcanized (cross-linked) by heating with sulfur or other agents. Vulcanization of rubber is a chemical reaction by which elastomer chains are linked together. The long chain molecules impart elasticity, and the cross-links give load-supporting strength.

Synthetic rubbers include elastomers that could be cross-linked such as polybutadiene, polyisoprene, and ethylene-propylene-diene terepolymer. It also includes TPEs that are not cross-linked and are adapted for special purposes such as automobile bumpers and wire and cable coatings. These materials could be scraped and reused. However, they cannot replace all traditional rubber as they do not have the wide temperature performance range of thermoset rubber.

The major use of rubber is for tire production. Nontire consumption includes hoses, footwear, molded and extruded materials, and plasticizers.

5.1 Butyl rubber

Butyl rubber is a copolymer of isobutylene (97.5%) and isoprene (2.5%). The polymerization is carried out at low temperature (below −95°C, <139°F) using aluminum chloride ($AlCl_3$) cocatalyzed with a small amount of water. The cocatalyst furnishes the protons needed for the cationic polymerization:

$$AlCl_3 + H_2O \rightarrow H^+(AlCl_3OH)^-$$

The product is a linear random copolymer that can be cured to a thermosetting polymer. This is made possible through the presence of some unsaturation from isoprene.

Butyl rubber vulcanizates have tensile strengths up to 2000 psi and are characterized by low permeability to air and a high resistance to many chemicals and to oxidation. These properties make it a suitable rubber for the production of tire inner tubes and inner liners of tubeless tires. The major use of butyl rubber is for inner tubes. Other uses include wire and cable insulation, steam hoses, mechanical goods, and adhesives. Chlorinated butyl is a low molecular weight polymer used as an adhesive and a sealant.

5.2 Ethylene propylene rubber

Ethylene propylene rubber (EPR) is a stereoregular copolymer of ethylene and propylene. Elastomers of this type do not possess the double bonds necessary for cross-linking. A third monomer, usually a mono-conjugated diene, is used to provide the residual double bonds needed for cross-linking. 1,4-Hexadiene and ethylidene norbornene are examples of these dienes. The main polymer chain is completely saturated while the unsaturated part is pending from the main chain. The product elastomer, termed ethylene-propylene terepolymer (EPT), can be cross-linked using sulfur. Cross-linking EPR is also possible without using a third component (a diene). This can be done with peroxides.

Important properties of vulcanized EPR and EPT include resistance to abrasion, oxidation, and heat and ozone; but they are susceptible to hydrocarbon derivatives. The main use of EPR is to produce automotive parts such as gaskets, mechanical goods, wire, and cable coating. It may also be used to produce tires.

5.3 Nitrile rubber

Nitrile rubber (NBR) is a copolymer of butadiene and acrylonitrile. It has the special property of being resistant to hydrocarbon liquids. The copolymerization occurs in an aqueous emulsion. When free radicals are used, a random copolymer is obtained. Alternating copolymers are produced when a Ziegler—Natta catalyst is employed. Molecular weight can be controlled by adding modifiers and inhibitors. When the polymerization reaches approximately 65%, the reaction mixture is vacuum distilled in presence of steam to recover the monomer. The ratio of acrylonitrile/butadiene could be adjusted to obtain a polymer with specific properties. Increasing the acrylonitrile ratio increases oil resistance of the rubber but decreases its plasticizer compatibility.

NBR is produced in different grades depending on the end use of the polymer. Low acrylonitrile rubber is flexible at low temperatures and is generally used in gaskets, 0-rings, and adhesives. The medium type is used in less flexible articles such as kitchen mats and shoe soles. High acrylonitrile polymers are more rigid and highly resistant to hydrocarbon derivatives and oils and are used in fuel tanks and hoses, hydraulic equipment, and gaskets.

5.4 Polychloroprene

Polychloroprene (neoprene rubber) is the oldest synthetic rubber. It is produced by the polymerization of 2-chloro-1,3-butadiene in a water emulsion with potassium sulfate as a catalyst. The product is a random polymer that is vulcanized with sulfur or with metal oxides (such as zinc oxide, ZnO, and magnesium oxide, MgO). Vulcanization with sulfur is very slow, and an accelerator is usually required.

Neoprene vulcanizates have a high tensile strength, excellent oil resistance (better than natural rubber), and heat resistance. Neoprene rubber could be used for tire production, but it is expensive. Major uses include cable coatings, mechanical goods, gaskets, conveyor belts, and cables.

5.5 Polyisoprene

Natural rubber is a stereoregular polymer composed of isoprene units attached in a cis configuration. This arrangement gives the rubber high resilience and strength. Isoprene can be polymerized using free radical initiators, but a random polymer is obtained. As with butadiene, polymerization of isoprene can produce a mixture of isomers. However, because the isoprene molecule is asymmetrical, the addition can occur in 1,2-, 1,4-, and 3,4- positions. Six tactic forms are possible from both 1,2- and 3,4- addition and two geometrical isomers from 1,4- addition (cis and trans).

Stereoregular polyisoprene is obtained when Ziegler−Natta catalysts or anionic initiators are used. The most important coordination catalyst is α-titanium trichloride cocatalyzed with aluminum alkyl derivatives. The polymerization rate and cis content depends on Al/Ti ratio, which should be greater than one. Lower ratios predominantly produce the trans structure.

Polyisoprene is a synthetic polymer (elastomer) that can be vulcanized by the addition of sulfur. Cis-polyisoprene has properties similar to that of natural rubber and is characterized by high tensile strength and insensitivity to temperature changes, but it has low abrasion resistance—it is attacked by oxygen and hydrocarbon derivatives. Trans-polyisoprene is similar to gutta percha, which is produced from the leaves and bark of the Sapotaceae tree but has different properties from the cis form and cannot be vulcanized. Important uses of cis-polyisoprene include the production of tires, specialized mechanical products, conveyor belts, footwear, and insulation.

5.6 Styrene-butadiene rubber

Styrene-butadiene rubber (SBR) is the most widely used synthetic rubber and can be produced by the copolymerization of butadiene (75%) and styrene (25%) using free radical initiators. A random copolymer is obtained. The microstructure of the polymer is 60%−68% trans, 14%−19% cis, and 17%−21% 1,2-. Wet methods are normally used to characterize polybutadiene polymers and copolymers. Solid-state **NMR** provides a more convenient way to determine the polymer microstructure.

Currently, more SBR is produced by copolymerizing the two monomers with anionic or coordination catalysts. The formed copolymer has better mechanical properties and a narrower molecular weight distribution. A random copolymer with ordered sequence can also be made in solution using butyl-lithium, provided that the two monomers are charged slowly. Block copolymers of butadiene and styrene may be produced in solution using coordination or anionic catalysts. Butadiene polymerizes first until it is consumed, then styrene starts to polymerize. SBR produced by coordination catalysts has better tensile strength than that produced by free radical initiators.

The main use of SBR is for tire production. Other uses include footwear, coatings, carpet backing, and adhesives.

6. Thermosetting plastics

This group of polymers includes many plastics produced by condensation polymerization. Among the important thermosets are the polyurethanes, epoxy resins, phenolic resins, and urea and melamine formaldehyde resins. Although the polymers covered in this section are not all hydrocarbon polymers, they are, however, for the most part based on hydrocarbon starting materials. It is for this reason that theses polymers are deemed worthy of mention in this text; without the hydrocarbon starting materials, there would be other time-consuming routes for that would have to be employed to produce the polymers. The same rationale is true for the inclusion and presentation of the polymers in the next section.

Plastics are available in the form of bars, tubes, sheets, coils, and blocks, and these can be fabricated to specification. However, plastic articles are commonly manufactured from plastic powders in which desired shapes are fashioned by compression, transfer, injection, or extrusion molding. In compression molding, materials are generally placed immediately in mold cavities, where the application of heat and pressure makes them first plastic, then hard. The transfer method, in which the compound is plasticized by outside heating and then poured into a mold to harden, is used for designs with intricate shapes and great variations in wall thickness. Injection molding machinery dissolves the plastic powder in a heating chamber and by plunger action forces it into cold molds, where the product sets. The operations take place at rigidly controlled temperatures and intervals. Extrusion molding employs a heating cylinder, pressure, and an extrusion die through which the molten plastic is sent and from which it exits in continuous form to be cut in lengths or coiled.

Thermoplastics are elastic and flexible above a glass transition temperature (Tg), which is specific to each plastic, specific for each one. Below a second, higher melting temperature, Tm, most thermoplastics have crystalline regions alternating with amorphous regions in which the chains approximate random coils. The amorphous regions contribute elasticity, and the crystalline regions contribute strength and rigidity. Above Tm,all crystalline structure disappears and the chains become randomly interdispersed. As the temperature increases above Tm, the viscosity gradually decreases without any distinct phase change.

Some thermoplastics normally do not crystallize—they are termed amorphous plastics and are useful at temperatures below the glass transition temperature. Generally, amorphous thermoplastics are less chemically resistant and can be subject to stress cracking. Thermoplastics will crystallize to a certain extent and are called semicrystalline. The speed and extent to which crystallization can occur depends in part on the flexibility of the polymer chain.

Semicrystalline thermoplastics are more resistant to solvents and other chemicals. If the crystallites are larger than the wavelength of light, the thermoplastic is hazy or opaque. Semicrystalline thermoplastics become less brittle above the glass transition temperature. If a plastic with otherwise desirable properties has too high a glass transition temperature, it can often be lowered by adding a low molecular weight plasticizer to the melt before forming and cooling. A similar result can sometimes be achieved by adding nonreactive side chains to the monomers before polymerization. Both methods make the polymer chains stand off a bit from one another. Another method of lowering the glass transition temperature (or raising the melting temperature) is to incorporate the original plastic into a copolymer (as with graft copolymers) of polystyrene.

Lowering the glass transition temperature is not the only way to reduce brittleness. Drawing (and similar processes that stretch or orient the molecules) or increasing the length of the polymer chains also decrease brittleness. Thermoplastics can go through melting/freezing cycles repeatedly, and the fact that they can be reshaped on reheating gives them their name. This quality makes thermoplastics recyclable. The processes required for recycling vary with the thermoplastic.

Although modestly vulcanized natural and synthetic rubbers are stretchy, they are elastomeric thermosets, not thermoplastics. Each has its own glass transition temperature and will crack and shatter when cold enough so that the cross-linked polymer chains can no longer move relative to one another. But they have no melting temperature and will decompose at high temperatures rather than melt.

6.1 Amino resins

Amino resins (aminoplasts) are condensation thermosetting polymers of formaldehyde with either urea or melamine. Melamine is a condensation product of three urea molecules. It is also prepared from cyanimide at high pressure and high temperature. The nucleophilic addition reaction of urea to formaldehyde produces mainly monomethylol urea and some dimethylol urea. When the mixture is heated in presence of an acid, condensation occurs, and water is released. This is accompanied by the formation of a cross-linked polymer. A similar reaction occurs between melamine and formaldehyde and produces methylolmelamine derivatives.

A variety of methylol derivatives are possible due to the availability of six hydrogens in melamine. As with urea formaldehyde resins, polymerization occurs by a condensation reaction and the release of water.

Amino resins are characterized by being more clear and harder (tensile strength) than phenol derivatives. However, their impact strength (breakability) and heat resistance are lower. Melamine resins have better heat and moisture resistance and better hardness than their urea analogs.

The most important use of amino resins is the production of adhesives for particleboard and hardwood plywood. Compression and injection molding are used with amino resins to produce articles such as radio cabinets, buttons, and cover plates. Because melamine resins have lower water absorption and better chemical and heat resistance than urea resins, they are used to produce dinnerware and laminates used to cover furniture. Almost all molded objects use fillers such as cellulose, asbestos, glass, wood flour, glass fiber, and paper.

6.2 Epoxy resins

Epoxy resins are produced by reacting epichlorohydrin and a diphenol. Bisphenol A is the diphenol generally used. The reaction, a ring opening polymerization of the epoxide ring, is catalyzed with strong bases such as sodium hydroxide. A nucleophilic attack of the phenoxy ion displaces a chloride ion and opens the ring. The linear polymer formed is cured by cross-linking either with an acid anhydride, which reacts with the —OH groups, or by an amine, which opens the terminal epoxide rings. Cresols and other bisphenols are also used for producing epoxy resins.

Epoxy resins have a wide range of molecular weights (approximately 1000 to 10,000). Those with higher molecular weight, termed phenoxy, are hydrolyzed to transparent resins that do not have the epoxide groups. These could be used in molding purposes or cross-linked by diisocyanate derivatives or by cyclic anhydride derivatives.

Important properties of epoxy resins include their ability to adhere strongly to metal surfaces, their resistance to chemicals, and their high dimensional stability. They can also withstand temperatures up to 500°C (930°F). Epoxy resins with improved stress cracking properties can be made by using toughening agents, such as carboxyl-terminated butadiene-acrylonitrile liquid polymers. The carboxyl group reacts with the terminal epoxy ring to form an ester. The ester, with its pendant hydroxyl groups, reacts with the remaining epoxide rings, thus more cross-linking occurs by forming ether linkages. This material is tougher than epoxy resins and suitable for encapsulating electrical units. Other uses of epoxy resins are coatings for appliance finishes, auto primers, adhesive, and in coatings for cans and drums. Interior coatings of drums used for chemicals and solvents manifest its chemical resistance.

6.3 Phenolformaldehyde resins

Phenol formaldehyde resins are the oldest thermosetting polymers. They are produced by a condensation reaction between phenol and formaldehyde.

Although many attempts were made to use the product and control the conditions for the acid-catalyzed reaction described by Bayer in 1872, there was no commercial production of the resin until the exhaustive work by Baekeland was published in 1909. In this paper, he describes the product as far superior to amber for pipe stem and similar articles, less flexible but more durable than celluloid, odorless, and fire-resistant.

The reaction between phenol and formaldehyde is either base or acid catalyzed, and the polymers are termed resols (for the base catalyzed) and novalacs (for the acid catalyzed). The first step in the base-catalyzed reaction is an attack by the phenoxide ion on the carbonyl carbon of formaldehyde, giving a mixture of ortho-substituted and para-substituted monomethylolphenol and di- and trisubstituted methylol phenol derivatives. The second step is the condensation reaction between the methylol phenol derivatives with the elimination of water and the formation of the polymer. Cross-linking occurs by a reaction between the methylol groups and results in the formation of ether bridges. It occurs also by the reaction of the methylol groups and the aromatic ring, which forms methylene bridges. The formed polymer is a three-dimensional network thermoset:

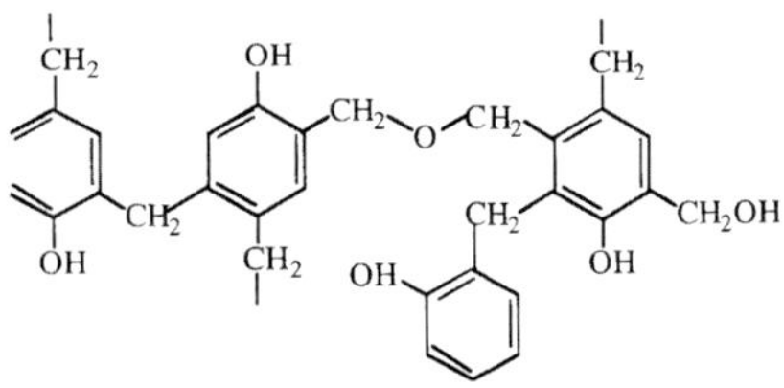

The acid-catalyzed reaction occurs by an electrophilic substitution where formaldehyde is the electrophile. Condensation between the methylol groups, and the benzene rings result in the formation of methylene bridges. Usually, the ratio of formaldehyde to phenol is kept less than unity to produce a linear fusible polymer in the first stage. Cross-linking of the formed polymer can occur by adding more formaldehyde and a small amount of hexamethylene tetramine (hexamine, $(CH_2)_6N_4$). Hexamine decomposes in the presence of traces of moisture to formaldehyde and ammonia. This results in cross-linking and formation of a thermoset resin:

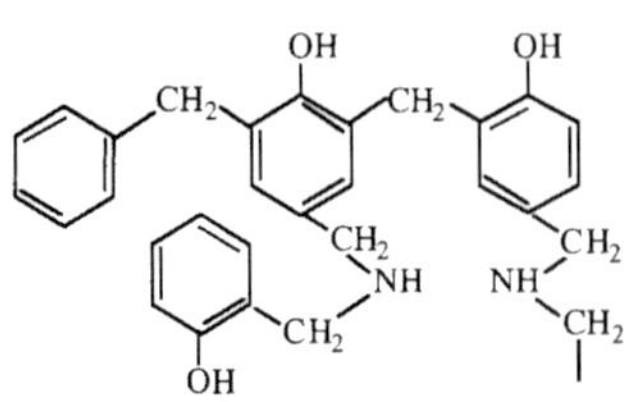

Important properties of phenolic resins are their hardness, corrosion resistance, rigidity, and resistance to water hydrolysis. They are also less expensive than many other polymers.

Many additives are used with phenolic resins such as wood flour, oils, asbestos, and fiberglass. Fiberglass piping made with phenolic resins can operate at 150°C and pressure up to 150 psi. Molding applications dominate the market of phenolic resins. Articles produced by injection molding have outstanding heat resistance and dimensional stability. Compression-molded glass-filled phenolic disk brake pistons are replacing the steel ones in many automobiles because of the low density (i.e., light weight) and corrosion resistance.

Phenol derivatives are also used in a variety of other applications such as adhesives, paints, laminates for building, automobile parts, and ion-exchange resins.

6.4 Polycyanurates

New polymer types that emerged as important materials for circuit boards are polycyanurate derivatives. The simplest monomer is the dicyanate ester of bisphenol A. When polymerized, it forms three-dimensional, densly cross-linked structures through three-way cyanuric acid (2,4,6-triazinetriol). The cyanurate ring is formed by the trimerization of the cyanate ester. Other monomers, such as hexaflurobisphenol A and tetramethyl bisphenol F, are also used. These polymers are characterized by high glass transition temperatures ranging between 192 and >350°C (377 to >660°F).

TPEs, as the name indicates, are plastic polymers with the physical properties of rubbers. They are soft, flexible, and possess the resilience needed of rubbers. However, they are processed like thermoplastics by extrusion and injection molding. TPEs are more economical to produce than traditional thermoset materials because fewer steps are required to manufacture them than to manufacture and vulcanize thermoset rubber. An important property of these polymers is that they are recyclable.

TPEs are multiphase composites, in which the phases are intimately depressed. In many cases, the phases are chemically bonded by block or graft copolymerization. At least one of the phases consists of a material that is hard at room temperature.

Currently, important TPEs include blends of semicrystalline thermoplastic polyolefin derivatives such as propylene copolymers, with EPT elastomer. Block copolymers of styrene with other monomers such as butadiene, isoprene, and ethylene or ethylene/propylene are the most widely used TPEs.

Polyurethane TPEs are relatively more expensive than other TPEs. However, they are noted for their flexibility, strength, toughness, and abrasion and chemical resistance. Blends of polyvinyl chloride with elastomers such as butyl are widely used in Japan. Random block copolymers such as polyesters (hard segments) and amorphous glycol soft segments, alloys of ethylene interpolymers, and chlorinated polyolefin derivatives are among the evolving TPEs.

Important properties of TPEs are the flexibility, softness, and resilience. However, compared with vulcanizable rubbers, they are inferior in resistance to deformation and solvents. Important markets for TPEs include shoe soles, pressure sensitive adhesives, insulation, and recyclable bumpers.

6.5 Polyurethanes

Polyurethanes are produced by the condensation reaction of a polyol and a diisocyanate:

$$\text{OCN-R-NCO} + \text{HO-R}^{\wedge\prime}\text{-OH} - \text{C}{-}\text{N-R-N-C-OR}^{\wedge\prime}\text{-O}$$

No by-product is formed from this reaction. Toluene diisocyanate (TDI) (Chapter 10) is a widely used monomer. TDI is an organic compound with the formula $CH_3C_6H_3(NCO)_2$:

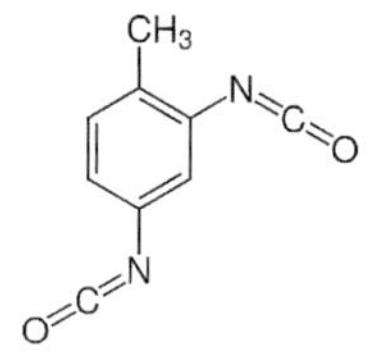

Two of the six possible isomers are commercially important: 2,4-TDI and 2,6-TDI.

Polyurethane polymers are either rigid or flexible, depending on the type of the polyol used. For example, triol derivatives derived from glycerol and propylene oxide are used for producing block slab foams. These polyols have moderate reactivity because the hydroxy groups are predominantly secondary. More reactive polyols (used to produce molding polyurethane foams) are formed by the reaction of polyglycols with ethylene oxide to give the more reactive primary group.

Other polyhydric compounds with higher functionality than glycerol (three OH) are commonly used. Examples are sorbitol (six OH) and sucrose (eight OH). Triethanolamine, with three OH groups, is also used. Diisocyanate derivatives generally employed with polyols to produce polyurethane derivatives are 2,4- and 2,6-TDI derivatives prepared from dinitro-toluene derivatives:

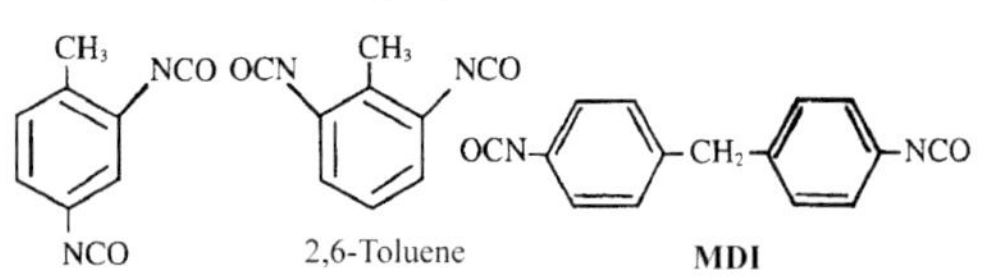

A different diisocyanate used in polyurethane synthesis is methylene diisocyanate (MDI), which is prepared from aniline and formaldehyde. The diamine product is reacted with phosgene to get MDI.

The physical properties of polyurethanes vary with the ratio of the polyol to the diisocyanate. For example, tensile strength can be adjusted within a range of 1200–600 psi and elongation between 150% and 800%. Improved polyurethanes can be produced by copolymerization. Block copolymers of polyurethanes connected with segments of isobutylene derivatives exhibit high-temperature properties, hydrolytic stability, and barrier characteristics. The hard segments of polyurethane block polymers consist of RNHCOOH, where R usually contains an aromatic moiety.

The major use of polyurethanes is to produce foam. The density as well as the mechanical strength of the rigid and the flexible types varies widely with polyol type and reaction conditions. For example, polyurethanes could have densities ranging between 1 and 6 lb/ft^3 for the flexible types and 1 and 50 lb/ft3 for the rigid types. Polyurethane foams have good abrasion resistance, low thermal conductance, and good load-bearing characteristics. However, they have moderate resistance to organic solvents and are attacked by strong acids. The ability of polyurethanes to acts as flame retardants can be improved by using special additives, spraying a coating material such as magnesium oxychloride, or by grafting a halogen phosphorous moiety to the polyol. Trichlorobutylene oxide is sometimes copolymerized with ethylene and propylene oxides to produce the polyol.

Major markets for polyurethanes are furniture, transportation, and building and construction. Other uses include carpet underlay, textural laminates and coatings, footwear, packaging, toys, and fibers. The largest use for rigid polyurethane is in construction and industrial insulation due to its high insulating property.

Molded urethanes are used in items such as bumpers, steering wheels, instrument panels, and body panels. Elastomers from polyurethanes are characterized by toughness and resistance to oils, oxidation, and abrasion. They are produced using short-chain polyols such as poly-tetramethylene glycol from 1,4-butanediol. Polyurethanes are also used to produce fibers. Spandex (trade name) is a copolymer of polyurethane (85%) and polyesters.

Polyurethane networks based on triisocyanate and diisocyanate connected by segments consisting of polyisobutylene are rubbery and exhibit high temperature properties, hydrolytic stability, and barrier characteristics.

6.6 Unsaturated polyesters

Unsaturated polyesters are a group of polymers and resins used in coatings or for castings with styrene. These polymers normally have maleic anhydride moiety or an unsaturated fatty acid to impart the required unsaturation. A typical example is the reaction between maleic anhydride and ethylene glycol.

Also, phthalic anhydride, a polyol, and an unsaturated fatty acid are usually copolymerized to unsaturated polyesters for coating purposes. Many other combinations in variable ratios are possible for preparing these resins.

7. Synthetic fibers

This group of polymers includes many plastics produced by condensation polymerization. Among the important thermosets are the polyurethanes, epoxy resins, phenolic resins, and urea and melamine formaldehyde resins. Although the polymers covered in this section are not all hydrocarbon polymers, they are, however, for the most part based on hydrocarbon starting materials. It is for this reason that theses polymers are deemed worthy of mention in this text—without the hydrocarbon starting materials, there would be other time-consuming routes for that would have to be employed to produce the polymers. The same rationale is true for the inclusion and presentation of some of the polymers in the previous section.

Briefly, and by way of explanation, a fiber is often is as a polymer with a length-to-diameter ratio of at least 100 (Browne and Work, 1983). Fibers (synthetic or natural) are polymers with high molecular symmetry and strong cohesive energies between chains that result usually from the presence of polar groups. Fibers possess a high degree of crystallinity characterized by the presence of stiffening groups in the polymer backbone and of intermolecular hydrogen bonds. Also, they are characterized by the absence of branching or irregularly space-dependent groups that will otherwise disrupt the crystalline formation. Fibers are normally linear and drawn in one direction to make them long, thin, and threadlike, with great strength along the fiber. These characteristics permit formation of this type of polymers into long fibers suitable for textile applications. Typical examples of fibers include polyesters, nylons, acrylic polymers such as polyacrylonitrile, and naturally occurring polymers such as cotton, wool, and silk.

Fibers fall into a class of materials that are continuous filaments or are in discrete elongated pieces, similar to lengths of thread. Fiber classification in reinforced plastics falls into two classes: (i) short fibers, also known as discontinuous fibers, with a general aspect ratio (defined as the ratio of fiber length to diameter) between 20 and 60, and (ii) long fibers, also known as continuous fibers, with a general aspect ratio between 200 and 500.

Polyethylene fiber properties depend markedly on the crystallinity or density of the polymer; although high-strength fibers can be made from linear polyethylene, resiliency properties are poor, tensile properties are highly time-dependent, and endurance under sustained loading is very poor. On the other hand, polypropylene fibers have good stress-endurance properties, excellent recovery from high extensions, and fair-to-good recovery properties at low strains; recovery at 'low strains is shown to depend on the extent of fiber orientation and annealing.

Anomalies in the change of the sonic modulus of polypropylene yarns during extension and relaxation are noted and interpreted in terms of structure changes in the crystalline phase. The high melting temperature of 235°C (455°F) for poly(4-methyl-1-pentene) appears to be due to its low entropy of melting, and fibers from this polymer are characterized by low tenacity when tested at elevated temperatures. Crystalline polystyrene fibers have relatively good retention of tenacity at elevated temperatures and are characterized by excellent resiliency at low strains, good wash-wear characteristics in cotton blends, and low abrasion resistance.

Thus, fibers are materials that are continuous filaments or discrete elongated pieces, similar to lengths of thread and are characterized by a high ratio of length to diameter. They are important for a variety of applications, including holding tissues together in both plants and animals. There are many different kinds of fibers including textile fiber, natural fibers, and synthetic or human-made fibers such as cellulose, mineral, polymer, and microfibers. Fibers can be manufactured from a natural origin such as silk, wool, and cotton or derived from a natural fiber such as rayon. They may also be synthesized from certain monomers by polymerization (synthetic fibers). In general, polymers with high melting points, high crystallinity, and moderate thermal stability and tensile strengths are suitable for fiber production.

Fibers can be spun into filaments, string, or rope; used as a component of composite material; or matted into sheets to make products such as paper and are often used in the manufacture of other materials. The strongest engineering materials are generally made as fibers, for example, carbon fiber and ultrahigh molecular weight polyethylene. Synthetic fibers can often be produced cheaply and in large amounts as compared with natural fibers, but natural fibers have benefit in some applications, especially for clothing.

Man-made fibers include, in addition to synthetic fibers, those derived from cellulose (cotton, wood) but modified by chemical treatment such as rayon, cellophane, and cellulose acetate. These are sometimes termed "regenerated cellulose fibers." Rayon and cellophane have shorter chains than the original cellulose due to degradation by alkaline treatment. Cellulose acetates produced by reacting cellulose with acetic acid and modified or regenerated fibers are excluded from this book because they are derived from a plant source. Fibers produced by drawing metals or glass (SiO_2) such as glass wool are also excluded.

Major fiber-making polymers are those of polyesters, polyamides (nylons), polyacrylic derivatives, and polyolefin derivatives. Polyesters and polyamides are produced by step polymerization reactions, while polyacrylic derivatives and polyolefin derivatives are synthesized by chain-addition polymerization.

7.1 Acrylic and modacrylic fibers

Acrylic fibers are a major synthetic fiber class and were developed at approximately the same time as polyesters. Modacrylic fibers are copolymers containing between 35% and 85% acrylonitrile. Acrylic fibers contain at least 85% acrylonitrile. Orlon is an acrylic fiber developed by DuPont in 1949; dynel is a modacrylic fiber developed by Union Carbide in 1951.

Polyacrylics are produced by copolymerizing acrylonitrile with other monomers such as vinyl acetate, vinyl chloride, and acrylamide. Solution polymerization may be used where water is the solvent in the presence of a redox catalyst. Free radical or anionic initiators may also be used. The produced polymer is insoluble in water and forms a precipitate. Copolymers of acrylonitrile are sensitive to heat, and melt spinning is not used. Solution spinning (wet or dry) is the preferred process where a polar solvent such as dimethyl formamide is used. In dry spinning, the solvent is evaporated and recovered.

Dynel, a modacrylic fiber, is produced by copolymerizing vinyl chloride with acrylonitrile. Solution spinning is also used where the polymer is dissolved in a solvent such as acetone. After the solvent is evaporated, the fibers are washed and subjected to stretching, which extends the fiber 4−10 times of the original length.

Acrylic fibers are characterized by having properties similar to wool and have replaced wool in many markets such as blankets, carpets, and sweaters. Important properties of acrylics are resistance to solvents and sunlight, resistance to creasing, and quick drying.

Acrylic fiber breaking strength ranges between 22,000 and 39,000 psi and they have water absorption of approximately 5%. Dynel, due to the presence of chlorine, is less flammable than many other synthetic fibers. Major uses of acrylic fibers are woven and knitted clothing fabrics (for apparel), carpets, and upholstery.

7.2 Graphite fibers

Carbon fibers are special reinforcement types having a carbon content of 92 99% w/w. They are prepared by controlled pyrolysis of organic materials in fibrous forms at temperatures ranging from 1000 to 3,000°C (1800 to 5,400°F). The commercial fibers are produced from rayon, polyacrylonitrile, and crude oil pitch. When acrylonitrile is heated in air at moderate temperatures (220°C, 430°F), hydrogen cyanide is emanated. Further heating above 1700°C (3,100°F) in the presence of nitrogen for a period of 24 hours produces carbon fiber. Carbon fibers are characterized by high strength, stiffness, low thermal expansion, and thermal and electrical conductivity, which makes them an attractive substitute for various metals and alloys.

7.3 Polyamides

Polyamide derivatives (particularly nylon fibers) are the second largest group of synthetic fibers after polyesters. Numbers that follow the word *nylon* denote the number of carbons present within a repeating unit and whether one or two monomers are being used in polymer formation.

Polyamides are produced by the reaction between a dicarboxylic acid (in many cases, terephthalic acid (TPA) produced from p-xylene) and a diamine (e.g., nylon 66), ring openings of a lactam, (e.g., nylon 6), or by the polymerization of ω-amino acids (e.g., nylon 11).

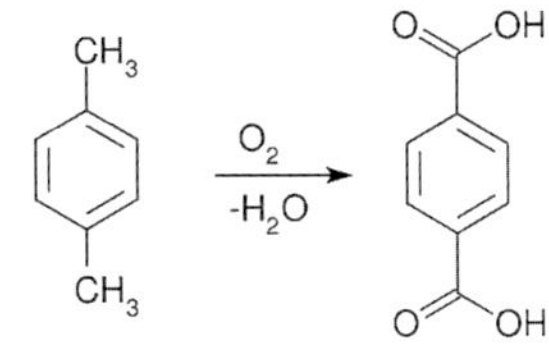

p-xylene terephthalic acid

The general reaction for the production nylon derivatives is

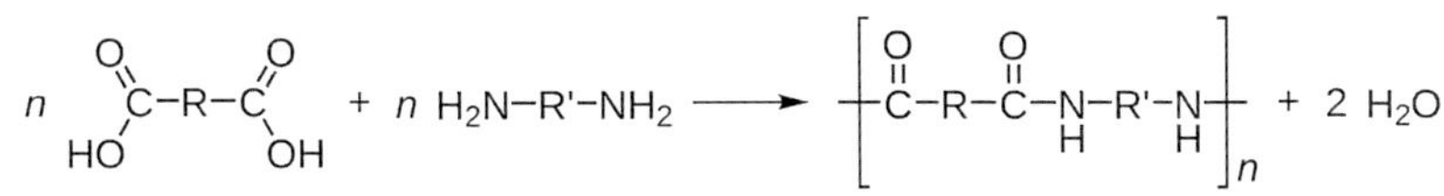

In this equation, R (i.e., R in HOOC-R-COOH) is typically TPA (1,4-$C_6H_4(COOH)_2$) and the diamine varies considerably.

For nylon derivatives using a single monomer such as nylon 6 or nylon 12, the numbers 6 and 12 denote the number of carbons in caprolactam and laurolactam, respectively.

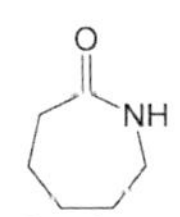

Caprolactam

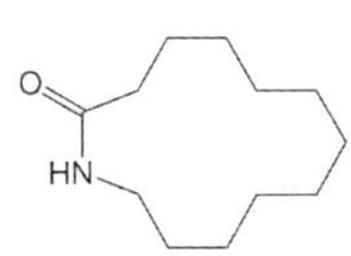

Laurolactam

On the other hand, for nylon derivatives using two monomers such as nylon 610, the first number, 6, indicates the number of carbons in the hexamethylene diamine and the other number, 10, is for the second monomer sebacic acid.

Hydrogen bonding in polyamides is fairly strong and has a pronounced effect on the physical properties of the polymer such as the crystallinity, melting point, and water absorption. For example, nylon 6, with six carbon atoms, has a melting point of 223°C (433°F), while it is only 190°C (374°F) for nylon 11, which reflects the higher hydrogen bonding in nylon 6 than in nylon 11.

Moisture absorption of nylons differs according to the distance between the amide groups. For example, nylon 4 has higher moisture absorption than most other nylons, and it is approximately similar to that of cotton. This is a result of the higher hydrophilic character of nylon. Nylons, however, are to some extent subject to deterioration by light. This has been explained on the basis of chain breaking and cross-linking. Nylons are liable to attack by mineral acids but are resistant to alkalis. They are difficult to ignite and are self-extinguishing.

In general, most nylons have remarkably similar properties, and the preference of using one nylon over the other is usually dictated by economic considerations except for specialized uses. Nylons have a variety of uses ranging from tire cord to carpet to hosiery. The most important application is cord followed by apparel. Nylon staple and filaments are extensively used in the carpet industry. Nylon fiber is also used for a variety of other articles such as seat belts, monofilament finishes, and knitwear. Because of its high tenacity and elasticity, it is a valuable fiber for ropes, parachutes, and underwear.

7.4 Polyester fibers

Polyesters are the most important class of synthetic fibers. In general, polyesters are produced by an esterification reaction of a diol and a diacid. Polyesters can be produced by an esterification of a dicarboxylic acid and a diol, a transesterification of an ester of a dicarboxylic acid and a diol, or by the reaction between an acid dichloride and a diol. Less important methods are the self-condensation of w-hydroxy acid and the ring opening of lactones and cyclic esters. In self-condensation of *w*-hydroxy acids, cyclization might compete seriously with linear polymerization, especially when the hydroxyl group is in a position to give five- or six membered-lactones.

PET is produced by esterifying TPA and ethylene glycol or, more commonly, by the transesterification of dimethyl terephthalate and ethylene glycol.

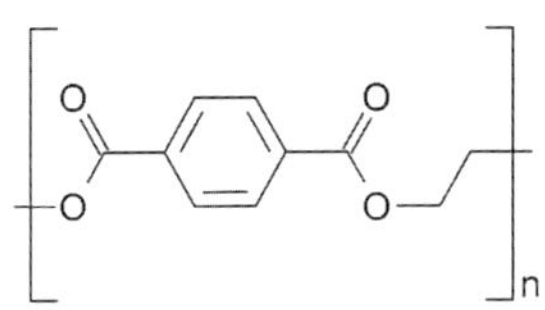

Polyethylene terephthalate

The reaction occurs in two stages: (i) in the first stage, methanol is released in the first stage at approximately 200°C (370°F) with the formation of bis(2-hydroxyethyl) terephthalate, and (ii) in the second stage, polycondensation occurs, and excess ethylene glycol is driven away at approximately 280°C (535°F) and at lower pressures.

Using excess ethylene glycol is the usual practice because it drives the equilibrium to near completion and terminates the acid end groups. This results in a polymer with terminal −OH. When the free acid is used (esterification), the reaction is self-catalyzed. However, an acid catalyst is used to compensate for the decrease in TPA as the esterification nears completion. In addition to the catalyst and terminator, other additives are used such as color improvers and dulling agents.

The molecular weight of the polymer is a function of the extent of polymerization and could be monitored through the melt viscosity. Batch polymerization is still used. However, most new processes use continuous polymerization and direct spinning. An alternative route to PET is by the direct reaction of TPA and ethylene oxide. The product bis(2-hydroxyethyl) terephthalate reacts in a second step with TPA to form a dimer and ethylene glycol, which is released under reduced pressure at approximately 300°C (570°F). This process differs from the direct esterification and the transesterification routes in that only ethylene glycol is released. In the former two routes, water or methanol is coproduced and the excess glycol released.

PET is an important thermoplastic. However, most PET is consumed in the production of fibers. Polyester fibers contain crystalline as well as noncrystalline regions. The degree of crystallinity and molecular orientation are important in determining the tensile strength of the fiber (between 18 and 22 denier) and its shrinkage. The degree of crystallinity and molecular orientation can be determined by X-ray diffraction techniques.

Important properties of polyesters are the relatively high melting temperatures (265°C, 510°F), high resistance to weather conditions and sunlight, and moderate tensile strength. Because of the hydrophobic nature of the fiber, sulfonated TPA may be used as a comonomer to provide anionic sites for cationic dyes. Small amounts of aliphatic di-acid derivatives such as adipic acid may also be used to increase the ability of the fibers to dyes by disturbing the crystallinity of the fiber.

Polyester fibers can be blended with natural fibers such as cotton and wool. The products have better qualities and are used for various types of clothing (for men and for women) wear, pillow cases, and bedspreads. Fiberfill, made from polyesters, is used in mattresses, pillows, and sleeping bags. High-tenacity polymers for tire cord reinforcement are equivalent in strength to nylon tire cords and are superior because they do not "flat spot." V-belts and fire hoses made from industrial filaments are another market for polyesters.

7.5 Polypropylene fibers

Polypropylene fibers represent a small percent of the total polypropylene production. (Most polypropylene is used as a thermoplastic.) The fibers are usually manufactured from isotactic polypropylene, the chain is arranged on one side of the backbone chain.

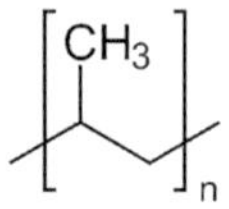

Polypropylene

By way of further clarification, atactic polymers are polymers where the side chain is arranged randomly along the backbone chain. Syndiotactic polymers are polymers where the side chain is arranged alternatively on both sides of the backbone chain.

Important characteristics of polypropylene are high abrasion resistance, strength, low static buildup, and resistance to chemicals. Crystallinity of fiber-grade polypropylene is moderate, and when heated, it starts to soften at approximately 145°C (293°F) and then melts at 170°C (340°F). The high melting points of polypropylene polymers are attributed to low entropy of fusion arising from stiffening of the chain.

Polyethylene fiber properties depend markedly on the crystallinity or density of the polymer; although high-strength fibers can be made from linear polyethylene, resiliency properties are poor, tensile properties are highly time-dependent, and endurance under sustained loading is very poor. On the other hand, polypropylene fibers have good stress-endurance properties, excellent recovery from high extensions, and fair-to-good recovery properties at low strains; recovery at ‘low strains is shown to depend on the extent of fiber orientation and annealing.

Anomalies in the change of the sonic modulus of polypropylene yarns during extension and relaxation are noted and interpreted in terms of structure changes in the crystalline phase. The high melting temperature of 235°C (455°F) for poly(4-methyl-1-pentene) appears to be due to its low entropy of melting, and fibers from this polymer are characterized by low tenacity when tested at elevated temperatures. Crystalline polystyrene fibers have relatively good retention of tenacity at elevated temperatures and are characterized by excellent resiliency at low strains, good wash-wear characteristics in cotton blends, and low abrasion resistance.

References

Abbas, A.S., Mohamed, F.A., 2015. Production and evaluation of liquid hydrocarbon fuel from thermal pyrolysis of virgin polyethylene plastics Iraqi. Journal of Chemical and Petroleum Engineering 16 (1), 21–33.

Ali, M.F., El Ali, B.M., Speight, J.G., 2005. Handbook of Industrial Chemistry: Organic Chemicals. McGraw-Hill, New York.

Austin, G.T., 1984. Shreve's Chemical Process Industries, fifth ed. McGraw- Hill, New York. Chapters 34, 35, and 36.

Braun, D., Cherdron, H., Ritter, H., 2001. Polymer Synthesis Theory and Practice: Fundamentals, Methods, Experiments. Springer-Verlag, Berlin, Germany.

Browne, C.L., Work, R.W., 1983. Man-made textile fibers. In: Kent, J.A. (Ed.), Riegel's Handbook of Industrial Chemistry, eighth ed. Van Nostrand Reinhold, New York (Chapter 1).

Carraher Jr., C.E., 2003. Polymer Chemistry 6th Edition: Revised and Expanded. Marcel Dekker Inc., New York.

Gary, J.G., Handwerk, G.E., Kaiser, M.J., 2007. Petroleum Refining: Technology and Economics, fifth ed. CRC Press, Taylor & Francis Group, Boca Raton, Florida.

Hsu, C.S., Robinson, P.R. (Eds.), 2017. Handbook of Petroleum Technology. Springer International Publishing AG, Cham, Switzerland.

Jones, R.W., Simon, R.H.M., 1983. Synthetic plastics. In: Kent, J.A. (Ed.), Riegel's Handbook of Industrial Chemistry, eighth ed. Van Nostrand Reinhold, New York (Chapter 10).

Lokensgard, E., 2010. Industrial Plastics: Theory and Applications. Delmar Cengage Learning, Clifton Park, New York.

Mannella, G.A., Pavia, F.C., La Carrubba, B., Brucato, V., 2016. Phase separation of polymer blends in solution: a case study. European polymer Jounral 79, 176–186.

Matar, S., Hatch, L.F., 2001. Chemistry of Petrochemical Process, second ed. Gulf Professional; Publishing, Elsevier BV, Amsterdam, Netherlands.

Odian, G., 2004. Principles of Polymerization, fourth ed. John Wiley 7 Sons Inc., New York.

Pundhir, S., Gagneja, A., 2016. Conversion of plastic to hydrocarbon. International Journal of Advances in Chemical Engineering & Biological Sciences (IJACEBS) 3 (1), 121–124.

Rudin, A., 1999. The Elements of Polymer Science and Engineering, second ed. Academic Press Inc., New York.

Sarker, M., Rashid, M.M., Molla, M., 2011. Waste plastic conversion into hydrocarbon fuel materials. Journal of Environmental Science and Engineering 5, 603–609.

Schroeder, E.E., 1983. Rubber. In: Kent, J.A. (Ed.), Riegel's Handbook of Industrial Chemistry, eighth ed. Van Nostrand Reinhold, New York (Chapter 9).

Speight, J.G., 2013. The Chemistry and Technology of Coal, third ed. CRC Press, Taylor and Francis Group, Boca Raton, Florida.

Speight, J.G., 2014 The Chemistry and Technology of Petroleum, fifth ed. CRC Press, Taylor and Francis Group, Boca Raton, Florida.

Speight, J.G., 2015. Handbook of Coal Analysis, second ed. John Wiley & Sons Inc., Hoboken, New Jersey.

Speight, J.G., 2017. Handbook of Petroleum Refining. CRC Press, Taylor & Francis Group, Boca Raton, Florida.

Speight, J.G., 2019. Handbook of Petrochemical Processes. CRC Press, Taylor & Francis Group, Boca Raton, Florida.

Chapter 15

Hydrocarbons in the environment

1. Introduction

Crude oil and natural gas are major sources of hydrocarbon derivatives, and experience over the past several decades has shown that it has been almost impossible to transport, store, and refine crude oil and natural gas without spills and losses resulting from damage to pipelines, transportation vehicles (including seagoing vessels), and storage containers, not only of crude oil and crude oil products but also from nonfossil fuel sources (Parkash, 2003; Gary et al., 2007; Speight, 2011, 2017; Lee et al., 2014; Hsu and Robinson, 2017). Typically, the soil suffers the most ecological damage in the damage areas of pipelines and spills of crude oils as there are often irreversible changes of the properties of the oil (Abha and Singh, 2012).

Large quantities of environmentally destructive crude oil and crude oil products are stored in (i) tank farms (multiple tanks), (ii) single aboveground storage tanks (ASTs), and (iii) semi-underground or underground storage tanks (USTs). Smaller quantities of materials may be stored in drums and containers of assorted compounds (such as lubricating oil, engine oil, other products for domestic supply). However, the above statements are not meant to diminish the potential effects on the various natural water ways and the atmosphere that can also suffer damage. The most affected soil properties by crude oil losses from pipelines are filtration, physical, and mechanical properties. These properties of the soil are important for maintaining the ecological equilibrium in the damaged area. For example, methane—the major constituents of natural gas—can drastically interfere with the workings of the atmosphere, as can the other hydrocarbon constituents of natural gas. Methane, itself a causative agent of global warming, can be oxidized into carbon dioxide (CO_2), increasing the amount of carbon dioxide in the atmosphere and adding to the greenhouse effect and global warming.

Thus, it is also necessary to consider (i) secondary containment of tanks and other storage areas and integrity of hard standing (without cracks, impervious surface) to prevent spills reaching the wider environment, also secondary containment of pipelines where appropriate, (ii) age, construction

Handbook of Industrial Hydrocarbon Processes. https://doi.org/10.1016/B978-0-12-809923-0.00015-1

details, and testing program of tanks, (iii) labeling and environmentally secure storage of drums (including waste storage), (iv) accident/fire precautions, emergency procedures, and (v) disposal/recycling of waste or "out-of-spec" oils and other materials.

There is a potential for significant soil and groundwater contamination to have arisen at crude oil refineries. Such contamination consists of (i) crude oil hydrocarbon derivatives including lower boiling, very mobile fractions (paraffins, cycloparaffins, and volatile aromatics such as benzene, toluene, ethylbenzene, and xylenes [BTEX]) typically associated with gasoline and similar boiling range distillates, (ii) middle distillate fractions (paraffins, cycloparaffins, and some polynuclear aromatics) associated with diesel, kerosene, and lower boiling fuel oil, which are also of significant mobility, (iii) higher boiling distillates (long chain paraffins, cycloparaffins, and polynuclear aromatics that are associated with lubricating oil and heavy fuel oil), (iv) various organic compounds associated with crude oil hydrocarbon derivatives or produced during the refining process, e.g., phenols, amines, amides, alcohols, organic acids, nitrogen, and sulfur containing compounds, (v) other organic additives, e.g., antifreeze (glycols), alcohols, detergents, and various proprietary compounds, and (vi) organic lead, associated with leaded gasoline and other heavy metals.

Key sources of such contamination at crude oil refineries are at (i) transfer and distribution points in tankage and process areas, also general loading and unloading areas, (ii) land farm areas, (iii) tank farms, (iv) individual ASTs and particularly individual USTs, (v) additive compounds, and (vi) pipelines, drainage areas, and on-site waste treatment facilities, impounding basins, and lagoons, especially if unlined.

While contamination may be associated with specific facilities, the contaminants are relatively highly mobile in nature and have the potential to migrate significant distances from the source in soil and groundwater. Crude oil contamination can take several forms: free-phase product, dissolved phase, emulsified phase, or vapor phase. Each form will require different methods of remediation so that cleanup may be complex and expensive. In addition, crude oil hydrocarbon derivatives include a number of compounds of significant toxicity, e.g., benzene and some polyaromatics are known carcinogens. Vapor-phase contamination can be of significance in terms of odor issues.

Because of the obvious risk of fire, refineries are equipped with sprinkler or spray systems that may draw on the main supply of water, water held in lagoons, or from reservoirs or neighboring water courses. Such water will be polluting and will require containment. Refining facilities require significant volumes of water for on-site processes (e.g., coolants, blowdowns, etc.) as well as for sanitary and potable use. Wastewater will derive from these sources (process water) and from storm water runoff. The latter could contain significant concentrations of crude oil product. Hydrocarbon derivatives, dissolved, emulsified, or occurring as free phase, will be the key constituents although wastewater may also contain significant concentrations of phenols, amines, amides, alcohols, ammonia, sulfide, heavy metals, and suspended solids.

Wastewaters may be collected in separate drainage systems (for process, sanitary, and storm water), although industrial and storm water systems may in some cases be combined. In addition, ballast water from bulk crude tankers may be pumped to receiving facilities at the refinery site before removal of floating oil in an interceptor and treatment as for other wastewater streams.

On-site treatment facilities may exist for wastewater, or treatment may take place at a public wastewater treatment plant. Storm water/process water is generally passed to a separator or interceptor before leaving the site that takes out free-phase oil (i.e., floating product) from the water before discharge or before further treatment, e.g., in settling lagoons. Other wastes that are typical of a refinery include (i) waste oils, process chemicals, still resides, (ii) non-specification chemicals and/or products, (iii) waste alkali (sodium hydroxide), (iv) waste oil sludge (from interceptors, tanks, and lagoons), and (v) solid wastes (cartons, rags, catalysts, and coke).

It is the purpose of this chapter to present the effects of hydrocarbon derivatives on the environment; although refineries have been cited above because they are the main source of hydrocarbon derivations and the ensuring pollution when spills occur, it is not the intent to lay all of the lame on refinery. Other sources of hydrocarbon derivatives are available, which must also shoulder some of the blame.

2. Release into the environment

Hydrocarbons are introduced into the environment through their extensive use as fuels and chemicals as well as through leaks or accidental spills during exploration, production, refining, or transport. Anthropogenic hydrocarbon contamination of soil is a serious global issue due to contaminant persistence and the negative impact on human health. However, to fully evaluate the environmental effects of natural gas, the general properties of the constituents (Table 15.1) must also be considered in addition to the effects of the combustion properties.

Currently, natural gas represents approximately one quarter of the energy consumed in the United States with increases in use projected for the next decade. These increases are expected because emissions of greenhouse gases are much lower with the consumption of natural gas relative to other fossil fuel consumption. For example, natural gas, when burned, emits lower quantities of greenhouse gases and criteria pollutants per unit of energy produced than other fossil fuels. This occurs in part because natural gas is fully combusted more easily and in part because natural gas contains fewer impurities than any other fossil fuel. However, the major constituent of natural gas, methane, also

TABLE 15.1 General Properties of the Constituents of Natural Gas up to n-Octane (C_8H_{18}), Including Benzene, Toluene, Ethylbenzene, and Xylene.

	Molecular weight	Specific gravity	Vapor density air = 1	Boiling point °C	Ignition temperature °C	Flash point °C
Methane	16	0.553	0.56	–160	537	–221
Ethane	30	0.572	1.04	–89	515	–135
Propane	44	0.504	1.50	–42	468	–104
Butane	58	0.601	2.11	–1	405	–60
Pentane	72	0.626	2.48	36	260	–40
Hexane	86	0.659	3.00	69	225	–23
Benzene	78	0.879	2.80	80	560	–11
Heptane	100	0.668	3.50	98	215	–4
Octane	114	0.707	3.90	126	220	13
Toluene	92	0.867	3.20	161	533	4
Ethylbenzene	106	0.867	3.70	136	432	15
Xylene	106	0.861	3.70	138	464	17

contributes directly to the greenhouse effect through venting or leaking of natural gas into the atmosphere (Speight, 2017).

Purified natural gas (methane) is the cleanest of all the fossil fuels. The main products of the combustion of natural gas are carbon dioxide and water vapor. Coal and crude oil release higher levels of harmful emissions, including a higher ratio of carbon emissions, nitrogen oxides (NO_x), and sulfur dioxide (SO_2). Coal and fuel oil also release ash particles into the environment, substances that do not burn but instead are carried into the atmosphere and contribute to pollution. The combustion of purified natural gas, on the other hand, releases very small amounts of sulfur dioxide and nitrogen oxides, virtually no ash or particulate matter, and lower levels of carbon dioxide, carbon monoxide, and other reactive hydrocarbon derivatives.

Coal mining operations are hazardous, and each year coal miners lose their lives or are seriously injured through the occurrence of roof falls, rock bursts, fires, and explosions. The latter results when flammable gases (such as methane) trapped in the coal are released during mining operations and accidently are ignited. Thus, provision of adequate ventilation is, amongst other aspects, an essential safety feature of underground coal mining. In some mines, the average weight of air passing daily through the coal mines may be many times the total daily weight of coal produced. Not all of this air is required to enable miners to work in comfort. Most of it is required to dilute the harmful gases, frequently termed damps (German *dampf*, vapor), produced during mining operations.

The gas, which occurs naturally in the coal seams, is methane (CH_4, *firedamp*) that is a highly flammable gas and forms explosive mixtures with air (5–14 volume percent methane). The explosion can then cause the combustion of the ensuing coal dust thereby increasing the extent of the hazard. To render the gas harmless, it is necessary to circulate large volumes of air to maintain the proportion of methane below the critical levels. Long boreholes may be drilled in the strata ahead of the working face, and the methane is drawn out of the workings and piped to the surface (*methane drainage*).

Carbon monoxide (CO, *whitedamp*) is a particularly harmful gas; as little as 1% in the air inhaled can cause death. It is often found after explosions and occurs in the gases evolved by explosives. Carbon dioxide (CO_2, *blackdamp*, *chokedamp*, or *stythe*) is found chiefly in old workings or badly ventilated headings. Hydrogen sulfide (H_2S, *stinkdamp*) is one of the first gases to be produced when coal is heated out of contact with air. It occasionally occurs in small quantities along with the methane given off by outbursts and is sometimes present in the fumes resulting from blasting. *Afterdamp* is the term applied to the mixture of gases found in a mine after an explosion or fire. The actual composition varies with the nature and amount of the materials consumed by the fire or with the extent to which firedamp or coal was involved in the explosion.

The continued inhalation of certain dusts is detrimental to health and may lead to reticulation of the lungs and eventually to fatal disease *pneumoconiosis* or *anthracosis* (*black lung disease*). Coal and silica dusts are particularly harmful, and methods that have been adopted to combat the dust hazard include the infusion of water under pressure into the coal before it is broken down; the spraying of water at all points where dust is likely to be formed; the installation of dust extraction units at strategic points; and the wearing of masks by miners operating drilling, cutting, and loading machinery.

Surface areas exposed during mining, as well as coal and rock waste (which were often dumped indiscriminately), weathered rapidly, producing abundant sediment and soluble chemical products such as sulfuric acid and iron sulfates. Nearby streams became clogged with sediment, iron oxides stained rocks, and *acid mine drainage* caused marked reductions in the numbers of plants and animals living in the vicinity. Potentially toxic elements, leached from the exposed coal and adjacent rocks, were released into the environment. Since the 1970s, however, stricter environmental laws have significantly reduced the environmental damage caused by coal mining. Once the coal has been extracted, it needs to be moved from the mine to the power plant or other place of use.

Over short distances, coal is generally transported by conveyor or truck, whereas trains, barges, ships, or pipelines are used for long distances. Preventative measures are taken at every stage during transport and storage to reduce potential environmental impacts. Dust can be controlled by using water sprays, compacting the coal and enclosing the stockpiles. Sealed systems, either pneumatic or covered conveyors, can be used to move the coal from the stockpiles to the combustion plant. Runoff of contaminated water is limited by appropriate design of coal storage facilities. All water is carefully treated before reuse or disposal.

In terms of the development of oil shale resources, the most serious environmental concerns are associated with the management and disposal of solid waste, especially the rock that remains after shale oil has been extracted. Oil shale comprises clastic, carbonate, organic and minor sulfide fractions, and also traces of some potentially toxic elements and, as a result, generates several types of environmentally harmful wastes. Shale (such as the Colorado shale) that contains a high proportion of dolomitic limestone (a mixture of calcium and magnesium carbonates) thermally deposes under the conditions of retorting and releases large volumes of carbon dioxide. This consumes energy and leads to the additional problem of sequestering the carbon dioxide to meet global climate change concerns.

Combustion of oil shale releases carbon dioxide (a greenhouse gas), derived from oxidation of organic matter and decomposition of carbonates. If carbonates are present in high proportions, this renders the oil shales inefficient in terms of energy per unit of carbon dioxide emitted. Furthermore, oil shale combustion emits acidic gases (nitrogen oxides, NOx, and sulfur dioxide, SO_2) derived both from inorganic sulfides and organically bound nitrogen and sulfur. Although the emissions of carbon dioxide, sulfur dioxide, and nitrogen oxides from combustion of oil shale are at the same level or lower than those from oil- or coal-based power plants with comparable capacity, the combustion of oil shales also yields particulate emissions (potentially enriched in a variety of metals, metalloids and organics) at a rate 20−50 times. The disposal of spent shale from the retort is also a problem that must be solved in economic fashion for the large-scale development of oil shale to proceed. Spent shale contains carbon as char, representing more than half of the original carbon values in the shale. The char is potentially pyrophoric and can burn if dumped into the open air while hot. The heating process results in a solid that occupies more volume than the fresh shale because of the problems of packing random particles.

One factor, which makes the extraction of oil from oil shale challenging, is that spent shale occupies 20%−30% percent greater volume after processing than raw shale due to a popcorn effect from the heating. This means that 50,000 barrels of oil per day oil shale plant will produce approximately 7500 cubic meters partially powdered rock waste per day in excess of that returned to the mine. Consequently, in the vicinity of oil shale operations, the environment will be altered, and costly environmental assessments of the impact on different ecological compartments have to be carried out parallel to developing the oil shale industry.

In situ processes avoid the spent shale disposal problems because the spent shale remains where it is created. In addition, ICP avoids carbon dioxide decomposition by operating at temperatures below approximately 350°C (650°F). On the other hand, the spent shale will contain uncollected liquids that can leach into groundwater, and vapors produced during retorting can potentially escape to the aquifer. Shell has gone to great efforts to design barrier methods for isolating its retorts to avoid these problems. In addition, there are also issues with the produced shale oil that also need resolution.

Shale oil is different to conventional crude oils, and several technologies have been developed to deal with this. The primary problems identified were arsenic, nitrogen, and the waxy nature of the crude. Nitrogen and wax problems were solved by Unocal and other companies using hydroprocessing approaches, essentially classical hydrocracking. In addtion, technologies that are focused on the production of high-quality lube stocks have been developed, which require that waxy materials be removed or isomerized. These technologies are well adapted for shale oils. However, the arsenic problem remains, no matter what the strategic significance of the oil shale reserves in the United States (US DOE, 2004a,b,c). However, arsenic can be removed from the oil by hydrotreating remains on the catalyst, generating a material that is a carcinogen, an acute poison, and a chronic poison. The catalyst must be removed and replaced when its capacity to hold arsenic is reached. Also, regulations require precautions to be taken when a reactor is opened to remove a catalyst.

Thus, several issues need to be resolved before an oil shale industry can be a viable option. These issues are not insurmountable but require the search for viable alternatives. For example, an alternative not much explored involves chemical treatment of shale to avoid the high-temperature process. The analogy with coal liquefaction here is striking: liquids can be generated from coal in two distinct ways: (i) by pyrolysis, creating a char coproduct or (ii) by dissolving the coal in a solvent in the presence of hydrogen. However, no similar *dissolution* approach to oil shale conversion is known because the chemistry of kerogen is markedly different from the chemistry of coal (Chapter 5).

As a first step in developing a direct route, some attempts were made in the 1970s to isolate kerogen from the oil shale by dissolving away the minerals. Acid treatment to dissolve the mineral carbonate followed by fluoride treatment to remove the aluminosilicate minerals might be considered. Such a scheme will only work if the kerogen is not chemically bonded to the inorganic matrix. However, if the kerogen is bonded to the inorganic matrix, the bonding arrangement must be defined for the scheme to be successful. Opportunities for circumventing the arsenic problem include development of an in-reactor process for regenerating the catalyst, collecting arsenic in a safe form away from the catalyst, and development of a catalyst or process where the removed arsenic exits the reactor in the gas or liquid phase to be scrubbed and confined elsewhere.

Shale oil produced by both aboveground and in situ techniques in the 1970s and 1980s was rich in organic nitrogen. Nitrogen compounds are catalyst poisons in many common refinery processes such as fluid catalytic cracking, hydrocracking, isomerization, naphtha reforming, and alkylation. The standard method for handling nitrogen poisoning is hydrodenitrogenation (HDN). HDN is a well-established high-pressure technology using nickel molybdenum catalysts. It can consume prodigious amounts of hydrogen, typically made by steam reforming of natural gas, with carbon dioxide as a by-product. Thus, after a decline of production since 1980 and the current scenarios that face a crude oil–based economy, the perspectives for oil shale can be viewed with a moderately positive outlook. This perspective is prompted by the rising demand for liquid fuels, the rising demand for electricity, and the change of price relationships between oil shale and conventional hydrocarbon derivatives.

In terms of innovative technologies, both conventional and in situ retorting processes result in inefficiencies that reduce the volume and quality of the produced shale oil. Depending on the efficiency of the process, a portion of the kerogen that does not yield liquid is either deposited as coke on the host mineral matter or is converted to hydrocarbon gases. For the purpose of producing shale oil, the optimal process is one that minimizes the regressive thermal and chemical reactions that form coke and hydrocarbon gases and maximizes the production of shale oil. Novel and advanced retorting and upgrading processes seek to modify the processing chemistry to improve recovery and/or create high-value by-products. Novel processes are being researched and tested in lab-scale environments. Some of these approaches include lower heating temperatures; higher heating rates; shorter residence time durations; introducing scavengers such as hydrogen (or hydrogen transfer/donor agents); and introducing solvents (Baldwin, 2002).

Finally, the development of western oil shale resources will require water for plant operations, supporting infrastructure, and the associated economic growth in the region. While some oil shale technologies may require reduced process water requirements, stable and secure sources of significant volumes of water may still be required for large-scale oil shale development. The largest demands for water are expected to be for land reclamation and to support the population and economic growth associated with oil shale activity. With consumption of fossil fuels allegedly outstripping discovery of new resources, it could be argued that oil shales may represent a viable energy and hydrocarbon-producing alternative for oil-poor countries, provided they are prepared for potential conflicts with international environmental agreements intended to regulate national emissions of greenhouse gases and thus to reduce the global emissions.

The fundamental problem with all oil shale technologies is the need to provide large amounts of heat energy to decompose the kerogen to liquid and gas products. More than 1 ton of shale must be heated to temperatures in the range 850−1,000F (425−525°C) for each barrel of oil generated, and the heat supplied must be of relatively high quality to reach retorting temperature. Once the reaction is complete, recovering sensible heat from the hot rock is very desirable for optimum process economics. This leads to three areas where new technology could improve the economics of oil recovery: (i) recovering heat from the spent shale, (ii) disposal of spent shale, especially if the shale is discharged at temperatures where the char can catch fire in the air, and (iii) concurrent generation of large volumes of carbon dioxide. The heat recovery from hot solids is generally not efficient, unless it is in the area of fluidized bed technology. However, applying fluidized bed technology to oil shale would require grinding the shale to sizes less than approximately 1 mm, an energy-intensive task that would result in an expensive disposal problem. However, such fine particles might be used in a lower temperature process for sequestering carbon dioxide.

The future development and expansion of the oil shale industry will be governed by the price of crude oil, unless oil shale−rich countries, such as the United States, decide to develop these resources to ensure a measure (yet to be defined) of energy and hydrocarbon security. Canada took this step in the early 1960s when various levels of government decided to join industry in the development of the Athabasca tar sand (oil sand) deposits. But how many politicians will be willing to tell their constituents that gasoline will increase in price by 50 to 100 percent (perhaps even more)? The fear of losing votes and of losing an elected position is strong! When the price of hydrocarbon products from oil shale is comparable to that of hydrocarbon derivatives from crude oil, and with an increasing number of countries experiencing decline in conventional oil production, hydrocarbon products from oil shale may find a place in the world energy mix.

The key is the development of efficient, economic, and environmentally friendly technology. Assuming that two-thirds of the remaining world oil resources will be produced in the Middle East and two-thirds of the resources of oil shale are located in North America, where the consumption of crude oil per capita is the greatest, one may wonder related to the geopolitical importance of shale oil in the future. Historically, energy sources have moved from wood to coal to oil and gas. Oil shale (via shale oil) has the potential to become the bridge between the impending shortage of crude oil in the coming decades and a transition to renewable energy sources and/or a hydrogen-based economy.

Natural gas has no known toxic or chronic physiological effects (that is, it is not poisonous), but it is a dangerous insofar as an atmosphere rich in natural gas will result in death to humans and animals. Exposure to a moderate concentration of natural gas may result in a headache or similar symptoms due to oxygen deprivation, but it is likely that the smell (through the presence of the odorant) would be detected well in advance of concentrations being high enough for this to occur.

In fact, in the natural gas and refining industries (Speight, 2014, 2019), as in other industries, air emissions include point and nonpoint sources. Point sources are emissions that exit stacks and flares and, thus, can be monitored and treated. Nonpoint sources are *fugitive emissions* that are difficult to locate and capture. Fugitive emissions occur throughout refineries and arise from, for example, the thousands of valves, pipe connections, seals in pumps and compressors, storage tanks, pressure relief valves, and flanged joints.

While individual leaks are typically small, the sum of all fugitive leaks at a gas processing plant can be one of its largest emission sources. These leaks can release methane and volatile constituents of natural gas into the air. Companies can minimize *fugitive emissions* by designing facilities with the fewest possible components and connections and avoiding components known to cause significant fugitive emissions. When companies quantify fugitive emissions, this provides them with important information they can then use to design the most effective leak repair program for their company. Directed inspection and maintenance programs are designed to identify the source of these leaks and to prioritize and plan their repair in a timely fashion. A reliable and effective directed inspection and maintenance plan for an individual facility will be composed of a number of components, including methods of leak detection, a definition of what constitutes a leak, set schedules and targeted devices for leak surveys, and allowable repair time.

A directed inspection and maintenance program begins with a baseline survey to identify and quantify leaks. Quantification of the leaks is critical because this information is used to determine which leaks are serious enough to justify their repair costs. Repairs are then made only to the leaking components that are cost effective to fix. Subsequent surveys are then scheduled and designed based on information collected from previous surveys, permitting operators to concentrate on the components that are more likely to leak. Some natural gas companies have demonstrated that directed inspection and maintenance programs can profitably eliminate as much as 95% of gas losses from equipment leaks.

There are also environmental issues related to the development of gas hydrate resources. As oil and gas exploration extends into progressively deeper waters, the potential hazard posed by gas hydrates to operations is gaining increasing recognition. Hazards can be considered as arising from two possible events: (i) the release of high-pressure gas trapped below the hydrate stability zone or (ii) the destabilization of in situ hydrates. A major issue is how gas hydrates alter the physical properties of sediment. The link between seafloor failure and gas hydrate destabilization has been well established, especially with respect to the previous glacial—interglacial eustatic sea-level changes. Sea slope failure, as a result of gas hydrate decomposition, is considered to pose a significant hazard to underwater installations, pipelines, and cables and, in extreme cases, to coastal populations through the generation of tsunamis. Furthermore, as a greenhouse gas, methane is roughly 10 times more potent than carbon dioxide. Over geologic time scales, there is evidence pointing to periodic, large releases of methane into the atmosphere. During formation of large polar ice sheets, the sea level falls, thereby reducing the pressure on the ocean margin gas hydrates.

In many pipelines, the temperature and the pressure conditions that are encountered place the flowing fluid well within the hydrate stability envelope. It is estimated that controlling and preventing hydrate formation (*flow assurance*) costs industry more than 100 million dollars per year (Bishnoi and Clarke, 2006). The problem is extremely severe in offshore pipelines. Conventional methods of preventing hydrate formation in pipelines are to process the crude oil fluids, typically by heating the fluid, water dew point control through moisture removal, or to inject thermodynamic inhibitors so that the operating conditions of the pipelines lie outside the hydrate stability envelopes (*hydrate avoidance methods*). More recently, kinetic methods of delaying hydrate formation and hydrate flow modifiers have been developed. The methods, based on modifying the flow characteristics of hydrates, seem to be gaining popularity with industry, especially for offshore applications.

Of the abovementioned techniques, thermodynamic inhibitors, which include alcohols, salts, and glycols, are by far the most prevalent. For example, adding methanol to a natural gas will shift the equilibrium conditions so that a higher pressure is required to form hydrates, at a given temperature. Kinetic inhibitors are typically water-soluble polymers or copolymers that delay hydrate nucleation and/or growth. An inhibitor molecule slows crystal growth by either adsorbing on to the growth sites on the crystal surface or by fitting into the crystal lattice. Antiagglomerants are designed to specifically interact with the growing hydrate crystal surface. These inhibitors permit hydrates to form but inhibit agglomeration, deposition, and plugging.

In the event of a deep-water well blowout, one of the environmental concerns is whether oil will surface and if so, where, when, and what will be the thickness of the oil slick. In the high-pressure and low-temperature conditions encountered in deep water, the gases are likely to form hydrates. As the density of hydrates is similar to that of oil, the conversion of gas into hydrates has a significant impact on the behavior of the jet plume due to the alteration of the buoyancy (Bishnoi and Clarke, 2006).

Contamination of the environment by hydrocarbon derivatives is a very serious problem whether it comes from crude oil, pesticides, or other toxic organic matter. Environmental pollution caused by crude oil is of great concern because crude oil hydrocarbon derivatives are toxic to all forms of life. Environmental contamination by crude oil is relatively common because of its widespread use and its associated disposal operations and accidental spills.

The term crude oil represents an extremely complex mixture of a wide variety of low molecular weight (low-boiling) and high molecular weight (high-boiling) hydrocarbon derivatives (Parkash, 2003; Gary et al., 2007; Speight, 2014, 2017; Hsu and Robinson, 2017). This complex mixture contains a variety of hydrocarbon derivatives such as saturated alkane derivatives, branched alkane derivatives, alkene derivatives, naphthene derivatives (homocyclic derivatives and heterocyclic derivatives), aromatic derivatives (including aromatics containing heteroatoms such as sulfur, oxygen, nitrogen, and other heavy metal complexes), naphthene aromatic derivatives, and higher molecular weight constituents such as resin constituents and asphaltene constituents (Speight, 2014, 2015). Along with the various hydrocarbon derivatives, crude oil also contains heavy metals, and much of the heavy metal content of crude oil is associated with pyrrolic structures known as porphyrins.

During refining operations, crude oil is separated into various hydrocarbon-containing fractions such as gases, naphtha, kerosene, fuel oil, lubricating oil, and wax (Parkash, 2003; Gary et al., 2007; Speight, 2014, 2017; Hsu and Robinson, 2017). The low-boiling fractions and the high-boiling fractions of crude oil have different hydrocarbon derivatives composition—the low-boiling fractions contain low molecular weight (predominantly) saturated hydrocarbon derivatives, unsaturated hydrocarbon derivatives, naphthene derivatives, and aromatic compounds while the high-boiling fractions consist of high molecular weight alkane derivatives, alkene derivatives, organometallic compounds, and high molecular weight aromatic compounds.

Principal sources of releases to air from refineries include (i) combustion plants, emitting sulfur dioxide, oxides of nitrogen, and particulate matter, (ii) refining operations, emitting sulfur dioxide, oxides of nitrogen, carbon monoxide, particulate matter, volatile organic compounds, hydrogen sulfide, mercaptans, and other sulfurous compounds, and (iii) bulk storage operations and handling of volatile organic compounds (various hydrocarbon derivatives). In light of this, it is necessary to consider (i) regulatory requirements—air emission permits stipulating limits for specific pollutants—and possibly health and hygiene permit requirements, (ii) requirement for monitoring program, and (iii) requirements to upgrade pollution abatement equipment.

Crude oil products released into the environment undergo weathering processes with time. These processes include evaporation, leaching (transfer to the aqueous phase) through solution and entrainment (physical transport along with the aqueous phase), chemical oxidation, and microbial degradation. The rate of weathering is highly dependent on environmental conditions. For example, gasoline, a volatile product, will evaporate readily in a surface spill while gasoline released below 10 feet of clay topped with asphalt will tend to evaporate slowly (weathering processes may not be detectable for years).

An understanding of weathering processes is valuable to environmental test laboratories. Weathering changes product composition and may affect testing results, the ability to bioremediate, and the toxicity of the spilled product. Unfortunately, the database available on the composition of weathered products is limited.

However, biodegradation processes, which influence the presence and the analysis of crude oil hydrocarbon at a particular site, can be very complex. The extent of biodegradation is dependent on many factors including the type of microorganisms present, environmental conditions (e.g., temperature, oxygen levels, and moisture), and the predominant hydrocarbon types. In fact, the primary factor controlling the extent of biodegradation is the molecular composition of the crude oil contaminant. Multiple ring cycloalkane derivatives are hard to degrade, while polynuclear aromatic hydrocarbon derivatives display varying degrees of degradation. Straight-chain alkane derivatives biodegrade rapidly with branched alkane derivatives and single saturated ring compounds degrading more slowly.

The primary processes determining the fate of crude oils and oil products after a spill are (i) dispersion, (ii) dissolution, (iii) emulsification, (iv) evaporation, (v) leaching, (vi) sedimentation, (vii) spreading, and (viii) wind. These processes are influenced by the spill characteristics, environmental conditions, and physicochemical properties of the spilled material.

2.1 Dispersion

The physical transport of oil droplets into the water column is referred to as dispersion. This is often a result of water surface turbulence and may also result from the application of chemical agents (dispersants). These droplets may remain in the water column or coalesce with other droplets and gain enough buoyancy to resurface. Dispersed oil tends to biodegrade and dissolve more rapidly than floating slicks because of high surface area relative to volume. Most of this process occurs from approximately half an hour to half a day after the spill.

2.2 Dissolution

Dissolution is the loss of individual oil compounds into the water. Many of the acutely toxic components of oils such as benzene, toluene, and xylene will readily dissolve into water. This process also occurs quickly after a discharge but tends to be less important than evaporation. In a typical marine discharge, generally less than 5% of the benzene is lost to dissolution while greater than 95% is lost to evaporation. For alkylated polynuclear aromatic compounds, solubility is inversely proportional to the number of rings and extent of alkylation. The dissolution process is thought to be much more important in rivers because natural containment may prevent spreading, reducing the surface area of the slick and thus retarding evaporation. At the same time, river turbulence increases the potential for mixing and dissolution. Most of this process occurs within the first hour of the spill.

Aromatics, and especially BTEX, tend to be the most water-soluble fraction (WSF) of crude oil. Crude oil—contaminated groundwater tends to be enriched in aromatics relative to other crude oil constituents. Relatively insoluble hydrocarbon derivatives may be entrained in water through adsorption into kaolinite particles suspended in the water or as an agglomeration of oil droplets (microemulsion). In cases where groundwater contains only dissolved hydrocarbon derivatives, it may not be possible to identify the original crude oil product because only a portion of the free product will be present in the dissolved phase. As the whole product floats on groundwater, the free product will gradually lose the water-soluble compounds. Groundwater containing entrained product will have a gas chromatographic fingerprint that is a combination of the free product chromatogram plus enhanced amounts of the soluble aromatics.

Generally, dissolved aromatics may be found quite far from the origin of a spill, but entrained hydrocarbon derivatives may be found in water close to the crude oil source. Oxygenates, such as methyl-t-butyl ether (MTBE), are even more water soluble than aromatics and are highly mobile in the environment.

2.3 Emulsification

Certain oils tend to form water-in-oil emulsions (where water is incorporated into oil) or "mousse" as weathering occurs. This process is significant because, for example, the apparent volume of the oil may increase dramatically, and the emulsification will slow the other weathering processes, especially evaporation. Under certain conditions, these emulsions may separate and release relatively fresh oil. Most of this process occurs from approximately half a day to 2 days after the spill.

2.4 Evaporation

Evaporative processes are very important in the weathering of volatile crude oil products and may be the dominant weathering process for gasoline. Automotive gasoline, aviation gasoline, and some grades of jet fuel (e.g., JP-4) contain 20%–99% highly volatile constituents (i.e., constituents with less than nine carbon atoms).

Evaporative processes begin immediately after oil is discharged into the environment. Some low-boiling products (like 1- to 2-ring aromatic hydrocarbon derivatives and/or low molecular weight alkane derivatives less than n-C15) may evaporate entirely; a significant fraction of heavy refined oils also may evaporate. For crude oils, the amount lost to evaporation can typically range from approximately 20 to 60%. The primary factors that control evaporation are the composition of the oil, slick thickness, temperature and solar radiation, wind speed, and wave height. While evaporation rates increase with temperature, this process is not restricted to warm climates. For the Exxon Valdez incident, which occurred in cold conditions (March 1989), it has been estimated that appreciable evaporation occurred even before all the oil escaped from the ship and that evaporation ultimately accounted for 20% of the oil. Most of this process occurs within the first few days after the spill.

It is not unusual for evaporative processes, however, to be working simultaneously with other processes to remove the volatile aromatics such as benzene and toluene.

2.5 Leaching

Leaching processes introduce hydrocarbon into the water phase by solubility and entrainment. Leaching processes of crude oil products in soils can have a variety of potential scenarios. Part of the aromatic fraction of a crude oil spill in soil may partition into water that has been in contact with the contamination.

2.6 Sedimentation or adsorption

As mentioned above, most oils are buoyant in water. However, in areas with high levels of suspended sediment, crude oil constituents may be transported to the river, lake, or ocean floor through the process of sedimentation. Oil may adsorb to sediments and sink or be ingested by zooplankton and excreted in fecal pellets that may settle to the bottom. Oil stranded on shorelines also may pick up sediments, float with the tide, and then sink. Most of this process occurs from approximately 2–7 days after the spill.

2.7 Spreading

As oil enters the environment, it begins to spread immediately. The viscosity of the oil, its pour point, and the ambient temperature will determine how rapidly the oil will spread, but low-boiling oils typically spread more rapidly than heavy oils. The rate of spreading and ultimate thickness of the oil slick will affect the rates of the other weathering processes. For example, discharges that occur in geographically contained areas (such as a pond or slow-moving stream) will evaporate more slowly than if the oil were allowed to spread. Most of this process occurs within the first week after the spill.

2.8 Wind

Wind (aeolian) transport (relocation by wind) can also occur and is particularly relevant when catalyst dust and coke dust are considered. Dust becomes airborne when winds traversing arid land with little vegetation cover pick up small particles such as catalyst dust, coke dust, and other refinery debris and send them skyward. Wind transport may occur through *suspension*, *saltation*, or *creep* of the particles.

3. Biodegradation

The biodegradability of crude oil and any crude oil constituent or product is a measure of the ability of that constituent to be metabolized (or cometabolized) by bacteria or other microorganisms through a series of biological process, which include ingestion by organisms as well as microbial degradation (Das and Chandran, 2011; Singh et al., 2012; Speight and Arjoon, 2012). The chemical characteristics of the contaminants influence biodegradability, and, in addition, the location and distribution of crude oil contamination in the subsurface can significantly influence the likelihood of success for bioremediation.

Briefly, bioremediation is the process of utilizing living organisms and microorganisms to degrade pollutants and contaminants from the environment and transform them into less toxic form. Bioremediation is based on the ability of a microorganism to degrade the hydrocarbons into components that can be taken up by other microorganisms as a nutrient source or can be safely returned to the environment. Degraded organic components are converted into water, carbon dioxide, and other inorganic compounds. Not only microbes but also plants help in biodegradation of hydrocarbons. An effective bioremediation requires enzymatic attack by microorganisms to convert pollutants into harmless products. Environmental parameters should be optimum to help the microorganisms to grow and degrade the pollutants at a rapid rate (Banerjee et al., 2016; De la Cueva et al., 2016).

Moreover, because biodegradation tends to favor the hydrocarbon constituents of a spill, the biodegradation of crude oil typically (i) raises the viscosity and decreases the API gravity, which adversely reduces the ability of the degraded product to flow, (ii) decreases the hydrocarbon content, thereby increasing the residuum content, (iii) increases the concentration of certain metals, (iv) increases the sulfur content, and (v) increases oil acidity and adds compounds such as carboxylic acids and phenols. All of these changes are seen in the product of the product relative to the unchanged (nonbiodegraded) crude oil.

The contamination of soils and aquifers by spilled crude oil is a persistent and widespread pollution problem, which causes ecological disturbances and the associated health implications (Okoh, 2006; Salam et al., 2011). Once crude oil is released and comes into contact with water, air, the necessary salts, and microorganisms present in the environment, the natural process of crude oil biodegradation begins (Antai, 1990; Davies et al., 2001). However, some of the crude oil—related pollutants are carcinogenic and mutagenic (Miller and Miller, 1981; Obayori et al., 2009).

The recognized mechanical and chemical methods for remediation of hydrocarbon-polluted environment are often expensive, technologically complex, and lack public acceptance (Speight, 2005; Speight and Lee, 2000; Vidali, 2001). Thus, bioremediation is often the method of choice for effective removal of hydrocarbon pollutants from a variety of ecosystems (Okoh and Trejo-Hernandez, 2006). In fact, crude oil and crude oil products are a rich source of carbon, and the reaction of the hydrocarbon derivatives contained therein with aerial oxygen (with the release of carbon dioxide) is promoted by a variety of microorganisms (Atlas, 1981; Atlas and Bartha, 1992; Steffan et al., 1997). However, the rate of microbial degradation of hydrocarbon derivatives in soils is affected by several physicochemical and biological parameters including the number and species of microorganisms present, the conditions for microbial degradation activity (e.g., presence of nutrient, oxygen, pH, and temperature), the quality, quantity, and bioavailability or bioaccessibility of the contaminants, and the soil characteristics such as particle size distribution (Atlas, 1991; Freijer et al., 1996; Margesin and Schinner, 1997a, 1997b; Speight and Lee, 2000; Dandie et al., 2010; Speight and Arjoon, 2012).

Hydrocarbon-degrading bacteria and fungi are mainly responsible for the mineralization (conversion of hydrocarbon derivatives to carbon dioxide and water) of crude oil—related pollutants and are distributed in diverse ecosystems (Leahy and Colwell, 1990). Furthermore, the population of microorganisms found in a polluted environment will degrade crude oil—related constituents differently and at a different rate than microorganisms in a relatively clean environment (Obire, 1990, 1993; Obire and Okudo, 1997; Obire and Nwaubeta, 2001).

However, it is uncommon to find organisms that could effectively degrade both aliphatic constituents and aromatic constituents possibly due to differences in metabolic routes and pathways for the degradation of the two classes of hydrocarbon derivatives. There are indications of the existence of bacterial species with propensities for simultaneous degradation of aliphatic hydrocarbon derivatives and aromatic hydrocarbon derivatives (Obayori et al., 2009). This rare ability may be as a result of long exposure of the organisms to different hydrocarbon pollutants resulting in genetic alteration and *acquisition* of the appropriate degradative genes.

Moreover, it is essential to recognize that the environmental impact of crude oil spills is dependent on previous hydrocarbon exposures and the adaptive status of the local microbiota (Greenwood et al., 2009). The different structural and functional response of microbial subgroups to different hydrocarbon derivatives confirms that the overall response of biota is sensitive to crude oil composition. This suggests that the preferred response to anticipated contaminants may be engineered by preexposure to representative substrates. The controlled adaption of microbes to a threatening contaminant is the basis of proactive bioremediation technology, including the augmentation of newly contaminated sites with locally remediated soil in which the biota had already been adapted (bioaugmentation).

Thus, establishing the chemical history of recently contaminated regions is an important aspect of environmental bioremediation, the premise being that microbial species adapted through a history of exposure to more bioavailable crude oil hydrocarbon derivatives is less severely impacted by a spill than microbes with no such preexposure or adaptation (Page et al., 1996; Peters et al., 2005). Indeed, the diversity of crude oil hydrocarbon degraders in most natural environments may be significant, but, in the absence of a previous pollution history, the numbers of the microbes may be low due to lack of prior stimulus (Swanell et al., 1996).

3.1 Specific constituents

Crude oil and crude oil products are mixtures of differing molecular species hydrocarbon derivatives, and the constituents of these molecular categories are present in varied proportions resulting in high variability in crude oil and crude oil products (Speight, 2014). In terms of bulk fractions (Chapter 3) (Speight, 2014, 2015), the resin constituents and the asphaltene constituents are of particular interest (or notoriety) because these constituents generally resist degradation. After a spill, the constituents of crude oil and crude oil products are subjected to physical and chemical processes such as evaporation or photochemical oxidation, which produce changes in the composition of the spilled material (Speight and Lee, 2000; Taghvaei Ganjali et al., 2007; Speight and Arjoon, 2012).

3.1.1 Alkanes

Conventional (light) crude oil contains 10%–40% w/w normal (straight-chain) alkane derivatives, but weathered and heavier oils may have only a fraction of a percent. Higher molecular weight alkane derivatives constitute 5%–20% w/w of light oils and up to 60% w/w of the more viscous oils and tar sand bitumen. Aromatic hydrocarbon derivatives are those characterized by the presence of at least one benzene (or substituted benzene) ring. The low molecular weight aromatic hydrocarbon derivatives are subject to evaporation and, although toxic to much marine life, are also relatively easily degraded. Conventional (light) crude oil typically contains between 2% and 20% w/w low-boiling aromatic compounds, whereas heavy oils contain less than 2% w/w aromatic compounds. As molecular weight and complexity increase, aromatics are less readily degraded. Thus, the degradation rate of polyaromatics is slower than that of monoaromatics.

Of these, the normal alkane series (straight-chain alkane series) is the most abundant and the most quickly degraded. Compounds with chains of up to 44 carbon atoms can be metabolized by microorganisms, but those having 10 to 24 carbon atoms (C_{10} to C_{24}) are usually the easiest to metabolize. Shorter chains (up to approximately C_8) also evaporate relatively easily. Only a few species can use C_1 to C_4 alkane derivatives, and C_5 to C_9 alkane derivatives are degradable by some microorganisms but toxic to others. Branched alkane derivatives are usually more resistant to biodegradation than normal alkane derivatives but less resistant than cycloalkane derivatives (naphthene derivatives)—those alkane derivatives having carbon atoms in ring-like central structures. Branched alkane derivatives are increasingly resistant to microbial attack as the number of branches increases. At low concentrations, cycloalkane derivatives may be degraded at moderate rates, but some highly condensed cycloalkane derivatives can persist for long periods after a spill.

Generally, with respect to the molecular composition of the aliphatic constituents of crude oil and crude oil–related products, microbial biodegradation attacks n-alkane derivatives and isoprenoid derivatives. The polycyclic alkane derivatives of sterane and triterpane type tend to be somewhat resistant to biodegradation. As this is the case even for naphthenic type crude oil (which is originally depleted in *n*-alkane derivatives), it has been concluded that the biodegradation of crude oil–type pollutants, under natural conditions, will be restricted to n-alkane derivatives and isoprenoid derivatives.

In aqueous systems, addition of acclimatized naturally occurring microorganisms (bioaugmentation) enhances the biodegradation of hydrocarbon derivatives. Because dissolved hydrocarbon derivatives are more available for microbiological degradation, application of dispersants and surfactants increases the bioavailability significantly and enhances oil degradation (Mohn and Stewart, 2000; Zahed et al., 2010). Other factors (such salinity and pH) have considerable effects on biodegradation of crude oil hydrocarbon derivatives in the marine environments as well.

3.1.2 Aromatic hydrocarbons

Spills of aromatic products such aromatic naphtha and leaks from underground fuel tanks contribute significantly to the contamination of groundwater by aromatic compounds. Nonoxygenated monoaromatic hydrocarbon derivatives, such as BTEX, are of particular concern. The high water solubility of the BTEX species enables them to migrate in the subsurface and contaminate drinking water.

Two thermophilic aerobic bacteria (*Thermus aquaticus* and *Thermus* sp.) have been reported to degrade BTEX fractions cometabolically (Chen and Taylor, 1995, 1997a). However, only small fractions of benzene and toluene were metabolized to carbon dioxide, and biodegradation was inhibited by higher BTEX concentrations but was enhanced if strains were pregrown on catechol and *o*-cresol—indicating that the preconditioning can enhance the performance of microbes (Chen and Taylor, 1995).

Two anaerobic consortia, consisting of unidentified bacterial cocci, could grow on all BTEX compounds as sole carbon and energy sources at 45–75°C (93–167°F), 50°C (122°F) being the optimal temperature (Chen and Taylor, 1997b). Only a small fraction of toluene was mineralized to carbon dioxide. Biodegradation was coupled by both consortia to sulfate reduction and to generation of hydrogen sulfide. No growth or BTEX metabolism occurred when sulfate was omitted. Thus, sulfate-reducing bacteria are most likely the principal species that carry out the biodegradation, while other thermophilic species may use the early water-soluble BTEX metabolites.

The bioremediation process is terminated by lowering the temperature below 40°C (104°F). Such an in situ follow-up treatment could also be applied to fuel-contaminated plumes subjected to thermally enhanced vapor stripping as a primary treatment method or as a stand-alone method, when the initial concentration of VOCs is low and the subsurface volume to be heated is small. In the latter case, thermophilic hydrocarbon degraders suspended in hot water are pumped into the subsurface.

The biodegradation of alkyltetralins derivatives has also been studied. However, tetralin has been shown to be biodegraded by both mixed cultures of microbes (Strawinski and Stone, 1940; Soli and Bens, 1972) and by some strains able to utilize the compound as sole carbon and energy source (Schreiber and Winkler, 1983; Sikkema and Bont, 1991).

3.1.3 Polynuclear aromatic hydrocarbons

Polynuclear aromatic hydrocarbon derivatives, in the current context, are persistent organic compounds with two or more aromatic rings in various structural configurations. Polynuclear aromatic hydrocarbon derivatives constitute a large and diverse class of organic compounds. However, derivatives such as tetralin (1,2,3,4-tetrahydronaphthalene) and decalin (decahydronaphthalene, bicyclo[4.4.0]decane) are not included in this group but are included in the alkane group because of the saturated ring.

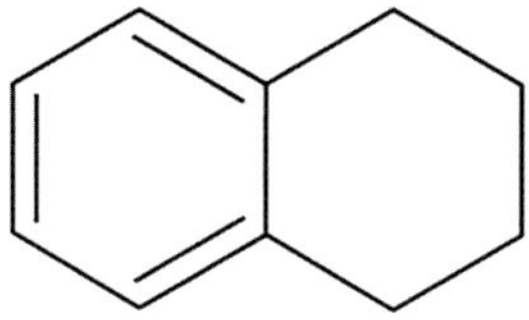

Tetralin

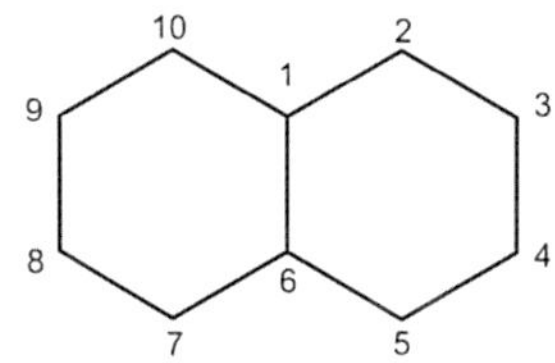

Decalin

The chemical properties, and hence the environmental fate, of polynuclear aromatic hydrocarbon derivatives are dependent in part on both molecular size (i.e., the number of aromatic rings and the pattern of ring linkage). Ring linkage patterns (also known as molecular topology) in polynuclear aromatic hydrocarbon derivatives may occur such that the tertiary carbon atoms are centers of two or three interlinked rings, as in the linear kata-condensed polynuclear aromatic hydrocarbon anthracene or the peri-condensed polynuclear aromatic hydrocarbon pyrene.

However, most polynuclear aromatic hydrocarbon derivatives occur as hybrids encompassing various structural components, such as in the polynuclear aromatic hydrocarbon benzo[*a*]pyrene.

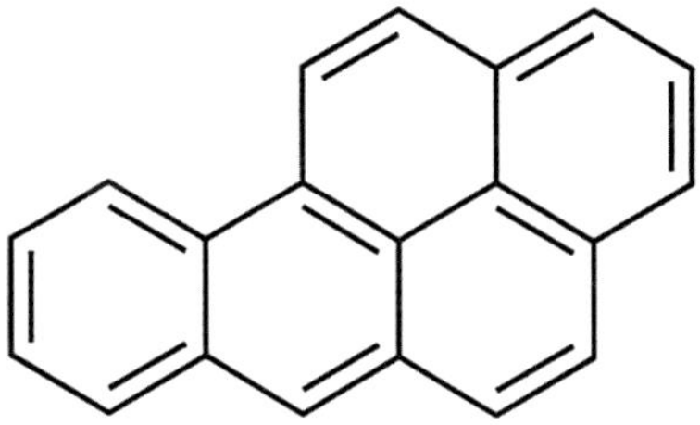

Benzo(a)pyrene

Generally, an increase in the size and angularity of a polynuclear aromatic hydrocarbon molecule results in a concomitant increase in hydrophobicity and electrochemical stability (Zander, 1983; Harvey, 1997). Polynuclear aromatic hydrocarbon molecule stability and hydrophobicity are two primary factors that contribute to their persistence of in the environment.

Polynuclear aromatic hydrocarbon derivatives are present as natural constituents in fossil fuels, are formed during the incomplete combustion of organic material, and are therefore present in relatively high concentrations in products of fossil fuel refining (Speight, 2013, 2014). Polynuclear aromatic hydrocarbon derivatives released into the environment may originate from crude oil products such as including gasoline, diesel fuel, and fuel oil (Pavlova and Ivanova, 2003). The concentration of polynuclear aromatic hydrocarbon derivatives in the environment varies widely, depending on the proximity of the contaminated site to the production source, the level of industrial development, and the mode(s) of polynuclear aromatic hydrocarbon transport.

Active bioremediation strategies (such as biostimulation) for application to polynuclear aromatic-contaminated soils can be used to supply nutrients, oxygen, and other amendments to the subsurface to enhance indigenous microbial activity and contaminant biodegradation (Bamforth and Singleton, 2005; Borchert et al., 1995; Mohan et al., 2006). The benefits of adding oxygen and/or nutrients on the biodegradation of polynuclear aromatics hydrocarbon derivatives have been reported for contaminated soils from various sites (Breedveld and Sparrevik, 2000; Eriksson et al., 2000; Liebeg and Cutright, 1999; Lundstedt et al., 2003; Talley et al., 2002). However, only a few studies have focused on the direct effects of biostimulation on the indigenous microbial community and polynuclear aromatic hydrocarbon−degrading bacteria (Ringelberg et al., 2001; Viñas et al., 2005).

Because of their lipophilic nature, polynuclear aromatic hydrocarbon derivatives have a high potential for bioconcentration (Clements et al., 1994). In addition to increases in environmental persistence with increasing polynuclear aromatic hydrocarbon molecular size, evidence suggests that in some cases, polynuclear aromatic hydrocarbon toxicity also increases with size, up to at least four or five fused benzene rings (Cerniglia, 1992). The relationship between polynuclear aromatic hydrocarbon environmental persistence and increasing numbers of benzene rings is consistent with the results of various studies correlating environmental biodegradation rates and polynuclear aromatic hydrocarbon molecule size (Shuttleworth and Cerniglia, 1995).

The biodegradation of naphthalene (the simplest polynuclear aromatic hydrocarbon) process was optimized with preliminary experiments in slurry aerobic microcosms (Bestetti et al., 2003). From soil samples collected on a contaminated site, a *Pseudomonas putida* strain (designated as M8), capable to degrade naphthalene, was selected. Microcosms were prepared with M8 strain by mixing noncontaminated soil and a mineral medium. Different experimental conditions were tested varying naphthalene concentration, soil/water ratio, and inoculum density. The disappearance of hydrocarbon, the production of carbon dioxide, and the ratio of total heterotrophic and naphthalene-degrading bacteria were monitored at different incubation times. The kinetic equation that best fitted the disappearance of contaminant with time was determined. The results showed that the isolated strain enhanced the biodegradation rate with respect to the natural biodegradation.

Of the four-ring polynuclear aromatic hydrocarbon derivatives, fluoranthene, pyrene, chrysene, and benz[*a*]anthracene have been investigated to various degrees.

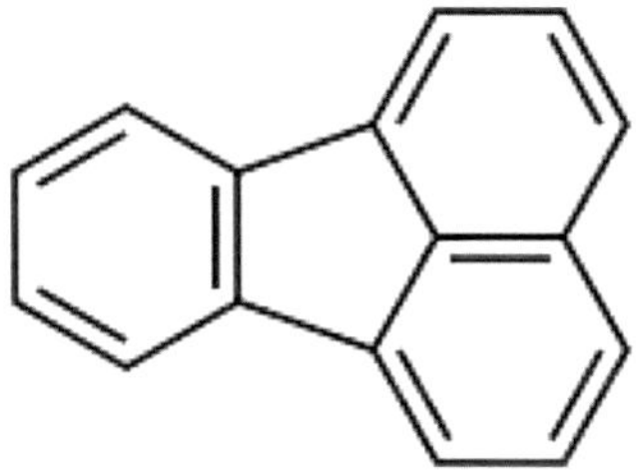

Fluoranthene

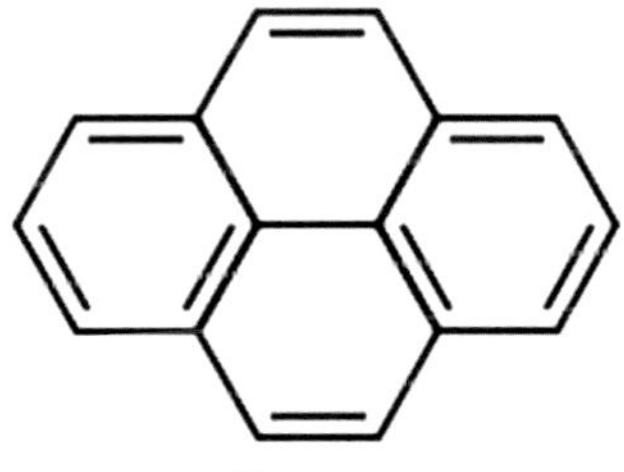

Pyrene

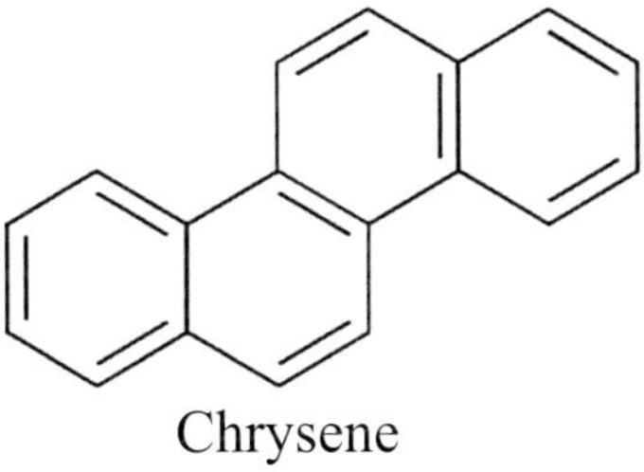

Chrysene

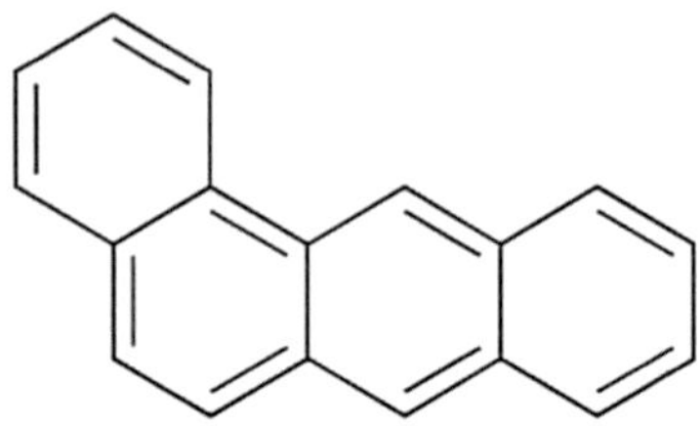

Benz(a)anthracene

Fluoranthene, a polynuclear aromatic hydrocarbon containing a five-membered ring, has been shown to be metabolized by a variety of bacteria, and pathways describing its biodegradation have been proposed (Mueller et al., 1990; Weissenfels et al., 1990, 1991; Ye et al., 1996). The bacterial degradation of pyrene, a peri-condensed polynuclear aromatic hydrocarbon, has been reported by a number of groups, and some have identified metabolites and proposed pathways (Cerniglia and Heitkamp, 1990). Sediment microcosms inoculated with the mycobacterium showed enhanced mineralization of various polynuclear aromatic hydrocarbon derivatives, including pyrene and benzo[*a*]pyrene (Heitkamp and Cerniglia, 1989).

Generally, aromatic constituents with five or more rings are not easily attacked and may persist in the environment for long periods. High molecular weight aromatics comprise 2%–10% w/w conventional (light) crude oil and up to 35% w/w of the more viscous crude oil. Increases in the understanding of the microbial ecology of polynuclear aromatic hydrocarbon–degrading communities and the mechanisms by which polynuclear aromatic hydrocarbon biodegradation occur will prove helpful for predicting the environmental fate of these compounds and for developing practical polynuclear aromatic hydrocarbon bioremediation strategies in the future (Okerentugba and Ezeronye, 2003).

3.2 Effects of biodegradation

These early stages of oil biodegradation (loss of n-paraffins followed by loss of acyclic isoprenoid hydrocarbon derivatives) can be readily detected by gas chromatography (GC) analysis of the crude oil. However, in heavily biodegraded crude oils, gas chromatographic analysis alone cannot distinguish differences in biodegradation due to interference of the unresolved complex mixture that dominates the gas chromatographic traces of heavily degraded crude oils. Among such crude oils, differences in the extent of biodegradation can be assessed using gas chromatography–mass spectrometry (GC-MS) to quantify the concentrations of biomarkers with differing resistances to biodegradation.

During biodegradation, the properties of the crude oil fluid changes because different classes of compounds in crude oil have different susceptibilities to biodegradation (Goodwin et al., 1983). The early stages of biodegradation (in addition to any evaporation effects) are characterized by the loss of n-paraffins (n-alkane derivatives or branched alkane derivatives) followed by loss of acyclic isoprenoid hydrocarbon derivatives (e.g., norpristane, pristane, and phytane). Compared with those compound groups, other compound classes (such as highly branched and cyclic saturated hydrocarbon derivatives as well as aromatic compounds) are more resistant to biodegradation. However, even the more resistant compound classes are eventually destroyed as biodegradation proceeds.

3.3 Effect of weathering

Weathered crude oil (i.e., crude oil and crude oil—related products) that has been exposed to air and oxidized and to other influence such as evaporation offer a differ challenge to bioremediation efforts. Oxygen is often the limiting factor in aerobic bioremediation at many sites. The degradation of crude oil hydrocarbon derivatives occurs much faster under aerobic conditions compared with anaerobic conditions. Therefore, the addition of oxygen can significantly increase the remediation rates. Oxygen addition is most frequently used to address dissolved phase contamination, such as total crude oil hydrocarbon derivatives and BTEX, as well as contamination in the capillary fringe zone. Oxygen can only be effective if the hydrocarbon derivatives are bioavailable and there is no nutrient limitation.

4. Analysis of hydrocarbons in the environment

The determination of the crude oil hydrocarbon derivatives in a sample is made by using several laboratory tests that are relatively inexpensive, relatively quick, sometimes ineffective, but not usually quantitative. The results are, thus, dependent on analysis of the medium in which the hydrocarbon derivatives are found.

There are several hundred individual hydrocarbon chemicals defined as crude oil—based and have been identified. Furthermore, each individual crude oil and each individual crude oil product has a specific mixture of the various constituents because of the variation in crude oil composition (Parkash, 2003; Gary et al., 2007; Speight, 2011, 2014, 2015, 2017; Lee et al., 2014; Hsu and Robinson, 2017), and this variation is reflected in the composition of the finished crude oil product.

Crude oil products, themselves, are the source of the many components, but knowing the composition of crude oil products does assist in defining the potential hydrocarbon derivatives that become environmental contaminants, but any ultimate exposure is determined also by how the product changes with use, by the nature of the release, and by the environmental fate of the released hydrocarbon derivatives. When crude oil products are released into the environment, changes that significantly affect their potential effects occur. Physical, chemical, and biological processes change the structure, concentration, and location of hydrocarbon derivatives at any particular site. Crude oil hydrocarbon derivatives are commonly found environmental contaminants, though they are not usually classified as hazardous waste. However, soil and groundwater contamination by crude oil hydrocarbon has spurred various analytical and site remediation developments, e.g., risk-based corrective actions.

The result of these processes is an alteration in the composition of the hydrocarbon discharged into the soil. Clearly, those hydrocarbon derivatives that are most strongly sorbed onto soil organic matter will be most resistant to loss or alteration by the other processes. Conversely, the more volatile/soluble hydrocarbon derivatives will be the most susceptible to change by volatilization/reaction/leaching/biodegradation. The ultimate result will be "weathering" of the hydrocarbon mixture discharged into the soil, with an accompanying change in its composition and a preferential transport of certain fractions to other environmental compartments. Moreover, interpretation of analytical results requires an understanding of how the determination was made (Miller, 2000; Dean, 2003).

The very volatile gases (compounds with four carbons or less), crude oil, and the solid asphaltic materials are not included in this discussion of analytical methods but are available elsewhere (Speight, 2015).

4.1 Environmental samples

Methods that use GC methods do provide some information related to the product type. Most of the methods involve a sample preparation procedure followed by analysis using gas chromatographic techniques. The gas chromatographic determination is based on selected components or the sum of all components detected within a given range. Frequently, the approach is to use two methods, one for the volatile range and another for the semivolatile range. Volatile constituents in water or solid samples are determined by purge-and-trap GC-flame ionization.

The analysis is often called the gasoline range organics (GRO) method. The semivolatile range is determined by analysis of an extract by GC-flame ionization and is referred to as diesel range organics (DRO).

In regard to releases from USTs, the most common method (EPA 418.1) is still used, but gas procedures have been developed to provide more specific information on hydrocarbon content of water and soil, as per local and/or regional legislation. These methods, coupled with specific extraction techniques, can provide information on product type by comparison of the chromatogram with standards. Quantitative estimates may be made for a boiling range or for a range of carbon numbers by summing peaks within a specific window. However, these methods do have limitations such as erroneous data caused by interferences, low recovery due to the standard selected, crude oil product changes caused by volatility, weathering, and microbial activity.

Another method (EPA 3611) that focus on the separation of groups or fractions with similar mobility in soils is based on the use of alumina and silica gel (EPA 3630) that are used to fractionate the hydrocarbon into aliphatic and aromatic fractions. A gas chromatograph equipped with a boiling point column (nonpolar capillary column) is used to analyze whole soil samples as well as the aliphatic and aromatic fractions to resolve and quantify the fate-and-transport fractions. The method is versatile and performance-based and, therefore, can be modified to accommodate data quality objectives.

Higher boiling hydrocarbon derivatives (C12–C26) are analyzed using an extraction procedure followed by a column separation using silica gel (EPA 3630 modified) of the aromatic and aliphatic groupings or fractions. The two fractions are then analyzed using GC-flame ionization. Polynuclear aromatic markers and n-alkane markers are used to divide the higher boiling aromatic and aliphatic fractions by carbon number, respectively.

4.1.1 Air

On the other hand, the complex mixture of crude oil hydrocarbon derivatives potentially present in an air sample can be minimized by separation of the sample into aliphatic and aromatic fractions, and then these two major fractions are separated into smaller fractions based on carbon number. Individual compounds (e.g., BTEX, MTBE, naphthalene) are also identified using this method. The range of compounds that can be identified includes C4 (1,3-butadiene) through C 12 (*n*-dodecane).

As a partial compromise between the use of on-site instrumental analysis and laboratory analysis, a passive sampler can be immersed into the soil (at a specified depth or at several depths) to collect the evolved gases that are adsorbed onto a solid-phase support. The sampler is then removed to the laboratory, where the gases are transferred by Curie point desorption, directly into the ion source of an interfaced quadrupole mass spectrometer. This procedure has its origin in the crude oil exploration industry, and the samplers can be used at a considerable range of depths (Einhorn et al., 1992).

A number of procedures, based on microanalysis of samples for known physical properties (Speight, 2014, 2015), have also been employed. For example, field screening, which uses infrared spectroscopy, employing a portable version of the laboratory procedure, has been used (Kasper et al., 1991). Field turbidometric methods favor the determination of high-boiling hydrocarbon derivatives and are of some use in delineating such pollution within soil (Kahrs et al., 1999). The fluorescence spectra exhibited by the aromatic components provide the basis for laser-induced fluorescence spectroscopy (Apitz et al., 1992; Löhmannsröben et al., 1999). They allow detection of polycyclic aromatic compounds and thus are able to take account of a fraction not measured by other field screening techniques.

4.1.2 Soils and sediments

Hydrocarbon species can enter the soil environment from a number of sources. The origin of the contaminants has a significant bearing on the species present and hence the analytical methodology to be used (Driscoll et al., 1992). Unlike other chemicals (notably pesticides), hydrocarbon derivatives were generally not applied to soils for a purpose, and thus hydrocarbon contamination results almost entirely from misadventure. The source that is probably most familiar to persons involved in the study of contaminated sites is leakage from USTs. This is particularly important at the site of former service stations, and the hydrocarbon derivatives involved are generally in the gasoline or diesel range. Other major sources include spillage during refueling and lubrication, the hydrocarbon derivatives being within the diesel and heavy oil range. Places in which transfer and handling of crude oils takes place (such as tanker terminals and oil refineries) are also potential places of contamination, the oil being largely of the heavier hydrocarbon type.

Thus, the relevant chemistry of hydrocarbon derivatives likely to be encountered at contaminated sites is briefly reviewed and the importance of hydrocarbon speciation noted in terms of a toxicological basis for risk assessment. Hydrocarbon interaction with soil contaminants is important both in terms of their toxicology and also their accessibility by analytical methods. There is no simple procedure that will give an overall picture of hydrocarbon derivatives present at contaminated sites. This is largely because the molecules are present in two separate categories—viz. volatile and semi- or nonvolatile. These two categories require significantly different sample collection, handling, and management techniques (Siegrist and Jenssen, 1990). Volatile hydrocarbon derivatives may be collected by zero headspace procedures or by immediate immersion of the soil into methanol.

The analysis involves gas chromatographic methods such as purge and trap, vacuum distillation, and headspace (Askari et al., 1996). On the other hand, samples for the determination of semi- and nonvolatile hydrocarbon derivatives need not be collected in such a rigorous manner. They require extraction by techniques such as solvent or supercritical fluid on arrival at the laboratory. Some cleanup of extracts is also necessary in most cases and the analytical finish is again by GC. Detectors used range from flame ionization to Fourier transform infrared and mass spectrometric, the latter types being necessary to achieve speciation of the component hydrocarbon derivatives.

The determination of hydrocarbon contaminants in soil is one of the most frequently performed analyses in the study of contaminated sites and is also one of the least standardized. Given the wide variety of hydrocarbon contaminants that can potentially enter and exist in the soil environment, a need exists for methods that satisfactorily quantify these chemicals. Formerly, the idea of total hydrocarbon determination in soil was seen as providing a satisfactory tool for assessing contaminated sites, but the nature of the method and the site specificity dictates a risk-based approach in data assessment. Quantitation of particular hydrocarbon species may be required.

There is a trend toward use of GC techniques in analysis of soils and sediments. One aspect of these methods is that *volatiles* and *semivolatiles* are determined separately. The volatile or GRO constituents are recovered using purge-and-trap or other stripping techniques. Semivolatiles are separated from the solid matrix by solvent extraction. Other extraction techniques have been developed to reduce the hazards and the cost of solvent use and to automate the process and techniques include supercritical fluid extraction, microwave extraction, Soxhlet extraction, sonication extraction, and solid phase extraction (EPA 3540C). Capillary column techniques have largely replaced the use of packed columns for analysis, as they provide resolution of a greater number of hydrocarbon compounds.

Because of the overall complexity of the problem and of the spectrum of hydrocarbon derivatives likely to be encountered, it is impossible to view the total crude oil hydrocarbon derivatives as a single entity. There have been many approaches to the problem, but the simplest and one most frequently used is the one based on the vapor pressure ranges of the relevant organic constituents. This also relates to the sampling methodology employed, and the approach consists of subdividing the hydrocarbon derivatives into the most volatile fraction (referred to as GRO) and the less volatile fraction. In the case of monitoring of storage tanks, a subfraction (known as DRO) is often distinguished among the semivolatile fraction.

Regarding contaminated soil, this type of analysis may not be possible because the various hydrocarbon derivatives cannot be extracted from the sample with equal efficiency. Volatile organic compounds require special procedures to achieve satisfactory recovery from the soil matrix. It thus becomes important to distinguish between those compounds that are considered to be volatile and those that rank as semivolatile compounds or nonvolatile compounds.

4.1.3 Water and wastewater

The overall method includes sample collection and storage, extraction, and analysis steps. Sampling strategy is an important step in the overall process. Care must be taken to assure that the samples collected are representative of the environmental medium and that they are collected without contamination. There is an extensive list of test methods for water analysis (ASTM, 2019) that includes numerous modifications of the original methods but most involve alternate extraction methods developed to improve overall method performance for the analysis. Solvent extraction methods with hexane are also in use.

4.2 Semi- and nonvolatile hydrocarbons

As mentioned above, the most usual analytical finish for hydrocarbon determination is GC (Speight, 2014, 2015). Depending on the degree of resolution and level of information required, a number of instrument configurations may be employed.

Because of the nature of the analytes (boiling point 170°C–430°C, 340–805°F), higher oven temperatures are required for chromatography of this fraction, compared with GRO compounds. Commonly, fused silica capillary columns are used, and the sample is generally introduced by direct injection. Temperatures of the injector and detector are maintained at 200°C (390°F) and 340°C (645°F), respectively, throughout the run and the column temperature ramped from 45 to 275°C (113–425°F). GC/FID may be used to simply fingerprint the components of a hydrocarbon pollution episode (Bruce and Schmidt, 1994), this strategy being most successful if the pollutant has only recently entered the soil environment.

Most frequently, however, some attempts are made to quantify the hydrocarbon fractions represented (Whittaker et al., 1995). It is possible to employ both external and internal standards in these determinations. When internal standards are used, they are generally compounds such as hexafluoro-2-propanol, hexafluoro-2-methyl-2-propanol, or 2-chloroacrylonitrile. Regarding determination of DRO compounds, regulatory authorities vary in terms of the prescribed range. Typically, the DRO range is considered to begin at C10 to C12 and end at C24 to C28.

More sophisticated detection methods for GC are also employed in the analysis of hydrocarbon derivatives, viz. GC-MS and GC-Fourier transform infrared spectroscopy. These procedures have a significant advantage in providing a better characterization of the contaminants and are thus of particular use where some environmental modification of the hydrocarbon derivatives have taken place subsequent to soil deposition.

For volatile organic compounds, the most significant process is through volatilization, resulting in a decrease of overall concentration with time. On the other hand, the higher molecular weight hydrocarbon derivatives are more prone to (chemical) modification through other processes, and it becomes necessary to identify the products of the various transformations. In addition, it is useful to obtain some index of overall weathering.

Such information cannot readily be obtained from simple GC-flame ionization profiles, and GC-MS has been used for such analyses. Electron impact ionization and chemical ionization procedures are available. The former procedure produces predominantly fragment ions, whereas the latter produces predominantly parent ions. With complex high molecular weight samples, CI can/will produce ambiguous results, as many of the analytes have identical parent ion peaks. Thus, GC–electron impact mass spectrometry has been the method of choice for analysis of most hydrocarbon studies (Speight, 2015).

The availability of this *piggyback* method GC/MS/MS has further enhanced the ability to examine environmental hydrocarbon samples for particular components.

Of particular significance in the study of crude oil weathering are the *biomarker* molecules (e.g., pristane, phytane, the hopanes, and steranes) that include the components of crude oils known as steranes. The biomarkers have historically been employed as crude oil signatures in prospecting and characterization. More recently, such molecules have also been employed in the environmental field, both for the determination of pollutant source and estimation of the degree of weathering.

The biomarker molecules are particularly resistant to microbial attack, and thus the ratio of other hydrocarbon components to the biomarker will decrease as the crude oil is biodegraded (Wang et al., 1994). In the case of an ongoing oil discharge into the soil, this ratio will be highest nearest the source and will decrease with increasing distance from the source. Thus, the ratio may be used to locate the source of the contaminant (Whittaker et al., 1995).

In a similar manner, expression of biodegradable hydrocarbon derivatives as a ratio to high molecular weight polynuclear aromatic hydrocarbon derivatives should have potential for fingerprinting purposes. The failure of some attempts to use polynuclear aromatic hydrocarbon derivatives for this purpose arises from the poor choice of molecules for comparison. Low molecular weight polynuclear aromatic hydrocarbon derivatives such as naphthalene or phenanthrene are often selected because of their abundance and relative ease of measurement, but these molecules are also the most prone to biodegradation as well as other forms of attenuation (Sadler and Connell, 2002).

5. Toxicity hazards

With few exceptions, the constituents of crude oil, crude oil products, and the various emissions are hazardous to the health. There are always exceptions that will be cited in opposition to such a statement, the most common exception being the liquid paraffin that is used medicinally to lubricate the alimentary tract. The use of such medication is common among miners who breathe and swallow coal dust every day during their work shifts.

Another approach is to consider crude oil constituents in terms of transportable materials, the character of which is determined by several chemical and physical properties (i.e., solubility, vapor pressure, and propensity to bind with soil and organic particles). These properties are the basis of measures of leachability and volatility of individual hydrocarbon derivatives. Thus, crude oil transport fractions can be considered by equivalent carbon number to be grouped into 13 different fractions. The analytical fractions are then set to match these transport fractions, using specific *n*-alkane derivatives to mark the analytical results for aliphatic compounds and selected aromatic compounds to delineate hydrocarbon derivatives containing benzene rings.

Although chemicals grouped by transport fraction generally have similar toxicological properties, this is not always the case. For example, benzene is a carcinogen, but many alkyl-substituted benzenes do not fall under this classification. However, it is more appropriate to group benzene with compounds that have similar environmental transport properties than to group it with other carcinogens such as benzo[a]pyrene that have very different environmental transport properties. Nevertheless, consultation of any reference work that lists the properties of chemicals will show the properties and hazardous nature of the types of chemicals that are found in crude oil. In addition, crude oil is used to make crude oil products, which can contaminate the environment.

Toxicity from hydrocarbon ingestion can affect many different organs, but the lungs are the most commonly affected organ. The chemical properties of the individual hydrocarbon determine the specific toxicity, while the dose and route of ingestion affect the organs exposed to the toxicity. The recreational use of inhaling hydrocarbon derivatives and other volatile solvents for the purposes of creating a euphoric state is becoming increasingly common. There are several methods for this misuse, including (i) *sniffing* (directly inhaling vapors), (ii) *huffing* (placing a hydrocarbon-saturated cloth over the mouth and nose and then inhaling), or (iii) *bagging* (inhaling through an opening in a plastic bag filled with hydrocarbon vapors).

The range of chemicals in crude oil and crude oil products is so vast that summarizing the properties and/or the toxicity or general hazard of crude oil in general or even for a specific crude oil is a difficult task. However, crude oil and some crude oil products, because of the hydrocarbon content, are at least theoretically biodegradable, but large-scale spills can overwhelm the ability of the ecosystem to break the oil down. The toxicological implications from crude oil occur primarily from exposure to or biological metabolism of aromatic structures. These implications change as the spill of crude oil, a crude oil product, or a hydrocarbon product ages or is weathered.

5.1 Lower boiling hydrocarbons

Exposure to hydrocarbon derivatives is common in modern society, and there are consequential effects (FR 2001). There is often a misconception of the effects and toxicity of hydrocarbon derivatives because of the perceived low chemical reactivity. For example, methane is not classed as a toxic chemical; however, it is an asphyxiant because it will displace oxygen in an enclosed space. While the gas is not in itself dangerous to humans, it causes a slow asphyxiation by displacing the oxygen normally present in the air in a closed environment. Asphyxia will result if the oxygen concentration is reduced to below 19.5% by displacement. Persons exposed over an extended period of time will suffer from lack of oxygen leading to brain damage or brain death, damage to other organs, and death. Symptoms of the low oxygen levels would be trouble focusing and sleepiness.

Methane is also highly flammable and will form explosive mixtures with air. Methane is violently reactive with oxidizers, halogens, and some halogen-containing compounds. The concentrations at which flammable or explosive mixtures form are much lower than the concentration at which asphyxiation risk is significant. When structures are built on or near landfills, methane off-gas can penetrate the buildings' interiors and expose occupants to significant levels of methane. Some buildings have specially engineered recovery systems below their basements to actively capture such fugitive off-gas and vent it away from the building. Similar, but often less drastic effects are evident when the lower molecular weight hydrocarbon gases are considered.

Hydrocarbon derivatives are easily accessible in products such as gasoline, turpentine, furniture polish, household cleansers, propellants, kerosene, and other fuels. Although hydrocarbon derivatives include all compounds composed predominantly of carbon and hydrogen, the compounds of interest are derived from crude oil and wood. Most of the dangerous hydrocarbon derivatives are derived from crude oil distillates and include aliphatic (straight-chain) hydrocarbon derivatives and aromatic (benzene-containing) hydrocarbon derivatives. Other hydrocarbon derivatives such as pine oil and turpentine are derived from wood.

Types of exposure include unintentional ingestion, intentional recreational abuse, unintentional inhalation, and dermal exposure or oral ingestion in a suicide attempt. The highest rates of morbidity and mortality result from accidental ingestion by children younger than 5 years. Aspiration pneumonitis is the most common complication of hydrocarbon ingestion, followed by CNS and cardiovascular complications.

The toxicity of hydrocarbon derivatives is directly related to their physical properties, specifically the viscosity, volatility, surface tension, and chemical activity of the side chains. The viscosity is a measure of resistance to flow and is measured in saybolt seconds universal (SSU). Substances with a lower viscosity (SSU <60, e.g., turpentine, gasoline, naphtha) are associated with a higher chance of aspiration. The surface tension is a cohesive force created by van der Waals forces between molecules and is a measure of the lability of the liquid to *creep*. Like the viscosity, the surface tension is also inversely related to aspiration risk; the lower the viscosity, the higher the risk of aspiration.

Volatility is the tendency for a liquid to change phases and become a gas. Hydrocarbon derivatives with a high volatility can vaporize and displace oxygen, which can lead to a transient state of hypoxia. Not surprisingly, the degree of volatility is directly related with the risk of aspiration. The amount of hydrocarbon ingested has not consistently been linked to the degree of aspiration, and hence pulmonary toxicity. In fact, toxicity from hydrocarbon exposure can be thought of as different syndromes, depending on which organ system is predominately involved. Organ systems that can be affected by hydrocarbon derivatives include the pulmonary, neurologic, cardiac, gastrointestinal, hepatic, renal, dermatologic, and hematologic systems.

Pulmonary complications, especially aspiration, are the most frequently reported adverse effect of hydrocarbon exposure. The lower the viscosity and higher the volatility, the greater the risk of pulmonary aspiration. The hydrophobic nature of hydrocarbon derivatives allows them to penetrate deep into the tracheobronchial tree, producing inflammation and bronchospasm. The volatile chemical may displace alveolar oxygen, leading to hypoxia. Direct contact with alveolar membranes can lead to hemorrhage, hyperemia, edema, surfactant inactivation, leukocyte infiltration, and vascular thrombosis. The result is poor oxygen exchange, atelectasis, and pneumonitis.

While the aliphatic hydrocarbon derivatives have little gastrointestinal absorption, aspiration frequently occurs, either initially or in a semidelayed fashion as the patient coughs or vomits, thereby resulting in pulmonary effects. Once aspirated, the hydrocarbon derivatives can create a severe pneumonitis.

Hydrocarbon pneumonitis results from a direct toxic affect by the hydrocarbon on the lung parenchyma. The type II pneumocytes are most affected, and as such, surfactant production and function are altered. The end result of hydrocarbon aspiration is interstitial inflammation, intraalveolar hemorrhage and edema, hyperemia, bronchial necrosis, and vascular necrosis. Pulmonary toxicity is the result of hydrocarbon aspiration. Respiratory symptoms generally begin in the first few hours after exposure and usually resolve in 2–8 days. Complications include hypoxia, barotrauma due to mechanical ventilation, and acute respiratory distress syndrome (ARDS). Prolonged hypoxia may result in encephalopathy, seizures, and death.

Hydrocarbon toxicity produces various effects to the central nervous system. Initial effects are similar to the disinhibition observed in patients with alcohol intoxication. Narcotic-like depression may also be observed. Euphoria may develop, as in alcohol or narcotic toxicity. Eventually, lethargy, headache, drowsiness, and coma may follow. Seizures are uncommon and are believed to be due to hypoxia.

Toxicity to the central nervous system can result from several mechanisms, including direct injury to the brain, or indirectly as a result of severe hypoxia or simple asphyxiation. Many of the hydrocarbon derivatives that affect the central nervous system directly are able to make their way across the blood–brain barrier because certain hydrocarbon derivatives are highly lipophilic. In addition for individuals who are *huffing* or *bagging*, the act of rebreathing can result in hypercarbia, which can contribute to decreased level of arousal. Prolonged abuse of hydrocarbon derivatives can result in white matter degeneration (leukoencephalopathy). In addition, prolonged exposure to certain hydrocarbon derivatives (e.g., n-hexane or methyl-n-butyl ketone) can result in peripheral neuropathy.

Cardiac effects are a major concern and include myocardial sensitization to catecholamines as well as direct myocardial damage. Sudden death has been reported as a result of coronary vasospasm due to hydrocarbon inhalation. Pulmonary toxicity is the major cause of morbidity and mortality from hydrocarbon poisoning, and long-term exposure may result in significant morbidity.

Chronic exposure to toluene, an aromatic hydrocarbon, can result in a distal renal tubular acidosis and present with an anion gap acidosis. A patient may have chronic exposure either via an occupational environment or by repeated recreational inhalation. Moreover, prolonged exposure to certain aromatic hydrocarbon derivatives (especially benzene) can lead to an increased risk of aplastic anemia, multiple myeloma, and acute myelogenous leukemia. In addition, hemolysis has been reported following the acute ingestion of various types of hydrocarbon derivatives. Hydrocarbon derivatives are also reported to cause bone marrow toxicity and hemolysis. Chlorinated hydrocarbon toxicity may cause hepatic and renal failure, and toluene toxicity may lead to renal tubular acidosis. Direct contact with the skin and mucous membranes may cause effects ranging from local irritation to extensive skin discoloration and dermatitis.

Inhalation injury due to hydrocarbon derivatives can occur as a result of either accidental or intentional exposure. Inhalant abuse, the deliberate inhalation of hydrocarbon derivatives as a form of recreational drug use, has become a significant health issue, and data indicate that, among adolescents, inhalants are the second most widely used class of illicit drugs. Death from intentional inhalation of hydrocarbon fumes is not uncommon and is usually due to sudden cardiac events or depression.

Deliberate inhalation of volatile hydrocarbon derivatives for their mood-altering effects is popular among adolescents. Their low cost, ready availability, and ease of use contribute to this problem. Volatile hydrocarbon derivatives are contained in glues, solvents, lighter fluid, gasoline, and paints. Most inhalants are composed of several compounds, and almost all pressurized aerosol products can be abused because the propellants are volatile hydrocarbon derivatives.

Recreational abuse of hydrocarbon derivatives by inhalation is accomplished in three ways: (i) sniffing, (ii) huffing, and (iii) bagging. *Sniffing*, the least potent delivery method, is the inhalation of the volatile substance through the nostrils. *Huffing* involves placing of a rag soaked with an inhalant such as gasoline or lighter fluid over the nose and mouth. *Bagging* involves repeated deep inhalations from a plastic or paper bag filled with a particular hydrocarbon such as spray paint or another propellant.

Two primary organ systems are affected by inhalation hydrocarbon toxicity: (i) the central nervous system and (ii) the cardiopulmonary system. Volatile hydrocarbon derivatives are highly lipid soluble and readily cross the blood−brain barrier. Rapid absorption occurs across the large surface area of the pulmonary vascular bed, and peak blood levels are noted approximately 15−30 minutes after inhalation. Confusion, disorientation, disinhibition, and euphoria are exhibited early. Speech becomes slurred, and motor function becomes impaired, with gait becoming staggered. Hallucinations are frequently described, followed by central nervous system depression, drowsiness, and sleep. Coma can occur with prolonged or repeated exposures—this is less likely to happen because intentional exposure ceases as the user becomes drowsy. With acute intoxication, death due to asphyxiation from a plastic bag over the head or from aspiration of stomach contents is not unusual. Also, trauma-related injury and motor vehicle accidents have been reported, resulting from disinhibition and disorientation following inhalation.

With long-term hydrocarbon inhalation, damage to the central nervous system occurs, including loss of cognitive functions, gait disturbances, and loss of coordination. Other, less common complications of long-term hydrocarbon inhalation include restrictive pulmonary disease, pulmonary hypertension, and reduced diffusion capacity. Pulmonary toxicity can occur as a result of hydrocarbon aspiration.

The common idea that solvent inhalation is innocuous undoubtedly contributes to solvent-inhalant abuse. The availability of organic hydrocarbon solvents in commonly used household products and in the pharmaceutical industry makes them readily accessible (Table 13.4). Commonly abused hydrocarbon products include the following: (i) liquids—model glue, gasoline, contact cement (rubber cement), lacquers, nail polish remover, and dry cleaning fluids, (ii) aerosols—spray paints, butane fuel, lighter fluid, cooking sprays, cosmetics, hairspray, toiletries, and deodorants. The hydrocarbon chemicals found in abused inhalants include (i) propane, (ii) butane, (iii) n-hexane, (iv) benzene, (v) toluene, and (vi) xylene.

Also, many of the gaseous and liquid constituents of the lower boiling fractions of crude oil and also in crude oil products fall into the class of chemicals that have one or more of the following characteristics that are considered to be hazardous by the Environmental Protection Agency: (i) ignitability and flammability, (ii) corrosivity, (iii) reactivity, and (iv) hazardous chemicals.

In terms of ignitability and flammability, a liquid that has a flash point of less than 60°C (140°F) is considered ignitable. Some examples are benzene, hexane, heptane, benzene, pentane, naphtha (low boiling, sometimes referred to as petroleum ether), toluene, and the xylene isomers.

An aqueous solution that has a pH of less than or equal to 2 or greater than or equal to 12.5 is considered corrosive. Most hydrocarbon derivatives and hydrocarbon products are not corrosive, but many of the chemicals used in refineries are corrosive. Corrosive materials also include substances such as sodium hydroxide and some other acids or bases.

Chemicals that react violently with air or water are considered hazardous. Examples are sodium metal, potassium metal, phosphorus, etc. Reactive materials also include strong oxidizers such as perchloric acid and chemicals capable of detonation when subjected to an initiating source, such as solid dry $<10\%$ H_2O picric acid, benzoyl peroxide, or sodium borohydride. Solutions of certain cyanide or sulfides that could generate toxic gases are also classified as reactive. The potential for finding such chemicals in a refinery is subject to the function and product slate of the refinery and/or the petrochemical complex. Moreover, the potential for finding such reactivity among hydrocarbon derivatives is low but is subject to the function and product slate of the refinery and/or the petrochemical complex.

Many chemicals have been shown in scientific studies to have toxic, carcinogenic, mutagenic, or teratogenic effects on humans or other life forms and are designated either as *acutely hazardous waste* or *toxic waste* by the Environmental Protection Agency. Substances found to be fatal to humans in low doses or, in the absence of data on human toxicity, have been shown in studies to have an oral LD_{50} toxicity (rat) of less than 2 mg per liter, or a dermal LD_{50} toxicity (rabbit) of less than 200 mg per kilogram, or is otherwise capable of causing or significantly contributing to an increase in serious irreversible, or incapacitating reversible illness are designated as acute hazardous waste. Materials containing any of the toxic constituents so listed are to be considered hazardous waste, unless, after considering the following factors, it can reasonably be concluded (by the Department of Environmental Health and Safety) that the waste is not capable of posing a substantial present or potential hazard to public health or the environment when improperly treated, stored, transported, or disposed of, or otherwise managed.

The issues to be held in consideration are (i) the nature of the toxicity presented by the constituent, (ii) the concentration of the constituent in the waste, (iii) the potential of the constituent or any toxic degradation product of the constituent to migrate from the waste into the environment under the types of improper management considered in item (g) below, (iv) the persistence of the constituent or any toxic degradation product of the constituent, (v) the potential for the constituent or any toxic degradation product of the constituent to degrade into nonharmful constituents and the rate of degradation, (vi) the degree to which the constituent or any degradation product of the constituent accumulates in an ecosystem, (vii) the plausible types of improper management to which the waste could be subjected, (viii) the quantities of the waste generated at individual generation sites or on a regional or national basis, (ix) the nature and severity of the public health threat and environmental damage that has occurred as a result of the improper management of wastes containing the constituent, and (x) actions taken by other governmental agencies or regulatory programs based on the health or environmental hazard posed by the waste or waste constituent. Other factors that may be appropriate may also be considered.

For the analysts, laboratories wishing to dispose of materials containing dilute concentrations of these constituents should contact the Department of Environmental Health and Safety for advice regarding the proper disposition of the materials. In addition, the list of such materials is not included here as it is subject to periodic updates. Furthermore, the list is not meant to be complete and may not include substances that have the hazardous characteristics as defined above. Omission of a chemical from this list does not mean it is without toxic properties or any other hazard.

More specifically, to crude oil and crude oil products, the alkane derivatives in gasoline and some other crude oil products are CNS depressants. In fact, gasoline was once evaluated as an anesthetic agent. However, sudden deaths, possibly as a result of irregular heartbeats, have been attributed to those inhaling vapors of hydrocarbon derivatives such as those in gasoline.

Alkane derivatives of various types of crude oils and various crude oil products were biodegraded faster than the *unresolved fractions*. Different types of crude oils and products biodegraded at different rates in the same environments. An oil product is a complex mixture of organic chemicals and contains within it less persistent and more persistent fractions. The range between these two extremes is greatest for crude oils. Because the many different substances in crude oil have different physical and chemical properties, summarizing the fate of crude oil in general (or even a particular oil) is very difficult. Solubility—fate relationships must be considered.

The relative proportion of hazardous constituents present in crude oil is typically quite variable. Therefore, contamination will vary from one site to another. In addition, the farther one progresses from low-boiling toward high-boiling constituents (the general progression from lower molecular weight to higher molecular weight constituents), the greater the percentage of polynuclear aromatic hydrocarbon derivatives and other semivolatile constituents or nonvolatile constituents (many of which are not so immediately toxic as the volatiles but which can result in long-term/chronic impacts). These higher molecular weight constituents thus need to be analyzed for the semivolatile compounds that typically pose the greatest long-term risk.

In addition to large spills, crude oil hydrocarbon derivatives are released into the aquatic environments from natural seeps as well as nonpoint source urban runoffs. Acute impacts from massive one-time spills are obvious and substantial. The impacts from small spills and chronic releases are the subject of much speculation and continued research. Clearly, these inputs of crude oil hydrocarbon derivatives have the potential for significant environmental impacts, but the effects of chronic low-level discharges can be minimized by the net assimilative capacities of many ecosystems, resulting in little detectable environmental harm.

Short-term (acute) hazards of lighter, more volatile, and water-soluble aromatic compounds (such as benzene, toluene, and the xylene isomers) include potential acute toxicity to aquatic life in the water column (especially, in relatively confined areas) as well as potential inhalation hazards. However, the compounds that pass through the water column often tend to do so in small concentrations and/or for short periods of time, and fish and other pelagic or generally mobile species can often swim away to avoid impacts from spilled oil in open waters. Most fish are mobile, and it is not known whether or not they can sense, and thus avoid, toxic concentrations of oil.

However, there are some potential effects of spilled oil on fish. The impacts to fish are primarily to the eggs and larvae, with limited effects on the adults. The sensitivity varies by species; pink salmon fry are affected by exposure to WSFs of crude oil, while pink salmon pink salmon eggs are very tolerant to benzene and water-soluble crude oil. The general effects are difficult to assess and quantitatively document due to the seasonal and natural variability of the species. Fish rapidly metabolize aromatic hydrocarbon derivatives due to their enzyme system. Long-term (chronic) potential hazards of lighter, more volatile, and water-soluble aromatic compounds include contamination of groundwater. Chronic effects of benzene, toluene, and xylene include changes in the liver and harmful effects on the kidneys, heart, lungs, and nervous system.

At the initial stages of a release, when the benzene-derived compounds are present at their highest concentrations, acute toxic effects are more common than later. These noncarcinogenic effects include subtle changes in detoxifying enzymes and liver damage. Generally, the relative aquatic acute toxicity of crude oil will be the result of the fractional toxicities of the different hydrocarbon derivatives present in the aqueous phase. Tests indicate that naphthalene-derived chemicals have a similar effect.

Except for short-term hazards from concentrated spills, BTEX compounds (benzene, toluene, ethylbenzene, and xylenes) have been more frequently associated with risk to humans than with risk to nonhuman species such as fish and wildlife. This is partly because plants, fish, and birds take up only very small amounts and because this volatile compound tends to evaporate into the atmosphere rather than persisting in surface waters or soils. However, volatiles such as this compound can pose a drinking water hazard when they accumulate in groundwater. See entry for BTEX in Table 15.1.

Crude oil is naturally weathered according to its physical and chemical properties, but during this process, living species within the local environment may be affected via one or more routes of exposure, including ingestion, inhalation, dermal contact, and, to a much lesser extent, bioconcentration through the food chain. Aromatic compounds of concern include alkylbenzenes, toluene, naphthalene derivatives, and polynuclear aromatic hydrocarbon derivatives (Abdel-Shafy and Mansour, 2015). Moreover, both atmospheric and hydrospheric impacts must be assessed when considering toxic implications from a crude oil release containing significant quantities of these single-ring aromatic compounds.

5.2 Higher boiling hydrocarbons

Naphthalene and its homologs are less acutely toxic than benzene but are more prevalent for a longer period during spills. The toxicity of different crude oils and refined oils depends on not only the total concentration of hydrocarbon derivatives but also the hydrocarbon composition in the WSF of crude oil, water solubility, concentrations of individual components, and toxicity of the components. The WSFs prepared from different oils will vary in these parameters. WSFs of refined oils (for example, No. 2 fuel oil and Bunker C oil) are more toxic than WSF of crude oil to several species of fish (killifish and salmon). Compounds with either more rings or methyl substitutions are more toxic than less substituted compounds but tend to be less water soluble and thus less plentiful in the WSF.

Among the polynuclear aromatic hydrocarbon derivatives, the toxicity of crude oil is a function of its di- and triaromatic hydrocarbon content. Like the single aromatic ring variations, including benzene, toluene, and the xylenes, all are relatively volatile compounds with varying degrees of water solubility.

There are indications that pure naphthalene (a constituent of mothballs that are, by definition, toxic to moths) and alkyl naphthalene derivatives are from 3 to 10 times more toxic to test animals than are benzene and alkylbenzenes. In addition, and because of the low water solubility of tricyclic and polycyclic (polynuclear) aromatic hydrocarbon derivatives (that is, those aromatic hydrocarbon derivatives heavier than naphthalene), these compounds are generally present at very low concentrations in the WSF of oil. Therefore, the results of this study and others conclude that the soluble aromatics of crude oil (such as BTEX isomers, and naphthalene derivatives) produce the majority of its toxic effects in the environment.

Once the acutely toxic lighter compounds have been left the aquatic environment through volatilization or degradation, the main concern is chronic effects from heavier and more alkylated polynuclear aromatic hydrocarbon derivatives. Bird species with water habitats are the species most commonly affected by crude oil spills and releases. Oil itself breaks down the protective waxes and oils in the feathers and fur of birds and animals and disrupts the fine strand structure of the feathers resulting in a loss of heat retention and buoyancy and possible hypothermia and death. Oiled birds often ingest crude oil while attempting to remove the crude oil from their feathers. The effects of ingested crude oil include anemia, pneumonia, kidney, and liver damage, decreased growth, altered blood chemistry, and decreased egg production and viability. Ingestion of the oil can also kill animals by interfering with their ability to digest food. Chicks may be exposed to crude oil by ingesting food regurgitated by impacted adults.

The dynamics of the oil-in-water dispersion (OWD) are complex and have relevance related to potential toxicity or hazard. In comparing the toxicities to marine animals of OWDs prepared from different oils, not only the amount of oil added but also the concentrations of oil in the aqueous phase and the composition and dispersion-forming characteristics of the parent oil must be taken into consideration. In comparing the potential impacts of spills of different oils on the marine biotic community, the amount of oil per unit water volume required to cause mortality is of greater importance than any other aspect of the crude oil behavior.

Several compounds in crude oil products are carcinogenic. The larger and higher molecular weight aromatic structures (with four to five aromatic rings), which are the more persistent in the environment, have the potential for chronic toxicological effects. Because these compounds are nonvolatile and are relatively insoluble in water, their main routes of exposure are through ingestion and epidermal contact. Some of the compounds in this classification are considered possible human carcinogens; these include benzo(a and e) pyrene, benzo(a)anthracene, benzo(b, j, and k)fluorene, benzo(ghi)perylene, chrysene, dibenzo(ah)anthracene, and pyrene (Bjorseth, 1983).

Mixtures of polynuclear aromatic hydrocarbon derivatives are often carcinogenic and possibly phototoxic. One way to approach site-specific risk assessments would be to collect the complex mixture of polynuclear aromatic hydrocarbon derivatives and other lipophilic contaminants in a semipermeable membrane device (also known as a *fat bag*) and then test the mixture for carcinogenicity, toxicity, and phototoxicity.

The solubility of hydrocarbon components in crude oil products is an important property when assessing toxicity. The water solubility of a substance determines the routes of exposure that are possible. Solubility is approximately inversely proportional to molecular weight; lighter hydrocarbon derivatives are more soluble in water than higher molecular weight compounds. Lower molecular weight hydrocarbon derivatives (C4 to C8, including the aromatic compounds) are relatively soluble, up to approximately 2000 ppm, while the higher molecular weight hydrocarbon derivatives are nearly insoluble. Usually, the most soluble components are also the most toxic.

Finally, the toxicity of crude oil may be affected by factors such as "weathering" time or the addition of oil dispersants. *Weathered* crude oil and *fresh* crude oil may have different toxicities, depending on oil type and weathering time.

5.3 Polynuclear aromatic hydrocarbons

Polynuclear aromatic hydrocarbon derivatives (another name is polyaromatic hydrocarbons, often written as PHAs) belong to a class of compounds found throughout the environment in the air, in the soil, and in the water. They are found naturally in crude oil as well as in the thermal products of coal such as coal tar and are constituents of emissions produced during the incomplete combustion of hydrocarbons such as coal, oil, gas, tobacco, and during forest fires. Emissions from anthropogenic activities predominate; nevertheless, some polynuclear aromatic hydrocarbon derivatives in the environment originate from natural sources such as open burning, natural losses or seepage of crude oil or coal deposits, and volcanic activities. Major anthropogenic sources of polynuclear aromatic hydrocarbon derivatives include residential heating, coal gasification and liquefying plants, carbon black, coal tar pitch and asphalt production, coke and aluminum production, catalytic cracking towers, and related activities in crude oil refineries as well as and motor vehicle exhaust. Polynuclear aromatic hydrocarbon derivatives are found in the ambient air in gas-phase and as sorbet to aerosols. Polynuclear aromatic hydrocarbon derivatives generally exist as colorless, pale yellow, or white solids (Bjorseth, 1983).

Atmospheric partitioning of polynuclear aromatic hydrocarbon derivatives between the particulate and the gaseous phases strongly influences their fate and transport in the atmosphere and the way they enter into the human body. The removal of polynuclear aromatic hydrocarbon derivatives from the atmosphere by dry and wet deposition processes are strongly influenced by their gas/particle partitioning. Atmospheric deposition is a major source for polynuclear aromatic hydrocarbon derivatives in soil.

Because they do not dissolve easily in water and generally do not burn, they can persist in the environment for months to years. It is difficult to isolate and analyze these hydrocarbon derivatives in the laboratory, due to the fact that they exist naturally as mixtures of many compounds. Polynuclear aromatic hydrocarbon derivatives are formed by incomplete combustion of hydrocarbon derivatives and can be isolated from the processing of fossil fuels. Many of these compounds have little use but there are some, however, that are important in the making of pharmaceuticals, dyes, plastics, and pesticides. Polynuclear aromatic hydrocarbon derivatives enter the body quickly and go to the fat containing tissues.

There is little data available for the toxic effects of naphthalene on humans. Short-term low exposure to naphthalene may cause eye and skin irritation. At slightly higher levels (above 10 ppm), headaches, fatigue, and nausea occur. If naphthalene is ingested, it has the potential to cause hemolytic anemia, a condition that involves the breakdown of red blood cells. Naphthalene is a suspected human carcinogen and has been proven to cause damage to the kidneys and to the liver. Chronic exposure can lead to reproductive defects including fatal damage and decreasing fertility. Higher incidences of lung and skin tumors have been reported for people who have been occupationally exposed to naphthalene and other polynuclear aromatic hydrocarbon derivatives. There are several hundred possible polynuclear aromatic hydrocarbon derivatives. As examples, and to illustrate the general properties of polynuclear aromatic hydrocarbon derivatives, naphthalene, anthracene, phenanthrene, and benzo[a]pyrene will be used here (Table 13.5).

Naphthalene is the smallest (lowest molecular weight) of the polynuclear aromatic hydrocarbon that contains two rings—synonyms for naphthalene include *mothballs*, *white tar*, *tar camphor*, and *albo carbon*. It is a white, crystalline solid that can be found in the form of scales, balls, powder, or cakes. Naphthalene has the strong aromatic odor that is associated with mothballs.

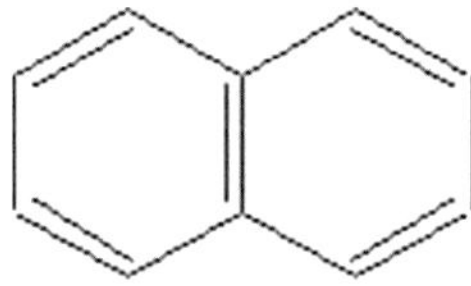

naphthalene

Naphthalene is the most abundant distillate of coal tar. Its most common use is as a household fumigant against moths (hence the name mothballs). Naphthalene is an important hydrocarbon raw material used in the manufacture of phthalic anhydride (used in dye making), of celluloid and hydronaphthalene derivatives (used in lubricants), and of motor fuels. At one time, it was used as an insecticide and vermicide. However, this use is decreasing due to the low toxicity of the vapor. Naphthalene is also of some use as an antiseptic and as a soil fumigant.

Anthracene (*anthracin*, *green oil*, and *paranaphthalene*) is common in the natural environment. It has a crystalline structure and is pale yellow in color. It exhibits a weak aromatic odor. It is a combustible solid.

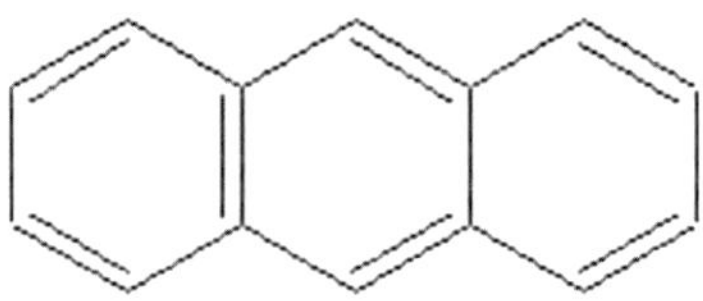

anthracene

Anthracene, like many polynuclear aromatic hydrocarbon derivatives, is used in the production of fast dyes as well as fibers and plastics. It is one of the most important feedstocks for the production of anthraquinone. Anthracene can also be used in insecticides and for a wood preservative. This compound is widely abundant. It is found in any type of coal or tar.

Anthracene exposure can cause skin and eye irritation, which can be aggravated by sunlight. Repeated exposure may cause alteration of skin pigments as well as cancerous growth, although there is no carcinogenic data for anthracene. Inhaled anthracene can cause bronchitis-like symptoms. There is limited information on human reproductive implications. Anthracene may cause genetic mutations in cells. Anthracene is not currently considered a toxic substance.

Phenanthrene is an isomer of anthracene, and, as a result, many of the physical properties of the two are very similar.

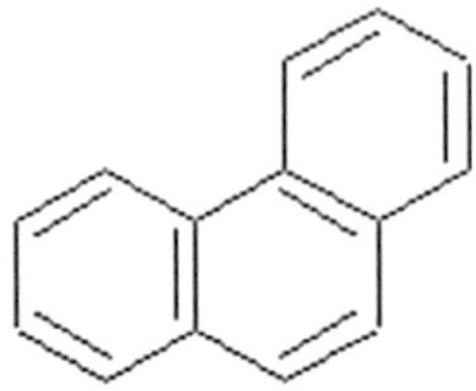

phenanthrene

The major differences between anthracene and phenanthrene lie in the melting point and the properties directly related to solubility (Table 13.5). Phenanthrene is purified as brown to white monoclinic crystals and also has the characteristic faint aromatic smell.

Like anthracene, phenanthrene is used in the production of dyes. It is also used in the manufacture of explosives and is an important starting material for phenanthrene-based drugs. This leads directly to use in biochemical research for the pharmaceutical industry. A mixture of phenanthrene and anthracene tar is used to coat water storage tanks to prevent rust. Phenanthrene is also a skin and eye irritant, with increasing effects in sunlight due to photosensitization. There is currently no data available for human oral and inhalation exposure. It is, however, a suspected carcinogen and, although there is no data for humans, it is best to err on the side of caution and suspect carcinogenic effects will be present.

Benzo[a]pyrene (3,4-benzpyrene) is the largest of the four compounds, with five rings. It also has the faint aromatic odor. Pure benzo[a]pyrene is pale yellow and is found as monoclinic or orthorhombic crystals. These can be separated from a mixture of polynuclear aromatic hydrocarbon derivatives using various standard separation techniques and recrystallized from benzene and methanol. Benzo[a]pyrene is the one of several compounds that is a known human carcinogen.

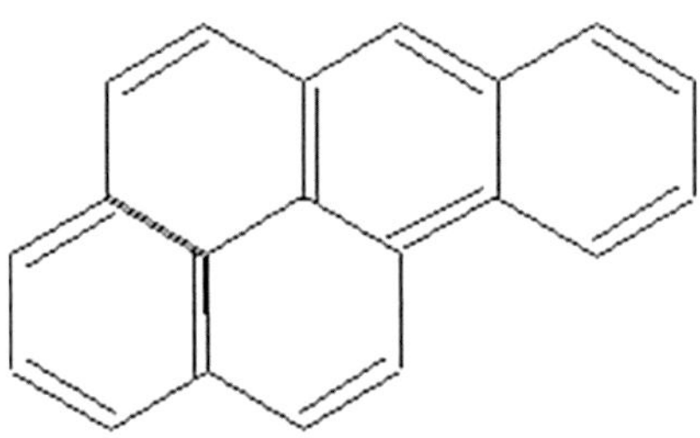

bnezo(a) pyrene

Benzo[a]pyrene is formed when gasoline, garbage, or any plant or animal materials are burned. So, it is usually present in soot and smoke. It is found in the coal tar pitch industry and is used to join electrical parts together. It is also found in creosote, a chemical used as a preservative for wood. It is used extensively as a positive control in a variety of laboratory mutagenic and carcinogenic short-term tests. Benzo[a]pyrene has been found in various types of cereals, vegetables, fruits, and meats. High-temperature cooking processes, such as charcoal grilling or charring, can increase the amount of benzo[a] pyrene found in food. It is also a component of tobacco products and is one of the cancer causing agents in cigarette smoke.

The greatest chance of high-level exposure to benzo[a]pyrene is likely to occur in the work place. People who work in coal tar production plants, coking plants, asphalt production plants, coal gasification sites, and smoke houses receive higher doses than the general population.

Benzo[a]pyrene is a known precursor to cancer causing metabolites in laboratory animals. Various studies have determined that this compound is toxic. It is converted by cytochrome P450 to a variety of oxides that react with DNA, making them highly mutagenic. The newborn animals of pregnant mice fed benzo[a]pyrene had other harmful effects including birth defects and low body weight. It is possible that similar effects could happen to humans exposed to benzo[a]pyrene. Thus, benzo[a]pyrene is a known carcinogen and is regulated along with substances known as coal *tar pitch volatiles*.

6. Remediation of hydrocarbon spills

When soil or water becomes contaminated with spilled hydrocarbon derivatives and/or hydrocarbon product, there are several methods of disposing or treating these wastes. Treatment occurs using methods such as adsorption, bioremediation (using microorganisms to convert the spilled hydrocarbon derivatives into less toxic forms), or thermal technologies (using high temperatures to reclaim or destroy hydrocarbon-contaminated material) (Speight and Lee, 2000; Speight, 2005).

Small spills of hydrocarbon derivatives and hydrocarbon products (such as lubricating oils, fuel oils, cooking oils, and radiator coolants) occur from time to time as a result of transportation and/or in industrial processes as well as a result of domestic and farming activities. Such spills are sometimes collected using sand, sawdust, and similar materials, but such materials have limited absorbent capacity and perform poorly if they become wet, often releasing much of the absorbed pollutant. Even though the method may have shown some success with small spills, the adsorption procedure is not always suitable or adequate to the task of cleaning up large spills.

When discharged into unlined pits, the hydrocarbon derivatives and hydrocarbon products can leach directly into the soil and may contaminate groundwater. Lined pits can also lead to pollution via ruptures in liners or by overflowing the pit area. These events can result in soil and water contamination, which can have a negative effect on both human and ecosystem health.

One of the most practical ways for treating soil contaminated by hydrocarbon derivatives is by some form of land surface treatment whereby crude oil contaminated soils are applied onto the soil surface and periodically turned over or tilled to aerate the contaminated soils to enhance the volatilization and biodegradation processes. To facilitate prompt remediation of the soils, they must be spread thinly and diced or rototilled regularly.

Bioremediation (*biological treatment*, *biotreatment*) of hydrocarbon derivatives may be accomplished using a variety of tillage and composting techniques. In all cases, the breakdown of hydrocarbon derivatives is maximized by providing the prime conditions for microbial activity, which requires a proper balance of moisture, and nutrients, as well as soil oxygen. Thus, it is often necessary to add water, nutrients, and additional soil and aerate the soil to enhance biodegradation of hydrocarbon derivatives. During periods of extended dry conditions, moisture control may also be needed to minimize dust. Application rates of water should be monitored and controlled to minimize the potential for runoff or leaching of contaminants.

Bioremediation uses microorganisms (bacteria and fungi) to biologically degrade hydrocarbon-contaminated waste into nontoxic residues. The objective of biotreatment is to accelerate the natural decomposition process by controlling oxygen, temperature, moisture, and nutrient parameters (McMillen et al., 2004).

Tilling the soils promotes volatilization (evaporation) of the lighter portions while the remaining compounds are immobilized within the soil mass and breakdown biologically. Naturally occurring soil microorganisms such as bacteria and fungi combine with sunlight, oxygen, and moisture to biodegrade (breakdown biologically) the crude oil products.

Composting is considered by some authorities to be a form adsorption but is still open to debate. In composting, wastes are mixed with bulking agents such as wood chips, straw, rice hulls, or husks to increase porosity and aeration potential for biological degradation. The bulking agents provide adequate porosity to allow aeration even when moisture levels are high. To increase the water-holding capacity of the waste–media mixture, and to increase trace nutrients, manure or agricultural wastes may be added. Adding nitrogen- and phosphorus-based fertilizers and trace minerals can also enhance microbial activity and reduce the time required to achieve the desired level of biodegradation.

Composting is similar to land treatment, but it can be more efficient. Also, with composting systems, treated waste is contained within the composting facility where its properties can be readily monitored. With composting, mixtures of the waste, soil (to provide indigenous bacteria), and other additives may be placed in piles to be tilled for aeration, or placed in containers or on platforms to allow air to be forced through the composting mixture. To optimize moisture conditions for biodegradation, the compost mixture is maintained at 40%–60% water by weight. Elevated temperatures (30–70°C,

86–158°F) in compost mixtures increase microbial metabolism. However, if temperatures exceed 70°, cell death can occur. Tilling the soil pile or forced aeration can help control temperature and oxygen levels. Composting in closed containers can control volatile emissions.

Some advantages of biological treatment are it is relatively environmentally benign; it generates few emissions; wastes are converted into products; and it requires minimal, if any, transportation. On occasion, bioremediation is used as an interim treatment or disposal step, which reduces the overall level of hydrocarbon contamination before final disposal. Bioremediation can create a drier, more stable material for land filling, thereby reducing the potential to generate leachate. Depending on the composition of the hydrocarbon components, the bioremediation environment, and the type of treatment utilized, bioremediation may be a fairly slow process and require months or years to reach the desired result.

Land spreading is typically a one-time application of wastes to an area of land. Wastes are spread on the land and incorporated into the upper soil zone (typically upper 6–8 inches of soil) to enhance hydrocarbon volatilization and biodegradation. It is important that other constituents, such as metals, salts, and acids, not be present at levels that will sterilize or permanently impair the soil system.

Bioreactors work according to the same aerobic biological reactions that occur in land treatment and composting, but the reactions occur in an open or closed vessel or impoundment. This environment accelerates the rate of biodegradation by allowing better control of the temperature and other conditions that affect the biodegradation rate. Bioreactor processes are typically operated as a batch or semicontinuous process. In a bioreactor, nutrients are added to a slurry of water and waste, and air sparging or intensive mechanical mixing of the reactor contents provides oxygen. This mechanical mixing results in significant contact between microorganisms and the waste components being degraded. To accelerate system start-up, introduction of microbes capable of degrading the organic constituents of the waste may be useful, although some companies have not had favorable experience with designer bugs. Many of the additives used for bioreactors are common agricultural products and plant or animal wastes.

After the desired treatment level has been reached, and depending on the constituents, liquids may be reused, transported to wastewater treatment facilities, injected, or discharged. Solids may be buried, applied to soils, used as fill, or treated further to stabilize components such as metals.

In tank-based bioreactors, operating conditions (temperature, nutrient concentration, pH, oxygen transport, and mixing) can be monitored and controlled easily. Optimized biological processes ensure the best rate of biodegradation and allow for reduced space requirements relative to land-based biological treatment processes.

Vermiculture is the process of using worms to decompose organic waste into a material capable of supplying necessary nutrients to help sustain plant growth. Worms can facilitate the rapid degradation of hydrocarbon-based products (Norman et al., 2002). Because worm cast (manure) has important fertilizer properties, the process may provide an alternative disposal method.

The feeding consists of applying the mixture as feedstock to windrows, which are covered to exclude light from the worm bed and protect it from becoming waterlogged. Controlled irrigation systems correct the moisture content during periods of low rainfall. The feedstock was applied to the windrows, generally once per week, at an average depth of 15–30 mm. The worms *work* the top of each windrow, consuming the applied material over 5–7 days. The resulting worm cast organic fertilizer is harvested and packaged for distribution and use as a beneficial fertilizer and soil conditioner.

Thermal technologies utilize high temperatures to destroy or remove hydrocarbon derivatives from waste materials. Depending on the final fate of the wastes, additional treatment may be needed to remove metals and salts. Thermal treatment technology generally occurs at a permanent or fixed facility, but some efforts are under way to develop mobile thermal treatment units.

There are two main types of thermal technologies: (i) incineration (e.g., rotary or cement kilns) destroys hydrocarbon derivatives by heating them to very high temperatures in the presence of air, and (ii) thermal desorption (e.g., indirect rotary kilns, thermal phase separation, thermal distillation) involves the application of heat to the wastes, to vaporize volatile and semivolatile hydrocarbon derivatives, which can be combusted to reduce the emission of toxic components or condensed and separated to recover higher molecular weight hydrocarbon derivatives.

References

Abdel-Shafy, H.I., Mansour, M.S.M., 2015. A review on polycyclic aromatic hydrocarbons: source, environmental impact, effect on human health and remediation. Egyptian Jounral of Petroleum 25 (1), 107–123.

Abha, S., Singh, C.S., 2012. Hydrocarbon pollution: effects on living organisms, remediation of contaminated environments, and effects of heavy metals Co-contamination on bioremediation. In: Romero-Zerón, L. (Ed.), Introduction to Enhanced Oil Recovery (EOR) Processes and Bioremediation of Oil-Contaminated Sites. InTech Europe, Rijeka, Croatia. http://cdn.intechopen.com/pdfs/37042/InTech-Hydrocarbon_pollution_effects_on_living_organisms_remediation_of_contaminated_environments_and_effects_of_heavy_metals_co_contamination_on_bioremediation.pdf.

Antai, S.P., 1990. Biodegradation of Bonny light crude oil by Bacillus sp and Pseudomonas sp. Waste Management 10, 61–64.

ASTM, 2019. Annual Book of Standards. ASTM International, West Conshohocken, Pennsylvania.

Atlas, R.M., 1981. Microbial degradation of petroleum hydrocarbons: an environmental perspective. Microbiological Reviews 45, 180–209.

Apitz, S.E., Borbridge, L.M., Theriault, G.A., Lieberman, S.H., 1992. Analysis 20, 461–474.

Askari, M.D., Masakarinec, M.P., Smith, S., Beam, P.M., Travis, C.C., 1996. Analytical Chemistry 68, 3431–3433.

Atlas, R.M., 1991. Microbial hydrocarbon degradation – bioremediation of oil spills. Journal of Chemical Technology Biotechnology 52, 149–156.

Atlas, R.M., Bartha, R., 1992. Hydrocarbon biodegradation and oil spill bioremediation. Advances in Microbial Ecology 12, 287–338.

Baldwin, R.M., July 2002. Oil Shale: A Brief Technical Overview. Colorado School of Mines, Golden, Colorado.

Bamforth, S.M., Singleton, I., 2005. Bioremediation of polycyclic aromatic hydrocarbons: current knowledge and future directions. Journal of Chemical Technology and Biotechnology 80, 723–736.

Banerjee, A., Roy, A., Dutta, S., Mondal, S., 2016. Bioremediation of hydrocarbons – a review. International Journal of Advanced Research 4 (6), 1303–1313.

Bestetti, G., Collina, E., Di Gennaro, P., Lasagni, M., Pitea, D., 2003. Kinetic study of naphthalene biodegradation in aerobic slurry phase microcosms for the optimization of the process. Water, Air, and Soil Pollution: Focus 3, 223–231.

Bishnoi, P.R., Clarke, M.A., 2006. Encyclopedia of Chemical Processing. CRC Press, Taylor & Francis Group, Boca Raton, Florida, pp. 1849–1863.

Bjorseth, A., 1983. Handbook of Polycyclic Aromatic Hydrocarbons. Marcel Dekker Inc, New York. Now: Taylor & Francie Inc., Boca Raton, Florida.

Borchert, S., Mueller, J., Alesi, E., Leins, C., Haninger, V., 1995. In situ bioremediation application strategies for soil and groundwater impacted by PAHs. Land Contamination and Reclamation 3, 6-1–6-4.

Breedveld, G.D., Sparrevik, M., 2000. Nutrient-limited biodegradation of PAH in various soil strata at a creosote contaminated site. Biodegradation 11, 391–399.

Bruce, L.G., Schmidt, G.W., 1994. The American Association of Petroleum Geologists Bulletin 78, 1692–1710.

Cerniglia, C.E., 1992. Biodegradation of polycyclic aromatic hydrocarbons. Biodegradation 3, 351–368.

Cerniglia, C.E., Heitkamp, M.A., 1990. Polycyclic aromatic hydrocarbon degradation by *Mycobacterium*. Methods in Enzymology 188, 148–153.

Chen, C.I., Taylor, R.T., 1995. Thermophilic biodegradation of BTEX by two *Thermus* species. Biotechnology and Bioengineering 48, 614–624.

Chen, C.I., Taylor, R.T., 1997a. Batch and fed-batch bioreactor cultivations of a *Thermus* species with thermophilic BTEX-degrading activity. Applied Microbiology and Biotechnology 47, 726–733.

Chen, C.I., Taylor, R.T., 1997b. Thermophilic biodegradation of BTEX by two consortia of anaerobic bacteria. Applied Microbiology and Biotechnology 48, 121–128.

Clements, W.H., Oris, J.T., Wissing, T.E., 1994. Accumulation and food chain transfer of fluoranthene and benzo[*a*]pyrene in *Chironomus riparius* and *Lepomis macrochirus*. Archives of Environmental Contamination and Toxicology 26, 261–266.

Dandie, C.E., Weber, J., Aleer, S., Adetutu, E.M., Ball, A.S., Juhasz, A.L., 2010. Assessment of five bioaccessibility assays for predicting the efficacy of petroleum hydrocarbon biodegradation in aged contaminated soils. Chemosphere 81, 1061–1068.

Das, N., Chandran, P., 2011. Microbial degradation of petroleum hydrocarbon contaminants: an overview. Biotechnology Research International, 94180. https://www.ncbi.nlm.nih.gov/pmc/articles/PMC3042690/.

Davies, L., Daniel, F., Swannell, R., Braddock, J., 2001. Biodegradability of Chemically-Dispersed Oil. Report No. AEAT/ENV/R0421. AEA Technology Environment, Abingdon, Oxfordshire, United Kingdom.

Dean, J.R., 2003. Methods for Environmental Trace Analysis. John Wiley 7 Sons Inc., Hoboken, New Jersey.

De la Cueva, S.C., Rodríguez, C.H., Cruz, N.O.S., Contreras, J.A.R., Miranda, J.L., 2016. Changes in bacterial populations during bioremediation of soil contaminated with petroleum hydrocarbons. Water, Air, & Soil Pollution 227 (3), 1−12.

Driscoll, J.N., Hanby, J., Panaro, J., 1992. Review of field screening methodology for analysis of soils. In: Kostecki, P.T., Calabrese, E.J., Bonazountas, M. (Eds.), Hydrocarbon Contaminated Soils, vol. II. Lewis Publishers, Boca Raton, pp. 153−171.

Einhorn, I.N., Sears, S.F., Hickey, J.C., Vielenave, J.H., Moore, G.S., 1992. Hydrocarbon Contaminated Soils 2, 89−143.

Eriksson, M., Dalhammar, G., Borg-Karlson, A.K., 2000. Biological degradation of selected hydrocarbons in an old PAH/creosote contaminated soil from a gas work site. Applied Microbiology and Biotechnology 53, 619−626.

Freijer, J.I., de Jonge, H., Bouten, W., Verstraten, J.M., 1996. Assessing mineralization rates of petroleum hydrocarbons in soils in relation to environmental factors and experimental scale. Biodegradation 7, 487−500.

Gary, J.G., Handwerk, G.E., Kaiser, M.J., 2007. Petroleum Refining: Technology and Economics, fifth ed. CRC Press, Taylor & Francis Group, Boca Raton, Florida.

Goodwin, N.S., Park, P.J.D., Rawlinson, A.P., 1983. Crude oil biodegradation under simulated and natural conditions. In: Bjorøy, M. (Ed.), Advances in Organic Geochemistry 1981. John Wiley & Sons Inc., New York, pp. 650−658.

Greenwood, P.F., Wibrow, S., George, S.J., Tibbett, M., 2009. Hydrocarbon biodegradation and soil microbial community response to repeated oil exposure. Organic Geochemistry 40, 293−300.

Harvey, R.G., 1997. Polycyclic Aromatic Hydrocarbons. John Wiley & Sons Inc.-VCH, New York.

Heitkamp, M.A., Cerniglia, C.E., 1989. Polycyclic aromatic hydrocarbon degradation by a *Mycobacterium* sp. In microcosms containing sediment and water from a pristine ecosystem. Applied and Environmental Microbiology 55, 1968−1973.

Hsu, C.S., Robinson, P.R. (Eds.), 2017. Handbook of Petroleum Technology. Springer International Publishing AG, Cham, Switzerland.

Kahrs, L.E., Horzempa, L.M., Peterson, D.M., 1999. Contaminated Soils 4, 221−232.

Kasper, K.D., Twoney, D.M., Dinsmore, D., 1991. Ground Water Management 8, 673−688.

Leahy, J.G., Colwell, R.R., 1990. Microbial degradation of hydrocarbons in the environment. Microbiological Reviews 54, 305−315.

Liebeg, E.W., Cutright, T.J., 1999. The investigation of enhanced bioremediation through the addition of macro and micro nutrients in a PAH contaminated soil. International Biodeterioration & Biodegradation 44, 55−64.

Lee, S., Speight, J.G., Loyalka, S., 2014. Handbook of Alternative Fuel Technologies, second ed. CRC Press, Taylor & Francis Group, Boca Raton, Florida.

Löhmannsröben, H.-G., Roch, T., Schulyzr, R.H., 1999. Polycyclic Aromatic Compounds 13, 165−174.

Lundstedt, S., Haglund, P., Oberg, L., 2003. Degradation and formation of polycyclic aromatic compounds during bioslurry treatment of an aged gasworks soil. Environmental Toxicology and Chemistry 22, 1413–1420.

Margesin, R., Schinner, F., 1997a. Efficiency of indigenous and inoculated cold-adapted soil microorganisms for biodegradation of diesel oil in alpine soils. Applied and Environmental Microbiology 63, 2660–2664.

Margesin, R., Schinner, F., 1997b. Bioremediation of diesel-oil-contaminated alpine soils at low temperatures. Applied Microbiology and Biotechnology 47, 462–468.

McMillen, S.J., Smart, R., Bernier, R., Hoffman, R.E., 2004. Biotreating E&P wastes: lessons learned from 1992–2003," SPE 86794. In: Proceedings. Seventh International Conference on Health, Safety, and Environment in Oil and Gas Exploration and Production, Calgary, Alberta, Canada, March 29–31.

Miller, E.C., Miller, J.A., 1981. Search for the ultimate chemical carcinogens and their reaction with cellular macromolecules. Cancer 47, 2327–2345.

Miller, M. (Ed.), 2000. Encyclopedia of Analytical Chemistry. John Wiley & Sons Inc., Hoboken, New Jersey.

Mohan, S.V., Kisa, T., Ohkuma, T., Kanaly, R.A., Shimizu, Y., 2006. Bioremediation technologies for treatment of PAH-contaminated soil and strategies to enhance process efficiency. Reviews in Environmental Science and Biotechnology 5, 347–374.

Mohn, W.W., Stewart, G.R., 2000. Limiting factors for hydrocarbon biodegradation at low temperature in arctic soils. Soil Biology and Biochemistry 32, 1161–1172.

Mueller, J.G., Chapman, P.J., Blattmann, B.O., Pritchard, P.H., 1990. Isolation and characterization of a fluoranthene-utilizing strain of *Pseudomonas paucimobilis*. Applied and Environmental Microbiology 56, 1079–1086.

Norman, M., Ross, S., McEwen, G., Getliff, J., 2002. Minimizing environmental impacts and maximizing hole stability - the significance of drilling with synthetic fluids in New Zealand. In: Proceedings. New Zealand Petroleum Conference Proceedings, February 24–27.

Obayori, O.S., Ilori, M.O., Adebusoye, S.A., Oyetibo, G.O., Omotayo, A.E., Amund, O.O., 2009. Degradation of hydrocarbons and biosurfactant production by *Pseudomonas sp.* strain LP1. World Journal of Microbiology and Biotechnology 25, 1615–1623.

Obire, O., 1990. Bacterial Degradation of three Different crude oils in Nigeria. Nigerian Journal of Botany 1, 81–90.

Obire, O., 1993. The suitability of various Nigerian petroleum fractions as substrate for bacterial growth. Discovery and Innovation 9, 25–32.

Obire, O., Okudo, I.V., 1997. Effects of crude oil on a freshwater stream in Nigeria. Discovery and Innovation 9, 25–32.

Obire, O., Nwaubeta, O., 2001. Biodegradation of refined petroleum hydrocarbons in soil. Journal of Applied Sciences and Environmental Management 5 (1), 43–46.

Okerentugba, P.O., Ezeronye, O.U., 2003. Petroleum degrading potentials of single and mixed microbial cultures isolated from rivers and refinery effluent in Nigeria. African Journal of Biotechnology 2 (9), 288–292.

Okoh, A.I., 2006. Biodegradation alternative in the cleanup of petroleum hydrocarbon pollutants. Biotechnology and Molecular Biology Reviews 1 (2), 38–50.

Okoh, A.I., Trejo-Hernandez, M.R., 2006. Remediation of petroleum hydrocarbon polluted systems: exploiting the bioremediation strategies. African Journal of Biotechnology 5, 2520–2525.

Page, D.S., Boehm, P.D., Douglas, G.S., Bence, A.E., Burns, W.A., Mankiewic, P.J., 1996. The natural petroleum hydrocarbon background in subtidal sediments of prince William sound Alaska, USA. Environmental Toxicology and Chemistry 15, 1266–1281.

Parkash, S., 2003. Refining Processes Handbook. Gulf Professional Publishing, Elsevier, Amsterdam, Netherlands.

Pavlova, A., Ivanova, R., 2003. Determination of petroleum hydrocarbons and polycyclic aromatic hydrocarbons in sludge from wastewater treatment basins. Journal of Environmental Monitoring 5, 319–323.

Peters, K.E., Walters, C.C., Moldowan, J.M., 2005. The Biomarker Guide – Edition II. Cambridge University Press, Cambridge, United Kingdom.

Ringelberg, D.B., Talley, J.W., Perkins, E.J., Tucker, S.G., Luthy, R.G., Bouwer, E.J., Fredrickson, H.L., 2001. Succession of phenotypic, genotypic, and metabolic community characteristics during in vitro bioslurry treatment of polycyclic aromatic hydrocarbon-contaminated sediments. Applied and Environmental Microbiology 67, 1542–1550.

Sadler, R., Connell, D., 2002. PAHs in the soil environment and their bioavailability. In: Kookana, R., Sadler, R., Sethunathan, N., Naidu, R. (Eds.), Environmental Protection and Risk Assessment of Organic Contaminants. Enfield: Science Publishers, pp. 27–43.

Salam, L.B., Obayori, O.S., Akashoro, O.S., Okogie, G.O., 2011. Biodegradation of bonny light crude oil by bacteria isolated from contaminated soil. International Journal of Agriculture and Biology 13 (2), 245–250.

Schreiber, A.F., Winkler, U.K., 1983. Transformation of tetralin by whole cells of *Pseudomonas stutzeri* As39. Applied Microbiology and Biotechnology 18 (1), 6–10.

Siegrist, R.L., Jenssen, P.D., 1990. Evaluation of sampling method effects on volatile organic compound measurements in contaminated soils. Environmental Science and Technology 24, 1387–1392.

Shuttleworth, K.L., Cerniglia, C.E., 1995. Environmental aspects of polynuclear aromatic hydrocarbon biodegradation. Applied Biochemistry and Biotechnology 54, 291–302.

Sikkema, J., Bont, J.A.M., 1991. Isolation and initial characterization of bacteria growing on tetralin. Biodegradation 2 (1), 15–23.

Singh, S.N., Kumari, B., Mishra, S., 2012. Microbial degradation of alkanes. In: Singh, S.N. (Ed.), Microbial Degradation of Xenobiotics. Springer-Verlag, Berlin, Germany.

Soli, G., Bens, E.M., 1972. Bacteria which attack petroleum hydrocarbons in a saline medium. Biotechnology and Bioengineering 14, 319–330.

Speight, J.G., Lee, S., 2000. Environmental Technology Handbook, second ed. Taylor & Francis, New York.

Speight, J.G., 2005. Environmental Analysis and Technology for the Refining Industry. John Wiley & Sons Inc., Hoboken, New Jersey.

Speight, J.G. (Ed.), 2011. The Biofuels Handbook. Royal Society of Chemistry, London, United Kingdom.

Speight, J.G., Arjoon, K.K., 2012. Bioremediation of Petroleum and Petroleum Products. Scrivener Publishing, Beverly, Massachusetts.

Speight, J.G., 2013. The Chemistry and Technology of Coal, third ed. CRC Press, Taylor & Francis Group, Boca Raton, Florida.

Speight, J.G., 2014. The Chemistry and Technology of Petroleum, fifth ed. CRC Press, Taylor & Francis Group, Boca Raton, Florida.

Speight, J.G., 2015. Handbook of Petroleum Product Analysis, second ed. John Wiley & Sons Inc., Hoboken, New Jersey.

Speight, J.G., 2017. Handbook of Petroleum Refining. CRC Press, Taylor & Francis Group, Boca Raton, Florida.

Speight, J.G., 2019. Natural Gas: A Basic Handbook, second ed. Gulf Publishing Company, Elsevier, Cambridge, Massachusetts.

Steffan, R.J., McCloy, K., Vainberg, S., Condee, C.W., Zhang, D., 1997. Biodegradation of the gasoline oxygenates methyl-tert-butyl ether, ethyl tert-butyl ether and tert − amyl methyl ether by propane-oxidizing bacteria. Applied and Environmental Microbiology 63 (11), 4216−4222.

Strawinski, R.J., Stone, R.W., 1940. The utilization of hydrocarbons by bacteria. Journal of Bacteriology 40 (3), 461.

Swanell, R.P.J., Lee, K., McDonagh, M., 1996. Field evaluation of marine oil spill bioremediation. Microbiological Reviews 342−365.

Taghvaei Ganjali, S., Nahri Niknafs, B., Khosravi, M., et al., 2007. Photo-oxidation of crude petroleum maltenic fraction I natural simulated conditions and structural elucidation of photoproducts. Iranian Journal of Environmental Health Science and Engineering 4 (1), 37−42.

Talley, J.W., Ghosh, U., Tucker, S.G., Furey, J.S., Luthy, R.G., 2002. Particle-scale understanding of the bioavailability of PAHs in sediment. Environmental Science and Technology 36, 477−483.

US DOE, 2004a. Strategic Significance of America's Oil Shale Reserves, I. Assessment of Strategic Issues. http://www.fe.doe.gov/programs/reserves/publications. http://www.fe.doe.gov/programs/reserves/publications.

US DOE, 2004b. Strategic Significance of America's Oil Shale Reserves, II. Oil Shale Resources, Technology, and Economics. http://www.fe.doe.gov/programs/reserves/publications. http://www.fe.doe.gov/programs/reserves/publications.

US DOE, 2004c. America's Oil Shale: A Roadmap for Federal Decision Making; USDOE Office of US Naval Petroleum and Oil Shale Reserves. http://www.fe.doe.gov/programs/reserves/publications. http://www.fe.doe.gov/programs/reserves/publications.

Vidali, M., 2001. Bioremediation: an overview. Pure and Applied Chemistry 73, 1163−1172.

Viñas, M., Sabate, J., Espuny, M.J., Solanas, A.M., 2005. Bacterial community dynamics and polycyclic aromatic hydrocarbon degradation during bioremediation of heavily creosote-contaminated soil. Applied and Environmental Microbiology 71, 7008−7018.

Wang, Z., Fingas, M., Sergy, Z., 1994. Environmental Science and Technology 28, 1733−1746.

Weissenfels, W.D., Beyer, M., Klein, J., 1990. Degradation of phenanthrene, fluorene and fluoranthene by pure bacterial cultures. Applied Microbiology and Biotechnology 32, 479−484.

Weissenfels, W.D., Beyer, M., Klein, J., Rehm, H.J., 1991. Microbial metabolism of fluoranthene: isolation and identification of ring fission products. Applied Microbiology and Biotechnology 34, 528−535.

Whittaker, M., Pollard, S.J.T., Fallick, T.E., 1995. Environmental Technology 16, 1009−1033.

Ye, D., Siddiqi, M.A., Maccubbin, A.E., Kumar, S., Sikka, H.C., 1996. Degradation of polynuclear aromatic hydrocarbons by *Sphingomonas paucimobilis*. Environmental Science and Technology 30, 136−142.

Zahed, M.A., Aziz, H.A., Isa, M.H., Mohajeri, L., 2010. Enhancement of biodegradation of n-alkanes from crude oil contaminated seawater. International Journal of Environmental Research 4 (4), 655−664.

Zander, M., 1983. Physical and chemical properties of polycyclic aromatic hydrocarbons. In: Bjørseth, A. (Ed.), Handbook of Polycyclic Aromatic Hydrocarbons. Marcel Dekker, Inc., New York, pp. 1−26.

Further reading

Björklöf, K., Salminen, J., Sainio, P., Jørgensen, K., 2008. Degradation rates of aged petroleum hydrocarbons are likely to be mass transfer dependent in the field. Environmental Geochemistry and Health 30 (2), 101–107.

Cerniglia, C.E., Heitkamp, M.A., 1989. Microbial degradation of polycyclic aromatic hydrocarbons in the aquatic environment. In: Varanasi, U. (Ed.), Metabolism of Polycyclic Aromatic Hydrocarbons in the Aquatic Environment. CRC Press, Inc., Boca Raton, Florida, pp. 41–68.

Irwin, R.J., VanMouwerik, M., Stevens, Seese, M.D., Basham, W., 1997. Environmental Contaminants Encyclopedia. National Park Service, Water Resources Division, Fort Collins, Colorado.

Conversion Factors

1. General

1 acre = 43,560 sq ft
1 acre-foot = 7758.0 bbl
1 atmosphere = 760 mm Hg = 14.696 psia = 29.91 in. Hg
1 atmosphere = 1.0133 bars = 33.899 ft. H_2O
1 barrel (oil) = 42 gal = 5.6146 *cu* ft
1 barrel (water) = 350 lb at 60°F
1 barrel per day = 1.84 cu cm per second
1 Btu = 778.26 ft-lb
1 centipoise x 2.42 = Ib mass/(ft) (hr), viscosity
1 centipoise x 0.000672 = Ib mass/(ft) (sec), viscosity
1 cubic foot = 28,317 cu cm = 7.4805 gal
Density of water at 60°F = 0.999 gram/cu cm = 62.367 Ib/cu ft = 8.337 Ib/gal
1 gallon = 231 cu in. = 3,785.4 cu cm = 0.13368 cu ft
1 horsepower-hour = 0.7457 kwhr = 2544.5 Btu
1 horsepower = 550 ft-lb/sec = 745.7 watts
1 inch = 2.54 cm
1 meter = 100 cm = 1,000 mm = 10 microns = 10 angstroms (A)
1 ounce = 28.35 grams
1 pound = 453.59 grams = 7,000 grains
1 square mile = 640 acres

2. Concentration Conversions

1 part per million (1 ppm) = 1 microgram per liter (1 μg/L)
1 microgram per liter (1 μg/L) = 1 milligram per kilogram (1 mg/kg)
1 microgram per liter (μg/L) x 6.243×10^8 = 1 lb per cubic foot (1 lb/ft^3)
1 microgram per liter (1 μg/L) x 10^{-3} = 1 milligram per liter (1 mg/L)
1 milligram per liter (1 mg/L) x 6.243×10^5 = 1 pound per cubic foot (1 lb/ft^3)
I gram mole per cubic meter (1 g mol/m^3) x 6.243×10^5 = 1 pound per cubic foot (1 lb/ft^3)
10,000 ppm = 1% w/w
1 ppm hydrocarbon in soil x 0.002 = 1 lb of hydrocarbons per ton of contaminated soil

3. Weight Conversion

1 ounce (1 oz) = 28.3495 grams (18.2495 g)
1 pound (1 lb) = 0.454 kilogram
1 pound (1 lb) = 454 grams (454 g)
1 kilogram (1 kg) = 2.20462 pounds (2.20462 lb)
1 stone (English, 1 st) = 14 pounds (14 lb)
1 ton (US; 1 short ton) = 2,000 lbs
1 ton (English; 1 long ton) = 2,240 lbs
1 metric ton = 2204.62262 pounds
1 tonne = 2204.62262 pounds

4. Temperature Conversions

$^oF = (^oC \times 1.8) + 32$
$^oC = (^oF - 32)/1.8$
$(^oF - 32) \times 0.555 = ^oC$
Absolute zero = −273.15°C
Absolute zero = −459.67°F

5. Area

1 square centimeter (1 cm^2) = 0.1550 square inches
1 square meter 1 (m^2) = 1.1960 square yards
1 hectare = 2.4711 acres
1 square kilometer (1 km^2) = 0.3861 square miles
1 square inch (1 $inch^2$) = 6.4516 square centimeters
1 square foot (1 ft^2) = 0.0929 square meters
1 square yard (1 yd^2) = 0.8361 square meters
1 acre = 4046.9 square meters
1 square mile (1 mi^2) = 2.59 square kilometers

6. Other Approximations

14.7 pounds per square inch (14.7 psi) = 1 atmosphere (1 atmos)
1 kilopascal (kPa) x 9.8692×10^{-3} = 14.7 pounds per square inch (14.7 psi)
1 yd^3 = 27 ft^3
1 US gallon of water = 8.34 lbs
1 imperial gallon of water = 10 lbs
1 yd^3 = 0.765 m^3
1 acre-inch of liquid = 27,150 gallons = 3.630 ft^3
1 ft depth in 1 acre (in situ) = 1,613 x (20−25 % excavation factor) = ~2,000 yd^3

1 yd^3 (clayey soils-excavated) = 1.1−1.2 tons (US)
1 yd^3 (sandy soils-excavated) = 1.2−1.3 tons (US)

SI METRIC CONVERSION FACTORS (**E = exponent; i.e., E + 03 = 10^3 and E − 03 = 10^{-3}**)

acre-foot × 1.233482	E + 03 = meters cubed
barrels × 1.589873	E − 01 = meters cubed
centipoise × 1.000000	E − 03 = pascal seconds
darcy × 9.869233	E − 01 = micrometers squared
feet × 3.048000	E − 01 = meters
pounds/acre-foot × 3.677332	E − 04 = kilograms/meters cubed
pounds/square inch × 6.894757	E + 00 = kilopascals
dyne/cm × 1.000000	E + 00 = mN/m
parts per million × 1.000000	E + 00 = milligrams/kilograms

Glossary

Absorber See Absorption tower.

Absorption gasoline Gasoline extracted from natural gas or refinery gas by contacting the absorbed gas with an oil and subsequently distilling the gasoline from the higher boiling components.

Absorption oil Oil used to separate the heavier components from a vapor mixture by absorption of the heavier components during intimate contacting of the oil and vapor; used to recover natural gasoline from wet gas.

Absorption plant A plant for recovering the condensable portion of natural or refinery gas, by absorbing the higher boiling hydrocarbons in an absorption oil, followed by separation and fractionation of the absorbed material.

Absorption tower A tower or column which promotes contact between a rising gas and a falling liquid so that part of the gas may be dissolved in the liquid.

Acetic acid An acid with the structure of CH_3COOH; acetyl groups are bound through an ester linkage to hemicellulose chains, especially xylan derivatives, in wood and other plants; the natural moisture present in plants hydrolyzes the acetyl groups to acetic acid, particularly at elevated temperatures.

Acetone–benzol process A dewaxing process in which acetone and benzol (benzene or aromatic naphtha) are used as solvents.

Acid catalyst A catalyst having acidic character; the aluminas are examples of such catalysts.

Acid deposition Acid rain; a form of pollution depletion in which pollutants, such as nitrogen oxides and sulfur oxides, are transferred from the atmosphere to soil or water; often referred to as atmospheric self-cleaning. The pollutants usually arise from the use of fossil fuels.

Acid drainage The runoff of acidic liquids from coal production waste piles. Such runoff can contaminate ground and surface waters.

Acid gas Hydrogen sulfide (H_2S) or carbon dioxide (CO_2).

Acid-insoluble lignin Lignin is mostly insoluble in mineral acids, and therefore can be analyzed gravimetrically after hydrolyzing the cellulose and hemicellulose fractions of the biomass with sulfuric acid.

Acidity The capacity of an acid to neutralize a base such as a hydroxyl ion (OH^-).

Acidizing A technique for improving the permeability of a reservoir by injecting acid.

Acid number A measure of the reactivity of petroleum with a caustic solution and given in terms of milligrams of potassium hydroxide that are neutralized by one gram of petroleum.

Acid rain The precipitation phenomenon that incorporates anthropogenic acids and other acidic chemicals from the atmosphere to the land and water (see Acid deposition).

Acid sludge The residue left after treating petroleum oil with sulfuric acid for the removal of impurities; a black, viscous substance containing the spent acid and impurities.

Acid treating A process in which unfinished petroleum products, such as gasoline, kerosene, and lubricating-oil stocks, are contacted with sulfuric acid to improve their color, odor, and other properties.

Acoustic log See Sonic log.

Acre-foot A measure of bulk rock volume where the area is one acre and the thickness is one foot.

Additive A material added to another (usually in small amounts) in order to enhance desirable properties or to suppress undesirable properties.

Add-on control methods The use of devices that remove refinery process emissions after they are generated but before they are discharged to the atmosphere.

Adsorption Transfer of a substance from a solution to the surface of a solid resulting in relatively high concentration of the substance at the place of contact; see also Chromatographic adsorption.

Adsorption gasoline Natural gasoline obtained by an adsorption process from wet gas.

Afterburn The combustion of carbon monoxide (CO) to carbon dioxide (CO_2); usually in the cyclones of a catalyst regenerator.

After flow Flow from the reservoir into the wellbore that continues for a period after the well has been shut in; after flow can complicate the analysis of a pressure transient test.

Agglomeration Formation of larger coal or ash particles by smaller particles sticking together.

Air-blown asphalt Asphalt produced by blowing air through residua at elevated temperatures.

Air injection An oil recovery technique using air to force oil from the reservoir into the wellbore.

Airlift Thermofor catalytic cracking A moving-bed continuous catalytic process for conversion of heavy gas oils into lighter products; the catalyst is moved by a stream of air.

Air pollution The discharge of toxic gases and particulate matter introduced into the atmosphere, principally as a result of human activity.

Air sweetening A process in which air or oxygen is used to oxidize lead mercaptides to disulfides instead of using elemental sulfur.

Air toxics Hazardous air pollutants.

Albertite A black, brittle, natural hydrocarbon possessing a conchoidal fracture and a specific gravity of approximately 1.1.

Alcohol The family name of a group of organic chemical compounds composed of carbon, hydrogen, and oxygen. The molecules in the series vary in chain length and are composed of a hydrocarbon plus a hydroxyl group. Alcohol includes methanol and ethanol.

Algae Primitive plants, usually aquatic, capable of synthesizing their own food by photosynthesis; currently being investigated as a possible feedstock for producing biodiesel.

Alicyclic hydrocarbon A compound containing carbon and hydrogen only which has a cyclic structure (e.g., cyclohexane); also collectively called naphthenes.

Aliphatic hydrocarbon A compound containing carbon and hydrogen only which has an open-chain structure (e.g., as ethane, butane, octane, butene) or a cyclic structure (e.g., cyclohexane).

Aliquot That quantity of material of proper size for measurement of the property of interest; test portions may be taken from the gross sample directly, but often preliminary operations such as mixing or further reduction in particle size are necessary.

Alkaline A high pH usually of an aqueous solution; aqueous solutions of sodium hydroxide, sodium orthosilicate, and sodium carbonate are typical alkaline materials used in enhanced oil recovery.

Alkaline flooding See EOR process.

Alkalinity The capacity of a base to neutralize the hydrogen ion (H^+).

Alkali treatment See Caustic wash.

Alkali wash See Caustic wash.

Alkanes Hydrocarbons that contain only single carbon–hydrogen bonds. The chemical name indicates the number of carbon atoms and ends with the suffix "ane."

Alkenes Hydrocarbons that contain carbon–carbon double bonds. The chemical name indicates the number of carbon atoms and ends with the suffix "ene."

Alkali lignin Lignin obtained by acidification of an alkaline extract of wood.

Alkylate The product of an alkylation process.

Alkylate bottoms Residua from fractionation of alkylate; the alkylate product which boils higher than the aviation gasoline range; sometimes called heavy alkylate or alkylate polymer.

Alkylation In the petroleum industry, a process by which an olefin (e.g., ethylene) is combined with a branched-chain hydrocarbon (e.g., *iso*-butane); alkylation may be accomplished as a thermal or as a catalytic reaction.

Alkyl groups A group of carbon and hydrogen atoms that branch from the main carbon chain or ring in a hydrocarbon molecule. The simplest alkyl group, a methyl group, is a carbon atom attached to three hydrogen atoms.

Alpha-scission The rupture of the aromatic carbon–aliphatic carbon bond that joins an alkyl group to an aromatic ring.

Alternative fuel As defined in the United States Energy Policy Act of 1992 (EPACT): methanol, denatured ethanol, and other alcohols, separately or in blends of at least 10 percent by volume with gasoline or other fuels; compressed natural gas; liquefied natural gas, liquefied propane gas, hydrogen, coal-derived liquid fuels, fuels other than alcohols derived from biological materials, electricity, biodiesel, and any other fuel deemed to be substantially not petroleum and yielding potential energy security benefits and substantial environmental benefits.

Alumina (Al_2O_3) Used in separation methods as an adsorbent and in refining as a catalyst.

American Society for Testing and Materials (ASTM) The official organization in the United States for designing standard tests for petroleum and other industrial products.

Amine washing A method of gas cleaning whereby acidic impurities such as hydrogen sulfide and carbon dioxide are removed from the gas stream by washing with an amine (usually an alkanolamine).

Anaerobic digestion Decomposition of biological wastes by microorganisms, usually under wet conditions, in the absence of air (oxygen), to produce a gas comprising mostly methane and carbon dioxide.

Analytical equivalence The acceptability of the results obtained from the different laboratories; a range of acceptable results.

Analyte The chemical for which a sample is tested or analyzed.

Antibody A molecule having chemically reactive sites specific for certain other molecules.

Aniline point The temperature, usually expressed in °F, above which equal volumes of a petroleum product are completely miscible; a qualitative indication of the relative proportions of paraffins in a petroleum product which are miscible with aniline only at higher temperatures; a high aniline point indicates low aromatics.

Anthracene oil The heaviest distillable coal tar fraction, with distillation range 270–400°C (520–750°F), containing creosote oil, anthracene, phenanthrene, carbazole, and so on.

Anthracite (hard coal) A hard, black, shiny coal, very high in fixed carbon and low in volatile matter, hydrogen, and oxygen.

Antiknock Resistance to detonation or pinging in spark-ignition engines.

Antiknock agent A chemical compound such as tetraethyl lead which, when added in small amount to the fuel charge of an internal combustion engine, tends to lessen knocking.

Antistripping agent An additive used in an asphaltic binder to overcome the natural affinity of an aggregate for water instead of asphalt.

API gravity A measure of the *lightness* or *heaviness* of petroleum which is related to density and specific gravity. $°API = (141.5/sp\ gr\ @\ 60°F) - 131.5$

Apparent bulk density The density of a catalyst as measured; usually loosely compacted in a container.

Apparent viscosity The viscosity of a fluid, or several fluids flowing simultaneously, measured in a porous medium (rock) and subject to both viscosity and permeability effects; also called effective viscosity.

Aquifer A subsurface rock interval that will produce water; often the underlay of a petroleum reservoir.

Areal sweep efficiency The fraction of the flood pattern area that is effectively swept by the injected fluids.

Aromatic hydrocarbon A hydrocarbon characterized by the presence of an aromatic ring or condensed aromatic rings; benzene and substituted benzene, naphthalene and substituted naphthalene, phenanthrene and substituted phenanthrene, as well as the higher condensed ring systems; compounds that are distinct from those of aliphatic compounds or alicyclic compounds.

Aromatics A range of hydrocarbons which have a distinctive sweet smell and include benzene and toluene, occur naturally in petroleum, and are also extracted as a petrochemical feedstock, as well as for use as solvents.

Aromatization The conversion of nonaromatic hydrocarbons to aromatic hydrocarbons by: (1) rearrangement of aliphatic (noncyclic) hydrocarbons into aromatic ring structures and (2) dehydrogenation of alicyclic hydrocarbons (naphthenes).

Arosorb process A process for the separation of aromatics from nonaromatics by adsorption on a gel from which they are recovered by desorption.

Ash The noncombustible residue remaining after complete coal combustion; the final form of the mineral matter present in coal.

Ash analysis Percentages of inorganic oxides present in an ash sample. Ash analyses are used for evaluation of the corrosion, slagging, and fouling potential of coal ash. The ash constituents of interest are silica (SiO_2) alumina (Al_2O_3), titania (TiO_2), ferric oxide (Fe_2O_3), lime (CaO), magnesia (MgO), potassium oxide (K_2O), sodium oxide (Na_2O), and sulfur trioxide (SO_3). An indication of ash behavior can be estimated from the relative percentages of each constituent.

Ash fusion temperatures A set of temperatures that characterize the behavior of ash as it is heated. These temperatures are determined by heating cones of ground, pressed ash in both oxidizing and reducing atmospheres.

Asphalt The nonvolatile product obtained by distillation and treatment of an asphaltic crude oil; a manufactured product.

Asphalt cement Asphalt especially prepared as to quality and consistency for direct use in the manufacture of bituminous pavements.

Asphalt emulsion An emulsion of asphalt cement in water containing a small amount of emulsifying agent.

Asphaltene (asphaltenes) The brown to black powdery material produced by treatment of petroleum, heavy oil, bitumen, or residuum with a low-boiling liquid hydrocarbon.

Asphalt flux An oil used to reduce the consistency or viscosity of hard asphalt to the point required for use.

Asphalt primer A liquid asphaltic material of low viscosity which, upon application to a nonbituminous surface, waterproofs the surface and prepares it for further construction.

Asphaltene (asphaltenes) The brown to black powdery material produced by treatment of petroleum, petroleum residua, or bituminous materials with a low-boiling liquid hydrocarbon, e.g. pentane or heptane; soluble in benzene (and other aromatic solvents), carbon disulfide, and chloroform (or other chlorinated hydrocarbon solvents).

Asphaltene association factor The number of individual asphaltene species which associate in nonpolar solvents as measured by molecular weight methods; the molecular weight of asphaltenes in toluene divided by the molecular weight in a polar nonassociating solvent, such as dichlorobenzene, pyridine, or nitrobenzene.

Asphaltic pyrobitumen See Asphaltoid.

Asphaltic road oil A thick, fluid solution of asphalt; usually a residual oil. See also Nonasphaltic road oil.

Asphaltite A variety of naturally occurring, dark brown to black, solid, nonvolatile bituminous material that is differentiated from bitumen primarily by a high content of material insoluble in n-pentane (asphaltene) or other liquid hydrocarbons.

Asphaltoid A group of brown to black, solid bituminous materials of which the members are differentiated from asphaltites by their infusibility and low solubility in carbon disulfide.

Asphaltum See Asphalt.

As-received moisture The moisture present in a coal sample when delivered.

Associated molecular weight The molecular weight of asphaltenes in an associating (nonpolar) solvent, such as toluene.

Atmospheric residuum A residuum obtained by distillation of a crude oil under atmospheric pressure and which boils above 350°C (660°F).

Atmospheric equivalent boiling point (AEBP) A mathematical method of estimating the boiling point at atmospheric pressure of nonvolatile fractions of petroleum.

Attainment area A geographical area that meets NAAQS for criteria air pollutants (See also Nonattainment area).

Attapulgus clay See Fuller's earth.

Attritus A microscopic coal constituent composed of macerated plant debris intimately mixed with mineral matter and coalified. US Bureau of Mines usage, viewed by transmitted light.

Auger mining Mining generally practiced but not restricted to hilly coal-bearing regions of the country that uses a machine designed on the principle of the auger, which bores into an exposed coal seam, conveying the coal to a storage pile or bin for loading and transporting. May be used alone or in combination with conventional surface mining. When used alone, a single cut is made sufficient to expose the coal seam and provide operating space for the machine. When used in combination with surface mining, the last cut pit provides the operating space.

Autofining A catalytic process for desulfurizing distillates.

Average particle size The weighted average particle diameter of a catalyst.

Aviation gasoline Any of the special grades of gasoline suitable for use in certain airplane engines.

Aviation turbine fuel See Jet fuel.

Backfill The operation of refilling an excavation. Also, the material placed in an excavation in the process of backfilling.

Back mixing The phenomenon observed when a catalyst travels at a slower rate in the riser pipe than the vapors.

BACT Best available control technology.

Baghouse A filter system for the removal of particulate matter from gas streams; so-called because of the similarity of the filters to coal bags.

Bank The concentration of oil (oil bank) in a reservoir that moves cohesively through the reservoir.

Bari-Sol process A dewaxing process which employs a mixture of ethylene dichloride and benzol as the solvent.

Barrel (bbl) The unit of measure used by the petroleum industry; equivalent to approximately 42 US gallons or approximately 34 (33.6) Imperial gallons or 159 liters; 7.2 barrels are equivalent to 1 tonne of oil (metric).

Barrel of oil equivalent (boe) The amount of energy contained in a barrel of crude oil, i.e., approximately 6.1 GJ (5.8 million Btu), equivalent to 1,700 kWh.

Base number The quantity of acid, expressed in milligrams of potassium hydroxide per gram of sample, that is required to titrate a sample to a specified end point.

Base stock A primary refined petroleum fraction into which other oils and additives are added (blended) to produce the finished product.

Basic nitrogen Nitrogen (in petroleum) which occurs in pyridine form

Basic sediment and water (bs&w, bsw) The material which collects in the bottom of storage tanks, usually composed of oil, water, and foreign matter; also called bottoms, bottom settlings.

Battery A series of stills or other refinery equipment operated as a unit.

Baumé gravity The specific gravity of liquids expressed as degrees on the Baum J (°Bé) scale; for liquids lighter than water:

$$\text{Sp gr } 60°\text{F} = 140/(130 + °\text{BJ})$$

For liquids heavier than water:

$$\text{Sp gr } 60°\text{F} = 145/(145 - °\text{BJ})$$

Bauxite Mineral matter used as a treating agent; hydrated aluminum oxide formed by the chemical weathering of igneous rock.

Bbl See Barrel.

Bedrock The rock material directly above and below the coal seam.

Beehive oven A dome-shaped oven not equipped to recover the by-product gas and liquids evolved during the coking process.

Bell cap A hemispherical or triangular cover placed over the riser in a (distillation) tower to direct the vapors through the liquid layer on the tray; see Bubble cap.

Belt feeder (feeder breaker) A crawler-mounted surge bin often equipped with a crusher or breaker and used in room-and-pillar sections positioned at the end of the section conveyor belt. It allows a quick discharge of the shuttle car. It sizes the coal, and a built-in conveyor feeds it at an appropriate rate onto the conveyor belt.

Bench The surface of an excavated area at some point between the material being mined and the original surface of the ground, on which equipment can sit, move, or operate. A working road or base below a high wall, as in contour stripping for coal.

Bender process A chemical treating process using a lead sulfide catalyst for sweetening light distillates by which mercaptans are converted to disulfides by oxidation.

Bentonite Montmorillonite (a magnesium–aluminum silicate); used as a treating agent.

Benzene A colorless aromatic liquid hydrocarbon (C_6H_6).

Benzin A refined light naphtha used for extraction purposes.

Benzine An obsolete term for light petroleum distillates covering the gasoline and naphtha range; see Ligroine.

Benzol The general term which refers to commercial or technical (not necessarily pure) benzene; also the term used for aromatic naphtha.

Beta-scission The rupture of a carbon–carbon bond, two bonds removed from an aromatic ring.

Billion 1×10^9

Biochemical conversion The use of fermentation or anaerobic digestion to produce fuels and chemicals from organic sources.

Biocide Any chemical capable of killing bacteria and biorganisms.

Biodiesel A fuel derived from biological sources that can be used in diesel engines instead of petroleum-derived diesel; through the process of transesterification, the triglycerides in the biologically derived oils are separated from the glycerin, creating a clean-burning, renewable fuel.

Biodiesel blend A blend of biodiesel and diesel fuels—the blend can be with Diesel #1, Diesel #2, or JP8; one standard blend that meets the minimum requirements of the federal EPA Clean Air Act criteria is B20. The number after "B" indicates the percentage of biodiesel included in the blend—in B20, there would be 20 percent biodiesel and 80 percent diesel in the fuel blend; a biodiesel blend can come in any mixture percentage, such as B2, B5, B50, and B85.

Bioenergy Useful, renewable energy produced from organic matter—the conversion of the complex carbohydrates in organic matter to energy; organic matter may either be used directly as a fuel, processed into liquids and gasses, or be a residual of processing and conversion.

Bioethanol Ethanol produced from biomass feedstocks; includes ethanol produced from the fermentation of crops, such as corn, as well as cellulosic ethanol produced from woody plants or grasses.

Biofuels A generic name for liquid or gaseous fuels that are not derived from petroleum-based fossils fuels or contain a proportion of non–fossil fuel; fuels produced from plants, crops such as sugar beet, rapeseed oil, or reprocessed vegetable oils or fuels made from gasified biomass; fuels made from renewable biological sources and include ethanol, methanol, and biodiesel; sources include, but are not limited to: corn, soybeans, flaxseed, rapeseed, sugarcane, palm oil, raw sewage, food scraps, animal parts, and rice.

Biogas A combustible gas derived from decomposing biological waste under anaerobic conditions. Biogas normally consists of 50–60 percent methane. See also landfill gas.

Biogenic Material derived from bacterial or vegetation sources.

Biological lipid Any biological fluid that is miscible with a nonpolar solvent. These materials include waxes, essential oils, chlorophyll, etc.

Biological oxidation The oxidative consumption of organic matter by bacteria by which the organic matter is converted into gases.

Biomass biological organic matter; any organic matter that is available on a renewable or recurring basis, including agricultural crops and trees, wood and wood residues, plants (including aquatic plants), grasses, animal manure, municipal residues, and other residue materials. Biomass is generally produced in a sustainable manner from water and carbon dioxide by photosynthesis. There are three main categories of biomass—primary, secondary, and tertiary.

Biopolymer A high-molecular-weight carbohydrate produced by bacteria.

Biopower The use of biomass feedstock to produce electric power or heat through direct combustion of the feedstock, through gasification and then combustion of the resultant gas, or through other thermal conversion processes. Power is generated with engines, turbines, fuel cells, or other equipment.

Biorefinery A facility that processes and converts biomass into value-added products. These products can range from biomaterials to fuels such as ethanol or important feedstocks for the production of chemicals and other materials.

Biomass to liquid (BTL) The process of converting biomass to liquid fuels. Hmm, that seems painfully obvious when you write it out.

Bioreactor A vessel in which a chemical process occurs, which usually involves organisms or biochemically active substances derived from such organisms.

Bitumen A semisolid to solid hydrocarbonaceous material found filling pores and crevices of sandstone, limestone, or argillaceous sediments; also, on occasion, referred to as native asphalt, and extra heavy oil; a naturally occurring material that has little or no mobility under reservoir conditions and which cannot be recovered through a well by conventional oil well production methods including currently used enhanced recovery techniques; current methods involve mining for bitumen recovery.

Bituminous Containing bitumen or constituting the source of bitumen.

Bituminous (soft) coal A relatively soft dark brown to black coal, lower in fixed carbon than anthracite but higher in volatile matter, hydrogen, and oxygen.

Bituminous rock See Bituminous sand.

Bituminous sand A formation in which the bituminous material (see Bitumen) is found as a filling in veins and fissures in fractured rock or impregnating relatively shallow sand, sandstone, and limestone strata; a sandstone reservoir that is impregnated with a heavy, viscous black petroleum-like material that cannot be retrieved through a well by conventional production techniques.

Black acid(s) A mixture of the sulfonates found in acid sludge which are insoluble in naphtha, benzene, and carbon tetrachloride; very soluble in water but insoluble in 30 per cent sulfuric acid; in the dry, oil-free state, the sodium soaps are black powders.

Black liquor Solution of lignin residue and the pulping chemicals used to extract lignin during the manufacture of paper.

Black oil Any of the dark-colored oils; a term now often applied to heavy oil.

Black soap See Black acid.

Black strap The black material (mainly lead sulfide) formed in the treatment of sour light oils with doctor solution and found at the interface between the oil and the solution.

Blown asphalt The asphalt prepared by air blowing a residuum or an asphalt.

Blue gas A mixture consisting chiefly of carbon monoxide and hydrogen formed by action of steam on hot coal or coke.

Bogging A condition that occurs in a coking reactor when the conversion to coke and light ends is too slow causing the coke particles to agglomerate.

Boghead coal Same as cannel coal except that algal remains can be seen under the microscope.

Boiling point A characteristic physical property of a liquid at which the vapor pressure is equal to that of the atmosphere and the liquid is converted to a gas.

Boiling range The range of temperature, usually determined at atmospheric pressure in standard laboratory apparatus, over which the distillation of an oil commences, proceeds, and finishes.

Bottled gas Usually butane or propane, or butane–propane mixtures, liquefied and stored under pressure for domestic use; see also Liquefied petroleum gas.
Bottoms The liquid which collects in the bottom of a vessel (tower bottoms, tank bottoms) either during distillation; also the deposit or sediment formed during storage of petroleum or a petroleum product; see also Residuum and Basic sediment and water.
Bright coal US Bureau of Mines term for a combination of clarain and vitrain with small amounts of fusain.
Bright stock Refined, high-viscosity lubricating oils usually made from residual stocks by processes such as a combination of acid treatment or solvent extraction with dewaxing or clay finishing.
Briquetting A process of applying pressure to coal fines, with or without a binder, to form a compact or agglomerate.
British thermal unit See Btu.
Bromine number The number of grams of bromine absorbed by 100 g of oil which indicates the percentage of double bonds in the material.
Brown acid Oil-soluble petroleum sulfonates found in acid sludge which can be recovered by extraction with naphtha solvent. Brown-acid sulfonates are somewhat similar to mahogany sulfonates but are more water soluble. In the dry, oil-free state, the sodium soaps are light-colored powders.
Brown soap See Brown acid.
Brønsted acid A chemical species which can act as a source of protons.
Brønsted base A chemical species which can accept protons.
BS&W See Basic sediment and water.
BTEX Benzene, toluene, ethylbenzene, and the xylene isomers.
Btu (British thermal unit) A nonmetric unit of heat, still widely used by engineers; one Btu is the heat energy needed to raise the temperature of one pound of water from 60°F to 61°F at one atmosphere pressure. 1 Btu = 1055 joules (1.055 kJ).
Bubble cap An inverted cup with a notched or slotted periphery to disperse the vapor in small bubbles beneath the surface of the liquid on the bubble plate in a distillation tower.
Bubble plate A tray in a distillation tower.
Bubble point The temperature at which incipient vaporization of a liquid in a liquid mixture occurs, corresponding with the equilibrium point of 0 per cent vaporization or 100 per cent condensation.
Bubble tower A fractionating tower so constructed that the vapors rising pass up through layers of condensate on a series of plates or trays (see Bubble plate); the vapor passes from one plate to the next above by bubbling under one or more caps (see Bubble cap) and out through the liquid on the plate where the less volatile portions of vapor condense in bubbling through the liquid on the plate, overflow to the next lower plate, and ultimately back into the reboiler thereby effecting fractionation.
Bubble tray A circular, perforated plates having the internal diameter of a bubble tower, set at specified distances in a tower to collect the various fractions produced during distillation.
Buckley–Leverett method A theoretical method of determining frontal advance rates and saturations from a fractional flow curve.
Bumping The knocking against the walls of a still occurring during distillation of petroleum or a petroleum product which usually contains water.
Bunker C oil See No. 6 Fuel oil.
Burner fuel oil Any petroleum liquid suitable for combustion.
Burning oil An illuminating oil, such as kerosene (kerosine), suitable for burning in a wick lamp.
Burning point See Fire point.
Burning-quality index An empirical numerical indication of the likely burning performance of a furnace or heater oil; derived from the distillation profile and the API gravity and generally recognizing the factors of paraffin character and volatility.

Burton process An older thermal cracking process in which oil was cracked in a pressure still and any condensation of the products of cracking also took place under pressure.

Butane dehydrogenation A process for removing hydrogen from butane to produce butenes and, on occasion, butadiene.

Butane vapor-phase isomerization A process for isomerizing n-butane to *iso*-butane using an aluminum chloride catalyst on a granular alumina support and with hydrogen chloride as a promoter.

Butanol Though generally produced from fossil fuels, this four-carbon alcohol can also be produced through bacterial fermentation of alcohol.

C_1, C_2, C_3, C_4, C_5 fractions A common way of representing fractions containing a preponderance of hydrocarbons having 1, 2, 3, 4, or 5 carbon atoms, respectively, and without reference to hydrocarbon type.

CAA Clean Air Act; this act is the foundation of air regulations in the United States.

Calcining Heating a metal oxide or an ore to decompose carbonates, hydrates, or other compounds often in a controlled atmosphere.

Cannel coal Predominately durain with lesser amounts of vitrain than splint coal and small quantities of fusain. Spores can be seen under the microscope.

Capillary forces Interfacial forces between immiscible fluid phases, resulting in pressure differences between the two phases.

Capillary number N_c, the ratio of viscous forces to capillary forces, and equal to viscosity times velocity divided by interfacial tension.

Carbene The pentane- or heptane-insoluble material that is insoluble in benzene or toluene but which is soluble in carbon disulfide (or pyridine); a type of rifle used for hunting bison.

Carboid The pentane- or heptane-insoluble material that is insoluble in benzene or toluene and which is also insoluble in carbon disulfide (or pyridine).

Carbonate washing Processing using a mild alkali (e.g. potassium carbonate) process for emission control by the removal of acid gases from gas streams.

Carbon dioxide (CO_2) A product of combustion that acts as a greenhouse gas in the Earth's atmosphere, trapping heat and contributing to climate change.

Carbon dioxide–augmented waterflooding Injection of carbonated water, or water and carbon dioxide, to increase waterflood efficiency; see immiscible carbon dioxide displacement.

Carbon dioxide–miscible flooding See EOR process.

Carbon-forming propensity See Carbon residue.

Carbonization The conversion of an organic compound into char or coke by heat in the substantial absence of air; often used in reference to the destructive distillation (with simultaneous removal of distillate) of coal; a process whereby coal is converted to coke by devolatilization.

Carbon monoxide (CO) A lethal gas produced by incomplete combustion of carbon-containing fuels in internal combustion engines. It is colorless, odorless, and tasteless.

Carbon–oxygen log Information about the relative abundance of elements such as carbon, oxygen, silicon, and calcium in a formation; usually derived from pulsed neutron equipment.

Carbon rejection Upgrading processes in which coke is produced, e.g. coking.

Carbon residue The amount of carbonaceous residue remaining after thermal decomposition of petroleum, a petroleum fraction, or a petroleum product in a limited amount of air; also called the *coke-* or *carbon-forming propensity*; often prefixed by the terms Conradson or Ramsbottom in reference to the inventor of the respective tests.

Carburetted blue gas See Water gas.

CAS Chemical Abstract Service.

Cascade tray A fractionating device consisting of a series of parallel troughs arranged on stair-step fashion in which liquid frown the tray above enters the uppermost trough and liquid thrown from this trough by vapor rising from the tray below impinges against a plate and a perforated baffle and liquid passing through the baffle enters the next longer of the troughs.

Casinghead gas Natural gas which issues from the casinghead (the mouth or opening) of an oil well.

Casinghead gasoline The liquid hydrocarbon product extracted from casinghead gas by one of three methods: compression, absorption, or refrigeration; see also Natural gasoline.

Catagenesis The alteration of organic matter during the formation of petroleum that may involve temperatures in the range 50°C (120°F) to 200°C (390°F); see also Diagenesis and Metagenesis.

Catalyst A chemical agent which, when added to a reaction (process), will enhance the conversion of a feedstock without being consumed in the process.

Catalyst selectivity The relative activity of a catalyst with respect to a particular compound in a mixture, or the relative rate in competing reactions of a single reactant.

Catalyst stripping The introduction of steam, at a point where the spent catalyst leaves the reactor, in order to strip, i.e., remove, deposits retained on the catalyst.

Catalytic activity The ratio of the space velocity of the catalyst under test to the space velocity required for the standard catalyst to give the same conversion as the catalyst being tested; usually multiplied by 100 before being reported.

Catalytic cracking The conversion of high-boiling feedstocks into lower boiling products by means of a catalyst which may be used in a fixed bed or fluid bed.

Cat cracking See Catalytic cracking.

Catalytic reforming Rearranging hydrocarbon molecules in a gasoline-boiling-range feedstock to produce other hydrocarbons having a higher antiknock quality; isomerization of paraffins, cyclization of paraffins to naphthenes, dehydrocyclization of paraffins to aromatics.

Catforming A process for reforming naphtha using a platinum-silica-alumina catalyst which permits relatively high space velocities and results in the production of high-purity hydrogen.

Caustic consumption The amount of caustic lost from reacting chemically with the minerals in the rock, the oil, and the brine.

Chemical flooding See EOR process.

Caustic wash The process of treating a product with a solution of caustic soda to remove minor impurities; often used in reference to the solution itself.

Ceresin A hard, brittle wax obtained by purifying ozokerite; see Microcrystralline wax and Ozokerite.

Cetane index An approximation of the cetane number calculated from the density and mid-boiling point temperature; see also Diesel index.

Cetane number A number indicating the ignition quality of diesel fuel; a high cetane number represents a short ignition delay time; the ignition quality of diesel fuel can also be estimated from the following formula: Diesel index = (aniline point (°F) x API gravity)100

CFR Code of Federal Regulations; Title 40 (40 CFR) contains the regulations for protection of the environment.

Characterization factor The UOP characterization factor K, defined as the ratio of the cube root of the molal average boiling point, T_B, in degrees Rankine (°R = °F + 460), to the specific gravity at 60°F/60°F: $K = (T_B)^{1/3}/\text{sp gr}$ The value ranges from 12.5 for paraffin stocks to 10.0 for the highly aromatic stocks; also called the Watson characterization factor.

Cheesebox still An early type of vertical cylindrical still designed with a vapor dome.

Chelating agents Complex-forming agents having the ability to solubilize heavy metals.

Chemical flooding See EOR process.

Chemical octane number The octane number added to gasoline by refinery processes or by the use of octane number improvers such as tetraethyl lead.

Chemical waste Any solid, liquid, or gaseous material discharged from a process and that may pose substantial hazards to human health and environment.

Chlorex process A process for extracting lubricating-oil stocks in which the solvent used is Chlorex (ß-ß-dichlorodiethyl ether).

Chromatographic adsorption Selective adsorption on materials such as activated carbon, alumina, or silica gel; liquid or gaseous mixtures of hydrocarbons are passed through the adsorbent in a stream of diluent, and certain components are preferentially adsorbed.

Chromatographic separation The separation of different species of compounds according to their size and interaction with the rock as they flow through a porous medium.

Chromatography A method of separation based on selective adsorption; see also Chromatographic adsorption.

Clarain A macroscopic coal constituent (lithotype) known as bright-banded coal, composed of alternating bands of vitrain and durain.

Clarified oil The heavy oil which has been taken from the bottom of a fractionator in a catalytic cracking process and from which the residual catalyst has been removed.

Clarifier Equipment for removing the color or cloudiness of an oil or water by separating the foreign material through mechanical or chemical means; may involve centrifugal action, filtration, heating, or treatment with acid or alkali.

Clay Silicate minerals that also usually contain aluminum and have particle sizes that are less than 0.002 micron; used in separation methods as an adsorbent and in refining as a catalyst.

Clay contact process See Contact filtration.

Clay refining A treating process in which vaporized gasoline or other light petroleum product is passed through a bed of granular clay such as fuller's earth.

Clay regeneration A process in which spent coarse-grained adsorbent clays from percolation processes are cleaned for reuse by deoiling them with naphtha, steaming out the excess naphtha, and then roasting in a stream of air to remove carbonaceous matter.

Clay treating See Gray clay treating.

Clay wash Light oil, such as kerosene (kerosine) or naphtha, used to clean fuller's earth after it has been used in a filter.

Clastic Composed of pieces of pre-existing rock.

Cleanup A preparatory step following extraction of a sample media designed to remove components that may interfere with subsequent analytical measurements.

Cloud point The temperature at which paraffin wax or other solid substances begin to crystallize or separate from the solution, imparting a cloudy appearance to the oil when the oil is chilled under prescribed conditions.

Coal An organic rock.

Coalescence The union of two or more droplets to form a larger droplet and, ultimately, a continuous phase.

Coal gas The mixture of volatile products (mainly hydrogen, methane, carbon monoxide, and nitrogen) remaining after removal of water and tar, obtained from carbonization of coal, having a heat content of 400–600 Btu/ft^3.

Coal gasification Production of gas from coal.

Coal liquefaction Conversion of coal to a liquid.

Coal seam A layer, vein, or deposit of coal. A stratigraphic part of the earth's surface containing coal.

Coal tar The condensable distillate containing light, middle, and heavy oils obtained by carbonization of coal. About 8 gal of tar is obtained from each ton of bituminous coal.

Coal tar pitch The specific name for the pitch produced from coal.

COFCAW An EOR process that combines forward combustion and waterflooding.

Cogeneration An energy conversion method by which electrical energy is produced along with steam generated for EOR use.

Coke A gray to black solid carbonaceous material produced from petroleum during thermal processing; characterized by having a high carbon content (95%+ by weight) and a honeycomb type of appearance and is insoluble in organic solvents; a porous, solid residue resulting from the incomplete combustion of coal, used primarily in the steelmaking process.

Coke drum A vessel in which coke is formed and which can cut oil from the process for cleaning.

Coke number Used, particularly in Great Britain, to report the results of the Ramsbottom carbon residue test, which is also referred to as a coke test.

Coke-oven gas A medium-Btu gas, typically 550 Btu/ft^3, produced as a by-product in the manufacture of coke by heating coal at moderate temperatures.

Coker The processing unit in which coking takes place.

Coking A process for the thermal conversion of petroleum in which gaseous, liquid, and solid (coke) products are formed.

Cold pressing The process of separating wax from oil by first chilling (to help form wax crystals) and then filtering under pressure in a plate and frame press.

Cold settling Processing for the removal of wax from high-viscosity stocks, wherein a naphtha solution of the waxy oil is chilled and the wax crystallizes out of the solution.

Color stability The resistance of a petroleum product to color change due to light, aging, etc.

Combustible liquid A liquid with a flash point in excess of 37.8°C (100°F) but below 93.3°C (200°F).

Combustion zone The volume of reservoir rock wherein petroleum is undergoing combustion during enhanced oil recovery.

Composition The general chemical make-up of petroleum.

Completion interval The portion of the reservoir formation placed in fluid communication with the well by selectively perforating the wellbore casing.

Composition map A means of illustrating the chemical make-up of petroleum using chemical and/or physical property data.

Con Carbon See Carbon residue.

Condensate A mixture of light hydrocarbon liquids obtained by condensation of hydrocarbon vapors: predominately butane, propane, and pentane with some heavier hydrocarbons and relatively little methane or ethane; see also Natural gas liquids.

Conductivity A measure of the ease of flow through a fracture, perforation, or pipe.

Conformance The uniformity with which a volume of the reservoir is swept by injection fluids in area and vertical directions.

Conradson carbon residue See Carbon residue.

Contact filtration A process in which finely divided adsorbent clay is used to remove color bodies from petroleum products.

Contaminant A substance that causes deviation from the normal composition of an environment.

Continuous contact coking A thermal conversion process in which petroleum-wetted coke particles move downward into the reactor in which cracking, coking, and drying take place to produce coke, gas, gasoline, and gas oil.

Continuous contact filtration A process to finish lubricants, waxes, or special oils after acid treating, solvent extraction, or distillation.

Conventional crude oil (conventional petroleum) Crude oil that is pumped from the ground and recovered using the energy inherent in the reservoir; also recoverable by application of secondary recovery techniques.

Conventional recovery Primary and/or secondary recovery.

Conversion The thermal treatment of petroleum which results in the formation of new products by the alteration of the original constituents.

Conversion cost The cost of changing a production well to an injection well, or some other change in the function of an oil field installation.

Conversion factor The percentage of feedstock converted to light ends, gasoline, other liquid fuels, and coke.

Copper sweetening processes involving the oxidation of mercaptans to disulfides by oxygen in the presence of cupric chloride.

Core floods Laboratory flow tests through samples (cores) of porous rock.

Cosurfactant A chemical compound, typically, alcohol that enhances the effectiveness of a surfactant.

Cp (centipoise) A unit of viscosity.

Craig-Geffen-Morse method A method for predicting oil recovery by waterflood.

Cracked residua Residua that have been subjected to temperatures above 350°C (660°F) during the distillation process.

Cracking A refining process that uses heat and/or a catalyst to break down high-molecular-weight chemical components into lower-molecular-weight products which can be used as blending components for fuels; the thermal processes by which the constituents of petroleum are converted to lower-molecular-weight products.

Cracking activity See Catalytic activity.

Cracking coil Equipment used for cracking heavy petroleum products consisting of a coil of heavy pipe running through a furnace so that the oil passing through it is subject to high temperature.

Cracking still The combined equipment furnace, reaction chamber, fractionator for the thermal conversion of heavier feedstocks to lighter products.

Cracking temperature The temperature (350°C; 660°F) at which the rate of thermal decomposition of petroleum constituents becomes significant.

Criteria air pollutants Air pollutants or classes of pollutants regulated by the Environmental Protection Agency; the air pollutants are (including VOCs): ozone, carbon monoxide, particulate matter, nitrogen oxides, sulfur dioxide, and lead.

Cross-linking Combining of two or polymer molecules by use of a chemical that mutually bonds with a part of the chemical structure of the polymer molecules.

Crude assay A procedure for determining the general distillation characteristics (e.g., distillation profile, *q.v.*) and other quality information of crude oil.

Crude oil See Petroleum.

Crude scale wax The wax product from the first sweating of the slack wax.

Crude still Distillation equipment in which crude oil is separated into various products.

Cumene A colorless liquid [$C_6H_5CH(CH_3)_2$] used as an aviation gasoline−blending component and as an intermediate in the manufacture of chemicals.

Cut point The boiling-temperature division between distillation fractions of petroleum.

Cutback The term applied to the products from blending heavier feedstocks or products with lighter oils to bring the heavier materials to the desired specifications.

Cutback asphalt Asphalt liquefied by the addition of a volatile liquid such as naphtha or kerosene which, after application and on exposure to the atmosphere, evaporates leaving the asphalt.

Cutting oil An oil to lubricate and cool metal-cutting tools; also called cutting fluid, cutting lubricant.

Cycle stock The product taken from some later stage of a process and recharged (recycled) to the process at some earlier stage.

Cyclic steams injection The alternating injection of steam and production of oil with condensed steam from the same well or wells.

Cyclization The process by which an open-chain hydrocarbon structure is converted to a ring structure, e.g., hexane to benzene.

Cyclone A device for extracting dust from industrial waste gases. It is in the form of an inverted cone into which the contaminated gas enters tangential from the top; the gas is propelled down a helical pathway, and the dust particles are deposited by means of centrifugal force onto the wall of the scrubber.

Cyclone collectors Equipment in which centrifugal force is used to separate particulates from a gas stream; see Cyclone.

Deactivation Reduction in catalyst activity by the deposition of contaminants (e.g., coke, metals) during a process.

Dealkylation The removal of an alkyl group from aromatic compounds.

Deasphaltened oil The fraction of petroleum after the asphaltene constituents have been removed.

Deasphaltening Removal of a solid powdery asphaltene fraction from petroleum by the addition of the low-boiling liquid hydrocarbons such as n-pentane or n-heptane under ambient conditions.

Deasphalting The removal of the asphaltene fraction from petroleum by the addition of a low-boiling hydrocarbon liquid such as n-pentane or n-heptane; more correctly the removal asphalt (tacky, semisolid) from petroleum (as occurs in a refinery asphalt plant) by the addition of liquid propane or liquid butane under pressure.

Debutanization Distillation to separate butane and lighter components from higher boiling components.

Decant oil The highest boiling product from a catalytic cracker; also referred to as slurry oil, clarified oil, or bottoms.

Decarbonizing A thermal conversion process designed to maximize coker gas oil production and minimize coke and gasoline yields; operated at essentially lower temperatures and pressures than delayed coking.

Decoking Removal of petroleum coke from equipment such as coking drums; hydraulic decoking uses high-velocity water streams.

Decolorizing Removal of suspended, colloidal, and dissolved impurities from liquid petroleum products by filtering, adsorption, chemical treatment, distillation, bleaching, etc.

De-ethanization Distillation to separate ethane and lighter components from propane and higher boiling components; also called de-ethanation.

Degradation The loss of desirable physical properties of EOR fluids, e.g., the loss of viscosity of polymer solutions.

Dehydrating agents Substances capable of removing water (drying, q.v.) or the elements of water from another substance.

Dehydrocyclization Any process by which both dehydrogenation and cyclization reactions occur.

Dehydrogenation The removal of hydrogen from a chemical compound; for example, the removal of two hydrogen atoms from butane to make butene(s) as well as the removal of additional hydrogen to produce butadiene.

Delayed coking A coking process in which the thermal reaction are allowed to proceed to completion to produce gaseous, liquid, and solid (coke) products.

Demethanization The process of distillation in which methane is separated from the higher boiling components; also called demethanation.

Density The mass (or weight) of a unit volume of any substance at a specified temperature; see also Specific gravity.

Deoiling Reduction in quantity of liquid oil entrained in solid wax by draining (sweating) or by a selective solvent; see MEK deoiling.

Depentanizer A fractionating column for the removal of pentane and lighter fractions from a mixture of hydrocarbons.

Depropanization Distillation in which lighter components are separated from butanes and higher boiling material; also called depropanation.

Desalting Removal of mineral salts (mostly chlorides) from crude oils.

Descending-bed system Gravity downflow of packed solids contacted with upwardly flowing gases. Sometimes referred to as "fixed-bed" or "moving-bed" system.

Desorption The reverse process of adsorption whereby adsorbed matter is removed from the adsorbent; also used as the reverse of absorption.

Desulfurization The removal of sulfur or sulfur compounds from a feedstock.

Detergent oil Lubricating oil possessing special sludge-dispersing properties for use in internal-combustion engines.

Dewaxing See Solvent dewaxing.

Devolatilization The removal of vaporizable material by the action of heat.

Devolatilized fuel Smokeless fuel; coke that has been reheated to remove all of the volatile material.

Dewatering The removal of water from coal by mechanical equipment such as a vibrating screen, filter, or centrifuge.

Diagenesis The concurrent and consecutive chemical reactions which commence the alteration of organic matter (at temperatures up to 50°C [120°F]) and ultimately result in the formation of petroleum from the marine sediment; see also Catagenesis and Metagenesis.

Diagenetic rock Rock formed by conversion through pressure or chemical reaction from a rock, e.g., sandstone is a diagenetic.

Diesel engine Named for the German engineer Rudolph Diesel, this internal-combustion, compression-ignition engine works by heating fuels and causing them to ignite; can use either petroleum or bio-derived fuel.

Diesel fuel Fuel used for internal combustion in diesel engines; usually that fraction which distills after kerosene.

Diesel cycle A repeated succession of operations representing the idealized working behavior of the fluids in a diesel engine.

Diesel index An approximation of the cetane number of diesel fuel calculated from the density and aniline point.

Diesel knock The result of a delayed period of ignition is long and the accumulated of diesel fuel in the engine.

Differential-strain analysis Measurement of thermal stress relaxation in a recently cut well.

Digester An airtight vessel or enclosure in which bacteria decomposes biomass in water to produce biogas.

Direct hydrogenation Hydrogenation of coal without use of a separate donor solvent hydrogenation step.

Direct-injection engine A diesel engine in which fuel is injected directly into the cylinder.

Dispersion A measure of the fluids flow in a reservoir.

Displacement efficiency The ratio of the amount of oil moved from the zone swept by the reprocess to the amount of oil present in the zone prior to start of the process.

Distillation A process for separating liquids with different boiling points.

Distillation curve See Distillation profile.

Distillation loss The difference, in a laboratory distillation, between the volume of liquid originally introduced into the distilling flask and the sum of the residue and the condensate recovered.

Distillation range The difference between the temperature at the initial boiling point and at the end point, as obtained by the distillation test.

Distillation profile The distillation characteristics of petroleum or petroleum products showing the temperature and the per cent distilled.

Distribution coefficient A coefficient that describes the distribution of a chemical in reservoir fluids, usually defined as the equilibrium concentrations in the aqueous phases.

Doctor solution A solution of sodium plumbite used to treat gasoline or other light petroleum distillates to remove mercaptan sulfur; see also Doctor test.

Doctor sweetening A process for sweetening gasoline, solvents, and kerosene by converting mercaptans to disulfides using sodium plumbite and sulfur.

Doctor test A test used for the detection of compounds in light petroleum distillates which react with sodium plumbite; see also Doctor solution.

Domestic heating oil See No. 2 Fuel Oil.

Donor solvent process A conversion process in which a hydrogen-donor solvent is used in place of or to augment hydrogen.

Downcomer A means of conveying liquid from one tray to the next below in a bubble tray column.

Downdraft gasifier A gasifier in which the product gases pass through a combustion zone at the bottom of the gasifier.

Downhole steam generator A generator installed downhole in an oil well to which oxygen-rich air, fuel, and water are supplied for the purposes of generating steam for it into the reservoir. Its major advantage over a surface steam generating facility is the losses to the wellbore and surrounding formation are eliminated.

Dry, ash-free (daf) basis A coal analysis basis calculated as if moisture and ash were removed.

Drying Removal of a solvent or water from a chemical substance; also referred to as the removal of solvent from a liquid or suspension; the removal of water from coal by thermal drying, screening, or centrifuging.

Dropping point The temperature at which grease passes from a semisolid to a liquid state under prescribed conditions.

Dry gas A gas which does not contain fractions that may easily condense under normal atmospheric conditions.

Dry point The temperature at which the last drop of petroleum fluid evaporates in a distillation test.

Dualayer distillate process A process for removing mercaptans and oxygenated compounds from distillate fuel oils and similar products, using a combination of treatment with concentrated caustic solution and electrical precipitation of the impurities.

Dualayer gasoline process A process for extracting mercaptans and other objectionable acidic compounds from petroleum distillates; see also Dualayer solution.

Dualayer solution A solution which consists of concentrated potassium or sodium hydroxide containing a solubilizer; see also Dualayer gasoline process.

Dubbs cracking An older continuous, liquid-phase thermal cracking process formerly used.

Dull coal Coal that absorbs rather than reflects light, containing mostly durain and fusain lithotypes.

Durain A macroscopic coal constituent (lithotype) that is hard and dull gray in color.

Dykstra–Parsons coefficient An index of reservoir heterogeneity arising from permeability variation and stratification.

E85 An alcohol fuel mixture containing 85 percent ethanol and 15 percent gasoline by volume, and the current alternative fuel of choice of the US government.

Ebullated bed A process in which the catalyst bed is in a suspended state in the reactor by means of a feedstock recirculation pump which pumps the feedstock upwards at sufficient speed to expand the catalyst bed at approximately 35% above the settled level.

Edeleanu process A process for refining oils at low temperature with liquid sulfur dioxide (SO_2), or with liquid sulfur dioxide and benzene; applicable to the recovery of aromatic concentrates from naphtha and heavier petroleum distillates.

Effective viscosity See Apparent viscosity.

Effluent Any contaminating substance, usually a liquid, which enters the environment via a domestic industrial, agricultural, or sewage plant outlet.

Electric desalting A continuous process to remove inorganic salts and other impurities from crude oil by settling out in an electrostatic field.

Electrical precipitation A process using an electrical field to improve the separation of hydrocarbon reagent dispersions. May be used in chemical treating processes on a wide variety of refinery stocks.

Electrofining A process for contacting a light hydrocarbon stream with a treating agent (acid, caustic, doctor, etc.), then assisting the action of separation of the chemical phase from the hydrocarbon phase by an electrostatic field.

Electrolytic mercaptan process A process in which aqueous caustic solution is used to extract mercaptans from refinery streams.

Electrostatic precipitation Separation of liquid or solid particles from a gas stream by the action of electrically charged wires and plates.

Electrostatic precipitators Devices used to trap fine dust particles (usually in the size range 30–60 microns) that operate on the principle of imparting an electric charge to particles in an incoming air stream and which are then collected on an oppositely charged plate across a high-voltage field.

Eluate The solutes, or analytes, moved through a chromatographic column (see *elution*).

Eluent Solvent used to elute sample.

Elution A process whereby a solute is moved through a chromatographic column by a solvent (liquid or gas) or eluent.

Emission control The use gas cleaning processes to reduce emissions.

Emissions Substances discharged into the air during combustion; waste substances released into the air or water.

Emission standard The maximum amount of a specific pollutant permitted to be discharged from a particular source in a given environment.

Emulsion A dispersion of very small drops of one liquid in an immiscible liquid, such as oil in water.

Emulsion breaking The settling or aggregation of colloidal-sized emulsions from suspension in a liquid medium.

Emulsion polymerization A polymerization process that is widely used to produce polymers in the form of emulsions, such as paints and floor polishes.

End-of-pipe emission control The use of specific emission control processes to clean gases after production of the gases.

Endothermic reaction A process in which heat is absorbed.

Energy The capacity of a body or system to do work, measured in joules (SI units); also the output of fuel sources.

Energy balance The difference between the energy produced by a fuel and the energy required to obtain it through agricultural processes, drilling, refining, and transportation.

Energy crops Crops grown specifically for their fuel value; include food crops such as corn and sugarcane, and nonfood crops such as poplar trees and switch grass.

Energy-efficiency ratio A number representing the energy stored in a fuel as compared to the energy required to produce, process, transport, and distribute that fuel.

Energy from biomass The production of energy from biomass.

Engler distillation A standard test for determining the volatility characteristics of a gasoline by measuring the per cent distilled at various specified temperatures.

Enhanced oil recovery (EOR) Petroleum recovery following recovery by conventional (i.e., primary and/or secondary) methods.

Enhanced oil recovery (EOR) process A method for recovering additional oil from a petroleum reservoir beyond that economically recoverable by conventional primary and secondary recovery methods. EOR methods are usually divided into three main categories: (1) *chemical flooding:* injection of water with added chemicals into a petroleum reservoir. The chemical processes include surfactant flooding, polymer flooding, and alkaline flooding, (2) *miscible flooding:* injection into a petroleum reservoir of a material that is miscible, or can become miscible, with the oil in the reservoir. Carbon dioxide, hydrocarbons, and nitrogen are used, (3) *thermal recovery:* injection of steam into a petroleum reservoir, or propagation of a combustion zone through a reservoir by air or oxygen-enriched air injection. The thermal processes include steam drive, cyclic steam injection, and in situ combustion.

Entrained bed A bed of solid particles suspended in a fluid (liquid or gas) at such a rate that some of the solid is carried over (entrained) by the fluid.

Entrained flow system Solids suspended in a moving gas stream and flowing with it.

EPA Environmental Protection Agency.

Equilibrium moisture The moisture capacity of coal at 30°C (86°F) in an atmosphere of 95% relative humidity.

Ester A compound formed by the reaction between an organic acid and an alcohol. Ethoxylated alcohols (i.e., alcohols having ethylene oxide functional groups attached to the alcohol molecule).

Ethanol See Ethyl alcohol.

Ethyl alcohol (ethanol, alcohol, or grain spirit) A clear, colorless, flammable oxygenated hydrocarbon (C_2H_5OH) formed during fermentation of sugars; used as a vehicle fuel by itself (E100 is 100% ethanol by volume), blended with gasoline (E85 is 85% ethanol by volume), or as a gasoline octane enhancer and oxygenate (10% by volume).

Evaporation A process for concentrating nonvolatile solids in a solution by boiling off the liquid portion of the waste stream.

Exinite A microscopic coal constituent (maceral) or maceral group containing spores and cuticles. Appears dark gray in reflected light.

Exothermic reaction A process in which heat is evolved.

Expanding clays Clays that expand or swell on contact with water, e.g., montmorillonite.

Explosive limits The limits of percentage composition of mixtures of gases and air within which an explosion takes place when the mixture is ignited.

Extract The portion of a sample preferentially dissolved by the solvent and recovered by physically separating the solvent.

Extractive distillation The separation of different components of mixtures which have similar vapor pressures by flowing a relatively high-boiling solvent, which is selective for one of the components in the feed, down a distillation column as the distillation proceeds; the selective solvent scrubs the soluble component from the vapor.

Fabric filters Filters made from fabric materials and used for removing particulate matter from gas streams (see Baghouse).

Facies One or more layers of rock that differs from other layers in composition, age, or content.

FAST Fracture-assisted steam flood technology.

Fat oil The bottom or enriched oil drawn from the absorber as opposed to lean oil.

Faujasite A naturally occurring silica–alumina ($SiO_2-Al_2O_3$) mineral.

FCC Fluid catalytic cracking.

FCCU Fluid catalytic cracking unit.

Feedstock Petroleum as it is fed to the refinery; a refinery product that is used as the raw material for another process; the term is also generally applied to raw materials used in other industrial processes; also the biomass used in the creation of a particular biofuel (e.g., corn or sugarcane for ethanol, soybeans, or rapeseed for biodiesel).

Fermentation Conversion of carbon-containing compounds by microorganisms for production of fuels and chemicals such as alcohols, acids, or energy-rich gases.

Ferrocyanide process A regenerative chemical treatment for mercaptan removal using caustic-sodium ferrocyanide reagent.

Fiber products Products derived from fibers of herbaceous and woody plant materials; examples include pulp, composition board products, and wood chips for export.

Field-scale The application of EOR processes to a significant portion of a field.

Filtration The use of an impassable barrier to collect solids but which allows liquids to pass.

Fingering The formation of finger-shaped irregularities at the leading edge of a displacing fluid in a porous medium which moves out ahead of the main body of fluid.

Firedamp An explosive mixture of carbonaceous gases, mainly methane, formed in coal mines by the decomposition of coal.

Fire point The lowest temperature at which, under specified conditions in standardized apparatus, a petroleum product vaporizes sufficiently rapidly to form above its surface an air–vapor mixture which burns continuously when ignited by a small flame.

First contact miscibility See miscibility.

Fischer–Tropsch process A process for synthesizing hydrocarbons and oxygenated chemicals from a mixture of hydrogen and carbon monoxide.

Fixed bed A stationary bed (of catalyst) to accomplish a process (see Fluid bed); see Descending-bed system.

Fixed carbon The combustible residue left after the volatile matter is driven off. In general, the fixed carbon represents that portion of the fuel that must be burned in the solid state.

Five-spot An arrangement or pattern of wells with four injection wells at the comers of a square and a producing well in the center of the square.

Flammability range The range of temperature over which a chemical is flammable.

Flammable A substance that will burn readily.

Flammable liquid A liquid having a flash point below 37.8°C (100°F).

Flammable solid A solid that can ignite from friction or from heat remaining from its manufacture, or which may cause a serious hazard if ignited.

Flash point The lowest temperature to which the product must be heated under specified conditions to give off sufficient vapor to form a mixture with air that can be ignited momentarily by a flame.

Flexible-fuel vehicle (flex-fuel vehicle) A vehicle that can run alternately on two or more sources of fuel; includes cars capable of running on gasoline and gasoline/ethanol mixtures, as well as cars that can run on both gasoline and natural gas.

Flexicoking A modification of the fluid coking process insofar as the process also includes a gasifier adjoining the burner/regenerator to convert excess coke to a clean fuel gas.

Flocculation threshold The point at which constituents of a solution (e.g. asphaltene constituents or coke precursors) will separate from the solution as a separate (solid) phase.

Floc point The temperature at which wax or solids separate as a definite floc.

Flood, flooding The process of displacing petroleum from a reservoir by the injection of fluids.

Flue gases The gaseous products of the combustion process mostly comprised of carbon dioxide, nitrogen, and water vapor.

Fluid bed Use of an agitated bed of inert granular material to accomplish a process in which the agitated bed resembles the motion of a fluid.

Flue gas desulfurization (FGD or scrubbing) The removal of sulfur oxides from stack gases of a coal-fired boiler.

Flue gas Gas from the combustion of fuel, the heating value of which has been substantially spent and which is, therefore, discarded to the flue or stack.

Fluid A reservoir gas or liquid.

Fluid bed A bed (of catalyst) that is agitated by an upward passing gas in such a manner that the particles of the bed simulate the movement of a fluid and has the characteristics associated with a true liquid; c.f. Fixed bed.

Fluid catalytic cracking Cracking in the presence of a fluidized bed of catalyst.

Fluid coking A continuous fluidized solids process that cracks feed thermally over heated coke particles in a reactor vessel to gas, liquid products, and coke.

Fluidized-bed boiler A large, refractory-lined vessel with an air distribution member or plate in the bottom, a hot gas outlet in or near the top, and some provisions for introducing fuel; the fluidized bed is formed by blowing air up through a layer of inert particles (such as sand or limestone) at a rate that causes the particles to go into suspension and continuous motion.

Fluidized-bed combustion A process used to burn low-quality solid fuels in a bed of small particles suspended by a gas stream (usually air that will lift the particles but not blow them out of the vessel). Rapid burning removes some of the offensive by-products of combustion from the gases and vapors that result from the combustion process; solids suspended in space by an upwardly moving gas stream.

Fly ash Particulate matter produced from mineral matter in coal that is converted during combustion to finely divided inorganic material and which emerges from the combustor in the gases.

Foots oil The oil sweated out of slack wax; named from the fact that the oil goes to the foot, or bottom, of the pan during the sweating operation.

Forest residues Material not harvested or removed from logging sites in commercial hardwood and softwood stands as well as material resulting from forest management operations such as precommercial thinnings and removal of dead and dying trees.

Formation An interval of rock with distinguishable geologic characteristics.

Formation volume factor The volume in barrel that one stock tank barrel occupies in the formation at reservoir temperature and with the solution gas that is held in the oil at reservoir pressure.

Fossil fuel resources A gaseous, liquid, or solid fuel material formed in the ground by chemical and physical changes (diagenesis, q.v.) in plant and animal residues over geological time; natural gas, petroleum, coal, and oil shale.

Fouling The accumulation of small, sticky molten particles of ash on a boiler surface.

Fractional composition The composition of petroleum as determined by fractionation (separation) methods.

Fractional distillation The separation of the components of a liquid mixture by vaporizing and collecting the fractions, or cuts, which condense in different temperature ranges.

Fractional flow The ratio of the volumetric flow rate of one fluid phase to the total fluid volumetric flow rate within a volume of rock.

Fractional flow curve The relationship between the fractional flow of one fluid and its saturator during simultaneous flow of fluids through rock.

Fracture A natural or man-made crack in a reservoir rock.

Fracturing The breaking apart of reservoir rock by applying very high fluid pressure at the rock face.

Fractionating column A column arranged to separate various fractions of petroleum by a single distillation and which may be tapped at different points along its length to separate various fractions in the order of their boiling points.

Fractionation The separation of petroleum into the constituent fractions using solvent or adsorbent methods; chemical agents such as sulfuric acid may also be used.

Frasch process A process formerly used for removing sulfur by distilling oil in the presence of copper oxide.

Free moisture (surface moisture) The part of coal moisture that is removed by air-drying under standard conditions approximating atmospheric equilibrium.

Free swelling index A measure of the agglomerating tendency of coal heated to 800°C (1470°F) in a crucible. Coals with a high index are referred to as coking coals; those with a low index are referred to as free-burning coal.

Fuel cell A device that converts the energy of a fuel directly to electricity and heat, without combustion.

Fuel cycle The series of steps required to produce electricity. The fuel cycle includes mining or otherwise acquiring the raw fuel source, processing and cleaning the fuel, transport, electricity generation, waste management, and plant decommissioning.

Fuel oil Also called heating oil is a distillate product that covers a wide range of properties; see also No. 1 to No. 4 Fuel oils.

Fuel wood (fuelwood) Wood used for conversion to some form of energy, primarily for residential use.

Fuller's earth A clay which has high adsorptive capacity for removing color from oils; attapulgus clay is a widely used fuller's earth.

Functional group The portion of a molecule that is characteristic of a family of compounds and determines the properties of these compounds.

Furfural extraction A single-solvent process in which furfural is used to remove aromatic, naphthene, olefin, and unstable hydrocarbons from a lubricating-oil charge stock.

Furnace oil A distillate fuel primarily intended for use in domestic heating equipment.

Fusain A black macroscopic coal constituent (lithotype) that resembles wood charcoal; extremely soft and friable. Also, US Bureau of Mines term for mineral charcoal seen by transmitted light microscopy.

Fusinite A microscopic coal constituent (maceral) with well-preserved cell structure and cell cavities empty or occupied by mineral matter.

Gas cap A part of a hydrocarbon reservoir at the top that will produce only gas.

Gaseous pollutants Gases released into the atmosphere that act as primary or secondary pollutants.

Gasification A chemical or heat process used to convert carbonaceous material (such as coal, petroleum, and biomass) into gaseous components such as carbon monoxide and hydrogen.

Gasifier A device for converting solid fuel into gaseous fuel; in biomass systems, the process is referred to as pyrolitic distillation.

Gasohol A mixture of 10% anhydrous ethanol and 90% gasoline by volume; 7.5% anhydrous ethanol and 92.5% gasoline by volume; or 5.5% anhydrous ethanol and 94.5% gasoline by volume.

Gas oil A petroleum distillate with a viscosity and boiling range between those of kerosine and lubricating oil.

Gas–oil ratio Ratio of the number of cubic feet of gas measured at atmospheric (standard) conditions to barrels of produced oil measured at stock tank conditions.

Gas oil sulfonate Sulfonate made from a specific refinery stream, in this case the gas oil stream.

Gasoline Fuel for the internal-combustion engine that is commonly, but improperly, referred to simply as gas.

Gas purification Gas treatment to remove contaminants such as fly ash, tars, oils, ammonia, and hydrogen sulfide.

Gas reversion A combination of thermal cracking or reforming of naphtha with thermal polymerization or alkylation of hydrocarbon gases carried out in the same reaction zone.

Gas to liquids (GTL) The process of refining natural gas and other hydrocarbons into longer chain hydrocarbons, which can be used to convert gaseous waste products into fuels.

Gel point The point at which a liquid fuel cools to the consistency of petroleum jelly.

Genetically modified organism (GMO) An organism whose genetic material has been modified through recombinant DNA technology, altering the phenotype of the organism to meet desired specifications.

Gilsonite An asphaltite that is >90% bitumen.

Girbotol process A continuous, regenerative process to separate hydrogen sulfide, carbon dioxide, and other acid impurities from natural gas, refinery gas, etc., using mono-, di-, or triethanolamine as the reagent.

Glance pitch An asphaltite.

Glycol-amine gas treating A continuous, regenerative process to simultaneously dehydrate and remove acid gases from natural gas or refinery gas.

Grahamite An asphaltite.

Gravity See API gravity.

Gravity drainage The movement of oil in a reservoir that results from the force of gravity.

Gravity segregation Partial separation of fluids in a reservoir caused by the gravity force acting on differences in density.

Gravity-stable displacement The displacement of oil from a reservoir by a fluid of a different density, where the density difference is utilized to prevent gravity segregation of the injected fluid.

Gray clay treating A fixed-bed, usually fuller's earth, vapor-phase treating process to selectively polymerize unsaturated gum-forming constituents (diolefins) in thermally cracked gasoline.

Grain alcohol See Ethyl alcohol.

Gravimetric Gravimetric methods weigh a residue.

Grease car A diesel-powered automobile rigged post production to run on used vegetable oil.

Greenhouse effect The effect of certain gases in the Earth's atmosphere in trapping heat from the sun.

Greenhouse gases Gases that trap the heat of the sun in the Earth's atmosphere, producing the greenhouse effect. The two major greenhouse gases are water vapor and carbon dioxide. Other greenhouse gases include methane, ozone, chlorofluorocarbons, and nitrous oxide.

Guard bed A bed of an adsorbent (such as, for example, bauxite) that protects a catalyst bed by adsorbing species detrimental to the catalyst.

Gulf HDS process A fixed-bed process for the catalytic hydrocracking of heavy stocks to lower boiling distillates with accompanying desulfurization.

Gulfining A catalytic hydrogen treating process for cracked and straight-run distillates and fuel oils, to reduce sulfur content; improve carbon residue, color, and general stability; and effect a slight increase in gravity.

Gum An insoluble tacky semisolid material formed as a result of the storage instability and/or the thermal instability of petroleum and petroleum products.

HAP(s) Hazardous air pollutant(s).

Hard coal Coal with a heat content greater than 10,260 Btu/lb. on a moist ash-free basis. Includes anthracite, bituminous, and the higher rank subbituminous coals.

Hardness The concentration of calcium and magnesium in brine.

Hardwoods Usually broad-leaved and deciduous trees.

HCPV Hydrocarbon pore volume.

Hearn method A method used in reservoir simulation for calculating a pseudo relative permeability curve that reflects reservoir stratification.

Headspace The vapor space above a sample into which volatile molecules evaporate. Certain methods sample this vapor.

Heating oil See Fuel oil.

Heating value The maximum amount of energy that is available from burning a substance.

Heavy crude See Heavy oil.

Heavy ends The highest boiling portion of a petroleum fraction; see also Light ends.

Heavy fuel oil Fuel oil having a high density and viscosity; generally residual fuel oil such as No. 5 and No 6. fuel oil

Heavy oil (heavy crude) Petroleum having an API gravity of less than 20°; a high boiling coal tar fraction with distillation range usually 250–300°C (480–570°F), containing naphthalene and coal tar bases; oil that is more viscous that conventional crude oil has a lower mobility in the reservoir but can be recovered through a well from the reservoir by the application of a secondary or enhanced recovery methods.

Hectare Common metric unit of area, equal to 2.47 acres. 100 hectares = 1 square kilometer.

Herbaceous Nonwoody type of vegetation, usually lacking permanent strong stems, such as grasses, cereals, and canola (rape).

Heavy petroleum See Heavy oil.

Heteroatom compounds Chemical compounds which contain nitrogen and/or oxygen and/or sulfur and/or metals bound within their molecular structure(s).

Heterogeneity Lack of uniformity in reservoir properties such as permeability.

HF alkylation An alkylation process whereby olefins (C_3, C_4, C_5) are combined with *iso*-butane in the presence of hydrofluoric acid catalyst.

Higgins–Leighton model Stream tube computer model used to simulate waterflood.

High-temperature tar The heavy distillate from the pyrolysis of coal at a temperature of about 800°C (1470°F).

Horton sphere A spherical pressure-type tank used to store a volatile liquid which prevents the excessive evaporation loss that occurs when such products are placed in conventional storage tanks.

Hot filtration test A test for the stability of a petroleum product.

Hot spot An area of a vessel or line wall appreciably above normal operating temperature, usually as a result of the deterioration of an internal insulating liner which exposes the line or vessel shell to the temperature of its contents.

Houdresid catalytic cracking A continuous moving–bed process for catalytically cracking reduced crude oil to produce high-octane gasoline and light distillate fuels.

Houdriflow catalytic cracking A continuous moving–bed catalytic cracking process employing an integrated single vessel for the reactor and regenerator kiln.

Houdriforming A continuous catalytic reforming process for producing aromatic concentrates and high-octane gasoline from low-octane straight naphtha.

Houdry butane dehydrogenation A catalytic process for dehydrogenating light hydrocarbons to their corresponding mono- or diolefins.

Houdry fixed-bed catalytic cracking A cyclic regenerable process for cracking of distillates.

Houdry hydrocracking A catalytic process combining cracking and desulfurization in the presence of hydrogen.

Huff-and-puff A cyclic EOR method in which steam or gas is injected into a production well; after a short shut-in period, oil and the injected fluid are produced through the same well.

Hydration The association of molecules of water with a substance.

Hydraulic fracturing The opening of fractures in a reservoir by high-pressure, high-volume injection of liquids through an injection well.

Hydrocarbonaceous material A material such as bitumen that is composed of carbon and hydrogen with other elements (heteroelements) such as nitrogen, oxygen, sulfur, and metals chemically combined within the structures of the constituents; even though carbon and hydrogen may be the predominant elements, there may be very few true hydrocarbons.

Hydrocarbon compounds Chemical compounds containing only carbon and hydrogen.

Hydrocarbon-producing resource A resource such as coal and oil shale (kerogen) which produces derived hydrocarbons by the application of conversion processes; the hydrocarbons so-produced are not naturally occurring materials.

Hydrocarbon resource Resources such as petroleum and natural gas which can produce naturally occurring hydrocarbons without the application of conversion processes.

Hydrocarbons Organic compounds containing only hydrogen and carbon.

Hydrolysis A chemical reaction in which water reacts with another substance to form one or more new substances.

Hydroconversion A term often applied to hydrocracking.

Hydrocracking A catalytic high-pressure high-temperature process for the conversion of petroleum feedstocks in the presence of fresh and recycled hydrogen; carbon- carbon bonds are cleaved in addition to the removal of heteroatomic species.

Hydrocracking catalyst A catalyst used for hydrocracking which typically contains separate hydrogenation and cracking functions.

Hydrocyclone Hydraulic device for separating suspended solid particles from liquids by centrifugal action. Cyclone action splits the inlet flow, a small part of which exits via the lower cone, the remainder overflowing the top of the cylindrical section. Particles are separated according to their densities, so that the denser particles exit via the cone underflow and less dense particles exit with the overflow.

Hydrodenitrogenation The removal of nitrogen by hydrotreating.

Hydrodesulfurization The removal of sulfur by hydrotreating.

Hydrofining A fixed-bed catalytic process to desulfurize and hydrogenate a wide range of charge stocks from gases through waxes.

Hydroforming A process in which naphtha is passed over a catalyst at elevated temperatures and moderate pressures, in the presence of added hydrogen or hydrogen-containing gases, to form high-octane motor fuel or aromatics.

Hydrogasification Reaction of carbonaceous material such as coal with hydrogen to produce methane.

Hydrogenation Chemical reaction in which hydrogen is added to a substance.

Hydrogen blistering Blistering of steel caused by trapped molecular hydrogen formed as atomic hydrogen during corrosion of steel by hydrogen sulfide.

Hydrogen addition An upgrading process in the presence of hydrogen, e.g., hydrocracking; see Hydrogenation.

Hydrogenation The chemical addition of hydrogen to a material. In nondestructive hydrogenation, hydrogen is added to a molecule only if, and where, unsaturation with respect to hydrogen exists.

Hydrogen transfer The transfer of inherent hydrogen within the feedstock constituents and products during processing.

Hydroprocesses Refinery processes designed to add hydrogen to various products of refining.

Hydroprocessing A term often equally applied to hydrotreating and to hydrocracking; also often collectively applied to both.

Hydrotreating The removal of heteroatomic (nitrogen, oxygen, and sulfur) species by treatment of a feedstock or product at relatively low temperatures in the presence of hydrogen.

Hydrovisbreaking A noncatalytic process, conducted under similar conditions to visbreaking, which involves treatment with hydrogen to reduce the viscosity of the feedstock and produce more stable products than is possible with visbreaking.

Hydropyrolysis A short residence time high-temperature process using hydrogen.

Hyperforming A catalytic hydrogenation process for improving the octane number of naphtha through removal of sulfur and nitrogen compounds.

Hypochlorite sweetening The oxidation of mercaptans in a sour stock by agitation with aqueous, alkaline hypochlorite solution; used where avoidance of free-sulfur addition is desired, because of a stringent copper strip requirements and minimum expense is not the primary object.

Ignitability Characteristic of liquids whose vapors are likely to ignite in the presence of ignition source; also characteristic of nonliquids that may catch fire from friction or contact with water and that burn vigorously.

Illuminating oil Oil used for lighting purposes.

Immiscible Two or more fluids that do not have complete mutual solubility and coexist as separate phases.

Immiscible carbon dioxide displacement Injection of carbon dioxide into an oil reservoir to effect oil displacement under conditions in which miscibility with reservoir oil is not obtained; see Carbon dioxide augmented waterflooding.

Immiscible displacement A displacement of oil by a fluid (gas or water) that is conducted under conditions so that interfaces exist between the driving fluid and the oil.

Immunoassay Portable tests that take advantage of an interaction between an antibody and a specific analyte. Immunoassay tests are semiquantitative and usually rely on color changes of varying intensities to indicate relative concentrations.

Incinerator Any device used to burn solid or liquid residues or wastes as a method of disposal.

Inclined grate A type of furnace in which fuel enters at the top part of a grate in a continuous ribbon, passes over the upper drying section where moisture is removed, and descends into the lower burning section. Ash is removed at the lower part of the grate.

Incompatibility The *immiscibility* of petroleum products and also of different crude oils and which is often reflected in the formation of a separate phase after mixing and/or storage.

Incremental ultimate recovery The difference between the quantity of oil that can be recovered by EOR methods and the quantity of oil that can be recovered by conventional recovery methods.

Indirect hydrogenation The process by which coal is first gasified to make a synthesis gas after which the gas is passed over a catalyst to produce methanol or paraffinic hydrocarbons.

Indirect-injection engine An older model of diesel engine in which fuel is injected into a pre-chamber, partly combusted, and then sent to the fuel-injection chamber.

Indirect liquefaction Conversion of biomass to a liquid fuel through a synthesis gas intermediate step.

Infill drilling Drilling additional wells within an established pattern.

Infrared spectroscopy An analytical technique that quantifies the vibration (stretching and bending) that occurs when a molecule absorbs (heat) energy in the infrared region of the electromagnetic spectrum.

Inhibitor A substance, the presence of which, in small amounts, in a petroleum product prevents or retards undesirable chemical changes from taking place in the product, or in the condition of the equipment in which the product is used.

Inhibitor sweetening A treating process to sweeten gasoline of low mercaptan content, using a phenylenediamine type of inhibitor, air, and caustic.

Initial boiling point The recorded temperature when the first drop of liquid falls from the end of the condenser.

Initial vapor pressure The vapor pressure of a liquid of a specified temperature and zero per cent evaporated.

Injection profile The vertical flow rate distribution of fluid flowing from the wellbore into a reservoir.

Injection well A well in an oil field used for injecting fluids into a reservoir.

Injectivity The relative ease with which a fluid is injected into a porous rock.

In situ In its original place; in the reservoir.

In situ combustion An EOR process consisting of injecting air or oxygen-enriched air into a reservoir under conditions that favor burning part of the in situ petroleum, advancing this burning zone and recovering oil heated from a nearby producing well.

Instability The inability of a petroleum product to exist for periods of time without change to the product.

Integrity Maintenance of a slug or bank at its preferred composition without too much dispersion or mixing.

Interface The thin surface area separating two immiscible fluids that are in contact with each other.

Interfacial film A thin layer of material at the interface between two fluids which differ in composition from the bulk fluids.

Interfacial polymerization A polymerization process that is mainly used in polycondensation reactions with very reactive monomers; one of the reactants, usually an acid chloride, dissolves in an organic solvent (such as benzene or toluene), and the second reactant, a diamine or a diacid, dissolves in water.

Interfacial tension The strength of the film separating two immiscible fluids, e.g., oil and water or microemulsion and oil; measured in dynes (force) per centimeter or millidynes per centimeter.

Interfacial viscosity The viscosity of the interfacial film between two immiscible liquids.

Interference testing A type of pressure transient test in which pressure is measured over time in a closed-in well while nearby wells are produced; flow and communication between wells can sometimes be deduced from an interference test.

Interphase mass transfer The net transfer of chemical compounds between two or more phases.

Iodine number A measure of the iodine absorption by oil under standard conditions; used to indicate the quantity of unsaturated compounds present; also called iodine value.

Ion exchange A means of removing cations or anions from solution onto a solid resin.

Ion exchange capacity A measure of the capacity of a mineral to exchange ions in amount of material per unit weight of solid.

Ions Chemical substances possessing positive or negative charges in solution.

Isocracking A hydrocracking process for conversion of hydrocarbons which operates at relatively low temperatures and pressures in the presence of hydrogen and a catalyst to produce more valuable, lower boiling products.

Isoforming A process in which olefinic naphtha is contacted with an alumina catalyst at high temperature and low pressure to produce isomers of higher octane number.

Iso-Kel process A fixed-bed, vapor-phase isomerization process using a precious metal catalyst and external hydrogen.

Isomate process A continuous, nonregenerative process for isomerizing C_5–C_8 normal paraffin hydrocarbons, using aluminum chloride–hydrocarbon catalyst with anhydrous hydrochloric acid as a promoter.

Isomerate process A fixed-bed isomerization process to convert pentane, heptane, and heptane to high-octane blending stocks.

Isomerization The conversion of a *normal* (straight-chain) paraffin hydrocarbon into an *iso* (branched-chain) paraffin hydrocarbon having the same atomic composition.

Isopach A line on a map designating points of equal formation thickness.

Iso-plus Houdriforming A combination process using a conventional Houdriformer operated at moderate severity, in conjunction with one of three possible alternatives—including the use of an aromatic recovery unit or a thermal reformer; see Houdriforming.

Jet fuel Fuel meeting the required properties for use in jet engines and aircraft turbine engines.

Joule Metric unit of energy, equivalent to the work done by a force of one Newton applied over distance of one meter (= 1 kg m2/s2). One joule (J) = 0.239 calories (1 calorie = 4.187 J).

Kaolinite A clay mineral formed by hydrothermal activity at the time of rock formation or by chemical weathering of rock with high feldspar content; usually associated with intrusive granite rock with high feldspar content.

Kata-condensed aromatic compounds Compounds based on linear condensed aromatic hydrocarbon systems, e.g., anthracene and naphthacene (tetracene).

Kauri butanol number A measurement of solvent strength for hydrocarbon solvents; the higher the kauri-butanol (KB) value, the stronger the solvency; the test method (ASTM D1133) is based on the principle that kauri resin is readily soluble in butyl alcohol but not in hydrocarbon solvents, and the resin solution will tolerate only a certain amount of dilution and is reflected as a cloudiness when the resin starts to come out of solution; solvents such as toluene can be added in a greater amount (and thus have a higher KB value) than weaker solvents like hexane.

Kerogen A complex carbonaceous (organic) material that occurs in sedimentary rock and shale; generally insoluble in common organic solvents.

Kerosene (kerosine) A fraction of petroleum that was initially sought as an illuminant in lamps; a precursor to diesel fuel; a light middle distillate that in various forms is used as aviation turbine fuel or for burning in heating boilers or as a solvent, such as white spirit.

K-factor See Characterization factor.

Kilowatt (kW) A measure of electrical power equal to 1,000 watts. 1 kW = 3412 Btu/hr = 1.341 horsepower.

Kilowatt hour (kWh) A measure of energy equivalent to the expenditure of one kilowatt for one hour. For example, 1 kWh will light a 100-watt light bulb for 10 hours. 1 kWh = 3412 Btu.

Kinematic viscosity The ratio of viscosity to density, both measured at the same temperature.

Knock The noise associated with self-ignition of a portion of the fuel–air mixture ahead of the advancing flame front.

Kriging A technique used in reservoir description for interpolation of reservoir parameters between wells based on random field theory.

LAER Lowest achievable emission rate; the required emission rate in nonattainment permits.

Lamp burning A test of burning oils in which the oil is burned in a standard lamp under specified conditions in order to observe the steadiness of the flame, the degree of encrustation of the wick, and the rate of consumption of the kerosene.

Lamp oil See Kerosene.

Landfill gas A type of biogas that is generated by decomposition of organic material at landfill disposal sites. Landfill gas is approximately 50 percent methane. See also biogas.

Leaded gasoline Gasoline containing tetraethyl lead or other organometallic lead antiknock compounds.

Lean gas The residual gas from the absorber after the condensable gasoline has been removed from the wet gas.

Lean oil Absorption oil from which gasoline fractions have been removed; oil leaving the stripper in a natural-gasoline plant.

Lewis acid A chemical species which can accept an electron pair from a base.

Lewis base A chemical species which can donate an electron pair.

Light ends The lower boiling components of a mixture of hydrocarbons; see also Heavy ends, Light hydrocarbons.

Light hydrocarbons Hydrocarbons with molecular weights less than that of heptane (C_7H_{16}).

Light oil The products distilled or processed from crude oil up to, but not including, the first lubricating-oil distillate; a coal tar and coal gas fraction with distillation range between 80 and 210°C (175−410°F) containing mainly benzene with smaller amounts of toluene and xylene.

Light petroleum Petroleum having an API gravity greater than 20°.

Lignin Structural constituent of wood and (to a lesser extent) other plant tissues, which encrusts the walls and cements the cells together.

Lignite A brownish-black woody-structured coal, lower in fixed carbon and higher in volatile matter and oxygen than either anthracite or bituminous coal. Similar to the "brown coal" of Europe and Australia.

Ligroine (Ligroin) A saturated petroleum naphtha boiling in the range of 20−135° C (68−275° F) and suitable for general use as a solvent; also called benzine or petroleum ether.

Linde copper sweetening A process for treating gasoline and distillates with a slurry of clay and cupric chloride.

Liquefaction The conversion of coal into nearly mineral-free hydrocarbon liquids or low-melting solids by a process of direct or indirect hydrogenation at elevated temperatures and pressures and separation of liquid products from residue by either filtration or distillation or both.

Liquefied petroleum gas (LPG) Propane, butane, or mixtures thereof, gaseous at atmospheric temperature and pressure, held in the liquid state by pressure to facilitate storage, transport, and handling; a mixture of propane and butane.

Liquid chromatography A chromatographic technique that employs a liquid mobile phase.

Liquid/liquid extraction An extraction technique in which one liquid is shaken with or contacted by an extraction solvent to transfer molecules of interest into the solvent phase.

Liquid petrolatum See White oil.

Liquid sulfur dioxide−benzene process A mixed-solvent process for treating lubricating-oil stocks to improve viscosity index; also used for dewaxing.

Lithology The geological characteristics of the reservoir rock.

Live cull A classification that includes live cull trees; when associated with volume, it is the net volume in live cull trees that are 5.0 inches in diameter and larger.

Live steam Steam coming directly from a boiler before being utilized for power or heat.

Liver The intermediate layer of dark-colored, oily material, insoluble in weak acid and in oil, which is formed when acid sludge is hydrolyzed.

Logging residues The unused portions of growing-stock and nongrowing-stock trees cut or killed logging and left in the woods.

Lorenz coefficient A permeability heterogeneity factor.

Low-Btu gas A nitrogen-rich gas with a heat content of 100−200 Btu/ft^3 produced in gasification processes using air as the oxygen source. The air-blown form of producer gas.

Lower phase microemulsion A microemulsion phase containing a high concentration of water that, when viewed in a test tube, resides near the bottom with oil phase on top.

LPG See Liquefied petroleum gas.

Lube See Lubricating oil.

Lube cut A fraction of crude oil of suitable boiling range and viscosity to yield lubricating oil when completely refined; also referred to as lube oil distillates or lube stock.

Lubricating oil A fluid lubricant used to reduce friction between bearing surfaces.

M85 An alcohol fuel mixture containing 85 percent methanol and 15 percent gasoline by volume. Methanol is typically made from natural gas but can also be derived from the fermentation of biomass.

MACT Maximum achievable control technology. Applies to major sources of hazardous air pollutants.

Mahogany acids Oil-soluble sulfonic acids formed by the action of sulfuric acid on petroleum distillates. They may be converted to their sodium soaps (mahogany soaps) and extracted from the oil with alcohol for use in the manufacture of soluble oils, rust preventives, and special greases. The calcium and barium soaps of these acids are used as detergent additives in motor oils; see also Brown acids and Sulfonic acids.

Major source A source that has a potential to emit for a regulated pollutant that is at or greater than an emission threshold set by regulations.

Maltenes That fraction of petroleum that is soluble in, for example, pentane or heptane; deasphaltened oil; also the term arbitrarily assigned to the pentane-soluble portion of petroleum that is relatively high boiling (>300°C, 760 mm) (see also Petrolenes).

Marine engine oil Oil used as a crankcase oil in marine engines.

Marine gasoline Fuel for motors in marine service.

Marine sediment The organic biomass from which petroleum is derived.

Marsh An area of spongy waterlogged ground with large numbers of surface water pools. Marshes usually result from: (1) an impermeable underlying bedrock; (2) surface deposits of glacial boulder clay; (3) a basin-like topography from which natural drainage is poor; (4) very heavy rainfall in conjunction with a correspondingly low evaporation rate; (5) low-lying land, particularly at estuarine sites at or below sea level.

Marx-Langenheim model Mathematical equations for calculating heat transfer in a hot water or steam flood.

Mass spectrometer An analytical technique that *fractures* organic compounds into characteristic "fragments" based on functional groups that have a specific mass-to-charge ratio.

Mayonnaise Low-temperature sludge; a black, brown, or gray deposit having a soft, mayonnaise-like consistency; not recommended as a food additive!

MCL Maximum contaminant level as dictated by regulations.

Medicinal oil Highly refined, colorless, tasteless, and odorless petroleum oil used as a medicine in the nature of an internal lubricant; sometimes called liquid paraffin.

Megawatt (MW) A measure of electrical power equal to one million watts (1,000 kW).

Membrane technology Gas separation processes utilizing membranes that permit different components of a gas to diffuse through the membrane at significantly different rates.

Methanation A process for catalytic conversion of 1 mole of carbon monoxide and 3 moles of hydrogen to 1 mole of methane and 1 mole of water.

Methanol A fuel typically derived from natural gas but which can be produced from the fermentation of sugars in biomass.

MDL See Method detection limit.

MEK-(methyl ethyl ketone) A colorless liquid ($CH_3COCH_2CH_3$) used as a solvent; as a chemical intermediate; and in the manufacture of lacquers, celluloid, and varnish removers.

MEK deoiling A wax-deoiling process in which the solvent is generally a mixture of methyl ethyl ketone and toluene.

MEK dewaxing A continuous solvent-dewaxing process in which the solvent is generally a mixture of methyl ethyl ketone and toluene.

MEOR Microbial-enhanced oil recovery.

Methanol See Methyl alcohol.

Method detection limit The smallest quantity or concentration of a substance that the instrument can measure.

Methyl t-butyl ether An ether added to gasoline to improve its octane rating and to decrease gaseous emissions; see Oxygenate.

Mercapsol process A regenerative process for extracting mercaptans, utilizing aqueous sodium (or potassium) hydroxide containing mixed cresols as solubility promoters.

Mercaptans Organic compounds having the general formula R-SH.

Metagenesis The alteration of organic matter during the formation of petroleum that may involve temperatures above 200°C (390°F); see also Catagenesis and Diagenesis.

Methyl alcohol (methanol; wood alcohol) A colorless, volatile, inflammable, and poisonous alcohol (CH_3OH) traditionally formed by destructive distillation of wood or, more recently, as a result of synthetic distillation in chemical plants.

Methyl t-butyl ether A highly refined high-octane light distillate used in the blending of petrol.

Methyl ethyl ketone See MEK.

Mica A complex aluminum silicate mineral that is transparent, tough, flexible, and elastic.

Micellar fluid (surfactant slug) An aqueous mixture of surfactants, cosurfactants, salts, and hydrocarbons. The term micellar is derived from the word micelle, which is a submicroscopic aggregate of surfactant molecules and associated fluid.

Micelle The structural entity by which asphaltene constituents are dispersed in petroleum.

Microcarbon residue The carbon residue determined using a themogravimetric method. See also Carbon residue.

Microcrystalline wax Wax extracted from certain petroleum residua and having a finer and less apparent crystalline structure than paraffin wax.

Microemulsion A stable, finely dispersed mixture of oil, water, and chemicals (surfactants and alcohols).

Microemulsion or micellar/emulsion flooding An augmented waterflooding technique in which a surfactant system is injected in order to enhance oil displacement toward producing wells.

Microorganisms Animals or plants of microscopic size, such as bacteria.

Microscopic displacement efficiency The efficiency with which an oil displacement process removes the oil from individual pores in the rock.

Mid-boiling point The temperature at which approximately 50 per cent of a material has distilled under specific conditions.

Middle distillate Distillate boiling between the kerosene and lubricating-oil fractions.

Middle oil (carbolic or creosote oil) A coal tar fraction with a distillation range of 200–270°C (390–520°F), containing mainly naphthalene, phenol, and cresols.

Middle-phase microemulsion A microemulsion phase containing a high concentration of both oil and water that, when viewed in a test tube, resides in the middle with the oil phase above it and the water phase below it.

Migration (primary) The movement of hydrocarbons (oil and natural gas) from mature, organic-rich source rocks to a point where the oil and gas can collect as droplets or as a continuous phase of liquid hydrocarbon.

Migration (secondary) The movement of the hydrocarbons as a single, continuous fluid phase through water-saturated rocks, fractures, or faults followed by accumulation of the oil and gas in sediments (traps, *q.v.*) from which further migration is prevented.

Million 1×10^6.

Mill residue Wood and bark residues produced in processing logs into lumber, plywood, and paper.

Mineral hydrocarbons Petroleum hydrocarbons, considered *mineral* because they come from the earth rather than from plants or animals.

Mineral oil The older term for petroleum; the term was introduced in the 19th century as a means of differentiating petroleum (rock oil) from whale oil which, at the time, was the predominant illuminant for oil lamps.

Minerals Naturally occurring inorganic solids with well-defined crystalline structures.

Mineral seal oil A distillate fraction boiling between kerosene and gas oil.

Mineral wax Yellow to dark brown, solid substances that occur naturally and are composed largely of paraffins; usually found associated with considerable mineral matter, as a filling in veins and fissures or as an interstitial material in porous rocks.

Minimum miscibility pressure (MMP) See Miscibility.

Miscibility An equilibrium condition, achieved after mixing two or more fluids, which is characterized by the absence of interfaces between the fluids: (1) *first-contact miscibility:* miscibility in the usual sense, whereby two fluids can be mixed in all proportions without any interfaces forming. Example: At room temperature and pressure, ethyl alcohol and water are first-contact miscible. (2) *multiple-contact miscibility (dynamic miscibility):* miscibility that is developed by repeated enrichment of one fluid phase with components from a second fluid phase with which it comes into contact. (3) *minimum miscibility* pressure: the minimum pressure above which two fluids become miscible at a given temperature, or can become miscible, by dynamic processes.

Miscible flooding See EOR process.

Miscible fluid displacement (miscible displacement) Is an oil displacement process in which an alcohol, a refined hydrocarbon, a condensed petroleum gas, carbon dioxide, liquefied natural gas, or even exhaust gas is injected into an oil reservoir, at pressure levels such that the injected gas or fluid and reservoir oil are miscible; the process may include the concurrent, alternating, or subsequent injection of water.

Mitigation Identification, evaluation, and cessation of potential impacts of a process product or by-product.

Mixed-phase cracking The thermal decomposition of higher boiling hydrocarbons to gasoline components.

Mobility A measure of the ease with which a fluid moves through reservoir rock; the ratio of rock permeability to apparent fluid viscosity.

Mobility buffer The bank that protects a chemical slug from water invasion and dilution and assures mobility control.

Mobility control Ensuring that the mobility of the displacing fluid or bank is equal to or less than that of the displaced fluid or bank.

Mobility ratio Ratio of mobility of an injection fluid to mobility of fluid being displaced.

Modified alkaline flooding The addition of a cosurfactant and/or polymer to the alkaline flooding process.

Modified diesel engine Traditional diesel engines must be modified to heat the oil before it reaches the fuel injectors in order to handle straight vegetable oil; a diesel engine can run on vegetable oil but without engine modification, the oil must first be converted to biodiesel.

Modified naphtha insolubles (MNI) An insoluble fraction obtained by adding naphtha to petroleum; usually the naphtha is modified by adding paraffin constituents; the fraction might be equated to asphaltenes *if* the naphtha is equivalent to n-heptane, but usually it is not.

Moisture The total moisture content of a sample customarily determined by adding the moisture loss obtained when air-drying the sample and the measured moisture content of the dried sample. Moisture does not represent all of the water present in coal, as water of decomposition (combined water) and hydration is not given off under standardized test conditions.

Moisture content (MC) The weight of the water contained in wood, usually expressed as a percentage of weight, either oven-dry or as received.

Moisture content, dry basis Moisture content expressed as a percentage of the weight of oven-wood, i.e., [(weight of wet sample - weight of dry sample)/weight of dry sample] x 100.

Moisture content, wet basis Moisture content expressed as a percentage of the weight of wood as-received, i.e., [(weight of wet sample - weight of dry sample)/weight of wet sample] x 100.

Molecular sieve A synthetic zeolite mineral having pores of uniform size; it is capable of separating molecules, on the basis of their size, structure, or both, by absorption or sieving.

Molten bath gasifier A reaction system in which coal and air or oxygen with steam are contacted underneath a pool of liquid iron, ash, or salt.

Motor Octane Method A test for determining the knock rating of fuels for use in spark-ignition engines; see also Research Octane Method.

Moving-bed catalytic cracking A cracking process in which the catalyst is continuously cycled between the reactor and the regenerator.

Moving-bed system See Descending-bed system.

MSDS Material safety data sheet.

MTBE See Methyl t-butyl ether.

Municipal wastes Residential, commercial, and institutional postconsumer wastes contain a significant proportion of plant-derived organic material that constitutes a renewable energy resource; waste paper, cardboard, construction and demolition wood waste, and yard wastes are examples of biomass resources in municipal wastes.

NAAQS National Ambient Air Quality Standards; standards exist for the pollutants known as the criteria air pollutants: nitrogen oxides (NO_x), sulfur oxides (SO_x), lead, ozone, particulate matter, less than 10 microns in diameter, and carbon monoxide (CO).

Naft Pre-Christian era (Greek) term for naphtha.

Napalm A thickened gasoline used as an incendiary medium that adheres to the surface it strikes.

Naphtha A generic term applied to refined, partly refined, or unrefined petroleum products and liquid products of natural gas, the majority of which distills below 240°C (464°F); the volatile fraction of petroleum which is used as a solvent or as a precursor to gasoline.

Naphthenes Cycloparaffins.

Native asphalt See Bitumen.

Natural asphalt See Bitumen.

Natural gas The naturally occurring gaseous constituents that are found in many petroleum reservoirs; also there are also those reservoirs in which natural gas may be the sole occupant; a naturally occurring gas with a heat content over 1000 Btu/ft^3, consisting mainly of methane but also containing smaller amounts of the C_2−C_4 hydrocarbons as well as nitrogen, carbon dioxide, and hydrogen sulfide.

Natural gas liquids (NGLs) The hydrocarbon liquids that condense during the processing of hydrocarbon gases that are produced from the oil or gas reservoir; see also Natural gasoline.

Natural gasoline A mixture of liquid hydrocarbons extracted from natural gas suitable for blending with refinery gasoline.

Natural gasoline plant A plant for the extraction of fluid hydrocarbon, such as gasoline and liquefied petroleum gas, from natural gas.

NESHAP National Emissions Standards for Hazardous Air Pollutants; emission standards for specific source categories that emit or have the potential to emit one or more hazardous air pollutants; the standards are modeled on the best practices and most effective emission reduction methodologies in use at the affected facilities.

Neutralization A process for reducing the acidity or alkalinity of a waste stream by mixing acids and bases to produce a neutral solution; also known as pH adjustment.

Neutral oil A distillate lubricating oil with viscosity usually not above 200 sec at 100° F.

Neutralization number The weight, in milligrams, of potassium hydroxide needed to neutralize the acid in 1 g of oil; an indication of the acidity of an oil.

Nitrogen oxides (NOx) Products of combustion that contribute to the formation of smog and ozone.

Nonasphaltic road oil Any of the nonhardening petroleum distillates or residual oils used as dust layers. They have sufficiently low viscosity to be applied without heating and, together with asphaltic road oils, are sometimes referred to as dust palliatives.

Nonattainment area A geographical area that does not meet NAAQS for criteria air pollutants (see also Attainment area).

Nonforest land Land that has never supported forests and lands formerly forested where use of timber management is precluded by development for other uses; if intermingled in forest areas, unimproved roads and nonforest strips must be more than 120 feet wide, and clearings, etc., must be more than 1 acre in area to qualify as nonforest land.

Nonionic surfactant A surfactant molecule containing no ionic charge.

Non-Newtonian A fluid that exhibits a change of viscosity with flow rate.

NOx Oxides of nitrogen; see Nitrogen oxides.

Nuclear magnetic resonance spectroscopy An analytical procedure that permits the identification of complex molecules based on the magnetic properties of the atoms they contain.

No. 1 Fuel oil Very similar to kerosene and is used in burners where vaporization before burning is usually required and a clean flame is specified.

No. 2 Fuel oil Also called domestic heating oil; has properties similar to diesel fuel and heavy jet fuel; used in burners where complete vaporization is not required before burning.

No. 4 Fuel oil A light industrial heating oil and is used where preheating is not required for handling or burning; there are two grades of No. 4 fuel oil, differing in safety (flash point) and flow (viscosity) properties.

No. 5 Fuel oil A heavy industrial fuel oil which requires preheating before burning.

No. 6 Fuel oil A heavy fuel oil and is more commonly known as Bunker C oil when it is used to fuel ocean-going vessels; preheating is always required for burning this oil.

Observation wells Wells that are completed and equipped to measure reservoir conditions and/or sample reservoir fluids, rather than to inject Dr produce reservoir fluids.

Octane barrel yield A measure used to evaluate fluid catalytic cracking processes; defined as (RON + MON)/2 times the gasoline yield, where RON is the research octane number and MON is the motor octane number.

Octane number A number indicating the antiknock characteristics of gasoline.

Oil bank See Bank.

Oil breakthrough (time) The time at which the oil−water bank arrives at the producing well.

Oil from tar sand Synthetic crude oil.

Oil mining Application of a mining method to the recovery of bitumen.

Oil originally in place (OOIP, oil originally in place, original oil in place) The quantity of petroleum existing in a reservoir before oil recovery operations begin.

Oils That portion of the maltenes that is not adsorbed by a surface-active material such as clay or alumina.

Oil sand See Tar sand.

Oil shale A fine-grained impervious sedimentary rock which contains an organic material called kerogen.

Olefin Synonymous with *alkene*.

OOIP See Oil originally in place.

Open-loop biomass Biomass that can be used to produce energy and bioproducts even though it was not grown specifically for this purpose; include agricultural livestock waste and residues from forest harvesting operations and crop harvesting.

Optimum salinity The salinity at which a middle-phase microemulsion containing equal concentrations of oil and water results from the mixture of a micellar fluid (surfactant slug) with oil.

Organic sedimentary rocks Rocks containing organic material such as residues of plant and animal remains/decay.

Overhead That portion of the feedstock which is vaporized and removed during distillation.

Override The gravity-induced flow of a lighter fluid in a reservoir above another heavier fluid.

Oxidation A process which can be used for the treatment of a variety of inorganic and organic substances.

Oxidized asphalt See Air-blown asphalt.

Oxygenate A substance which, when added to gasoline, increases the amount of oxygen in that gasoline blend; includes fuel ethanol, methanol, and methyl tertiary butyl ether (MTBE).

Oxygenated fuels Ethanol is an oxygenate, meaning that it adds oxygen to the fuel mixture—more oxygen helps the fuel burn more completely thereby reducing the amount of harmful emissions from the tailpipe; a fuel such as ethanol-blended gasoline that contains a high oxygen content is called *oxygenated*.

Ozokerite (Ozocerite) A naturally occurring wax; when refined also known as ceresin.

Oxygenate An oxygen-containing compound that is blended into gasoline to improve its octane number and to decrease gaseous emissions; includes fuel ethanol, methanol, and methyl tertiary butyl ether (MTBE).

Oxygenated gasoline Gasoline with added ethers or alcohols, formulated according to the Federal Clean Air Act to reduce carbon monoxide emissions during winter months.

Oxygen scavenger A chemical which reacts with oxygen in injection water, used to prevent degradation of polymer.

Pale oil A lubricating oil or a process oil refined until its color, by transmitted light, is straw to pale yellow.

Palm oil A form of vegetable oil obtained from the fruit of the oil palm tree; widely used feedstock for traditional biodiesel production; the palm oil and palm kernel oil are composed of fatty acids, esterified with glycerol just like any ordinary fat.

Paraffinum liquidum See Liquid petrolatum.

Paraffin wax The colorless, translucent, highly crystalline material obtained from the light lubricating fractions of paraffin crude oils (wax distillates).

Particle density The density of solid particles.

Particulate A small, discrete mass of solid or liquid matter that remains individually dispersed in gas or liquid emissions.

Particulate emissions Particles of a solid or liquid suspended in a gas or the fine particles of carbonaceous soot and other organic molecules discharged into the air during combustion.

Particulate matter (particulates) Particles in the atmosphere or on a gas stream that may be organic or inorganic and originate from a wide variety of sources and processes.

Particle size distribution The particle size distribution (of a catalyst sample) expressed as a percent of the whole.

Partitioning In chromatography, the physical act of a solute having different affinities for the stationary and mobile phases.

Partition ratios, K The ratio of total analytical concentration of a solute in the stationary phase, CS, to its concentration in the mobile phase, CM.

Pattern The areal pattern of injection and producing wells selected for a secondary or enhanced recovery project.

Pattern life The length of time a flood pattern participates in oil recovery.

Peat Partially carbonized plant matter, formed by slow decay in water.

Penex process A continuous, nonregenerative process for isomerization of C_5 and/or C_6 fractions in the presence of hydrogen (from reforming) and a platinum catalyst.

Pentafining A pentane isomerization process using a regenerable platinum catalyst on a silica–alumina support and requiring outside hydrogen.

Pepper sludge The fine particles of sludge produced in acid treating which may remain in suspension.

Pericondensed aromatic compounds Compounds based on angular condensed aromatic hydrocarbon systems, e.g., phenanthrene, chrysene, picene, etc.

Permeability The ease of flow of the water through the rock.

Petrol A term commonly used in some countries for gasoline.

Petrolatum A semisolid product, ranging from white to yellow in color, produced during refining of residual stocks; see Petroleum jelly.

Petrolenes The term applied to that part of the pentane-soluble or heptane-soluble material that is low boiling (<300°C, <570°F, 760 mm) and can be distilled without thermal decomposition (see also Maltenes).

Petroleum (crude oil) A naturally occurring mixture of gaseous, liquid, and solid hydrocarbon compounds usually found trapped deep underground beneath impermeable cap rock and above a lower dome of sedimentary rock such as shale; most petroleum reservoirs occur in sedimentary rocks of marine, deltaic, or estuarine origin.

Petroleum asphalt See Asphalt.

Petroleum ether See Ligroine.

Petroleum jelly A translucent, yellowish to amber or white, hydrocarbon substance (melting point: 38–54°C) having almost no odor or taste, derived from petroleum and used principally in medicine and pharmacy as a protective dressing and as a substitute for fats in ointments and cosmetics; also used in many types of polishes and in lubricating greases, rust preventives, and modeling clay; obtained by dewaxing heavy lubricating-oil stocks.

Petroleum refinery See Refinery.

Petroleum refining A complex sequence of events that result in the production of a variety of products.

Petroleum sulfonate A surfactant used in chemical flooding prepared by sulfonating selected crude oil fractions.

Petroporphyrins See Porphyrins.

Phase A separate fluid that coexists with other fluids; gas, oil, water, and other stable fluids such as microemulsions are all called phases in EOR research.

Phase behavior The tendency of a fluid system to form phases as a result of changing temperature, pressure, or the bulk composition of the fluids or of individual fluid phases.

Phase diagram A graph of phase behavior. In chemical flooding a graph showing the relative volume of oil, brine, and sometimes one or more microemulsion phases. In carbon dioxide flooding, conditions for formation of various liquid, vapor, and solid phases.

Phase properties Types of fluids, compositions, densities, viscosities, and relative amounts of oil, microemulsion, or solvent, and water formed when a micellar fluid (surfactant slug) or miscible solvent (e.g., CO2) is mixed with oil.

Phase separation The formation of a separate phase that is usually the prelude to coke formation during a thermal process; the formation of a separate phase as a result of the instability/incompatibility of petroleum and petroleum products.

pH adjustment Neutralization.

Phosphoric acid polymerization A process using a phosphoric acid catalyst to convert propene, butene, or both, to gasoline or petrochemical polymers.

Photoionization A gas chromatographic detection system that utilizes a *detector (PID)* ultraviolet lamp as an ionization source for analyte detection. It is usually used as a selective detector by changing the photon energy of the ionization source.

Photosynthesis Process by which chlorophyll-containing cells in green plants concert incident light to chemical energy, capturing carbon dioxide in the form of carbohydrates.

PINA analysis A method of analysis for paraffins, *iso*-paraffins, naphthenes, and aromatics.

PIONA analysis A method of analysis for paraffins, *iso*-paraffins, olefins, naphthenes, and aromatics.

Pipeline gas A methane-rich gas with a heat content of 950–1050 Btu/ft^3 compressed to 1000 psi.

Pipestill A still in which heat is applied to the oil while being pumped through a coil or pipe arranged in a suitable firebox.

Pipestill gas The most volatile fraction that contains most of the gases that are generally dissolved in the crude. Also known as pipestill light ends.

Pipestill light ends See *Pipestill gas.*

Pitch The nonvolatile, brown to black, semisolid to solid viscous product from the destructive distillation of many bituminous or other organic materials, especially coal.

Platforming A reforming process using a platinum-containing catalyst on an alumina base.

PNA A polynuclear aromatic compound.

NA A polynuclear aromatic compound.

PNA analysis A method of analysis for paraffins, naphthenes, and aromatics.

Polar aromatics Resins; the constituents of petroleum that are predominantly aromatic in character and contain polar (nitrogen, oxygen, and sulfur) functions in their molecular structure(s).

Pollutant A chemical (or chemicals) introduced into the land water and air systems of that is (are) not indigenous to these systems; also an indigenous chemical (or chemicals) introduced into the land water and air systems in amounts greater than the natural abundance.

Pollution The introduction into the land water and air systems of a chemical or chemicals that are not indigenous to these systems or the introduction into the land water and air systems of indigenous chemicals in greater-than-natural amounts.

Polyacrylamide Very-high-molecular-weight material used in polymer flooding.

Polycyclic aromatic hydrocarbons (PAHs) Polycyclic aromatic hydrocarbons are a suite of compounds comprised of two or more condensed aromatic rings. They are found in many petroleum mixtures, and they are predominantly introduced to the environment through natural and anthropogenic combustion processes.

Polyforming A process charging both C_3 and C_4 gases with naphtha or gas oil under thermal conditions to produce gasoline.

Polymer In EOR, any very-high-molecular weight material that is added to water to increase viscosity for polymer flooding.

Polymer augmented waterflooding Waterflooding in which organic polymers are injected with the water to improve areal and vertical sweep efficiency.

Polymer gasoline The product of polymerization of gaseous hydrocarbons to hydrocarbons boiling in the gasoline range.

Polymerization The combination of two olefin molecules to form a higher molecular weight paraffin.

Polymer A large molecule (macromolecule) composed of repeating structural units (monomers) typically connected by covalent chemical bonds; polymers are classified by the characteristics of the reactions by which they are formed.

Polymer stability The ability of a polymer to resist degradation and maintain its original properties.

Polynuclear aromatic compound An aromatic compound having two or more fused benzene rings, e.g. naphthalene, phenanthrene.

Polysulfide treating A chemical treatment used to remove elemental sulfur from refinery liquids by contacting them with a nonregenerable solution of sodium polysulfide.

PONA analysis A method of analysis for paraffins (P), olefins (O), naphthenes (N), and aromatics (A).

Pore diameter The average pore size of a solid material, e.g., catalyst.

Pore space A small hole in reservoir rock that contains fluid or fluids; a 4-inch cube of reservoir rock may contain millions of interconnected pore spaces.

Pore volume Total volume of all pores and fractures in a reservoir or part of a reservoir; also applied to catalyst samples.

Porosity The percentage of rock volume available to contain water or other fluid.

Porphyrins Organometallic constituents of petroleum that contain vanadium or nickel; the degradation products of chlorophyll that became included in the protopetroleum.

Positive bias A result that is incorrect and too high.

Possible reserves Reserves where there is an even greater degree of uncertainty but about which there is some information.

Potential reserves Reserves based upon geological information about the types of sediments where such resources are likely to occur and they are considered to represent an educated guess.

Pour point The lowest temperature at which oil will pour or flow when it is chilled without disturbance under definite conditions.

Powerforming A fixed-bed naphtha-reforming process using a regenerable platinum catalyst.

Power-law exponent An exponent used to model the degree of viscosity change of some non-Newtonian liquids.

Precipitation number The number of milliliters of precipitate formed when 10 ml of lubricating oil is mixed with 90 ml of petroleum naphtha of a definite quality and centrifuged under definitely prescribed conditions.

Preflush A conditioning slug injected into a reservoir as the first step of an EOR process.

Pressure cores Cores cut into a special coring barrel that maintains reservoir pressure when brought to the surface; this prevents the loss of reservoir fluids that usually accompanies a drop in pressure from reservoir to atmospheric conditions.

Pressure gradient Rate of change of pressure with distance.

Pressure maintenance Augmenting the pressure (and energy) in a reservoir by injecting gas and/or water through one or more wells.

Pressure pulse test A technique for determining reservoir characteristics by injecting a sharp pulse of pressure in one well and detecting it surrounding wells.

Pressure transient testing Measuring the effect of changes in pressure at one well on other well in a field.

Primary oil recovery Oil recovery utilizing only naturally occurring forces; recovery of crude oil from the reservoir using the inherent reservoir energy.

Primary wood–using mill A mill that converts round wood products into other wood products; common examples are sawmills that convert saw logs into lumber and pulp mills that convert pulpwood round wood into wood pulp.

Primary structure The chemical sequence of atoms in a molecule.

Primary tracer A chemical that, when injected into a test well, reacts with reservoir fluids form a detectable chemical compound.

Probable reserves Mineral reserves that are nearly certain but about which a slight doubt exists.

Process heat Heat used in an industrial process rather than for space heating or other housekeeping purposes.

Producer gas Mainly carbon monoxide with smaller amounts of hydrogen, methane, and variable nitrogen, obtained from partial combustion of coal or coke in air or oxygen, having a heat content of 110–160 Btu/ft^3 (air combustion) or 400-500 Btu/ft^3 (oxygen combustion).

Producibility The rate at which oil or gas can be produced from a reservoir through a wellbore.

Producing well A well in an oil field used for removing fluids from a reservoir.

Propane asphalt See Solvent asphalt.

Propane deasphalting Solvent deasphalting using propane as the solvent.

Propane decarbonizing A solvent extraction process used to recover catalytic cracking feed from heavy fuel residues

Propane dewaxing A process for dewaxing lubricating oils in which propane serves as a solvent.

Propane fractionation A continuous extraction process employing liquid propane as the solvent; a variant of propane deasphalting.

Protopetroleum A generic term used to indicate the initial changes have occurred to the precursors of petroleum.

Proved reserves Mineral reserves that have been positively identified as recoverable with current technology.

PSD Prevention of significant deterioration.

PTE Potential to emit; the maximum capacity of a source to emit a pollutant, given its physical or operation design and considering certain controls and limitations.

Pulpwood Round wood, whole-tree chips, or wood residues that are used for the production of wood pulp.

Pulse-echo ultrasonic borehole televiewer Well-logging system wherein a pulsed, narrow acoustic beam scans the well as the tool is pulled up the borehole; the amplitude of the reflecting beam is displayed on a cathode-ray tube resulting in a pictorial representation of wellbore.

Purge and trap A chromatographic sample introduction technique in volatile components that are purged from a liquid medium by bubbling gas through it. The components are then concentrated by "trapping" them on a short intermediate column, which is subsequently heated to drive the components on to the analytical column for separation.

Purge gas Typically helium or nitrogen used to remove analytes from the sample matrix in purge/trap extractions.

Pyrobitumen See Asphaltoid.

Pyrolysis Exposure of a feedstock to high temperatures in an oxygen-poor environment; the thermal decomposition of biomass at high temperatures (greater than 400°F, or 200°C) in the absence of air; the end product of pyrolysis is a mixture of solids (char), liquids (oxygenated oils), and gases (methane, carbon monoxide, and carbon dioxide) with proportions determined by operating temperature, pressure, oxygen content, and other conditions.

Pyrolysis oil A bio-oil produced by fast pyrolysis of biomass; typically a dark brown, mobile liquid containing much of the energy content of the original biomass, with a heating value about half that of conventional fuel oil; conversion of raw biomass to pyrolysis oil represents a considerable increase in energy density and it can thus represent a more efficient form in which to transport it.

Pyrophoric Substances that catch fire spontaneously in air without an ignition source.

Quad One quadrillion Btu (10^{15} Btu) = 1.055 exajoules (EJ), or approximately 172 million barrels of oil equivalent.

Quadrillion 1×10^{15}.

Quench The sudden cooling of hot material discharging from a thermal reactor.

RACT Reasonably Available Control Technology standards; implemented in areas of nonattainment to reduce emissions of volatile organic compounds and nitrogen oxides.

Raffinate That portion of the oil which remains undissolved in a solvent refining process.

Ramsbottom carbon residue See Carbon residue

Rank A complex property of coals that is descriptive of their degree of coalification (i.e., the stage of metamorphosis of the original vegetal material in the increasing sequence peat, lignite, subbituminous, bituminous, and anthracite).

Rapeseed (*Brassica napus*; rape, oilseed rape or canola) A bright yellow flowering member of the family *Brassicaceae* (mustard or cabbage family); a traditional feedstock used for biodiesel production; canola is a name taken from *Canada oil* due to the fact that much of the development of the oil was performed in Canada; see Colza.

Rapeseed oil Food-grade oil produced from rapeseed.

Raw materials Minerals extracted from the earth prior to any refining or treating.

Recoverable reserves Reserves that can be removed by current technology, considering economic, legal, political, and social variables.

Recovery boiler A pulp mill boiler in which lignin and spent cooking liquor (black liquor) is burned to generate steam.

Recycle ratio the volume of recycle stock per volume of fresh feed; often expressed as the volume of recycle divided by the total charge.

Recycle stock The portion of a feedstock which has passed through a refining process and is recirculated through the process.

Recycling The use or reuse of chemical waste as an effective substitute for a commercial products or as an ingredient or feedstock in an industrial process.

Reduced crude A residual product remaining after the removal, by distillation or other means, of an appreciable quantity of the more volatile components of crude oil.

Refinery A series of integrated unit processes by which petroleum can be converted to a slate of useful (salable) products.

Refinery gas A gas (or a gaseous mixture) produced as a result of refining operations.

Refining The processes by which petroleum is distilled and/or converted by application of a physical and chemical processes to form a variety of products are generated.

Reformate The liquid product of a reforming process.

Reformed gasoline Gasoline made by a reforming process.

Reforming The conversion of hydrocarbons with low-octane numbers into hydrocarbons having higher octane numbers; e.g., the conversion of an n-paraffin into an iso-paraffin.

Reformulated gasoline (RFG) Gasoline designed to mitigate smog production and to improve air quality by limiting the emission levels of certain chemical compounds such as benzene and other aromatic derivatives; often contains oxygenates.

Refractory lining A lining, usually of ceramic, capable of resisting and maintaining high temperatures.

Refuse-derived fuel (RDF) Fuel prepared from municipal solid waste; noncombustible materials such as rocks, glass, and metals are removed, and the remaining combustible portion of the solid waste is chopped or shredded.

Reid vapor pressure A measure of the volatility of liquid fuels, especially gasoline.

Regeneration The reactivation of a catalyst by burning off the coke deposits.

Regenerator A reactor for catalyst reactivation.

Relative permeability The permeability of rock to gas, oil, or water, when any two or more are present, expressed as a fraction of the sir phase permeability of the rock.

Renewable energy sources Solar, wind, and other nonfossil fuel energy sources.

Rerunning The distillation of an oil which has already been distilled.

Research Octane Method A test for determining the knock rating, in terms octane numbers of fuels for use in spark-ignition engines; see also Motor Octane Method.

Reserves Well-identified resources that can be profitably extracted and utilized with existing technology.

Reservoir A rock formation below the earth's surface containing petroleum or natural gas; a domain where a pollutant may reside for an indeterminate time.

Reservoir simulation Analysis and prediction of reservoir performance with a computer model.

Residual asphalt See Straight-run asphalt.

Residual fuel oil Obtained by blending the residual product(s) from various refining processes with suitable diluent(s) (usually middle distillates) to obtain the required fuel oil grades.

Residual oil See Residuum; petroleum remaining in situ after oil recovery.

Residual resistance factor The reduction in permeability of rock to water caused by the adsorption of polymer.

Residues Bark and woody materials that are generated in primary wood−using mills when round wood products are converted to other products.

Residuum (resid; *pl.* residua) The residue obtained from petroleum after nondestructive distillation has removed all the volatile materials from crude oil, e.g., an atmospheric (345°C, 650°F+) residuum.

Resins That portion of the maltenes that is adsorbed by a surface-active material such as clay or alumina; the fraction of deasphaltened oil that is insoluble in liquid propane but soluble in n-heptane.

Resistance factor A measure of resistance to flow of a polymer solution relative to the resistance to flow of water.

Resource The total amount of a commodity (usually a mineral but can include nonminerals such as water and petroleum) that has been estimated to be ultimately available.

Retention The loss of chemical components due to adsorption onto the rock's surface, precipitation, or to trapping within the reservoir.

Retention time The time it takes for an eluate to move through a chromatographic system and reach the detector. Retention times are reproducible and can therefore be compared to a standard for analyte identification.

Rexforming A process combining platforming with aromatics extraction, wherein low-octane raffinate is recycled to the platformer.

Rich oil Absorption oil containing dissolved natural gasoline fractions.

Riser The part of the bubble-plate assembly which channels the vapor and causes it to flow downward to escape through the liquid; also the vertical pipe where fluid catalytic cracking reactions occur.

Rock asphalt Bitumen which occurs in formations that have a limiting ratio of bitumen-to-rock matrix.

Rock matrix The granular structure of a rock or porous medium.

Room-and-pillar mining A mining method in which a designated area is divided into regular-shaped coal pillars through the parallel development of entries and cross-cuts. After the area is so developed, the remaining pillars are mined by slicing them into smaller pillars.

Rotation Period of years between establishment of a stand of timber and the time when it is considered ready for final harvest and regeneration.

Round wood products Logs and other round timber generated from harvesting trees for industrial or consumer use.

Run-of-the-river reservoirs Reservoirs with a large rate of flow-through compared to their volume.

Salinity The concentration of salt in water.

Sand A course granular mineral mainly comprising quartz grains that are derived from the chemical and physical weathering of rocks rich in quartz, notably sandstone and granite.

Sand face The cylindrical wall of the wellbore through which the fluids must flow to or from the reservoir.

Sandstone A sedimentary rock formed by compaction and cementation of sand grains; can be classified according to the mineral composition of the sand and cement.

SARA analysis A method of fractionation by which petroleum is separated into saturates, aromatics, resins, and asphaltene fractions.

SARA separation See SARA analysis.

Saturated steam Steam at boiling temperature for a given pressure.

Saturates Paraffins and cycloparaffins (naphthenes).

Saturation The ratio of the volume of a single fluid in the pores to pore volume, expressed as a percent and applied to water, oil, or gas separately; the sum of the saturations of each fluid in a pore volume is 100 percent.

Saybolt Furol viscosity The time, in seconds (Saybolt Furol Seconds, SFS), for 60 ml of fluid to flow through a capillary tube in a Saybolt Furol viscometer at specified temperatures between 70 and 210° F; the method is appropriate for high-viscosity oils such as transmission, gear, and heavy fuel oils.

Saybolt Universal viscosity The time, in seconds (Saybolt Universal Seconds, SUS), for 60 ml of fluid to flow through a capillary tube in a Saybolt Universal viscometer at a given temperature.

Scale wax The paraffin derived by removing the greater part of the oil from slack wax by sweating or solvent deoiling.

Screen factor A simple measure of the viscoelastic properties of polymer solutions.

Screening guide A list of reservoir rock and fluid properties critical to an EOR process.

Scrubber A device that uses water and chemicals to clean air pollutants from combustion exhaust; see Flue gas desulfurization.

Scrubbing Purifying a gas by washing with water or chemical; less frequently, the removal of entrained materials.

Seam Underground layer of coal or other mineral of any thickness.

Secondary oil recovery Application of energy (e.g., waterflooding) to recovery of crude oil from a reservoir after the yield of crude oil from primary recovery diminishes.

Secondary pollutants A pollutant (chemical species) produced by interaction of a primary pollutant with another chemical or by dissociation of a primary pollutant or by other effects within a particular ecosystem.

Secondary recovery Oil recovery resulting from injection of water, or an immiscible gas at moderate pressure, into a petroleum reservoir after primary depletion.

Secondary structure The ordering of the atoms of a molecule in space relative to each other.

Secondary tracer The product of the chemical reaction between reservoir fluids and an injected primary tracer.

Secondary wood–processing mills A mill that uses primary wood products in the manufacture of finished wood products, such as cabinets, moldings, and furniture.

Sediment An insoluble solid formed as a result of the storage instability and/or the thermal instability of petroleum and petroleum products.

Sedimentary Formed by or from deposits of sediments, especially from sand grains or silts transported from their source and deposited in water, as sandstone and shale; or from calcareous remains of organisms, as limestone.

Sedimentary strata Typically consist of mixtures of clay, silt, sand, organic matter, and various minerals; formed by or from deposits of sediments, especially from sand grains or silts transported from their source and deposited in water, such as sandstone and shale; or from calcareous remains of organisms, such as limestone.

Selective solvent A solvent which, at certain temperatures and ratios, will preferentially dissolve more of one component of a mixture than of another and thereby permit partial separation.

Separation process An upgrading process in which the constituents of petroleum are separated, usually without thermal decomposition, e.g., distillation and deasphalting.

Separator-Nobel dewaxing A solvent (tricholoethylene) dewaxing process.

Separatory funnel Glassware shaped like a funnel with a stoppered rounded top and a valve at the tapered bottom, used for liquid/liquid separations.

Shear Mechanical deformation or distortion, or partial destruction of a polymer molecule as it flows at a high rate.

Shear rate A measure of the rate of deformation of a liquid under mechanical stress.

Shear-thinning The characteristic of a fluid whose viscosity decreases as the shear rate increases.

Shell fluid catalytic cracking A two-stage fluid catalytic cracking process in which the catalyst is regenerated.

Shell still A still formerly used in which the oil was charged into a closed, cylindrical shell and the heat required for distillation was applied to the outside of the bottom from a firebox.

Sidestream A liquid stream taken from any one of the intermediate plates of a bubble tower.

Sidestream stripper A device used to perform further distillation on a liquid stream from any one of the plates of a bubble tower, usually by the use of steam.

Single well tracer A technique for determining residual oil saturation by injecting an ester, allowing it to hydrolyze; following dissolution of some of the reaction product in residual oil the injected solutions produced back and analyzed.

Slack wax The soft, oily crude wax obtained from the pressing of paraffin distillate or wax distillate.

Slime A name used for petroleum in ancient texts.

Slim tube testing Laboratory procedure for the determination of minimum miscibility pressure using long, small-diameter, sand-packed, oil-saturated, stainless steel tube.

Sludge A semisolid to solid product which results from the storage instability and/or the thermal instability of petroleum and petroleum products.

Slug A quantity of fluid injected into a reservoir during enhanced oil recovery.

Slurry A mixture of pulverized insoluble material and water.

Slurry hydroconversion process A process in which the feedstock is contacted with hydrogen under pressure in the presence of a catalytic coke-inhibiting additive.

Slurry phase reactors Tanks into which wastes, nutrients, and microorganisms are placed.

Slurry pipeline A pipeline that can transport a coal–water mixture for long distances.

Smoke point A measure of the burning cleanliness of jet fuel and kerosine.

Sodium hydroxide treatment See Caustic wash.

Sodium plumbite A solution prepared from a mixture of sodium hydroxide, lead oxide, and distilled water; used in making the doctor test for light oils such as gasoline and kerosine.

Softwood Generally, one of the botanical groups of trees that in most cases have needle-like or scale-like leaves; the conifers; also the wood produced by such trees; the term has no reference to the actual softness of the wood; the botanical name for softwoods is gymnosperms.

Solubility parameter A measure of the solvent power and polarity of a solvent.

Solution polymerization A polymerization process in which an organic solvent dissolves the monomer.

Solutizer-steam regenerative process A chemical treating process for extracting mercaptans from gasoline or naphtha, using solutizers (potassium isobutyrate, potassium alkyl phenolate) in strong potassium hydroxide solution.

Solvent A liquid in which certain kinds of molecules dissolve. While they typically are liquids with low boiling points, they may include high-boiling liquids, supercritical fluids, or gases.

Solvent asphalt The asphalt produced by solvent extraction of residua or by light hydrocarbon (propane) treatment of a residuum or an asphaltic crude oil.

Solvent deasphalting A process for removing asphaltic and resinous materials from reduced crude oils, lubricating-oil stocks, gas oils, or middle distillates through the extraction or precipitant action of low-molecular-weight hydrocarbon solvents; see also Propane deasphalting.

Solvent decarbonizing See Propane decarbonizing.

Solvent deresining See Solvent deasphalting.

Solvent dewaxing A process for removing wax from oils by means of solvents usually by chilling a mixture of solvent and waxy oil, filtration or by centrifuging the wax which precipitates, and solvent recovery.

Solvent extraction A process for separating liquids by mixing the stream with a solvent that is immiscible with part of the waste but that will extract certain components of the waste stream.

Solvent gas An injected gaseous fluid that becomes miscible with oil under reservoir conditions and improves oil displacement.

Sonic log A well log based on the time required for sound to travel through rock, useful in determining porosity.

Solvent naphtha A refined naphtha of restricted boiling range used as a solvent; also called petroleum naphtha; petroleum spirits.

Solvent refining See Solvent extraction.

Sonication A physical technique employing ultrasound to intensely vibrate a sample media in extracting solvent and to maximize solvent/analyte interactions.

Sour crude oil Crude oil containing an abnormally large amount of sulfur compounds; see also Sweet crude oil.

SOx Oxides of sulfur.

Soxhlet extraction An extraction technique for solids in which the sample is repeatedly contacted with solvent over several hours, increasing extraction efficiency.

Soy (soy oil) A vegetable oil pressed from soybeans.

Soy diesel A general term for biodiesel which accentuates the renewable nature of biodiesel; popular in soy-producing regions.

Spontaneous ignition Ignition of a fuel, such as coal, under normal atmospheric conditions; usually induced by climatic conditions.

Specific gravity The mass (or weight) of a unit volume of any substance at a specified temperature compared to the mass of an equal volume of pure water at a standard temperature; see also Density.

Spent catalyst Catalyst that has lost much of its activity due to the deposition of coke and metals.

Stabilization The removal of volatile constituents from a higher boiling fraction or product (q.v. stripping); the production of a product which, to all intents and purposes, does not undergo any further reaction when exposed to the air.

Stabilizer A fractionating tower for removing light hydrocarbons from an oil to reduce vapor pressure particularly applied to gasoline.

Stack gas The product gas evolved during complete combustion of a fuel.

Stand (of trees) A tree community that possesses sufficient uniformity in composition, constitution, age, spatial arrangement, or condition to be distinguishable from adjacent communities.

Standpipe The pipe by which catalyst is conveyed between the reactor and the regenerator.

Starch A molecule composed of long chains of a-glucose molecules linked together (repeating unit $C_{12}H_{16}O_5$); these linkages occur in chains of alpha-1,4 linkages with branches formed as a result of alpha-1,6 linkages; widely distributed in the vegetable kingdom and is stored in all grains and tubers (swollen underground plant stems); this polymer is highly amorphous, making it more readily attacked by human and animal enzyme systems and broken down into glucose; gross heat of combustion: Qv(gross)=7560 Btu/lb.

Stationary phase In chromatography, the porous solid or liquid phase through which an introduced sample passes. The different affinities the stationary phase has for a sample allow the components in the sample to be separated or resolved.

Steam cracking A conversion process in which the feedstock is treated with superheated steam.

Steam distillation Distillation in which vaporization of the volatile constituents is effected at a lower temperature by introduction of steam (open steam) directly into the charge.

Steam drive injection (steam injection) EOR process in which steam is continuously injected into one set of wells (injection wells) or other injection source to effect oil displacement toward and production from a second set of wells (production wells); steam stimulation of production wells is *direct steam stimulation* whereas steam drive by steam injection to increase production from other wells is *indirect steam stimulation*.

Steam stimulation Injection of steam into a well and the subsequent production of oil from the same well.

Steam turbine A device for converting energy of high-pressure steam (produced in a boiler) into mechanical power which can then be used to generate electricity.

Stiles method A simple approximate method for calculating oil recovery by waterflood that assumes separate layers (stratified reservoirs) for the permeability distribution.

Storage stability (or storage instability) The ability (inability) of a liquid to remain in storage over extended periods of time without appreciable deterioration as measured by gum formation and the depositions of insoluble material (sediment).

Straight-run A fraction distilled from crude oil without thermal alteration.

Straight-run asphalt The asphalt produced by the nondestructive distillation of asphaltic crude oil.

Straight-run products Obtained from a distillation unit and used without further treatment.

Straight vegetable oil (SVO) Any vegetable oil that has not been optimized through the process of transesterification.

Strata Layers including the solid iron-rich inner core, molten outer core, mantle, and crust of the earth.

Straw oil Pale paraffin oil of straw color used for many process applications.

Stripper well A well that produces (strips from the reservoir) oil or gas.

Stripping A means of separating volatile components from less volatile ones in a liquid mixture by the partitioning of the more volatile materials to a gas phase of air or steam (q.v. stabilization).

Subbituminous coal A glossy-black weathering and nonagglomerating coal that is lower in fixed carbon than bituminous coal, with more volatile matter and oxygen.

Substitute natural gas See Synthetic natural gas.

Sulfonic acids Acids obtained by petroleum or a petroleum product with strong sulfuric acid.

Sulfuric acid alkylation An alkylation process in which olefins (C_3, C_4, and C_5) combine with *iso*-butane in the presence of a catalyst (sulfuric acid) to form branched chain hydrocarbons used especially in gasoline blending stock.

Supercritical fluid An extraction method where the extraction fluid is present at a pressure and temperature above its critical point.

Superheated steam Steam which is hotter than boiling temperature for a given pressure.

Surface active material A chemical compound, molecule, or aggregate of molecules with physical properties that cause it to adsorb at the interface between *two* immiscible liquids, resulting in a reduction of interfacial tension or the formation of a microemulsion.

Surfactant A type of chemical, characterized as one that reduces interfacial resistance to mixing between oil and water or changes the degree to which water wets reservoir rock.

Suspension polymerization A polymerization process in which the monomer is first dispersed in a liquid, such as water, and mechanical agitation keeps the monomer dispersed.

Suspensoid catalytic cracking A nonregenerative cracking process in which cracking stock is mixed with slurry of catalyst (usually clay) and cycle oil and passed through the coils of a heater.

Sustainable An ecosystem condition in which biodiversity, renewability, and resource productivity are maintained over time.

SW-846 An EPA multivolume publication entitled *Test Methods for Evaluating Solid Waste, Physical/Chemical Methods*; the official compendium of analytical and sampling methods that have been evaluated and approved for use in complying with the RCRA regulations and that functions primarily as a guidance document setting forth acceptable, although not required, methods for the regulated and regulatory communities to use in responding to RCRA-related sampling and analysis requirements. SW-846 changes over time as new information and data are developed.

Sweated wax A crude wax freed from oil by having been passed through a sweater.

Sweating The separation of paraffin oil and low-melting wax from paraffin wax.

Sweep efficiency The ratio of the pore volume of reservoir rock contacted by injected fluids to the total pore volume of reservoir rock in the project area. *(See also* areal sweep efficiency *and* vertical sweep efficiency.)

Sweet crude oil Crude oil containing little sulfur; see also Sour crude oil.

Sweetened gas Gas from which acid (sour) gases such as H_2 S and CO_2 have been removed.

Sweetening The process by which petroleum products are improved in odor and color by oxidizing or removing the sulfur-containing and unsaturated compounds.

Swelling Increase in the volume of crude oil caused by absorption of EOR fluids, especially carbon dioxide. Also increase in volume of clays when exposed to brine.

Swept zone The volume of rock that is effectively swept by injected fluids.

Syncrude Synthetic crude oil produced by pyrolysis or hydrogenation of coal or coal extracts.

Syngas See Synthesis gas.

Synthesis gas (syngas) Approximately 2:1 molar mixture of hydrogen and carbon monoxide with varying amounts of carbon dioxide.

Synthetic crude oil (syncrude) A hydrocarbon product produced by the conversion of coal, oil shale, or tar sand bitumen that resembles conventional crude oil; can be refined in a petroleum refinery.

Synthetic ethanol Ethanol produced from ethylene, a petroleum by-product.

Synthetic natural gas (substitute natural gas) Pipeline-quality gas that is interchangeable with natural gas (mainly methane).

Tail gas Residual gas leaving a process.

Tar The volatile, brown to black, oily, viscous product from the destructive distillation of many bituminous or other organic materials, especially coal; a name used for petroleum in ancient texts.

Target analyte Target analytes are compounds that are required analytes in U S EPA analytical methods. BTEX and PAHs are examples of petroleum-related compounds that are target analytes in U S EPA Methods.

Tar sand A formation in which the bituminous material (bitumen) is found as a filling in veins and fissures in fractured rocks or impregnating relatively shallow sand, sandstone, and limestone strata; a sandstone reservoir that is impregnated with a heavy, extremely viscous, black hydrocarbonaceous, petroleum-like material that cannot be retrieved through a well by conventional or enhanced oil recovery techniques; (FE 76-4): The several rock types that contain an extremely viscous hydrocarbon which is not recoverable in its natural state by conventional oil well production methods including currently used enhanced recovery techniques; see also Bituminous sand.

Tertiary structure The three-dimensional structure of a molecule.

Tetraethyl lead (TEL) an organic compound of lead, $Pb(CH_3)_4$, which, when added in small amounts, increases the antiknock quality of gasoline.

Thermal coke the carbonaceous residue formed as a result of a noncatalytic thermal process; the Conradson carbon residue; the Ramsbottom carbon residue.

Thermal cracking A process which decomposes, rearranges, or combines hydrocarbon molecules by the application of heat, without the aid of catalysts.

Thermal polymerization A thermal process to convert light hydrocarbon gases into liquid fuels.

Thermal process Any refining process which utilizes heat, without the aid of a catalyst.

Thermal recovery See EOR process.

Thermal reforming A process using heat (but no catalyst) to effect molecular rearrangement of low-octane naphtha into gasoline of higher antiknock quality.

Thermal stability (thermal instability) The ability (inability) of a liquid to withstand relatively high temperatures for short periods of time without the formation of carbonaceous deposits (sediment or coke).

Thermochemical conversion Use of heat to chemically change substances from one state to another, e.g., to make useful energy products.

Thermofor catalytic cracking A continuous, moving-bed catalytic cracking process.

Thermofor catalytic reforming A reforming process in which the synthetic, bead-type catalyst of coprecipitated chromia (Cr_2O_3) and alumina (Al_2O_3) flows down through the reactor concurrent with the feedstock.

Thermofor continuous percolation A continuous clay treating process to stabilize and decolorize lubricants or waxes.

Thief zone Any geologic stratum not intended to receive injected fluids in which significant amounts of injected fluids are lost; fluids may reach the thief zone due to an improper completion or a faulty cement job.

Thin layer chromatography (TLC) A chromatographic technique employing a porous medium of glass coated with a stationary phase. An extract is spotted near the bottom of the medium and placed in a chamber with solvent (mobile phase). The solvent moves up the medium and separates the components of the extract, based on affinities for the medium and solvent.

Timberland Forest land that is producing or is capable of producing crops of industrial wood, and that is not withdrawn from timber utilization by statute or administrative regulation.

Time-lapse logging The repeated use of calibrated well logs to quantitatively observe changes in measurable reservoir properties over time.

Ton (short ton) 2,000 pounds.

Tonne (imperial ton, long ton, shipping ton) 2,240 pounds; equivalent to 1,000 kilograms or in crude oil terms about 7.5 barrels of oil.

Topped crude Petroleum that has had volatile constituents removed up to a certain temperature, e.g., 250°C+ (480°F+) topped crude; not always the same as a residuum.

Topping The distillation of crude oil to remove light fractions only

Topping cycle A cogeneration system in which electric power is produced first. The reject heat from power production is then used to produce useful process heat.

Topping and back pressure turbines Turbines which operate at exhaust pressure considerably higher than atmospheric (noncondensing turbines); often multistage with relatively high efficiency.

Total petroleum hydrocarbons (TPHs) The family of several hundred chemical compounds that originally come from petroleum.

Tower Equipment for increasing the degree of separation obtained during the distillation of oil in a still.

Town gas A gaseous mixture of coal gas and carbureted water gas manufactured from coal with a heat content of 600 Btu/ft^3.

TPH E Gas chromatographic test for TPH extractable organic compounds.

TPH V Gas chromatographic test for TPH volatile organic compounds.

TPH-D(DRO) Gas chromatographic test for TPH diesel-range organics.

TPH-G(GRO) Gas chromatographic test for TPH gasoline-range organics.

Trace element Those elements that occur at very low levels in a given system.

Tracer test A technique for determining fluid flow paths in a reservoir by adding small quantities of easily detected material (often radioactive) to the flowing fluid and monitoring their appearance at production wells. Also used in cyclic injection to appraise oil saturation.

Transesterification The chemical process in which an alcohol reacts with the triglycerides in vegetable oil or animal fats, separating the glycerin and producing biodiesel.

Transmissibility (transmissivity) An index of producibility of a reservoir or zone, the product of permeability and layer thickness.

Traps Sediments in which oil and gas accumulate from which further migration is prevented.

Traveling grate A type of furnace in which assembled links of grates are joined together in a perpetual belt arrangement. Fuel is fed in at one end and ash is discharged at the other.

Treatment Any method, technique, or process that changes the physical and/or chemical character of petroleum.

Triaxial borehole seismic survey A technique for detecting the orientation of hydraulically induced fractures, wherein a tool holding three mutually seismic detectors is clamped in the borehole during fracturing; fracture orientation is deduced through analysis of the detected microseismic perpendicular events that are generated by the fracturing process.

Trickle hydrodesulfurization A fixed-bed process for desulfurizing middle distillates.

Trillion 1×10^{12}.

True boiling point (true boiling range) The boiling point (boiling range) of a crude oil fraction or a crude oil product under standard conditions of temperature and pressure.

Tube-and-tank cracking An older liquid-phase thermal cracking process.

Tumbling-bed gasifier An apparatus in which coal is lifted vertically in a revolving cylinder and dropped through an axially flowing stream of oxygen and steam.

Turbine A machine for converting the heat energy in steam or high-temperature gas into mechanical energy. In a turbine, a high velocity flow of steam or gas passes through successive rows of radial blades fastened to a central shaft.

Turn down ratio The lowest load at which a boiler will operate efficiently as compared to the boiler's maximum design load.

Two-stage gasification Partial gasification or pyrolysis in a first step followed by essentially complete gasification of the resultant char in a second step.

Ultimate analysis Elemental composition; the analytical percentage by weight of coal carbon, hydrogen, nitrogen, sulfur, oxygen, and ash.

Ultimate recovery The cumulative quantity of oil that will be recovered when revenues from further production no longer justify the costs of the additional production.

Ultrafining A fixed-bed catalytic hydrogenation process to desulfurize naphtha and upgrade distillates by essentially removing sulfur, nitrogen, and other materials.

Ultraforming A low-pressure naphtha-reforming process employing onstream regeneration of a platinum-on-alumina catalyst and producing high yields of hydrogen and high-octane-number reformate.

Unassociated molecular weight The molecular weight of asphaltenes in a nonassociating (polar) solvent, such as dichlorobenzene, pyridine, or nitrobenzene.

Unconformity A surface of erosion that separates younger strata from older rocks.

Unifining A fixed-bed catalytic process to desulfurize and hydrogenate refinery distillates.

Unisol process A chemical process for extracting mercaptan sulfur and certain nitrogen compounds from sour gasoline or distillates using regenerable aqueous solutions of sodium or potassium hydroxide containing methanol.

Universal viscosity See Saybolt Universal viscosity.

Unresolved complex The thousands of compounds that a gas chromatograph *mixture (UCM)* is unable to fully separate.

Unstable Usually refers to a petroleum product that has more volatile constituents present or refers to the presence of olefin and other unsaturated constituents.

UOP alkylation A process using hydrofluoric acid (which can be regenerated) as a catalyst to unite olefins with *iso*-butane.

UOP copper sweetening A fixed-bed process for sweetening gasoline by converting mercaptans to disulfides by contact with ammonium chloride and copper sulfate in a bed.

UOP fluid catalytic cracking A fluid process of using a reactor-over-regenerator design.

Upgrading The conversion of petroleum to value-added salable products.

Upper-phase microemulsion A microemulsion phase containing a high concentration of oil that, when viewed in a test tube, resides on top of a water phase.

Urea dewaxing A continuous dewaxing process for producing low-pour-point oils, and using urea which forms a solid complex (adduct) with the straight-chain wax paraffins in the stock; the complex is readily separated by filtration.

Vacuum distillation Distillation under reduced pressure; a secondary distillation process which uses a partial vacuum to lower the boiling point of residues from primary distillation and extract further blending components.

Vacuum residuum A residuum obtained by distillation of a crude oil under vacuum (reduced pressure); that portion of petroleum which boils above a selected temperature such as 510°C (950°F) or 565°C (1050°F).

Vapor-phase cracking A high-temperature, low-pressure conversion process.

Vapor-phase hydrodesulfurization A fixed-bed process for desulfurization and hydrogenation of naphtha.

Vertical sweep efficiency The fraction of the layers or vertically distributed zones of a reservoir that are effectively contacted by displacing fluids.

Visbreaking A process for reducing the viscosity of heavy feedstocks by controlled thermal decomposition.

Viscosity A measure of the ability of a liquid to flow or a measure of its resistance to flow; the force required to move a plane surface of area 1 meter^2 over another parallel plane surface 1 meter away at a rate of 1 meter/sec when both surfaces are immersed in the fluid.

VGC (viscosity-gravity constant) An index of the chemical composition of crude oil defined by the general relation between specific gravity, sg, at 60° F and Saybolt Universal viscosity, SUV, at 100° F: a = 10sg - 1.0752 log (SUV - 38)/10sg - log (SUV - 38) The constant, a, is low for the paraffin crude oils and high for the naphthenic crude oils.

VI (Viscosity index) An arbitrary scale used to show the magnitude of viscosity changes in lubricating oils with changes in temperature.

Viscosity-gravity constant See VGC.

Viscosity index See VI.

VOC (VOCs) Volatile organic compound(s); volatile organic compounds are regulated because they are precursors to ozone; carbon-containing gases, and vapors from incomplete gasoline combustion and from the evaporation of solvents.

Volatile compounds A relative term that may mean (1) any compound that will purge, (2) any compound that will elute before the solvent peak (usually those < C6), or (3) any compound that will not evaporate during a solvent removal step.

Volatile matter Hydrogen, carbon monoxide, methane, tar, other hydrocarbons, carbon dioxide, and water obtained on coal pyrolysis.

Volumetric sweep The fraction of the total reservoir volume within a flood pattern that is effectively contacted by injected fluids.

VSP Vertical seismic profiling, a method of conducting seismic surveys in the borehole for detailed subsurface information.

Waste streams Unused solid or liquid by-products of a process.

Waste vegetable oil (WVO) Grease from the nearest fryer which is filtered and used in modified diesel engines or converted to biodiesel through the process of transesterification and used in any ol' diesel car.

Water-cooled vibrating grate A boiler grate made up of a tuyere grate surface mounted on a grid of water tubes interconnected with the boiler circulation system for positive cooling; the structure is supported by flexing plates allowing the grid and grate to move in a vibrating action; ash is automatically discharged.

Waterflood Injection of water to displace oil from a reservoir (usually a secondary recovery process).

Waterflood mobility ratio Mobility ratio of water displacing oil during waterflooding. (See *also* mobility ratio.)

Waterflood residual The waterflood residual oil saturation; the saturation of oil remaining after waterflooding in those regions of the reservoir that have been thoroughly contacted by water.

Water gas (carbureted blue gas) A mixture of carbon monoxide and hydrogen formed by the action of air and then steam on hot coal or coke and enriched with hydrocarbon gases from the pyrolysis of oils.

Watershed The drainage basin contributing water, organic matter, dissolved nutrients, and sediments to a stream or lake.

Watson characterization factor See Characterization factor.

Watt The common base unit of power in the metric system; one watt equals one joule per second, or the power developed in a circuit by a current of one ampere flowing through a potential difference of one volt. One Watt = 3.412 Btu/hr.

Wax See Mineral wax and Paraffin wax.

Wax distillate A neutral distillate containing a high percentage of crystallizable paraffin wax, obtained on the distillation of paraffin or mixed-base crude, and on reducing neutral lubricating stocks.

Wax fractionation A continuous process for producing waxes of low oil content from wax concentrates; see also MEK deoiling.

Wax manufacturing A process for producing oil-free waxes.

Weathered crude oil Crude oil which, due to natural causes during storage and handling, has lost an appreciable quantity of its more volatile components; also indicates uptake of oxygen.

Wellbore The hole in the earth comprising a well.

Well completion The complete outfitting of an oil well for either oil production or fluid injection; also the technique used to control fluid communication with the reservoir.

Wellhead That portion of an oil well above the surface of the ground.

Wet gas Gas containing a relatively high proportion of hydrocarbons which are recoverable as liquids; see also Lean gas.

Wet scrubbers Devices in which a counter-current spray liquid is used to remove impurities and particulate matter from a gas stream.

Wettability The relative degree to which a fluid will spread on (or coat) a solid surface in the presence of other immiscible fluids.

Wettability number A measure of the degree to which a reservoir rock is water-wet or oil-wet, based on capillary pressure curves.

Wettability reversal The reversal of the preferred fluid wettability of a rock, e.g., from water-wet to oil-wet, or vice versa.

White oil A generic tame applied to highly refined, colorless hydrocarbon oils of low volatility, and covering a wide range of viscosity.

Whole-tree harvesting A harvesting method in which the whole tree (above the stump) is removed.

Wobbe Index (or Wobbe Number) The calorific value of a gas divided by the specific gravity.

Wood alcohol See Methyl alcohol.

Xylan A polymer of xylose with a repeating unit of $C_5H_8O_4$, found in the hemicellulose fraction of biomass—can be hydrolyzed to xylose.

Xylose A five-carbon sugar $C_5H_{10}O_5$; a product of hydrolysis of xylan found in the hemicellulose fraction of biomass.

Yarding The initial movement of logs from the point of felling to a central loading area or landing.

Zeolite A crystalline aluminosilicate used as a catalyst and having a particular chemical and physical structure.

Ziegler–Natta catalysts Catalysts for polymerization processes that are prepared by reacting specific transition metal halides with organometallic reagents such as alkyl aluminum, lithium, and zinc reagents.

Yeast Any of various single-cell fungi capable of fermenting carbohydrates; bioethanol is produced by fermenting sugars with yeast.

Yellow grease A term from the rendering industry—usually means used frying oils from deep fryers and restaurants' grease traps; can also refer to lower quality grades of tallow from rendering plants.

Index

'*Note*: Page numbers followed by "f" indicate figures and "t" indicate tables.'

B

C

E

F

G

H

I

J

K

M

P

T

U

V

X

Y

Z

FSC
www.fsc.org
MIX
Papier | Fördert
gute Waldnutzung
FSC® C083411

Kolibri 360 GmbH
Ferdinand-Jühlke-Straße 7
99095 Erfurt, Deutschland
produktsicherheit@kolibri360.de